Agriculture and Environment Series

Jack E. Rechcigl
Editor-in-Chief

Agriculture is an essential part of our economy on which we all depend for food, feed, and fiber. With the increased agricultural productivity in this country, as well as abroad, the general public has taken agriculture for granted while voicing its concern and dismay over possible adverse effects of agriculture on the environment. The public debate that has ensued on the subject has been brought about, in part, by the indiscriminate use of agricultural chemicals and, in part, by disinformation, based largely on anecdotal evidence.

At the national level, recommendations have been made for increased research in this area by such bodies as the Office of Technology Assessment, the National Academy of Sciences, and the Carnegie Commission on Science, Technology, and Government. Specific issues identified for attention include contamination of surface and groundwater by natural and chemical fertilizers, pesticides and sediment, the continued abuse of fragile and nutrient-poor soils, and suitable disposal of industrial and agricultural waste.

Although a number of publications have recently appeared on specific environmental effects of some agricultural practice, no attempt has been made to approach the subject in a systematic, comprehensive manner. The aim of this series is to fill the gap by providing the synthesis and critical analysis of the state of the art in different areas of agriculture bearing on environment and vice versa. Efforts will also be made to review research in progress and comment on perspectives for the future. From time to time methodological treatises as well as compendia of important data in handbook form will also be included. The emphasis throughout the series will be on comprehensiveness, comparative aspects, alternative approaches, innovation, and worldwide orientation.

Specific topics will be selected by the Editor-in-Chief with the council of an international advisory board. Imaginative and timely suggestions for the inclusion in the series from individual scientists will be given serious consideration.

Published Titles

Soil Amendments and Environmental Quality
Soil Amendments: Impacts on Biotic Systems

Forthcoming Title

Environmentally Safe Approaches to Crop Disease

Agriculture and Environment Series

SOIL AMENDMENTS and ENVIRONMENTAL QUALITY

Edited by Jack E. Rechcigl

To Nancy, Gregory, Kevin, and Lindsey, with love.

Agriculture and Environment Series

SOIL AMENDMENTS and ENVIRONMENTAL QUALITY

Edited by Jack E. Rechcigl
University of Florida
Soil and Water Science Department
Research and Education Center
Ona, Florida

LEWIS PUBLISHERS

Boca Raton New York London Tokyo

Library of Congress Cataloging-in-Publication Data

Rechcigl, Jack E.
Soil amendments and environmental quality / edited by Jack E. Rechcigl.
p. cm. — (Agriculture and environment series)
Includes bibliographical references and index.
ISBN 0-87371-859-3
1. Fertilizers—Environmental aspects. 2. Soil amendments—Environmental aspects. I. Title. II. Series.
TD196.F47R43 1995
631.8—dc20

95-7403
CIP

Direct all inquiries to CRC Press, Inc., 2000 Corporate Blvd., N.W., Boca Raton, Florida 33431.

Lewis Publishers is an imprint of CRC Press

International Standard Book Number 0-87371-859-3
Library of Congress Card Number 95-7403
Printed in the United States of America 1 2 3 4 5 6 7 8 9 0
Printed on acid-free paper

The Editor

Jack E. Rechcigl is an Associate Professor of Soil and Environmental Sciences at the University of Florida located at the Research and Education Center in Ona. He received his B.S. degree (1982) in Plant Science from the University of Delaware, Newark, and his M.S. (1983) and Ph.D. (1986) degrees in Soil Science from Virginia Polytechnic Institute and State University, Blacksburg. He joined the faculty of the University of Florida in 1986 as Assistant Professor and in 1991 was promoted to Associate Professor.

Dr. Rechcigl has authored over 100 publications including contributions to books, monographs, and articles in periodicals in the fields of soil fertility, environmental quality, and water pollution. His research has been supported by grants totaling over $2 million from both private sources and government agencies. Dr. Rechcigl has been a frequent speaker at national and international workshops and conferences and has consulted in various countries including Brazil, Nicaragua, Venezuela, Australia, Taiwan, Philippines, and Czechoslovakia.

He is currently an Associate Editor for the *Journal of Environmental Quality* and is Editor-in-Chief of the **Agriculture and Environment Book Series**. He is also serving as an invitational reviewer for manuscripts and grant proposals for scientific journals and granting agencies.

Dr. Rechcigl is a member of the Soil Science Society of America, American Society of Agronomy, International Soil Science Society, Czechoslovak Society of Arts and Sciences, various trade organizations, and the honorary societies of Sigma Xi, Gamma Sigma Delta, Phi Sigma, and Gamma Beta Phi.

Preface

For years fertilizers have been the key to successful agriculture. Many soils worldwide are deficient in one or more plant nutrients which makes them unproductive. Ever since the pioneering work of Julius von Liebig on the "Law of the Minimum", which demonstrated that crop yield was determined by the minimum amount of the nutrient in the soil, it has been a standard practice to supply soil with the missing or deficient plant nutrient in order to achieve maximal yield. The missing nutrients have usually been supplied in the form of chemical fertilizers.

In the U.S. 30 to 40% of crop production is attributable to fertilizers, while in some developing countries figures as high as 50 to 75% have been reported.

As a result of the role agricultural chemicals have played in the Green Revolution, some agricultural practitioners have developed the mistaken belief that more fertilizer is better than less and that overfertilization would guarantee them maximum profits. They have clearly forgotten the long established principle of "Law of Diminishing Returns" which states that with each additional increment of a fertilizer the increase in yield becomes smaller and smaller.

The practice of overfertilization is not only unnecessary and inordinately costly, but may also be detrimental to the crop itself and even more so to the surrounding environment.

Although literally hundreds of publications of various types exist on soil fertility and the use of fertilizers, no monograph exists on the environmental aspects of fertilizer use.

The aim of this publication is to present a comprehensive and balanced overview of the influences of soil amendments on the environment, addressing both positive and negative aspects. This book deals specifically with the environmental aspects of chemical fertilizers that supply nitrogen, phosphorus, potassium, sulfur, lime, micronutrients, and trace metals as well as organic wastes, including manures, compost, sludge, and industrial by-products, such as fly ash, phosphogypsum, and other by-product gypsums. The final two chapters deal with alternative agronomic practices and biotechnology that may ameliorate or minimize the possible adverse effects resulting from fertilizer use. This book should be of interest to researchers as well as to practitioners in the field of agriculture and environmental sciences. Moreover, it should be of benefit to policymakers and the general public concerned with environmental issues.

The editor wishes to thank the individual contributors for the time and effort they put into the preparation of their chapters. In addition, special thanks are due to Lewis Publishers and CRC Press staff and Editorial Board.

Jack E. Rechcigl

Contributors

Domy C. Adriano
Savannah River Ecology Laboratory
University of Georgia
Aiken, South Carolina

Isabelo S. Alcordo
Research and Education Center
University of Florida
Ona, Florida

A. K. Alva
Citrus Research and Education Center
University of Florida
Lake Alfred, Florida

J. S. Angle
Department of Agronomy
University of Maryland
College Park, Maryland

Allen V. Barker
Department of Plant and Soil Sciences
University of Massachusetts
Amherst, Massachusetts

J. J. Bilski
Citrus Research and Education Center
University of Florida
Lake Alfred, Florida

James J. Camberato
Pee Dee Research and Education Center
Clemson University
Florence, South Carolina

Thanh H. Dao
Grazinglands Research Laboratory
USDA/ARS
El Reno, Oklahoma

H. V. Eck
Conservation and Production Research Laboratory
USDA/ARS
Bushland, Texas

R. L. Hill
Department of Agronomy
University of Maryland
College Park, Maryland

N. V. Hue
Department of Agronomy and Soil Science
College of Tropical Agriculture and Human Resources
University of Hawaii at Manoa
Honolulu, Hawaii

B. R. James
Department of Agronomy
University of Maryland
College Park, Maryland

Alina Kabata-Pendias
Institute of Soils Science and Plant Cultivation
Osada Palacowa Pulawy, Poland

Robert L. Mikkelsen
Department of Soil Science
North Carolina State University
Raleigh, North Carolina

Rosa M. C. Muchovej
Southwest Florida Research and Education Center
University of Florida
Immokalee, Florida

Jack E. Rechcigl
Research and Education Center
University of Florida
Ona, Florida

K. S. Sajwan
School of Science and Technology
Savannah State College
Savannah, Georgia

Andrew Sharpley
Pasture Systems and Watershed Research Laboratory
USDA/ARS
University Park, Pennsylvania

B. A. Stewart
Conservation and Production Research Laboratory
USDA/ARS
Bushland, Texas

Margie Lynn Stratton
Research and Education Center
University of Florida
Ona, Florida

Paul W. Unger
Conservation and Production Research Laboratory
USDA/ARS
Bushland, Texas

J. L. Wiebers
Buchart-Horn, Inc.
Eschborn, Germany

Paul J. Withers
ADAS
Bridgets Research Centre
Martyr Worthy
Winchester, Hampshire, England

Table of Contents

CHAPTER 1

Nitrogen Fertilizers

Rosa M. C. Muchovej and Jack E. Rechcigl

0-87371-859-3/95/$0.00+$.50

I. INTRODUCTION

A. CONSUMPTION AND PRODUCTION OF NITROGEN FERTILIZERS

Agricultural systems rely heavily on nitrogen (N) fertilizer inputs to meet plant requirements. An exception to this may be the leguminous crops that form associations with bacteria species of the genera *Bradyrhizobium* and *Rhizobium* and are able to obtain part or most of the necessary N from biological N_2 fixation. Nitrogenous fertilizers may be either organic or inorganic compounds. Due to the low available N content and the high cost per unit of N supplied, organic N fertilizers are not extensively used for crop production. (Organic fertilizers are discussed in Chapters 5, 6, and 7, which address manure, sewage sludge, compost, and other N-containing organic amendments.)

Application of N fertilizers has been increasing worldwide in order to enhance food and fiber production to meet the demands of a growing global population. From the period of 1967 to late 1987, there was an increase in world N fertilizer utilization from 22.1 to 72.7 million metric tons.[1] In 1991, this amount increased to 77 million metric tons.[2] It has been estimated that, by the end of 1993, the world consumption of N fertilizer will exceed 83.5 million metric tons,[3] and that the developing countries will be consuming as much N fertilizer as the developed countries. In developed countries, annual N fertilizer use has increased about 25%, while in developing countries this increase has been 160%, in a period of 10 years, i.e., from about 1979 to 1989.[4] Although the U.S. is one of the major consumers of N fertilizers in the world (Table 1),[5] annual consumption in the nation has reached a plateau since 1975.[4] The same trend of stabilizing application is true for many other developed countries, mainly because of excess crop production and reduced area of agricultural use. However, in developing countries that are expanding crop production to reduce food imports, consumption as well as production of N fertilizers has been increasing continuously.[6] It has been estimated that, by the year 2000, production and consumption of N fertilizers worldwide will exceed 130 million metric tons.[7]

B. FORMS AND METHODS OF APPLICATION

The global distribution of N varies greatly (Table 2).[8] Nitrogen (N_2) constitutes approximately 78% of the earth's atmosphere. However, N_2 can be utilized directly only by a few prokaryotic microbes. Combination of N_2 with H_2 or O_2 is necessary to produce forms usable by higher plants and animals. This conversion of N_2 can be accomplished biologically by N_2-fixing bacteria, either free living or associated with plants, by atmospheric discharges as oxides of N, and by industrial processes for the manufac-

Table 1 Annual N Fertilizer Consumption for Selected Countries in 1983/1984

	N Fertilizers (Mg × 10^3)					
Country	**AP**	**AS**	**AN**	**Urea**	**Other**	**Total**
U.S.	55	149	687	1124	8151	10111
El Salvador	—	38	—	4	17	59
Honduras	0.7	0.7	0.7	15	2	19
Mexico	54	302	45	370	239	1010
Brazil	96	169	85	244	48	642
Peru	3	5	12	30	3	53
Tanzania	—	7	5	1	2	15
Israel	—	10	11	8	15	44
Japan	—	207	12	113	370	702
Pakistan	59	15	79	673	88	914
Philippines	23	29	—	157	22	231
India	428	95	91	3791	232	4637
Indonesia	—	75	—	969	5	1049
China	—	—	—	—	—	13678
Germany (West)	—	66	858	—	454	1378
Hungary	—	1	353	180	91	625
Italy	—	75	203	370	348	996
Netherlands	0.2	0.4	402	2	73	478
Poland	61	158	812	255	35	1321
USSR (former)	—	—	—	—	—	10292
New Zealand	10	9	2	11	—	32

Note: AP = Ammonium phosphate; AS = Ammonium sulfate; AN = Ammonium nitrate; total = all nitrogenous fertilizers.

From Peterson, G.A. and Frye, W.W., *Nitrogen Management and Ground Water Protection*, Follett, R.F., Ed., Elsevier, Amsterdam. With permission.

Table 2 Estimates of the Global Distribution of N

Environment	**N (Mg)**
Atmosphere	3.9×10^{15}
Soil (nonliving)	1.5×10^{11}
Microbes in soil	6×10^{9}
Plants	1.5×10^{10}
Animals (land)	2×10^{8}
People	1×10^{7}
Sea (various)	2.4×10^{13}

From Addiscott, T.M., Whitmore, A.P., and Powlson, D.S., *Farming, Fertilizers and the Nitrate Problem*, C.A.B. International Wallington, Oxon, 1991. With permission.

ture of commercial N fertilizers. A brief discussion of commercially produced fertilizers, some of which are listed in Table 3,[9] will follow.

Nitrogen fertilizer manufacturing has changed little over the years. Ammonia is the principal commercial N compound and it is the building block for most synthetic N fertilizers. Anhydrous ammonia production is dependent upon a supply of natural gas, the largest variable cost component of manufacturers. Fuel oil, naphtha, and coal are also substitutes for natural gas. In the process, N_2 gas is reacted with H_2 in the presence of a catalyst, normally magnetite (Fe_3O_4), with additions of potassium (K), alumina, and calcium (Ca), at temperatures close to 1200°C and high pressure. Several derivatives of

Table 3 Sources of N Fertilizer, Grade, Chemical, and Physical Form and Recommended Method of Application

Fertilizer	Grade (N-P_2O_5-K_2O)	Physical Form	Method of Recommended Application
Anhydrous ammonia (NH_3)	82-0-0	High-pressure liquid	Must be injected 15 to 20 cm deep in friable, moist soil
Aqua ammonia ($NH_3 + H_2O$)	20-0-0 to 24-0-0	Low-pressure liquid	Must be injected 5 to 8 cm deep in friable, moist soil
Low-pressure N solutions ($NH_3 + H_2O$)	37-0-0 to 41-0-0	Low-pressure liquids	Must be injected 5 to 8 cm deep (NH_4NO_3) in friable, moist soil
Pressureless N solutions (NH_4NO_3 + urea + H_2O)	28-0-0 to 32-0-0	Pressureless liquids	Spray on surface or sidedress; incorporate surface application to prevent volatilization loss of NH_3 from the urea
Ammonium nitrate (NH_4NO_3)	33.5-0-0 to 34-0-0	Dry prills, granules	Broadcast or sidedress; can be left on the soil surface
Ammonium sulfate $(NH_4)_2SO_4$	20-0-0	Dry granules	Broadcast or sidedress; can be left on the soil surface
Urea ($NH_2 \cdot CO \cdot NH_2$)	45-0-0	Dry prills, granules	Broadcast or sidedress; incorporate surface application to prevent volatilization loss of NH_3 from the urea
Sodium nitrate ($NaNO_3$)	16-0-0	Dry granules	Broadcast or sidedress; can be left on the soil surface
Calcium nitrate $Ca(NO_3)_2$	15.5-0-0	Dry granules	Broadcast or sidedress; can be left on the soil surface
Ammonium nitrate of lime ($NH_4NO_3 + CaCO_3$)	20.5-0-0	Dry granules	Broadcast or sidedress; can be left on the soil surface.
Diammonium phosphate $(NH_4)_2HPO_4$	18-46-0	Dry granules	Broadcast or apply in the row; can be left on the soil surface
Potassium nitrate (KNO_3)	13-0-44	Dry granules	Broadcast or apply in the row; can be left on the soil surface
Ammonium phosphate	10-34-0 to 11-37-0	Pressureless liquids	Spray on surface or sidedress

From Walsh, L.M., Soil and Applied Nitrogen, Ext. Fact Sheet A 2519, University of Wisconsin, Madison, 1973.

ammonia can be made, including ammonium nitrate, urea (or carbamide), ammonium chloride, ammonium sulfate, calcium cyanamide, calcium nitrate, oxamide, and sodium nitrate.

Ammonia is reacted with O_2 in the presence of a platinum catalyst at 800°C to produce nitric acid, which is then used to manufacture ammonium nitrate and other nitrate (NO_3) fertilizers. Ammonium phosphate, ammonium sulfate, and ammonium nitrate are manufactured by neutralization of the respective acids with ammonia. Ammonium nitrate is produced in liquid and solid form, depending on whether it is to be applied directly or incorporated into the soil. Ammonium phosphates can be produced as many grades, depending on the proportions of the primary ingredients (anhydrous ammonia, phosphoric acid, and sulfuric acid) used in the mixture.

The cyanamide process consists of reacting pure N_2 gas with calcium carbide at high temperatures (1100°C) to form calcium cyanamide ($CaCN_2$). The use of this process for N_2 fixation is not widespread throughout the world.

Urea is produced commercially by reacting ammonia with carbon dioxide (CO_2), under very high pressure (1.38×10^7 Pa) in the presence of a catalyst. The end product is an aqueous solution containing approximately 75% urea and can be used to produce urea-ammonium nitrate fertilizer solutions as well as solid urea or granular fertilizers.

Nitrogen solutions are aqueous mixtures of ammonia, ammonium nitrate, and urea, separate or in combination. They may be applied directly to the soil, in the irrigation water, sprayed on leaves, or may be used to manufacture other N fertilizers. Free anhydrous ammonia solutions must be stored under pressure to prevent volatilization. Urea is used in both solid and liquid forms. Liquid urea is used extensively in N solutions, and the solid product can be applied directly or incorporated into mixed and bulk blend fertilizers.[10]

The process of N_2 fixation by oxidation involves the passage of electric sparks through a mixture of N_2 and O_2 gases. The process requires large amounts of electricity and does not compete with the less-costly ammonia fixation process.

Nitrogen from urea sources can be lost as free NH_3 or undergo hydrolysis by the enzyme urease and be transformed to NO_3 and ammonium (NH_4) by microbial processes. Sulfur-coated urea (SCU) was developed by the Tennessee Valley Authority to control dissolution and to inhibit microbial degradation of urea, thus providing slow release of N to crops.[11] Slow-release N compounds extend the residence time of the N fertilizer in soil. Some examples of such compounds are urea-formaldehyde complexes (ureaform), crotonylidene diurea (CDU), isobutylidene urea (IBDU), and magnesium ammonium sulfate.

Nitrification inhibitors are toxic to nitrifying organisms and temporarily prevent the production of NO_3. These materials are blended with the N fertilizer or used as coatings on the fertilizer particles to keep N as NH_4 and decrease rate of leaching and possible loss by denitrification. The only nitrification inhibitors available commercially in the U.S. are nitrapyrin [2-chloro-6-(trichloromethyl)pyridine] and etridiazol [5-ethoxy-3-(trichloromethyl)-1,2,4-thiadiazole].[5]

The method of application and placement of the N fertilizer in relation to the plant (Table 3)[10] will influence the efficiency of its use, and will depend on the crop and on the amount of fertilizer to be applied. For instance, for row crops the fertilizer may be placed in the hill, slightly below and on the side or even both sides of the seed. It may also be placed in a narrow band on one or both sides of the row, away from and below the seed level. Application of large quantities of fertilizer is commonly done by broadcasting part of it and working it into the soil before planting. For small grains, the drill is equipped with a fertilizer distributor. If the rate of fertilizer application is high, seed injury may occur. Ornamental or forest trees are treated individually and the fertilizer is usually worked into the soil around the tree. For pastures and lawns, fertilizer is usually applied at the time of seeding and top-dressing is performed in subsequent years.

Several reviews have been published on N fertilizer types, sources, and forms of application,[5,12–14] and the reader is referred to them for more detailed descriptions. The review by Peterson and Frye[5] provides a summary of advantages and disadvantages of the various N carriers.

II. SOIL NITROGEN

A. THE NITROGEN CYCLE

Nitrogen is present in the atmosphere mainly as the free, inert dinitrogen (N_2) gas, and is not used as such by most life systems. The soil N is mostly in the form of organic matter (OM), originating from inputs from animal and green manures, decomposition of crop residues, composts and sludges, seasonal renewal of grass root systems, and soil biota and microbiota. The organic forms are not immediately available to plants. Inorganic N forms, largely ammonium-N (NH_4-N), and nitrate-N (NO_3-N), comprise the remaining significantly smaller portion of soil N and are directly available to plants.

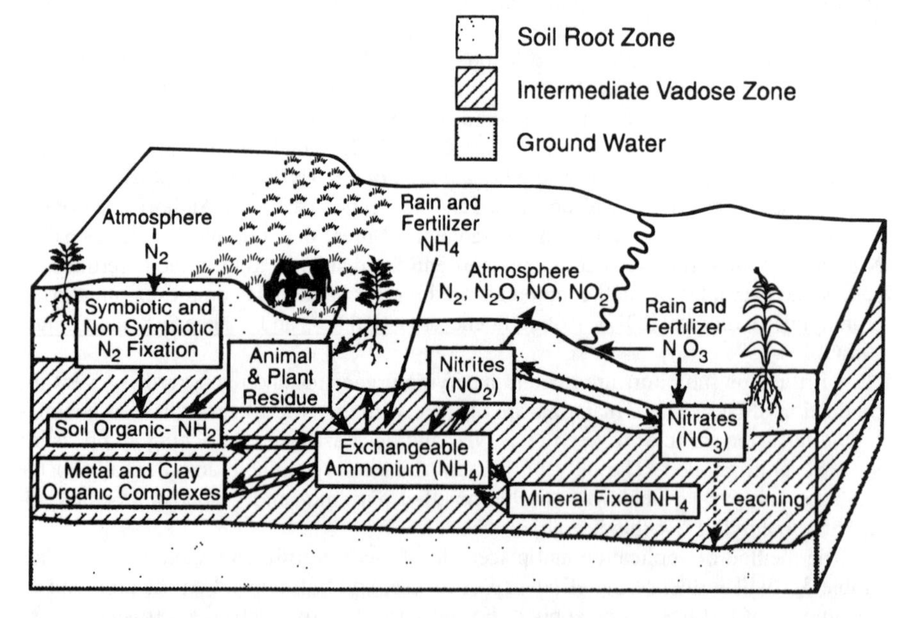

Figure 1 Nitrogen processes occurring in the soil-plant system. (From Follett, R.F. and Walker, D.J., *Nitrogen Management and Ground Water Protection*, Follett, R.F., Ed., Elsevier, Amsterdam, 1989, 1. With permission.)

Nitrogen is a dynamic element and is transformed to various forms by processes that constitute the N cycle (Figure 1). Microbes play a major role in the soil-N cycle, transforming N compounds into available or unavailable forms, mobile or immobile forms, oxidized or reduced forms and gaseous or soluble forms.

The major divisions of the soil-N cycle are immobilization (removal of N from readily available forms) and mineralization (release of N). Nitrogen removal by crops, animals, and microbial populations are forms of N immobilization and, in a way, are considered as a type of loss in the soil-N cycle since N is now less available for plant growth. Depending on the conditions prevailing in the specific environment, mineralized N is susceptible to further transformations, such as volatilization, leaching, or denitrification, resulting in losses of inorganic N. The major environmental impacts of N fertilizers concern losses of inorganic N to surface and groundwater and/or the atmosphere. More detailed information on the N cycle may be obtained in several published reviews.[15–20]

B. TRANSFORMATIONS OF NITROGEN FERTILIZERS IN SOILS

Several physical, chemical, and biological factors that interact with each other and with the environment determine the fate of N fertilizers in the soil-plant system.[13,21] Ammonia can be lost rapidly from calcareous soils (pH >7), under conditions of increasing soil and air temperatures, in soils with low cation exchange capacity, such as sands, or when high organic-N substances, such as manures, are placed on the soil surface.[17] When quantities of anhydrous NH_3 applied exceed the sorption capacity of the soil and the gas is injected superficially, considerable losses (as high as 70%) of this fertilizer can occur in the environment through volatilization.[15] Ammonia volatilization from fertilizers is normally reduced by application at depths below 10 cm.[22]

Ammonia added as anhydrous NH_3 or produced in soil through microbial activity reacts readily with soil water and H ions to form the positively charged NH_4 ion.[13] The distribution between the NH_3 and NH_4 forms is highly influenced by pH (Figure 2); at neutral or acid pH, the predominant form is NH_4. The NH_4 ion can be absorbed, adsorbed to cation exchange sites, transformed further by biological processes, or enter into fixation reactions with certain clay minerals, most frequently vermiculite and illite. The mechanism of NH_4 fixation is presumed similar to that of K fixation. Apparently additions of K fertilizer prior to or simultaneously with the NH_4 fertilizer reduce fixation.[17] Ammonia (NH_3) can also be fixed by reactions with lignin and fractions of OM, and may not be immediately usable by plants, although eventually it will be released by the process of mineralization.[17] This NH_3 fixation, not to be confused with NH_4 fixation, has been used for decades to convert organic materials, such as sawdust, peat, high-lignin materials, and coal products, into nitrogenous fertilizers.[17]

Ammonium fertilizers may be transformed biologically to NO_3 (nitrification). The process, carried out by autotrophic bacteria of the genera *Nitrosomonas* and *Nitrobacter,* is slowed at pH below neutrality and nitrification rates approach 0 at pH values lower than 5 or under conditions of high soil moisture. Because the cation NH_4 undergoes adsorption and retention reactions with soil colloids, fertilizers containing NH_4-N may be applied months before planting to fine-textured soils without considerable loss by leaching, provided temperatures are low (<10°C). However, N application to sandy, low-exchange capacity soils will probably result in appreciable movement of the NH_4 through the profile.

The NO_3 forms of N, whether supplied as fertilizers or produced by nitrification of NH_4, are quite susceptible to leaching. Nitrate compounds are readily soluble in soil so-

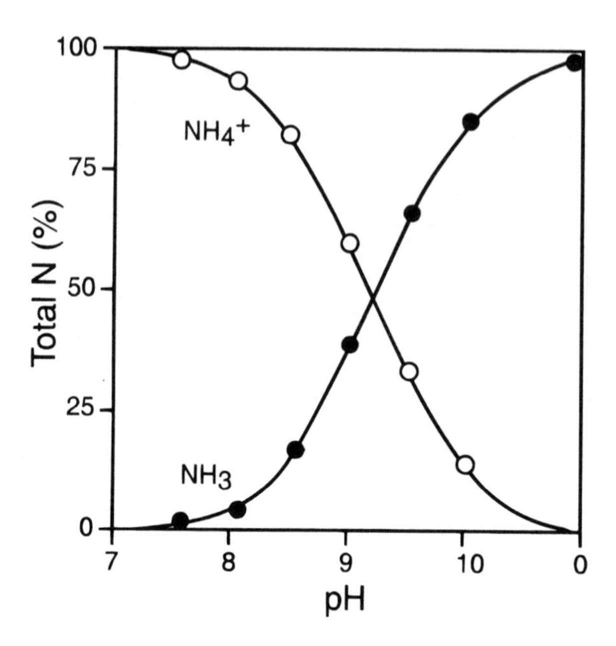

Figure 2 Percent of total N as NH_3 and NH_4 at various solution pH levels. (From Boswell, F.C., Meisinger, J.J., and Case, N.L., *Fertilizer Technology and Use*, Engelstad, O.P., Ed., SSSA, Madison, WI, 1985, 229. With permission.)

lution and the NO_3 is not held tightly by the anion exchange sites. Thus, NO_3 is very mobile in soils, i.e., moves with soil water, upward due to capillary forces, during extremely dry weather, and downward, under conditions of excessive precipitation or irrigation. However, on soils having an anion exchange capacity, such as the ando soils which are positively charged, NO_3 is adsorbed and NH_4 may be leached. Nitrogenous oxides may also undergo biological and/or chemical denitrification in soil. The biological process is carried out only by bacteria and the principal products are N_2 and nitrous oxide (N_2O). The denitrifying bacteria are aerobic, although in a reduced environment they are capable of using NO_3 as an electron acceptor instead of O_2. Abiological denitrification, also known as chemodenitrification, is important especially in acid soils (pH <6), in the presence of nitrous oxide (HNO_2).[23] It is difficult to distinguish chemical from biological denitrification since the former depends on the biological reduction of NO_3 to NO_2.

III. BENEFICIAL EFFECTS OF NITROGEN

A. SOIL FERTILITY

Insufficient N is a common soil fertility problem, limiting the production of most crops. Most of the soil N is in organic forms that are used by microbes as a source of energy and food. Complex combinations of organic and inorganic substances comprise the soil humic fraction, which is a very important constituent of a fertile soil. OM in various amounts and stages of decomposition is normally present, as well as some inorganic forms such as nitrates and ammoniates. Plowing can hasten OM decomposition and the release of available soil N. Reserves of soil N are also diminished by cropping without replenishing removed N. Declines in the original reserves of N in soils of the North Central Region of the U.S. are estimated at 25 to 59% during the first 75 years of cultivation.[24]

Processes such as erosion, leaching, denitrification, and fixation by clays also result in decreases in available soil N and, consequently, in reduced fertility of a particular soil. Thus, adoption of management measures that inhibit or prevent losses or excessive removal of N from agricultural soils contributes significantly to maintenance of soil fertility.

B. CROP PRODUCTION

Of all essential elements, N is required in the largest amount, except for O_2, H, and C. Nitrogen concentrations in the vegetative tissues of most plants vary between 1 and 5%, by weight. In the seed or storage organs of protein-rich crops, especially legumes, seed N concentrations may even exceed 10 to 12% N, by weight.

Nitrogen is a vital constituent of the chlorophyll molecule and of all amino acids. Nitrogen is a structural component of many compounds essential for plant growth, including many coenzymes and metabolites such as nucleic acids, adenosine nucleotides (ATP, ADP, AMP), nicotinamide- or flavin-containing dinucleotides (NADH, NADPH, FAD), porphyrins, diamines, and various secondary metabolites and is even found in cell walls. Nitrogen is required for carbohydrate utilization, and in most plants, it regulates the uptake and utilization of K, P, and other constituents.

Plants receiving an insufficient supply of N are stunted and have chlorotic leaves, i.e., leaves that are yellow to yellow-green, often fail to flower or set seed, and may have as little as 0.5 to 0.6% N in vegetative tissues. When applied as a chemical fertilizer, N appears to have the quickest and most pronounced effect on plants, as compared with other mineral nutrients. Nitrogen is absorbed by plants usually in the form of NO_3; however, uptake of NH_4 ions and urea occurs in some plants.

Ample literature stresses the economic importance and the beneficial effects of N fertilizers for crop growth and production. Crops such as corn (*Zea mays* L.), sorghum [(*Sorghum bicolor* L. (Moench)], wheat (*Triticum aestivum* L.), and cotton (*Gossypium hirsutum* L.) have higher yields following N fertilizer application.[25] Supplemental N is also beneficial for a wide range of crops such as sugarcane (*Saccharum officinarum*), barley (*Hordeum vulgare* L.), rice (*Oryza sativa* L.), and rapeseed (*Brassica napus*). In terms of requirement and plant response, N is the most important nutrient for pasture crops. For instance, a positive response in forage yield of bahiagrass (*Paspalum notatum* Flugge) was obtained due to fertilization with NH_4NO_3.[26]

C. LIVESTOCK PRODUCTION

Approximately 50% of the productivity in grassland systems is due to N-fertilizer application.[27] Crude protein concentration, palatability, and vigor are increased substantially by addition of N in small grains and grass forages.[28] Addition of N fertilizer significantly increased protein content of early and late-season wheat forage, used for grazing by livestock in the Great Plains.[29]

Grasslands, in the form of native meadows, are some of the most protective land-use patterns regarding downward movement of N fertilizers.[30] Native and extensively managed grasslands are usually deficient in N and little NO_3 leaching occurs. Permanent grass swards generally have a high absorption capacity for NO_3, even when annual fertilization is performed.[31-34] Nitrate does not generally accumulate under native grasslands during the growing season because it is utilized as it is formed by nitrification.[35] The prolific root mass of perennial grasses is capable of absorption of N present in excess of crop demands.[25] The amount of vegetative growth of grasslands is closely related to the moisture supply and the amount of available N. In grassland systems, N recovery, as harvested yield, ranges from 50 and 65% of the applied N fertilizer in the year of application.[25] Application of large quantities of N fertilizer and concentrates contribute significantly to current increased productivity of grassland and livestock on intensively managed grassland farms.[36,37]

IV. POTENTIAL ADVERSE EFFECTS OF NITROGEN FERTILIZERS

A. PLANT SYSTEMS

For a given crop, area, climate, and year, N-response curves are normally curvilinear, with a gradual decrease in yield increment per unit N.[38] Addition of N fertilizer beyond the requirement for maximum yield usually does not increase crop yields, but will likely produce plants with undesirable characteristics and increase potential hazards to the environment.

An oversupply of N alters the nutrient balance and disrupts the physiology of the plant; leaves become dark green and there is excessive vegetative growth and succulence. As a consequence of luxury consumption of N, lodging and a weakening of fibers, reduced sugar content, delayed crop maturity, and enhanced disease and insect susceptibility usually occur.

Nitrate concentrations as high as 3000 mg kg^{-1} have been verified in some vegetable, forage, and cereal crops as a result of high soil NO_3 levels, low light intensity during maturation, reduced soil moisture, and inadequate supply of plant nutrients.[17] Other causes of NO_3 accumulation include high temperature, low CO_2 concentration, and high air humidity; and these factors are discussed by van Diest.[39] Elevated NO_3 levels constitute health hazards to both humans and animals.

B. AQUATIC ECOSYSTEMS

Pollutants that reach bodies of water originate from either point or nonpoint (diffuse) sources. The point sources are usually caused by discharges from industrial plants, urban sewers, or sewage treatment works, whereas the diffuse sources arise from locations such as the land surface or the atmosphere. Generally the diffuse sources are more difficult to identify, measure, and control.

Intensification of agriculture together with the high use of synthetic fertilizers and animal wastes have aggravated problems associated with eutrophication of surface waters, such as lakes, and increased NO_3 concentrations in drinking water sources. Most soluble N from agricultural land that reaches lakes, rivers, and aquifers is in the form of NO_3. During the past 30 years, NO_3 concentrations in ground and surface waters have been rising rapidly in many locations, and either closely approach or exceed international permissible standards for drinking water. For certain areas in high latitudes, eutrophication of surface waters may become the major problem. In most other areas, the main concern is contamination of aquifers.

The magnitude of the NO_3 problem originates primarily from the following: aquifers provide most of the drinking water for most communities. Rivers receive water draining from aquifers, agricultural runoff, and urban and industrial waste waters which may contain significant amounts of NO_3. Increasing fertilizer use and aquifer contamination will eventually result in poor water quality.[40] While nutrients and sediment are the major non-point pollution problems for surface waters, the primary concern for groundwater supplies is NO_3 contamination. The present NO_3 concentrations in waters are considered to reflect fertilization rates of several years ago.[40] Sources other than fertilizers and soil N may also contribute to high levels of NO_3 encountered in certain areas. According to Henry and Menely,[41] a 1948 report by Robertson and Riddell clearly established NO_3 problems in Saskatchewan farm wells before any significant use of N fertilizers was being made. Livestock feeding, barnyards, septic tanks, and animal and human contamination are frequently associated with elevated NO_3 in well waters.[17] Techniques to remove NO_3 from water are costly and normally are not employed in conventional drinking water treatment; thus, possible health hazards to large populations may be considerable even though individual risk is small.[40]

The impact of fertilizer use on the environment has been a major concern since the 1960s, but at that time there was a scarcity of scientific evidence to support the contention that there was harm from fertilizer applications. Although the benefits of N for plant production are well recognized, excessive use of N fertilizers has often been indicated as a major source of NO_3 in ground and surface water in rural and suburban areas.[42–44] Nitrate leaching is a potential problem in urban and rural areas that are characterized by the predominance of sandy soils, abundant rainfall, frequent shallow water tables, high fertilization rates, and reliance on groundwater as a source of drinking water. Reports from Florida, dating as early as 1916,[45] mentioned that $(NH_4)_2SO_4$, rather than $NaNO_3$, should be the preferred N fertilizer form in the wet season to prevent leaching losses. Excessive irrigation (>10 cm) or heavy rainfall is likely to move NO_3 below the root zone of many crops grown on sandy soils in Florida.[46,47] In humid regions and under irrigation, leaching loss can account for up to 50% of the N input.[19,22]

1. *Groundwater*

In the past, research and surveys on groundwater degradation were justified by potential adverse effects on human and animal health. Presently, there has been an increase in public awareness of environmental quality, especially groundwater, since groundwater is the major component of baseflow in streams and rivers.[48] In the U.S., groundwater is the major source of drinking water for 50% of the urban and 85% of the rural population.[49] In Great Britain, drinking water originates from surface waters and from Cretaceous chalk and Triassic sandstone.[50] In natural systems, NO_3-N rarely occurs in groundwater in concentrations greater than 3 mg L^{-1}.[51] Elevated NO_3 levels, greater than 10 mg NO_3-N L^{-1}, have been found in areas of recharge, but, in areas where the aquifers move under clay cover, levels were lower.[52] In Israel, the danger of groundwater pollution from various possible sources, such as outflow from urban and agricultural areas, is acute. Nitrate-N concentrations in excess of 10 mg L^{-1} were detected in large portions of the shallow

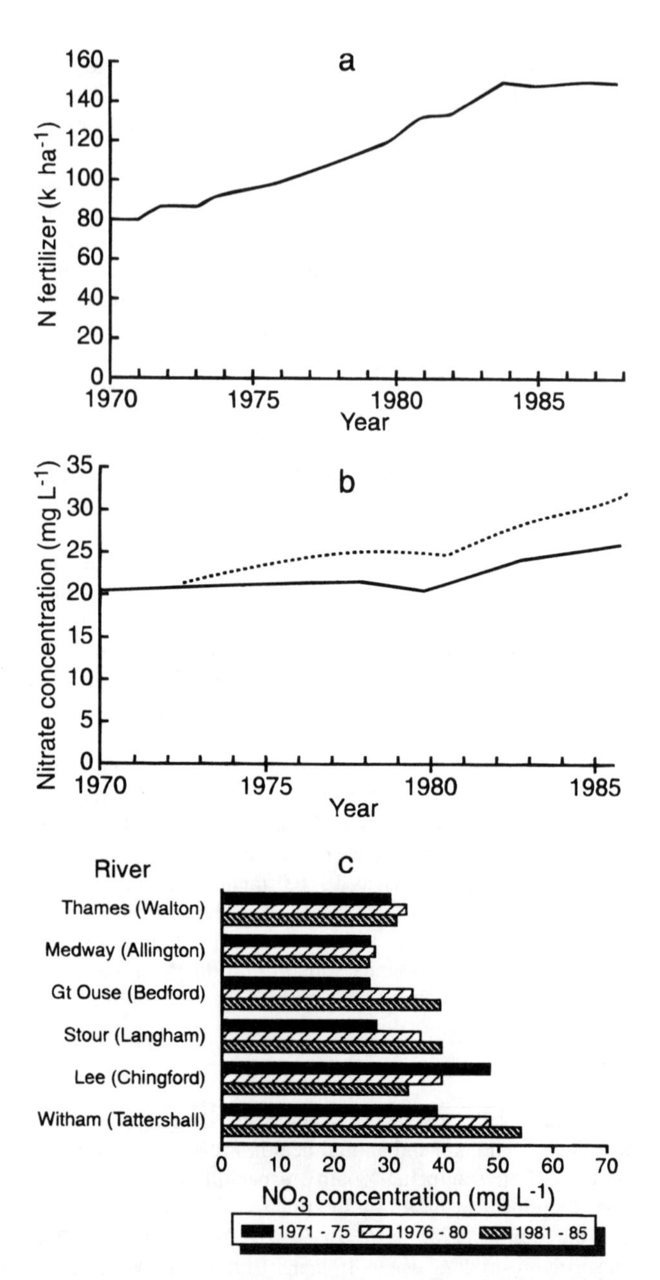

Figure 3 (a) Increase in the amount of N fertilizer (kg ha^{-1}) applied to all crops and grass 1970–1988 (survey of fertilizer practice); (b) increase in the concentration of NO_3 in water drawn from two bore holes between 1970 and 1986; Mill Meece (-), Hatton (…); (c) change in the concentration of NO_3 in water drawn from some U.K. rivers between 1971 and 1985. (From Addiscott, T.M., Whitmore, A.P., and Powlson, D.S., *Farming, Fertilizers and the Nitrate Problem*, C.A.B. International Wallingford, Oxon, 1991. With permission.)

coastal aquifer, and the concentration is steadily rising at the rate of 1 to 2 mg L^{-1} per year.[53] In 10% of the wells, NO_3-N concentration exceeded 22 mg L^{-1}. Over 90% of available water resources is already being utilized, and about 30% of this is obtained from the coastal aquifer.

Identification of sources of NO_3 to groundwater is difficult due to the complexity of the N cycle and the multitude of sources. Nevertheless, intensive agricultural activities have caused increases in groundwater NO_3. Application of excessive amounts of N fertilizers has frequently been suggested as a primary source of NO_3 contamination in groundwater.[43,44] Nitrate levels in natural waters have been increasing over the years as a result of the increase in N fertilizer use (Figure 3). The NO_3 levels in the Alliston region of Ontario exceeded 10 mg L^{-1}, especially in areas of high N fertilizer use for potato (*Solanum tuberosum* L.) production, but values were low in areas under forests or permanent pasture.[54]

There have been several statewide surveys of NO_3 concentrations in rural wells in the U.S. A combination of N fertilizer and irrigation runoff was determined to be responsible for increase in groundwater NO_3 in Merrick County, Nebraska.[55] A later study, involving three counties, indicated N fertilizer to be the primary source of contamination in most wells sampled.[56] Schepers et al.[25] verified that excess N fertilization (exceeding 100 kg N ha^{-1} above the recommendation) came from 14% of the area planted with corn in Nebraska, resulting in increased NO_3 concentrations in groundwater. As a consequence of these findings, efforts to reduce NO_3 leaching in Nebraska involve some water management practices that are now mandatory. Hallberg[57] verified that NO_3 concentrations in Iowa groundwater had increased from the 1950s, and that by 1983 values that were below 3 mg L^{-1} had increased to 10.1 mg L^{-1}. The high NO_3 concentration was correlated with fertilizer use and an increase in area of corn production. A survey in Montana[58] revealed that 5% of 1300 private well samples had more than 10 mg L^{-1}. Equal contributions to the high NO_3 occurred from summer fallow and point sources.

There is increasing concern with NO_3 enrichment of the groundwater[51,59] and estuaries[60] in the Mid-Atlantic states. In southern Delaware, 8 to 32% of the wells sampled were determined to have greater than 10 mg NO_3-N L^{-1} and NO_3 contaminated areas were linked to intensive broiler chicken production.[61] High levels of NO_3-N, up to 50 mg L^{-1}, have been reported in several counties in Florida.[62] In 20 states in the U.S., 41 aquifers were shown to have areas, considered as "hot spots", with NO_3-N concentrations above 10 mg L^{-1}, and some of this contamination was linked to agricultural activities.[63]

The first comprehensive evaluation of the areal distribution of NO_3 in groundwater in the U.S. was accomplished by Madison and Brunett[64] who mapped close to 124,000 wells. For that survey, they used data stored in the U.S. Geological Survey (USGS) WATSTORE database and obtained information for identification of regions with NO_3 problems (Figure 4). It was determined that 6.4% of the wells presented NO_3-N values in excess of 10 mg L^{-1}. Values above 3 mg L^{-1}, which is the concentration indicative of some human activity, were found in 13.2% of the wells. By utilizing Madison's and Brunett's data,[64] Lee and Nielsen[65] delineated areas with potential for NO_3 contamination of groundwater. A 5-year program of systematic sampling of 564 community wells and 783 rural wells showed that only 1.2 and 2.4% of the respective wells contained concentrations of NO_3-N above 10 mg L^{-1}.[66]

Spalding and Exner[48] recently published a review of federal, state, and local surveys totaling more than 200,000 NO_3-N data points in U.S. groundwaters. Nitrate-N levels in water apparently decreased with depth and the shallow wells (<8 m) presented the

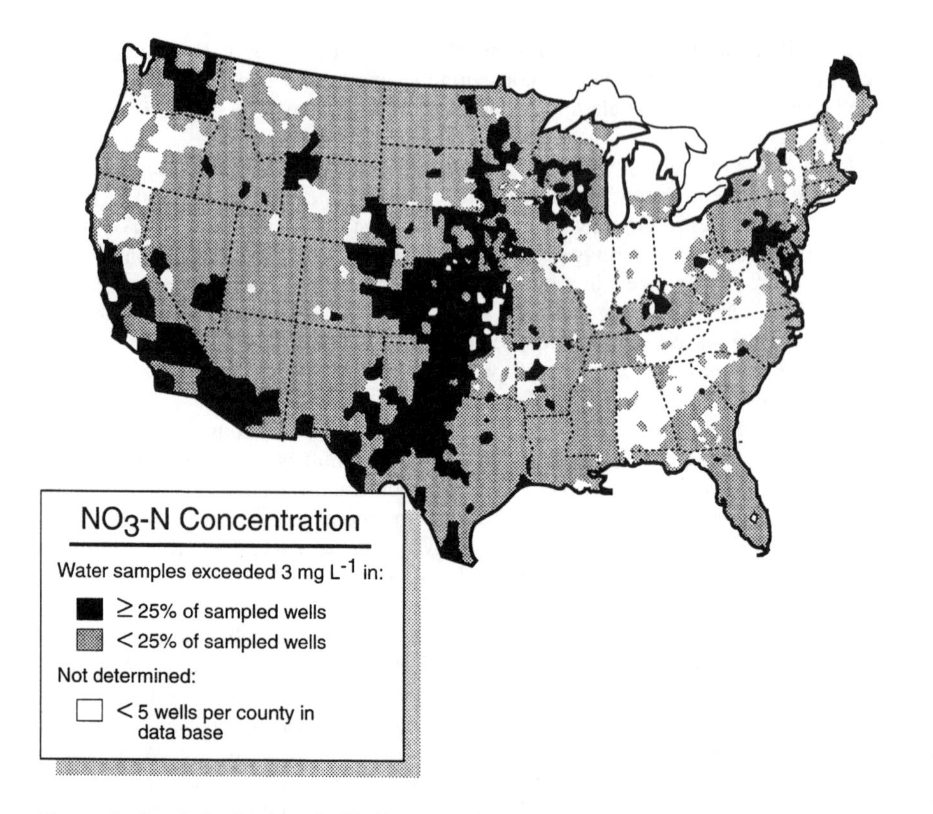

Figure 4 Areal distribution of NO_3-N concentrations in groundwater in the contiguous U.S. (From Madison, R.J. and Brunett, J.O., National, Water Summary 1984, U.S. Geol. Surv. Water Supply paper, 2275, 1985, 93.)

higher levels. Significantly higher NO_3 concentrations were also detected in water from older and poorly constructed wells. Regions under irrigated cropland and with well-drained soils, such as those west of the Missouri River, are more prone to have large areas that exceed the maximum level of NO_3 for drinking water (10 mg NO_3-N L^{-1}). According to the survey, groundwater in southeastern areas in the U.S. is apparently not contaminated. This lack of contamination was attributed to areas of warm C-rich environment and to natural remediation of NO_3 by plant uptake and denitrification. Much of the groundwater in the Corn Belt states has been protected from major NO_3 problems by extensive tile drainage. In poorly drained soils, NO_3 pollution of groundwater is not a serious problem, especially if seepage to deep aquifers is small. Denitrification in the subsoil is normally an intense process of N removal under these conditions.[67]

Although most of the surveys indicate that NO_3-contaminated areas are small in proportion to the total land use,[48] nevertheless protective and preventive measures can and must be taken to avoid more serious problems in the future. On farms, water sampling and analysis for NO_3 must be done on the household well, as well as all other water sources and supplies, including dug and drilled wells, cisterns, farm ponds, springs, streams, drainage ways, tile outlets, and pits, for an overall assessment of the NO_3 situation.[68] These samplings preferably should be done in the spring and in the autumn, before establishment of the new crop and after harvest.

Transport time for NO_3 in the soil zone to cross the unsaturated zone of the aquifer and then to reach the water table may be several years or decades, depending on the water table depth, the permeability, and fissuring of the unsaturated zone and the rate of recharge from the surface.[40] Aquifers, similarly to lakes and reservoirs, show small response in variability of inputs through mixing as compared with rivers and consequently NO_3 concentrations in groundwater sources generally show little seasonal variation.

Naturally occurring environmental processes, such as heating and cooling, exposure to sunlight, microbial transformations, and oxidation that chemically transform pollutants, aid in the remediation of surface water pollution. However, groundwater is not subject to these processes and thus recovery is expected to be slow, if it occurs at all.

2. *Surface Runoff*

Groundwater moves in the permeable geologic formations called aquifers, and from there can recharge deeper aquifers or discharge to streams, lakes, and other bodies of water[69] and may contribute to surface water contamination.[70] In humid environments, groundwater discharges to streams, lakes, and wetlands provide the generally observed perennial flow; in arid regions, surface water runoff may recharge groundwater. Excellent reviews on NO_3 movement and transformation in aquifers have been published[69,71] and the reader should refer to them for a better understanding of these systems.

Drainage water or runoff from agricultural lands may contribute significantly to NO_3 pollution of surface waters.[67,72] However, surface runoff generally contains only small quantities of soluble-N, unless high rates of surface applied N fertilizer are followed by heavy rains.[22] There is increasing evidence that agricultural activities are largely responsible for higher concentrations of NO_3 in shallow aquifers underlying major crop-producing areas in the U.S.[19,49] Analysis of surface water performed by the EPA[73] and summarized by Hallberg[70] indicated increasing levels of NO_3 for more than 50% of the Iowa sites evaluated between 1979 and 1984 (Table 4). Normally less NO_3 is lost to drainage water from poorly drained soils with high water tables than from well-drained soils.[74] Nitrate movement in soils is enhanced when the soils are drained, even in the absence of additional fertilizer N. In certain soils, such as histosols, N-mineralization rates may be higher than the crop uptake and, therefore, water draining from these soils may have a high NO_3 content, even from unfertilized areas.[75,76] Normally surface runoff from rains or snowmelts into streams is low in NO_3, although the concentration of organic N may be high. Nitrate moves into the soil mainly through infiltration. Therefore, a high NO_3 concentration, verified frequently in streams draining agricultural lands, originates mostly from groundwater.

Table 4 Trends in NH_4 and NO_3 ($+NO_2$) N Concentrations in Iowa Streams from 1979 to 1984

	Trend of Decreasing Concentration	No Trend	Trend of Increasing Concentration
		(%)	
NH_4-N	21	73	6
NO_3-N	0	46	54

From Hallberg, G.R., *Nitrogen Management and Groundwater Protection*, Follett, R.F., Ed., Elsevier, Amsterdam, 1989, 35. With permission.

The vegetation on site or nearby affects the quality of the surface water. For cultivated crops, losses may occur by both surface and subsurface drainage. As summarized by Gilliam et al.,[77] N losses via runoff are small, especially for inorganic N, even from fertilized fields, unless surface application is followed by an intense rainstorm. Whenever N fertilizer application greatly exceeds N absorption by the crop, large losses may occur through tile outlets and thus excess N may enter streams or simply seep into nearby lakes.[77]

About 20 to 50 times more NO_3 per unit area is lost from an agricultural watershed as compared with a forested watershed.[78] In an upland coastal watershed, 99% of the applied N was lost via leaching and subsurface flow.[79] Losses of N are very low, in the range of 1 to 2 kg N ha^{-1} for both fertilized and unfertilized forests.[77] Much of the N from forest fertilization enters the streams directly and appears to result from fallout during aerial application. The predominant N species in streams in agricultural areas is NO_3, whereas in natural or even in forested environments organic N predominates (Figure 5). Nitrogen losses from pasture crops to surface water are quite small and application of N fertilizers does not usually have a large influence on N movement from pastures.[77] Nevertheless, the intensification of grassland farms can lead to potentially large emissions of N to the environment[36] and leaching and surface runoff loss of N.[20,37] Although a major fraction of the N lost by surface runoff is organic N, which is associated with sediments,[80,81] data obtained from studies on NO_3 leaching from grasslands in the U.K. suggest that substantial loss of N as NO_3 can occur in leachate and surface runoff, particularly in swards grazed throughout the year.[82] Efficiency of N use for intensive grassland production may be improved through management practices that will be discussed in another section.

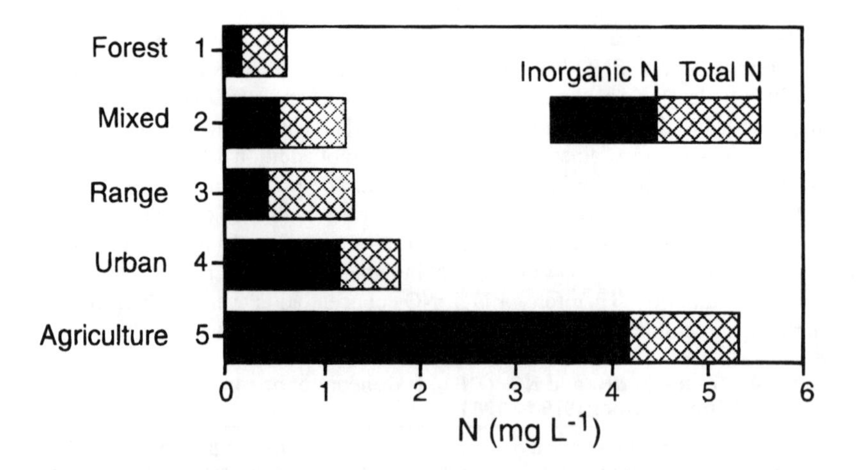

Figure 5 Land use and mean inorganic and total N concentrations from stream sample data from 904 nonpoint source-type watersheds distributed throughout the U.S. Land use categories: (1) ≥90% forest (n = 68); (2) mixed (n = 103); (3) ≥75% range (n = 17); (4) ≥40% urban (n = 11); (5) ≥90% agriculture (n = 74). (After Hallberg, G.R., *Nitrogen Management and Groundwater Protection*, Follett, R.F., Ed., Elsevier, Amsterdam, 1989, 35. With permission.)

Since the early 1970s, water management practices that enhance denitrification and reduce the amount of NO_3 entering surface and groundwater have been suggested.[83] A decrease of approximately 50% in NO_3 outflow to drainage waters was achieved in field studies in North Carolina, as a result of controlled drainage, which forced water to leave the fields by deep seepage through horizons where denitrification occurred.[384]

In many areas, such as in Kentucky, North Carolina, and Colorado, little change in NO_3 occurred since the mid-1930s or 1940s, regardless of the increase in fertilizer N applied.[85–87] However, increases in N content in streams have been observed in other areas.[88–91] Stream flow normally presents lower and more stable NO_3 concentrations, due to the mixing of groundwater discharge and surface runoff from various sources.[71]

The intermediate vadose zone (IVZ) is the subsurface material bounded by the root zone and water table.[71] The thickness and depth of the IVZ affects the rate of NO_3 transmission. In deeper IVZ, denitrification is insignificant; however, in a shallow vadose zone considerable denitrification can occur, provided O_2 depletion occurs. Denitrification and N uptake by deeply rooted vegetation are more likely to remove NO_3 entering shallow aquifer systems than the deep, unconfined or confined aquifers, which are less biologically active.[71]

Impacts of riparian ecosystems on streamflow patterns, quality, and quantity have been studied intensively in the past few years.[92–97] Forested riparian ecosystems effectively reduce nutrient and sediment levels in streamflow.[69] Nitrate-N removal occurs primarily via denitrification although uptake by vegetation is also significant.[71] Riparian forest communities have higher N contents when they are in association with agricultural watersheds.[98]

Considerable reduction in point-source loads, such as waste waters from sewage treatment or industrial plants, has occurred in the U.S. since the establishment of the Clean Water Act of 1972; however, non-point source pollution, due to agricultural practices, continues to degrade water quality in many areas.[70] Surface water contamination may be minimized by numerous management practices.[99] Physical alteration of the landscape, contouring, terraces, settling basins and impoundments, grass filter strips, and tile drains have been used successfully to control sediment discharge.[100]

Prevention of aquifer contamination is more effective and a less-costly strategy than is aquifer restoration. This is particularly the case for NO_3 contamination originating from agricultural activities since the areas affected may be very extensive.[69] Physical containment and *in situ* chemical/biological treatment are some of the methods proposed for aquifer restoration.[69] The high structural and energy costs inherent to physical containment limit its utilization. Biodegradation of contaminants may be accomplished by enhancing the native microbial communities through injection of nutrients and O_2 or inoculation with acclimated or even genetically engineered populations.[101] The same principle may be applied to remove NO_3 from waters.

3. Lakes, Reservoirs, Watersheds, and Wetlands

It is estimated that 5 billion tons of soil reaches U.S. waterways and surface water impoundments annually.[102] Erosion accounts for 99% of the total suspended solids in these waterways. In semi-arid regions, soil erosion is frequently severe due to sparse vegetation cover and high-intensity storms. A study conducted in the Loess Plateau in North China demonstrated that 98% of the N losses were associated with sediments and that most of the N removed from both cultivated and grazed areas was exported by erosion.[103]

Excessive increases in nutrient concentration in surface water normally lead to the proliferation of algae and aquatic plants. The ecological impacts of eutrophication can be dramatic since the decomposition of the excessive biomass leads to deoxygenation of water and the decline and disappearance of fish. Eutrophication is also a source of problems for drinking water, since filters get clogged rapidly, unpleasant tastes, odors, and colorations may result, and the risk of increased bacterial growth exists. As a consequence, eutrophied water is usually highly chlorinated during treatment, which is also not very desirable, due to the health implications of the Cl. The sources of N that lead to eutrophication are both point and diffuse. Nitrogen is distributed predominantly by water, but may occur from local air spreading of powdered fertilizers. Eutrophication of several important lakes in New Zealand[104] and other countries has been attributed to increased N inputs in agriculture.

Since algae require ten times as much N as P by weight and in most water bodies this ratio is exceeded, P is very often the nutrient that controls algal growth[40] as demonstrated in northern Europe and other parts of the world. With increasing NO_3 concentrations, this P limitation is expected to become more general. The limiting nutrient for plant and algal growth in temperate freshwaters is usually P, whereas N limits in tropical lakes and marine ecosystems. A P limitation results in decreased algal growth, whereas N limitation normally leads to a shift in the species composition, since some species are N_2-fixers and less dependent on N in water. For instance, NO_3 enrichment of Lough Neagh in Northern Ireland was accompanied by a shift from a population dominated by the N_2-fixing blue-green alga *Anabaena flos aquae*, which forms surface blooms, to dominance by *Oscillatoria agardii*, a non-N_2 fixer, that is more evenly distributed in the water column.[27] Since these two algae provide differing surface cover, NO_3 leaching may result in qualitative, as well as quantitative, effects on the trophic and biological status of surface waters.

Previously eutrophication prevention was directed toward reducing nutrient concentration, by controlling the inflow to affected water body. The new era of eutrophication prevention is also concerned with understanding the production processes in the water bodies, since the aquatic environment presents a high assimilative capacity for the critically limiting nutrients, rather than for the nonlimiting ones.[105] Thus, novel approaches for eutrophication control include the reduction of the nutrient concentration inside, as well as outside the water body in question.

Nutrients move from agricultural soils to water bodies by surface or subsurface routes. In upland coastal plain, watershed losses of NO_3 are mainly by subsurface flow and can account for 79% of the total runoff loss.[79] A study of N outputs from an agricultural watershed on the Georgia coastal plain revealed that N movement in subsurface flow was 97% NO_3 and about ten times that in surface runoff.[106] The author concluded that the low levels of N lost were unlikely to be a threat agronomically or as output to the environment. In a previous study[107] denitrification and vegetation uptake removed considerable amounts of NO_3 from groundwater in those riparian ecosystems. Riparian forests and riparian buffer strips located between croplands and streams are important in the maintenance of stream water quality on agricultural watersheds. They act as a sink for NO_3 from shallow groundwater.[92,96,108]

The Great Lakes contain approximately 20% of the world's fresh-water supply. A study conducted from 1975 to 1977 on 11 agricultural watersheds in the Canadian Great Lakes basin indicated that only 12% of the 3000 runoff samples analyzed had NO_3-N concentrations below 0.3 mg L^{-1}, while 3% exceeded the 10 mg L^{-1} Ontario drinking water standard.[109] Nitrate loads correlated with total N addition to the watershed and with the percent of watershed under cultivation.

Aquatic plants have the inherent ability to remove nutrients from waters and have received much attention for that reason.[110–116] The rational use of these plants for removal of pollutants in ponds, lakes, and estuaries is a promising possibility for alleviating water pollution problems.

Wetlands are some of the most productive ecosystems known.[117] Wetlands have the ability to improve water quality in a dual manner: by removing nutrients before they enter the lakes and streams through production of biomass and by providing area for sediment accumulation.[118] Wetland plants are able to absorb large quantities of nutrients in their biomass.[119–121] However, once these plants die and decompose, nutrients are liberated to the environment again. Thus, sediment deposition may provide a more permanent solution to the maintenance of groundwater quality than plant uptake. Some estimates of the magnitude of N removal by wetland soil deposition have been made. DeLaune et al.[122] have calculated a rate of 210 kg N ha^{-1} $year^{-1}$ streamside removal in a tidal marsh. Johnston et al.[118] determined that the annual accumulation of N in runoff water from a Wisconsin lake was 12.8 g N m^{-2} $year^{-1}$, and concluded that soil mechanisms were more important than vegetative uptake for long-term nutrient and sediment retention.

Forest ecosystems are essentially closed systems for N cycling since leaching and gaseous losses are minimal due to low inputs and mineralization.[22] There is a balance between nutrient inputs and uptake in forested areas which results in virtually no escape of NO_3 beyond the root zone. Results of a 20-year study of N utilization regimes in Douglas-fir (*Pseudostuga taxifolia* L.) forests of the Pacific Northwest indicated that 85% of the total N was in the soil, 10% in the vegetation, and 5% in the forest floor litter.[123] As in the Douglas fir ecosystem, recycling was responsible for maintaining a small labile pool of N within the biomass of a hardwood forest on a northeastern U.S. watershed.[124] In forested land in Europe, NO_3 leaching has also been reported as being low.[30] Disturbance of climax forests leads to larger losses of NO_3 to the groundwater.[125] However, losses are small compared to those originating from agricultural practices.

C. HEALTH EFFECTS OF NITRATES

Usually drinking water sources are not developed in areas where there is risk of pollution from domestic or industrial wastes. However, many regions due to their topography or population distribution rely on water drainage from areas where agriculture has intensified. In these regions, NO_3 concentrations in the drinking water have been continuously increasing over the past 20 years. Problems of high NO_3 concentrations in water sources appear to be magnified in areas of intense horticultural activity. Vegetables receiving high N-fertilizer applications can also accumulate NO_3 in their leaves and roots; thus, a dual source to dietary intake of NO_3 is provided in these areas. Contamination of natural waters by NO_3 in Europe has shown marked increase since the early 1960s with the areas affected the most being usually near the most intensively farmed lands.[40]

The World Health Organization (WHO) established the first European Standards for NO_3 content in drinking water in 1970, with a "recommended" limit of 50 mg NO_3 L^{-1} (11.3 mg NO_3-N L^{-1}) and a maximum "acceptable" limit of 100 mg NO_3 L^{-1} (22.5 mg

NO_3-N L^{-1}). Upon consideration of higher human fluid intake and bacteria concentration in some non-European tropical countries, in 1971 the WHO International Standards made 44.5 mg NO_3 L^{-1} or 10 mg NO_3-N L^{-1} the limit.[40] Due to the increasing intake from other dietary sources and potential risk of cancer, the European Community (EC) Directive on the Quality of Water for Human Consumption (1980, effective 1985) established 11.3 mg NO_3-N L^{-1} as the maximum acceptable concentration and proposed 5.6 mg NO_3-N L^{-1} as a guide level.[126,127] The difference between the WHO and EC standards is due to use of rounded numbers.[127] Croll and Hayes[52] reported that, as of 1972 in Great Britain, public water supplies should not be used if NO_3-N concentration was greater than 22.5 mg L^{-1} and that public health and medical authorities must be informed when the level is above 11.3 mg L^{-1}. The U.S. Public Health Service Standards specify that NO_3-N content of drinking water should not exceed 10 mg L^{-1}.[128,129] Presently, 10 mg NO_3-N L^{-1} is accepted as the limit for drinking water in most countries, including Canada.[41]

Due to the delay in response of groundwater to changes in the N cycle in the soil, some endangered aquifers may not yet show the expected increases in NO_3 concentrations. Once these aquifers reach the polluted status, they will remain that way for decades, even after NO_3 leaching from the surface has decreased considerably. In areas where aquifers feed rivers, the NO_3 load will also be higher and an "all-year" NO_3 problem will likely arise.

The health effects of nitrates have been discussed in a number of reviews.[35,49,126,130–135] High NO_3 concentrations in drinking water are considered a health risk because they may cause methemoglobinemia, or blue baby syndrome. This disease is associated with a reduction in oxygen-carrying capacity of the blood. Methemoglobinemia can also occur at zero NO_3 levels, provided coliforms are present in water in elevated numbers. Usually the symptoms of methemoglobinemia are easily recognized and can be treated.[136]

Nitrate itself is not considered harmful, since it is normally absorbed high in the gastrointestinal tract and excreted rapidly via the kidneys. However, upon reduction to NO_2

Table 5 Some Cases of Methemoglobinemia in Humans and Associated NO_3 Concentrations in Water

Country (Date)	Cases	NO_3 Concentration (mg L^{-1})
U.S. (1945)	2	388, 619
	Anecdotal	283–620
U.S. (1980)	1	1200
U.S. (1982)	1	545
U.S. (1987)	1	665
U.K. (1951)	2	200,[a] 95[b]
U.K. (1985)	14	>100
Hungary (1985)	95	40–100
	1258	>100
Denmark (1985)	1	200[b]

Note: All of these cases were associated with water from wells. The wells were dug privately in 98% of the cases.

[a] Known fatal case.

[b] Water known to have been bacterially polluted.

From Addiscott, T.M., Whitmore, A.P., and Powlson, D.S., Farming, Fertilizers and the Nitrate Problem, C.A.B. International Wallingford, Oxon, 1991. With permission.

by bacteria present in the upper gastrointestinal tract and absorption into the bloodstream, hemoglobin is converted to methemoglobin, which is incapable of transporting O_2. The condition is reversible in normal adults but death may occur if more than 40% of the hemoglobin is converted.[127] Young infants in particular are more susceptible to methemoglobinemia due to low gastric acidity which allows NO_3-reducing bacteria to proliferate in their intestine, due to their relatively high fluid intake in proportion to body weight, and easier conversion of fetal hemoglobin to methemoglobin than adult hemoglobin. The number of documented cases of methemoglobinemia worldwide is small and some of the reported cases are presented (Table 5). Apparently, the most recently reported case of death resulting from methemoglobinemia was that of an infant girl in South Dakota.[137]

Nitrites can react with amines and amides to form nitrosamines and nitrosamides, and experiments have proved these compounds to be carcinogenic in animals.[40] The possibility of carcinogenic nitrosamine formation from ingested amines and nitrite is considered another health hazard from exposure to high levels of NO_3.[127] Epidemiological studies have indicated a positive correlation between NO_3 consumption in drinking water and the incidence of gastric cancer,[138] although only limited evidence linking stomach cancer with NO_3 in water has been found by several authors.[8,127] There exists some indication of other health disorders, such as non-Hodgkin's lymphoma,[139] increased infant mortality,[140] central nervous system birth defects,[141] and even hypertension.[142]

Nitrate poisoning of animals was recorded in livestock farming, where lethal consequences were verified in cattle ingesting drought-affected forage.[39] Nitrate accumulation in animals from water sources is not as frequent, but has been reported in Saskatchewan, under conditions where well water contained over 600 mg L^{-1} NO_3-N.[143] Various clinical effects have been reported in cattle and other animals, as related to high NO_3 intake through feeds, forages, and/or water. The disorders include deficient growth/health of young animals, increased abortion rates, and depressed yields and milk quality.[40] Since N fertilization of grassland and forage is expected to increase in European countries and due to higher NO_3 verified in the water, NO_3 intake by farm animals is likely to increase.[40] In Canada loss of cattle has been caused by ingestion of oat (*Avena sativa*) hay or oat straw containing high NO_3 levels.[41] Plants can accumulate excessive amounts of NO_3 especially by conditions of drought and by most conditions where photosynthesis is depressed but NO_3 uptake still continues.[39]

D. ATMOSPHERE

Gaseous losses of N from soils occur primarily as nitrous oxide (N_2O) and N_2 gases formed during denitrification (reduction) and nitrification (oxidation). Volatilization of free NH_3 and chemical decomposition of NO_2 to form N_2, NO plus NO_2, and smaller quantities of N_2O also contribute to gaseous losses of N to the atmosphere.

Ammonia gas constitutes a direct health hazard and can damage human tissue and cause injuries to eyes, ears, and the respiratory tract.[144] Ammonia may be present in the atmosphere in particularly high concentrations in areas such as agricultural sites, feedlots, and close to fertilizer manufacturing plants. Ammonia is lost by volatilization, and losses in the magnitude of 10 to 15 kg N ha^{-1} $year^{-1}$ may occur, particularly if animal manures are applied to the surface.[40] Substantial amounts of NH_4-N can be transferred from these localized sites to nearby areas and contribute to the N content of soil and water.[22] A portion of the NH_3 in the atmosphere originates from the soil, due to transformations that take place in the soil environment. Anhydrous NH_3, a popular fertilizer because of its 82% N content, is usually injected in bands at a depth of 10 to 15 cm. Some NH_3 gas escapes to the atmosphere through the injection slits and soil cracks. This emission continues for some time after injection but at a decreasing rate.[145] Photorespiration in C_3 plants also release NH_3 to the atmosphere and up to 10% of the plant total N may be emitted in this manner. The soil is capable of adsorbing NH_3 gas from the atmosphere, independently of that added in rainfall. However, precipitation from areas of high NH_3 concentrations, such as cattle feedlots, may contribute significantly to N enrichment of water bodies.[146]

There is increasing concern about deleterious effects on the environment from release of soil N_2O to the atmosphere. As N_2O moves into the stratosphere, it participates in a series of catalytic reactions that result in the destruction of the ozone (O_3) layer that shields the earth's surface from harmful ultraviolet radiation.[147] An excellent review of the stratospheric reactions of N_2O was written by Crutzen.[148]

Accurate determination of N_2O evolution from agricultural lands is difficult,[149] since evolution from soils is variable, both spatially and temporally. Studies conducted in Colorado indicated that losses due to N_2O emissions from an irrigated corn field were small and correlated strongly with soil water but not with NO_3 content.[150] Similar results were reported for barley and the values were lower than the ones projected by models developed to describe effects of agricultural fertilizers on stratospheric O_3.[151] Addition of manure in a reduced tillage system resulted in increased N_2O fluxes in Wisconsin.[152] There are indications that N_2O emissions from anhydrous NH_3 application to soils greatly exceed those induced by other N fertilizers.[153] This aspect of atmosphere pollution has not been extensively investigated, however.

Eichner[154] summarized data from 104 field experiments on N_2O emissions from fertilized soils worldwide. She concluded that a correlation exists between emissions and quantity of fertilizer applied and that recent global estimates for the year 2000 are fairly accurate. Hence, if 100 Tg N fertilizer are consumed annually worldwide, global release of fertilizer emissions will probably not exceed 3 Tg N_2O-N. Agricultural practices and policies as well as trends among types of fertilizers used and natural processes will affect the real emission rates. More systematic research is necessary to determine contributions of N fertilizers to atmospheric N_2O emissions.

The process of denitrification also affects atmospheric composition through the production and consumption of N_2O and thus it has an impact on climate. Denitrification intermediates, such as NO and NO_2, are toxic and can lead to carcinogenic nitrosamines, also produced during the process.[155] There is considerable variation as to the extent to which N fertilizer is lost from the soil to the atmosphere by denitrification. This variation has been reported as 0 to 70% of the N applied[156] and depends on the concentration of NO_3, soil porosity, water tension, pH, temperature, and content of mineralizable OM. Addition of leguminous green manure or residues to soils low in available C may enhance denitrifying activity. Denitrification in soils under continuous grassland is about five times greater than that in soils cultivated annually for long-term cropping. Lower denitrification rates in some soils under grassland have been attributed mainly to their greater porosity and capacity to quickly return to a relatively high moisture tension in the event of rain.[157]

V. REDUCING ADVERSE EFFECTS OF NITROGEN FERTILIZERS

A. ENHANCED AWARENESS OF QUALITY OF THE ENVIRONMENT

Nitrogen fertilizers are applied to correct deficiencies, increase the nutrition levels, maintain soil fertility conditions, and improve crop yield and quality. Adequate information, training, and motivation of farmers should be promoted by various state and federal government agencies for best management of soil and fertilizer N. The training should address both agricultural development and pollution prevention and should take into account optimum yield and also soil, water, and health protection criteria. Public concern about environmental quality pressures the agricultural sector to take preventive measures.

In the agricultural sector, short-, medium-, and long-term water management measures, in addition to the necessary policies for source pollution control, should aim at guaranteeing a continuous supply of low-NO_3 water. If NO_3 pollution from agriculture is reduced, drinking water quality will be improved and other problems, such as eutrophication, will be alleviated. In regions where NO_3 is a serious problem, rational allocation and distribution of available low NO_3 waters can frequently reduce the problem to more manageable proportions. The highest quality waters should be allocated primarily as drinking water supplies whereas industry, irrigation, and municipal cleaning could use multipurpose disinfected water, not necessarily low in NO_3 concentration.[40] Appropriate authorities worldwide should also update regulations and standards to maintain allowable NO_3 levels as low as possible and should promote management practices that reduce N fertilizer losses, to limit overall NO_3 intake through water and food. The cost of synthetic fertilizers has been maintained at a relatively low level in a number of the Organization for Economic Cooperation and Development (OECD) countries, through total or partial tax reliefs.[40] The elimination, even if gradual, of these tax exemptions would foster fertilizer conservation and stimulate a more rational NO_3 pollution policy in those countries.

B. BEST MANAGEMENT PRACTICES

In 1977, the U.S. Congress recognized "Best Management Practices" (BMPs) as the standard for controlling non-point pollution sources (NPSs).[72] In 1987, Section 319 was added to the Clean Water Act, which included a requirement for the states to submit an assessment report to the Environmental Protection Agency (EPA), describing processes for identification of BMPs for non-point source problems.[72] Many NO_3 increases in groundwater are a direct result of low efficiency of N fertilizer. The main goal in a BMP approach is the utilization of fertilizer N in the most effective manner. This approach should be employed by every grower for every crop on every field. Problems with N control include (1) maintenance of an adequate supply in the soil; (2) regulation of N turnover, to assure availability for plants; and (3) practices that minimize losses such as leaching and volatilization.

Plant growth is often severely restricted by insufficient N supply, partly due to the high cost of increasing soil N levels and also from the difficulty in predicting adequate N fertilizer for the plant. Upon determining the amount of fertilizer to apply, farmers must consider what was left from previous applications and the contributions from decomposing plants in the field. When excessive amounts of N are applied, as fertilizer or

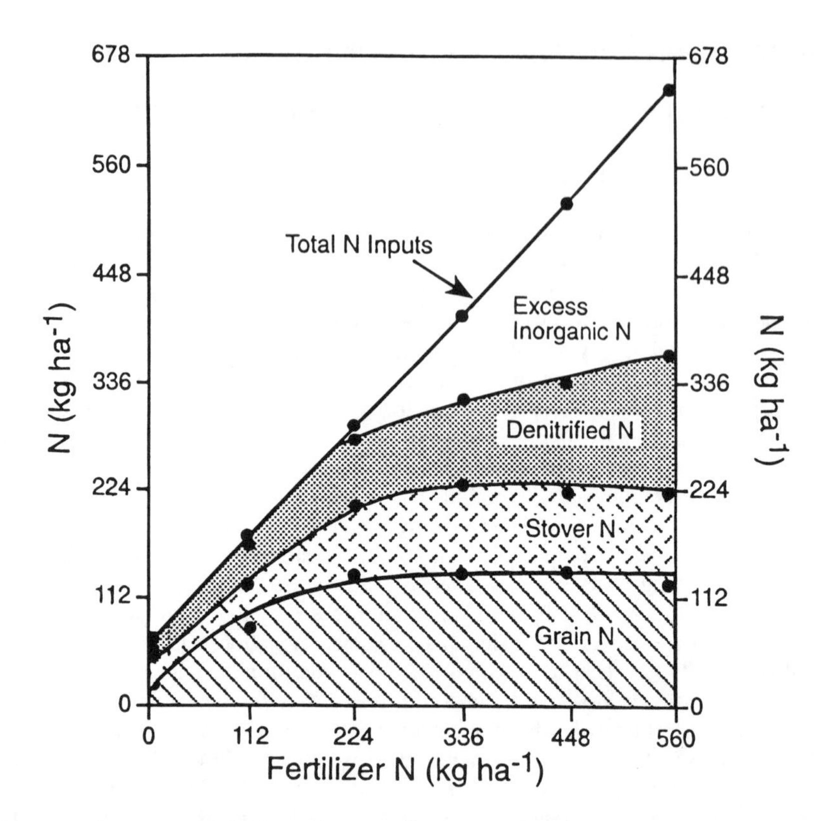

Figure 6 Compartmentalization of N in a corn plant-soil system as influenced by N fertilizer rates. (From Boswell, F.C., Meisinger, J.J., and Case, N.L., *Fertilizer Technology and Use*, Engelstad, O.P., Ed., SSSA, Madison, WI, 1985, 229. With permission.)

as organic manures, yields will not be increased (Figure 6) and NO_3 concentration in drainage water may be elevated to levels that cause environmental degradation.

Adoption of proper farm management practices such as (1) fertilization (rate, type, form, and time of application); (2) irrigation and drainage; (3) erosion control and tillage; (4) cover crops; (5) plant species; (6) crop geometry; and (7) cropping strategies by farmers will minimize the possibility of NO_3 leaching into the groundwater, both in humid and drier regions.[68,158]

1. Fertilization

Rates, types, and placement of a N fertilizer are selected as a function of the crop, soil characteristics, and climate. The desired crop yield, fertilizer cost, and impact of the fertilizer upon the environment must also be considered. Nitrogen recommendations based on soil-plant mass balances may aid in lowering fertilizer use and environmental consequences of excess fertilizer without affecting economic returns.[159] The quantity of NO_3 derived from fertilizer that leaches below the root zone is subject to some degree

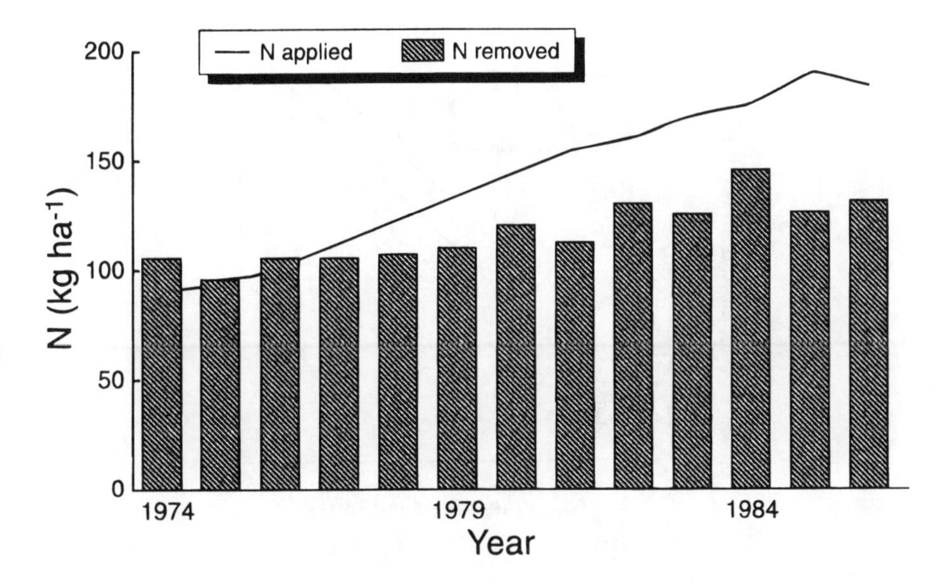

Figure 7 Amounts of N fertilizer applied to winter wheat in the U.K. and amount of N removed in grain during the period of 1974 to 1986. (From Addiscott, T.M., Whitmore, A.P., and Powlson, D.S., *Farming, Fertilizers and the Nitrate Problem*, C.A.B. International Wallingford, Oxon, 1991. With permission.)

of control by fertilizer management practices such as rate and timing of N fertilizer application.[13]

Rate of Fertilizer — Fertilizer recommendations have been established for each crop for a range of typical soils and climates. If applied rates are at or slightly below crop assimilative capacity, N will be absorbed readily and the risk of loss will be reduced (Figure 6). In general, crops can assimilate between 30 and 70% of the fertilizer N applied.[13] Researchers in Europe find that for many arable crops nearly 50% of the applied N is recovered in the harvested crop.[160] Grasses recover more mineral N than field crops.[22] For corn grain, evaluations of long-term experiments have indicated that N fertilizer removal is usually less than 40% at economically optimum yields.[161,162] For grasslands, N recovery rate may be over 80% and thus less is available for leaching. In the U.K., NO_3 concentrations in the soil profile reflect these differences in leaching losses.[40] Crop N recovery decreases as rates of applied N fertilizer increase (Figure 7). Consequently NO_3 leaching losses become more significant.[163]

Contributions from nonfertilizer N inputs, such as those from residual soil N, from OM mineralization, that contained in irrigation water, from legume crops grown in rotation with cereal crops, and from animal manures should be accounted for when making N fertilizer recommendations.[25,38] It has long been recognized by agronomists that a great contribution toward improving N-use efficiency could be obtained from a soil-N test. However, a practical test to assess soil supplying factors was established only in 1984.[164] The test, termed *presidedress soil NO_3 test* (PSNT),[165] measures soil NO_3 at a

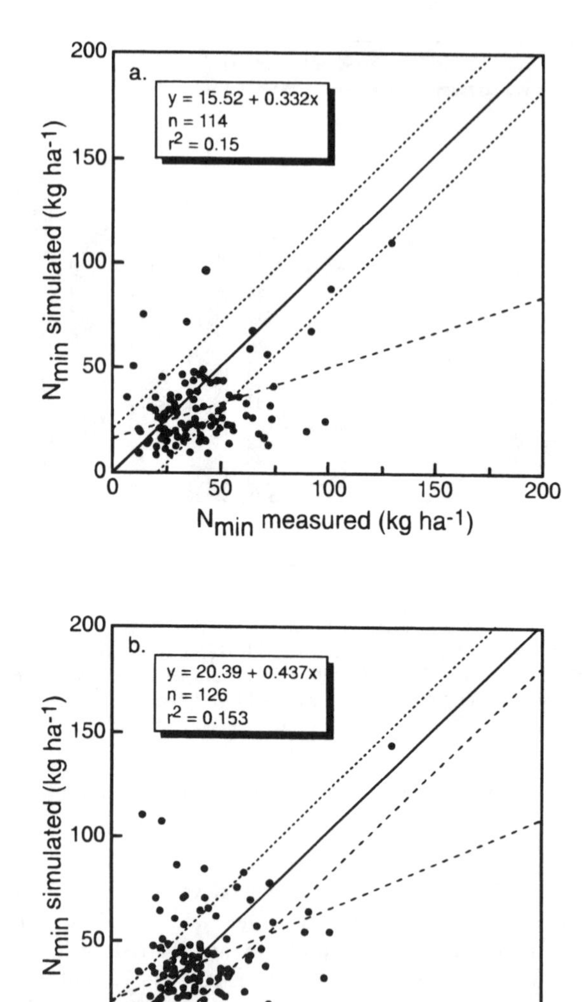

Figure 8 (a) Comparison between measured and simulated mineral N in the soil profile (0 to 90 cm) under cereal crops, in the spring of 1988. Test runs with the model of Whitmore. Dotted lines represent a 20 kg N ha^{-1} deviation from the solid 1:1 line; the regression function is depicted in a broken line; (b) Comparison between measured and simulated mineral N in the soil profile (0 to 90 cm) under cereal crops, in the spring of 1988. Test runs with the model of Kersebaum. Dotted lines represent a 20 kg N ha^{-1} deviation from the solid 1:1 line; the regression function is depicted in a broken line; (c) Comparison between measured and simulated mineral N in the soil profile (0 to 90 cm) under cereal crops, in the spring of 1988. Test runs with the model of Groot. Dotted lines represent a 20 kg N ha^{-1} deviation from the solid 1:1 line; the regression function is depicted in a broken line; (d) Comparison

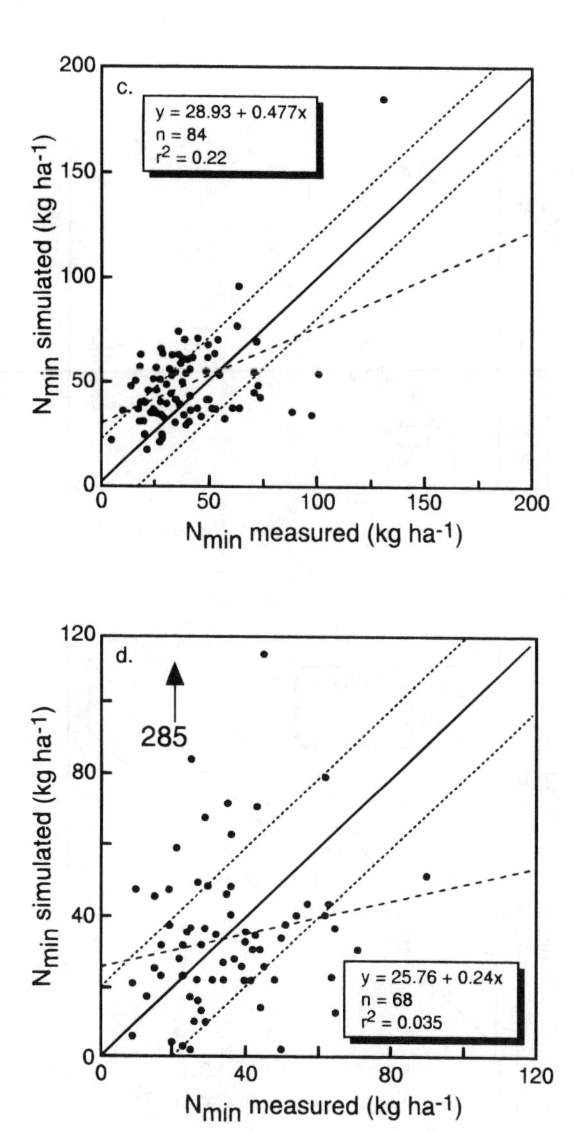

Figure 8 (*Continued*)

between measured and simulated mineral N in the soil profile (0 to 90 cm) under cereal crops, in the spring of 1989. Test runs with the model of Whitmore. Dotted lines represent a 20 kg N ha^{-1} deviation from the solid 1:1 line; the regression function is depicted in a broken line; (e) Comparison between measured and simulated mineral N in the soil profile (0 to 90 cm) under cereal crops, in the spring of 1989. Test runs with the model of Kersebaum. Dotted lines represent a 20 kg N ha^{-1} deviation from the solid 1:1 line; the regression function is depicted in a broken line; (f) Comparison between measured and simulated mineral N in the soil profile (0 to 90 cm) under cereal crops, in the spring of 1989. Test runs with the model of Groot.

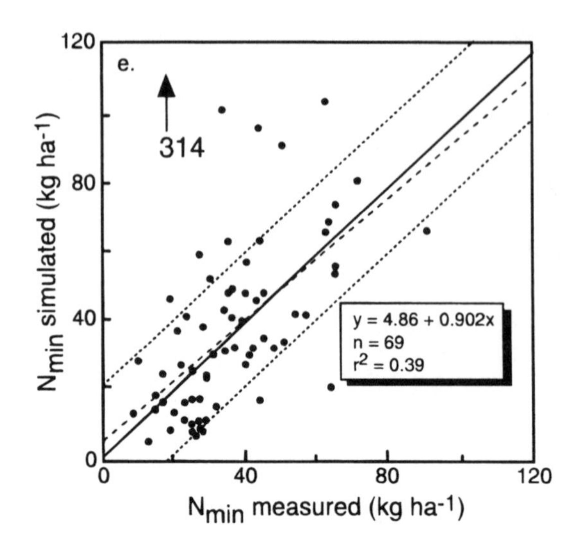

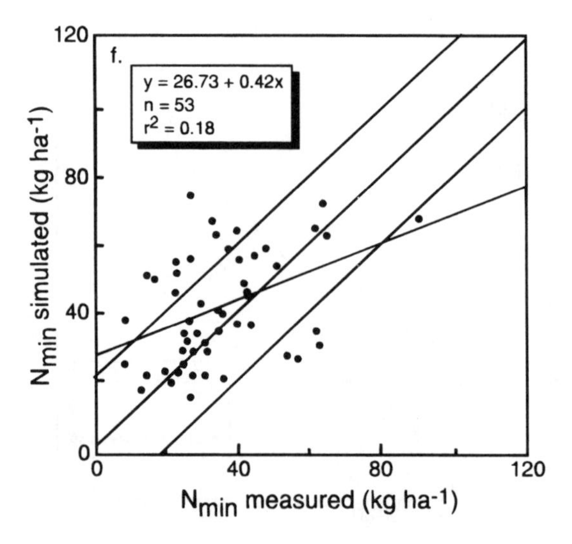

Figure 8 (*Continued*)

Dotted lines represent a 20 kg N ha^{-1} deviation from the solid 1:1 line; the regression function is depicted in a broken line. (From Otter-Nacke, S. and Kulmann, H., *Fertilizer Res.*, 27, 341, 1991. Reprinted by permission of Kluwer Academic Publishers.)

crucial time, when NO_3 concentrations are likely to be high, but before significant N uptake has begun.[166] The PSNT aims to help farmers conserve fertilizer N and reduce NO_3 losses to the environment. The test can successfully identify N-sufficient sites across a range of textures, drainage classes, and years, thus, minimizing the application of "insurance N".[164] The PSNT provided a satisfactory indication of corn response to sidedress

Table 6 Nitrogen Leaching Following Use of 200 kg N ha^{-1} Slow-Release Fertilizers (Lysimeter Trial over 10 years for Sugar Beets, Potatoes, and Carrots on Sandy Soil)

	N Leached (kg N ha^{-1})		
Type of Fertilizer	**Summer Half-Year**	**Winter Half-Year**	**Total**
No N applied	8.7	19.4	28.1
NH_4NO_3	40.4	56.3	96.7
Gold-N	19.5	44.7	64.2
Isobutylidene diurea	14.5	32.8	47.3
Ureaform	16.3	30.7	47.0

After Dressel and Jung,[175] as presented by Juergens-Gschwind.[30]

N in Pennsylvannia,[167] Iowa,[168] and, more recently, in Maryland.[164] The underlying principles of the PSNT reveal that it is primarily an *in situ* N-mineralization test, well suited for warm-season crops and for fine- to medium-textured soils.[169]

Simple models have been developed to predict the amount of N for various crops by using easily available soil and meteorological data. The performance of three models was compared[170] with data from 2 years and several locations (Figure 8). It was concluded that the results were too variable for any of the models to be used with confidence. However, once proper adjustments are made, models can be valuable and aid in making accurate predictions.

A qualitative N-screening model, the *long-term potentially leachable nitrogen* (LPLN), was proposed with the intent to estimate a field-scale N budget.[171] Based on the estimated excess N, a given field can be classified into a potential N-leaching risk group. It is suggested that such a classification would lead to a field-sampling program to verify the existence of a potential environmental N problem. Recently Meisinger et al.[169] suggested the necessity of a team approach to evaluate NO_3 behavior and to develop management practices such as the use of presidedress tests to decrease N fertilizer losses.

Type of Fertilizer — As mentioned previously, in most soils NH_4 fertilizers are less subject to leaching losses due to their positive charge. A comparative study of NO_3 and NH_4 fertilizers in the shallow-stony and free-deep ferrallitic soils of Mauritius demonstrated the movement of N in the form of NO_3 from those soils into the groundwater.[172] This movement was more pronounced when the fertilizer was in the NO_3 rather than the NH_4 form. Loss of N by leaching increased with intensity of rainfall but decreased with an increase in clay content and cation exchange capacity of the soil. The mobility of the applied N governed N-use efficiency by sugarcane. Although NO_3 was more efficient in low-rainfall areas, NH_4 was best in superhumid areas.

Nitrification inhibitors, such as dicyandiamide (DCD) and nitrapyrin, reduce leaching and denitrification losses of fertilizer N from the root zone. The use of nitrification inhibitors requires the synchronization of soil N availability and crop N uptake to prevent N deficiencies.[173] Blending of urea with neem (*Azadirachta indica*) cake, which has nitrification inhibition properties, can prevent some NO_3 loss in fields planted with wheat.[174]

Slow-release N fertilizers also generally result in less leaching than mineral fertilizers (Table 6).[30,175] As with nitrification inhibitors, there may be either slight deficit or excess in N during the growth period, if release rates do not correspond with crop

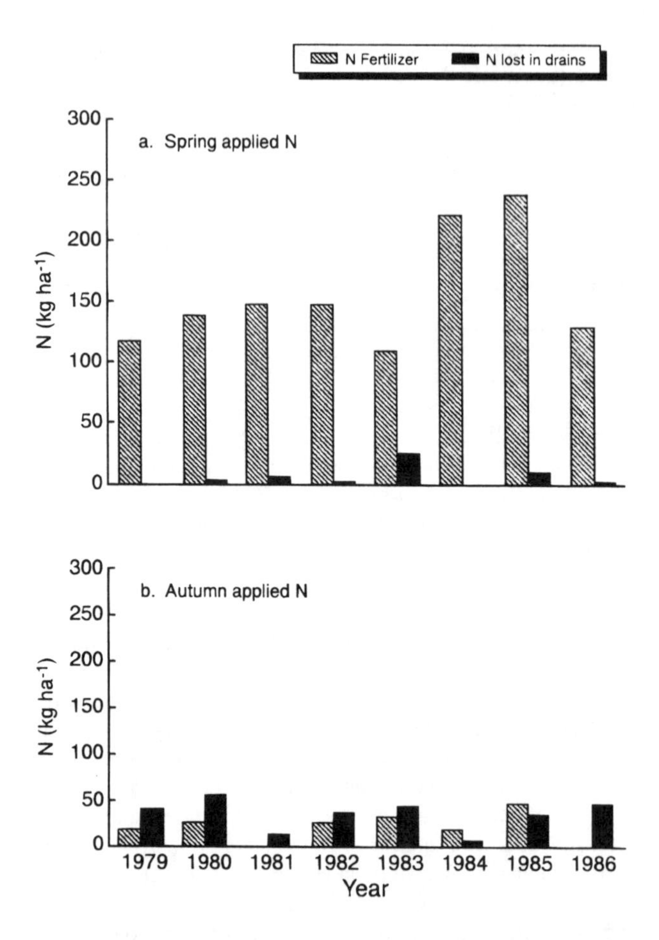

Figure 9 Losses of N in water draining from (a) fertilizer applied in spring and (b) applied in autumn. (From Addiscott, T.M., Whitmore, A.P., and Powlson, D.S., *Farming, Fertilizers and the Nitrate Problem,* C.A.B. International Wallingford, Oxon, 1991. With permission.)

requirements. Sulfur-coated urea fertilizers have been shown to increase yields of various crops, mainly by maintaining N in the soil longer and preventing leaching.[176] Recent studies in northeast Florida indicated that nitrification inhibitors were not suitable for potato production on irrigated sandy soils.[116] On the other hand, the use of IBDU, a slow-release N fertilizer, resulted in greater tuber yields than did the nitrification inhibitors tested.

Form of Application — Nitrogen losses with surface runoff increase when N fertilizers are surface applied, especially if heavy rains occur right after application. These losses are diminished by incorporation of the fertilizer into the soil.[68] The most efficient form of application of liquid or gaseous forms of NH_3 is usually by injection. This method minimizes leaching by increasing NO_3 availability in the root zone and promoting early plant absorption.

There is great variation in the movement of soluble N, both laterally and vertically in the soil. When NO_3 is present or applied below the water level in the furrows, movement is lateral and downward.[22] If, however, the NO_3 is above the water level, it will concentrate in the ridges and at the surface, and crop N recovery may be undesirably low.

Chemigation or fertigation consists of application of fertilizer in irrigation water and may be a more efficient mode of application, since the N fertilizer may be applied when needed by the crop.[68]

Time of Application — Application of N fertilizer should be done at the time when both water and N uptake by the crop are high.[177] Losses of N in drainage water are usually small for spring-applied N but these losses can be considerable for autumn application (Figure 9).

Split applications are desirable in all soils, particularly on well-drained, coarse-textured soils that are more susceptible to NO_3 leaching. For crops with relatively short growing periods, such as cereals, two applications may be sufficient to limit leaching.[68,177] Split applications of the recommended rate in order to satisfy immediate crop needs and reduce surplus in the early stages of growth is an important BMP for N fertilization.

2. Drainage and Irrigation

Water management practices are designed to increase infiltration and storage of precipitation and snowmelt, while reducing runoff and soil water evaporation. Drainage and irrigation are the major forms of soil water management,[67] and these practices vary with crop, soil, location, and climate. Drainage of moist soils improves aeration and crop growth by lowering the water table. However, in some instances, the result may be increased leaching since NO_3 generally moves with the drainage water.

In nonirrigated conditions, water availability is the most important factor for crop production.[178] Efficient water and N use is the first step in development of sustainable production since water is the vehicle for NO_3 movement down the soil profile.[178] A combination of water and N management is more effective on reducing NO_3 accumulation than either approach alone.[67]

Table 7 Correlations between NO_3 Content of Nebraska Groundwater and Several Factors

Independent Variable	r Value
1. Overlying soil clay content	–0.49
2. Irrigation well density	0.43
3. Total fertilizer use	0.28
4. Irrigation well depth	–0.28
5. Water pH	–0.23
6. Cattle density	0.18
7. Human density	0.06

Note: Individual well water NO_3 level related to site characteristics 1, 4, and 5 above and to average county-wide statistics for characteristics 2, 3, 6, and 7. Water sampled from 480 wells, 1971–1972.

From Follett, R.F. and Walker, D.J., *Nitrogen Management and Ground Water Protection*, Follett, R.F., Ed., Elsevier, Amsterdam, 1989, 1. With permission.

Maintenance of or an increase in the OM content in the form of animal or green manures and crop residues is a practice used to reduce leaching. Another practice that discourages leaching is to prevent overirrigation, especially after application of the fertilizer, since movement of water and NO_3 beyond the root system is likely to occur in that situation. Storage of high NO_3 water before it reaches water sources may be performed, provided that appropriate drainage systems exist. In that case, water can be used for irrigation at a later time or can be allowed to denitrify before discharge.

In the U.S., most of the irrigated crops are located in the western arid, semiarid, and subhumid regions[72] and occupy 15% of the crop land.[63] In Europe, the amount of irrigated land and the rate of irrigation have increased considerably in the last few years.[40] The soils in most arid and semiarid regions are frequently the most susceptible to leaching. Furthermore, in arid regions, the rooting zone must be periodically leached or flushed to prevent salinity problems. Moderate and efficient irrigation in the case of drought or arid regions will normally decrease chances of NO_3 leaching by increasing plant growth and NO_3 uptake. Irrigation water should be supplied when needed by the crop and in amounts just sufficient for storage and use within the crop's rooting zone.[68] Marked increases in NO_3 movement from urea occurred under overirrigated wheat.[179] Excessive irrigation of many crops invariably causes NO_3-leaching.[40,180]

Nitrate content in waters correlates positively with irrigation and with fertilizer use and negatively with clay content and irrigation well depth (Table 7). Research performed at the Bet Dagan Experimental Station, Israel, investigated the effects of different irrigation and Ammonium sulfate $[(NH_4)_2SO_4]$-fertilizer regimes on yield of Rhodes grass (*Chloris gayana*) and movement of NO_3 residues through the soil. The results indicated that in finer-textured soils leaching did not occur during the irrigation season, while on sandy soils some leaching was found even when no deliberate water excess was applied.[53] Under Mediterranean climatic conditions (characterized by a definite rainy season during the winter when the crop is dormant), amounts of excess fertilizer that leached from the root zone annually depended primarily on fertilizer practices, since winter rains of approximately 50 cm leached out all residues left in the upper 200 cm of soil, regardless of their vertical distribution. Forcing the crop to scavenge during the last portion of the growth cycle by decreasing the last fertilizer application by half minimized NO_3 excess and maintained satisfactory yields.

Table 8 Lysimeter NO_3-N Concentrations in Lysimeter Drainage Water as Influenced by Tillage and N Source

	Average NO_3-N Concentration (mg L^{-1})			
	Conventional Tillage		No-Tillage	
Time Period	Fertilizer N	Legume N	Fertilizer N	Legume N
2/83–5/83[a]	0.26	1.95	0.50	2.04
2/84–5/84[a]	0.04	1.02	0.25	0.87
11/83–1/84	4.75	7.70	3.30	3.02
2/83–5/84	2.33	3.76	1.67	1.96

[a] Significant effect of N sources at $p = 0.05$.

After Groffman et al.,[189] reprinted by permission of Kluwer Academic Publishers; Russelle and Hargrove.[158]

Some crops, especially shallow-rooted species such as potatoes, are difficult to manage for fertilizer and water use.[20] Intensive potato production has resulted in increases in NO_3 concentration in shallow groundwater.[19] Controlled irrigation and N fertilization, based on the plant demand and crop rotation with shallow- followed by deep-rooted plants, reduce the potential for groundwater pollution from N fertilizers.[181] Leaching of NO_3 is also reduced when the soil is kept under continual cropping.

Another approach to reduce leaching losses is to create a soil environment more conducive to storage of autumn and spring precipitation. For instance, in irrigated areas, a deficit may be imposed at the end of the growing season and autumn cover crops that grow rapidly and survive low winter temperatures and thus readily utilize spring precipitation may also be planted.[173]

Under irrigated crops, NO_3 leaching to groundwater can be reduced by proper irrigation design and schedule, applying just enough fertilizer to meet yield goals, timing the application and use of nitrification inhibitors and slow-release fertilizers.[67,182]

3. *Erosion Control and Tillage*

The use of sound erosion control practices such as no-tillage planting, conservation (minimum tillage), contouring or terracing, contour strip cropping, grassed waterways, and outlets and sod-based crop rotations are important aspects of conserving N, thus reducing water pollution from agricultural inputs.[100]

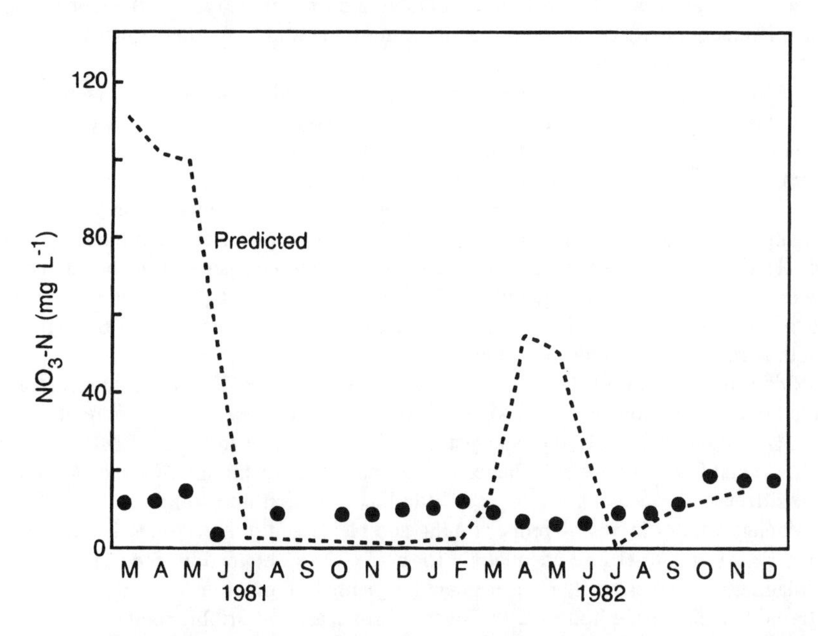

Figure 10 Nitrate in groundwater under a Georgia cornfield (circles) and predicted NO_3 in groundwater (dashed line). (From Thomas, G.W., Smith, M.S., and Phillips, R.E., *Nitrogen Management and Ground Water Protection*, Follett, R.F., Ed., Elsevier, Amsterdam, 1989, 247. With permission.)

The adoption of practices that slow water runoff in soils increases infiltration rates. Although runoff is reduced, there is a greater chance for water to move through the root zone, since infiltration is maximized.[177] Residue or crop canopy cover are involved in maintenance of water infiltration rate in a manner that surface runoff, from irrigation or rainfall, is lessened.[177] The amount of effective coverage depends on the slope and, for the steeper slopes, coverages of as much as 50% may be necessary. Contouring is effective in controlling erosion on slopes of less than 5%.

The effects of conservation or reduced-tillage on NO_3 leaching are not as clear as those for surface runoff. Smith et al.[100] concluded that both no-tillage and conventional tillage have been shown to cause increases and decreases in downward NO_3 movement. Several reports show evidence of reduced N runoff loss under no-tillage practices.[183–185] However, since infiltration rates are higher under these systems, the potential exists for increased NO_3 leaching out of the root zone into the groundwater.[186,187]

Tillage has a profound effect on soil N dynamics and soil water relations and promotes OM mineralization. Therefore, by increasing ammonification, with subsequent nitrification, tillage can affect the potential for groundwater contamination by NO_3. These losses sometimes occur even in the absence of fertilizer N application.[188] Uptake of N by summer crops is greater under conventional tillage, whereas for winter crops, more N is taken up under no tillage systems (Table 8).[189]

Cultivation systems, e.g., conventional and minimum tillage, which change the pore size distribution will alter the pattern of leaching. A small increase in the number of large channels will reduce the modal depth of NO_3 leaching. There is a tendency for water to flow down preferred paths, and these paths predominate in soils that are less altered by tillage. Data obtained under a Georgia corn field (Figure 10) suggest a sporadic movement of NO_3 from soil to groundwater rather than a gradual downward movement.[190]

The potential for denitrification, particularly in humid climates, is enhanced by no-tillage due to increased soil water content and decreased air-filled pore spaces, and thus decreased O_2 diffusion.[191] Denitrifying conditions reduce NO_3 leaching; however, in cultivated soils this would be a very inefficient way of controlling N leaching.[190]

Angle et al.[186] reported that the concentration of NO_3 in groundwater under no-tillage cultivation was equal or less than the NO_3 concentration under conventional tillage for corn. Results from a subsequent study in which infiltration capacity of the no-tillage was greater than conventional tillage indicated that the use of no-tillage cultivation decreased soil-NO_3 concentrations.[192] They suggested that no-tillage cultivation is a BMP relative to both surface and groundwater quality.

Ridge-tillage and no-tillage corn production systems have lower NO_3 concentration in tile drainage and lower total N losses.[187] Since NO_3-N concentrations were above the 10 mg L^{-1} limit in their study, they suggested implementation of additional BMPs. Potential for NH_3 volatilization may be enhanced under no-tillage or minimum tillage systems if the fertilizer is broadcast, since there is little incorporation into soil.[22]

Management of losses by proper timing and placement of N fertilizer is required for no-tillage systems to ensure adequate N for crops. Data obtained by Stecker et al.[193] for no-tillage corn suggested that knife injection of urea ammonium nitrate (UAN) is the preferred placement method over the surface broadcast and dribble methods to prevent N loss, regardless of application time.

Maintenance of crop residues on the soil surface may influence available N levels in percolate and/or surface water runoff.[194,195] Surface placement of crop residue was

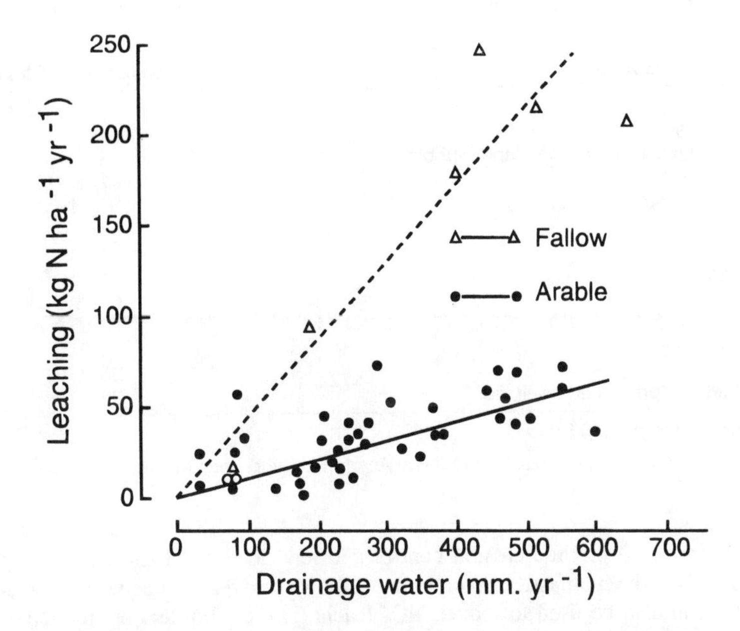

Figure 11 Leaching of N under fallow and arable land as a function of amount of drainage water in a Maschhaupt lysimeter (after Kolenbrander,[197] as presented by Juergens-Gschwind[30]).

shown to reduce residue N availability, compared with soil incorporation.[196] However, since differences were small, the authors concluded that crop residue placement may have only a minor impact on N availability in terms of environment.

4. Cover Crops

Temporary immobilization by the microbes occurs when crop residues are plowed into the soil. Mineralization occurs subsequently. Bare soils release more NO_3 than cropped soils (Figure 11).[197] If the field is barren through the autumn and winter, the NO_3 released is liable to be leached before the spring crop usage. Leguminous cover crops increase both organic and inorganic soil N[198] and supply a substantial part of the N needed for optimum yields of nonleguminous summer crops. They also conserve water, by improving soil structure and increasing water infiltration, and decrease runoff and evaporation losses, if left as a surface mulch. Groffman et al.[189] verified that concentrations of NO_3 under clover were higher than under rye (*Secale cereale* L.) in the spring (Table 8). Nevertheless, the potential for NO_3 leaching in a legume-based production system appears to be small compared with a fertilizer-based system. Research conducted by Muller[199] indicated that only 2% of the N input was leached, even though there was a release of 65% of the total N from the clover tissue (Table 9).

Fallowing during the growing season may increase the potential for NO_3 leaching. Sandy soils should not be summer fallowed, regardless of tillage method, if annual precipitation exceeds 25 cm year^{-1}; otherwise, severe NO_3 leaching may occur.[177] Cover

Table 9 Balance for Clover-Derived ^{15}N 1 Year after Burial of the Clover Tissue in Soil

Source	Recovery (% of Input)
Vegetation	
Barley shoots	4
Clover residues + barley roots and stubble	3
Total	7 ± 1[a]
Soil (inorganic + organic fractions)	
0–27 cm	67
27–46 cm	1
46–70 cm	<1
Total	68 ± 7
Leaching	2 ± 1
Total recovery	75 ± 6
Total loss	25 ± 6
Unaccounted for (presumed denitrified)	23

[a] Confidence limits $p < 0.05$ (n = 5).

From Muller[199]; reprinted by permission of Kluwer Academic Publishers; as presented by Russelle and Hargrove.[158]

crops on fallow land provide erosion control and tend to retard NO_3 leaching by absorption of N into the biomass. Strip cropping (fallow or cover crop between crop under production) can also be used to reduce NO_3 leaching. Field borders or riparian zones and grassed waterways filter sediments from runoff and therefore reduce sediment load in streams and rivers, acting as scrubbers of the subsurface flow. Deep-rooted trees and grasses are also able to intercept the NO_3 in laterally moving water.[173]

Cover crops may be regarded both as sinks for residual N and as tools to reduce surface runoff.[198] Thus, cover crops preserve soil N for use by subsequent crops and, with proper management, improve water quality.[200]

5. Plant Species

Factors such as plant variety or hybrid, plant parts removed at harvest, and root morphological and physiological characteristics need to be considered when managing N to prevent losses to the environment. With knowledge of the root characteristics of the particular plant, the most effective placement of the fertilizer can be determined. For example, the root systems of plants such as corn are capable of exploiting the soil more thoroughly than those of potatoes. Plants with more restricted root systems have a greater dependence on fertilizer. As larger amounts of fertilizer are applied to supply needs, risks of losses increase. Movement of NO_3 below the root zone of shallow-rooted crops such as potatoes has been verified.[201] Shallower-rooted plants and those that produce fewer fine roots are also more dependent on microbial associations such as mycorrhizae, which are known to increase uptake of nutrients, especially N and P. High levels of available inorganic N inhibit mycorrhizal development[202,203] and colonization.[204] The beneficial effects of these symbiotic structures may be reduced considerably under conditions of excess N fertilizers. This symbiosis has been largely overlooked as a N management tool.

Deep-rooted crops have greater capability to intercept and use water and residual NO_3 in soil.[8,177] Thus, deep-rooted crops such as alfalfa (*Medicago sativa* L.) may be used to remove NO_3 accumulated after several years of residual N in soils under abandoned feedlots.[173] When roots have access to the entire soil solution, NO_3 is not leached unless N fertilization and irrigation are excessive.[180] Selection of superior cultivars and modification of plant N metabolic processes through genetic manipulation to improve N efficiency use are among areas requiring additional attention for future research.[100]

6. Crop Geometry

Row crops occupy a major share of the land area in the U.S. Of the approximately 760 million hectares in the continental U.S., 24% is defined as cropland, mostly cropped with corn, wheat, cotton, and soybean.[20] A large area of land receives application of fertilizers and is bare for a portion of the season. These areas can constitute more of an erosion hazard and inefficient N use for part of the year.

Management to minimize NO_3 leaching losses for row crops is based on avoiding excess NO_3 in the root zone, especially when the soil is subject to irrigation or excess rainfall.[136] This management involves genetic selection to improve N-use efficiency, well-controlled irrigation, multiple fertilizer applications, use of cover crops or deep-rooted crops, nitrification inhibitors, and slow-release fertilizers.

Row spacing varies depending on plant characteristics and the required machinery. Manipulation of inter- and intrarow spacing appears to be the easiest component of crop geometry to alter.[205] Normally, N extraction efficiency of broadcast fertilizer is greater with narrower row spacing, which results in increased radiation interception and increased yields, especially if adequate soil water is available. Nitrogen uptake, soil N utilization, and water depletion appear to be faster under narrow-row and higher-density plantings. Erosion control is also promoted by denser plant canopies due to enhanced infiltration rates. A combination of improved planting geometries with precise fertilizer placement increases fertilizer recovery and reduces potential for groundwater contamination.[205]

7. Cropping Strategy

Soil N is not evenly distributed in time and space. Mineralized N is usually more readily available in the spring. This N may be efficiently utilized by a sorghum crop which converts it into protein, whereas if the crop is winter wheat that N may be lost since the crop's uptake period would precede rapid mineralization.[206] The rooting habit also influences the N-use efficiency in a multiple-cropping system. Nitrogen movement below shallow roots of lettuce (*Lactuca sativa* L.), for example, leads to a low capture efficiency. However, if subsequently it is captured by a deeper-rooted crop such as corn, the efficiency of that multiple-crop sequence may be high[206] and NO_3 leaching losses would be reduced. Depending on the region's climate, decomposition of plant residues may occur all year, and multiple cropping can allow for capture and conversion of the N that could be lost during the off-season.[206] Inclusion of legumes in multiple-cropping systems increases N comparable to application of fertilizer N. In Kentucky, a hairy vetch (*Vicia villosa* Roth) winter cover crop supplied over 90 kg N ha^{-1} to a succeeding corn crop.[207]

Intercropping systems that include a legume crop transfer some of the fixed N to the associated nonlegume crop.[208,209] Use and capture of this N fixed by the associated crop can increase total N utilization efficiency and lower the fertilizer input.

Rotations such as wheat-fallow promote NO_3 accumulation during the fallow period and increase N for the following wheat crop.[188] The potential negative impacts of this NO_3 accumulation are increased leaching losses, groundwater contamination, and reduced symbiotic N_2 fixation by following legume crops. Crop rotation also presents the potential to reduce chemical fertilizer inputs in agricultural production. Rotations generally increase soil organic N or decrease the rate of loss of soil organic N. Some or, in some instances, all of the N requirement of the succeeding crop can be provided through crop rotation.[210] The type of residue from the preceding crop affects the N mineralization. Leguminous crops residues release N more rapidly than grasses such as sorghum or wheat; however, N mineralization of wheat residue is initially faster than that of sorghum residue. In either case, addition of N to counteract immobilization would be beneficial in terms of release of residue N for the succeeding crop.[210] Water-use efficiency by plants and efficient water use through management are potentially increased in rotation, leading to greater and more efficient N uptake and reduced leaching losses.[173,210]

Alfalfa[211] and soybean (*Glycine max* L. Merr.)[212] can utilize much of the inorganic N remaining after removal of a crop. These scavenger species have the potential to alleviate NO_3 leaching hazards.

8. Nitrate Leaching under Agricultural Crops

A compilation of selected studies from various countries is presented in this section without intention of exhaustion of all existing literature on the subject. Surprisingly, very little information is available on developing countries, where increases in N fertilizer use have been in the order of 160% in the last 10 years.

Soil texture and OM content can have a major influence on leaching of NO_3. In Sweden, Bergstrom and Johanson[213] verified that leaching was greater in a sandy soil with low OM content, intermediate losses occurred in loamy soils, and the smallest losses were observed in a clay soil and in another sandy soil rich in OM. As mentioned previously, it has been indicated that sandy soils should not be summer fallowed, when annual precipitation exceeds 25 cm $year^{-1}$.[177]

Under Swedish conditions, leaching of NO_3 from arable lands occurs mainly in the autumn, which is characterized by high precipitation and low evapotranspiration.[214] During that time, levels of inorganic N are often high and the soils are commonly bare; thus, the supply of soluble N may be immobilized if a cover crop is planted.

Results from a field study in which labeled ^{15}N- $(NH_4)_2SO_4$ was applied to a bean crop at the rate of 100 kg N ha^{-1} on a typical dark red latosol (alfisol) indicated that NO_3 leaching below the depth of 120 cm was equivalent to approximately 16 kg ha^{-1}, of which less than 10% was derived from the labeled fertilizer.[215] Recovery of the labeled fertilizer was low (30%) by a first harvest and even less by a second crop, suggesting a high immobilization of the soil-bean (*Phaseolus vulgaris* L.) plant-residue system. Nearly all the N_2 fixed in legume-rhizobia systems is utilized by the plant, with little excretion by nodules and roots into the soil. Only moderate quantities of NO_3 leach from the root zone of N_2-fixing soybeans.[216]

In Yugoslavia, however, results from a field experiment involving application of ^{15}N-labeled NH_4NO_3 to corn, also at the rate of 100 kg N ha^{-1}, indicated downward movement of the labeled fertilizer below the 100-cm depth.[217] The leaching process was apparently retarded by application of OM in the form of ground straw and oats. Still in a different field study with corn in Yugoslavia,[218] evidence was found that, after harvest, mineral N moved to a depth of 2 m in a chernozem, and that only 10% was derived from labeled fertilizer. Addition of OM also tended to reduce leaching of NO_3.

In India, rising NO_3 concentrations in certain ground and surface waters have been associated with the intensive use of fertilizer.[219] The fate of ^{15}N-labeled fertilizer, applied to corn only, in a multiple-cropping system for two crop rotations of corn-wheat-moong was studied. The authors concluded that leaching losses of fertilizer N were heavy under monsoon conditions, and they suggested split applications of fertilizer for more efficient crop utilization, as well as use of slow-release N fertilizers. The use of deep-rooted fodder crop intercropped with a grain crop, such as wheat, were also suggested as scavengers of NO_3 in order to reduce the quantities that might contribute to groundwater pollution as a result of continuous leaching.

Rice is the major cereal crop in India and N is the most important single nutrient determining yield in that country.[220] The fate of an initial pulse of ^{15}N-urea (100 kg N ha^{-1}) was followed under flooded rice conditions over a sequence of three crops and two intercrop fallows. The soil was a clay vertisol. Ammonia volatilization and leaching losses amounted to 9.7 and 7.5%, respectively. Major losses of fertilizer N occurred during the crop season, immediately following its application. At the end of the experiment, 26.5% of the N fertilizer was recovered in the root zone, and 0.9% was clay-fixed, nonexchangeable NH_4-N. The maximum contribution of the ^{15}N pulse to the NO_3 content of the groundwater, approximately 2%, occurred in the first crop season and had declined below 0.2% in the third crop season. The maximum concentration of NO_3-N in the groundwater was 3.2%, well within the recommended 10 mg L^{-1}.

A field investigation of water and NO_3 movement in a Yolo soil from Davis, California, revealed that, when evapotranspiration exceeded irrigation, movement of soil NO_3 was upward, whether derived from native soil mineralization or added fertilizer.[221] When irrigation was applied, annual cumulative seepage of NO_3 at 300-cm soil depth varied from 30 kg N ha^{-1} $year^{-1}$ with no fertilizer addition to 140 kg, depending on the rate of irrigation and fertilizer application. However, even though labeled N fertilizer seepage was detected at a 300-cm depth, N derived from fertilizer in the seeping NO_3 was always a relatively small fraction. Results from ^{15}N-labeled N fertilizer applied to corn over a 3-year period in Romania showed that application of N fertilizer during the vegetation period contributed to NO_3 available in the top soil for the next crop and reduced NO_3 leaching potential.[181] Most of the ^{15}N-NO_3 movement occurred in the 0- to 160-cm profile after the first year of application in both irrigated and nonirrigated plots. Excessive precipitation, especially during the first part of the vegetation period, favored NO_3 percolation, when N fertilizer was applied at planting.

In a comparative leaching study of Cl and NO_3 in a well-drained uncovered sandy soil, Cameron et al.[222] determined that both anions showed similar distribution patterns and that Cl/NO_3 ratios indicated that leaching rather than denitrification was responsible for NO_3 losses from 0- to 75-cm depth.

The major inputs of N into soils and surface water of England and Wales are N_2 fixation and rainfall, followed by animal and human wastes, and then by inorganic fertilizers.[91] Most leaching occurs in the winter when crop growth and transpiration are very low. During this period, any NO_3 not taken up by the crop fertilizer applications or from mineralization of soil OM is likely to be leached, as will be NO_3 mineralized in the postharvest period.

Wild and Cameron[91] suggested that the importance of minimizing losses of N from agricultural land should be seen primarily as a step toward more productive farming rather than an answer to environmental pollution. They concluded that, with the exception of young infants, there seems to be a negligible risk of any damage to health if concentrations in drinking water rise to 20 mg NO_3-N L^{-1}. A plea is made by the same authors for a balance to be achieved between the relatively small health hazard (in economic terms) and the substantially higher cost of keeping NO_3 concentration low.

C. NITRATE REMOVAL FROM CONTAMINATED WATERS

The conventional treatment for drinking water removes practically no NO_3. The cost of treatment will vary with the amount of NO_3 that must be removed. Wild[223] pointed out that threshold limits set for safe concentrations of NO_3 in drinking water are critical, as an unnecessarily high limit would increase the cost of water purification. The most feasible processes for NO_3 removal from contaminated waters are ion exchange and biological denitrification. Ion exchange is applied especially for borehole sites, and biological denitrification, for treatment of concentrated NO_3 sources and rivers.[40,91]

Ion exchange is an expensive process,[224] especially because it involves the basic costs of the exchange resins and those involved in disposal of spent regenerant. Ion exchange can be technically used to remove NO_3 by using a strong base exchange resin. Chloride ions are exchanged for NO_3 and sulfate in the treated water. This type of treatment has been prohibited in France for drinking water since organic compounds may leach from the resin.[40] Reverse osmosis has been the subject of experimentation, but it is also a very expensive process.

Reduction of NO_3 by biological denitrification is performed by naturally occurring bacteria. It is a rather inexpensive process, but may be inadequate at borehole sites because it requires expert attention and is a slow process.[223] For large bodies of water denitrification by storage is an option for NO_3 removal, although significant effects are not obtained on a short-term basis.

Under anaerobic conditions, NO_3 can be denitrified in a relatively short time, especially if enough energy is available for the denitrifying bacteria in the form of OM. Losses are generally negligible at moisture levels below two thirds of the water-holding capacity of the soil and, above this value, they correlate directly with moisture.[15] Denitrification rates depend strongly on NO_3 concentration, C supply, content of ferrous material in the environment, moisture level, and temperature. Denitrification in poorly drained soils is normally reduced by soil drainage.[225]

The process of denitrification has been researched in the past mostly in an effort to understand and minimize loss of N, especially in flooded soil conditions. The role of denitrification has shifted from a negative process to a positive process with the potential to prevent contamination of aquifers by NO_3.[41] Reduction of nitrates in groundwater in North Carolina,[226] in Ontario,[227] in clay-confined aquifers,[228] and vadose zone in the Chalk aquifer[229] have been attributed to denitrification. Denitrification in riparian zones has been reported to be responsible for removal of up to 98% of the NO_3 from groundwater.[230] They further suggest that, by controlling *in situ* denitrification rates, it becomes possible to enhance NO_3 removal by proper management of riparian zones.

Both processes, ion exchange and denitrification, may give rise to secondary contamination of drinking water. Methanol residues, bacterial proliferation, organic metabolites, and organochlorinated compounds have been associated with denitrification, while release of undesirable organic compounds may occur with the use of resins in ion exchange.

Costs are high either to remove NO_3 from water or to prevent contamination, but both aspects should be considered for establishment of a rational long-term policy.[40] A favorable option is the development of new good-quality sources such as removal of water from deeper aquifers or the construction of reservoirs. Policies of "strategic water reserves" for the future could be systematically applied, such as Italy's "Aqueduct Plan".[40]

Long-term planning and preventive measures are essential for any water supply policy, especially in the case of NO_3 contamination of aquifers where the lag time may be in the order of decades for both verification of contamination and restoration. Short-term solutions, although attractive and politically easier to implement, may eventually increase environmental impacts.[40] Organizational problems with the interdisciplinary character of agricultural-environmental problems are well recognized.[231] The major obstacle to combining knowledge from various disciplines is the lack of an integrated database that can support both high-quality disciplinary research and informed interdisciplinary policy evaluation.[232]

D. SITE-SPECIFIC EVALUATION FOR POTENTIAL NITRATE LEACHING

Several models for evaluating potential NO_3 leaching of a site have been developed. The DRASTIC Index was developed by the EPA[233] to measure NO_3 leaching risk. It evaluates the vulnerability of aquifers to contamination by taking into account and assigning values to each of seven factors: depth to water, net recharge, aquifer media, soil media, topography, impact on the unsaturated zone, and conductivity of the aquifer. These factors, collectively, determine site potential for NO_3 leaching. For instance, some aquifers are more susceptible to contamination, depending on the hydrogeological condition, the flow system, and the climate. Shallow, permeable, unconfined aquifers, with high recharge rates, are generally more susceptible to contamination.[234] Lack of correlation between the DRASTIC model and data obtained from field studies has been reported by some authors.[100]

An indication of the relative susceptibility of a site to contamination can sometimes be obtained through the use of detailed, accurate, and local hydrologic data.[224] The quantity of water percolating through the rooting zone is also important for determining NO_3 leaching.[235] The soil, crop, and climatic factors that collectively affect water percolation are well discussed by these authors. The erosion-productivity impact calculator (EPIC) model,[236] used to characterize average percolation and the variation in percolation, has been a very useful tool for evaluation of leaching potential at various agricultural regions in the U.S. The model addresses hydrology, erosion, temperature, weather data, plant growth, nutrient status, and tillage. Additional testing for N is necessary, although field tests for NO_3 leaching have been conducted.[235]

Another method, the N-leaching index (LI),[235] simpler than EPIC, has been used to identify problem areas. Estimates for annual percolation obtained with this index agree closely with those obtained by use of EPIC.

E. MODELING THE FATE OF NITROGEN FERTILIZERS

The agricultural community currently shows substantial interest in the use of models to guide the application of chemicals and water to soils and crops and to predict the fate of these in the environment. A variety of approaches have been developed to describe the movement of water and solutes in field soils. Several new models have been proposed and they vary widely in their conceptual approach, complexity, and resolution. These models are also strongly influenced by the environment and many have been produced as the result of research into the basic physics and chemistry of N transport and transformation in agricultural soils. Models, however, have not been extensively used for N management purposes.

Computer models of fertilizer use are designed in three categories: screening models, for preliminary assessment; research models, which provide quantitative estimates of water and solute movement and the fate of nutrients; and planning or management models, used in planning, regulation, and management. Computer models are especially helpful in estimating use and fate of fertilizers, due to the necessity to integrate a number of variables involved in the particular ecosystem. For instance, effectiveness of a fertilizer application depends on rainfall, soil conditions, chemical form of the fertilizer, etc.

Some models are very specific and concentrate on a particular process, whereas others try to cover a wide field. Models differ considerably in their underlying concepts, due primarily to differences in emphasis on the role of water regime and soil OM. Some models have been developed to provide a better understanding of the processes whereas others may concentrate on prediction of available mineral N or on overall management. Management models are more generalized and less quantitative in their ability to predict water and solute fate under changing field conditions. However, this type of approach may be more easily adapted for varying soil and environmental conditions. The most widely used modeling approach in soil science over the last 30 years has been the use of deterministic, mechanistic models of solute movement which are based on the miscible displacement theory.[221] Several models for the behavior of N in soil and uptake by plants will be presented and the reader may refer to Frissel and van Veen,[237] Follett,[238,239] and Smith et al.[100] for more details on some of the models.

The research applied to national needs (UCD-RANN) model[240,241] simulates corn field, irrigated and fertilized, one-dimensional transient flow and Mediterranean climate, simulating different irrigation schedules and leaching patterns in the root zone. It comprises a water flow submodel and a N flow submodel, involves physical and microbiological processes, and also considers N uptake by plants.

A model for water and NO_3 movement through soil was developed by Burns.[242] Rainfall is assumed to increase the water content of any soil layer until field capacity is attained. Nitrate leaching is correlated with the water percolated and the amount of NO_3 leached below the root zone is calculated, under the assumption that the soil solution of each layer is in equilibrium with the water percolated. The model does not consider the N transformation processes and the main goal is calculating NO_3 leaching under rainfall and irrigation conditions for a wide range of fallow or cropped soils. Frissel and van Veen[237] suggest that the model may be a useful submodel for a more complex model.

Unsaturated and saturated flows, water table fluctuations, tile flow of water, N transformation and movement processes, and crop growth are considered in the simulation model presented by Duffy et al.,[243] where the main inputs are N fertilizers and rainfall. This model is applicable to any typical field in the Corn Belt and allows for prediction of NO_3 concentrations in the tile effluent as a function of farm management and climatic conditions.

A digital model was developed to calculate the occurrence, movement, and dissipation of NO_3 within the soil profile of fertilized agricultural lands by Saxton et al.[244] It considers the vertical water and NO_3 ion movement in the soil, fertilizer, and rainfall additions and plant uptake of N, nitrification, and mineralization. The model targets at obtaining a better understanding of how to maintain optimum NO_3 in the soil for crop use and to minimize potential environment pollution and can be applied to agricultural watersheds with permeable soils and high-yielding corn.

A steady-state model for optimization of fertilizer application with respect to economic returns and permissible N releases was devised by Singh et al.[245] In the model, grain yield and N uptake for wheat and corn are described as a quadratic response function of the N fertilizer application. Another steady-state model presented by Tanji et al.[246] aimed to predict N leaching losses on croplands under intensive production, located in humid regions or in areas where irrigation has been practiced for a long period. It consists of a hydrologic submodel and a N submodel and, upon input of data such as N in irrigation water, N in precipitation, applied fertilizer, N_2 fixation, irrigation water-flux, and precipitation, allows for estimation of N emissions from cropped lands.

A one-dimensional multicompartment model, NFLUX, was designed to be used in the description of transient upward or downward fluxes of N within the soil profile.[247] The model was structured to transport and transform urea, NH_4 and NO_3. No mineralization of OM is modeled but irrigation is included.

Nitrogen dynamics in soil (NDS) is a model that calculates N leaching below depths of 2 m from soil surface.[248] It simulates the behavior of N under different possible fertilizer treatments and makes it possible to estimate N fertilizer requirements for various soil and climatic conditions. Climatic soil physical and biochemical data and many empirical correlation factors are required.

van Veen and Frissel[249] developed a model, named M3, primarily for intensive agricultural systems of northwestern Europe characterized by a temperate climate and high N fertilizer use, aiming to optimize fertilizer use. Emphasis was placed on the role of microbes and the model describes the impact of the soil C cycle on N transformations. Leaching can be calculated and a program version is available which enables the user to follow the fate of ^{15}N within soil OM. The main driving variables in the model are manure, fresh OM, and N fertilizer application; however, several other data are required, some of which are difficult to measure.

One of the most extended models is PHOENIX, which includes microbiological, plant physiological, and physical processes.[250] The objective of this model is the exploration, through simulation modeling, of the relationships among plant processes and microbial processes and their effect on N cycling, plant production, and microbial secondary production. Emphasis is placed on mineral N, soil OM, and microbes.

A model with a dual objective (applied and scientific)[251] was developed which is applicable to a field of annual pasture or a small-grain crop growing in semiarid environment. The production of arid pastures limited by rainfall and nitrogen (PAPRAN) models one growing season with a time of resolution of one day. Leaching of N can be calculated since climatic, soil physical, and biochemical data are included.

Addiscott[252] reported SL3 and SLD3 as models and programs for modeling of NO_3 leaching on plot or field scale in structured soils. The objective is simulation of downward movement or loss of NO_3 and other nonadsorbed solutes to provide predictions of solute concentrations in the profile and in drainage. Leaching is calculated from the vertical water flux and allowance is made for mobile and retained waterphases.

Specific processes or transformations are addressed in some models. For instance, dynamics of NO_3 concentrations in soil with prediction of N transformations and movement from manure application are considered in the model presented by Walter et al.[253] This model can be applied to lands with heavy applications of livestock waste in early spring. Ammonification, oxidation of NH_4, in soil and chemical decomposition of NO_2 are the main processes studied in the model by Laudelot et al.[254] in which the main goal is prediction of N loss from soil as a result of chemical decomposition of NO_2.

Smith[255] presented a model with the coined name AMAZON (a model of the extent of anaerobic zones in soil and the effects on N loss by denitrification). Parameters describing the soil aggregates, physical constants to describe O_2 diffusion, oxygen-consumption rate of soil, concentrations and coefficients for NO_3, and the gaseous products of denitrification are the input data.

Selim and Iskandar[256] described the model WASTEN to describe N behavior in soils irrigated with liquid waste as affected by physical and microbiological processes, environmental conditions taking also into account water and N uptake by plants. The model may be used to study the fate of fertilizer N applied to agricultural areas.

The development of NITROSIM was aimed at simulating the transport and plant uptake of N and water and transformations of N and C in the root zone for agricultural fields receiving animal manure, plant residue applications, and fertilizer N.[257] The main inputs for this model are climatic, soil physical, and biochemical data in the soil.

An NH_3 volatilization model, NLOS, which considers physical processes, such as gas flux, ammonium adsorption and exchange, ammonium volatilization, and hydrolysis of urea, was developed by Parton et al.,[258] with the objective of quantifying losses of volatile NH_3, especially from urine patches, as a function of environmental conditions such as soil temperature, texture, pH, evaporation rate, and moisture.

Use of tracer and computer simulation techniques to assess mineralization and immobilization of soil N (TRAMIN)[259] was developed under laboratory conditions and can be used to obtain a better understanding of the N transformations in soil. Soil organic C is divided into seven pools, each having its own availability as substrate for microbes. A program version that describes the fate of ^{15}N and ^{14}C besides ^{14}N and ^{12}C within the soil OM is available.

The models listed below have been used as management tools for NO_3 leaching studies.

The leaching estimation and chemistry model (LEACHM), developed by Wagenet and Hutson,[260,261] includes only the processes that are likely to be of general interest to soil scientists such as expanded range of N pools and transformations in the N version, kinetic, two-site sorption processes in the pesticide version, and a new version that describes the growth of two microbial populations, either competing for the same substrate or acting as a predator-prey system. Five versions of the simulation model describe the water regime and the chemistry and transport of solutes in unsaturated or partially saturated soil to a depth of about 2 m. The model is intended to be applied to laboratory and field situations. The version LEACHN describes N transport and transformations, including NH_3 volatilization and N uptake by plants. The authors of LEACHM indicate that many important processes that influence the fate of chemicals in the root zone are represented in the model in at least a semiquantitative way. This model has been successfully used to describe nutrient and pesticide movement in field soils in the U.S. and other countries.[262]

Nitrate leaching and economic analysis package (NLEAP)[263] is a screening model recently developed to assess the potential for NO_3 leaching and considers the amount of N added as fertilizer and calculates the amounts added by precipitation or in irrigation water, produced in the soil from OM decomposition, the amount removed through plant absorption, lost in biological activity and runoff, and that still present in the soil.

Research models are complex and simulate the physical processes of soil water movement, nutrient transformation, adsorption to soil particles, etc. They aim to determine how long it takes the nutrient to dissipate and help to design measures to reduce the movement of these to the groundwater.

The root zone water quality model (RZWQM)[264] simulates the movement of water and solutes from layer to layer in the soil profile. Nitrogen cycling is simulated in the soil by using factors such as moisture levels, temperature, bacteria populations, degree of acidity, and type and quantity of OM. A plant growth submodel is included that simulates root and shoot growth and grain production.

Planning and management models are used in regulation and planning as guides to farming practice, where they are used to select the type of fertilizers, pesticides, and timing and to fit alternative practices to specific soil properties. This type of model has been used by regulatory agencies such as the EPA. Some of these models incorporate economics and costs of farming practices. Agricultural nonpoint source pollution model (AGNPS),[265] CREAMS (chemicals, runoff and erosion from agricultural management systems),[266] PRMS (precipitation-runoff modeling system),[267] HSPF (hydrologic simulation program — Fortam),[268] and ANSWERS (aerial nonpoint source watershed environment response simulation[269] are examples of such models. AGNPS and CREAMS were developed by ARS to look at non-point source pollution and the environmental consequences of alternative agricultural practices, respectively. They are regularly used by the Soil Conservation Service. HSPF and PRMS are mainly hydrologic models. ANSWERS simulates pollutant movement for small watersheds and was developed at Purdue University, West Lafayette, Indiana.

Groundwater leaching effects of agricultural management systems (GLEAMS),[270] an outgrowth of the CREAMS model, was developed by ARS and incorporates many of its original process descriptions, adding improvements especially in the nutrient and pesticide subsystems. The model aims to evaluate effects of agricultural management systems on the movement of agricultural chemicals within and through the plant root zone. It includes a hydrologic component to simulate infiltration and soil water and solute movement through soil layers and a soil erosion component, for estimation of soil loss by particle size is included.

A multiparameter model for the behavior of N compounds in soil was presented by Frissel and van Veen,[271] in which mineralization and immobilization of N are thought to be controlled by the biomass and only NO_3 is considered to be leachable. The model proposed contains 23 parameters and 19 functions and was tested in a fallow soil, rich in OM. Some adjustments were made and the respective authors reported that the simulation model explains observed amounts of N fairly well, but that more research to determine essential parameters was necessary.

Environmental models are increasingly being used to predict the effects of human activities on the land and water resources. Models such as CREAMS and AGNPS are being used to describe and predict soil loss in a field, chemical and sediment runoff in a field, and pesticide and nutrient runoff in a watershed, as well as erosion's impact on agricultural productivity. These models rely heavily on large amounts of geographic information and geographic analyses. A useful computerized information management tool has been emerging as a tool to support resource management applications: GIS (geographic information systems). The GIS allows for input, management, analysis, manipulation, and display of geographic data. Using traditional tabular data and database management system techniques, as well as geographic databases and powerful analytical software programs that model space in two or three dimensions, the system handles large spatial databases. The results are displayed in a way that makes relationships, trends, changes, and proposed solutions easily understood. GIS can be used for such broad applications as land use planning, economic analysis, urban planning, social and cultural evaluations, environmental analysis, and natural resource planning. In the Soil Conservation Service (SCS), GIS technology is being used to support project planning, water quality analysis, planning for resource conservation and development, among others. The farmer's field would be displayed, along with the soils, roads, and streams. Several analyses would describe erosion loss or nutrient runoff, predicting the probability of it reaching the stream or groundwater. It is being used in land and water resource management research to determine the spatial variability of surface and groundwater resources, and to study the transport of pesticides and other chemicals into the groundwater.

In several occasions, modeling the transport and fate of agricultural chemicals is of limited success due to factors such as soil heterogeneity, variable water inputs, localized nonlinear transport processes, and influences of temperature, plants, and microbes. There is no doubt that mathematical and computer capabilities are available for describing behavior of N in soil; however, the present knowledge on the various processes that affect this behavior is still far from complete. As the models are field tested more extensively and modified as necessary, better predictions may be possible; however, a certain degree of empiricism and estimation will remain.[100]

F. PROGRAMS AND LEGISLATION FOR WATER PROTECTION

Programs and legislation at all government levels have been established to promote the protection of ground and surface water from agricultural contaminants.[72,272,273]

1. Federal Water Quality Legislation

The Clean Water Act — The Clean Water Act, under Section 303 of 101 (a) United States Code, Vol. 331251 (a) and The Federal Water Pollution Control Act Amendments of 1972 (PL 92-500) with subsequent changes in 1977, 1982, and 1987 have been the guidelines for water pollution control in the U.S. The 1987 amendment (clean lakes — section 314; nonpoint sources — section 319, and national estuaries — section 320) delegates regulatory authority for control of point and nonpoint sources of pollution. The Clean Water Act is directed at maintaining and restoring the quality of waters of the U.S. in a broad sense, including wetlands. The Act is enforced by the United States, which has established standards for maximum amounts of pollutants that may be released. Individual states are authorized to establish their own standards for allowable levels of N as long as these are at least as strict as those established by the USEPA.

The Coastal Zone Management Act — Originally passed in 1972, amended in 1980, and reauthorized in 1990, this act requires that each state, with a Federal Coastal Zone Management Program, develop a nonpoint pollution control protection program to curb physical and biological degradation of the coastal environment.

The Safe Drinking Water Act (SDWA) — Initially passed in 1974, the SDWA has undergone several amendments. As amended in 1987, it regulates drinking water standards for all public water supplies with the establishment of maximum contaminant levels (MCLs) for contaminants that may adversely affect health. Among the areas of SDWA coverage, regulation and issuing permits for injection wells probably have the most effect on agriculture.

In practically all states, the USEPA serves only as the supervisor of the state programs. However, it has the authority to enforce the act if the state does not take action within 30 days of receiving notice from the USEPA of violation of water quality standards.

2. Federal Programs

The United States Environmental Protection Agency (USEPA) — The primary federal agency responsible for protection of the water resources from pollution implements regulatory and nonregulatory programs, among these the Drinking Water Programs, under the SDWA.

The United States Department of Agriculture (USDA) — The USDA established in 1990 its new Policy for Water Quality Protection and has assigned several of its agencies, including the Soil Conservation Service (SCS), Extension Service (ES), Agricultural Research Service (ARS), Cooperative State Research Service (CSRS), and the Economic Research Service (ERS) responsibilities for water quality protection.

The President's Water Quality Initiative (WQI) is an interagency, interdepartmental program to develop, test, and deliver information to farmers on crop and livestock management systems that reduce the risk of agricultural chemicals, commercial fertilizers, and wastes from reaching water supplies, particularly groundwater. It uses the combined efforts of the USDA, USGS, USEPA, and National Oceanic and Atmospheric Administration (NOAA).

The 1990 Farm Bill, or the Food, Agriculture, Conservation and Trade Act of 1990 (FACT), established WQI projects through four programs. The Conservation Reserve Program (CRP) directs an enrollment of over 2.4 million new hectares of environmentally sensitive lands. The Wetlands Reserve Program will enroll up to 1 million acres of wetlands into 30-year or permanent easements out of the total CRP acreage. The SCS, under the Water Quality Incentives Program (WQIP), will provide incentive payments, up to $3500 per year, per farm for up to 5 years, to farmers adopting BMPs to protect water quality. The Environmental Easement Program provides for permanent easements of lands that pose a significant environmental threat.

The Integrated Crop Management (ICM) Program helps farmers develop an overall crop management system that promotes efficient use of agricultural inputs in an environmentally sound and profitable manner. This program is promoted by the Agricultural Stabilization and Conservation Service (ASCS), as part of the Agricultural Conservation Program (ACP).

The USDA's agencies (ARS, CSRS, and ERS) also conduct research on various aspects of agriculture and water quality.

The United States Geological Survey (USGS) — The USGS monitors ground and surface water, assesses quality through the National Water Quality Assessment, State Water Resources Research Institutes, and Information Dissemination Programs.

There are several state/regional/local programs that consist of efforts initiated solely at the state, regional, or local levels and at the federal level, and some of these are reported in CAST.[72] The public is well aware of the necessity of protecting the environment. Surveys conducted in the U.S. and other countries on NO_3 in groundwater have been useful for detection of potential problem areas. With knowledge of vulnerable regions, prevention of NO_3 pollution, using various strategies mentioned in previous sections, we can reduce the source of contamination and reduce the risks to human health and the environment.

VI. SUMMARY AND CONCLUSIONS

Concerns about the environment and prevention of its degradation are increasing worldwide. Nonetheless, bountiful crops worldwide have been produced thanks, to a major extent, to fertilization with N. The use of N fertilizers in most non-leguminous crop production is a necessity particularly in nutrient-poor or intensively cropped soils. A drastic reduction in N fertilizer application would invariably result in a reduction in production, leading to harvests that will be insufficient to supply the demands of the growing population. Thus, solutions to adverse effects of N rely heavily on BMPs, based on site-specific conditions of soil, climate, and crop. The use of simulation models may also aid in minimizing N losses, once these are properly tested under field conditions.

The environmental fate of N fertilizers may be altered by several farming strategies, including tillage, rate, form, and time of N fertilizer application, cropping system, irrigation practice, erosion control, and grazing management which are controllable by man. The development of more effective fertilizing techniques, continuous and integrated crop cycles capable of best utilization of available soil N throughout the year, and more efficient plant varieties are also among the promising approaches for minimizing losses while maintaining high agricultural production. Also of utmost importance is that the agricultural sector complies with the existing laws and regulations for the specific area in question. In this manner, N fertilizers may be used more efficiently and "safely" on agricultural lands, without further degradation of soil, atmosphere, and especially ground and surface water. If degradation has already occurred, several measures can and must be taken to alleviate the problems, although it is a difficult task.

The severity of adverse health and environmental aspects of NO_3 contamination is still questioned by many, as evidenced in some of the literature presented in this chapter. However, as stated by Hallberg,[70] "there is clearly no benefit or redeeming value to such contamination." Preventive and protective measures can and should be adopted since restoration of a degraded environment, especially water quality, is not an easy solution. Once contaminated, restoration of the environment to the original conditions is difficult. Just meeting the safety standards may take many years and prove to be a major expense.

ACKNOWLEDGMENTS

The authors wish to express their gratitude to Drs. Raymond S. Pacovsky (USDA-Albany), David C. Martens (Virginia Polytechnic Institute and State University), Paul T. Rygiewicz (EPA-Corvallis), Vivien G. Allen (Virginia Polytechnic Institute and State University), Sam J. Smith (USDA-Durant), Jerry B. Sartain (University of Florida), and Isabelo S. Alcordo (UF), for critically reviewing this chapter. They also thank Ms. Amie Smith, for drawing the figures, Ms. Stephanie Haas, for help in conducting library searches, Mrs. Christina Markham, for assistance with typing, and James B. Platt and Stephen J. and Sarah C. Muchovej, for assistance in indexing the references.

REFERENCES

1. FAO (Food and Agriculture Organization of the United Nations), *FAO Fertilizer Yearbooks*, FAO, Rome, 1962–1987.
2. Fertilizer Institute, *Fertilizer Facts and Figures, Washington*, D.C., 1993.
3. FAO (Food and Agriculture Organization of the United Nations), *Current World Fertilizer Situation and Outlook 1986/87–1992/93*, FAO, Rome, 1988.
4. Follett, R.F. and Walker, D.J., Ground water quality concerns about nitrogen, in *Nitrogen Management and Ground Water Protection*, Follett, R.F., Ed., Elsevier, Amsterdam, 1989, 1.
5. Peterson, G.A. and Frye, W.W., Fertilizer nitrogen management, in *Nitrogen Management and Ground Water Protection*, Follett, R.F., Ed., Elsevier, Amsterdam, 1989, 183.
6. Harre, E.A. and Bridges, J.D., Importance of urea fertilizers, in *Ammonia Volatilization from Urea Fertilizers*, Bock, B.R. and Kissel D.E., Eds., National Fertilizer Development Center, Bull. Y206, Tennessee Valley Authority, Muscle Shoals, AL, 1988, 1.
7. Newbould, P., The use of nitrogen fertiliser in agriculture. Where do we go practically and ecologically?, *Plant Soil*, 115, 297, 1989.
8. Addiscott, T.M., Whitmore, A.P., and Powlson, D.S., *Farming, Fertilizers and the Nitrate Problem*, C.A.B. International Wallingford, Oxon, U.K., 1991.
9. Walsh, L.M., Soil and Applied Nitrogen, Ext. Fact Sheet A 2519, University of Wisconsin, Madison, 1973.
10. Paul, D. and Kilmer, R.L., *The Manufacturing and Marketing of Nitrogen Fertilizers in the United States*, National Economic Analysis Division, Economic Research Service, U.S. Department of Agriculture, Agricultural Economic Report No. 390, Washington, D.C., 1977.
11. Blowin, G.M. and Rindt, D.W., Sulfur-coated fertilizer pellet having controlled dissolution rate and inhibited against microbial decomposition, U.S. Patents 3, 342, 577, 1967.
12. Follett, R.H., Murphy, L.S., and Donahue, R.L., Nitrogen fertilizers, in *Fertilizers and Soil Amendments*, Prentice-Hall, Inc., Englewood Cliffs, NJ, 1981, 23.
13. Boswell, F.C., Meisinger, J.J., and Case, N.L., Production, marketing, and use of nitrogen fertilizers, in *Fertilizer Technology and Use*, Engelstad, O.P., Ed., Soil Science Society of America, Madison, WI, 1985, 229.
14. Tisdale, S.L., Nelson, W.L., and Beaton, J.D., *Soil Fertility and Fertilizers*, Macmillan, New York, 1985.
15. Stevenson, F.J., Soil nitrogen, in *Fertilizer Nitrogen: Its Chemistry and Technology*, Sauchelli, V., Ed., Reinhold, New York, 1964, 18.

16. Stevenson, F.J., Origin and distribution of nitrogen in soil, in *Nitrogen in Agricultural Soils*, Stevenson, F.J., Ed., Agronomy Monograph no. 22, ASA, CSSA, SSSA, Madison, WI, 1982, 1.
17. Stevenson, F.J., The internal cycle of nitrogen in soil, in *Cycles of Soil: Carbon, Nitrogen Phosphorus, Sulfur, Micronutrients*, John Wiley & Sons, New York, 1986, 155.
18. Keeney, D.R., Transformations and transport of nitrogen, in *Agricultural Management and Water Quality*, Schaller, F.W. and Bailey, G.W., Eds., Iowa State University Press, Ames, 1983, 48.
19. Keeney, D.R., Sources of nitrate to ground water, *Crit. Rev. Environ. Control*, 16, 257, 1986.
20. Keeney, D.R., Sources of nitrate to ground water, in *Nitrogen Management and Ground Water Protection*, Follett, R.F., Ed., Elsevier, Amsterdam, 1989, 23.
21. Nelson, D.W., Gaseous losses of nitrogen other than through denitrification, in *Nitrogen in Agricultural Soils, Stevenson*, F.J., Ed., Agronomy Monograph no. 22, ASA, CSSA, SSSA, Madison, WI, 1982, 327.
22. Legg, J.O. and Meisinger, J.J., Soil nitrogen budgets, in *Nitrogen in Agricultural Soils*, Stevenson, F.J., Ed., Agronomy Monograph no. 22, 1982, ASA, CSSA, SSSA, Madison, WI, 503.
23. Tiedje, J.M., Denitrification, in *Methods of Soil Analysis Part 2: Chemical and Microbiological Properties*, 2nd ed., Agronomy Monograph no. 9, ASA, SSSA, Madison, WI, 1982, 1011.
24. Sauchelli, V., Nitrogen: chemical and physical properties, in *Fertilizer Nitrogen: Its Chemistry and Technology*, Sauchelli, V., Ed., Reinhold, New York, 1964, 10.
25. Schepers, J.S. and Mosier, A.R., Accounting for nitrogen in nonequillibrium soil-crop systems, in *Managing Nitrogen for Groundwater Quality and Farm Profitability*, Follett, R.F., Keeney, D.R., and Cruse, R.M., Eds., SSSA, Madison, WI, 1991, 125.
26. Mata, A. and Blue, W.G., Fertilizer nitrogen distribution in Pensacola bahiagrass sod during the first year of development on an Aeric Haplaquod, *Soil Crop Sci. Soc. Fla. Proc.*, 33, 209, 1973.
27. Garrett, M.K., Nitrogen losses from grassland systems under temperate climatic conditions, in *Chemistry, Agriculture and the Environment*, Richardson, M.L., Ed., The Royal Society of Chemistry, Thomas Graham House, Science Park, Cambridge, 1991, 121.
28. Blaser, R.E., Symposium of forage utilization: effects of fertility levels and stage of maturity on forage nutritive value, *J. Anim. Sci.*, 23, 246, 1964.
29. Freyh, R.L. and Lamond, R.E., Sulfur and nitrogen fertilization, *J. Prod. Agric.*, 5, 488, 1992.
30. Juergens-Gschwind, S., Ground water nitrates in other developed countries (Europe) — Relationships to land use patterns, in *Nitrogen Management and Ground Water Protection*, Follett, R.F., Ed., Elsevier, Amsterdam, 1989, 75.
31. Larson, K.L., Carter, J.F., and Vasey, E.H., Nitrate-nitrogen accumulation under bromegrass sod fertilized annually at six levels of nitrogen for fifteen years, *Agron. J.*, 63, 527, 1971.
32. Jaakkola, A., Leaching losses of nitrogen from a clay soil under grass and cereal crops in Finland, *Plant Soil*, 76, 59, 1984.
33. Barraclough, D., Geens, E.L., and Maggs, J.M., Fate of fertilizer nitrogen applied to grassland. II. Nitrogen-15 leaching results, *J. Soil Sci.*, 35, 191, 1984.

34. Standley, J., Thomas, G.A., Hunter, H.M., Webb, A.A., and Berthelsen, S., Decreases over seven years in subsoil nitrate in a vertisol with grain sorghum and grass, *Plant Soil*, 125, 1, 1990.
35. Fuleky, G., Leaching of nutrient elements from fertilizers into deeper soil zones, in *Chemistry, Agriculture, and the Environment*, Richardson, M.L., Ed., The Royal Society of Chemistry, Cambridge, 1991, 185.
36. van der Meer, H.G. and van Uum-van Lohuyzen, M.G., The relationship between inputs and outputs of nitrogen in intensive grassland systems, in *Nitrogen Fluxes in Intensive Grassland Systems*, van der Meer, H.G., Ryden, J.C., and Ennik, G.C., Eds., Martinus Nijhoff, Dordrecht, 1986, 1.
37. Steenvoorden, J.H.A.M., Fonck, H., and Oosteron, H.P., Losses of nitrogen from intensive grassland systems by leaching and surface runoff, in *Nitrogen Fluxes in Intensive Grassland Systems*, van der Meer, H.G., Ryden, J.C., and Ennik, G.C., Eds., Martinus Nijhoff, Dordrecht, 1986, 85.
38. Bock, B.R. and Hergert, G.W., Fertilizer nitrogen management, in *Managing Nitrogen for Groundwater Quality and Farm Profitability*, Follett, R.F., Keeney, D.R., and Cruse, R.M., Eds., SSSA, Madison, WI, 1991, 139.
39. van Diest, A., Accumulation of nitrate in higher plants — its causes and prevention, in *Nitrogen in Higher Plants*, Abrol, Y.P., Ed., Research Studies Press, Taunton, Somerset, 1990, 441.
40. OECD (Organization for Economic Co-operation and Development), *Water Pollution by Fertilizers and Pesticides*, OECD Publications, Paris, 1986.
41. Henry, J.L. and Menely, W.A., Nitrates in Western Canadian Groundwater, Western Canada Fertilizer Association, Saskatchewan, 1993, 1.
42. DeRoo, H.C., Nitrate Fluctuation in Groundwater as Influenced by Use of Fertilizer, Connecticut Agricultural Experimental Station, Bull. 779, University of Connecticut, New Haven, 1980.
43. Spalding, R.F., Exner, M.E., Lindau, C.W., and Eaton, D.W., Investigation of sources of groundwater nitrate contamination in the Burbank-Wallula area of Washington, U.S.A., *J. Hydrol.*, 58, 307, 1982.
44. Schepers, J.S., Frank, K.D., and Watts, D.G., Influence of irrigation and nitrogen fertilization on groundwater quality, Proc. Int. Union of Geodesy and Geophysics, Hamburg, Germany, 1984, 21.
45. Collison, S.E. and Walker, S.S., Loss of fertilizers by leaching, Bull. 132, University of Florida, Agric. Exp. Sta., Gainsville, 1916.
46. Graetz, D.A., Hammond, L.C., and Davidson, J.M., Nitrate movement in a Eutis fine sand planted to millet, *Soil Crop Sci. Soc. Fla. Proc.*, 33, 157, 1973.
47. Endelman, F.J., Keeney, D.R., Gilmour, J.T., and Saffira, P.G., Nitrate and chloride movement in the Plainfield loamy sand under intensive irrigation, *J. Environ. Qual.*, 3, 295, 1974.
48. Spalding, R.F. and Exner, M.E., Occurrence of nitrate in groundwater — a review, *J. Environ. Qual.*, 22, 392, 1993.
49. CAST (Council for Agricultural Sciences and Technology), *Agriculture and Groundwater Quality*, Council for Agricultural Sciences and Technology Reports, Ames, IA, 103, 1985, 62.
50. Foster, S.S.D., Cripps, A.C., and Smith Carington, A., Nitrate leaching to groundwater, *Phil. Trans. 2, Soc. Land*, 296, 477, 1982.

51. Bachman, J.L., Nitrate in the Columbia Aquifer, Central Delmarva Peninsula, Maryland, U.S. Geol. Surv. Water-Resour. Invest. Rep. 84-4322, U.S. Geological Survey, Towson, MD, 1984.
52. Croll, B.T. and Hayes, C.R., Nitrate and water supplies in the United Kingdom, *Environ. Pollut.*, 50, 163, 1988.
53. Rawitz, E., Burns, S., Etkin, H., Hardiman, R., Hillel, D., and Terkeltoub, R., Fate of fertilizer nitrogen in irrigated fields under semi-arid conditions, in *Soil Nitrogen as Fertilizer or Pollutant*, International Atomic Energy Agency, Vienna, 1980, 195.
54. Hill, A.R., Nitrate distribution in the groundwater of the Alliston region of Ontario Canada, *Ground Water*, 20, 696, 1982.
55. Spalding, R.F., Gormly, J.R., Curtiss, B.H., and Exner, M.E., Nonpoint nitrate contamination of groundwater in Merrick County, Nebraska. *Ground Water*, 26, 86, 1978.
56. Gormly, J.R. and Spalding, R.F., Sources and concentrations of nitrate-nitrogen in ground water of the central Platte region, Nebraska, *Ground Water*, 17, 291, 1979.
57. Hallberg, G.R., Nitrates in Iowa groundwater, in *Rural Groundwater Contamination*, D'Itri, F.M. and Wolfson, L.G., Eds., Lewis Publishers, Chelsea, MI, 1988, 23.
58. Bauder, J.W., White, B.A., and Inskeep, W.P, Montana extension initiative focuses on private well quality, *J. Soil Water Cons.*, 46, 69, 1991.
59. Sharpe, W.E., Mooney, D.M., and Adams, R.S., An analysis of groundwater quality data obtained from private individual water systems in Pennsylvania, *Northeast Environ. Sci.*, 4, 155, 1985.
60. D'elia, C.F., Sanders, J.G., and Boynton, W.R., Nutrient enrichment studies in a coastal plain estuary: phytoplankton growth in large-scale, continuous cultures, *Can. J. Fish Aquatic Sci.*, 43, 397, 1986.
61. Ritter, W.F. and Chirnside, A.E.M., Impact of land use on groundwater quality in southern Delaware, *Ground Water*, 22, 38, 1984.
62. Jones, G.W., Dehaven, E.C., Clark, L.F., Rauch, J.T., Rasmussen, J.R., and Guillen, C.G., Groundwater Quality Sampling Results from Wells in Southwest Florida Water Management District, Ambient Groundwater Quality Monitoring Program, SFWMD, and Florida Dept. of Environmental Regulations, West Palm Beach, FL, 1990.
63. USDA (United States Department of Agriculture), *Agricultural Statistics, 1991*. U.S. Government Printing Office, Washington, D.C., 1991.
64. Madison, R.J. and Brunett, J.O., Overview of the occurrence of nitrate in ground water of the United States, in *National Water Summary (1984)*, U.S. Geol. Surv. Water Supply paper, Washington, D.C., 1985, 93.
65. Lee, L.K. and Nielsen, E.G., Farm Chemicals and Groundwater Çontamination, in *Agriculture and Groundwater Quality — Examining the Issue*, Nelson, J.R. and McTerman, E.M., Eds., University Center for Water Research, Oklahoma State University, Stillwater, 1989, 2.
66. USEPA (United States Environmental Protection Agency), *National Pesticide Survey: Summary Results of EPA's National Survey of Pesticides in Drinking Water Wells*, USEPA Office of Water, Washington, D.C., 16, 1990.
67. Martin, D.L., Gilley, J.R., and Skaggs, R.W., Soil water balance and management, in *Managing Nitrogen for Groundwater Quality and Farm Profitability*, Follett, R.F., Keeney, D.R., and Cruse, R.M., Eds., SSSA, Madison, WI, 1991, 199.
68. Smith, S.J. and Cassel, D.K., Estimating nitrate leaching in soil materials, in *Managing Nitrogen for Groundwater Quality and Farm Profitability*, Follett, R.F., Keeney, D.R., and Cruse, R.M., Eds., SSSA, Madison, WI, 1991, 165.

69. Lowrance, R.R. and Pionke, H.B., Transformations and movement of nitrate in aquifer systems, in *Nitrogen Management and Groundwater Protection*, Follett, R.F., Ed., Elsevier, Amsterdam, 1989, 373.
70. Hallberg, G.R., Nitrate in groundwater in the United States, in *Nitrogen Management and Groundwater Protection*, Follett, R.F., Ed., Elsevier, Amsterdam, 1989, 35.
71. Pionke, H.B. and Lowrance, R.R., Fate of nitrate in subsurface drainage water, in *Managing Nitrogen for Groundwater Quality and Farm Profitability*, Follett, R.F., Keeney, D.R., and Cruse, R.M., Eds., SSSA, Madison, WI, 1991, 237.
72. CAST (Council for Agricultural Sciences and Technology), Water Quality: Agriculture's Role. Task Force Rep. 120, CAST, Ames, IA, 1992.
73. Holloway, T.T., Evaluation of Ambient Surface Water Quality in the State of Iowa; Based on Monitoring Data from Years 1982–1984. EPA 90/9-85-003, U.S. EPA, Kansas City, KS, 1985.
74. Gambrell, R.P., Gilliam, J.W., and Weed, S.B., Nitrogen losses from soils of the North Carolina Coastal Plains, *J. Environ. Qual.*, 4, 317, 1975.
75. Duxbury, J.M. and Peverly, J.H., Nitrogen and phosphorus losses from organic soils, *J. Environ. Qual.*, 7, 566, 1978.
76. Miller, M.H., Contribution of nitrogen and phosphorus to subsurface drainage water from intensively cropped mineral and organic soils in Ontario, *J. Environ. Qual.*, 8, 42, 1979.
77. Gilliam, J.W., Logan, T.J., and Broadbent, F.E., Fertilizer use in relation to the environment, in *Fertilizer Technology and Use*, Engelstad, O.P., Ed., SSSA, Madison, WI, 1985, 56.
78. Procházková, L., Blazka, P., and Brandl, Z., The output of NO_3-N and other elements from small homogeneous watersheds, in *Land Use Impacts on Aquatic Systems*, Zolánkai, G. and Roberts, G., Eds., Programme on Man and Biosphere (MAB), Budapest, 1983, 291.
79. Hubbard, R.K. and Sheridan, J.M., Water and nitrate-nitrogen losses from a small, upland, coastal plain watershed, *J. Environ. Qual.*, 12, 291, 1983.
80. Schuman, G.E., Burwell, R.E., Piest, R.F., and Spomer, R.G., Nitrogen losses in surface runoff from agricultural watersheds on Missouri Valley loess, *J. Environ. Qual.*, 2, 299, 1973.
81. Kissel, D.E., Richardson, C.W., and Burnett, E., Losses of nitrogen in surface runoff in the blackland prairie of Texas, *J. Environ. Qual.*, 5, 288, 1976.
82. Garwood, E. and Ryden, J.C., Nitrate loss through leaching and surface runoff from grassland: effects of water supply, soil type and management, in *Nitrogen Fluxes in Intensive Grassland Systems*, van der Meer, H.G., Ryden, J.C., and Ennik, G.C., Eds., Martinus Nijhoff, Dordrecht, 1986, 99.
83. Raveh, A. and Avnimelich, Y., Minimizing nitrate seepage from the Hula Valley into Lake Kinneret (Sea of Galilee). I. Enhancement of nitrate reduction by sprinkling and flooding, *J. Environ. Qual.*, 2, 455, 1973.
84. Gilliam, J.W. Skaggs, R.W., and Weed, S.B., Drainage control to diminish nitrate loss from agricultural fields, *J. Environ. Qual.*, 8, 137, 1979.
85. Thomas, G.W. and Crutchfield, J.D., Nitrate-nitrogen and phosphorus contents of streams draining small agricultural watersheds in Kentucky, *J. Environ. Qual.*, 3, 46, 1974.
86. Gilliam, J.W. and Terry, D.L., Potential for water pollution from fertilizer use in North Carolina, North Carolina Agric. Ext. Circ., Raleigh, NC, 1973, 550.

87. Bower, C.A. and Wilcox, L.V., Nitrate content of the upper Rio Grande as influenced by nitrogen fertilization of adjacent irrigated lands, *Soil Sci. Soc. Am. Proc.*, 33, 971, 1969.
88. Budd, T., Are we polluting steams?, *Prairie Farmer*, 144, 16, 1972.
89. Johnson, A.H., Bouldin, D.R., Goyette, E.A., and Hedges, A.M., Nitrate dynamics in Fall Creek, N.Y., *J. Environ. Qual.*, 5, 386, 1976.
90. Skogerboe, G.V., Agricultural impact on water quality in western rivers, in *Environmental Impact on Rivers*, Shen, H.W., Ed., Hsieh Wen Shen Publisher, Fort Collins, CO, 1974, 1.
91. Wild, A. and Cameron, K.C., Nitrate leaching through soils and environmental considerations — with special reference to recent work in the United Kingdom, in *Soil Nitrogen as Fertilizer or Pollutant*, International Atomic Energy Agency, Panel Proceedings Series, Vienna, 1980, 289.
92. Jacobs, T.C. and Gilliam, J.W., Riparian losses of nitrate from agricultural drainage waters, *J. Environ. Qual.*, 14, 472, 1985.
93. Karr, J.R. and Schlosser, I.J., Water resources and the land-water interface, *Science*, 201, 229, 1978.
94. Lowrance, R.R., Todd, R.L., and Asmussen, L.E., Nutrient cycling in an agricultural watershed. II. Streamflow and artificial drainage, *J. Environ. Qual.*, 13, 27, 1984.
95. Lowrance, R.R., Leonard, R.A., and Sheridan, J.M., Managing riparian ecosystems to control non-point pollution, *J. Soil Water Conserv.*, 40, 87, 1985.
96. Peterjohn, W.T. and Cornell, D.L., Nutrient dynamics in an agricultural watershed: observations on the role of a riparian forest, *Ecology*, 65, 1146, 1984.
97. Peterjohn, W.T. and Cornell, D.L., The effect of riparian forest on the volume and chemical composition of baseflow in an agricultural watershed, in *Watershed Research Perspectives*, Correll, D.L., Ed., Smithsonian Institution Washington, D.C., 1986, 224.
98. Fail, J.L., Jr., Haines, B.L., and Todd, R.L., Riparian forest communities and their role in nutrient conservation in an agricultural watershed, *Am. J. Alter. Agric.*, 2, 114, 1987.
99. Novotny, V. and Chesters, G., *Handbook of Non-Point Pollution: Sources and Management*, van Nostrand Reinhold, New York, 1981.
100. Smith, S.J., Schepers, J.S., and Porter, L.K., Assessing and managing agricultural nitrogen losses to the environment, in *Advances in Soil Science*, Vol. 14, Springer-Verlag, New York 1990, 1.
101. Knox, R.C., Canter, L.W., Kincannon, D.E., Stover, E.L., and Ward, C.H., *Aquifer Restoration, State of the Art*, Pollution Technology. Review No. 131. Noyes Publications, Park Ridge, NJ, 1986.
102. USDA (United States Department of Agriculture), *The Second RCA Appraisal. Soil, Water, and Related Resources on Non-Federal Land in the U.S. Analysis of Condition and Trends*, U.S. Government Printing Office, Washington, D.C., 1989.
103. Hamilton, H. and Luk, S.-H., Nitrogen transfers in a rapidly eroding agroecosystem: Loess Plateau China, *J. Environ. Qual.*, 22, 133, 1993.
104. Viner, A.B. and White, E., Phytoplankton growth, in *Inland Waters of New Zealand*, Viner, A.B., Ed., DSIR Science Information Publishing Centre, Wellington, NZ, 1987, 191.
105. Straskraba, M., Ecotechnological measures against eutrophication, *Limnologica*, 17, 237, 1986.
106. Lowrance, R., Nitrogen outputs from a field-size agricultural watershed, *J. Environ. Qual.*, 21, 602, 1992.

107. Lowrance, R.R. Todd, R.L., and Asmussen, L.E., Waterborne nutrient budgets for the riparian zone of an agricultural watershed, *Agric. Ecosyst. Environ.*, 10, 371, 1983.
108. Lowrance, R.R., Todd, R.L., and Asmussen, L.E., Nutrient cycling in an agricultural watershed. I. Phreatic movement, *J. Environ. Qual.*, 13, 22, 1984.
109. Nielsen, G.H., Culley, J.L.B., and Cameron, D.R., Agriculture and water quality in the Canadian Great Lakes Basin. IV. Nitrogen, *J. Environ. Qual.*, 11, 493, 1982.
110. Cornwell, D.A., Zoltek, J., Jr., Patrinely, C.D., des Furman, T., and Kim, J.I. Nutrient removal by waterhyacinths, *J. Water Pollut. Control Fed.*, 49, 57, 1977.
111. Dinges, R., Upgrading stabilization pond effluents by water hyacinth culture, *J. Water Pollut. Control Fed.*, 50, 833, 1978.
112. McDonald, R.C. and Wolverton, B.C., Comparative study of waste water lagoon with and without water hyacinth, *Econ. Bot.*, 34, 101, 1980.
113. Reddy, K.R., Campbell, K.L., Graetz, D.A., and Portier, K.M., Use of biological filters for agricultural drainage water treatment. *J. Environ. Qual.*, 11, 591, 1982.
114. Hauser, J.R., Use of water hyacinth aquatic treatment systems for ammonia control and effluent polishing, *J. Water Pollut. Control Fed.*, 56, 219, 1984.
115. Moorhead, K.K., Reddy, K.R., and Graetz, D.A., Nitrogen transformations in a water hyacinth-based water treatment system, *J. Environ. Qual.*, 17, 71, 1988.
116. Martin, H.W., Graetz, D.A., Locascio, S.J., and Hensel, D.R., Nitrification inhibitor influences on potato, *Agron. J.*, 85, 651, 1993.
117. Moore, P.D. and Bellamy, D.J., *Peatlands*, Springer-Verlag, New York, 1974.
118. Johnston, C.A., Bubenzer, G.D., Lee, G.B., Madison, F.W., and McHenry, J.R., Nutrient trapping by sediment deposition in a seasonally flooded lakeside wetland. *J. Environ. Qual.*, 13, 283, 1984.
119. Bernard, J.M. and Solsky, B.A., Nutrient cycling in a *Carex lacustris* wetland, *Can. J. Bot.*, 55, 630, 1977.
120. Davis, C.B. and van der Valk, A.G., Litter decomposition in prairie glacial marshes, in *Freshwater Wetlands*, Good, R.E., Whigham, D.F., and Simpson, R.L., Eds., Academic Press, New York, 1978, 89.
121. Whigham, D.F. and Simpon, R.L., The potential use of freshwater tidal marshes in the management of water quality in the Delaware River, in *Biological Control of Water Pollution*, Tourbier, J. and Pierson, R.W., Eds., University of Pennsylvania Press, Philadelphia, 1976, 173.
122. DeLaune, R.D., Patrick, W.H., Jr., and Buresh, R.J., Sedimentation rates determined by Cs-137 dating in a rapidly accreting salt marsh, *Nature*, 275, 532, 1978.
123. Gessel, S.P., Cole, D.W., and Steinberger, E.P., Nitrogen balances in forest ecosystems of the Pacific Northwest, *Soil Biol. Biochem.*, 5, 19, 1973.
124. Bormann, F.H., Likens, G.E., and Melillo, J.M., Nitrogen budget for an aggrading northern hardwood forest ecosystem, *Science*, 196, 981, 1977.
125. Keeney, D.R., Prediction of soil nitrogen availability in forest ecosystems: a literature review, *Forest Sci.*, 26, 159, 1980.
126. Fraser, P. and Chilvers, C., Health aspects of nitrate in drinking water, *Sci. Total Environ.*, 18, 103, 1981.
127. Bockman, O.C. and Granli, T., Human health aspects of nitrate intake from food and water, in *Chemistry, Agriculture, and the Environment*, Richardson, M.L., Ed., The Royal Society of Chemistry, Cambridge, 1991, 373.
128. USEPA (United States Environment Protection Agency), *Quality Criteria for Water*, U.S. Government Printing Office, Washington, D.C., 1976.

129. Stevenson, F.J., Impact of nitrogen on health and the environment, in *Cycles of Soil: Carbon, Nitrogen Phosphorus, Sulfur, Micronutrients*, John Wiley & Sons, New York, 1986, 216.
130. Aldrich, S.R., Nitrogen management to minimize adverse effects on the environment, in *Nitrogen in Crop Production*, Hauck, R.D., Ed., ASA, Madison, WI, 1984, chap. 45.
131. Brezonik, P.L., *Nitrates: An Environment Assessment*, National Academy of Sciences, Washington, D.C., 1978.
132. Keeney, D.R., Nitrogen management for maximum efficiency and minimum pollution, in *Nitrogen in Agricultural Soils*, Stevenson, F.J., Ed., ASA, CSSA, SSSA, Madison, WI, 1982, 605.
133. Viets, F.G., Jr. and Hageman, R.H., *Factors Affecting the Accumulation of Nitrate in Soil, Water and Plants*, U.S. Department of Agriculture, Agricultural Handbook, Washington, D.C., 413, 1971, 63.
134. Deeb, B.S. and Sloan, K.W., Nitrates, Nitrites, and Health, Illinois Agric. Exp. Sta. Bull. 750, Urbana, 1975.
135. Magee, P.N., Nitrogen as a potential health hazard, *Phil. Trans. R. Soc. Land.*, B296, 543, 1982.
136. Keeney, D.R. and Follett, R.F., Managing nitrogen for groundwater quality and farm profitability: overview and introduction, in *Managing Nitrogen for Groundwater Quality and Farm Profitability*, Follett, R.F., Keeney, D.R., and Cruse, R.M., Eds., SSSA, Madison, WI, 1991, 1.
137. Johnson, C.J., Bonrud, P.A., Dosch, T.L., Kilness, A.W., Senger, K.A., Busch, D.C., and Meyer, M.R., Fatal outcome of methemoglobinemia in an infant, *JAMA*, 257, 2796, 1987.
138. Hill, M.J., Hawksworth, G., and Tattersall, G., Bacteria, nitrosamines and cancer of stomach, *Br. J. Cancer*, 28, 562, 1973.
139. Weisenburger, D.D., Potential health consequences of groundwater contaminations by nitrates in Nebraska, in *Nitrate Contamination: Exposure, Consequence, and Control*, Bogardi, I. and Kuzelka, R.D., Eds., NATO ASI Ser. G: Ecological Sciences 30, Springer-Verlag, Berlin, 1991, 309.
140. Super, M., Heese, H., MacKenzie, D., Dempster, W.S., duPless, J., and Ferreira, J.J., An epidemiologic study of well-water nitrates in a group of southwest African Namibian infants, *Water Res.*, 15, 1265, 1981.
141. Dorsche, M.M., Scragg, R.K.R., McMichael, A.J., Baghurst, P.A., and Dyer, K.F., Congenital malformations and maternal drinking water supply in rural south Australia: a case control study, *Am. J. Epidemiol.*, 119, 473, 1984.
142. Malberg, J.W., Savage, E.P., and Osteryoung, J., Nitrates in drinking water and the early onset of hypertension, *Environ. Pollut.*, 15, 155, 1978.
143. Campbell, J.B., Davis, A.N., and Myhr, P.J., Methemoglobinemia of livestock caused by high nitrate contents of well water, *Can. J. Camp. Med.*, 18, 93, 1954.
144. Johnston, L., The effects of anhydrous ammonia on the environment and its direct effect on humans, *J. Environ. Health*, 37, 462, 1975.
145. Denmead, O.T., Simpson, J.R., and Freney, J.R., A direct, field measurement of ammonia emission after injection of anhydrous ammonia, *Soil Sci. Soc. Am. J.*, 41, 1001, 1977.
146. Hutchinson, G.L. and Viets, F.G., Nitrogen enrichment of surface water by absorption of ammonia volatilization from cattle feed lots, *Science*, 166, 514, 1969.
147. Crutzen, P.J. and Ehhalt, D.H., Effects of nitrogen oxides on the atmospheric ozone content, *Q. J. R. Meterol. Soc.*, 96, 320, 1970.

148. Crutzen, P.J., Atmospheric chemical processes of the oxides of nitrogen, including nitrous oxide, in *Denitrification, Nitrification and Atmospheric Nitrous Oxide*, Delwieke, C.C., Ed., John Wiley & Sons, New York, 1981, 17.
149. Gilliam, J.W., Dasberg, S., Lund, L.J., and Focht, D.D., Denitrification in four California soils: effect of soil profile characteristics, *Soil Sci. Soc. Am. J.*, 42, 61, 1978.
150. Mosier, A.R. and Hutchinson, G.L., Nitrous oxide emission from cropped fields, *J. Environ. Qual.*, 10, 169, 1981.
151. Mosier, A.R., Hutchinson, G.L., Sabey, B.R., and Baxter, J., Nitrous oxide emissions from barley plots treated with ammonium nitrate or sewage sludge, *J. Environ. Qual.*, 11, 78, 1982.
152. Goodroad, L.L., Keeney, D.R., and Peterson, L.A., Nitrous oxide emissions from agricultural soils in Wisconsin, *J. Environ. Qual.*, 13, 557, 1984.
153. Bremner, J.M., Breitenbeck, G.A., and Blackmer, A.M., Effect of anhydrous ammonia fertilization on emissions of nitrous oxide from soils, *J. Environ. Qual.*, 10, 77, 1981.
154. Eichner, M.J., Nitrous oxide emissions from fertilized soils: summary of available data, *J. Environ. Qual.*, 19, 272, 1990.
155. Tiedje, J.M., Ecology of denitrification and dissimilatory nitrate reduction to ammonium, in *Biology of Anaerobic Microorganisms*, Zehnder, A.J.B., Ed., John Wiley & Sons, New York, 1988, 179.
156. Firestone, M.K., Biological denitrification, in *Nitrogen in Agricultural Soils*, Stevenson, F.J., Ed., ASA, Madison, 1982, 289.
157. Bijay-Singh, Ryden, J.C., and Whitehead, D.C., Denitrification potential and actual rates of denitrification in soils under long-term grassland and arable cropping, *Soil Biol. Biochem.*, 21, 897, 1989.
158. Russelle, M.P. and Hargrove, W.L., Cropping systems: ecology and management, in *Nitrogen Management and Groundwater Protection*, Follett, R.F., Ed., Elsevier, Amsterdam, 1989, 277.
159. Meisinger, J.J., Evaluating plant-available nitrogen in soil-crop systems, in *Nitrogen in Crop Production*, Hauck, R.D., Ed., ASA, CSSA, and SSSA, Madison, WI, 1984, 391.
160. Kolenbrander, G.J., Nitrogen in organic matter and fertilizers as a source of pollution, *Prog. Water Technol.*, 8, 67, 1977.
161. Oberle, S.L. and Keeney, D.R., Factors influencing corn fertilizer N requirements in the Northern U.S. Corn Belt, *J. Prod. Agric.*, 3, 527, 1990.
162. Oberle, S.L. and Keeney, D.R., Soil type, precipitation, and fertilizer N effects on corn yields, *J. Prod. Agric.*, 3, 522, 1990.
163. Roth, L.W. and Fox, R.H., Soil nitrate accumulations following nitrogen-fertilized corn in Pennsylvania, *J. Environ. Qual.*, 19, 243, 1990.
164. Meisinger, J.J., Bandel, V.A., Angle, J.S., O'Keefe, B.E., and Reynolds, C.M., Presidedress soil nitrate test evaluation in Maryland, *Soil Sci. Soc. Am. J.*, 56, 1527, 1992.
165. Magdoff, F.R., Ross, D., and Amadon, J., A soil test for nitrogen availability to corn, *Soil Sci. Soc. Am. J.*, 48, 1301, 1984.
166. Magdoff, F.R., Understanding the Magdoff pre-sidedness nitrate test for corn, *J. Prod. Agric.*, 4, 297, 1991.
167. Fox, R.H., Roth, G.W., Iverson, K.V., and Piekielek, W.P., Soil and tissue nitrate tests compared for predicting soil nitrogen availability to corn, *Agron. J.*, 81, 971, 1989.
168. Blackmer, A.M., Pottker, D., Cerrato, M.E., and Webb, J., Correlations between soil nitrate concentrations in late spring and corn yields in Iowa, *J. Prod. Agric.*, 2, 103, 1989.

169. Meisinger, J.J., Magdoff, F.R., and Schepers, J.S., Predicting N fertilizer needs for corn in humid regions: underlying principles, in *Predicting N Fertilizer Needs for Corn in Humid Regions*, Bock, B.R. and Kelley, K.R., Eds., TVA, National Fertilizer and Environmental Research Center, Muscle Shoals, AL, 1992, 7.
170. Otter-Nacke, S. and Kuhlmann, H., A comparison of the performance of N simulation models in the prediction of N_{min} on farmers' fields in the spring, *Fertilizer Res.*, 27, 341, 1991.
171. Meisinger, J.J. and Randall, G.W., Estimating nitrogen budgets for soil-crop systems, in *Managing Nitrogen for Ground Water Quality and Farm Profitability*, Follett, R.F., Keeney, D.R., and Cruse, R.M., Eds., SSSA, Madison, WI, 1991. 85.
172. Wong, Yon Cheong, Y., NG Kee Kwong, K.F., and Cavalot, P.C., Comparative study of ammonium and nitrate fertilizers in two soils of Mauritius cropped with sugarcane, in *Soil Nitrogen as Fertilizer on Pollutant*, International Atomic Energy Agency, Panel Proceedings Series, Vienna, 1980, 351.
173. Schepers, J.S., Role of cropping systems in environmental quality: ground water nitrogen, in *Cropping Strategies for Efficient Use of Water and Nitrogen*, Hargrove, W.L., Ed., ASA, CSSA, SSSA, Special Publ. No. 51, Madison, WI, 1988, 167.
174. Singh, M. and Singh, T.A., Leaching losses of nitrogen from urea as affected by application of neem cake, *J. Indian Soc. Soil Sci.*, 34, 767, 1986.
175. Dressel, J. and Jung, J., Naehrstoffverlagerung in einem Sandboden in Abhaengigkeit von der Bepflanzung und Stickstoffduengung (Lysimeterversuche), *Landwirtsch. Forsch.*, 36, 363, 1984.
176. Locascio, S.J., Fiskell, J.G.A., and Elmstrom, G.W., Comparison of sulfur-coated and uncoated urea for watermelons, *Proc. Soil Crop Sci. Soc. Florida*, 37, 197, 1978.
177. Peterson, G.A. and Power, J.F., Soil, crop, and water management, in *Managing Nitrogen for Groundwater Quality and Farm Profitability*, Follett, R.F., Keeney, D.R., and Cruse, R.M., Eds., Soil Science Society of America, Madison, WI, 1991, 189.
178. Hargrove, W.L., Black, A.L., and Mannering, J.V., Cropping strategies for efficient use of water and nitrogen: introduction, in *Cropping Strategies for Efficient Use of Water and Nitrogen*, Hargrove, W.L., Ed., Special Publ. No. 51, ASA, CSSA, SSSA, Madison, WI, 1988, 1.
179. Bauder, J.W. and Schneider, R.P., Nitrate-nitrogen leaching following urea fertilization and irrigation, *Soil Sci. Soc. Am. J.*, 43, 348, 1979.
180. Pratt, P.F., Nitrogen use and nitrate leaching in irrigation agriculture, in *Nitrogen in Crop Production*, Hauck, R.D., Ed., American Society of America, Madison, WI, 1984, 319.
181. Paltineanu, I.C., Hera, C., Paltineanu, R., Idriceanu, A., Eliade, Gh., Suteu, Gh., Bologa, M., Canarache, A., Postolache, T., and Apostol, I., Irrigation water and N fertilizer application efficiencies for reduction of water and N losses and for water pollution control, in *Soil Nitrogen as Fertilizer or Pollutant*, International Atomic Energy Agency, Panel Proceedings Series, Vienna, 1980, 169.
182. Ritter, W.F., Nitrate leaching under irrigation in the United States — a review, *J. Environ. Sci. Health*, A24, 349, 1989.
183. McDowell, L.L. and McGregor, K.G., Nitrogen and phosphorus losses in runoff from no-till soybeans, *Trans. Am. Soc. Agric. Eng.*, 23, 643, 1980.
184. Angle, J.S., McClung, G., McIntosh, M. S., Thomas, P.M., and Wolf, D.C., Nutrient losses in runoff from conventional and no-till watersheds, *J. Environ. Qual.*, 13, 431, 1984.

185. Alberts, E.E. and Spomer, R.A., Dissolved nitrogen in runoff from watersheds in conservation and conventional tillage, *J. Soil Water Conserv.*, 40, 153, 1985.
186. Angle, J.S., Gross, C.M., and McIntosh, M.S., Nitrate concentrations in percolate and groundwater under conventional and no-till *Zea mays* watersheds, *Agric. Ecosystems Environ.*, 25, 279, 1989.
187. Drury, C.F., McKenney, D.J., Findlay, W.I., and Gaynor, J.D., Influence of tillage on nitrate loss in surface runoff and tile drainage, *Soil Sci. Soc. Am. J.*, 57, 797, 1993.
188. Lamb, J.A., Peterson, G.A., and Fenster, C.R., Wheat fallow tillage systems — effect on a newly cultivated grassland soil's nitrogen budget, *Soil Sci. Soc. Am. J.*, 49, 352, 1985.
189. Groffman, P.M., Hendrix, P.F., and Crossley, D.A., Jr., Nitrogen dynamics in conventional and no-tillage agroecosystems with inorganic fertilizer or legume nitrogen inputs, *Plant Soil*, 97, 315, 1987.
190. Thomas, G.W., Smith, M.S., and Phillips, R.E., Impact of soil management practices on nitrogen leaching, in *Nitrogen Management and Ground Water Protection*, Follett, R.F., Ed., Elsevier, Amsterdam, 1989, 247.
191. Power, J.F., and Doran, J.W., Role of crop residue management in nitrogen cycling and use, in *Cropping Strategies for Efficient Use of Water and Nitrogen*, Hargrove, W.L., Ed., Special Publ. No. 51, ASA, CSSA, SSSA, Madison, WI, 1988, 101.
192. Angle, J.S., Gross, C.M., Hill, R.L., and McIntosh, M.S., Soil nitrate concentrations under corn as affected by tillage, manure, and fertilizer applications, *J. Environ. Qual.*, 22, 141, 1993.
193. Stecker, J.A., Buchholz, D.D., Hanson, R.G., Wollenhaupt, N.C., and McVay, K.A., Application placement and timing of nitrogen solution for no-till corn, *Agron. J.*, 85, 645, 1993.
194. Sharpley, A.N. and Smith, S.J., Effects of cover crops on surface water quality, in *Cover Crops for Clean Water*, Hargrove, W.L., Ed., Soil and Water Conserv. Soc. Publ., Ankeny, IA, 1991. 41.
195. Smith, S.J., Sharpley, A.N., Naney, J.W., Berg, W. A., and Jones, O.R., Water quality impacts associated with wheat culture in the Southern Plains, *J. Environ. Qual.*, 20, 244, 1991.
196. Smith, S.J. and Sharpley, A.N., Nitrogen availability from surface-applied and soil-incorporated crop residues, *Agron. J.*, 85, 776, 1993.
197. Kolenbrander, G.J., Leaching of nitrogen in agriculture, in *Nitrogen Losses and Surface Run-Off*, Brogan, J. C., Ed., Nijhoff/Junk, The Netherlands, 1981, 199.
198. Frye, W.W., Blevins, R.L., Smith, M.S., and Corak, S.J., Role of annual legume cover crops in efficient use of water and nitrogen, in *Cropping Strategies for Efficient Use of Water and Nitrogen*, Hargrove, W.L., Ed., Special Publ. No. 51, ASA, CSSA, SSSA, Madison, WI, 1988, 129.
199. Muller, M.M., Leaching of subterranean clover-derived N from a loam soil, *Plant Soil*, 102, 185, 1987.
200. Wagger, M.G. and Mengel, D.B., The role of non-leguminous crops in the efficient use of water and nitrogen, in *Cropping Strategies for Efficient Use of Water and Nitrogen*, Hargrove, W.L., Ed., Special Publ. No. 1, ASA, CSSA, SSSA, Madison, WI, 1988, 115.
201. Singh, B. and Sekhan, G.S., Some measures of reducing leaching loss of nitrates beyond potential rooting zone. III. Proper crop rotation, *Plant Soil*, 47, 585, 1977.

202. Hayman, D.S., *Endogone* spore numbers in soil and vesicular-arbuscular mycorrhiza in wheat as influenced by season and soil treatment, *Trans. Br. Mycol. Soc.*, 54, 53, 1970.
203. Piché, Y. and Fortin, J.A., Development of mycorrhizal extramatricial mycelium and sclerotia on *Pinus strobus* seedlings, *New Phytol.*, 91, 211, 1982.
204. Abbott, L.K. and Robson, A.D., Growth stimulation of subterranean clover with vesicular arbuscular mycorrhizas, *Aust. J. Agric. Res.*, 28, 539, 1977.
205. Sojka, R.E., Karlen, D.L., and Sadler, E.J., Planting geometries and the efficient use of water and nutrients, in *Cropping Strategies for Efficient Use of Water and Nitrogen*, Hargrove, W.L., Ed., Special Publ. No. 51, ASA, CSSA, SSSA, Madison, WI, 1988, 43.
206. Hook, J.E. and Gascho, G.J., Multiple cropping for efficient use of water and nitrogen, in *Cropping Strategies for Efficient Use of Water and Nutrients*, Hargrove, W.L., Ed., Special Publ. No. 51, ASA, CSSA, SSSA, Madison, WI, 1998, 7.
207. Ebelhar, S.A., Frye, W.W., and Blevins, R.L., Nitrogen for legume cover crops for no tillage corn, *Agron. J.*, 76, 51, 1984.
208. Ismaili, M. and Weaver, R.W., Competition between Siratro and Kleingrass for ^{15}N labeled mineralized nitrogen, *Plant Soil*, 96, 327, 1986.
209. Ta, T.C. and Ferris, M.A., Effects of alfalfa proportions and clipping frequencies on timothy-alfalfa mixtures. II. Nitrogen fixation and transfer, *Agron. J.*, 79, 820, 1987.
210. Pierce, F.J. and Rice, C.W., Crop rotation and its impact on efficiency of water and nitrogen use, in *Cropping Strategies for Efficient Use of Water and Nitrogen*, Hargrove, W.L., Ed., Special Publ. No. 51, ASA, CSSA, SSSA, Madison, WI, 1988, 21.
211. Robbins, C.W. and Carter, D.L., Nitrate-nitrogen leached below the root zone during and following alfalfa, *J. Environ. Qual.*, 9, 447, 1980.
212. Johnson, J.W., Welch, L.F., and Kurtz, L.T., Environmental implications of N-fixation by soybeans, *J. Environ. Qual.*, 4, 303, 1975.
213. Bergstrom, L. and Johansson, R., Leaching of nitrate from monolith lysimeters of different types of agricultural soils, *J. Environ. Qual.*, 20, 801, 1991.
214. Bergstrom, L. and Brink, N., Effects of differential applications of fertilizer N on leaching losses and distribution of inorganic N in the soil, *Plant Soil*, 93, 333, 1986.
215. Cervellini, A., Ruschel, A.P., Matsui, E., Salati, E., Zagatto, E.A.G., Ferreyra, H.F.F., Krug, F.J., Bergamin, H., Reichardt, K., Meirelles, N.M.F., Libardi, P.L., Victoria, R., Saito, S.M.T., and Nascimento, V.F., Fate of ^{15}N applied as ammonium sulfate to a bean crop, in *Soil Nitrogen as Fertilizer or Pollutant*, International Atomic Energy Agency, Panel Proceedings Series, Vienna, 1980, 23.
216. Angle, J.S., Nitrate leaching losses from soybeans (*Glycine max* L. Merr.), *Agric. Ecosystems Environ.*, 31, 91, 1990.
217. Filipović, R. and Stevanović, D., Soil and water nitrate levels in relation to fertilizer utilization in Yugoslavia, in *Soil Nitrogen as Fertilizer or Pollutant*, International Atomic Energy Agency, Panel Proceedings Series, Vienna, 1980, 37.
218. Filipović R., Fertilizer-nitrogen residues: useful conservation and pollutant potential under maize, in *Soil Nitrogen as Fertilizer or Pollutant*, International Atomic Energy Agency, Panel Proceedings Series, Vienna, 1980, 47.
219. Arora, R.P., Sachdev, M.S., Sud, Y.K., Luthra, U.K., and Subbiah, B.V., Fate of fertilizer nitrogen in a multiple cropping system, in *Soil Nitrogen as Fertilizer or Pollutant*, International Atomic Energy Agency, Panel Proceedings Series, Vienna, 1980, 3.
220. Krishnappa, A.M. and Shinde, J.E., Fate of ^{15}N labelled urea fertilizer under conditions of tropical flooded-rice culture, in *Soil Nitrogen As Fertilizer or Pollutant*, International Atomic Energy Agency, Panel Proceedings Series, Vienna, 1980, 127.

221. Nielsen, D.R., Biggar, J.W., MacIntyre, J., and Tanji, K.K., Field investigation of water and nitrate-nitrogen movement in yolo soil, in *Soil Nitrogen as Fertilizer or Pollutant*, International Atomic Energy Agency, Panel Proceedings Series, Vienna, 1980, 145.
222. Cameron, D.R., Kowalenko, C.G., and Ivarson, K.C., Nitrogen and chloride leaching in a sandy field plot, *Soil Sci.*, 126, 174, 1978.
223. Wild, A., Nitrate in drinking water: health hazard unlikely, *Nature*, (London), 268, 197, 1977.
224. Gauntlet, R.B., Nitrate removal from water by ion exchange, *Water Treat. Exam.*, 24, 172, 1975.
225. Gambrell, R.P., Gilliam, J. W., and Weed, S. B., Denitrification in subsoils of the North Carolina Coastal Plains as affected by soil drainage, *J. Environ. Qual.*, 4, 311, 1975.
226. Gilliam, J.W., Daniels, R.B., and Lutz, J.F., Nitrogen content of shallow groundwater in the North Carolina Coastal Plain, *J. Environ. Qual.*, 147, 3, 1974.
227. Egboka, B.C.E., Nitrate contamination of shallow groundwaters in Ontario, Canada, *Sci. Total Environ.*, 35, 53, 1984.
228. Foster, S.S.D., Kelley, D.P., and James, R., The evidence for zones of biodenitrification in British aquifers, in *Planetary Ecology*, Caldwell, D. E., Brierly, J. A., and Brierly, C. L., Eds., Van Nostrand Reinhold, New York, 1985, 356.
229. Whitelaw, K. and Rees, J.F., Nitrate-reducing and ammonium-oxidizing bacteria in the vadose zone of the chalk aquifer of England, *Geomicrobiol J.*, 2, 179, 1980.
230. Schipper, L.A., Cooper, A.B., and Dyck, W.J., Mitigating nonpoint-source nitrate pollution by riparian-zone denitrification, in *Nitrate Contamination: Exposure Consequence and Control*, Bogárdi, I. and Kuzelka, R. D., Eds., NATO ASI Series, Vol. G30, Springer-Verlag, Berlin, 1991, 401.
231. Capalbo, S.M. and Antle, J.M., Incorporating social costs in the returns to agricultural research, *Am. J. Agric. Econ.*, 71, 458, 1989.
232. Antle, J.M. and Just, R.E., Conceptual and empirical foundations for agricultural-environmental policy analysis, *J. Environ. Qual.*, 21, 307, 1992.
233. Allen, L., Bennett, T., Lehr, J.H., and Petty, R.J., DRASTIC — A Standardized System for Evaluating Ground-Water Pollution Potential Using Hydrogeologic Settings, EPA 600/2-85-028, U.S. Government Printing Office, Washington, D.C., 1985, 1.
234. Knox, E. and Moody, D.W., Influence of hydrology, soil properties and agricultural land use on nitrogen in groundwater, in *Managing Nitrogen for Groundwater Quality and Farm Profitability*, Follett, R.F., Keeney, D.R., and Cruse R. M., Eds., Soil Science Society of America, Madison, WI, 1991, 19.
235. Williams, J.R. and Kissel, D.E., Water percolation: an indicator of nitrogen-leaching potential, in *Managing Nitrogen for Groundwater Quality and Farm Profitability*, Follett, R.F., Keeney, D.R., and Cruse, R.M., Eds., Soil Science Society of America, Madison, WI, 1991, 59.
236. Williams, J.R., Jones, C.A., and Dyke, P.T., A modeling approach to determining relationship between erosion and soil productivity, *Trans. Am. Soc. Agric. Eng.*, 27, 129, 1984.
237. Frissel, M.J. and van Veen, J.A., Eds., *Simulation of Nitrogen Behaviour of Soil-Plant Systems*, Pudoc, Centre for Agricultural Publishing and Documentation, Wageningen, 1981.
238. Follett, R.F., Ed., *Nitrogen Management and Ground Water Protection*, Developments in Agricultural and Managed-Forest Ecology 21, Elsevier, Amsterdam, 1989.

239. Follett, R.F., Keeney, D.R., and Cruse, R.M., Eds., *Managing Nitrogen for Ground Water Quality and Profitability*, Soil Science Society of America, Madison, WI, 1991.
240. Tanji, K.K. and Mehran, M., Conceptual and dynamic models for nitrogen in irrigated croplands, in *Nitrate in Effluents from Irrigated Agriculture*, Pratt, P.F., Ed., Final Report to NSF, University of California, Davis, 16, 555, 1979.
241. Tanji, K.K., Mehran, M., and Gupta, S.K., Water and nitrogen fluxes in the root zone of irrigated maize, in *Simulation of Nitrogen Behaviour of Soil-Plant Systems*, Frissel, M.J. and van Veen, J.A., Eds., Pudoc, Centre for Agricultural Publishing and Documentation, Wageningen, 1981, 51.
242. Burns, I.G., An equation to predict the leaching of surface-applied nitrate, *J. Agric. Sci.*, 85, 443, 1975.
243. Duffy, J., Chung, C., Boast, C., and Franklin, M., A simulation model of biophysiochemical transformation of nitrogen in tile-drained corn belt soil, *J. Environ. Qual.*, 4, 477, 1975.
244. Saxton, K.E., Schuman, G.E., and Burwell, R.E., Modeling nitrate movement and dissipation in fertilized soils, *Soil Sci. Soc. Am. J.*, 41, 265, 1977.
245. Singh, B., Biswas, C.R., and Sekhon, G.S., A rational approach for optimizing application rates of fertilizer nitrogen to reduce potential nitrate pollution of natural water, *Agric. Environ.*, 4, 57, 1978.
246. Tanji, K.K., Fried, M., and van de Pol, R.M., A steady-state conceptual nitrogen model for estimating nitrogen emissions from cropped lands, *J. Environ. Qual.*, 6, 155, 1977.
247. Wagenet, R.J., Simulation of soil-water and nitrogen movement, in *Simulation of Nitrogen Behaviour of Soil-Plant Systems*, Frissel, M. J. and van Veen, J.A., Eds., Pudoc, Centre for Agricultural Publishing and Documentation, Wageningen, 1981, 67.
248. Kruh, G. and Segall, E., Nitrogen dynamics in soil, in *Simulation of Nitrogen Behaviour of Soil-Plant Systems*, Frissel, M. J. and van Veen, J. A., Eds., Pudoc, Centre for Agricultural Publishing and Documentation, Wageningen, 1981, 109.
249. van Veen, J.A. and Frissel, M.J., Simulation model of the behaviour of N in soil, in *Simulation of Nitrogen Behaviour of Soil-Plant Systems*, Frissel, M.J. and van Veen, J. A., Eds., Pudoc, Centre for Agricultural Publishing and Documentation, Wageningen, 1981, 126.
250. McGill, W.B., Hunt, H.W., Woodmansee, R.G., Reuss, J.O., and Paustian, K.H., Formulation, process controls, parameters and performance of PHOENIX: a model of carbon and nitrogen dynamics in grassland soils, in *Simulation of Nitrogen Behaviour of Soil-Plant Systems*, Frissel, M.J. and van Veen, J.A., Eds., Pudoc, Centre for Agricultural Publishing and Documentation, Wageningen, 1981, 171.
251. Seligman, N.G. and van Keulen, H., PAPRAN: a simulation model of annual pasture production limited by rainfall and nitrogen, in *Simulation of Nitrogen Behaviour of Soil-Plant Systems*, Frissel, M.J. and van Veen, J.A., Eds., Pudoc, Centre for Agricultural Publishing and Documentation, Wageningen, 1981, 192.
252. Addiscott, T. M., Leaching of nitrate in structured soils, in *Simulation of Nitrogen Behaviour of Soil-Plant Systems*, Frissel, M.J. and van Veen, J.A., Eds., Pudoc, Centre for Agricultural Publishing and Documentation, Wageningen, 1981, 245.
253. Walter, M.F., Bubenzer, G.D., and Converse, J.C., Predicting vertical movement of manurial nitrogen in soil, *Trans. Am. Soc. Agr. Eng.*, 18, 100, 1975.
254. Laudelot, H., Germain, L., Chabalier, P.F., and Chiang, C.N., Computer simulation of loss of fertilizer nitrogen through chemical decomposition of nitrite, *J. Soil Sci.*, 28, 329, 1977.

255. Smith, K.A., A model of denitrification in aggregated soils, in *Simulation of Nitrogen Behaviour of Soil-Plant System*, Frissel, M.J. and van Veen, J.A., Eds., Pudoc, Centre for Agricultural Publishing and Documentation, Wageningen, 1981, 259.
256. Selim, H.M. and Iskandar, I.K., WASTEN: a model for nitrogen behaviour in soils irrigated with liquid waste, in *Simulation of Nitrogen Behaviour of Soil-Plant Systems*, Frissel, M.J. and van Veen, J.A., Eds., Pudoc, Centre for Agricultural Publishing and Documentation, Wageningen, 1981, 96.
257. Rao, P.S.C., Davidson, J.M., and Jessup, R.E., Simulation of nitrogen behaviour in the root zone of cropped land areas receiving organic wastes, in *Simulation of Nitrogen Behaviour of Soil Plant Systems*, Frissel, M.J. and van Veen, J.A., Eds., Pudoc, Centre for Agricultural Publishing and Documentation, Wageningen, 1981, 81.
258. Parton, W.J., Gould, W.D., Adamsen, F.J., Torbit, S., and Woodmansee, R.G., NH_3 volatilization model, in *Simulation of Nitrogen Behaviour of Soil-Plant Systems*, Frissel, M.J. and van Veen, J.A., Eds., Pudoc, Centre for Agricultural Publishing and Documentation, Wageningen, 1981, 233.
259. Juma, N.G. and Paul, E.A., Use of tracers and computer simulation techniques to assess mineralization and immobilization of soil nitrogen, in *Simulation of Nitrogen Behaviour of Soil-Plant Systems*, Frissel, M.J. and van Veen J.A., Eds., Pudoc, Centre of Agricultural Publishing and Documentation, Wageningen, 1981, 145.
260. Wagenet, R.J. and Hutson, J.L., *LEACHM: Leaching Estimates and Chemistry Model*, Vol. 2, Water Res. Inst., Cornell University, Ithaca, NY, 1987.
261. Wagenet, R.J. and Hutson, J.L., *LEACHM: Leaching Estimation and Chemistry Model: A Process Based Model of Water and Solute Movement Transformation, Plant Uptake and Chemical Reactions in the Unsaturated Zone*, Vol. 2, Version 2, Water Resources Inst., Cornell University, Ithaca, NY, 1989.
262. Hutson, J.L. and Wagenet, R.J., *LEACHM: Leaching Estimation and Chemistry Model: A Process-Based Model of Water and Solute Movement, Transformation, Plant Uptake and Chemical Reactions in the Unsaturated Zone*, Version 3, Cornell University, Ithaca, NY, 1992.
263. Shaffer, M.J., Halvorson, A.D., and Pierce, F.J., Nitrate leaching and economic analysis package (NLEAP): model description and application, in *Managing Nitrogen for Ground Water Quality and Farm Profitability*, Follett, R.F., Keeney, D.R., and Cruse, R.M., Eds., Soil Science Society of America, Madison, WI, 1991, 285.
264. DeCoursey, D.G. and Rojas, K.W., RZWQM: a model for simulating the movement of water and solutes in the root zone, in *Proc. Int. Symp. Water Quality Modeling of Agricultural Non-Point Sources*, Parts 1 and 2, DeCorsey, D.G., Ed., Utah State University, Logan, June 19–23, 1988, USDA, ARS-81, 1990.
265. Young, R.A., Onstad, C.A., Bosh, D.B., and Anderson, W.P., AGNPS, Agricultural Nonpoint-source Pollution Model. USDA, Agric. Res. Ser. Conservation Research Report 35, U.S. Government Printing Office, Washington, D.C., 1987.
266. Knisel, W.G., *CREAMS: A Field-Scale for Chemicals, Runoff, and Erosion from Agricultural Management Systems*, U.S. Department of Agriculture, Agric. Res. Ser. Conservation Research Report 26, U.S. Government Printing Office, Washington, D.C., 1980.
267. Leavesley, G.H., Lichty, R.W., Troutman, B.M., and Saindon, L.G., Precipitation-Runoff Modeling System: User's manual, Water Resources Investigations, Report 83-4238, U.S. Geol. Surv., Washington, D.C., 1983.

268. Johanson, R.C., Imhoff, J.C., Kittle, J.L., Jr., and Donigian, A.S., Jr., Hydrological Simulation Program Fortran (HSPF): User's Manual for Release 8.0 EPA 600/3-84-66, U.S. EPA, Washington D.C., 1984.
269. Beasley, D.B., Huggins, L.F., and Monke, E.J., Planning for water quality using the ANSWERS approach, in *Proc. Hydrological Transport Modeling Symposium*, New Orleans, December 10–11, 1979, Am. Soc. Agric. Eng., 1979, 21.
270. Leonard, R.A., Knisel, W.G., and Still, D.A., GLEAMS: groundwater loading effects of agricultural management systems, *Trans. Am. Soc. Agric. Eng.*, 30, 1403, 1987.
271. Frissel, M.J. and van Veen, J.A., Soil-nitrogen transformations in relation to leaching, in *Soil Nitrogen as Fertilizer or Pollutant*, International Atomic Energy Agency, Panel Proceedings Series, Vienna, 1980, 61.
272. Cooper, L.I., Law — part 1: Clean Water Act, Coastal Zone Management Act, and the Safe Drinking Water Act, *Res. J. Water Pollut. Control Fed.*, 63, 302, 1991.
273. Nichols, A.B., USDA moves to clean up the nation's water, *Water Environ. Technol.*, 3, 50, 1991.

CHAPTER 2

Phosphorus Fertilizers

Paul J. Withers and Andrew N. Sharpley

0-87371-859-3/95/$0.00+$.50

I. INTRODUCTION

Phosphorus (P) fertilizers have had an enormous impact on agricultural production worldwide and are an essential input in the conversion of virgin and impoverished land to agricultural use.[1] Large increases in crop yields are obtained when inorganic and organic P fertilizers are applied to P-deficient soil and maintaining adequate soil P fertility is considered essential to optimize the utilization of other plant nutrients, particularly nitrogen (N).[2]

The manufacture of inorganic P fertilizers expanded dramatically after the Second World War in response to the demand for increased agricultural output. In the developed countries, large processing plants were built to convert imported rock phosphate into a variety of water-soluble and partially water-soluble P fertilizer products. Basic slag, a by-product from the steel industry, also became widely used. As a consequence, soil P fertility was improved and, in many areas, is now at a level at which the need for P fertilizer is less. In addition, there is increasing awareness among farmers that P is a comparatively expensive plant nutrient and its application needs to be cost effective.

From an environmental viewpoint, the mining of phosphate rock can be considered an undesirable intervention in the cycling of P within natural landscapes. Since P is not recycled in the atmosphere in any quantity, there is a small but continuous loss of P, by natural erosion and leaching, from the terrestrial environment to the oceans.[3,4] This removal of P from the terrestrial cycle has been accelerated by the increased use of manufactured derivatives of phosphate rock and their ultimate disposal, in domestic and industrial effluents, to surface waters. Maybeck[5] estimated that over 20 million tonnes of P are transported annually to the oceans by world rivers with consequent risks of eutrophication.

The accumulation of P in soil from the unbalanced use of organic and inorganic P fertilizers has also raised concerns over the agricultural contribution to eutrophication of inland waters and the marine environment.[6,7] Restrictions on P inputs to land are already in operation in some European countries, notably, The Netherlands. Harmful emissions of fluoride,[8] the accumulation of by-product gypsums,[9] and the uneven redistribution of heavy metals[10] are other environmental problems associated with either the manufacture or use of P products derived from rock phosphate. In the developed countries, reclamation of land after mining may also be of local concern.

With these potential environmental problems, economic considerations, and the lower demand for P, the manufacture of inorganic P fertilizers in the developed countries has declined. From 1990 to 1992, world production of phosphate rock decreased by about 10%.[11] Phosphorus fertilizers are now manufactured more in the countries where phosphate rock deposits occur and lower-quality products are more commonly produced. In Eastern Europe, Russia, and the developing countries, the demand for P fertilizers remains large and in line with the need to grow more food. However, actual consumption is often less than required for economic reasons.[11,12]

A. CURRENT SOURCES AND SUPPLIES

World reserves of *in situ* phosphate rock have been estimated at nearly 13 billion tonnes of P, although only about 3.5 billion tonnes are currently minable.[11] About 8.5 billion tonnes are potentially available with further investment in the industry and improvements in mining technology. Offshore deposits of rock phosphate are also poten-

Table 1 Reserves, Production, and Consumption of Phosphorus (P) in the Major Continents

			Consumption[c]		
Continent	Reserves[a]	Production[b]	Total	Straights[d]	Compounds[e]
		(Million tonnes P)		(% of total)	
Asia	907	2.7	5.2	50.1	49.9
Oceania	3	0.2	0.4	6.8	93.2
Middle East	145	1.5	0.8	48.9	51.1
Africa	1961	4.8	0.4	54.9	45.1
North America	161	6.2	1.9	26.7	73.3
Latin America	92	0.5	1.0	23.7	76.3
Europe	6	0.1	2.6	26.3	73.7
Russia	110	2.9	3.4	18.4	81.6
World total	3385	18.9	15.7	33.4	66.6

[a] Estimates of currently minable P reserves.
[b] Actual production in 1992.
[c] Total P deliveries in 1990/1991. About 10 to 15% is used in livestock feeds and the detergent industries.
[d] Includes basic slag, ground rock phosphate, superphosphate, triple superphosphate (TSP), and other elemental P fertilizers.
[e] Includes ammonium phosphates and other NP, PK, and NPK compound fertilizers.

Data adapted from References 11 and 13.

tially available. The reserves of rock phosphate can therefore be considered sufficient to cope with current demand (Table 1). The main areas of production are the U.S., Africa (particularly Morocco, Tunisia, and South Africa), Russia, China, and the Middle East (particularly Israel and Jordan). The source of the rock phosphate affects both its chemical composition and physical characteristics.

B. USE, FORMS, AND MODES OF APPLICATION

1. Inorganic Phosphorus Fertilizers

The main forms of inorganic P fertilizers available on the world market, the processes involved in their manufacture, and their relative availabilities for plant uptake are summarized in Figure 1. Their agronomic effectiveness depends on the speed of dissolution in the soil, which is usually chemically defined by their solubility in either water, 2% citric acid, ammonium citrate, or 2% formic acid, and crop requirements. Potatoes, for example, have a large demand for water-soluble P at planting. These factors largely determine the method of application and the farming systems to which different P fertilizers are most suited. Application rates for selected crops in selected countries are shown in Table 2.

Water Insoluble — Water-insoluble P fertilizers are characterized by their variable quality and their application in powdered or mini-granular form. Ground rock phosphate (GRP) is the main water-insoluble straight P fertilizer used on agricultural land but represents only about 3% of world consumption.[13] Citric-soluble dicalcium phosphate (DCP) is more usually a constituent of compound fertilizers, although it can be applied straight, as a by-product of the detergent industry.

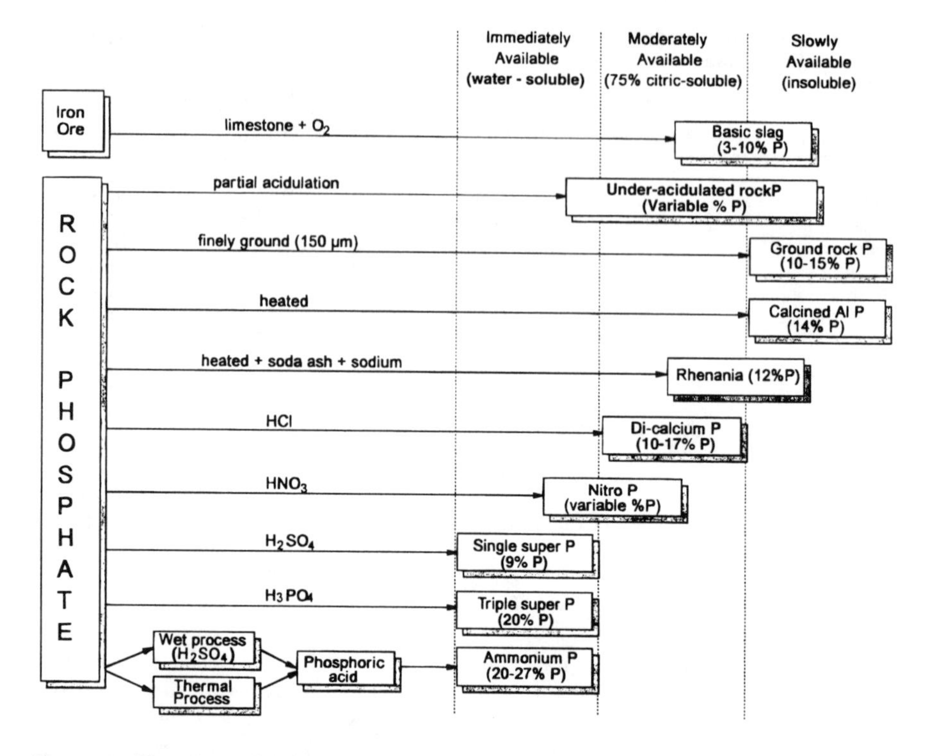

Figure 1 Phosphorus fertilizers, their manufacture, and relative plant availabilities.

Table 2 Phosphorus Fertilizer Application Rates to Selected Crops in Selected Countries (kg P ha^{-1})

Country	Wheat	Maize	Potatoes	Soybean	Sugar Beet	Tobacco	Grass
Australia	10	NA[a]	66[b]	NA	35[c]	70	9[d]
Brazil	20	NA	112	16	17[c]	NA	NA
China	22	9	20	17	13	13	NA
Russia	18	31	47	22	73	NA	12[e]
South Africa	13	14	59	29	20[c]	97	24[e]
Turkey	12	14	16	17	41	7	56[f]
U.K.	29	21	90	NA	28	NA	19[e]
U.S.	16	31	73	23	18	59	1[d]

[a] NA denotes the crop is either not grown or data are not available.
[b] All vegetables.
[c] Sugar cane.
[d] Pasture and fodder.
[e] Temporary or intensive grass.
[f] Clover mixtures.

Data adapted from Reference 14.

Table 3 Agronomic Effectiveness of Gafsa GRP at Two Degrees of Fineness Compared to SSP on Limed and Unlimed Grassland Sites

	Relative Increases in Dry Matter Yield[a]							
	Unlimed Sites (year[b])				Limed sites (year[b])			
	0	1	2	3	0	1	2	3
Normally ground (60–80% > 150 μm)	25	82	136	154	0	4	28	51
Hyperphosphate (90% > 63 μm)	25	100	143	154	21	52	80	70

[a] Yield response as a percentage of that obtained from SSP when applied at a rate of 146 kg P ha^{-1}.
[b] Year 0 is the year of application. Years 1, 2, and 3 are the first, second, and third residual years, respectively. Application was in winter or early spring.
Data from Davies, G.R., *J. Sci. Food Agric.*, 35, 265, 1984. With permission.

The quality and reactivity of GRP fertilizers is variable and depends on the amount of carbonate present in the rock. Chemical extraction with 2% formic acid is considered the most appropriate measure of chemical reactivity, although repeated extraction with 2% citric acid is recommended for carbonate-rich rocks.[15] GRP is not considered a suitable fertilizer in the U.K. unless it is "soft", i.e., 55% soluble in 2% formic acid. Moroccan "Gafsa" and North Carolina GRPs usually meet these criteria.

Davies[16] found GRP to be usually less than 60% as effective as water-soluble P fertilizers in the first year of application, but equally or more effective over a number of years for grassland on slightly acidic soils (Table 3). Under current EU fertilizer regulations, GRP products must be ground such that 80% passes a 150-μm sieve. GRP is certainly slower acting when granulated, but its agronomic effectiveness is not greatly increased by grinding finer than 150 μm, except where dissolution is inhibited by free calcium ions in the soil solution (Table 3). GRP is therefore best suited to acidic soils[17,18] and is not recommended for P-demanding arable crops on soils at or above pH 6.5 or where lime has recently been applied.

Other forms of water-insoluble straight P fertilizers are used only in small quantities. Basic slag, once widely applied to grass and forage crops, is now in short supply because of changes in the method of steel manufacture, although it is still applied in France and Germany. Basic slag is more citric-soluble than GRP and has been found to be more quickly available in the first year of application.[16] It also has a liming value. Various mixtures of GRP and basic slag have been marketed and their agronomic effectiveness depends on which is in most abundance in the mixture. Calcined aluminum phosphate (CAP) is slower acting than GRP in acid and neutral soils, although its reactivity is less affected by free calcium carbonate than GRP. Rhenania P appears equally or slightly more effective than GRP,[16] contains sodium, and has a liming value.

Water Soluble — Granulated water-soluble straight P fertilizers are produced by reaction of phosphate rock with either sulfuric, nitric, or phosphoric acids (Figure 1). Phosphoric acid may also be used directly in liquid fertilizers. Single superphosphate (SSP), which contains a mixture of monocalcium phosphate and calcium sulfate, was first manufactured in 1840 and dominated the world P fertilizer market until about 1955.[12] It is still widely used but now represents only 17% of world consumption.[13]

Triple superphosphate (TSP), which contains mainly mono-calcium phosphate, has increased and now represents 10% of world consumption.[13] Commercial use TSP fertilizers may not be 100% water soluble because of impurities [iron (Fe) and aluminum (Al) phosphates] formed during the manufacturing process. Mullins and Evans[19] found solubilities ranging from 80 to 94%, although this did not limit their agronomic effectiveness.

Granulated ammonium phosphates and other NP, potassium (K), and NPK compounds are the most widely used P fertilizers, representing two thirds of world consumption (Table 1). Mono- and diammonium phosphates produced from phosphoric acid are considered to be totally water soluble, but nitrophosphates contain a mixture of ammonium phosphate and DCP, which is citrate soluble but not water soluble. Compounds with <50% of the P as water soluble must be considered less effective and therefore best suited for maintenance applications to soils of high fertility.

Rock phosphates that are only partially treated with acid (termed underacidulated) contain mixtures of monocalcium phosphate, tricalcium phosphate, calcium sulfate, and GRP. These materials are up to 50% water soluble and therefore considered to be intermediate in effectiveness between GRP and SSP or TSP.[16] The presence of water-soluble monocalcium phosphate alongside GRP may improve dissolution of GRP because of the localized acidity that develops whem monocalcium phosphate dissolves.[20] Underacidulated rock phosphate products have been advocated for use in the tropics by Hammond et al.,[21] who argued that the greater dissolution of GRP would not only reduce P fixation by Al and Fe on highly weathered acid soils but also improve the utilization of a locally mined resource. Blends of granulated GRP and SSP or TSP have also been marketed to useful effect in developed countries.

Water-soluble P fertilizers are characterized by their more uniform quality and ease of application, and are best applied to P-deficient soils, where a rapid availability of P is required, especially for P-demanding crops such as potatoes. Numerous experiments have shown that, in such soils, the placement of P either with or close to the germinating seed gives a yield advantage compared to broadcast application and/or incorporation in the seedbed.[22,23] For cereals on soils low in P, yield benefits are obtained by drilling the seed and fertilizer together. For other crops, fertilizer P is best placed 5 cm to the side and 2.5 cm below the seed, because of the risk of damage to the germination of small seeds or development of potato sprouts from the highly acidic and salt-rich environment surrounding fertilizer granules. Placement of ammonium phosphate fertilizer may give additional yield benefits on neutral and alkaline soils due to enhanced P uptake from the acidifying effect around plant roots when ammonium ions are absorbed.[24] Deep placement of P in the upper subsoil has also been used to increase total P availability to deep-rooting crops.

2. *Organic Phosphorus Fertilizers*

Primary sources of organic P include livestock manures and sewage sludge. Bonemeal (or flour) and various miscellaneous by-products produced from the sea are only of minor importance. The P content and composition of fresh livestock manures is variable (Table 4) depending on the type and age of the animal, the type of bedding material used, and the composition and digestibility of the diet, particularly where mineral supplements (DCP) are used. Haynes and Williams[27] found that the proportion of inorganic P (mostly tri-calcium phosphate) in sheep feces increased with increasing rates of super phosphate

Table 4 Average P, N, and K Contents (Dry Weight Basis) of Animal Manures (g kg^{-1})

	Nitrogen	Phosphorus	Potassium
Beef	32.5	9.6	20.8
Dairy	39.6	6.7	31.6
Poultry layers	49.0	20.8	20.8
Poultry broilers	40.0	16.9	19.0
Sheep	44.4	10.3	30.5
Swine	76.2	17.6	26.2
Turkeys	59.6	16.5	19.4
Sewage sludge	33.0	23.0	3.0

Data adapted from References 25 and 26.

applied to pasture. Upon storage, the inorganic P component in feces is further increased by microbial mineralization of the organic P fraction.[28,29] In stored manures, inorganic P represents 80 to 90% of the total P content and is largely composed of calcium phosphates of varying solubility. The P contained in livestock manures is more slowly available than in water-soluble P fertilizers but is generally considered to become totally available in the longer term.[30,31]

Sewage sludge is considered a useful source of P for agricultural crops but its P content and potential plant availability can also be variable depending on the source and method of treatment. Sommers[26] found total P concentrations in the dry matter of between 11 and 55 g kg^{-1} in aerobically treated (i.e., primary or secondary treatment only) sludges but between 5 and 143 g kg^{-1} in anaerobically digested sludges. However, median concentrations of P in the two types of sludge were very similar (Table 4). Sewage sludge P contents are larger where P is chemically precipitated at sewage treatment works discharging to sensitive waters.

Phosphorus contained in anaerobically digested sludge largely occurs as inorganic precipitates of calcium (Ca), iron (Fe), or aluminum (Al).[32] Sludges produced after only primary or secondary treatment have a higher content of organic P.[33] As with manures, sewage sludge is generally considered to be about 50% available in the first year of application,[34] although anaerobically digested and chemically precipitated or heat-treated sludges tend to have lower plant availability than aerobically treated sludges.[35,36]

II. MOVEMENT AND REACTION IN SOIL

A. FORMS OF SOIL PHOSPHORUS

Soil P exists in inorganic and organic forms (Figure 2). Inorganic P forms are dominated by hydrous sesquioxides, amorphous and crystalline Al and Fe compounds in acidic, non-calcareous soils, and by Ca compounds in alkaline, calcareous soils. Organic P forms include relatively labile phospholipids, inositols, and fulvic acids, while more resistant forms are comprised of humic acids (Figure 2). Dynamic transformations between the different P forms occur continuously to maintain an equilibrium P concentration in the soil solution.

In most soils, the P content of surface horizons is greater than of subsoil due to the adsorption of added P, greater biological activity, and accumulation of organic material

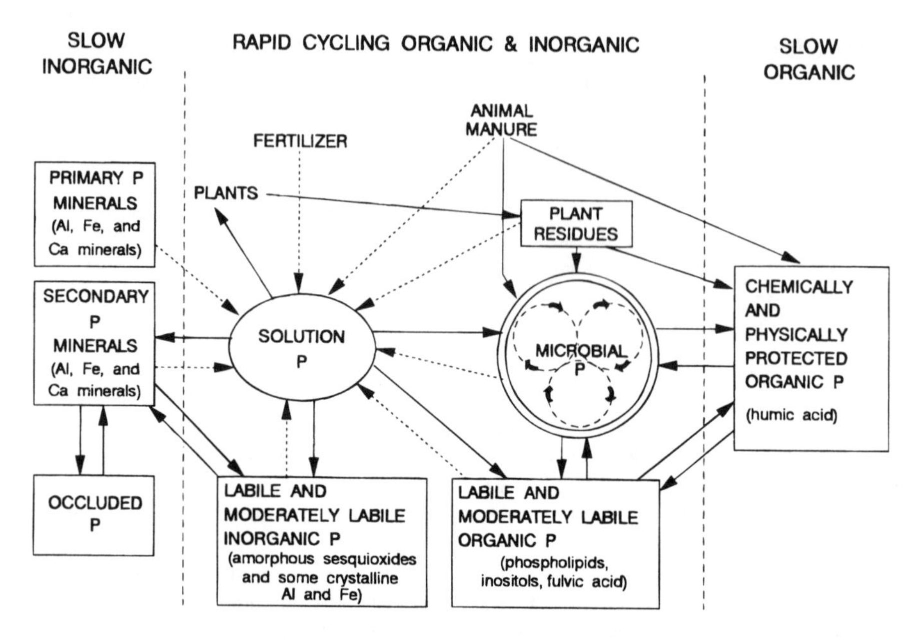

Figure 2 The soil P cycle: its components and measurable fractions. (Adapted from Reference 37.)

in surface layers. Soil P content varies with parent material, texture, and management factors, such as rate and type of P applied, and whether the soil is regularly cultivated or conservation tilled. These factors also influence the relative amounts of inorganic and organic P. In most agricultural soils, 50 to 75% of the P is inorganic, although this fraction can vary from 10 to 90%. In acid upland soils, dissolved organic P can form the major portion of total P in the soil solution, even where water-soluble inorganic P fertilizers have been applied.[38]

B. TRANSFORMATIONS

Phosphorus amendments, either in organic or inorganic form, are needed to maintain adequate available soil P for plant uptake in modern agricultural systems. The level of these amendments varies with both soil and plant type.[34,39] Once applied, P is either taken up by the crop, incorporated into organic P,[40] or becomes weakly (physisorption) or strongly (chemisorption) adsorbed onto Al, Fe, and Ca surfaces.[41] After the initial adsorption reaction, there is a gradual fixation (absorption) of added P that renders a proportion of the adsorbed P unavailable for plant uptake (Figure 3). Organic P compounds may also become resistant to hydrolysis by phosphatase through complexation with Al and Fe.[42]

Sharpley et al.[43,44] showed that the portion of P remaining as plant available 6 months after fertilizer P application decreased from 47 to 27% as clay, organic C, Fe, Al, and calcium carbonate ($CaCO_3$) content increased for over 200 soils in different degrees of weathering. Although once considered to be largely irreversible, fixed P can be slowly

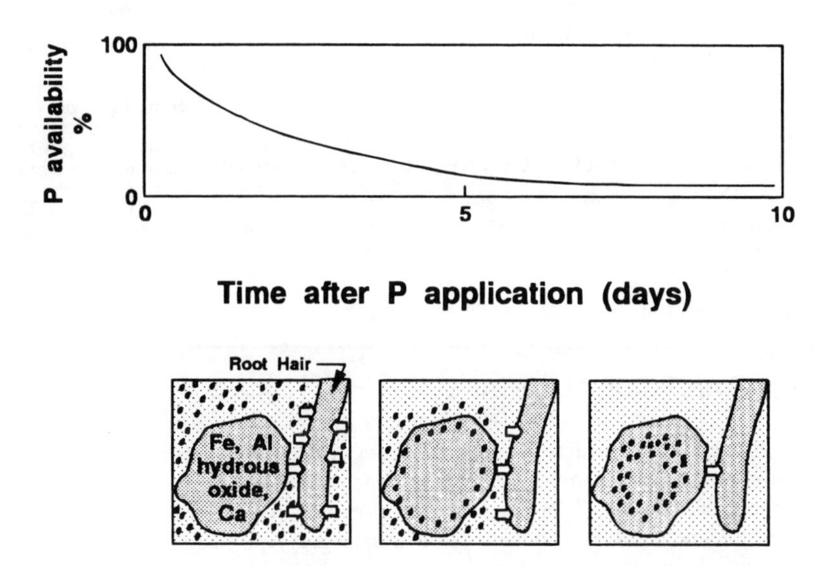

Figure 3 Plant availability of P with time after application.

released back into the soil solution when reserves of less strongly held P are exhausted.[2,45] However, the release of fixed P is very slow and, on impoverished or highly weathered, acidic soils, annual application of fertilizer P is necessary to maintain optimum available P in the soil. With regular fertilization, the importance of fixation processes is diminished as the soil P sorption capacity becomes slowly saturated and a larger concentration of P is maintained in the soil solution.

Even though inorganic P has generally been considered the major source of plant available P in soils, the mineralization of labile organic P has also been shown to be important in both low- and high-fertility soils.[46,47] The amounts of P mineralized may be sufficient for plant needs without the need for supplementation with inorganic P fertilizers,[48] although crops receiving organic P amendments often only show irregular growth and uneven ripening. Amounts of P mineralized in temperate dryland soils range from 5 to 20 kg P ha^{-1} $year^{-1}$.[37] Mineralization of soil organic P tends to be higher in the tropics (67 to 157 kg P ha^{-1} $year^{-1}$), where distinct wet and dry seasons and higher soil temperatures enhance microbial activity.

The role of microbial biomass P as a dynamic intermediary between organic and inorganic forms is evident from Figure 2. In a study of P cycling through soil microbial biomass in England, Brookes et al.[49] measured annual P fluxes of 5 and 23 kg P ha^{-1} $year^{-1}$ in soils under continuous wheat and permanent grass, respectively. Although biomass P flux under continuous wheat was less than P uptake by the crop (20 kg P ha^{-1} $year^{-1}$), annual P flux in the grassland soils was much greater than P uptake by the grass (12 kg P ha^{-1} $year^{-1}$). Clearly, microbial P plays an important intermediary role in the short-term dynamics of organic P transformations and suggests that management practices maximizing the buildup of organic matter during autumn and winter may reduce the need for P fertilizers for plant growth during the following spring and early summer.

Table 5 Phosphorus Balance (Kg P ha^{-1} $year^{-1}$) and Efficiency of Plant and Animal Uptake of P for the U.S. and Selected European Countries

				Efficiency (%) of		
	Input	Output	Surplus	Plant Uptake	Animal Uptake	Total Uptake
U.S.	39	13	26	56	15	33
The Netherlands	143	55	88	69	24	38
Germany (East)	79	8	71	59	10	11
Germany (West)	84	29	55	76	34	35

Data adapted from References 55 and 56.

C. PLANT AND ANIMAL UPTAKE

The efficiency with which P amendments are utilized by plants in the year of application is usually less than 20%,[50,51] but dependent on a number of environmental[52–54] and management factors.[23] These include temperature, soil moisture, soil aeration, soil pH, type, and amount of clay content, soil nutrient status, and microbial activity. When soil temperatures are low during early plant growth or where soil moisture is limiting during drought periods, P uptake is reduced. Soil compaction inhibits P uptake by reducing pore space and the availability of water and oxygen. The availability of P to plants is reduced by complexation with Ca at high pH and by Fe and Al at low pH. Soils with high clay content tend to fix more P than sandy soils with a low clay content and therefore require more P to raise the soil test level.

Liming can increase P availability in soils by stimulating mineralization of organic P or may decrease P availability by the formation of insoluble calcium phosphates. High concentrations of ammonium-N in the soil with fertilizer P may interfere with and delay normal P fixation reactions, prolonging the availability of fertilizer P. Differences in pH and biological activity within the rhizosphere, including vesicular-arbuscular mycorrhizal associations with plant roots, can considerably enhance P uptake, especially on impoverished soils. In accessing subsoil P, deep-rooted plants naturally recycle P to the surface when crop residues are left on the surface. Placement of fertilizer P can help overcome low P utilization due to low soil temperatures or resticted rooting volume, thereby reducing application rates.

Uptake of P by animals from feed materials can influence the fate and transport of P in soils. A generalized P balance and efficiency of plant and animal uptake of P for the U.S. and several European countries indicates the potential for P accumulation in agricultural systems (Table 5). Although the magnitude of P input and output varies between countries, the relative proportions of P removal in plant and animal products is similar. In spite of the relatively efficient recycling of P in crop production of 56 to 76%, total P recovery by agriculture is only 11 to 38% (Table 5). In contrast to crop production, the recycling of P in animal production is only 10 to 34%. Thus, the efficiency of P recycling in agriculture is dominated by that of animal production, as 76 to 94% of the total crop production is fed to animals (in addition to P additives). Animal-specific studies of P excretion rates substantiate this poor retention of P with values of 70 to 80% measured for dairy cows,[57] sheep,[27] feeder pigs,[56] and 87% for poultry.[56] Clearly, agricultural systems that include confined animal operations can determine the overall efficiency of P recycling in agriculture and thereby the magnitude of P surpluses or potential soil P accumulations.

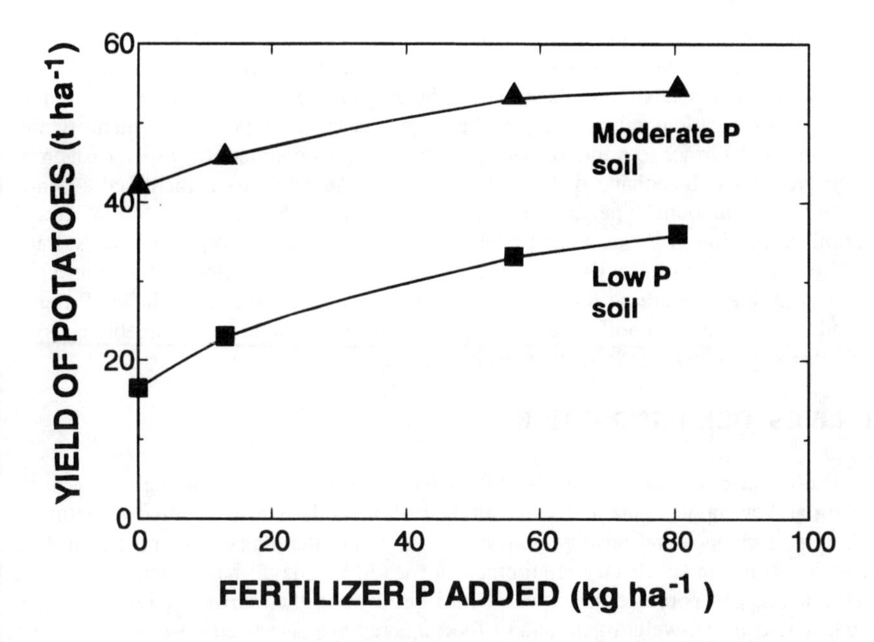

Figure 4 Yield response of potatoes to fertilizer P in the presence or absence of residual soil P. (Adapted from Reference 60.)

III. BENEFICIAL EFFECTS

A. SOIL FERTILITY AND CROP PRODUCTION

Phosphorus fertilizers have dramatically increased levels of food production worldwide and are essential to maximize agricultural output where land is in short supply. Economic responses in yield to applications of fresh P have been recorded for a number of crops.[58,59] Yield responses are largest on soils with a low level of available P and diminish as soil P fertility is increased. Residues of P in the soil, built up from previous fertilization, also confer a unique yield advantage that cannot be compensated for by fresh P, whatever quantity is applied (Figure 4).

At any given level of soil P, the amount of fertilizer P required depends on the crop being grown. Potatoes give the largest response to both fresh and residual fertilizer P and require larger amounts of P to achieve that response (Table 2). Differences between crops reflect differences in rooting patterns, and cereals that quickly develop a vigorous root system require only small amounts of P even on soils of low P fertility.[61] In the U.K., threshold soil P concentrations or target indices above which there is no economic yield response to freshly applied fertilizer P have been identified for individual crops.[34,59] Once target soil P levels have been achieved, recommended fertilizer policies are designed to balance crop offtake of P over a rotation and fertilizer P need not be applied annually to most crops.[2] This flexibility in fertilizing operations is a major advantage on large farms and in seasons where there is a limited time span in which to prepare a seedbed and drill the crop.

In establishing a vegetative cover, inorganic and organic P amendments can be considered beneficial in terms of reducing erosion and the transport of nutrients from agricultural land. In conservation tillage, where the crop residue is left in place to minimize soil water evaporation and erosion, or where cover crops are killed before maturity and left on the soil surface to minimize water, light, and weed competition with a cash crop, soil P cycling can be enhanced. The extent to which P availability is increased depends on the residue amount, type, and degree of incorporation. Sharpley and Smith[62] found that mineralization of P was greater when the residue of several crop types was surface applied compared to incorporated. However, such "secondary" flows of P from crop residues may also increase the amount, concentration, and bioavailability of P transported in agricultural runoff,[63–65] and must be taken into account in sustainable ecosystems that seek to balance P inputs with P outputs.

B. LIVESTOCK PRODUCTION

The maintenance of a satisfactory P content in livestock feeds through crop fertilization and/or supplementation with inorganic P minerals is also required to optimize animal performance and fertility. Kincaid et al.[66] found that dairy cows receiving 90% of their recommended dietary requirement for P (70 g P day^{-1}) had lower milk yields, feed intakes, and poorer efficiencies of milk production than cows receiving 120 g P day^{-1}. Reduced live-weight gain and/or food intake have also been observed in sheep[67] and beef cattle[68] on diets containing inadequate P.

Minimum dietary phosphorus requirements are dictated by animal live weight and performance. Diets containing P in excess of minimum dietary requirements do not improve animal performance but serve only to increase the amount of P excreted.[69,70] For non-ruminants, the level of phytase in the diet can have a considerable impact on the utilization and excretion of dietary P intake.[71] Carefully balancing the amount of P supplied with the animals requirements can, therefore, reduce excreted P and minimize many adverse environmental impacts.[72] The transfer of P in hay and grain from the area of production to the area of animal confinement is a secondary flow that also must be taken into account in systems management in sustainable agriculture.

IV. ADVERSE EFFECTS

A. SOIL ENVIRONMENT

1. Phosphorus Accumulation

The continual long-term application of P in fertilizer and manure at levels exceeding crop requirements can raise soil test P levels above those required for optimum crop yields (Figure 5). Once soil test P levels become excessive, the potential for P loss in runoff and drainage water is greater than any agronomic benefits of further P applications. In areas of intensive agricultural and livestock production, there is often a high proportion of soils with unnecessarily high soil test P (Figure 6). Often these soils are located near sensitive water bodies, such as Florida lakes, the Great Lakes, New England lakes, and Chesapeake Bay (Figure 6). In England and Wales, >50% of arable soils have

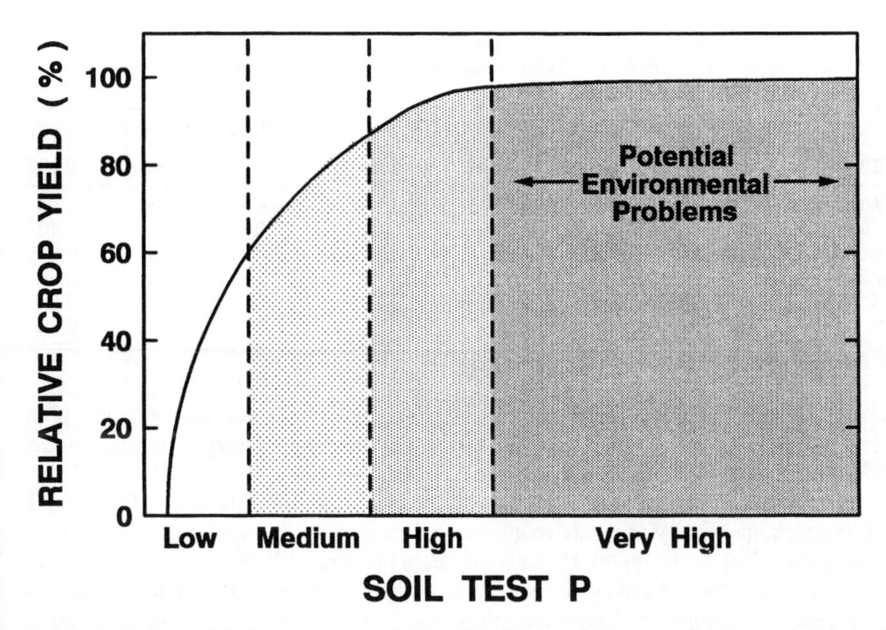

Figure 5 Relationship between crop yield, soil test P, and the potential for environmental problems.

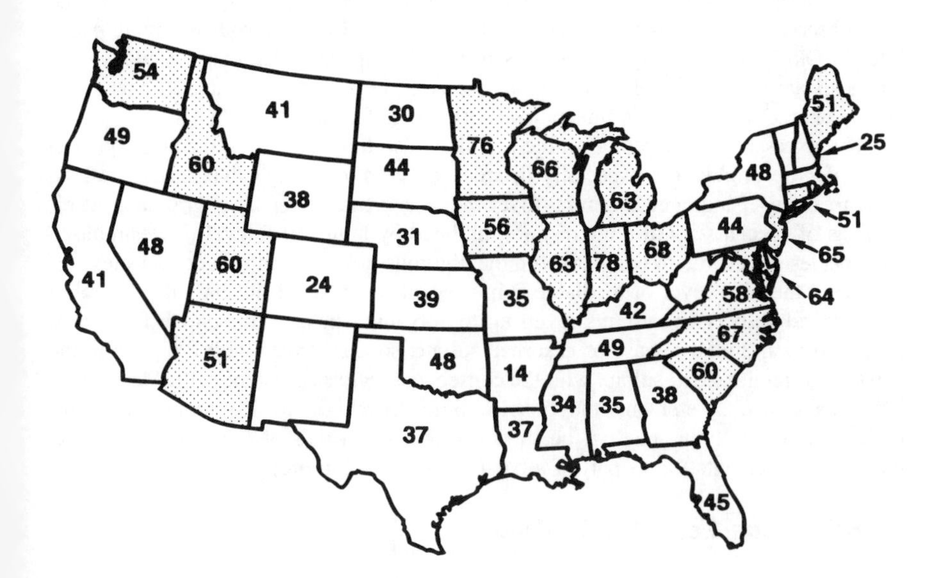

Figure 6 Percent of soil samples testing high or above high for P in 1989. Highlighted states have 50% or greater of soil samples testing in the high or above range. (Adapted from References 73 and 74.)

Table 6 Concentrations of Orthophosphate in the Soil Solution at Increasing Soil Depth of P-rich Sandy Soils Receiving Regular Additions of Organic Manures in the U.K.

			Orthophosphate P (mg L^{-1})		
Site	**Manure Type**	**Soil P[a] (mg kg^{-1})**	**30 cm**	**60 cm**	**90 cm**
Walesby	Poultry	232	14.68	8.46	4.94
Stoke Bardolph	Sewage sludge	185	3.04	0.32	0.13
Rowde	Dairy	114	2.99	1.56	0.85
Watton	Swine	82	1.74	1.83	2.33
Kennyhill	Poultry	61	0.54	0.04	0.01
Ringwood	Dairy	35	0.54	0.11	0.08

[a] Determined with 0.5 *M* Na HCO_3 at pH 8.5.

above-average levels of P in the soil.[75] Clearly high soil test P levels are a regional problem, and there are many areas, e.g., the Great Plains, where fertilizer P is still required for optimum crop yields (Figure 6).

The risk of P accumulation is greatest where there is a limited land area for recycling of livestock manures. Manures have a lower N:P ratio than is removed in crops, and manure applications based on the N requirement lead to a rapid buildup of P. For example, Graetz and Nair[76] found soil test P levels (double acid) of 453 mg P kg^{-1} in the surface soils of several dairy farms that have been in operation for up to 32 years. Soils not receiving manure had soil test P levels of only 3 mg P kg^{-1}, while P fertilizer is not recommended at levels above 66 mg P kg^{-1} on these Florida soils. Further, these authors clearly demonstrated the environmental impact of P contained in these manured soils by calculating that about 4000 kg P ha^{-1} would be available for transport from these areas. Also, soil test P levels (Bray-I) of up to 200 mg P kg^{-1} in soils receiving long-term applications of dairy manure were observed in Wisconsin[77] and up to 280 mg P kg^{-1} (Bray-I) in soils receiving poultry litter in Oklahoma.[78] Recent experiments in the U.K. indicate that concentrations of orthophosphate in the soil solution of fields receiving large regular organic P amendments greatly exceed those found in mineral soils of average fertility (Table 6).

The excessive buildup of P in the soil may also have adverse effects for other aspects of crop production, especially in areas where more sustainable agricultural systems are being encouraged. Attempts to regenerate a greater diversity of grassland species in areas of special scientific interest are inhibited by high soil P levels.[79] Desirable native species are unable to compete with the vigorous growth of introduced species stimulated by the high level of available P in the soil. High soil P levels will also inhibit mycorrhizal activity in systems reverting to "organic" status. Although P fertilization may lead to an initial release of adsorbed sulfate, soils saturated with P have a limited capacity to retain added sulfate with the consequent increased risk of sulfur deficiency.[80] The uptake of trace elements may also be limited by high soil P, although agronomically these interactions may not be sufficient to affect yields of crops receiving trace-element supplementation as part of a balanced fertilization program.[23]

2. Hazardous Metal Accumulation

Phosphate rock and its manufactured fertilizer derivatives are a potential source of toxic or hazardous metal contamination of soils. The elemental composition of a number of phosphate rocks of varying reactivity has been studied by Syers et al.[15] The results, which were broadly typical of previously published data, showed large variation in the concentrations of zinc (Zn, 57 to 1010 mg kg^{-1}); arsenic (As, <2 to 23 mg kg^{-1}); cadmium (Cd, 2 to 100 mg kg^{-1}); and uranium (U, 64 to 153 mg kg^{-1}). Of particular concern is the accumulation of As, Cd, and U which can be potentially toxic to man at low concentrations. Undesirable levels of Cd in the offal tissue of sheep over 8 months old was attributed by Bramley[10] to the accumulation of Cd in the pasture soil. The soil had received annual applications of super phosphate made from Cd-rich Nauru rock phosphate for 20 years.

Williams and David[81] found Cd concentrations ranging from 18 to 91 mg kg^{-1} in 21 samples of straight and compound fertilizers, suggesting that much of the Cd contained in phosphate rock is retained during manufacture. Cd concentrations in 24 samples of TSP collected by Charter et al.[82] ranged from 7 to 47 mg kg^{-1} but in a similar number of monoammonium P and diammonium P fertilizers Cd concentrations were, with one exception, in the range 5 to 9 mg kg^{-1}. Similarly, in a recent survey of U.K. fertilizers, lower median Cd, and other metal, concentrations were obtained in compound fertilizers largely based on ammonium phosphates than in straight P fertilizers derived directly from phosphate rock.[83] Organic fertilizers based on recycled pig or poultry manure contained the lowest concentrations of Cd but highest concentrations of Zn and Cu.[83] Thus, although levels of Cd and other metals generally increase with increasing fertilizer P content, the exact metal concentrations and their loading to agricultural soils may vary according to the manufacturing process.

The production of phosphoric acid removes radioactive contaminants such as U but this element is a constitutent of SSP and TSP and can accumulate where these fertilizers are continuously used. U concentrations of between 24 and 50 mg kg^{-1} have been reported for SSPs[84] and up to 180 mg kg^{-1} for TSP.[85] Spalding and Sackett[85] attributed significant increases in the U content of rivers flowing into the Gulf of Mexico from intensively managed farmland to increased use of P fertilizers over a 20-year period. Goody et al.[86] report increased concentrations of U in the unsaturated zone of the Upper Chalk in England beneath a lagoon used for storing dairy manure for 18 years.

The significance of potentially harmful elements in P fertilizers must be assessed in relation to the inputs of these elements from atmospheric sources and their background levels in uncontaminated soils. Nutrient budget studies have shown that inputs of Cd from the atmosphere are of the same magnitude as those in P fertilizers.[87,88] Levels of Zn and As in uncontaminated soils can be relatively high compared to the inputs of these elements in P fertilizers. Since metals applied in P fertilizers are largely retained within the cultivated layer, they may be transported to surface waters in suspended sediments eroding from fields.[89]

B. AQUATIC ENVIRONMENT

1. Sources of Phosphorus Contamination

Phosphorus contamination of the aquatic environment from agriculture arises from point and nonpoint sources. Point sources of P loss from agriculture include consented discharge from livestock farms, accidental spillage from overflowing manure stores, or direct contamination during broadcast applications of fertilizer and manure. Nonpoint agricultural P loss occurs as surface flow (surface runoff and erosion) and subsurface flow (leaching and drainflow). The main nonpoint sources contributing to the P load of water bodies include erosion, fertilizer, manure, sewage sludge, and septage. Amounts of P transported in surface flow from uncultivated or pristine land and deposited in rainfall are considered the background loading, which cannot be reduced. These sources determine the natural status of a lake.

Although the atmospheric input of P to agricultural land is generally small compared to fertilizer and manure inputs, rainwater P entering aquatic systems directly may be sufficient to enhance algal growth in certain situations, for example, in Clear Lake, Ontario,[90] and Wisconsin lakes.[91] Elder[92] estimated that rainfall P may account for up to 50% of the P entering Lake Superior. Since 25 to 50% of the total phosphorus (TP) in rainfall is soluble, it is directly available to organisms in the lake. Ryding and Rast[93] report that the amounts of P deposited from the atmosphere were dependent on the land use of the surrounding area, emphasizing the often particulate nature of atmospheric P.

There is often little information on background loadings of P for watersheds and it is consequently difficult to quantify any increase in P loss due to intensification of agricultural activities or assess the overall impact on the waterbody. Water quality monitoring is expensive and any measured effects are difficult to replicate because of site specificity and the large spatial and temporal variation in measured values. However, a combination of selective water monitoring and desk studies of P nutrient budgets enable some assessment of the agricultural contribution.[94,95]

The loss of P from several mixed land-use watersheds in the British Isles is generally less than 1 kg P ha^{-1} $year^{-1}$.[96] The estimated contribution of non-point sources from agriculture to these losses is variable (5 to 90%) but can be the major contributor. Similar results have been found from surveys of U.S. watersheds, where P loss in runoff increases as the portion of the watershed under forest decreases and agriculture increases (generally <1 kg P ha^{-1} $year^{-1}$).[97] The loss of P from forested land tends to be similar to that found in subsurface or base flow from agricultural land.[98] In general, forested watersheds conserve P, with P input in rainfall usually exceeding outputs in stream flow.[99,100] As a result forested areas are often utilized as buffer or riparian zones around streams or water bodies to reduce P inputs from agricultural land.[101] The main factors controlling these non-point P losses from agricultural land include the relative importance of surface and subsurface runoff in a watershed, fertilizer and manure amendments, and runoff and erosion potential as influenced by land management.

2. Nonpoint Phosphorus Loss

Transport Mechanisms — Phosphorus is transported in dissolved (DP) and particulate (PP) forms (Figure 7). PP includes P sorbed by soil particles and organic matter eroded during flow events and constitutes the major proportion of P transported from cultivated land (75 to 90%). Runoff from grass or forest land carries little sediment and is, therefore, generally dominated by the dissolved form. While DP is, for the most part, immediately available for biological uptake,[103,104] PP can provide a long-term source of P for aquatic biota.[105,106] The bioavailability of PP can vary from 10 to 90%, depending on the nature of the eroding soil.[89] Together DP and bioavailable PP constitute bioavailable P or P available for uptake by aquatic biota.

During the transport of P from the edge of the field to the receiving lake or ocean, DP and PP fractions continuously change as a result of in-stream processes. These processes include uptake of DP by aquatic biota, transformations between PP and DP caused by changes in the equilibrium stream DP concentration, deposition of suspended PP, and

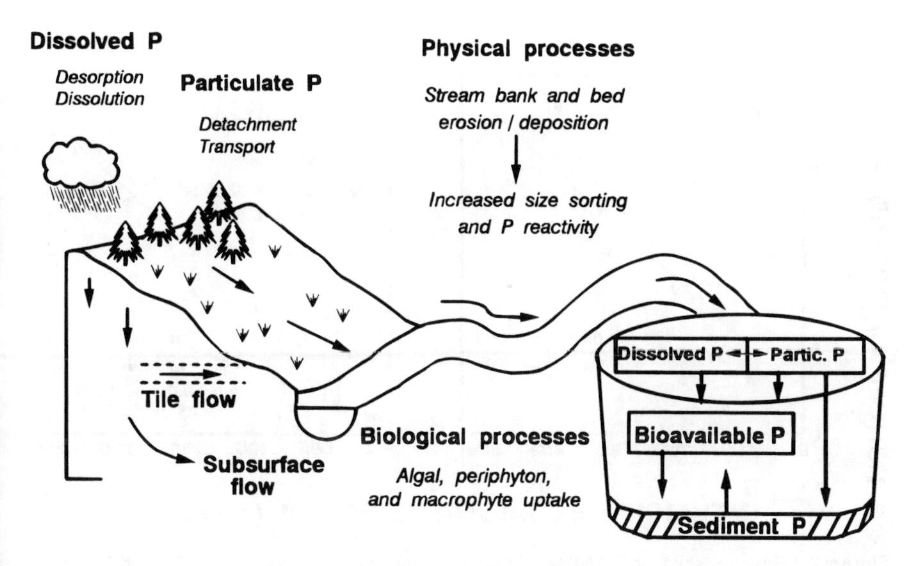

Figure 7 Processes involved in the transport and bioavailability of P from agricultural land. (Adapted from Reference 102.)

resuspension of streambed or streambank PP.[89,107,108] The direction and extent of these P transformations during transport depends on the time of year, the relative amounts of P entering from different sources, and, in particular, the rate of flow. In watersheds where point sources of P are dominant, DP concentrations in water generally decrease with increasing flow due to dilution. In agricultural watersheds with little point source P input, DP concentrations increase during storm events as a result of the transport of suspended sediment from fields.

On arrival at the receiving lake, further exchanges of P at the sediment-water interface affect the amount of P available for biological productivity.[109,110] For example, Theis and McCabe[111] found that the DP concentration in the lakewater of two shallow hypereutrophic lakes in Indiana was reduced by sorption during aerobic periods and increased by release of sediment-bound P during anaerobic periods. The transformations of P within water bodies must be considered in assessing the impact of P transported from agricultural land on the potential biological productivity and eutrophication risk of the receiving lake.

Runoff and Erosion — The transport of DP in runoff is initiated by the desorption, dissolution, and extraction of P from soil and plant material (Figure 7). These processes occur as a portion of rainfall interacts with a thin layer of surface soil (1 to 5 cm) before leaving the field as runoff.[112,113] Although this depth is difficult to quantify in the field, it is expected to be highly dynamic due to variations in rainfall intensity, slope, soil tilth, and vegetative cover. Several studies have reported that the loss of DP in runoff is dependent on the soil P content of surface soil. For example, a highly significant linear relationship was obtained between the DP concentration of runoff and soil P content (Mehlich 3) of surface soil (5 cm) from cropped and grassed watersheds in Arkansas and Oklahoma (Figure 8). A similar dependence of the DP concentration of runoff on

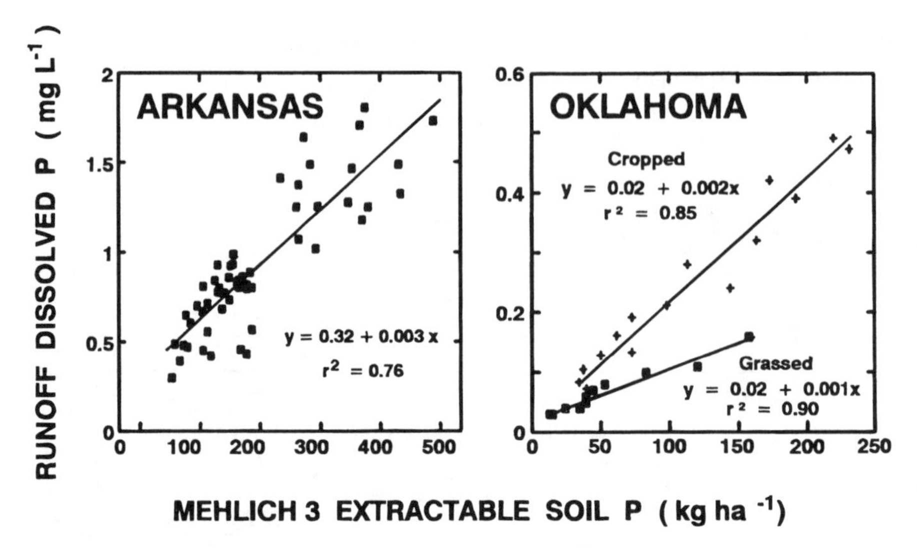

Figure 8 Effect of soil test P (Mehlich 3) on the dissolved P concentration of runoff from several Arkansas and Oklahoma watersheds.

Bray-1 P was found by Romkens and Nelson[114] for a silt loam in Illinois ($r^2 = 0.81$) and on water-extractable soil P ($r^2 = 0.61$) of 17 Mississippi watersheds by Schreiber[115] and 11 Oklahoma watersheds ($r^2 = 0.88$) by Olness et al.[116]

As the sources of PP in streams include eroding surface soil, streambanks, and channel beds, processes determining soil erosion also control PP transport (Figure 7). In general, the P content and reactivity of eroded particulate material is greater than that of source soil, due to preferential transport of clay-sized material (<2 μm). Sharpley[117] observed that the plant available P content of runoff sediment was an average 3-fold greater than of source soil and 1.5-fold greater for total, inorganic, and organic forms.

Increases in P loss in surface runoff have been measured after fertilizer (Table 7) and manure amendments (Table 8). Phosphorus losses are influenced by the rate, time, and method of fertilizer and manure application, form of P applied, amount and time of rainfall after application, and vegetative cover. The portion of fertilizer P transported in runoff for the studies reported in Table 7 was generally greater from conventional compared to conservation-tilled watersheds. However, McDowell and McGregor[120] found that fertilizer P application to no-till corn reduced PP transport, probably due to an increased vegetative cover afforded by fertilization.

Concentrations of DP in runoff can be larger from manure-amended soils than from soils receiving inorganic P fertilizer. Sherwood and Fanning[134] recorded DP concentrations of up to 30 mg L^{-1} in runoff soon after application of pig and cattle slurry to grassland in Ireland. DP concentrations, which represented the majority of the total P present, were still larger than 1 mg L^{-1}, 6 weeks after their application. Although the number of runoff events and period of study varied, no one type of manure appeared to be more susceptible to loss in runoff (Table 8). Larger runoff total P and PP losses occur where stocking density is high.[135–137]

Although of limited agronomic value, the loss of P can be of environmental importance. Although it is difficult to distinguish between losses of fertilizer or manure, and

Table 7 Effect of Fertilizer Phosphorus (P) Application on the Loss of P in Surface Runoff

Land Use	P Applied (kg ha^{-1} year^{-1})	Concentration (mg L^{-1})		Amount (kg ha^{-1} year^{-1})		Fertilizer Loss (%)		Ref.
		Soluble	Particulate	Soluble	Particulate	Soluble	Particulate	
Contour corn	40	0.19	0.71	0.12	0.45	—	—	118
	66	0.25	1.27	0.15	0.76	0.1	1.2	
Grass	0	0.01	0.06	0.01	0.20	—	—	119
	75	0.03	0.14	0.04	0.29	0.04	0.1	
No-till corn silage	0	0.23	0.43	0.70	1.30	—	—	120
	30	0.39	0.49	0.80	1.00	0.3	+23.1[a]	
No-till corn grain	0	0.23	0.46	1.10	2.20	—	—	
	30	0.57	0.51	1.80	1.60	2.3	+27.3[a]	
Conventional corn	15	0.07	3.57	0.30	15.10	—	—	
	30	0.11	9.71	0.20	17.50	+3.3[a]	16.0	
Wheat, summer fallow	0	0.30	1.80	0.20	1.40	—	—	121
	54	3.70	7.40	1.20	2.90	1.9	2.8	
Bahiagrass	0	0.69	0.32	0.88	0.41	—	—	122
	12	1.10	0.05	1.10	0.05	1.8	+3.0[a]	
	48	2.00	0.43	2.36	0.51	3.1	0.2	
Grass	0	0.18	0.24	0.50	0.67	—	—	123
	50	0.98	0.96	2.80	2.74	4.6	4.1	

[a] Percent decrease in P loss from fertilized compared to check treatment.

Table 8 Proportion of P Added in Manure Transported in Surface Runoff

Crop	Amount P Added	Study Period	Phosphorus Loss Amount (kg ha^{-1} year^{-1})	Phosphorus Loss Percent	Ref.
		Dairy Manure			
Corn	108	3 months	8.7	8.1	124
Fescue	142	4 events	1.8	1.3	125
Corn	100	2 events	6.2	6.2	126
Fescue, dry[a]	104	8 events	8.2	7.9	127
Fescue, slurry[a]	112	8 events	13.6	12.1	127
Alfalfa, spring[b]	21	1 year	2.5	11.9	128
Alfalfa, fall[b]	55	1 year	7.4	13.5	128
Corn, spring[b]	21	1 year	0.3	1.3	128
Corn, fall[b]	55	1 year	1.4	2.5	128
		Poultry Litter			
Fescue	54	1 event	1.2	2.2	129
	108	1 event	2.5	2.3	129
Fescue	150	1 year	2.9	1.9	130
Fallow	165	1 event	31.4	19.0	131
		Poultry Manure			
Fescue	76	1 event	2.1	2.8	132
	304	1 event	9.7	3.2	132
Fescue	85	4 events	2.0	2.4	125
Fallow	95	1 event	12.4	13.0	131
		Swine Manure			
Fescue	19	1 event	1.5	7.9	133
	38	1 event	3.3	8.5	133

[a] Applied as dry manure or as slurry.
[b] Manure applied in the spring and fall.

native soil P, without the use of expensive and hazardous radio tracers, losses of applied P in runoff are generally less than 5% of that applied, unless rainfall immediately follows application or where runoff has occurred on steeply sloping, poorly drained, and/or frozen soils. Also, a major portion (>75%) of annual watershed runoff can occur in one or two severe events.[138,139] These few events can contribute over 90% of annual P loads.

Leaching and Drainflow — The P content of water percolating through the soil profile is generally low (<0.1 mg L^{-1}) due to sorption of P by P-deficient subsoils. Exceptions may occur in organic or peaty soils, where organic matter may accelerate the downward movement of P together with organic acids and Fe and Al.[140,141] Similarly, P is more susceptible to movement through sandy soils with low P sorption capacities[142] and in soils that have become waterlogged, where conversion of Fe (III) to Fe (II) content and organic P mineralization occurs.[143]

Because of the variable path and time of water flow through a soil with subsurface drainage, factors controlling DP in subsurface waters are more complex than for surface runoff. However, soil P content has been shown to influence the loss of P in drainage water as well as surface runoff. For example, Sharpley et al.[144] found that the amount of P extracted by 0.1 *M* sodium chloride (NaCl) from soil at the tile drain depth (40 to

50 cm) was related (r^2 = 0.86) to the DP loss in tile drainage during storm events. A similar dependence of DP concentration in tile drainage on the P sorption-desorption properties of subsoil material was found for Histosols in Florida,[145] New York,[146] Ontario,[147] and for Haploquolls in Ontario and Michigan.[148] The flow of PP in drainflow from fissured soils has also been noted.[149,150]

The amounts of P loss in subsurface flow are appreciably lower than that in surface flow, because P is sorbed from infiltrating water as it moves through the soil profile (Table 9). Where tile or more drains are present, the flow of P through the soil is accelerated. McAllister[153] recorded significant increases in the P concentration of drainflow (up to 10 mg L^{-1}) within a half hour of application of diluted livestock manures to land intensively drained with permeable backfill. In general, P concentrations and losses in natural subsurface runoff are lower than in drainflow (Table 9) due to the longer contact time with the subsoil. Increased sorption of P from percolating water accounted for lower TP losses from 1-m- (0.50 kg ha^{-1} $year^{-1}$) than 0.6-m- (1.07 kg ha^{-1} $year^{-1}$) deep tiles draining a Brookston clay soil under alfalfa.[148] For the shallower drains, TP losses were about 1% of fertilizer P applied, whereas 1-m-deep tiles exported about 0.6% of that applied (60 kg P ha^{-1} $year^{-1}$).

3. *Impacts on the Aquatic Environment*

Advanced eutrophication of surface water leads to problems with its use for fisheries, recreation, industry, or drinking, due to the increased growth of undesirable algae and aquatic weeds and oxygen shortages caused by their senescence and decomposition. Also, many drinking water supplies throughout the world experience periodic massive surface blooms of cyanobacteria.[154] These blooms contribute to a wide range of water-related problems including summer fish kills, unpalatability of drinking water, and formation of trihalomethane during water chlorination.[155,156] Consumption of cyanobacterial blooms or water-soluble neuro- and hepatoxins released when these blooms die can kill livestock and may pose a serious health hazard to humans.[157,158]

From the fisherman's point of view, advanced eutrophication of lakes can increase the population of rough fish compared to desirable game fish. This has a negative impact on the recreational value of lakes. However, fishery management often recommends a higher productivity to maintain an adequate phytoplankton-zooplankton-fish food chain for optimum commercial fish production. This food chain may be manipulated by stocking water with certain fish species in addition to P-load reductions, in efforts to reduce the incidence of algal blooms and improve overall water quality. For example, stocking lakes with predatory game fish at the top of the food chain (piscivore fish—bass, pike, trout) can reduce the number of planktivore or coarser fish (yellow perch, crappies) on which they feed. Zooplankton should then thrive, which in turn will reduce phytoplankton populations, improving water quality.

Clearly, lake use has an impact on desirable water quality goals, which will require differing management. Watershed management often becomes more complex with multiple-use lakes and streams, which tend to dominate U.S. waters. For example, a lake may have been built for water supply, hydropower, and/or flood control and, although not a primary purpose, recreation is often considered a benefit, with aesthetic enhancement (including property value) an additional fringe benefit.

Table 9 Effect of Fertilizer P Application on the Loss of P in Subsurface Runoff

Land Use	P Applied (kg ha^{-1} year^{-1})	Concentration (mg L^{-1})		Amount (kg ha^{-1} year^{-1})		Fertilizer Loss (%)		Ref.
		Soluble	Particulate	Soluble	Particulate	Soluble	Particulate	
Alfalfa (tile drainage)	0	0.180	—	0.12	0	—	—	151
	29	0.210	—	0.19	—	1.0	—	
Continuous corn	40	0.007	—	0.03	—	—	—	118
	66	0.009	—	0.04	—	—	—	
Terraced corn	67	0.028	—	0.17	—	—	—	
Bromegrass	40	0.005	—	0.03	—	—	—	
Continuous corn	0	0.020	0.100	0.13	0.29			148
(tile drainage)	30	0.110	0.360	0.20	0.42	0.2	0.4	
Bluegrass sod	0	0.02	0.15	0.06	0.09	—	—	
(tile drainage)	30	1.01	3.29	0.16	0.21	0.3	0.4	
Oats (tile drainage)	0	0.02	0.09	0.10	0.19	—	—	
	30	0.42	1.10	0.20	0.30	0.3	0.4	
Alfalfa (tile drainage)	0	0.02	0.11	0.12	0.20	—	—	
	30	0.37	1.03	0.20	0.31	0.3	0.3	
Corn (tile drainage)	17	0.018	0.043	0.005	0.02	—	—	152
	42	0.000	0.000	0.000	0.00	—	—	
	44	0.004	0.024	0.004	0.04	—	—	
Bahiagrass	0	0.074	0.058	—	—	—	—	122
	12	0.077	0.142	—	—	—	—	
	48	0.133	0.291	—	—	—	—	
Grass	0	0.020	0.022	0.04	0.44	—	—	123
	50	0.033	0.019	0.12	0.07	0.2	+7.4[a]	
Grass (tile drainage)	0	0.064	0.072	0.08	0.09	—	—	
	50	0.190	0.161	0.44	0.37	0.7	0.6	

[a] Percent decrease in P loss from fertilized compared to check treatment.

V. MINIMIZING ADVERSE EFFECTS

Agriculture can make a significant contribution to eutrophication problems in inland and coastal waters. There is clearly a need to minimize the adverse effects of P buildup in agricultural soils and the transport of soil and applied P to aquatic systems. While the symptoms of eutrophication in lakes and reservoirs can be overcome in the short term by water management practices (manipulation of aquatic biota populations, control of P release from bottom sediments), there is a longer-term and more fundamental requirement to control the sources of P loss to aquatic systems.

For agricultural systems, source and transport control strategies include balancing P inputs and P outputs and adopting appropriate land management practices. Strategies to minimize P loss transported in runoff will be most effective if sensitive or source areas within a watershed are identified, rather than implementing widespread and costly general strategies over a broad area. It is also clear from the extent of soils with P in excess of levels sufficient for optimum crop yields (Figure 6) that more attention should be paid to avoiding soil P buildup via P-source management. However, before cost-effective control measures can be targeted, critical or unacceptable P concentrations in runoff must be established, considering local climatic, topographic, and agronomic factors and the presence and proximity of P-sensitive waters to edge-of-field losses. Then soil P testing and vulnerability to P loss in runoff can assess under what conditions the critical runoff P levels will be exceeded.

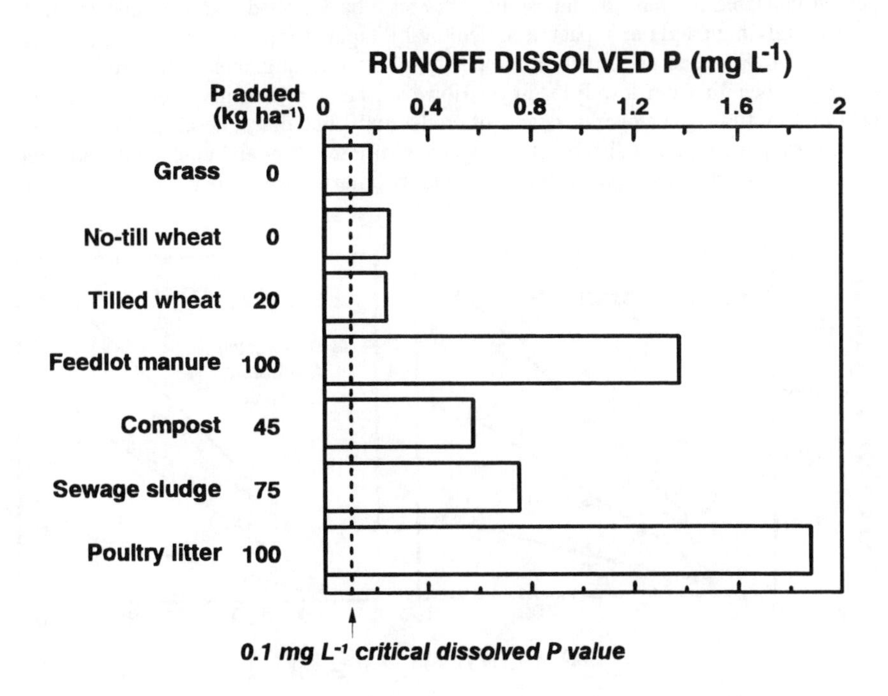

Figure 9 The concentration of dissolved P in runoff as a function of watershed and manure management in the Southern Plains relative to critical values associated with accelerated eutrophication. (Adapted from References 130 and 163.)

A. REALISTIC WATER QUALITY CRITERIA

Critical DP and TP concentrations above which the eutrophication of lake waters may be accelerated were proposed by Vollenweider[159] and Sawyer[160] as 0.01 and 0.02 mg L^{-1}, respectively. Foy and Withers[161] considered a critical TP concentration of 0.1 mg L^{-1} was more appropriate for lakes and rivers and this limit has been used in the restoration of the Norfolk Broads, England.[94] U.S. state agencies adopt values ranging from 0.05 to 0.1 mg L^{-1} for DP and total P in lake inputs.[162]. For example, Florida recently identified 0.05 mg L^{-1} as the concentration of DP allowable in drainage water entering the Everglades. By the year 2000, the state hopes to be able to reduce this concentration to 0.01 mg L^{-1}.

However, these criteria are too stringent with respect to agricultural P loss in surface runoff and drainage water. For example, the mean annual DP concentration of runoff from several wheat watersheds in the Southern Plains receiving mineral fertilizer P (20 kg P ha^{-1} $year^{-1}$) and grassed watersheds receiving various types of manure all exceed a "generous" critical DP value of 0.1 mg L^{-1} (Figure 9). This is also the case for unfertilized native grass watersheds. Thus, it is unlikely that any form of fertilizer or manure management involving continuous P inputs to balance P outputs will reduce DP in runoff to below critical DP concentrations of 0.1 mg L^{-1} or less.

A more flexible approach considers the water P concentration in relation to the geological and physical characteristics of the watershed and the use to which the water is put. In glaciated regions and limestone-rich areas, background water P concentrations are naturally high and the impact of agricultural P inputs (or point source P inputs) may be less.[164] Lakes used principally for water supply, swimming, and multipurpose recreation will benefit from low P loadings. However, lakes principally used for fish production benefit from a moderate degree of productivity and, thus, tolerate higher P inputs.

Clearly, environmentally sound management of P fertilizer and manure amendments must be based on realistic, workable water quality criteria, which are acceptable to farm-

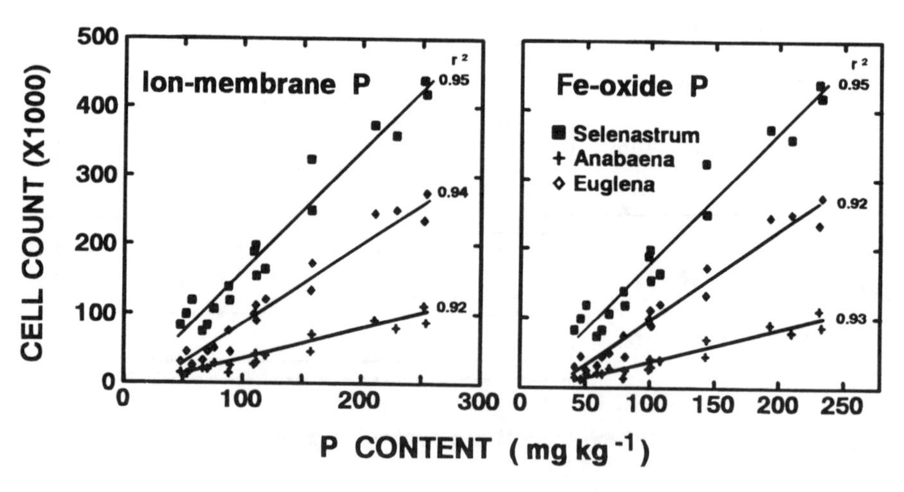

Figure 10 Relationship between the growth of P-starved algae during a 29-day incubation and bioavailable P content if runoff sediment determined by ion exchange membranes and Fe-oxide strips. (Adapted from Reference 168.)

Table 10 Soil P Interpretations and Management Guidelines

State	Critical Value	Management Recommendation	Rationale[a]
Arkansas	150 mg kg^{-1} Mehlich 3 P	At or above 150 mg kg^{-1} soil P: 1. Apply no P from any source 2. Provide buffers next to streams 3. Overseed pastures with legumes to aid P removal 4. Provide constant soil cover to minimize erosion	CV: Ohio sewage sludge data MR: reduce soil P and minimize movement of P from field
Delaware	120 mg kg^{-1} Mehlich 1 P	Above 120 mg kg^{-1} soil P: Apply no P from any source until soil P is significantly reduced	CV: greater P loss potential from high P soils MR: protect water quality by minimizing further soil P accumulations
Ohio	150 mg kg^{-1} Bray P1	Above 150 mg kg^{-1} soil P: 1. Institute practices to reduce erosion 2. Reduce or eliminate P additions	CV: greater P loss potential from high P soils as well as role of high soil P in zinc deficiency MR: protect water quality by minimizing further soil P accumulations
Oklahoma	130 mg kg^{-1} Mehlich 3 P	30 to 130 mg kg^{-1} soil P: Half P rate on >8% slopes 130 to 200 mg kg^{-1} soil P: Half P rate on all soils and institute practices to reduce runoff and erosion Above 200 mg kg^{-1} soil P: P rate not to exceed crop removal	CV: greater P loss potential from high P soils MR: protect water quality, minimize further soil P accumulation, and maintain economic viability
Michigan	75 mg kg^{-1} Bray P1	Above 75 mg kg^{-1} soil P: P application must not exceed crop removal Above 150 mg kg^{-1} soil P: Apply no P from any source	CV: minimize P loss by erosion or leaching in sandy soils MR: protect water quality and encourage wider distribution of manures
Texas	200 mg kg^{-1} Bray P1 or Texas A&M P	Above 200 mg kg^{-1} soil P: P addition not to exceed crop removal	CV: greater P loss potential from high P soils MR: Protect water quality by minimizing further soil P accumulations
Wisconsin	75 mg kg^{-1} Bray P1	Above 75 mg kg^{-1} soil P: 1. Rotate to P-demanding crops 2. Reduce manure application rates Above 150 mg kg^{-1} soil P: Discontinue manure applications	CV: at that level, soils will remain nonresponsive to applied P for 2 to 3 years MR: Minimize further soil P accumulations

[a] CV represents critical value rationale and MR, management recommendation rationale.

Adapted from References 165 and 166.

ers without creating economic hardship within rural communities. Such criteria must take into account both the concentration and amount of P leaving the field and the likely impact on the receiving water. Fertilizer and manure P amendment restrictions have been successfully imposed on farming communities in Europe, which recognize the need for such measures and are receptive to suggestions for their implementation.

B. SOIL TESTING

In areas of P-related water quality problems, recommendations for fertilizer and manure P applications need to take account of the buildup of P in the soil as well as the crop P requirement. Critical values of soil P above which the potential for unacceptable P loss in runoff is high have been proposed in states with a high proportion of high P soils (Table 10). However, the use of soil test P values as indicators of P loss in runoff has been controversial. Often the relationships between soil test P and runoff P have been based on very limited data which may not be applicable to local conditions and land area restrictions on manure application have imposed significant economic problems. In many areas dominated by animal-based agriculture, there is simply no economically viable alternative to land application.

While soil testing procedures were developed to estimate the amount of plant-available P in soil, recent research has shown that soil test P is well correlated with several parameters needed to assess nonpoint source pollution.[74,167] Thus, routine soil tests can usefully pinpoint high P loss risk soils for further environmental assessment. For an environmental assessment, the bioavailability of soil (or sediment) P to aquatic organisms should be determined. Amounts of soil and sediment P extracted by P-sink approaches such as ion exchange membranes or Fe-oxide impregnated paper strips have been shown to be closely related to the growth of P-starved algae in bioassays (Figure 10). These approaches are quicker than bioassays and theoretically more sound than chemical extraction for estimating algal availability of runoff P.[169] However, a lack of field calibration and quantification of critical P-sink levels restricts their applicability for environmental management recommendations.

In the Netherlands, the national strategy is to limit P entry into both surface and groundwater. One of the fundamental mechanisms for accomplishing this goal is the identification of a soil P saturation level above which P application should not exceed crop removal rates.[170,171] The P saturation approach is based on the fact that soil P desorption increases as sorbed P accumulates in soil following P additions, which can lead to increased P loss via runoff or leaching.

To determine the critical level of soil P accumulation, Dutch regulators have set a critical limit of 0.1 mg L^{-1} as DP tolerated in groundwater at a given soil depth (mean highest water level).[171] The degree of accumulation is related to the P sorption capacity, or degree of P saturation (%):

$$\text{P sorption saturation} = \frac{\text{Extractable soil P}}{\text{P sorption capacity}} \times 100 \qquad (1)$$

where the units of extractable soil P and P sorption capacity are unit mass of DP for a given depth of soil (kg P ha^{-1}). In the Netherlands, extractable soil P is estimated by oxalate extraction (Al- and Fe-bound P) and P sorption capacity from the oxalate-extractable Al and Fe content of noncalcareous soils.[171] A P sorption saturation of 25% has been established as the critical value above which the potential for P movement in surface and groundwaters becomes unacceptable.[170]

The relationship between the P sorption saturation of surface soil and DP concentration of runoff was investigated for 56 plots of a Captina silt loam under fescue in Arkansas which had received different amounts of P (as manure) to adjust soil P and P saturation. Phosphorus sorption saturation was calculated from Equation 1, where extractable soil P was represented by surface soil water-soluble P content of each plot (1 g soil shaken with 25 mL water for 1 h). Water-extractable soil P was used to more closely reflect the release of soil P to rainfall/runoff. Phosphorus sorption maximum was calculated from a Langmiur sorption isotherm for an unfertilized Captina silt loam.[172]

Phosphorus sorption saturation was related ($p < 0.001$) to the DP concentration of runoff from fescue (Figure 11). With an increase in P saturation and release of soil P, the DP concentration of runoff increased. Using the relationship between soil P saturation and runoff DP shown in Figure 11, we can determine either a P saturation that will support an "acceptable" DP concentration of runoff or vice versa, a DP concentration that could be expected in runoff from a soil of given P saturation.

The added complexity of this approach in terms of obtaining a reliable estimate of soil P sorption capacity, compared to standard soil test methods, may limit its acceptability at the present time. However, as soil properties affecting P sorption and desorption are accounted for, by this approach, it may be independent of soil type. Thus, the P saturation approach may provide a greater degree of flexibility across soil types than soil test P alone, in estimating the potential for P loss in runoff from a given site. This approach could then be used on soils that have already been identified as being vulnerable to P loss, as a result of high soil P and erodibility or leachability.

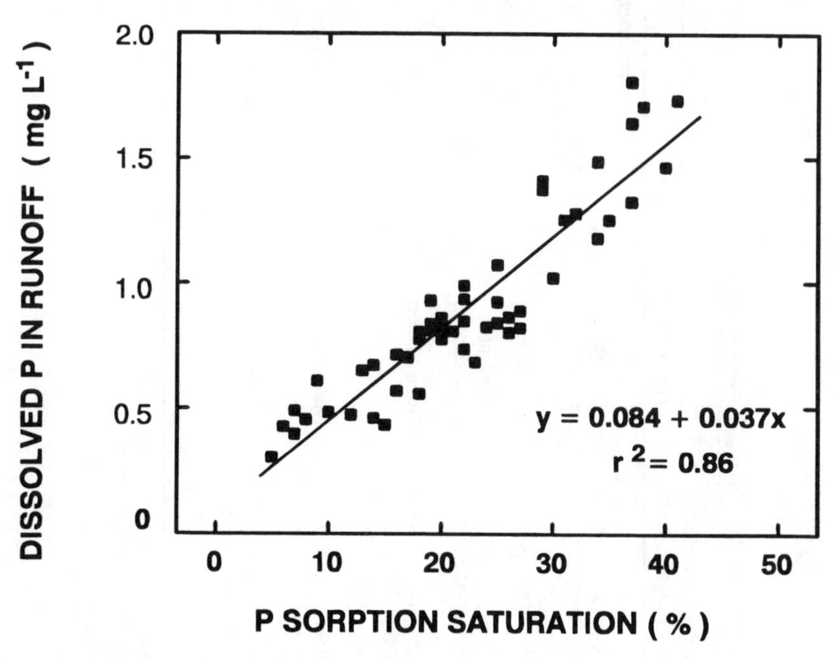

Figure 11 Relationship between P-sorption saturation of surface soil (0 to 5 cm) and dissolved P concentration of runoff from fescue in Arkansas.

Table 11 The P Indexing System to Rate the Potential P Loss Runoff from Site Characteristics

Site Characteristic (Weight)	Phosphorus Loss Potential (Value)				
	None (0)	Low (1)	Medium (2)	High (4)	Very High (8)
			Transport Factors		
Soil erosion (1.5)[a]	Negligible	<10	10–20	20–30	>30
Runoff class (0.5)	Negligible or low	Very low	Medium	High	Very high
			Phosphorus Source Factors		
Soil P test (1.0)	Negligible	Low	Medium	High	Excessive
P fertilizer application rate (0.75)[b]	None applied	1–15	16–45	46–75	>76
P fertilizer application method (0.5)	None appled	Placed with planter deeper than 5 cm before crop	Incorporated immediately applied <3 months before crop	Incorporated >3 months before crop or surface	Surface applied >3 months before crop
Organic P source application rate (0.5)>[b]	None applied	1–15	16–30	30–45	>45
Organic P source application method (1.0)	None	Injected deeper than 5 cm before crop	Incorporated immediately applied <3 months before crop	Incorporated >3 months before crop or surface	Surface applied >3 months before crop

[a] Units for soil erosion are mg ha^{-1}.
[b] Units for P application are kg P ha^{-1}.

Adapted from Reference 173.

Table 12 Site Vulnerability to P Loss as a Function of Total Weighted Rating Values from the Index Matrix

Site Vulnerability	Total Index Rating Value
Low	<10
Medium	10–18
High	19–36
Very high	>36

An added advantage of the P saturation approach is that it not only describes the potential for P release from soil but also indicates how close the P sorption sites of a soil are to being saturated. In other words, measuring P saturation both describes the potential of a soil to enrich runoff with DP (high degree of P saturation) and also helps to predict how much of the P added in fertilizers and manures will be retained by the soil in a form that is relatively resistant to loss in runoff (low degree of P saturation).

C. SITE ASSESSMENT

Soil testing alone cannot assess the potential for agricultural P loss from an individual site or watershed. Any environmental soil P test must be linked to site assessment of drainage, runoff, and erosion potential and management factors affecting the vulnerability for P transport from a site. While land-use modeling of P loss from catchments can help identify broad land uses with a high export coefficient,[95] an assessment of P loss risk from individual fields is often required. For example, adjacent fields having a similar soil test P level but differing susceptibilities to runoff and erosion, due to contrasting topography or management, should not have similar P recommendations.

Thus, a P indexing system was developed to identify soils vulnerable to P loss in runoff (Tables 11 and 12). Each site characteristic affecting P loss is assigned a weighting, assuming that certain characteristics have a relatively greater effect on potential P loss than others. The P loss potential is given a value (Table 11), although each user must establish a range of values for different geographic areas. An assessment of site vulnerability to P loss in runoff is made by selecting the rating value for each site characteristic from the P index (Table 11). Each rating is multiplied by the appropriate weighting factor. Weighted values of all site characteristics are summed and site vulnerability obtained from Table 12.

The index is intended for use as a tool for field personnel to easily identify agricultural areas or practices that have the greatest potential to accelerate eutrophication. It is intended that the index will identify management options available to land users that will allow them flexibility in developing remedial strategies. Based on site vulnerability to P loss in runoff using the P index, Sims[174] proposed management options to minimize nonpoint source pollution of surface waters by soil P (Table 13).

D. REMEDIAL STRATEGIES

1. Source Management

Source management includes determining appropriate P application rates based on environmental soil tests and suitable methods of application. Balancing P inputs with P outputs within the watershed is a primary consideration, especially in areas with con-

Table 13 Soil Management Options Based on the P Index

Phosphorus Index	Management Options to Minimize Nonpoint Source Pollution of Surface Waters by Soil P
<10 (low)	*Soil testing*: have soils tested for P at least every 3 years to monitor buildup or decline in soil P *Soil conservation*: follow good soil conservation practices; consider effects of changes in tillage practices or land use on potential for increased transport of P from site *Nutrient management*: consider effects of any major changes in agricultural practices on P losses *before* implementing them on the farm; examples include increasing the number of animal units on a farm or changing to crops with a high demand for fertilizer P
10–18 (medium)	*Soil testing*: for areas with low P index values, have soils tested for P at least every 3 years to monitor buildup or decline in soil P; conduct a more comprehensive soil testing program in areas that have been identified by the P index as being most sensitive to P loss by erosion, runoff, or drainage *Soil conservation*: implement practices that control P losses via erosion, runoff, or drainage in the most sensitive fields; examples include reduced tillage, wider field border strips, grassed waterways, and improved irrigation and drainage management *Nutrient management*: any changes in agricultural practices may affect P loss: carefully consider the sensitivity of fields to P loss before implementing any activity that will increase soil P; avoid broadcast applications of P fertilizers and apply manures only to fields with low P index values
19–36 (high)	*Soil testing*: a comprehensive soil testing program should be conducted on the entire farm to determine fields that are most suitable for further additions of P; for fields that are excessive in P, estimates of the time required to deplete soil P to optimum levels should be made for use in long-range planning *Soil conservation*: implement practices to control P losses via erosion, runoff, or drainage; examples are reduced tillage, wider field border strips, grassed waterways, and improved irrigation and drainage management; consider using crops with high P removal capacities in fields with high P index values *Nutrient management*: in most situations fertilizer P, other than a small amount used in starter fertilizers, will not be needed; manure may be in excess on the farm and should only be applied to fields with low P index values; a long-term P management plan should be considered
>36 (very high)	*Soil testing*: for fields that are excessive in P, estimates of the time required to deplete soil P to optimum levels should be made for use in long-range planning. Consider the use of new soil testing methods that may provide more information on environmental impact of soil P *Soil conservation*: implement practices that control P losses via erosion, runoff, or drainage; examples include reduced tillage, wider field border strips, grassed waterways, and improved irrigation or drainage management; consider using crops with high P removal in fields with high P index values *Nutrient management*: fertilizer and manure P will not be required for at least 3 years and perhaps longer; a comprehensive, long-term P management plan must be developed and implemented

Adapted from Reference 174.

fined animal operations. Basing manure application on soil P and crop removal of P mitigates the excessive buildup of soil P and at the same time lowers the risk for nitrate leaching to groundwater.

However, basing manure applications on P rather than N management could present several problems to many landowners. A soil test P-based strategy could eliminate much of the land area with a history of continual manure application from further manure additions, as several years are required for significant depletion of high soil P levels. McCollum[175] estimates that, without further P addition, 16 to 18 years of cropping corn (*Zea mays* L.) or soybean (*Glycine max* L. Merr.) would be needed to deplete the soil test P content (Mehlich III) of a Portsmouth soil from 100 mg P kg^{-1} to the threshold agronomic level of 20 mg P kg^{-1}. This would force landowners to identify larger areas of land to utilize the generated manure, further exacerbating the problem of local land area limitations.

There may also be a limited time period over which fertilizer P can be omitted on soils with high test P without affecting yield. Withers et al.[176] found that, after 3 years of withholding P fertilizer on P-rich calcareous soils in England, crop yields of cereals were consistently reduced. It may therefore be more appropriate to reduce P application rates on certain soils rather than omit P altogether. Small amounts of placed starter fertilizer for vegetable crops have successfully reduced the need for much larger P applications broadcast,[177] and a similar strategy (e.g., foliar P applications) may be appropiate for other crops.

Methods of P application are also important. Rotational applications of P designed to streamline fertilizer operations may leave large amounts of available P in the surface and should be avoided in high-risk areas. Efficient management of P amendments on soils susceptible to P loss involves the subsurface placement of fertilizer and manure away from the zone of removal in surface runoff, and the periodic plowing of no-till soils to redistribute surface P accumulations throughout the root zone. Both practices may indirectly reduce the loss of P by increasing crop uptake of P and yield, which affords a greater vegetative protection of surface soil from erosion.

Subsurface application or knifing of P fertilizer or manure may be recommended to minimize P loss in runoff. However, this may be unacceptable if it reduces residue cover below 30%, a Soil Conservation Service BMP guideline to reduce erosion risk. Clearly, assessments of priorities and greater compatability between different recommended management plans are needed.

It may be possible that by utilizing residual soil P careful crop selection will reduce the amount of nutrients potentially available to be transferred to surface waters.[39] "Scavenger" crops that have a higher affinity or requirement for P may thereby reduce the amount of P amendment needed and reduce soil nutrient stratification. Alfalfa, for example, has reduced subsoil nitrate accumulations[178] and may reduce soil accumulations.

Manipulation of dietary P intake by livestock is another aspect of source management that is receiving increasing attention. In the Netherlands, the concentration of P in manure decreased temporarily during the Second World War when concentrates and fertilizers were less available, and reductions in concentrate P are now being similarily implemented to help reduce the amounts of P excreted to land.[179] Morse et al.[70] recorded a 23% reduction in excretion of P in feces and a 17% reduction in total P excretion when dairy cows reduced their daily P intake from 82 to 60 g day^{-1}. Increasing the dietary P intake from 82 to 112 g day^{-1} increased excretion of P in feces by 49% and total P excretion by 37%. Thus, there is a clear indication that amounts of excreted P can be reduced by carefully matching dietary P inputs to the animal's requirements, especially as P intakes above minimum dietary requirements do not seem to confer any milk yield advantage.[69,70]

2. Transport Management

Loss by erosion and runoff may be reduced by increasing vegetative cover through conservation tillage, although this may increase the amount of N leaching that occurs because of the larger volumes of water draining through the soil profile.[96] Also, DP and bioavailable P losses can be greater from no-till than from conventional till practices, since accumulation of crop residues and added P at the soil surface provides a source of P to runoff that would be decreased during tillage. Such water quality tradeoffs must be weighed against the potential benefits of conservation measures in assessing their effectiveness. In some watersheds or fields within watersheds, the risk of N contamination of groundwater may override the need for widespread measures for controlling potential P contamination of surface waters.

Additional specific measures to minimize P loss by erosion and runoff include riparian zones, terracing, contour tillage, cover crops, and impoundments or small reservoirs. However, these practices are generally more efficient at reducing PP than DP. For example, several studies have indicated little decrease in lake productivity with reduced P inputs following implementation of conservation measures.[180,181] The lack of biological response was attributed to an increased bioavailability of P entering the lakes as well as internal recycling. Clearly, effective remedial strategies must address the management of P sources and applications as well as erosion and runoff control.

3. Targeting Remedial Strategies

Remedial strategies for controlling P loss from agricultural land must be carefully targeted if they are to produce cost-effective improvements in water quality. While most freshwaters are P limited, there are notable exceptions where P controls will have marginal to no benefit; for example, high elevation lakes in the western U.S. are N limited. In some lakes and in many streams, plant productivity is limited by high turbidity either from anthropogenic or natural sources. In lakes where eutrophication is sustained by internal P recycling mechanisms within the water body,[182–184] controls on external P sources will have limited effect without some form of in-lake management to reduce aquatic bioproductivity, such as the removal of bottom sediments or the introduction of specific fish species. Reducing agricultural P inputs to lakes may not always achieve the desired or even expected water quality improvements, due to the continued contribution of P from other external sources, i.e., point source P inputs and rainfall.

To optimize control activities there is therefore often a need to prioritize management actions to those watersheds where the control of P will provide the greatest benefit. Management agencies are also often required to further target limited financial and human resources to those P-sensitive lakes having the highest public or ecosystem value. Phosphorus-sensitive lakes are generally those that are greater than 10 ha in size, stratify during the summer, and have water flushing rates of less than six times per year.

Several states are adopting a watershed approach to target nonpoint source control strategies.[162] The Wisconsin Department of Natural Resources[185] targets priority lakes and watersheds by considering the threat to the water quality and the practicability of alleviating the threat, the practicability of achieving a significant reduction in P inputs, water use, and unique or endangered environmental resources.

VI. CONCLUSIONS

The fate of P in soil is governed by dynamic climatic, edaphic, and agronomic variables. Up to 80% of soil P amendments react with Al, Fe, and Ca to form complexes that are temporarily unavailable for plant uptake. This P can, however, be transported from the site of application by runoff and erosion. Unless added P is incorporated into the soil, it usually accumulates in the surface 10 cm of soil, increasing the potential for its transport in runoff.

Although we have been successful in reducing P inputs to aquatic systems via point sources, municipal and urban discharge, and detergents, less success has been achieved in minimizing nonpoint agricultural inputs. This is exacerbated where P inputs in manure from confined animal operations often exceed local crop removal rates. The subsequent accumulation of P in soil is of environmental rather than agronomic concern in many cases. As many years are required to bring about a significant reduction in soil P levels by crop removal, time is not on our side. Also, once lake eutrophication is accelerated, it is usually not cost effective to treat the lake and internal recycling of sedimentary P can support the growth of aquatic biota even if external inputs could be stopped.

Consequently, efforts to minimize P transport from terrestrial to aquatic environments and slow down freshwater eutrophication must identify critical source areas of P in a watershed that present a greater risk to P-sensitive lakes, in order to target cost-effective remedial strategies. In areas of confined animal operations, the development and adoption of innovative measures to transport manure to greater distances and to find alternative end uses must be encouraged. Finally, and perhaps most crucial to any water quality improvement strategy, is efficient transfer of research technology to the land user. Effective implementation will involve education programs to overcome the perception by end users of water that it is often much cheaper to treat the symptoms of eutrophication rather than control the nonpoint sources.

REFERENCES

1. Viets, F.G., The environmental impact of fertilizers, *Crit. Rev. Environ. Control*, 5, 423, 1975.
2. Johnston, A.E., Phosphorus cycling in intensive arable agriculture, in *Proc. SCOPE/UNEP Workshop, Phosphorus Cycles in Terrestrial and Aquatic Ecosystems, Regional Workshop 1: Europe*, Czerniejewo Poland, Tiessen, H., Ed., Saskatchewan Institute of Pedology, Saskatoon, Canada, 1989, 123.
3. Walker, T.W. and Syers, J.K., The fate of phosphorus during pedogenesis, *Geoderma*, 15, 1, 1976.
4. Stevenson, F.J., The phosphorus cycle, in *Cycles of Soil Carbon, Nitrogen, Phosphorus, Sulfur, Micronutrients*, John Wiley & Sons, New York, 1986, chap. 7.
5. Maybeck, M., Carbon, nitrogen and phosphorus transport by world rivers, *Am. J. Sci.*, 401, 1982.

6. Vighi, M. and Chiaudani, G., Eutrophication in Europe: the role of agricultural activities, *Rev. Env. Toxicol.*, 3, 213, 1987.
7. Uunk, E.J.B., Eutrophication of surface waters and the contribution of agriculture, *Proc. Fert. Soc. No. 303*, The Fertiliser Society, Peterborough, England, 1991.
8. Sidhu, S.S., Fluoride levels in air, vegetation and soil in the vicinity of a phosphorus plant, *J. Air Pollut. Control Assoc.*, 29, 1069, 1979.
9. Alcordo, I.S. and Rechcigl, J.E., Phosphogypsum and other by-product gypsums, in *Soil Amendments and Environmental Quality*, Rechcigl, J. E., Ed., Lewis Publishers, Boca Raton, FL, 1995, chap. 9.
10. Bramley, R.G.V., Cadmium in New Zealand agriculture, *N. Z. J. Agric. Res.*, 33, 505, 1990.
11. Louis, P.L., Availability of fertiliser raw materials, *Proc. Fert. Soc. No. 336*, The Fertiliser Society, Peterborough, England, 1993, 13.
12. Kaarstad, O., Phosphorus fertilisers in Europe, in *Proc. SCOPE/UNEP Workshop, Phosphorus Cycles in Terrestrial and Aquatic Ecosystems, Regional Workshop 1: Europe*, Czerniejewo, Poland, Tiessen, H., Ed., Saskatchewan Institute of Pedology, Saskatoon, Canada, 1989, 207.
13. IFA, *World Fertilizer Consumption Statistics No. 24*, International Fertilizer Industry Association, Paris, 1992.
14. FAO, *Fertilizer Use by Crop*, Food and Agriculture Organisation of the United Nations, Rome, 1992.
15. Syers, J.K., Mackay, A.D., Brown, M.W., and Currie, L.D., Chemical and physical characteristics of phosphate rock materials of varying reactivity, *J. Sci. Food Agric.*, 37, 1057, 1986.
16. Davies, G.R., Comparison of water-insoluble phosphate fertilizers with superphosphate — a review, *J. Sci. Food Agric.*, 35, 265, 1984.
17. Kucey, R.M.N. and Bole, J.B., Availability of phosphorus from 17 rock phosphates in moderately and weakly acidic soils as determined by P dilution, A value and total P uptake methods, *Soil Sci.*, 138, 180, 1984.
18. Chien, S.H., Reactions of phosphate rocks with acid soils of the humid tropics, in *Proc. Workshop on Phosphate Sources for Acid Soils in the Humid Tropics of Asia*, Malaysian Society of Soil Science, Kuala Lumpur, Malaysia, 1992, 18.
19. Mullins, G.L. and Evans, C.E., Field evaluation of commercial triplesuperphosphate fertilizers, *Fert. Res.*, 25, 101, 1990.
20. Golden, D.C., White, R.E., Tillman, R.W., and Stewart, R.B., Partially acidulated reactive phosphate rock (PAPR) fertilizer and its reactions in soil, *Fert. Res.*, 28, 181, 1991.
21. Hammond, L.L., Chien, S.H., and Mokwunye, A.U., Agronomic value of unacidulated and partially acidulated phosphate rocks indigenous to the tropics, *Adv. Agron.*, 40, 89, 1986.
22. Peterson, G.A., Sander, D.H., Grabouski, P.H., and Hooker, M.L., A new look at raw and broadcast phosphate recommendations for winter wheat, *Agron. J.*, 73, 13, 1981.
23. Dibb, D.W., Fixen, P.E., and Murphy, L.S., Balanced fertilization with particular reference to phosphates: interaction of phosphorus with other inputs and management practices, *Fert. Res.*, 26, 29, 1990.
24. Gahoonia, T.S., Claasen, N., and Jungk, A., Mobilization of phosphate in different soils by ryegrass supplied with ammonium or nitrate, *Plant Soil*, 140, 241, 1993.

25. Gilbertson, C.B., Norstadt, F.A., Mathers, A.C., Holt, R.F., Barnett, A.P., McCalla, T.M., Onstad, C.A., and Young, R.A., Animal waste utilization on cropland and pastureland, *USDA Utilization Research Report No. 6*, USDA, Washington, D.C., 1979.
26. Sommers, L.E., Chemical composition of sewage sludges and analysis of their potential use as fertilizers, *J. Environ. Qual.*, 6, 225, 1977.
27. Haynes, R.J. and Williams, P.H., Long-term effects of superphosphate on accumulation of soil phosphorus and exchangeable cations on a grazed, irrigated pasture site, *Plant Soil*, 142, 123, 1992.
28. Gerritse, R.G. and Vriesma, R., Phosphate distribution in animal waste slurries, *J. Agric. Sci. Cambridge*, 102, 159, 1974.
29. Gerritse, R.G. and Zugec, I., The phosphorus cycle in pig slurry measured from 32 PO_4 distribution rates, *J. Agric. Sci. Cambridge*, 88, 101, 1977.
30. Goss, D.W. and Stewart, B.A., Efficiency of phosphorus utilization by alfalfa from manure and superphosphate, *Soil Sci. Soc. Am. J.*, 43, 523, 1979.
31. Smith, K.A. and van Dijk, T.A., Utilization of phosphorus and potassium from animal manures on grassland and forage crops, in *Animal Manure on Grassland and Fodder Crops*, H. G. van de Meer et al., Eds., Martinus Nijhoff, Dordrecht, The Netherlands, 1987, 88.
32. Sommers, L.E., Nelson, D.W., and Jost, K.J., Variable nature of chemical composition of sewage sludge, *J. Environ. Qual.*, 5, 303, 1976.
33. Hinedi, Z.R., Chang, A.C., and Lee, R.W.K., Characterization of phosphorus in sewage sludge extracts using phosphorus-31 nuclear magnetic resonance spectroscopy, *J. Environ. Qual.*, 18, 323, 1989.
34. MAFF (Ministry of Agriculture, Fisheries, and Food), Fertilizer recommendations for agricultural and horticultural crops, *Min. Agric. Fish. Food. Ref. Book 209*, HMSO, London, 1994.
35. Chaussod, R., Gupta, S.K., Hall, J.E., Pommel, B., and Williams, J.H., Nitrogen and phosphorus value of sewage sludge, in *Concerted Action on Treatment and Use of Sewage Sludge*, Hall, J.E. and Williams, J.H., Eds., Commission of the European Communities, Brussels, 1985, 25.
36. Zhang, L.M., Morel, J.L., and Frossard, E., Phosphorus availability in sewage sludge, in *Proc. 1st ESA Congress, Paris*, Session 5, Scaife, A., Ed., European Society of Agronomy, Paris, 1990, 32.
37. Stewart, J.W.B. and A.N. Sharpley, Controls on dynamics of soil and fertilizer phosphorus and sulfur, in *Soil Fertility and Organic Matter as Critical Components of Production*, Follett, R.F., Stewart, J.W.B. and Cole, C.V., Eds., SSSA Spec. Publ. 19, American Society of Agronomists, Madison, WI, 1987, 101.
38. Edwards, A.C., Factors influencing plant availability of P from acid soils, in *Phosphorus in Agriculture*, Vol. 3, C.A.B. International, Wallingford, U.K., 1993.
39. Pierzynski, G.M. and Logan, T.J., Crop, soil, and management effects on phosphorus soil test levels, *J. Prod. Agric.*, 6, 513, Wallingford, U.K., 1993.
40. McLaughlin, M.J., Alston, A.M., and Martin, J.K., Phosphorus cycling in wheat-pasture rotations. III. Organic phosphorus turnover and phosphorus cycling, *Aust. J. Soil Res.*, 26, 343, 1988.
41. Syers, J.K. and Curtin, D., Inorganic reactions controlling phosphorus cycling, in *Proc. SCOPE/UNEP Workshop, Phosphorus Cycles in Terrestrial and Aquatic Ecosystems, Regional Workshop 1: Europe*, Czerniejewo, Poland, Tiessen, H., Ed., Saskatchewan Institute of Pedology, Saskatoon, Canada, 1989, 17.

42. Tate, K.R., The biological transformations of P in soil, *Plant Soil*, 76, 245, 1984.
43. Sharpley, A.N., Jones, C.A., Gray, C., and Cole, C.V., A simplified soil and plant phosphorus model. II. Prediction of labile, organic, and sorbed phosphorus, *Soil Sci. Soc. Am. J.*, 48, 805, 1984.
44. Sharpley, A.N., Singh, U., Uehara, G., and Kimble, J., Modelling soil and plant phosphorus dynamics in calcareous and highly weathered soils, *Soil Sci. Soc. Am. J.*, 53, 153, 1989.
45. Tiessen, H., Abetcoe, M.K., Salcedo, I.H., and Owusu-Bennoah, E., Reversibility of phosphorus soption by ferruginous nodules, *Plant Soil*, 153, 113, 1993.
46. Stewart, J.W.B. and Tiessen, H., Dynamics of soil organic phosphorus, *Biogeochemistry*, 4, 41, 1987.
47. Tate, K.R., Spier, T.W., Ross, D.J., Parfitt, R.L., Whale, K.N., and Cowling, J.C., Temporal variations in some plant and soil phosphorus pools in two pasture soils of different phosphorus fertility status, *Plant Soil*, 132, 219, 1991.
48. Doerge, T. and Gardner, E.H., Soil testing for available P in southwest Oregon, in *Proc. 29th Annu. Northwest Fertilizer Conference*, Oregon State University, Corvallis 1978, 143.
49. Brookes, P.C., Powlson, D.S., and Jenkinson, D.S., Measurement of microbial biomass phosphorus in soil, *Soil Biol. Biochem.*, 14, 319, 1982.
50. Widdowson, F.V., Penny, A., and Hewitt, M.V., Results from the Woburn Reference experiment. III. Yields of the crops and the recoveries of N, P, K and Mg from manures and soil, 1975–1979. *Rothamsted Experimental Station Report for 1981*, Part 2, 5, 1982.
51. Paynter, R.M. and Dampney, P.M.R., The effect of rate and timing of phosphorus fertilizer on the yield and phosphate offtake of grass grown for silage at moderate to high levels of soil phosphorus. *Grass Forage Sci.*, 46, 131, 1991.
52. Munson, R.D. and Murphy, L.S., Factors affecting crop response to phosphorus, in *Phosphorus for Agriculture, A Situation Analysis*, Potash and Phosphate Institute, Atlanta, 1986, 9.
53. Sumner, M.E. and Farina, M.P.W., Phosphorus interactions with other nutrients and lime in field cropping systems, *Adv. Soil Sci.*, 5, 201, 1986.
54. Kucey, R.M.N., Janzen, H.H., and Leggett M.E., Microbially mediated increases in plant-available phosphorus, *Adv. Agron.*, 42, 199, 1989.
55. National Research Council, *Soil and Water Quality: An Agenda for Agriculture*, National Academy Press, Washington, D.C., 1993.
56. Iserman, K., Share of agriculture in nitrogen and phosphorus emissions into the surface waters of Western Europe against the background of their eutrophication, *Fert. Res.*, 26, 253, 1990.
57. Aarts, H.F.M., Biewinga, E.E., and van Keulen, H., Dairy farming systems based on efficient nutrient management, *Neth. J. Agric. Sci.*, 40, 285, 1993.
58. Johnston, A.E., Mattingley, G.E.G., and Poulton, P.R., Effect of phosphate residues on soil P values and crop yields. I. Experiments on barley, potatoes and sugar beet on sandy loam soils at Woburn, *Rothamsted Exp. St. Rep. 1975*, Part 2, 5, 1976.
59. Johnston, A.E., Lane, P.W., Mattingly, G.E.G., Poulton, P.R., and Hewitt, M.V., Effects of soil and fertilizer P on yields of potatoes, sugar beet, barley and winter wheat on a sandy clay loam at Saxmundham, *J. Agric. Sci. Cambridge*, 106, 155, 1986.
60. Archer, J., *Crop Nutrition and Fertiliser Use*, Farming Press, Ipswich, England, 1985, chap. 5.

61. Arnold, P.W. and Shepherd, M.A., Phosphorus and potassium requirement of cereals, *HGCA Res. Rev. 16*, Home-Grown Cereals Authority, London, 1990.
62. Sharpley, A.N. and Smith, S.J., Mineralization and leaching of phosphorus from soil incubated with surface-applied and incorporated crop residues, *J. Environ. Qual.*, 18, 101, 1989.
63. Langdale, G.W., Leonard, R.A., and Thomas, A.N., Conservation practice effects on phosphorus losses from Southern Piedmont watersheds, *J. Soil Water Conserv.*, 40, 157, 1985.
64. Seta, A.K., Blevins, R.L., Frye, W.W., and Barfield, B.J., Reducing soil erosion and agricultural chemical losses with conservation tillage, *J. Environ. Qual.*, 22, 661, 1993.
65. Sharpley, A.N., Smith, S.J., Jones, O.R., Berg, W.A., and Coleman, G.A., The transport of bioavailable phosphorus in agricultural runoff, *J. Environ. Qual.*, 21, 30, 1992.
66. Kincaid, R.L., Hilliers, J.K., and Cronrath, J.D., Calcium and phosphorus supplementation of rations for lactating dairy cows, *J. Dairy Sci.*, 64, 754, 1981.
67. Fishwick, G. and Hemingway, R.G., Urea phosphate and mono-ammonium phosphate as dietary supplements for sheep fed diets inadequate in phosphorus and nitrogen, *J. Agric. Sci. Cambridge*, 81, 139, 1973.
68. Bass, J.M., Fishwick, G., Hemingway, R.G., Parkins, J.J., and Ritchie, N.S., The effects of supplementary phosphorus on the voluntary consumption and digestibility of a low phosphorus straw-based diet given to beef cows during pregnancy and early lactation, *J. Agric. Sci. Cambridge*, 97, 365, 1981.
69. Brodison, J.A., Goodall, E.A., Armstrong, J.D., Givens, D.I., Gordon, F.J., McCaughey, W.J., and Todd, J.R., Influence of dietary phosphorus on the performance of lactating dairy cattle, *J. Agric. Sci. Cambridge*, 112, 303, 1989.
70. Morse, D., Head, H.H., Wilcox, C.J., van Hern, H.H., Hissem, C.D., and Harris, B., Jr., Effects of concentration of dietary phosphorus on amount and route of excretion, *J. Dairy Sci.*, 75, 3039, 1992.
71. Lee, P.A., Reducing the phosphorus content of pig slurry, Unpublished ADAS Internal Report, London, 1992.
72. van Horn, H.H., Achieving environmental balance of nutrient flow through animal production systems, *Prof. Anim. Sci.*, 7, 22, 1991.
73. Potash and Phosphate Institute, Soil test summaries: phosphorus, potassium, and pH, *Better Crops*, 74, 16, 1990.
74. Sims, J.T., Environmental soil testing for phosphorus, *J. Prod. Agric.*, 6, 501, 1993a.
75. Skinner, R.J., Church, B.M., and Kershaw, C.D., Recent trends in soil pH and nutrient status in England and Wales, *Soil Use Manag.*, 8, 16, 1992.
76. Graetz, D.A. and Nair, V.D., Fate of phosphorus in Florida spodosols contaminated with cattle manure, *Ecol. Eng.*, in press.
77. Motshall, R.M. and Daniel, T.C., A soil sampling method to identify critical manure management areas, *Trans. Am. Soc. Agric. Eng.*, 25, 1641, 1982.
78. Sharpley, A.N., Smith, S.J., and Bain, W.R., Nitrogen and phosphorus fate from long-term poultry litter applications to Oklahoma soils, *Soil Sci. Soc. Am. J.*, 57, 1131, 1993.
79. Tallowin, J.R.B., Mountford, J.O., Kirkham, F.W., Smith, R.E., and Lathani, K.H., The effect of inorganic fertilizer on a species rich grassland—implications for nature conservation, in *Proc. 15th General Meeting of the European Grassland Federation*, Mannatje, L.'t. and Frame, J., Eds., Wageningen, The Netherlands, 6–9 June, 1994, 332.

80. Syers, J.K., Skinner R.J., and Curtin, D., Soil and fertilizer sulphur in U.K. agriculture, *Proc. Fert. Soc. No. 264*, The Fertiliser Society, London, 1987.
81. Williams, C.H. and David, D.J., The effect of superphosphate on the cadmium content of soils and plants, *Aust. J. Soil Res.*, 11, 43, 1973.
82. Charter, R.A., Tabatabai, M.A., and Schafer, J.W., Metal contents of fertilizers marketed in Iowa, *Commun. Soil Sci. Plant Anal.*, 24(9 and 10), 961, 1993.
83. Marks, M.J., Survey of the trace contaminants in phosphatic fertilizers, Unpublished Report to the Misistry of Agriculture, Fisheries, and Food, 1994.
84. Rothbaum, H.P., McGaveston, D.A., Wall, T., Johnston, A.E., and Mattingley, G.E.C., Uranium accumulation in soils from long-continued applications of superphosphate, *J. Soil Sci.*, 30, 147, 1979.
85. Spalding, R.F. and Sackett, W.M., Uranium in run-off from the Gulf of Mexico distributive province: anomalous concentrations, *Science*, 175, 629, 1972.
86. Goody, D.C., Chilton, P.J., Smith, K.L., Kinniburgh, D.G., and Bridge, L.R., Unsaturated zone drilling beneath a slurry pit on the chalk at Bridgets Farm, Winchester, British Geological Survey, BGS Tech. Rep. WD/94/16C, Wallingford, England, 1994.
87. Hutton, M. and Symon, C., Sources of cadmium discharge to the UK environment, in *Pollutant Transport and Fate in Ecosystems*, Coughtrey, P.J., Martin, M.H., and Unsworth, M.H., Eds., Blackwell Scientific, London, 1987, 223.
88. Merry, R.H. and Tiller, K.G., Distribution and budget of cadmium and lead in an agricultural region near Adelaide, South Australia, *Water, Air, Soil Pollut.*
89. Sharpley, A.N. and Menzel, R.G., The impact of soil and fertilizer phosphorus on the environment, *Adv. Agron.*, 41, 297, 1987.
90. Schindler, D.W. and Nighswander, J.E., Nutrient supply and primary production in Clear Lake, eastern Ontario, *J. Fish. Res. Bd. Can.*, 27, 260, 1970.
91. Lee, G.F., Role of phosphorus in eutrophication and diffuse source control, *Water Res.*, 7, 111, 1973.
92. Elder, F.C., *International Joint Commission Program for Atmospheric Loading of the Upper Great Lakes*, Second Interagency Committee on Marine Science and Engineering Conference on the Great Lakes, Argonne, IL, 1975.
93. Ryding, S.O. and Rast, W., The control of eutrophication of lakes and reservoirs, *Man and the Biosphere Series*, Vol. 1, UNESCO, Parthenon Publishing Group, Paris, 1989.
94. Moss, B., Balls, H., Booker, I., Manson, K., and Timms, M., Problems in the construction of a nutrient budget for the R. Bore and its Broads (Norfolk) prior to its restoration from agriculture, in *Algae and the Aquatic Environment*, Round, F.E., Ed., Biopress, Bristol, 1988, 326.
95. Heathwaite, A.L. and Johnes, P.J., Modelling the impact on water quality of land use change in two agricultural catchments, *Hydrol. Processes*, in press.
96. Sharpley, A.N. and Withers, P.J.A., The environmentally sound management of agricultural phosphorus, *Fert. Res.*, 39, 133, 1994.
97. Omernik, J.M., Nonpoint source — stream nutrient level relationships: a nationwide study, EPA-600/3-77-105, Corvallis, OR, 1977.
98. Ryden, J.C., Syers, J.K., and Harris, R.F., Phosphorus in runoff and streams, *Adv. Agron.*, 25, 1, 1973.
99. Taylor, A.W., Edwards, W.M., and Simpson, E.C., Nutrients in streams draining woodland and farmland near Coshocton, Ohio, *Water Resour. Res.*, 7, 81, 1971.

100. Schreiber, J.D., Duffy, P.D., and McClurkin, D.C., Dissolved nutrient losses in storm runoff from five southern pine watersheds, *J. Environ. Qual.*, 5, 201, 1976.
101. Lowrance, R.R., Todd, R.L., Fail, J., Hendrickson, O., Jr., Leonard, R., Jr., and Asmussen, L., Riparian forests as nutrient filters in agricultural watersheds, *BioScience*, 34, 374, 1984.
102. Sharpley, A.N., Daniel, T.C., and Edwards, D.R., Phosphorus movement in the landscape, *J. Prod. Agric.*, 6, 492, 1993.
103. Walton, C.P. and Lee, G.F., A biological evaluation of the molybdenum blue method for orthophosphate analysis, *Tech. Int. Ver. Limnol.*, 18, 6765, 1972.
104. Nurnberg, G.K. and Peters, R.H., Biological availability of soluble reactive phosphorus in anoxic and oxic freshwaters, *Can. J. Fish. Aquat. Sci.*, 41, 757, 1984.
105. Wildung, R.E., Schmidt, R.L., and Gahler, A.R., The phosphorus status of eutrophic lake sediments as related to changes in limnological conditions — total, inorganic, and organic phosphorus, *J. Environ. Qual.*, 3, 133, 1974.
106. Carignan, R. and Kalff, J., Phosphorus sources for aquatic weeds: water or sediments?, *Science*, 207, 987, 1980.
107. Meyer, J.L., The role of sediments and bryophites in phosphorus dynamics in a headwater stream ecosystem, *Limnol. Oceanog.*, 24, 365, 1979.
108. House, W.A. and Casey, H., Transport of phosphorus in rivers, in *Proc. SCOPE/UNEP Workshop, Phosphorus Cycles in Terrestrial and Aquatic Ecosystems, Regional Workshop 1: Europe*, Czerniejewo, Poland, Tiessen, H., Ed., Saskatchewan Institute of Pedology, Saskatoon, Canada, 1989, 253.
109. Syers, J.K., Harris, R.F., and Armstrong, D.E., Phosphate chemistry in lake sediments, *J. Environ. Qual.*, 2, 1, 1973.
110. Nurnberg, G.K., Dillon, P.J., and McQueen, D.J., Internal phosphorus load in an oligotrophic precambrian shield lake with an anoxic hypolimnion, *Can. J. Fish. Aquat. Sci.*, 43, 574, 1986.
111. Theis, T.L. and McCabe, P.J., Phosphorus dynamics in hypereutrophic lake sediments, *Water Res.*, 12, 677, 1978.
112. Sharpley, A.N., Depth of surface soil-runoff interaction as affected by rainfall, soil slope and management, *Soil Sci. Soc. Am. J.*, 49, 1010, 1985.
113. Ahuja, L.R., Characterization and modelling of chemical transfer to run-off, *Adv. Soil Sci.*, 4, 149, 1986.
114. Romkens, M.J.M. and Nelson, D.W., Phosphorus relationships in runoff from fertilized soil, *J. Environ. Qual.*, 3, 10, 1974.
115. Schreiber, J.D., Estimating soluble phosphorus (PO_4-P) in agricultural runoff, *J. Miss. Acad. Sci.*, 33, 1, 1988.
116. Olness, A.E., Smith, S.J., Rhoades, E.D., and Menzel, R.G., Nutrient and sediment discharge from agricultural watersheds in Oklahoma, *J. Environ. Qual.*, 4, 331, 1975.
117. Sharpley, A.N., The selective erosion of plant nutrients in runoff, *Soil Sci. Soc. Am. J.*, 49, 1527, 1985.
118. Burwell, R.E., Schuman, G.E., Heinemann, H.G., and Spomer, R.G., Nitrogen and phosphorus movement from agricultural watersheds, *J. Soil Water Conserv.*, 32, 226, 1977.
119. McColl, R.H.S., White, E., and Gibson, A.R., Phosphorus and nitrate runoff in hill pasture and forest catchments, Taita, New Zealand, *N.Z. J. Mar. Freshwater Res.*, 11, 729, 1977.

120. McDowell, L.L. and McGregor, K.C., Plant nutrient losses in runoff from conservation tillage corn, *Soil Tillage Res.*, 4, 79, 1984.
121. Nicholaichuk, W. and Read, D.W.L., Nutrient runoff from fertilized and unfertilized fields in western Canada, *J. Environ. Qual.*, 7, 542, 1978.
122. Rechcigl, J.E., Payne, G.G., Bottcher, A.B., and Porter, P.S., Development of fertilization practices for beef cattle pastures to minimize nutrient loss in runoff, South Florida Water Management District Report, South Florida Water Management District, West Palm Beach, FL, 1990.
123. Sharpley, A.N. and Syers, J.K., Phosphorus inputs into a stream draining an agricultural watershed. II. Amounts and relative significance of runoff types, *Water, Air Soil Pollut.*, 11, 417, 1979.
124. Klausner, S.D., Zwerman, P.J., and Ellis, D.F., Nitrogen and phosphorus losses from winter disposal of dairy manure, *J. Environ. Qual.*, 5, 47, 1976.
125. McLeod, R.V. and Hegg, R.O., Pasture runoff water quality from application of inorganic and organic nitrogen sources, *J. Environ. Qual.*, 13, 122, 1984.
126. Mueller, D.H., Wendt, R.C., and Daniel, T.C., Phosphorus losses as affected by tillage and manure application, *Soil Sci. Soc. Am. J.*, 48, 901, 1984.
127. Reese, L.E., Hegg, R.O., and Gantt, R.E., Runoff water quality from dairy pastures in the Piedmont region, *Trans. Am. Soc. Agric. Eng.*, 25, 697, 1982.
128. Young, R.A. and Mutchler, C.K., Pollution potential of manure spread on frozen ground, *J. Environ. Qual.*, 5, 174, 1976.
129. Edwards, D.R. and Daniel, T.C., Effects of poultry litter application rate and rainfall intensity on quality of runoff from fescuegrass plots, *J. Environ. Qual.*, 23, 361, 1993.
130. Heathman, G.C., Sharpley, A.N., and Robinson, J.S., Poultry litter applciation and water quality in Oklahoma, *Fert. Res.*, in press.
131. Westerman, P.W., Donnely, T.L., and Overcash, M.R., Erosion of soil and poultry manure — a laboratory study, *Trans. Am. Soc. Agric. Eng.*, 26, 1070, 1983.
132. Edwards, D.R. and Daniel, T.C., Potential runoff quality effects of poultry manure slurry applied to fescue plots, *Trans. Am. Soc. Agric. Eng.*, 35, 1827, 1992.
133. Edwards, D.R. and Daniel, T.C., Runoff quality impacts of swine manure applied to fescue plots, *Trans. Am. Soc. Agric. Eng.*, 36, 81, 1993.
134. Sherwood, M.T. and Fanning, A., Nutrient content of surface run-off from land treated with animal wastes, in *Nitrogen Losses and Surface Runoff from Land Spreading of Manures*, Brogan, J.C., Ed., Martinus Nijhoff, The Hague, 1981, 5.
135. Duda, A.M. and Finan, D.S., Influence of livestock on non-point nutrient levels of streams, *Trans. Am. Soc. Agric. Eng.*, 26, 1710, 1882.
136. Schepers, J.S., Hackes, B.L., and Francis, D.D., Chemical water quality runoff from grazing land in Nebraska. II. Contributing factors, *J. Environ. Qual.*, 11, 355, 1982.
137. Heathwaite, A.L., Burt, T.P., and Trudgill, S.T., The effect of land use on nitrogen, phosphorus and suspended sediment delivery to streams in a small catchment in South-West England, in *Vegetation and Erosion*, Thornes, J.B., Ed., John Wiley & Sons, Chichester, England, 1990, 161.
138. Edwards, W.M. and Owens, L.B., Large storm effects on total soil erosion, *J. Soil Water Conserv.*, 46, 75, 1991.
139. Smith, S.J., Sharpley, A.N., Williams, J.R., Berg, W.A., and Coleman, G.A., Sediment-nutrient transport during severe storms, in *Fifth Interagency Sedimentation Conference*, Fan, S.S. and Kuo, Y.H., Eds., Federal Energy Regulatory Commission, Washington, D.C., 1991b, 48.

140. Duxbury, J.M. and Peverly, J.H., Nitrogen and phosphorus losses from organic soils, *J. Environ. Qual.*, 7, 566, 1978.
141. Miller, M.H., Contribution of nitrogen and phosphorus to subsurface drainage water form intensively cropped mineral and organic soils in Ontario, *J. Environ. Qual.*, 8, 42, 1979.
142. Ozanne, P.G., Kirton, D.J., and Shaw, T.C., The loss of phosphorus from sandy soils, *Aust. J. Agric. Res.*, 12, 409, 1961.
143. Gotol, S. and Patrick, W.H., Jr., Transformations of iron in a waterlogged soil as influenced by redox potential and pH, *Soil Sci. Soc. Am. Proc.*, 38, 66, 1974.
144. Sharpley, A.N., Tillman, R.W., and Syers, J.K., Use of laboratory extraction data to predict losses of dissolved inorganic phosphate in surface runoff and tile drainage, *J. Environ. Qual.*, 6, 33, 1977.
145. Horstenstine, C.C. and Forbes, R.B., Concentration of nitrogen, phosphorus, potassium and total soluble salts in soil solution samples from fertilized and unfertilized histosols, *J. Environ. Qual.*, 1, 446, 1972.
146. Cogger, G. and Duxbury, J.M., Factors affecting phosphorus losses from cultivated organic soils, *J. Environ. Qual.*, 13, 111, 1984.
147. Nicholls, K.H. and MacCrimmon, H.R., Nutrients in subsurface and runoff waters of the Holland Marsh, Ontario, *J. Environ. Qual.*, 3, 31, 1974.
148. Culley, J.L.B., Bolton, E.F., and Bernyk, V., Suspended solids and phosphorus loads from a clay soil. I. Plot studies, *J. Environ. Qual.*, 12, 493, 1983.
149. Williams, R.J.B., The chemical composition of rain, land drainage and borehole water from Rothamsted, Brooms Barn, Saxmundham and Woburn experimental stations, in *Agriculture and Water Quality*, MAFF Tech. Bull. 32, HMSO, London, 1976, H4.
150. Johnes, P.J., An Investigation of the Effects of Land Use upon Water Quality in the Windrush Catchment, Ph.D. thesis University of Oxford, Oxford, 1990.
151. Bolton, E.F., Aylesworth, J.W., and Hove, F.R., Nutrient losses through tile drainage under three cropping systems and two fertility levels on a Brookston clay soil, *Can. J. Soil Sci.*, 50, 272, 1970.
152. Hanway, J.J. and Laflen, J.M., Plant nutrient losses from tile outlet terraces, *J. Environ. Qual.*, 7, 208, 1974.
153. McAllister, J.S.V., Studies in Northern Ireland on problems related to the disposal of slurry, in *Agriculture and Water Quality*, MAFF Tech. Bull. 32, HMSO, London, 1976, 418.
154. Kotak, B.G., Kenefick, S.L., Fritz, D.L., Rousseaux, C.G., Prepas, E.E., and Hrudey, S.E., Occurrence and toxicological evaluation of cyanobacterial toxins in Alberta lakes and farm dugouts, *Water Res.*, 27, 495, 1993.
155. Palmstrom, N.S., Carlson, R.E., and Cooke, G.D., Potential links between eutrophication and formation of carcinogens in drinking water, *Lake Reserv. Manage.*, 4, 1, 1988.
156. Kotak, B.G., Prepas, E.E., and Hrudey, S.E., Blue green algal toxins in drinking water supplies: research in Alberta, *Lake Line*, 14, 37, 1994.
157. Lawton, L.A. and Codd, G.A., Cyanobacterial (blue-green algae) toxins and their significance in UK and European waters, *J. Inst. Water Environ. Manage.*, 5, 460, 1991.
158. Martin, A. and Cooke, G.D., Health risks in eutrophic water supplies, *Lake Line*, 14, 24, 1994.

159. Vollenweider, R.A., Scientific fundamentals of the eutrophication of lakes and flowing waters with particular reference to nitrogen and phosphorus, OECD Rep. DAS/CSI/68.27, Paris, 1968.
160. Sawyer, C.N., Fertilization of lakes by agricultural and urban drainage, *J. N. Engl. Water Works Assoc.*, 61, 109, 1947.
161. Foy, R.H. and Withers, P.J.A., The contribution of agricultural phosphorus to eutrophication, *Proc. Fert. Soc. No. 365,* The Fertiliser Society, London, 1995.
162. U.S. Environmental Protection Agency, *Quality Criteria for Water*, U.S. Government Printing Office, Washington, D.C., 1976.
163. Jones, O.R., Willis, W.M., Smith, S.J., and Stewart, B.A., Nutrient cycling of cattle feedlot manure and composted manure applied to Southern High Plains drylands, in *Impact of Animal Manure and the Land-Water Interface*, Steele, K., Ed., Lewis Publishers, Chelsea, MI, 1994.
164. Lund, J.W.G. and Moss, B., Eutrophication in the United Kingdom — Trends in the 1980's, Report to the Soap and Detergent Industry Association, Hayes, England, 1989.
165. Sharpley, A.N., Chapra, S.C., Wedepohl, R., Sims, J.T., Daniel, T.C., and Reddy, K.R., Managing agricultural phosphorus for protection of surface waters: issues and options, *J. Environ. Qual.*, 23, 437, 1994.
166. Gartley, K.L. and Sims, J.T., Phosphorus soil testing: environmental uses and implications, *Commun. Soil Sci. Plant Anal.*, 25, 1565, 1994.
167. Wolf, A.M., Baker, D.E., Pionke, H.B., and Kunishi, H.M., Soil tests for estimating labile, soluble, and algae-available phosphorus in agricultural soils, *J. Environ. Qual.*, 14, 341, 1985.
168. Sharpley, A.N., An innovative approach to estimate bioavailable phosphorus in agricultural runoff using iron oxide-impregnanted paper, *J. Environ. Qual.*, 22, 597, 1993.
169. Dorich, R.A., Nelson, D.W., and Sommers, L.E., Estimating algal available phosphorus in suspended sediments by chemical extraction, *J. Environ. Qual.*, 14, 400, 1985.
170. Van der Zee, SEA.T.M., Fokkink, L.G.J., and van Riemsdjkik, W.H., A new technique for assessment of reversibly adsorbed phosphate, *Soil Sci. Soc. Am. J.*, 51, 599, 1987.
171. Breeuwsman, A. and Silva, S., Phosphorus fertilization and environmental effects in The Netherlands and the Po region (Italy), Rep. 57. Agric. Res. Dep., The Winand Staring Centre for Integrated Land, Soil and Water Research, Wageningen, The Netherlands, 1992.
172. Nair, P.S., Logan, T.J., Sharpley, A.N., Sommers, L.E., Tabatabai, M., and Yuan, T.L., Interlaboratory comparison of a standardized phosphorus adsorption procedure, *J. Environ. Qual.*, 13, 591, 1984.
173. Lemunyon, J.L. and Gilbert, R.G., Concept and need for a phosphorus assessment tool, *J. Prod. Agric.*, 6, 483, 1993.
174. Sims, J.T., The phosphorus index: a phosphorus management strategy for Delaware's agricultural soils, Department of Plant and Soil Science, University of Delaware, Newark, 1993.
175. McCollum, R.E., Buildup and decline in soil phosphorus: 30-year trends on a Typic Umprabuult, *Agron. J.*, 83, 77, 1991.
176. Withers, P.J.A., Unwin, R.J., Grylls, J.P., and Kane, R., Effects of withholding phosphate and potash fertilizer on grain yield of cereals and on plant-available phosphorus and potassium in calcareous soils, *Eur. J. Agron.*, 3, 1, 1994.

177. Costigan, P., The placement of starter fertilizers to improve early growth of drilled and transplanted vegetables, *Proc. Fert. Soc. No. 274*, The Fertiliser Society, Peterborough, England, 1988.
178. Mathers, A.C., Stewart, B.A., and Blair, B., Nitrate removal from soil profiles by alfalfa, *J. Environ. Qual.*, 4, 403, 1975.
179. Wadman, W.P., Sluijsmans, C.M.J., and De La Lande Cremer, L.C.N., Value of animal manures: changes in perception, in *Animal Manure on Grassland and Fodder crops*, H.G.van der Meer, et al., Eds., Martinus Nijhoff, Dordrecht, The Netherlands, 1987.
180. Young, T.C. and DePinto, J.V., Algal — availability of particulate phosphorus from diffuse and point sources in the lower Great Lakes basin, *Hydrobiologia*, 91, 111, 1982.
181. Gray, C.B.J. and Kirkland, R.A., Suspended sediment phosphorus composition in tributaries of the Okanagan Lakes, BC, *Water Res.*, 20, 1193, 1986.
182. Ahlgren, I., Role of sediments in the process of recovery of a eutrophicated lake, in *Interactions between Sediments And Fresh Water*, Golterman, H.L., Ed., Dr. W. Junk, The Hague, 1977.
183. Larsen, D.P., Schults, D.W., and Malueg, K.W., Summer internal phosphorus supplies in Shagawa Lake, Minnesota, *Limnol. Oceanogr.*, 26, 740, 1981.
184. Jacoby, J.M., Lynch, D.D., Welch, E.B., and Perkins, M.S., Internal phosphorus loading in a shallow eutrophic lake, *Water Res.*, 16, 911, 1982.
185. Wisconsin Department of Natural Resources, Nonpoint Source Pollution Abatement Program, Wisconsin Administrative Code NR 120, Madison, WI, 1986.

CHAPTER 3

Potassium, Sulfur, Lime, and Micronutrient Fertilizers

Robert L. Mikkelsen and James J. Camberato

0-87371-859-3/95/$0.00+$.50

I. INTRODUCTION

The impact of soil amendments on the environment is of concern to both agriculturalists and the general public. Although potassium (K), sulfur (S), lime, and essential plant micronutrients do not receive as much attention as N and P, they can have significant direct and indirect effects on agricultural ecosystems. Since the published literature in these topics is far too great to comprehensively review in a single chapter, the reader is directed to further details that can be obtained by consulting the listed references.

II. POTASSIUM

Potassium has an essential role in both plant and animal nutrition. It is found in all living organisms, and is thus widely distributed throughout the biosphere. Potassium is typically found in greater concentrations in plants than any other soil-derived nutrient, except perhaps N. For example, when grown in soils with abundant fertilization, K may comprise as much as 8% of the leaf dry weight, although 1 to 4% is a more typical range. However, some plants contain such low concentrations of K that they cannot sustain growth.

For animals, K plays an essential role as a cofactor for enzymes and in maintaining the proper distribution of fluids and electrolytes in tissues, muscles, and nerves. As an essential nutrient for humans and animals, there are many food sources that contain high concentrations of K. For example, orange juice (*Citrus sinensis*), bananas (*Musa* spp.), potatoes (*Solanum tuberosum*), and milk all contain relatively high concentrations of K. For animals, soybean (*Glycine max*) and cottonseed meal (*Gossypium hirsutum*), alfalfa (*Medicago sativa*), and cane molasses (*Sorghum vulgare*) are good sources of K in feed.

Although it does not appear that losses of K will directly create any environmental problems,[1] maintenance of adequate levels of K in soil is essential for efficient crop production. When soil K becomes insufficient for vigorous plant growth, many indirect effects may occur that can result in undesirable environmental impacts. The goal for users of all fertilizer and nutrients should be complete recovery by the crop, with no loss from the soil/plant system. In some situations the recovery of applied K fertilizer may approach this efficiency goal.

A. SOIL REACTION

Soil K exists in four phases: water soluble, exchangeable, nonexchangeable, and mineral (Figure 1). The rate of movement between these phases depends on a combination of environmental factors (such as temperature and rainfall), the soil parent material, the plant species present, and fertilization practices.[2]

Potassium in unfertilized soil originates from the weathering and decomposition of K-bearing rocks and minerals. Although the earth crustal rock contains an average of 2.1% K, it is not uniformly distributed throughout the world. Important minerals that are primary sources of soil K include feldspars such as orthoclase [$KAlSi_3O_8$], muscovite [$KAl_2(AlSi_3O_{10})(OH)_2$], and biotite [$K(Mg,Fe)_3(AlSi_3O_{10})(OH)_2$]. The K feldspars comprise a large potential K reserve in many soils, while K-bearing minerals may be completely absent in highly weathered soils. However, K found in soil minerals is released very slowly and is not of immediate benefit for plant nutrition until weathering and dissolution occurs.[3]

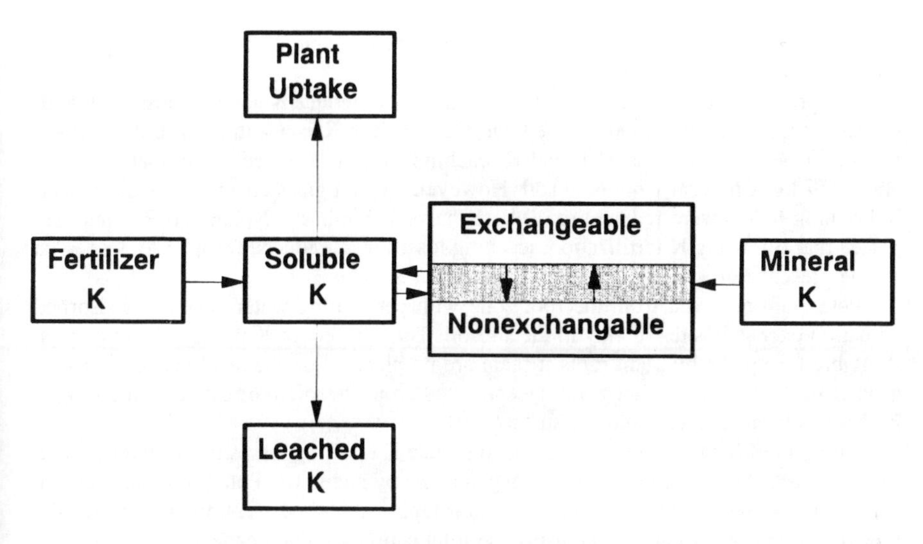

Figure 1 Phases of soil K and potential inputs and outputs in plants, soil, and minerals.

As soil minerals weather and dissolution products enter the soil solution, K is subject to many reversible reactions. Like other cations, K may be held by electrostatic attraction to soil colloids (on negatively charged cation exchange sites) in an exchangeable form where it can replace or be replaced by other dissolved cations. The degree to which K is held on soil exchange sites is determined by the concentration and properties of other ions present in the soil solution and the nature of the cation exchange sites. In soils containing clays such as illite, vermiculite, or montmorillonite, K may be found in soils in a "fixed" form, unavailable for plant uptake. This nonexchangeable K is trapped between clay layers and only slowly released to the soil solution again. The extent of K fixation will vary considerably, depending on the amount and type of clay present in the soil, but it can be extremely high in certain conditions. For example, Cassman et al.[4] found that, shortly following an application of 1440 kg K/ha to a vermiculitic soil, 86% of the added K was found in the fixed form, unavailable for plant uptake.

B. POTENTIAL LOSSES

Potassium does not move to a great extent in soil (<1 to 2 cm/year) if the texture is a loam or finer, since it is held by clay minerals on negatively charged cation exchange sites. In such soils added fertilizer K is slowly recovered by later crops and not appreciably lost by leaching.[5] Additionally, increased clay content of soil not only results in a greater cation exchange capacity (CEC), but also a greater moisture-holding capacity, both of which can retard K movement. In such soils, leaching losses may be as low as 1 kg K/ha/year.[6]

In sandy soils with a low CEC, the magnitude of K leaching can be much greater. For example, when a sandy Norfolk soil was fertilized with 150 kg K/ha and leached with 40 cm water, over 90% of the added K was removed via leaching.[7] Similarly, Blue et al.[8] found that following 40 years of K fertilization, no K accumulated in the surface 0.75 m of a sandy soil due to leaching losses. To reduce leaching losses from sandy soils, K fertilizer should be applied close to the time of planting if possible or split into multiple smaller applications.

The presence of growing plants has been shown to reduce K leaching losses by both reducing the volume of water leached and reducing the K concentration in the soil solution. Nielsen and Stevenson[9] found K leaching from a fertilized sandy loam soil averaged 57 kg K/ha/year from bare soil. However, when a grass cover crop was present, K leaching losses were reduced to 30 kg K/ha/year. Similarly, Nolan and Prichett[10] reported that following K fertilization, leaching losses were reduced from 51 to 3 kg K/ha by the presence of a growing millet (*Pennisetum glaucum*) crop.

Potassium removed from the root zone with percolating water is often readsorbed on cation exchange sites of clay in the subsoil. This buildup of K in the subsoil may be available for uptake by plant roots if there are no physical or chemical barriers for root growth in this lower soil horizon.[11] Deep-rooted crops are often effective in recovering K that has been leached from the surface soil.

Addition of lime to acid soils may also reduce K leaching losses in two ways.[3] First, liming causes the replacement of exchangeable Al by added Ca. Potassium may replace Ca on soil exchange sites more readily than it replaces Al. Second, liming increases the CEC of many soils, thereby increasing the total cation retention capacity.

The leaching of all soil cations, including K, depends on the type and amount of soluble anion present since cations and anions must leach together to maintain electrical neutrality. Potassium leaching losses are generally greater from KCl and KNO_3 than from K_2SO_4.[12]

Direct removal of K from the surface of the soil may be another significant pathway of loss.[13] Surface runoff water may contain considerable quantities of clay and silt, resulting in a relatively high loss of K via suspended soil solids. Losses of soluble K in surface runoff water are usually even lower than losses from leaching. In a review of published literature, Barrows and Kilmer[14] reported that annual losses of dissolved K directly from the soil surface generally range from 0.8 to 8 kg K/ha, exchangeable K losses range from 0.7 to 23 kg K/ha, and total K losses range from 2.8 to 1110 kg K/ha, depending on soil properties, management practices, and the fertilization history.

C. EFFECTS ON PLANT GROWTH

Potassium is generally the most abundant cation in plants; therefore, it is absorbed by the root in relatively large quantities. Potassium primarily moves through the soil to the root surface by diffusion. However, diffusion of K through the soil to the root generally only occurs over short distances (1 to 4 mm); therefore, factors that affect diffusion will regulate the K-supplying power of the soil. Potassium diffusion is governed by properties such as the soil moisture content, soil texture, and the soil density. It is also governed by the concentration of K in the soil solution. Therefore, soils containing abundant K will generally have greater diffusion rates and greater nutrient availability than soils with a low concentration of soluble K. Plants with an abundance of fine roots and a large root surface area for absorption will have a greater potential for nutrient acquisition.[15]

Within the plant, K is involved in many essential physiological functions. Perhaps its most important role is the regulation of the water status in plant tissue, where it is involved in cell expansion and stomatal movement. Additionally, K is involved in the activation of more that 60 enzymes and is required for efficient protein synthesis.[16] When K is lacking, soluble N compounds within the plant (e.g., amino acids, amides, nitrate, etc.) cannot be efficiently assimilated into proteins.[17] Potassium also plays an essential role in photosynthesis, phloem transport, and maintaining the appropriate anion/cation balance within the plant.[18]

Legume growth and N accumulation are frequently increased following fertilization with K.[19] Plant deficiencies of K may reduce the net N_2 fixation by symbiotic bacteria such as *Rhizobia*. The stimulation of N_2 fixation often observed following K fertilization is likely due to increased plant photosynthesis which improves the supply of photosynthate to the legume nodule, rather than a direct effect of K on the nodule or bacteria itself. Improving the efficiency of N_2 fixation may decrease the need for purchased N fertilizer for subsequent crops and, if plant residues are properly managed, may also improve the physical properties of the soil.[20]

Potassium has a direct effect on the disease resistance of many plants.[21] Adequate K has been shown to result in increased thickness of plant support tissues (such as sclerenchyma) in many plants, compared with K-deficient plants. This typically results in reduced incidence of stalk rot, stalk breakage, and lodging for many grain crops when K is sufficient. Resistance to pathogens, such as some leaf spots, mildews, blights, rusts, and wilts, has been linked to an abundant K supply. Forest trees are more resistant to fungal diseases when well supplied with K. Decreased insect damage may also occur when plants contain sufficient K. The exact nature of the enhanced disease and insect resistance resulting from K fertilization is not clear, but it may result from the increased tissue hardness that occurs. The need for pesticides may be decreased when crops are well supplied with essential nutrients, especially K.

Adequate K nutrition is essential for efficient utilization of all applied nutrients. When plant growth is limited by a specific nutrient deficiency, other nutrients present in the soil will not be efficiently utilized. This was well illustrated in an experiment where crops were grown in soil where K and P were insufficient for vigorous growth.[22] In this experiment, considerable amounts of residual NO_3 remained in the soil following crop harvest (Figure 2). In the treatments that received P and K fertilization, very little leachable NO_3 remained following harvest. Soil NO_3 that leaches below the root zone is a potential contaminant of groundwater. This illustrates the requirement for balanced plant nutrition and potential problems resulting from the deficiency of an essential nutrient.

Potassium has been identified as contributing to crop "quality" for a variety of crops. When K is deficient, plant metabolism (e.g., photosynthesis, translocation, water relations, and enzyme systems) are disrupted. This may result in decreased plant growth and a reduction in the quality of the harvested portion of the plant. Although this subject is beyond the scope of this chapter, the relationship between K nutrition and crop quality was comprehensively reviewed by Usherwood.[23]

Potassium is commonly added as a fertilizer to stimulate plant growth and increase crop yields by overcoming conditions of low soil K. High-yielding crops remove large quantities of K in the harvested product. For example, it is not uncommon for grain crops such as corn (*Zea mays*), soybeans, and sorghum to remove 100 kg K/ha each season. Forage crops, such as alfalfa, may remove more than 500 kg K/ha in a season. Proper fertilization and replacement of K removed by the crop is therefore essential for maintaining a sustainable and productive agriculture and to avoid nutrient depletion of soil resources.

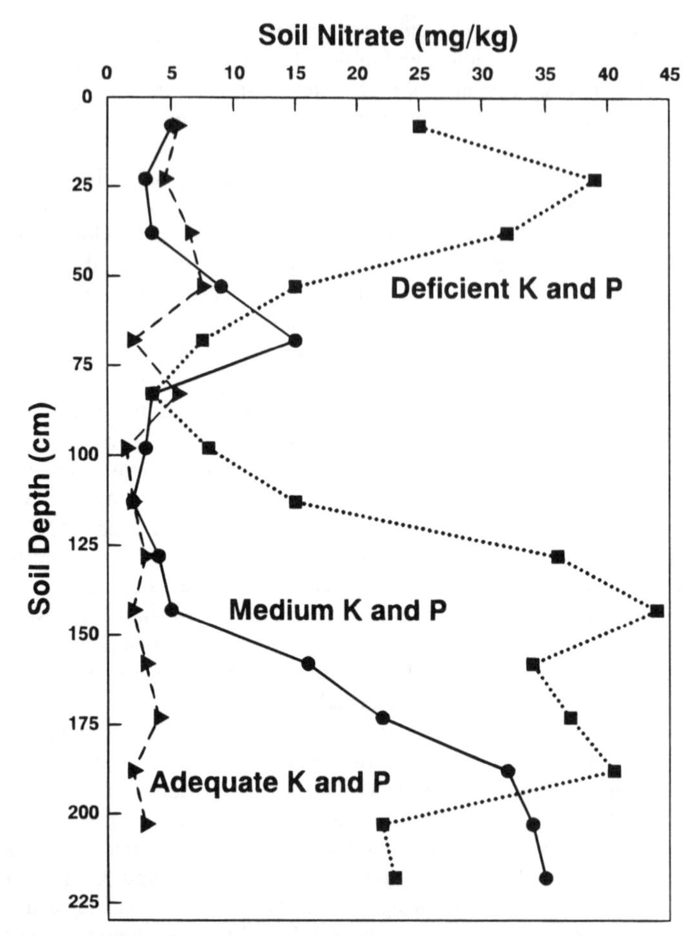

Figure 2 Residual NO_3 remaining in the soil profile following crop harvest where both P and K are (a) severely deficient, (b) moderately deficient, or (c) adequately supplied. (Modified from Reference 22.)

D. POTENTIAL ADVERSE EFFECTS

Potassium fertilization has not been reported to cause direct adverse environmental impacts. It is generally not the limiting nutrient in aquatic ecosystems; therefore, additions resulting from leaching or runoff do not cause direct environmental damage. Excessive environmental disturbance can occur in the vicinity of K mining and processing operations if careful management is not followed, but this is not a widespread problem.

E. SOURCES AND CURRENT USAGE

Fertilizer K is commonly referred to as potash, a name originating from early methods whereby a K-containing solution was obtained by leaching wood ash. The collected solution was evaporated to an ash-like crystalline residue [primarily potassium carbonate (K_2CO_3)] in large iron pots: hence, the name potash. This method of making potash is registered as U.S. Patent number 1.

Although K is found in minerals and rocks throughout the world, commercial sources are primarily obtained from ancient deposits left following the evaporation of oceans and seas.[24] Depending on the conditions during the evaporation period, a variety of soluble evaporite minerals may be found, including calcium carbonate ($CaCO_3$), gypsum ($CaSO_4$), and chloride salts of sodium (Na), magnesium (Mg), and K. These marine salt deposits were often covered with overburden as higher ground eroded. Many of the commercially important K deposits are now found as deep as 1000 m beneath the present land surface.

Production of K fertilizer is widespread throughout the world. Major production areas include the former U.S.S.R., Canada, the Dead Sea region, the western U.S., and western Europe. Other smaller commercial sources are also found in other parts of the world.[25]

The K salts are obtained from these deposits by one of three methods:

1. Conventional shaft mining with heavy boring equipment
2. Solution mining, where water is injected into the K-bearing ore and the solution returned to the surface where it is evaporated and purified
3. Evaporation from naturally occurring brines (such as the Great Salt Lake and the Dead Sea)

Commercial production of K fertilizer is primarily as muriate of potash (KCl), which accounts for 90 to 95% of the world production of K fertilizers. Smaller quantities of K_2SO_4 are made throughout the world. Potassium sulfate is processed directly or formed by reacting KCl with a soluble SO_4 source, such as H_2SO_4. Potassium nitrate is also a common fertilizer source that supplies both K and N to plants. The choice of fertilizer used to supply K is largely based on the price of the various sources and other nutrients required by the crop. For example, S-deficient soils are widespread in certain parts of the world. In these areas, use of K_2SO_4 may be the best choice of a K source.[26] Certain crops, such as tobacco (*Nicotiana glauca*), may be sensitive to high concentrations of chloride (Cl) and the use of Cl-containing fertilizers should be avoided. Other crops, notably sugar beets (*Beta vulgaris*) and coconuts (*Cocos nucifera*), respond well to additions of Cl. Salt damage to seedlings resulting from the placement of fertilizers too close to the seed should be avoided for all soluble K fertilizers.

III. SULFUR

All organisms contain sulfur (S) in the form of the amino acids, cysteine, and methionine, which are used to build proteins. Transformation of inorganic sulfate (SO_4) into organic S compounds is done largely by plants that synthesize cysteine and methionine from SO_4 accumulated by roots from the soil. Animals cannot synthesize S-containing amino acids directly from SO_4. Sulfur is also a component of the vitamins and coenzymes, thiamin, biotin, coenzyme A, and lipoic acid, which participate in many of the important enzymatic reactions in living organisms.

A. PLANT NUTRITION

Sulfur deficiencies of crops have been reported in many countries throughout the world.[27] In the U.S., approximately half of the states have confirmed S deficiencies. Widely grown field crops such as corn, wheat, and soybeans accumulate approximately 15 to 20 kg S/ha. Pastures and crucifers (such as canola, *Brassica napus*, and cabbage, *B. oleracea*) accumulate greater quantities of S (25 to 35 kg S/ha). The average tissue S concentration in healthy crop plants ranges from 0.08 to 0.30%. The N:S ratio of field crops is often used as a diagnostic criteria of S deficiency, because N and S occur in plant proteins in a constant proportion. For example, the typical ratio of N to S in corn is approximately 15:1.[28] A N:S ratio greater than 15:1 indicates S may be deficient and limiting protein formation, while narrower ratios indicate an accumulation of S or a deficiency of N.

Visual symptoms of S and N deficiency are often confused because both are essential to the formation of the chlorophyll molecule and low concentrations result in plant tissue of pale green color. Sulfur deficiency symptoms can generally be separated from N deficiency symptoms because the symptoms are more pronounced in the younger tissue, whereas those of N are more severe in the oldest tissue. The symptoms are more pronounced in the newly formed leaves in S-deficient plants because of slow redistribution of S within the plant.

B. SULFUR IN RELATION TO ANIMAL NUTRITION

Although monogastric animals (e.g., man, swine, poultry) can synthesize cysteine from methionine, and some of the requirement for methionine can be replaced by cysteine, they cannot synthesize these amino acids directly from SO_4.[29] Cysteine and methionine must be ingested and then adsorbed by the intestine into the bloodstream to be utilized in these animals. In contrast, the cysteine and methionine requirements of ruminant animals (e.g., cows, goats, sheep) can be met by feeding inorganic S.[29] The animal itself cannot directly assimilate SO_4, but microorganisms in the rumen reduce the SO_4 to sulfide and then other microbes assimilate the sulfide into amino acids. Transfer and digestion of these microbial cells in the abomasum thereby provides cysteine and methionine to the ruminant animal.

Forages grown on S-deficient soil generally have a lower protein content and higher NO_3 concentration than when supplied with adequate S. Both conditions are detrimental to animal performance and growth. When S is added to the soil to produce higher quality forage or the animals are fed a S supplement, greater average daily weight gains occur.[30]

Sulfur toxicity can occur in ruminant and nonruminant animals.[31] Supplementation of the diet with excessive amounts of ammonium sulfate or gypsum may result in S toxicity in ruminants, whereas excessive feeding of methionine is responsible for toxicity in nonruminants. Chronic S toxicity is expressed in poor animal performance, depressed weight gains, and poor feed conversion, but more severe toxicities can result in metabolic disorders.

Sulfur deficiencies not only limit crop yields and limit food production, but the quality and nutritive value of the harvested crop may be affected as well. For example, the baking quality of wheat flour for bread is, in part, dependent on the S content of the grain.[32,33] The appearance, nutritional value, and flavor of several vegetables is also affected by S supply.[34] A more detailed review of the effects of S on crop quality for human and livestock consumption has been published by Rendig.[35]

C. SOIL REACTIONS WITH SULFUR

The predominance of S accumulated by plants is taken up from the soil as SO_4. Sulfur occurs in aerated soils primarily in three intraconvertible forms: S-containing organic compounds, soil-adsorbed SO_4, and soluble SO_4. In anaerobic soils, sulfide may also be present. Although organically bound S is the largest constituent of the three forms present in noncalcareous surface soils (as much as 95% of the total S[36]), the total amount of organically bound S present in the soil relates poorly to the amount of S available for plant uptake.[37] The S nutrition of crops is dependent on the concentration of SO_4 in the soil solution and adsorbed onto soil constituents.[38] The mineralization and oxidation of organically bound S to SO_4 and the soil's ability to retain adsorbed SO_4 and maintain the concentration of SO_4 in soil solution determines the S supply for crops grown on that soil.

The nature of organically bound S in soils is not well understood because of difficulties in extracting S compounds without altering their form.[36] Much of the organically bound S is contained in the biomass of soil organisms and in plant tissues in various stages of decomposition. Compounds identified in the soil itself include the S-containing amino acids, sulfated polysaccharides, lipid S, sulfonated esters, and C-bonded S. Since the mineralization of organic S is dependent on microbial activity, factors that affect this activity, such as temperature, moisture, pH, and availability of C and N substrates, will determine the rate of S release.[39]

Adsorption of SO_4 occurs on positively charged exchange sites existing at crystal edges of clays, in soil organic matter, and at the surface of hydrous iron and aluminum oxides.[40] The potential number of negative adsorption sites in a given soil is a function of pH and the clay, organic matter, and Fe and Al oxide content of the soil. The SO_4 adsorption capacity of a soil is not only dependent upon the potential number of adsorption sites, but also the charge expression of these sites. As the soil pH increases, the number of positively charged sites available for SO_4 retention is reduced.[41] Sulfate adsorption by soils decreases with increasing pH and is negligible in the pH range suitable for most crops.[42] High concentrations of orthophosphate anions in the soil solution also reduce SO_4 adsorption by the soil, as the phosphate anion is preferentially adsorbed to the positively charged sites.[41,42] The concentration of SO_4 in the soil solution, that which is immediately accessible to plants, is dependent upon the amount of SO_4 released from adsorption sites and the rate of S mineralization from organic matter.

In many soils, the greatest retention of SO_4 occurs in the subsoil, not at the soil surface, because subsoils usually have greater clay contents, lower pH, and a lower concentration of P, so SO_4 adsorption is optimized.[43] The factors governing the availability of subsoil S to plants are toxic accumulations of Al (see discussion on lime), physical barriers to root growth, and the depth at which the SO_4 accumulates. Oates and Kamprath[44] reported that, for annual crop species, subsoils within approximately 30 to 40 cm of the soil surface can fulfill the S needs of the crop if they contain adequate accumulations of SO_4, and if no chemical or physical barriers to root growth and proliferation exist.

D. ADDITIONS AND LOSSES OF S

There has been considerable interest regarding the entry of S into the atmosphere. Excessive S emissions are undesirable because they have the potential to adversely affect the environment. Of the 120 million tonnes S/year released from natural sources into the atmosphere, nearly 60% arises from microbial activity in soil and plant materials.[45] Other significant natural S emissions arise from oceans, coastal wetlands, volcanoes, and wildfires. Volatile losses occur primarily as hydrogen sulfide (H_2S) and dimethyl sulfide [$(CH_3)_2S$]. The primary anthropogenic source of S results from the combustion of S-containing fossil fuels which emit primarily sulfur dioxide (SO_2). Together with emissions from the smelting of S-bearing metal ores and the emissions of reduced S gas from pulp mills, the contribution of human activity releases another 65 million tonnes S/year to the global atmosphere. Hydrogen sulfide is rapidly oxidized to SO_2 and to sulfurous acid (H_2SO_3) and sulfuric acid (H_2SO_4) by a complex series of reactions.[46]

Upon senescence and decomposition of crop tissues, S may be released into the atmosphere.[47] Estimates for crop plants range from 0.08 to 3.0 kg S/ha/year, with most in the range of 0.1 to 1.0 kg S/ha/year. Comparable losses of S from bare soil surfaces occur (0.03 to 3.3 kg S/ha/year) primarily as H_2S.

Crop plants can assimilate gaseous S compounds directly from the atmosphere or accumulate SO_4 deposited in precipitation or as dry deposition to the soil. In the eastern and midwestern U.S., atmospheric deposition of S is large enough to supply much of the crop requirement of S.[48,49] Sulfur deposition in areas of the western U.S. far from industrial areas is generally quite low (<6 kg S/ha/year), but can be as high as 200 kg S/ha/year in areas of concentrated industry.[50] Kamprath and Jones[49] concluded that annual atmospheric accretions of S in the southeastern U.S. were approximately 20 to 28 kg S/ha/year, an amount equivalent to the annual requirement of high-yielding crops.

Crop plants grown in the vicinity of point sources of S, such as coal-fired power plants and smelters, can be adversely affected by high atmospheric levels of SO_2 when it is absorbed as a gas.[51] Documented yield decreases in field situations due directly to atmospheric S toxicity, although few, typically range from 5 to 15%.[45]

Acidification of soils by atmospheric deposition of S is not considered a major agronomic problem except in close proximity to smelters.[52] However, the long-term implications of acidification of soils may eventually lead to a serious decline in soil pH, resulting in problems such as Al or Mn toxicity for plants.

An accurate assessment of the economic importance of atmospheric S deposition on crop production, both beneficial and detrimental, is difficult because of uncertainties in the S-supplying capacity of soils relative to crop requirements, the quantification of S deposition, and measuring the degree and extent of crop injury. Sulfate is readily leached from coarse-textured surfaces of cultivated soils;[53] therefore, annual additions of S to satisfy the current crop's requirements are generally suggested in these conditions. Although leaching of S from the crop root zone is documented, there has been little research examining loss from deeper soil zones.

E. SULFUR FERTILIZERS, AMENDMENTS, AND THEIR USES

Sulfur is often added to soils concomitantly with N because of the recognized need for both N and S in plant nutrition and the availability of fertilizers containing N and S. The basic source used in many fertilizers containing N and S is $(NH_4)_2SO_4$ (21% N and

24% S), which can be used in solid or liquid fertilizers. However, when added to supply all the N needs of a crop this material provides much more S than is generally needed. Therefore, $(NH_4)_2SO_4$ is often mixed with solutions of urea, ammonium nitrate, and ammonium phosphate to create fertilizer materials of wider N:S ratios. Ammonium bisulfite (32% S) and thiosulfate (26% S) can also be used to make liquid fertilizers containing N and S, although the cost of these materials is usually greater than that of $(NH_4)_2SO_4$ and the S must be oxidized before it is available for plant uptake.

Potassium magnesium sulfate (23% S) is a solid fertilizer material that is frequently used to make granular and blended fertilizers. In addition to supplying S it also contains two other essential plant nutrients — K and Mg. Potassium sulfate (18% S) is another nutrient source used to provide K and S to crop plants. Although ordinary superphosphate (12% S) is used to some extent as a source of P and S in solid fertilizers, this fertilizer is not as common as it was before the development of concentrated superphosphate and ammoniated P fertilizers that contain no S. Gypsum (17% S) is often used as a source of Ca and as a soil conditioner, but is seldom added to soils solely as a source of S. However, when used for these alternative purposes, significant quantities of S are provided to crops.

Elemental S, and to a much lesser degree aluminum and iron sulfates and H_2SO_4, are frequently used in alkaline and calcareous soils to reduce soil pH, reclaim sodic soils, and improve the solubility of elements such as Fe, Cu, Mn, and Zn.[54] With elemental S, these purposes cannot be effected until the S has been oxidized to SO_4. The primary oxidizers of S in soil are several diverse autotrophic bacteria of the *Thiobacillus* genus. The reaction proceeds as follows:

$$2S + 3O_2 + 2H_2O \rightarrow 2H_2SO_4 \qquad (1)$$

with two H^+ ions produced for every atom of S oxidized. This reaction occurs over a wide range of soil conditions because several species of *Thiobacillus*, with diverse environmental requirements and tolerances, will oxidize S. The amount of S needed to lower soil pH is dependent on the same factors considered when increasing soil pH with lime: the initial pH, the desired pH, and the acid-buffering capacity of the soil. The rate of oxidation is greatly affected by the coarseness of the S material, with smaller particles of high surface area being oxidized more rapidly than large particles.

Elemental S is also proposed as a good S source on extremely sandy soils because the SO_4 is slowly released following oxidation and is less susceptible to leaching loss. The utility of this "slow-release" form of S must be weighed against the increased need for liming as acidity is generated by the oxidation reaction. Indeed, for this reason the use of elemental S to provide SO_4 to plants growing on nonalkaline soils is quite limited.

IV. LIME

The leaching of basic cations such as Ca, Mg, and K from the soil, the hydrolysis of trivalent aluminum (Al^{+3}), organic acid production associated with organic matter decomposition, and the microbial oxidation of S- and N-containing compounds generates acidity that continually lowers soil pH.[55] In many crop production areas of the U.S., the addition of liming materials is needed on a regular basis to ameliorate the detrimental effects associated with soil acidity.

Table 1 Common Liming Materials Used to Adjust Soil pH

Lime Material	Chemical Formula	Neutralizing Value (%)
Calcitic	$CaCO_3$	100
Dolomitic	$CaCO_3 \cdot MgCO_3$	110
Hydrated	$Ca(OH)_2$	120–135
Burned	CaO	150–175
Wood ash		20–140

Note: Neutralizing value of various liming materials is relative to pure calcium carbonate (100%).

A. LIME REACTIONS IN SOIL

Lime materials, containing various proportions of carbonates, hydroxides, and oxides of Ca and Mg, have been used for centuries to increase the pH of agricultural soils. The relative neutralizing value of a number of common liming materials is listed in Table 1. The relative neutralizing value of a material expresses its ability to neutralize acidity compared with pure $CaCO_3$.

The rapidity and extent of the reaction of lime and soil are dependent on the lime source and its neutralizing value, the particle size, and the degree of mixing into the soil. The most commonly used liming materials are calcitic and dolomitic limestone. The reaction of calcitic lime in soil forms Ca bicarbonate:

$$Ca(CO_3) + H_2CO_3 \rightarrow Ca(HCO_3)_2 \quad (2)$$

Further reaction of the Ca bicarbonate results in the production of two OH^- ions that can then react with two H^+ ions to form H_2O:

$$Ca(HCO_3)_2 \rightarrow Ca^{2+} + 2OH^- + 2CO_2 \quad (3)$$

Lime sources also react with exchangeable Al^{3+} and other partially hydrolyzed Al compounds to form less-soluble Al compounds:

$$Al^{3+} + 3OH^- \rightarrow Al(OH)_3 \quad (4)$$

The pH of the soil is thereby increased as H^+ and Al species are removed from the soil solution.

B. SOIL ACIDITY

The amount of lime added to soils on a routine basis is dependent not only on the neutralizing value of the lime but on the initial and target pH and the pH-buffering capacity of the soil. The target pH is determined by the species of plant to be grown and the soil type. Plants differ in their tolerance to soil acidity. For example, some plants such as blueberries (*Vaccinium corymbosum*) and azaleas (*Rhododendron indicum*) grow well at low soil pH (4.5 to 5.2), whereas the majority of important agricultural plants, [including corn, soybeans, and wheat (*Triticum aestivum*)] grow well at moderate acidity levels (pH 5.8 to 6.5).[56] Others species, such as alfalfa and sweet clover (*Melilotus alba*), grow optimally in soils near neutral pH levels.[56] The optimum soil pH needed to grow a given plant species decreases as the organic matter content of the soil increases.[57] This occurs because the organic matter complexes polycationic Al ions that are in most instances the phytotoxic agent in acid soils.[58]

Soil acidity may indirectly decrease plant growth due to mineral toxicities or deficiencies. Toxicities of Al and/or Mn occur in many acid soils.[59] In soils receiving metal additions from pesticides and sludges, or developed in mine spoils, metal toxicities (zinc, ZN; copper, Cu; nickel, Ni) may also occur.[60] Nutritional deficiencies of Ca, Mg, P, and molybdenum (Mo) are also commonly associated with low-pH soils.[59]

The most important agent of phytotoxicity in acid soils are polycationic Al species,[61] the prevalence and solubility of which is greatly increased in mineral soils at pH <5.[59] High concentrations of Al^{3+} in the soil solution greatly limit root growth in most crop species, thereby reducing the availability of water and plant nutrients and resulting in poor plant growth. Also, soluble Al may directly distort plant growth through restriction of cell division[62] and DNA synthesis[63] in the root tip and alteration of cytokinin production.[64]

The incidence of Mn toxicity in plants is less frequent than that of Al, but still may be important. The Mn concentration in the soil solution is a good indicator of Mn toxicity. In instances of Mn toxicity, increasing soil pH to 5.5 through liming is generally sufficient to precipitate soluble Mn and reduce the concentrations below toxicity levels, except in soils conducive to reducing conditions where toxicities can occur at higher pH.

Calcium deficiencies do not occur often in soils, but when they do it is typically associated with low pH, occurring commonly in sandy soils, with a low CEC.[59] Magnesium deficiencies are generally associated with soil pH values <5, although plant deficiencies can occur at high pH or following excessive K fertilization.

C. BENEFITS OF LIMING

Liming has variable effects on K availability to plants. By increasing the CEC through liming (pH-dependent charge), K losses may be retarded from surface soils, thereby increasing its availability to the crop.[65,66] However, liming has been reported to occasionally reduce available K, although a definitive causal mechanism for decreased K supply has not been identified.[67,68]

Soil pH affects nitrification, the conversion of NH_4 to NO_3 in soils.[69,70] At pH <5.4, the rate of nitrification is considerably less than that at 6.3, the optimum pH for this process. Therefore, liming to increase soil pH into an acceptable range for the growth of many crop plants also enhances nitrification. Nitrate is susceptible to leaching losses, whereas NH_4 is attracted to the negative charges of soil particles and does not leach to a great extent. When the greater the proportion of N is present as NO_3, the greater the probability exists for N loss by leaching.

The relative rates of NH_4 and NO_3 uptake by plants are also affected by soil pH. It is generally accepted that the uptake of NO_3 is decreased and the uptake of NH_4 is increased as the soil solution pH is increased.[71] The relative amounts of NH_4 and NO_3 taken up by the plant also affect the pH of soil immediately adjacent to the root.[72] The decrease in rhizosphere pH may be as great as 2.4 pH units in plants supplied with NH_4[73] which can result in toxicity and plant death in some instances.[74]

Liming increases P availability for plants by neutralizing exchangeable Al^{3+} and solubilizing unavailable Al-P compounds, as well as by promoting vigorous root exploration in a larger volume of soil.[59] Considerably more P fertilizer is needed when pH is low than when a soil is limed to an acceptable range. For example, only half as much P was required for optimum growth of millet when exchangeable Al^{3+} was neutralized with lime compared with nonlimed soil.[75]

Liming soils generally decreases the availability of SO_4 for plants. Sulfate adsorption is decreased as soil pH is increased.[42] When not adsorbed, SO_4 is readily leached with water and may be removed from the root zone.[53]

The concentration of a hydrogen ion (H^+) found in most mineral soils is insufficient to have a detrimental effect on plant growth. An exception to this are the "cat-clay" soils of coastal marshes that may contain substantial quantities of reduced S compounds.[56] Upon drainage of these soils and oxidation of S to H_2SO_4, the soil pH may drop into the range of pH 1 to 2. Most plants do not grow at this high a concentration of H.[+76]

The plant availability of Mo is largely governed by soil pH. The solubility of Mo is greatly enhanced by raising the soil pH, thus increasing its availability to plants.[77,78] In one report, increasing the soil pH with lime additions completely eliminated the need to supply fertilizer Mo.[79]

For several other plant micronutrients (B, Mn Cu, Zn, and Fe), increasing soil pH with lime additions will decrease their solubility and may induce deficiencies of these elements. On soils prone to micronutrient deficiency, the target pH for lime recommendations could be adjusted downward to eliminate the possibility of inducing a nutritional deficiency following lime application.

Lime is often used to increase soil pH for the detoxification of heavy metals in soils. This practice may occur in agricultural situations, in sites receiving application of industrial and municipal wastes, or in the revegetation of disturbed lands. For example, it was noted that, when peach (*Prunus persica*) orchards were returned to row crop production, cotton and soybeans often showed Zn toxicity symptoms and reduced yields.[80,81] Peanuts (*Arachis hypogaea*) grown on these same soils also exhibited distinct Zn toxicity symptoms. The orchards had received substantial amounts of Zn oxysulfate and elemental S as fungicides; thus, the soils had large accumulations of Zn and had a low pH. When soil pH was increased with lime to pH >5.5 to 6 these effects were eliminated.[82]

For the same reason, the U.S. Environmental Protection Agency[83] suggests all soils receiving wastes be maintained at pH >6.5 to reduce the availability and movement of heavy metals contained in the waste materials (see Chapter X for further discussion).

The revegetation of disturbed lands often requires the amelioration of pH to allow plants to grow. In coal-mining regions, mine refuse consisting of waste coal, rock pyrites, and other materials is often highly acidic (a result of oxidation of the sulfide compounds), making revegetation difficult. For example, on a mine-land reclamation site with a soil of pH of 2.8, there was no growth of bromegrass (*Bromus secalinus*) or alfalfa over a 5-year period even though additional nutrients were provided from sewage sludge.[84,85] However, when the soil pH was increased to 5 with lime, an 89% vegetative cover was achieved by the 5th year.

V. MICRONUTRIENTS

The behavior of plant micronutrients (Mn, Mo, Zn, B, Cu, Fe, and Cl) in soil is not closely related to the total quantity of the element present in the soil, but is largely determined by the interaction of complex biological, chemical, and environmental factors.[86]

Although micronutrient deficiencies limit plant growth in many parts of the world, there are also circumstances where excessively high concentrations of these elements can cause potentially serious environmental effects on soils, plants, and animals.[87] In the following section, the function and behavior of each micronutrient is briefly reviewed with additional references provided for further information.

A. MANGANESE

Manganese is generally well distributed in soils, with an average soil concentration of 500 to 600 μg Mn/g soil. It is released from the weathering of various primary and secondary minerals, the most common being the oxides, carbonates, and silicates. Manganese ore, mined from sources throughout the world, is used extensively in the metallurgical industry for alloys of steel, copper, and aluminum as well as many other industrial applications.

Although Mn can be found in several oxidation states in soils, the dominant forms are Mn^{2+} and Mn^{4+}. These forms of Mn are in equilibrium with one another and differ in their solubility and plant availability. The solubility of all plant nutrients is important since the plant supply is obtained from the soil solution. Predicting Mn solubility in soil is complicated due to the many chemical and biochemical processes that occur, but it is primarily controlled by soil pH, redox potential, and organic matter complexation.[88]

Soil pH is one of the most important factors controlling Mn solubility in soils. The solubility of soil Mn increases 10^4 times with each unit decrease in pH the range of 7 to 4. In extremely acid soils, Mn toxicity to plants can become important, along with toxicity from Al, which also becomes soluble at low soil pH. Manganese toxicity occurring in acid soils can be alleviated by application of lime to raise the pH above 5.5.

When a soil becomes flooded or oxygen becomes limited (e.g., due to soil compaction or anaerobic microsites), tetravalent manganese (Mn^{4+}) may be microbially or chemically reduced to the more soluble form, divalent manganese (Mn^{2+}). Flooding a soil with water may initially increase Mn solubility, but, since flooding may also increase the pH of an acid soil, an enhanced Mn accumulation by the plant may not always occur.[89]

Adding organic matter to the soil can also increase the plant availability of Mn by complexation or chelation. The added organic matter may also reduce the soil pH, thereby increasing Mn solubility. Soil oxygen may also be temporarily depleted following heavy organic matter additions, resulting in a reduction of soil Mn^{4+} to more-soluble Mn^{2+}. The strong influence of the soil environment on Mn availability to plants and its rapidly changing solubility can make it difficult to predict soils where Mn deficiency or toxicity may occur.

In plants, Mn is involved with many critical biochemical processes. Due to its ability to easily change oxidation states, Mn is involved in many reduction/oxidation processes within the plant, including electron transport in photosynthesis (Hill reaction), detoxification of free radicals, and many enzyme reactions.[90]

Both Mn deficiencies and toxicities are commonly found in plants throughout the world. Deficiencies occur most commonly on shallow soils containing substantial amounts of organic matter and a high pH (>6.5). In organic soils, Mn deficiencies may occur when the pH exceeds 5.5 to 6. Manganese deficiencies are also found in very sandy, mineral soils that are low in native Mn.[91]

The relative sensitivity to low Mn concentrations varies considerably among plant species. Soybeans, oats (*Avena sativa*), and peas (*Pisum sativum*) are common crops that are especially sensitive to Mn deficiency. Typical visual symptoms of Mn deficiency initially include interveinal chlorosis, followed by the appearance of brown or gray spots on the leaves and a loss of the developing leaves. Deficiency symptoms of Mn appear on many plants when tissue concentrations drop below 15 to 25 mg Mn/kg.[92]

When Mn is deficient for crop growth, it is frequently added to soils as an inorganic salt such as $MnSO_4 \cdot 4H_2O$ or as $MnCl_2$ at rates between 5 and 25 kg Mn/ha, depending on the crop and the soil conditions. Foliar applications of Mn can also effectively supply nutrients to crops. [93,94]

Plant toxicities from excessive Mn are most often found when the soil pH is below 5. However, toxic Mn thresholds for plants are highly variable and depend on soil, plant, and environmental factors.[95,96] Manganese toxicity commonly occurs in association with Al toxicity, as Al solubility also increases in acid soil conditions. Toxicity problems with Mn are frequently managed by additions of lime to increase the soil pH. Waste products such as sewage sludge and coal ash may also supply Mn when applied to soil, but, when the soil pH is maintained at the recommended level, Mn is not considered an environmental concern.[97]

When Mn is orally ingested, it is relatively nontoxic, but cases of human toxicity have been reported as a result of inhalation of airborne Mn in factories and mines.

B. MOLYBDENUM

In soils, soluble Mo exists primarily as molybdic acid, with MoO_4^{2-} the predominant species ranging in concentration from 0.5 to 6.0 mg Mo/kg soil.[98] The primary mineral molbdenide (MoS_2) is the most abundant source of soil Mo. However, the availability of Mo for plants is dependent on the primary parent material, its degree of weathering, and the presence of organic matter. In the U.S., Mo deficiencies commonly occur on sandy, acid soils of the Southeast and in coarse-textured soils surrounding the Great Lakes. Soils of the western U.S. derived from shale and granite sources are often high in Mo.

While Mo is required only in trace amounts by plants (0.1 to 1.0 mg Mo/kg plant tissue), where it functions in electron transfer reactions, most plants are capable of accumulating high concentrations (200 to 1000 mg/kg) without exhibiting toxicity symptoms. In plants, Mo is essential in Mo-containing enzyme systems such as nitrate reductase and sulfite oxidase. The nitrate-reductase mediated conversion of NO_3 to NO_2 occurs as the first step in amino acid and protein formation within the plant. A deficiency of Mo may cause symptoms similar to those of N deficiency. Some plants exhibit specific symptoms of Mo deficiency with reduced or irregular leaf formation known as "whiptail" with cauliflower (*Brassica oleracea*) and "yellow spot" of citrus.

Plants supplied primarily with NO_3 have a much greater requirement for Mo than do plants supplied predominantly with NH_4.[90] This close relationship between Mo supply and N use suggests that Mo deficiency may lead to inefficient utilization of applied NO_3 fertilizer with associated risk of NO_3 leaching into groundwater. It is important that all of the essential plant nutrients be present in adequate supply to minimize residual fertilizer remaining in the soil due to poor plant growth.

The biological fixation of atmospheric N_2 by microorganisms requires the presence of the Mo-containing enzyme nitrogenase. Because of this additional need for Mo in N_2

fixation, the requirement for Mo by plants with N_2-fixing nodules is greater than for non-nodulated plants.

In soil, Mo is strongly adsorbed by Fe and Al oxides and organic matter. The equilibrium concentration established with Mo and these adsorbing surfaces is strongly influenced by soil pH. Unlike most of the other essential micronutrients, Mo solubility increases as soil pH increases. This means that raising the pH of acid soils may increase Mo solubility and replace the need for direct fertilization in many cases. For example, plant uptake of soil Mo can increase by two to three times for every unit increase of pH between 5.0 and 7.0.[99] Likewise, Mo solubility can be decreased by the addition of acid or acid-forming fertilizers.

The requirement for Mo varies with plant species. For example, legumes generally contain higher Mo concentrations (0.75 to 2.5 mg/kg) than grasses do (0.25 to 1.5 mg/kg). Plant species of the *Brassica* family are frequently observed to have a greater Mo requirement than other crops, such as grains. When Mo supplementation is needed, several sources of fertilization have been successfully used such as sodium or ammonium molybdate. Application rates of 0.01 to 0.5 kg Mo/ha to soil will generally correct Mo deficiencies. Although Mo may be added to soil, foliage, or to seed to correct deficiencies, caution needs to be taken to avoid overapplication since Mo in forage crops may be toxic for animals, even at low concentrations (>2.5 mg Mo/kg).[92,100]

Excessive accumulation of Mo in forages consumed by grazing ruminant animals may lead to a Mo-induced Cu deficiency (molybdenosis). This most commonly occurs with vegetation growing in poorly drained soils where the parent material or organic matter was initially high in Mo. Molybdenosis is generally treated by addition of Cu to the animal diet, decreasing the soil pH, or improving soil drainage.[101]

Plants growing in Mo-contaminated soils may accumulate high concentrations of Mo. For example, Hornick et al. reported that plants growing on a Mo-polluted soil near a Mo processing plant contained between 124 and 1060 mg Mo/kg.[102] Schalscha et al. also reported high tissue Mo concentrations in plants grown near a Mo ore processing plant.[103] Plants grown in soil amended with municipal sludge ash also contained elevated concentrations of Mo.[104] However, Mo toxicity to plants rarely occurs in the field.

C. ZINC

Naturally occurring levels of Zn in soil depend primarily on the type of parent material, the presence of organic matter, and soil pH. Estimates of average Zn concentrations in soil worldwide range from 50 to 100 mg Zn/kg soil. Soil Zn may be found in the soil solution, on soil cation exchange sites, or adsorbed on organic matter and other soil constituents.[105]

Zinc-bearing minerals (especially zinc sulfide, ZnS) are mined from many locations throughout the world. The primary industrial use of Zn is for preventing corrosion of metals (such as galvanized steel) and in alloys such as brass and bronze. Zinc is used in many household and chemical applications. It is also a common micronutrient fertilizer and constituent in pesticides.

Deficiencies of Zn are common worldwide for both plants and animals. Inadequate Zn for plants is typically associated with high pH soils or coarse-textured, highly leached, acid soils. Animal Zn deficiency (especially with swine and poultry) is often associated with other nutritional imbalances (such as excess Ca supply). For humans, low dietary intake may result in Zn deficiency, but this is not common.[106]

In plants Zn is found exclusively as Zn^{2+} and is not subject to changes in valence state as are other nutrients such as Fe and Mn. Zinc is found in several important enzyme systems within plants, including alcohol dehydrogenase, superoxide dismutase, carbonic anhydrase, and RNA polymerase. Zinc is also required as an activator of many essential enzyme systems.[90]

Zinc deficiency is the most common micronutrient deficiency for plants worldwide. In the U.S., Zn deficiency is known to occur in 39 states.[107] Similar widespread Zn deficiencies are common on every continent. Zinc deficiencies are most likely to occur in highly weathered acid soils, organic soils, and in calcareous soils with high pH. In high-pH soils, adequate amounts of Zn may be present in the soil, but its solubility and plant availability is limited. For example, in one study the concentration of soluble Zn decreased approximately 100 times as the soil pH was raised from 4.4 to 7.5.[108] Liming acid soils commonly reduces the availability of Zn to crops and may lead to deficiencies in sensitive crops.

Symptoms of Zn deficiency are typically manifested in dicotyledonous plants by shortened stem internodes and decreased leaf size. In monocots, deficiency symptoms often appear as chlorotic stripes along the leaf. For many plants, Zn deficiency occurs when tissue concentrations drop below 10 to 20 mg Zn/kg.[109]

When Zn is present in the soil in high concentrations, toxicity may result in non-tolerant plants resulting in important regional environmental problems. Zinc toxicity for plants is usually associated with Zn smelting or the addition of unusually high amounts of Zn to the soil (such as bacterial sprays or galvanized steel).[80] Naturally high concentrations of Zn may occur occasionally as well.[110] Although the physiological mechanism responsible for Zn toxicity is complicated, it is usually manifested initially by inhibition of root growth. Increasing the soil pH through liming is effective in reducing excessive Zn concentrations and minimizing toxicity in plants. Plant species and genotypes vary widely in their tolerance to Zn deficiencies and toxicities.[105] Most plant species have a great tolerance to excessive Zn, which is not considered highly phytotoxic. However, some sensitive species may be damaged when tissue concentrations exceed 150 to 200 mg Zn/kg.

Commercial Zn fertilizers are routinely applied to soils to eliminate plant nutrient deficiencies. The most common inorganic sources are $ZnSO_4$ and ZnO, although many other sources have been successfully utilized. When chelated ZnEDTA is used in soil to correct Zn deficiencies, much less total Zn is needed to meet the plant demand than when inorganic Zn salts are used.[111]

D. BORON

Among the plant micronutrients, B is unique in that it is a nonmetallic element and its chemistry and behavior is unlike the previously discussed metals. Boron is fairly rare in nature and nutritional deficiencies are widespread. It commonly exists in soils in the borate form (in compounds such as borax and boric acid). A commercially important source of B is from evaporite minerals (such as kernite) found in the Mohave Desert of California. In most soils, the total content of B is not relevant for plant nutrition because this element is often found in very insoluble forms, such as minerals in the tourmaline group. When these B-containing minerals are solubilized, it is generally in the form of borate, such as nonionized boric acid (H_3BO_3). This soluble form of B is readily available for plant uptake.[112]

Soil B can be found dissolved in the soil solution, adsorbed to clays and Fe and Al oxides, in association with organic matter, or in the mineral phase itself. Among micronutrients, B is considered to be one of the most mobile and is easily leached from many soils. The B concentration in the soil solution is generally the best indicator of plant availability. Consequently, a hot-water extract of the soil is often used to predict the B availability to plants.

The precise function of B within the plant is not as well known as the role of the other mineral nutrients. Most of the information gained regarding the physiological function of B comes from observing the behavior of plants when B is withheld and then resupplied. Boron is unique in that it does not appear to participate in any enzymatic functions within the plant. The functions of B appear to be primarily extracellular, especially related to lignification and xylem development. Much of the total B in the plant is found in cell wall materials and membranes, associated with sugar transport, and is essential for pollen development.[113]

Boron is immobile within the plant and once absorbed it cannot be redistributed among plant parts. As a result, deficiencies will first appear on new leaves or buds on the plant and for normal growth a continual supply of B is needed throughout the growing period. Inadequate soil B may be manifested as misshapen leaves, shortened internodes, or excessive flower drop, especially in rapidly growing plants. Boron deficiencies begin to appear for many plants when tissue concentrations drop below 5 to 30 mg B/kg.

When soil B concentrations are insufficient for plant growth, borax and sodium tetraborate are the most commonly used B fertilizers. Other soluble B sources may also be effectively used. Typical soil application rates range from 0.5 to 4 kg B ha^{-1}. Foliar application of B is also effective in correcting deficiencies.

When the soil concentration of soluble B exceeds 1 to 3 mg B L^{-1}, plant toxicity may occur.[114] The range between plant deficiency and toxicity is quite narrow for B. The most common symptom of excessive B accumulation is necrosis along leaf margins and at the growing points, reflecting the buildup of B following the transpiration stream. Although toxicity from excessive soluble B sometimes occurs in some arid soils and in geologically young soils, in most soils toxicity does not occur unless B has been added in excessive amounts. Maas[114] presented a useful summary of the relative sensitivity of various crop and ornamental plants to B toxicity.

In many arid and semi-arid regions, the B concentration of surface and groundwater used for irrigation may be sufficiently high in soluble B to cause a toxic accumulation within the root zone. Sewage effluent and sludge may also contain relatively high B concentrations due to the use of B in many detergents and bleaches. High disposal rates of sewage wastes on soil could cause a potential problem for sensitive plants. The addition of coal ash to soil may also contribute large amounts of B. Coal contains an average of 50 to 60 mg B kg^{-1};[115] however, following combustion, the B may be concentrated in the ash to concentrations as high as 600 mg B kg^{-1}.[116] A heavy application of B-enriched coal ash to agricultural soil may cause plant toxicity for a period of time following ash addition.

Soils containing high B concentrations can be reclaimed if adequate drainage is present to allow leaching. Boron removal from the root zone is possible by passing large quantities of water through the soil to move the soluble B to soil zones beneath the root zone and ultimately into drainage conveyances. If the irrigation water has a high B concentration, an alternative water supply might be needed to sufficiently reduce the soil B concentration.

E. COPPER

Copper in soils is found primarily as Cu^{2+}, where it is strongly fixed by organic matter, clays, and oxide minerals.[117] Since Cu has such a high affinity for complexation by organic compounds and by mineral components in the soil, it is the least mobile of the plant micronutrients. Total Cu in noncontaminated soils typically averages approximately 30 mg Cu/kg, but the total quantity of Cu does not accurately reflect its availability to a growing plant. Extracting a soil with water, a weak acid, or a chelate typically correlates well with plant availability.

Although the requirement for Cu by higher plants is relatively small, it is involved in many essential reactions. Copper is a direct and indirect constituent of many important enzyme systems. It is involved in various Cu-containing proteins, in carbohydrate and N metabolism, lignification, and pollen formation. When plant tissue concentrations fall below 2 to 5 mg Cu kg^{-1}, deficiencies may occur. Visual symptoms of Cu deficiency typically include leaf chlorosis, leaf distortion, and dieback of the young shoots. Deficiencies of Cu may result in poor lignification, which can be manifested as wilting, bent shoots, and lodging.

Deficiency of Cu in plants occurs in many parts of the world where organic soils are used for crop production. Peat and muck soils commonly have low concentrations of plant-available Cu and respond well to Cu fertilizer additions. A Cu deficiency occurring following development of acid, organic soils is commonly called "reclamation disease". Plant species differ in their sensitivity to low concentrations of soil Cu and it may be possible to select certain species and varieties to overcome this limitation to plant growth without fertilization.

Crop removal of Cu is negligible (20 to 40 g Cu/ha/year) when compared to the total Cu content in soil; however, Cu fertilization is essential in many soils. When required, many inorganic and organic compounds have been used to correct Cu deficiencies in plants.[118] The most common Cu fertilizer source is $CuSO_4$, which is popular due to its relatively low price, high solubility, and its commercial availability. Synthetic chelates containing Cu (such as $Na_2CuEDTA$) and natural organic complexes are also effective in maintaining Cu in a soluble and plant-available form. Prior to adding Cu fertilizer to soil, it is necessary to know the likely behavior of the applied Cu. On some sandy, Cu-deficient soils, application of 0.5 kg Cu ha^{-1} is sufficient to eliminate plant deficiencies. In these situations, application of 2.5 kg Cu ha^{-1} can cause toxicity. However on highly organic soils, Cu applications as high as 70 kg Cu ha^{-1} have been recommended. Copper may also be delivered effectively to deficient plants through foliar sprays of $CuSO_4$ or chelated Cu.[118]

Due to the extremely low mobility of Cu in soil, knowledge of the residual fertilizer effects is important for the long-term management of Cu-fertilized soils. In many soils, the residual effects of a light Cu application (1 to 10 kg Cu ha^{-1}) may persist for many years.[119,120] Excessive application of Cu should be avoided by careful soil and plant analysis.

The application of waste products such as sludges, composts, and animal wastes may also contribute considerable amounts of Cu to soil and plants. For example, swine feed is commonly enriched with $CuSO_4$ (125 to 250 mg Cu/kg feed) to increase feed efficiency, weight gains, and prevent dysentery. Poultry feed is also commonly enriched in Cu. Long-term soil application of animal wastes containing high Cu concentrations may lead to an enrichment of soils and sometimes to plant toxicity. The Cu concentration of other organic wastes, such as sewage sludge, varies widely and may lead to a gradual increase in Cu accumulation by plants.

Copper has been extensively used as a fungicide where a mixture of $CuSO_4$ and $CaCO_3$ is sprayed on plants (Bordeaux mixture). Copper-containing sprays are effective for disease control in many crops and are still used extensively in many parts of the world. While use of these sprays may prevent Cu deficiencies in some regions, there is also a danger of over-application. In some soils repeated application of Cu-containing sprays can lead to a buildup of Cu to potentially toxic concentrations.[121] For example, Cu toxicity has been reported in grape-growing (*Vitis vinifera*) regions of France and Cu-toxicity problems have been reported in coffee-growing (*Coffea arabica*) areas in Kenya.[122] Excessively high Cu concentrations may also have deleterious effects on microbial, earthworm, and insect populations and hinder normal biological processes in soil.[123]

Soil contamination by Cu compounds has also been reported from industrial emissions and mining activities. Soil and water in the vicinity of old mining areas and metal processing may contain elevated Cu concentrations. Copper contamination from such activities has been documented in many parts of the world (such as Japan, Wales, the U.S., and the lower Rhine Valley). When sediments are dredged from rivers, canals, and harbors, crops grown in the sediments may contain elevated concentrations of Cu and other metals.[123]

F. IRON

Iron is one of the most common elements in rocks and soils. The total iron content of soils average 3 to 4% where it is found in primary soil minerals, clays, oxides, and hydroxides. Soil color is often used as an indicator of the form and concentration of Fe found through the profile. Although abundant Fe may be present in most soils, plant deficiencies are very common throughout the world. Most Fe-containing compounds in soil are extremely insoluble and the concentration of soluble ferrous iron (Fe^{2+}) is generally extremely low.[124]

Iron solubility in soil is controlled largely by the soil pH and the redox potential. For example, every unit decrease in soil pH increases the concentration of ferric iron (Fe^{3+}) 1000 times and the concentration of Fe^{2+} increases 100 times. In soils where O_2 diffusion is limited (such as flooded soils), the insoluble Fe^{3+} becomes reduced to the more-soluble Fe^{2+} form, which results in greatly increased Fe availability for plants.

Plant deficiencies of Fe are common in areas with alkaline and calcareous soils where Fe solubility is extremely low. An Fe deficiency occurring in these soils is called "lime-induced chlorosis". The presence of bicarbonate (HCO_3^-) in calcareous soils may also aggravate Fe deficiencies by interfering with Fe uptake and translocation within the plant.[125] Iron deficiencies may also occur on sandy, highly leached soils where there is an absolute lack of Fe in the soil.

In most well-aerated soils, Fe^{3+} is the dominant form, although Fe^{2+} is the species taken up by the plant. Therefore, Fe^{3+} must be reduced at the surface of the root before transport to the shoots. This reduction may involve complexation by organic root exudates released in response to the soil Fe concentration.[126]

Within the plant Fe has many important roles due to its relative ease in changing oxidation state and its ability to form coordination complexes with many groups. For example, Fe is found in many plant proteins, in chlorophyll, and in many electron transfer reactions. When Fe is deficient, the first visual evidence is interveinal chlorosis of young leaves due to a decrease in chlorophyll concentration. The leaves may continue to develop in size temporarily until the chlorosis can be alleviated with soil or foliage treatment.

Many organic and inorganic sources of Fe have been used as fertilizer sources for deficient plants. Commercial chelates containing Fe (such as FeEDDHA) are generally the most effective sources, but the cost of treatment may be prohibitive in many circumstances. Soil application of inorganic Fe sources (such as $FeSO_4$) is not generally effective in improving Fe availability in aerated soils because it is rapidly oxidized to insoluble Fe^{3+}. Foliar applications of either inorganic and organic Fe compounds can correct deficiency symptoms, although repeated applications may be needed through the growing season.[127]

Iron toxicity caused by application of fertilizers or waste products is not common because most soils (except organic soils) already contain large quantities of Fe. Additional Fe applications do not greatly increase the concentration. In most well-aerated soils, the added Fe rapidly precipitates to form very insoluble compounds. Iron toxicity can occur in highly acidic, reduced soils where Fe solubility may be very high. Plants adapted to flooded conditions are generally more tolerant to elevated Fe concentrations than plants grown in well-aerated soils. Increasing soil pH through the use of lime would also alleviate Fe toxicity associated with extreme soil acidity.[109]

G. CHLORIDE

Chloride is widely distributed throughout soils of the world. Since it can be distributed in the atmosphere from volcanic emissions and marine aerosols, Cl concentrations are often greater near oceans and seas. For example, near the Pacific Ocean of the U.S., the Cl^- concentration may reach 2 mg Cl L^{-1}in rainwater, while the Cl concentration drops to <0.2 mg Cl L^{-1} in the central U.S. Common Cl salts are quite water soluble and Cl is readily leached through the soil profile with drainage water at virtually the same speed as the water itself. In humid regions, Cl is leached from the profile and transported with water, while in drier climates Cl often accumulates in the surface horizons.

Plant deficiencies of Cl are quite rare in field-grown plants. In fact, it is difficult to induce Cl deficiencies except under carefully controlled conditions due to atmospheric deposition and contamination. Chloride is involved with some specific reactions within the plant, but the major physiological role appears to be as a mobile counter-ion for the transport of cations within the plant. Positive crop responses to Cl have been reported in unusual circumstances. Fertilization with Cl has also been linked with suppression of some foliar and root diseases.[128]

Concerns regarding excess Cl are much greater than are deficiency problems. In most soils, Cl is present as Cl^- in association with various soluble salts such as sodium chloride (NaCl), potassium chloride (KCl), magnesium chloride ($MgCl_2$), and calcium chloride ($CaCl_2$) and not subject to significant retention by soil minerals and organic matter. When soluble salts accumulate in the root zone, Cl toxicity may occur. This Cl accumulation may occur from many sources including natural salt deposits, salt added with irrigation water, road-deicing agents, and waste materials applied to land.

Excessively high concentrations of Cl and soluble salts increase the soil water potential and lower the availability of water to growing plants. Chloride toxicity thresholds vary among plant species, but symptoms are typically manifested by necrosis of the leaf margins, yellowing, and premature leaf drop. If adequate water and drainage is available, Cl can be leached and removed from the soil.[129]

The presence of Cl in soil may also have an indirect effect on the behavior and solubility of other cations. For example, Hahne and Kroontje reported that the solubility of Zn and Pb is significantly greater in the presence of high Cl concentrations.[130] Therefore the interactions of various anions and cations within the soil cannot be ignored.

REFERENCES

1. Cooke, G.W., The present use and efficiency of fertilisers and their future potential in agricultural production systems, in *Proc. Environment and Chemicals in Agriculture, Winteringham*, P.W., Ed., Elsevier, London, 1985, 163.
2. Sparks, D.L. and Huang, P.M., Physical chemistry of soil potassium, in *Potassium in Agriculture*, Munson, R.D., et al., Eds., American Society of Agronomy, Madison, WI, 1985, 201.
3. Goulding, K.W.T., Thermodynamics and potassium exchange in soils and clay minerals, *Adv. Agron.*, 36, 215, 1983.
4. Cassman, K.G., Roberts, B.A., Kerby, T.A., Bryant, D.C., and Higashi, S.L., Soil potassium balance and cumulative cotton response to annual potassium additions on a vermiculitic soil, *Soil Sci. Soc. Am. J.*, 53, 805, 1989.
5. Vomel, A., Nutrient balance in various lysimeter soils. I. Water leaching and nutrient balance, *Zeit. Acker. Pflanzenbau.*, 123, 155, 1966.
6. Stauffer, R.S., Runoff, percolate, and leaching losses from some Illinois soils, *J. Am. Soc. Agron.*, 34, 830, 1942.
7. Volk, G.M. and Bell, C.E., Some major factors in the leaching of calcium, potassium, sulfur, and nitrogen from sandy soils: a lysimeter study, *Florida Agric. Exp. Sta. Bull. No. 416*, 1945.
8. Blue, W.G., Eno, C.F., and Westgate, P.J., Influence of soil profile characteristics and nutrient concentrations on fungi and bacteria in Leon fine sands, *Soil Sci.*, 80, 303, 1955.
9. Nielsen, G.H. and Stevenson, D.S., Leaching of soil calcium, magnesium, and potassium inirrigated orchard lysimeters, *Soil Sci. Soc. Am. J.*, 47, 692, 1983.
10. Nolan, C.W. and Prichett, W.C., Certain factors affecting the leaching of potassium from sandy soils, *Proc. Soil Crop Sci. Fla.*, 20, 139, 1960.
11. Sparks, D.L., Chemistry of soil potassium in Atlantic Coastal Plain soils: a review, *Commun. Soil Sci. Plant Anal.*, 11, 435, 1980.
12. Cooke, G.W., The fate of fertilizers, in *The Chemistry of Soil Processes*, Greenland, D.J. and Hayes, M.H.B., Eds., John Wiley & Sons, Chichester, England, 1981, 563.
13. Munson, R.D. and Nelson, W.L., Movement of applied potassium in soils, *J. Agric. Food Chem.*, 11, 193, 1963.
14. Barrows, H.L. and Kilmer, V.J., Plant nutrient losses from soils by water erosion, *Adv. Agron.*, 15, 303, 1963.
15. Mengel, K. and Kirkby, E.A., Potassium in crop production, *Adv. Agron.*, 33, 59, 1980.
16. Suelter, C.H., Role of potassium in enzyme catalysis, in *Potassium in Agriculture*, Munson, R.D. et al., Eds., American Society of Agronomy, Madison, WI, 1985, 337.

17. Blevins, D.G., Role of potassium in protein metabolism in plants, in Potassium in Agriculture, Munson, R.D. et al., Eds., American Society of Agronomy, Madison, WI, 1985, 413.
18. Mengel, K., Potassium movement within plants and its importance to assimilate transport, in *Potassium in Agriculture*, Munson, R.D. et al., Eds., American Society of Agronomy, Madison, WI, 1985, 397.
19. Duke, S.H. and Collins, M., Role of potassium in legume dinitrogen fixation, in *Potassium in Agriculture*, Munson, R.D. et al., Eds., American Society of Agronomy, Madison, WI, 1985, 443.
20. Meisinger, J.J., Hargrove, W.L., Mikkelsen, R.L., Williams, J.R., and Benson, V.W., Effect of cover crops on groundwater quality, in *Proc. International Conf. Cover Crops for Clean Water*, Jackson, TN, 1991, 57.
21. Huber, D.M. and Arny, D.C., Interactions of potassium with plant disease, in *Potassium in Agriculture*, Munson, R.D. et al., Eds., American Society of Agronomy, Madison, WI, 1985, 467.
22. Singh, B. and Sekhon, G.S., Some measures of reducing leaching loss of nitrates beyond potential rooting zone. II. Balanced fertilization, *Plant Soil*, 44, 391, 1976.
23. Usherwood, N.R., The role of potassium in crop quality, in *Potassium in Agriculture*, Munson, R.D. et al., Eds., American Society of Agronomy, Madison, WI, 1985, 489.
24. Nelson, L.B., *History of the U.S. Fertilizer Industry*, Tennessee Valley Authority, Muscle Shoals, AL, 1990.
25. Barber, S.A., Munson, R.D., and Dancy, W.B., Production, marketing, and use of potassium fertilizers, in *Fertilizer Technology and Use*, Engelsadt, O.P., Ed., Soil Science Society of America, Madison, WI, 1985, 377.
26. Zehler, E., Kreipe, H., and Gethring, P.A., *Potassium Sulphate and Potassium Chloride: Their Influence on the Yield and Quality of Cultivated Plants*, International Potash Institute Research Topics No. 9, International Potash Institute. Bern, Switzerland, 1981.
27. Bixby, D.W. and Beaton, J.D., *Sulphur-Containing Fertilizers: Properties and Applications*, Tech. Bull. No. 17, The Sulphur Institute, Washington, D.C., 1970.
28. Gaines, T.P. and Phatak, S.C., Sulfur fertilization effects on the constancy of the protein N:S ratio in low and high sulfur accumulating crops, *Agron. J.*, 74, 415, 1982.
29. Anderson, J.W., *Sulphur in Biology*, University Park Press, Baltimore, 1978.
30. Anon., *The Fourth Major Nutrient*, The Sulphur Institute, Washington, D.C., 1982.
31. Goodrich, R.D. and Garrett, J.E., Sulfur in livestock nutrition, in *Sulfur in Agriculture*, Tabatabai, M.A., Ed., Soil Science Society of America, Madison, WI, 1986, 617.
32. Haneklaus, S., Evans, E., and Schnug, E., Baking quality and sulphur content of wheat. I. Influence of grain sulphur and protein concentrations on loaf volume, *Sulphur Agric.*, 16, 31, 1992.
33. Lehane, L., Sulfur deficiency and wheat quality, *Rural Res.*, 111, 11, 1981.
34. Schnug, E., Sulphur nutrition and quality of vegetables, *Sulphur Agric.*, 14, 3, 1990.
35. Rendig, V.V., Sulfur and crop quality, in *Sulfur in Agriculture*, Tabatabai, M.A., Ed., Soil Science Society of America, Madison, WI, 1986, 635.
36. Freney, J.R., Forms and reactions of organic sulfur compounds in soils, in *Sulfur in Agriculture*, Tabatabai, M.A., Ed., Soil Science Society of America, Madison, WI, 1986, 207.
37. Ensminger, L.E. and Freney, J.R., Diagnostic techniques for determining sulfur deficiencies in crops and soils, *Soil Sci.*, 101, 283, 1966.

38. Spencer, K. and Freney, J.R., A comparison of several procedures for estimating the sulphur status of soils, *Aust. J. Agric. Res.*, 11, 948, 1960.
39. Stevenson, F.J., *Cycles of Soil*, John Wiley & Sons, New York, 1986.
40. Chao, T.T., Harward, M.E., and Fang, S.C., Soil constituents and properties in the adsorption of sulfate ions, *Soil Sci.*, 94, 276, 1962.
41. Harward, M.E. and Reisenauer, H.M., Reactions and movement of inorganic soil sulfur, *Soil Sci.*, 101, 326, 1966.
42. Kamprath, E.J., Nelson, W.L., and Fitts, J.W., The effect of pH, sulfate and phosphate concentration on the adsorption of sulfate by soils, *Soil Sci. Soc. Am. Proc.*, 20, 463, 1956.
43. Camberato, J.J. and Kamprath, E.J., Solubility of adsorbed sulfate in Coastal Plain soils, *Soil Sci.*, 142, 211, 1986.
44. Oates, K.M. and Kamprath, E.J., Sulfur fertilization of winter wheat grown on deep sandy soils, *Soil Sci. Soc. Am. J.*, 49, 925, 1985.
45. Noggle, J.C., Meagher, J.F., and Jones, U.S., Sulfur in the atmosphere and its effects on plant growth, in *Sulfur in Agriculture*, Tabatabai, M.A., Ed., Soil Science Society of America, Madison, WI, 1986, 251.
46. Kennedy, I.R., *Acid Soil and Acid Rain*, 2nd ed., Research Studies Press, Taunton, Somerset, England, 1992.
47. Aneja, V.P. and Cooper, W.J., Biogenic sulfur emissions: a review, in *Biogenic Sulfur in the Environment*, Saltzman, E.S. and Cooper, W.J., Eds., American Chemical Society, Washington, D.C., 1989, 2.
48. Hoeft, R.G. and Fox, R.H., Plant response to sulfur in the Midwest and Northeastern United States, in *Sulfur in Agriculture*, Tabatabai, M.A., Ed., Soil Science Society of America, Madison, WI, 1986, 345.
49. Kamprath, E.J. and Jones, U.S., Plant response to sulfur in the Southeastern United States, in *Sulfur in Agriculture*, Tabatabai, M.A., Ed., Soil Science Society of America, Madison, WI, 1986, 323.
50. Rasmussen, P.E. and Kresge, P.O., Plant response to sulfur in the Western United States, in *Sulfur in Agriculture*, Tabatabai, M.A., Ed., Soil Science Society of America, Madison, WI, 1986, 357.
51. Heck, W.W., Heagle, A.S., and Shriver, D.S., Effects on vegetation: native crops and forests, in *Air Pollution*, Stern, A.S., Ed., Academic Press, New York, 6, 247, 1986.
52. Terman, G.L., *Atmospheric Sulphur — The Agronomic Aspects*. Tech. Bull. No. 23. The Sulphur Institute, Washington, D.C., 1978.
53. Rhue, R.D. and Kamprath, E.J., Leaching losses of sulfur during winter months when applied as gypsum, elemental S or prilled S, *Agron. J.*, 65, 603, 1973.
54. Stromberg, L.K. and Tisdale, S.L., *Treating Irrigated Arid-Land Soils with Acid-Forming Sulphur Compounds*, Tech. Bull. No. 17, The Sulphur Institute, Washington, D.C., 1979.
55. Tisdale, S.L. and Nelson, W.L., *Soil Fertility and Fertilizers*, 3rd ed., Macmillan, New York, 1975.
56. Brady, N.C., *The Nature and Properties of Soils*, 8th ed., Macmillan, New York, 1974.
57. Evans, C.E. and Kamprath, E.J., Lime response as related to percent Al saturation, solution Al, and organic matter content, *Soil Sci. Soc. Am. Proc.*, 34, 893, 1970.
58. Marschner, H., Mechanisms of adaptation of plants to acid soils, in *Plant-Soil Interactions at Low pH*, Wright, R.J., Baligar, V.C., and Murrman, R.P., Eds., Kluwer Academic, Dordrecht, The Netherlands, 1991, 683.

59. Kamprath, E.J. and Foy, C.D., Lime-fertilizer-plant interactions in acid soils, in *Fertilizer Technology and Use*, Engelstad, O.P., Ed., Soil Science Society of America, Madison, WI, 1985, 91.
60. Massey, H.F., pH and soluble Cu, Ni, and Zn in eastern Kentucky coal mine spoil materials, *Soil Sci.*, 114, 217, 1972.
61. Kinraide, T.B., Identity of the rhizotoxic aluminum species, in *Plant-Soil Interactions at Low pH*, Wright, R.J., Baligar, V.C., and Murrman, R.P., Eds., Kluwer Academic, Dordrecht, The Netherlands, 1991, 717.
62. Clarkson, D.T., The effect of aluminum and some other trivalent metal cations on cell division of root apices of *Allium cepa, Ann. Bot.*, 29, 309, 1965.
63. Wallace, S.U. and Anderson, I.C., Aluminum toxicity and DNA synthesis in wheat roots, *Agron. J.*, 76, 5, 1984.
64. Pan, W.L., Hopkins, A.G., and Jackson, W.A., Aluminum-inhibited shoot development in soybean: a possible consequence of impaired cytokinin supply, *Commun. Soil Sci. Plant Anal.*, 19, 1143, 1988.
65. Krause, H.H., Effect of pH on leaching losses of potassium applied to forest nursery soils, *Soil Sci. Soc. Am. Proc.*, 29, 613, 1965.
66. Mahilum, B.C., Fox, R.L., and Silva, J.A., Residual effects of liming volcanic ash soils in the humid tropics, *Soil Sci.*, 109, 102, 1970.
67. MacLeod, L.B., Bishop, R.F., and Calder, F.W., Effect of various rates of liming and fertilization on certain chemical properties of a strongly acid soil on the establishment, yield, botanical and chemical composition of a forage mixture, *Can. J. Soil Sci.*, 44, 237, 1964.
68. Powell, A.J. and Hutcheson, T.B., Effect of lime and potassium additions on soil potassium reactions and plant response, *Soil Sci. Soc. Am. Proc.*, 29, 76, 1965.
69. Morrill, L.G. and Dawson, J.E., Patterns observed for the oxidation of ammonium to nitrate by soil organisms, *Soil Sci. Soc. Am. Proc.*, 31, 757, 1967.
70. Dancer, W.S., Peterson, L.A., and Chesters, G., Ammonification and nitrification of N as influenced by soil pH and previous N treatments, *Soil Sci. Soc. Am. Proc.*, 37, 67, 1973.
71. Munn, D.A. and Jackson, W.A., Nitrate and ammonium uptake by rooted cuttings of sweet potato, *Agron. J.*, 70, 312, 1978.
72. Marschner, H., Romheld, V., and Ossenberg-Neuhaus, H., Rapid method for measuring changes in pH and reducing processes along roots of intact plants, *Z. Pflanzenphysiol. Bd.*, 105, 407, 1982.
73. Grinsted, M.J., Hedley, M.J., White, R.E., and Nye, P.H., Plant-induced changes in the rhizosphere of rape (*Brassica napus* var. Emerald) seedlings, *New Phytol.*, 91, 19, 1982.
74. Gijsman, A.J. and Van Noordwijk, M., Critical ammonium:nitrate uptake ratios for Douglas-fir determining rhizosphere pH and tree mortality, in *Plant-Soil Interactions at Low pH*, Wright, R.J., Baligar, V.C., and Murrman, R.P., Eds., Kluwer Academic, Dordrecht, The Netherlands, 1991, 181.
75. Woodruff, J.R. and Kamprath, E.J., Phosphorus adsorption maximum as measured by the Langmuir isotherm and its relationship to phosphorus availability, *Soil Sci. Soc. Am. Proc.*, 29, 148, 1965.
76. Islam, A.K.M.S., Edwards, D.G., and Asher, C.J., pH optima for crop growth: results of a flowing culture: experiment with six species, *Plant Soil*, 54, 339, 1980.

77. Reisenauer, H.M., Tabilh, A.A., and Stout, P.R., Molybdenum reactions with soils and the hydrous oxides of iron, aluminum and titanium, *Soil Sci. Soc. Am. Proc.*, 26, 23, 1962.
78. Barshad, I., Factors affecting the molybdenum content of pasture plants. I. Nature of soil molybdenum, growth of plants, and soil pH, *Soil Sci.*, 71, 297, 1951.
79. Parker, M.B. and Harris, H.B., Soybean response to molybdenum and lime and relationship between yield and chemical composition, *Agron. J.*, 54, 480, 1962.
80. Lee, C.R. and Page, N.R., Soil factors influencing the growth of cotton following peach orchards, *Agron. J.*, 59, 237, 1967.
81. Lee, C.R. and Craddock, G.R., Factors affecting plant growth in high-zinc medium. II. Influence of soil treatments on growth of soybeans on strongly acid soil containing zinc from peach sprays, *Agron J.*, 61, 565, 1969.
82. Keisling, T.C., Lauer, D.A., Walker, M.E., and Henning, R.J., Visual, tissue, and soil factors associated with Zn toxicity of peanuts, *Agron. J.*, 69, 765, 1977.
83. U.S. Environmental Protection Agency, *Process Design Manual — Land Aplication of Municipal Sludge*, USEPA Rep. 625/1-83-016, U.S. Government Printing Office, Washington, D.C., 1983.
84. Pietz, R.I., Carlson, C.R., Jr., Peterson, J.R., Zenz, D.R., and Lue-Hing, C., Application of sewage sludge and other amendments to coal refuse material. I. Effects on chemical composition, *J. Environ. Qual.*, 18, 164, 1989.
85. Pietz, R.I., Carlson, C.R., Jr., Peterson, J.R., Zenz, D.R., and Lue-Hing, C., Application of sewage sludge and other amendments to coal refuse material. II. Effects on revegetation, *J. Environ. Qual.*, 18, 169, 1989.
86. Sposito, G., Distribution of potentially hazardous trace metals, in *Metal Ions in Biological Systems*, Vol. 20, Sigel, H., Ed., Marcel-Dekker, Basel, Switzerland, 1986, 1.
87. Adriano, D.C., *Trace Elements in the Terrestrial Environment*, Springer-Verlag, New York, 1986.
88. Stahlberg, S. and Sombatpanit, S., Manganese relationships of soil and plant. I. Investigation and classification of Swedish manganese deficient soils, *Acta Agric. Scand.*, 24, 179, 1974.
89. Norvell, W.A., Inorganic reactions of manganese in soils, in *Manganese in Soils and Plants*, Graham, R.D. et al., Eds., Kluwer Academic, Dordrecht, The Netherlands, 1988, 37.
90. Romheld, V. and Marschner, H., Function of micronutrients in plants, in *Micronutrients in Agriculture*, Mortvedt, J.J. et al., Eds., Soil Science Society of America, Madison, WI, 1991, 297.
91. Labanauskas, C.K., Manganese, in *Diagnostic Criteria for Plants and Soils*, Chapman, H.D., Ed., Quality Printing Co., Abilene, TX, 1965, 264.
92. Kabata-Pendias, A. and Pendias, H., *Trace Elements in Soils and Plants*, 2nd ed., CRC Press, Boca Raton, FL, 1992.
93. Mascagni, H.J., Jr. and Cox, F.R., Effective rates of fertilization for correcting manganese deficiency in soybeans, *Agron. J.*, 77, 363, 1985.
94. Mascagni, H.J., Jr. and Cox, F.R., Evaluation of inorganic and organic manganese fertilizer sources, *Soil Sci. Soc. Am. J.*, 49, 458, 1985.
95. Horst, W.J., The physiology of manganese toxicity, in *Manganese in Soils and Plants*, Graham, R.D. et al., Eds., Kluwer Academic, Dordrecht, The Netherlands, 1988, 175.
96. Mikkelsen, R.L. and Jarrell, W.M., Injection of urea phosphate and urea sulfate to drip-irrigated tomatoes grown in calcareous soil, *Soil Sci. Soc. Am. J.*, 51, 464, 1987.

97. Council for Agricultural Science and Technology (CAST), Application of Sewage Sludge to Cropland: Appraisal of Potential Hazards of the Heavy Metals to Plants and Animals, Report No. 64, Ames, IA, 1976.
98. Kubota, J., Molybdenum status of United States soils and plants, in *Molybdenum in the Environment*, Chappell, W.R. and Petersen, K.K., Eds., Marcel Decker, New York, 1977, 555.
99. Davies, B.E. and Jones, L.H.P., Micronutrients and toxic elements, in *Russell's Soil Conditions and Plant Growth*, Wild, A., Ed., Longman, Essex, England, 1988, 780.
100. Gupta, U.C. and Lipsett, J., Molybdenum in soils, plants, and animals, *Adv. Agron.*, 34, 73, 1981.
101. Miller, E.R., Lei, X., and Ullery, D.E., Trace elements in animal nutrition, in *Micronutrients in Agriculture*, Mortvedt, J.J. et al., Eds., Soil Science Society of America, Madison, WI, 1991, 593.
102. Hornick, S.B., Baker, D.E., and Guss, S.B., Crop production and animal health problems associated with high molybdenum soils, *Symp. Molybdenum in the Environment*, June 12, Denver, 1975, 12.
103. Schalscha, E.B., Morales, M., and Pratt, P.F., Lead and molybdenum in soils and forage near an atmospheric source, *J. Environ. Qual.*, 16, 313, 1987.
104. Furr, A.K., Parkinson, T.F., Bache, C.A., Gutenmann, W.H., Pakkala, I., and Lisk, D.J., Multielement absorption by crops grown on soils amnded with municipal sludges ashes, *J. Agric. Food Chem.*, 28, 660, 1980.
105. Bingham, F.T., Peryea, F.J., and Jarrell, W.M., in *Metal Ions in Biological Systems*, Sigel, H., Ed., Marcel Dekker, New York, 1986, 119.
106. Van Campen, D.R., Trace elements in human nutrition, in *Micronutrients in Agriculture*, Mortvedt, J.J. et al., Eds., Soil Science Society of America, Madison, WI, 1991, 663.
107. Sparr, M.C., Micronutrient needs — which, where, or what — in the United States, *Commun. Soil Sci. Plant Anal.*, 1, 241, 1970.
108. Jeffery, J.J. and Uren, N.C., Copper and zinc species in the soil solution and the effects of soil pH, *Aust. J. Soil Res.*, 21, 479, 1983.
109. Mikkelsen, D.S. and Kuo, S., *Zinc Fertilization and Behavior in Flooded Soils*, Special Publ. No. 5, Commonwealth Agriculture Bureaux, Farnham Royal, U.K., 1977.
110. Staker, E.V. and Cummings, R.W., The influence of zinc on the productivity of certain New York peat soils, *Soil Sci. Soc. Am. Proc.*, 6, 207, 1941.
111. Martens, D.C. and Westermann, D.T., Fertilizer applications for correcting micronutrient deficiencies, in *Micronutrients in Agriculture*, Mortvedt, J.J. et al., Eds., Soil Science Society of America, Madison, WI, 1991, 549.
112. Keren, R. and Bingham, F.T., Boron in water, soils, and plants, *Adv. Soil Sci.*, 1, 229, 1985.
113. Gupta, U.C., Boron nutrition of crops, *Adv. Agron.*, 31, 273, 1979.
114. Maas, E.V., Salt tolerance of plants, *Appl. Agric. Res.*, 1, 12, 1986.
115. Swanson, V.E. Medlin, J.M., Hatch, J.R., Coleman, S.L., Wood, G.H., Jr., Woodruff, S.D., and Hildebrand, R.T., U.S. Geol. Survey Report 76-468, Washington, D.C., 1979, 503.
116. Page, A.L., Elseewi, A.A., and Straughan, I.R., Physical and chemical properties from coal-fired power plants with reference to environmental impacts, *Residue Rev.*, 71, 83, 1979.
117. McBride, M.B., Forms and distribution of copper in solid and solution phases of soil, in *Copper in Soils and Plants*, Loneragan, J.F. et al., Eds., Academic Press, Sydney, 1981, 24.

118. Gilkes, R.J., Behaviour of Cu additives — fertilisers, in *Copper in Soils and Plants*, Lonerragan, J.F. et al., Eds., Academic Press, Sydney, 1981, 97.
119. Reith, J.W.S., Copper deficiency in crops in north-east Scotland, *J. Agric. Sci.*, 70, 39, 1968.
120. Cox, F.R., Residual value of copper fertilization, *Commun. Soil Sci. Plant Anal.*, 23, 101, 1992.
121. Hirst, J.M., LeRiche, H.H., and Bascomb, C.K., Copper accumulations in the soils of apple orchards near Wisbech, *Plant Pathol.*, 10, 105, 1961.
122. Lepp, N.W., Dickinson, N.M., and Ormand, K.L., Distribution of fungicide-derived copper in soils, litter and vegetation of different aged stands of coffee (*Coffea arabica* L.) in Kenya, *Plant Soil*, 77, 263, 1984.
123. Tiller, K.G. and Merry, R.H., Copper pollution of agricultural soils, in *Copper in Soils and Plants*, Loneragan, J.F. et al., Eds., Academic Press, Sydney, 1981, 119.
124. Lindsay, W.L., Inorganic equilibria affecting micronutrients in soils, in *Micronutrients in Agriculture*, Mortvedt, J.J. et al., Eds., Soil Science Society of America, Madison, WI, 1991, 89.
125. Mengel, K., Scherrer, H.W., and Malissiovas, N., Chlorosis from the aspect of soil chemistry and vine nutrition, *Mitt. Klosterneuburg*, 29, 151, 1979.
126. Brown, J.C. and Jolley, V.D., Plant metabolic responses to iron-deficiency stress, *Bioscience*, 38, 546, 1989.
127. Mortvedt, J.J., Iron sources and management practices for correcting iron chlorosis problems, *J. Plant Nutr.*, 9, 961, 1986.
128. Christensen, N.W., Taylor, R.G., Jackson, T.L., and Mitchell, B.L., Chloride effects on water potential of wheat infected with take-all root rot, *Agron. J.*, 73, 1053, 1987.
129. Eaton, F.M., Chlorine, in *Diagnostic Criteria for Plants and Soils*, Chapman, H.D., Ed., Quality Printing Co., Abilene, TX, 1966, 98.
130. Hahne, H.C.H. and Kroontje, W., Significance of pH and chloride concentration on behavior of heavy metal pollutants: mercury (II), cadmium (II), zinc (II), and lead (II), *J. Environ. Qual.*, 2, 444, 1973.

CHAPTER 4

Trace Metals

Alina Kabata-Pendias and Domy C. Adriano

I. INTRODUCTION

Soil, as a part of the terrestrial ecosystem, plays a crucial role in elemental cycling. It has important functions as a storage, buffer, filter, and transformation compartment, supporting a homeostatic interrelationship between the biotic and abiotic components.

The chemical composition of soils is diverse and, although governed by many different factors, climatic condition and parent material usually predominate. Although trace elements (in this chapter, trace metals in both cationic and anionic forms will be adopted) are minor components of the soil solid phase, they play an important role in soil bioactivity and fertility. A knowledge of the association of trace metals with particular soil phases and their affinity to each soil constituent is a key to a better understanding of the principles governing their behavior in soils. The speciation and spatial distribution of trace metals in soils are closely related to their behavior in multiphase and dynamic soil systems. Surface reactions at soil-solution-root interfaces play a role in the determination of the amounts of mobile and bioavailable trace metals.

0-87371-859-3/95/$0.00+$.50

The behavior of trace metals in soils is related also to their origin and source. The indigenous trace metal content of soils is inherited from the parent material (i.e., termed lithogenic) and altered due to weathering and soil-forming processes (i.e., termed pedogenic). Trace metals deposited into soils from human activities (i.e., termed anthropogenic) may greatly vary in behavior from metals of another origin. Metals are a major group of pollutants that are known to have caused environmental problems in agricultural and natural areas.

Certain trace metals should be at sufficient levels in agricultural soils because they are essential in plant nutrition. These are the commonly known "micronutrients" which include the metals iron, manganese, copper, zinc, molybdenum, and cobalt. It should be pointed out that the usual concern for these essential trace metals is their deficiency in plant nutrition rather than toxicity. Zinc deficiency is the most widespread form of micronutrient disorder in economic crops on a worldwide basis, although to a much lesser extent, deficiency of manganese, copper, and iron are also frequently reported in agricultural and horticultural crops.[1] Other trace metals that become of concern are those that are supplied with waste by-products, such as municipal sewage sludge, livestock manures, and now the likelihood of coal combustion residues being applied on the land. These by-products contain a myriad of trace metals, some of which could be of concern in the food chain transfer.

The assessment of soil pollution and adoption of acceptable standards for permissible trace metal levels in soils could be instrumental in protecting a well-balanced ecological function in soils and for fostering sustainable agriculture.

II. ORIGIN OF METALS IN SOILS

The natural pool of trace metals in surface soils arose from the net effect of geological and soil-forming processes, as well as of geochemical properties of the elements. Under the usual soil-forming conditions, trace metal status of soil originates from the parent rock. Thus, an understanding of the distribution of trace metals in various rocks is the first step in predicting the natural metal contents of soils.

The distribution of trace metals in magmatic rocks (Table 1) is a manifestation of the net effect of properties of the rock on the elements' concentration. Some trace metals are expected to occur in highest amounts in ultramafic and mafic rocks (e.g., Cd, Co, Cr, Cu, Mn, Ni), while others tend to accumulate in acid and intermediate rocks (e.g., Ba, Mo, Pb, Sn, U, W, Zn).

The status of trace metals in sedimentary rocks is controlled by geological and geochemical processes of which weathering, transport with water, sorption, and precipitation are probably the most significant. Because of the differing influence of these processes, the distribution of trace metals in sedimentary rocks is generally variable but exhibits the tendency that most metals are accumulated in argillaceous sediments (Table 2). Only some metals, e.g., Mn, Sr, and U, are also infrequently concentrated in calcareous rocks. Sandstones and other sandy deposits and many calcareous rocks (limestones, dolomites) usually contain much lower amounts of trace metals than rocks enriched in fine grain fractions. This general tendency allows soil scientists to predict the natural content (i.e., termed baseline or background) of trace metals in soils. Expected values should always be compared with calculated values based on a broad database[3] (Table 3). Metal content of main soil categories shows a somewhat rising trend with increasing content of fine

Table 1 Commonly Reported Values for Trace Elements in Magmatic Rocks (mg kg^{-1})

	Element	Rocks				
	1	Ultramafic 2	Mafic 3	Intermediate 4	Acid 5	Acid (volcanic) 6
Ag	Silver	0.05–0.06	0.1	0.05–0.07	0.04	0.05
As	Arsenic	0.5–1	0.6–2	1–2.5	1–2.6	1.5–2.5
Au	Gold	0.005	0.0005–0.003	0.0032	0.0012–0.0018	0.0015
Ba	Barium	0.5–25	250–400	600–1000	400–850	600–1200
Be	Beryllim	0.2	0.3–1	1–1.8	2–5	5–6.5
Bi	Bismuth	0.001–0.02	0.01–0.15	0.01–0.1	0.01–0.12	0.01–0.12
Cd	Cadmium	0.03–0.05	0.13–0.22	0.13	0.09–0.2	0.05–0.2
Co	Cobalt	100–200	35–50	1–10	1–7	15
Cr	Chromium	1600–3400	170–200	15–50	4–25	4–16
Cu	Copper	10–40	60–120	15–80	15–30	5–20
Hg	Mercury	0.0X	0.0X	0.0X	0.08	0.0X
Li	Lithium	0.5–15	6–20	20–28	25–40	15–45
Mn	Manganese	850–1500	1200–2000	500–1200	350–600	600–1200
Mo	Molibdenum	0.2–0.3	1–1.5	0.6–1	1–2	2
Ni	Nickel	1400–2000	130–160	5–55	5–15	20
Pb	Lead	0.1–1	3–8	12–15	15–25	10–20
Sb	Antimony	0.1	0.2–0.1	0.X	0.2	0.2
Se	Selenium	0.02–0.05	0.01–0.05	0.02–0.05	0.01–0.05	0.02–0.05
Sn	Tin	0.35–0.5	0.9–1.5	1.3–1.5	1.5–3.6	2–3
Tl	Thallium	0.05–0.2	0.1–0.4	0.5–1.4	0.6–2.3	0.5–1.8
U	Uranium	0.003–0.01	0.3–1	1.4–3	2.5–6	5
V	Vanadium	40–100	200–250	30–100	40–90	70
W	Tungsten	0.1–0.77	0.36–1.1	1–1.9	1.3–2.4	2
Zn	Zinc	40–60	80–120	40–100	40–60	40–100

Adapted from Reference 2.

Table 2 Commonly Reported Values for Trace Elements in Sedimentary Rocks

Element	Argillaceous Sediments 2	Shales 3	Sandstones 4	Limestones, Dolomites 5
Ag	0.07	0.07–0.01	0.05–0.25	0.1–0.15
As	13	5–13	1–1.2	1–2.4
Au	0.003–0.004	0.0025–0.004	0.003–0.007	0.002–0.006
Ba	500–800	500–800	100–320	50–200
Be	2–6	2–5	0.2–1	0.2–2
Bi	0.05–0.4	0.05–0.5	0.1–0.2	0.1–0.2
Cd	0.3	0.22–0.3	0.05	0.035
Co	14–20	11–20	0.3–10	0.1–3
Cr	80–120	60–100	20–40	5–16
Cu	40–60	40	5–30	2–10
Hg	0.2–0.4	0.18–0.4	0.04–0.1	0.04–0.05
Li	60	50–75	10–40	5–20
Mn	400–800	500–850	100–500	200–1000
Mo	2–2.6	0.7–2.6	0.2–0.8	0.14–0.4
Ni	40–90	50–70	5–20	7–20
Pb	20–40	18–25	5–10	3–10
Sb	1.2–2	0.8–1.5	0.05	0.03
Se	0.4–0.6	0.6	0.05–0.08	0.03–0.1
Sn	6–10	6	0.5	0.5
Tl	0.5–1.5	0.5–2	0.4–1	0.01–0.14
U	3–4	3–4.1	0.45–0.59	2.2–2.5
V	80–130	100–130	10–60	10–45
W	1.8–2	1.8–2	1–2	0.4–0.6
Zn	80–120	80–120	15–30	10–25

Adapted from Reference 2.

soil fraction (clay) which in turn is positively correlated with the soil cation exchange capacity (CEC) values. However, the amount and quality of soil organic matter (SOM) and various concretions (mainly oxides and carbonates of Fe, Mn, and Ca) are likely to change this pattern in some soil. The distribution of certain elements, e.g., the lanthanides (also called rare earth elements), in soil shows a more discernible impact by the parent rocks and generally follows the geochemical rule that their abundance decreases with increasing atomic number in both even- and odd-numbered sequences.

Mean trace metal contents of top soils are, in most cases, lower than the average values (i.e., Clarke's values) for rocks of the earth's crust (Figure 1). Mafic rocks (granites and basalts) predominate in the earth's crust, and soils are developed most commonly on sedimentary deposits. Perhaps a more reliable method to compare the relative abundance of a chemical element in geologic material with a soil sample is to normalize a particular element to an element that is universally abundant and relatively more stable in geochemical environments. Aluminum is commonly used as the reference element.[5] Many elements are somewhat enriched in surface soils compared with their abundance in bedrocks when calculated on the basis of Al content (Figure 2). The significant enrichment of Cd, Pb, and As in topsoil is obvious when compared to their occurrence in the earth crust. According to Harrison and Rahn,[6] enriched elements in soils are of chalcophilic nature and have small mass-median diameters. This phenomenon may be explained by a hypothesis that the enriched trace metals arose from high-temperature combustion sources, either indigenous or anthropogenic.

Table 3 Ranges and Means[a] of Total Contents of Trace Metals in Surface Soils[b] Calculated on the World Scale (mg kg^{-1})

	Podzols (Sandy Soils)		Cambisoils		Rendzinas		Chernozems, Kastanozems		Histosols	
Metal	Range	Mean	Range	Mean	Range	Mean	Range	Mean	Range	Mean
As	<0.1–30	4.4	1.3–27	8.4	—	—	1.9–23	8.5	<0.1–66.5	9.3
Ba	20–1,500	330	19–1,500	520	150–1,500	520	100–1,000	520	10–700	175
Cd	0.01–2.7	0.37	0.08–1.61	0.45	0.38–0.84	0.62	0.18–0.71	0.44	0.19–2.2	0.78
Co	0.1–65	5.5	3–58	10	1–70	12	0.5–50	7.5	0.2–49	4.5
Cr	1.4–530	47	4–1,100	51	5–500	83	11–195	77	1–100	12
Cu	1–70	13	4–100	23	6.8–70	23	6.5–140	24	1–113	16
Hg	0.008–0.7	0.05	0.01–1.1	0.1	0.01–0.5	0.05	0.02–0.53	0.1	0.04–1.11	0.26
Li	<5–72	22	1.4–130	46	6–105	56	9–175	53	0.01–32	1.3
Mn	7–2,000	270	45–9,200	525	50–7,750	445	100–3,907	480	7–2,200	465
Mo	0.17–3.7	1.3	0.1–7.2	2.8	0.3–7.4	1.5	0.4–6.9	2	0.3–3.2	1.5
Ni	1–110	13	3–110	26	2–450	34	6–61	25	0.2–119	12
Pb	2.3–70	22	1.5–70	28	10–50	26	8–70	23	1.5–176	44
Sr	5–1,000	87	15–1,000	210	15–1,000	195	70–500	145	5–300	100
Ti	200–17,000	2,600	500–24,000	3,300	400–10,000	4,800	700–7,000	3500	80–6,700	2,300
V	10–260	67	15–330	76	10–500	115	25–150	78	6.3–150	18
Zn	3.5–220	45	9–362	60	10–570	100	20–770	65	5–250	50

[a] Ranges of common abundance in topsoils and arithmetic means.
[b] Soil units as given in the FAO/UNESCO system.[4]

From Kabata-Pendias, A. and Pendias, H., *Biogeochemistry of Trace Elements*, PWN, Warsaw, 1993, 364. With permission.

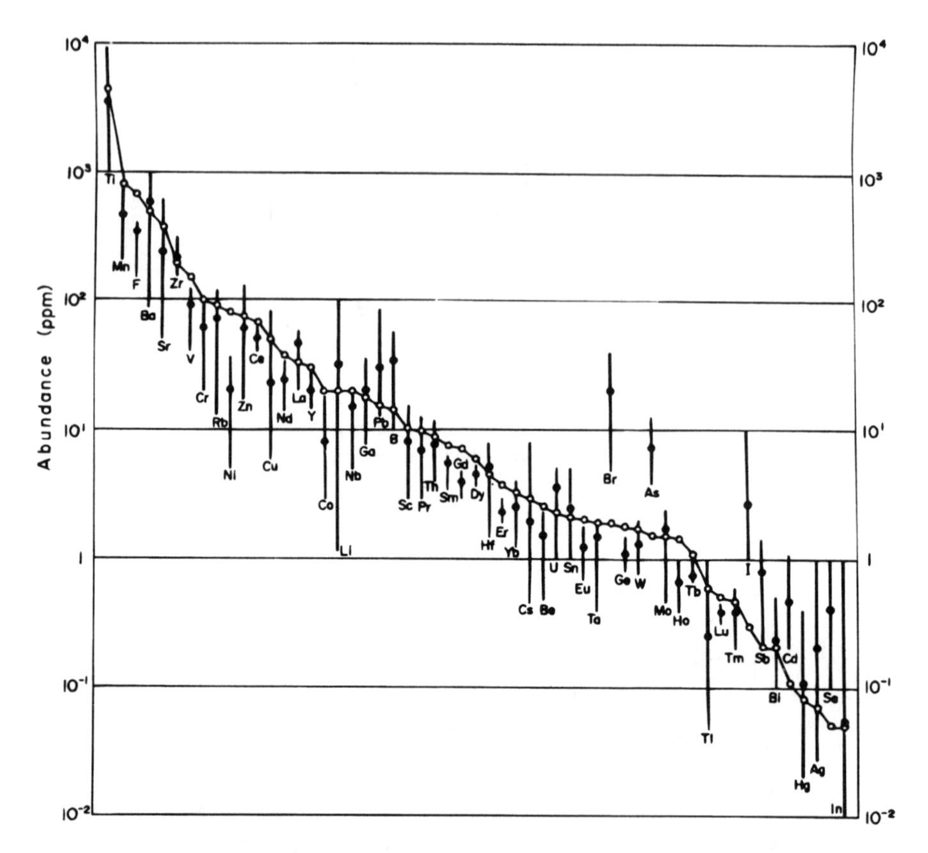

Figure 1 Trace elements in soils compared to their abundance in the earth crust.[2] Open circles, mean content in the earth crust; black circles, mean content of the soils; vertical lines, values commonly found in topsoils.

A. LITHOGENIC METALS

The trace metal composition of soils is primarily inherited from the parent material. The mobility of these elements during weathering is determined, first, by the stability of the host minerals and, second, by their electrochemical properties. Because lithogenic and pedogenic metals are of the same origin from the parent material they are difficult to distinguish.

In the dynamic chemical system of the soil, mineral transformation is continuous. The degree of weathering of parent material influences the metal content of the soil. Other parameters also affect the trace metal status in the soil.[7,8] Their importance varies, depending upon their intensity; however, under moderate soil conditions, pH, CEC, and redox potential are likely to exert the greatest influence.

In most soils, quartz is the predominant constituent (up to 70%) of the sand and silt fractions (20 to 200 μm). Feldspars comprise about 20% and the heavy minerals <5%. Trace metals associated with these constituents are the most stable. It is still an open question for scientists in the matter of how trace metals evolved from their parent ma-

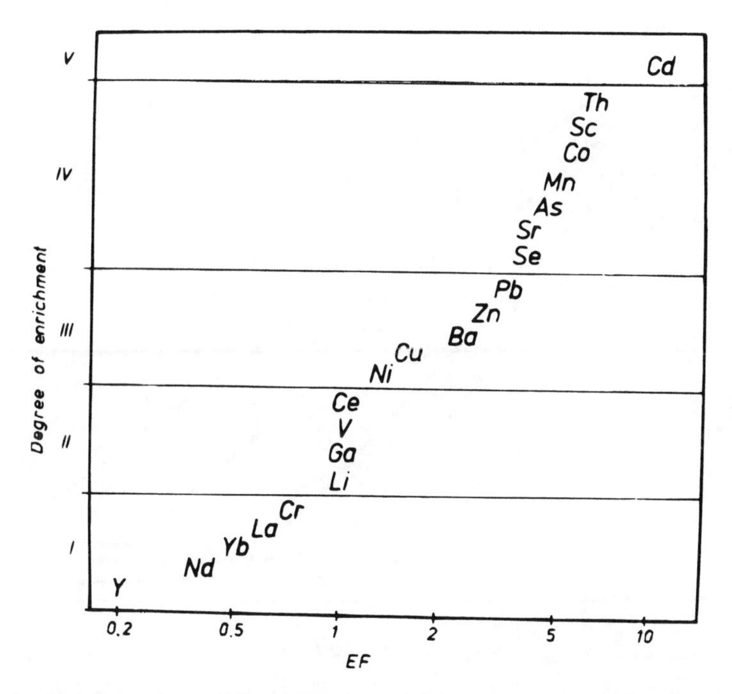

Figure 2 Enrichment factor (EF) values for trace elements in surface sandy soils of Poland as compared with mean concentrations in sandstones, and normalized to Al content.[2] Degrees of trace elements enrichment: I = decrease, II relative stability, III = slight increase, IV = moderate increase, V = strong increase.

terials and their eventual formation of compounds. It seems more likely that these metals are bound onto soil minerals by isomorphic substitution or by fixation on free structural sites.[9] The adsorption capacity by certain soil components for metals is high and, therefore, considerable amounts of metal could be unavailable for their compound formation, such as phosphates, carbonates, sulfites, sulfates, and others.

Lithogenic metals form a pool of relatively immobile elements in soil. However, they are subject to transformation into more mobile species under certain soil conditions and especially under the influence of microbial activity and of plant root exudates.

B. PEDOGENIC METALS

Soil is a heterogeneous mixture of various organic and organo-mineral substances, clay minerals, oxides of Al, Fe, and Mn, and other solid components, as well as of a variety of soluble substances and microbiota. The binding mechanisms are therefore complex and vary with the composition of the soil, redox potential, and soil pH (Figure 3). However, the influence of soil pH on the mobility of trace metals depends upon the geochemical properties of the metal as well as on several other soil variables. For example, the activity of trace metals in relation to soil pH is known to be modified by the amount and type of organic matter (Figure 4). Several metal-organic complexes are fairly soluble even at pH values above 6.[11,12]

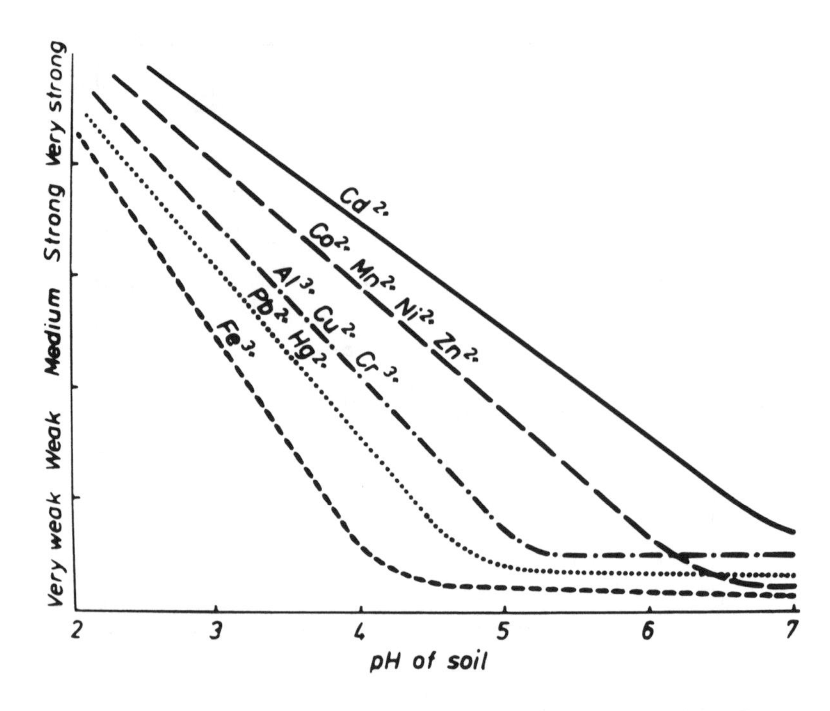

Figure 3 Schematic trends in the mobility of metals as influenced by soil pH. (Data for light mineral soil.)[2]

The complexity of the soil properties in its natural state is a result of the following processes: (1) dissolution, (2) sorption-desorption, (3) complexation, (4) migration, (5) precipitation, (6) occlusion, (7) diffusion, (8) binding by organic substances, (9) absorption and sorption by microbiota, and (10) volatilization. Biological, chemical, and physical characteristics of the natural soil system are variable with depth and space because of the influence by the various soil factors.[2,13,17] Therefore, the distribution of trace metals within the soil profile, as well as their speciation, is not stable.

Both the distribution and mobility of pedogenic metals are influenced by the specific adsorption of metals on various soil constituents, in particular the hydrous oxides of Fe and Mn. The pedogenic formation of these oxides seems to be a significant factor governing the distribution of metals on the solid phase of the soil. The fixation of trace metals by Fe and Mn oxides (in the form of nodules, concretions, or layers) is most often responsible for their deposition in the soil profile. Usually this form of metals is sparingly mobile, and probably mostly unavailable to plant roots. Such localized immobilization of metals has been observed at the root-soil interface where these oxides are precipitated under certain soil conditions.[14,15]

Various analytical procedures involving successive (or sequential) extraction have been developed to measure the speciation of metals in soil. The distribution patterns of metal species in soils vary widely, and are comparable among soils only when the same extraction procedure is applied. However, depending upon the variability in physicochemical characteristics of metals, their affinity to soil components may govern their speciation (Figure 5). More mobile metals (e.g., Cd and Zn) may exist mainly as inor-

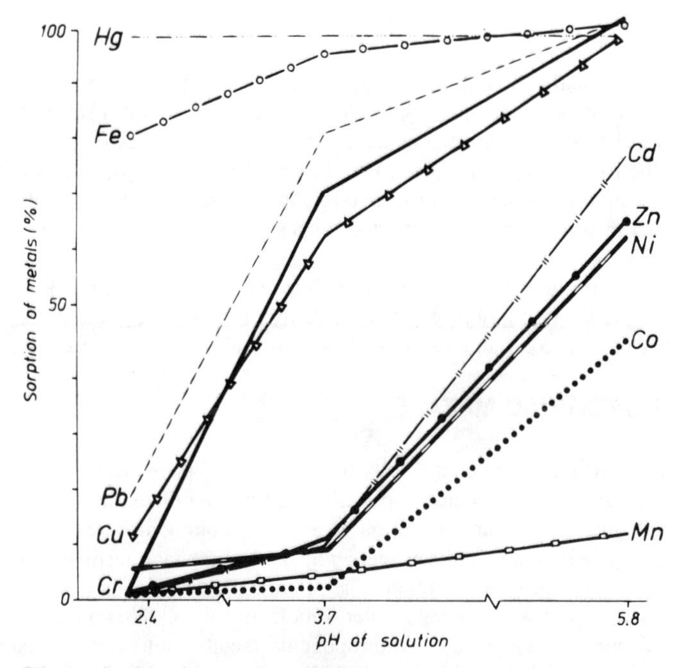

Figure 4 Effects of pH on the sorption of metals on humic acid. Sorption is given in percent of initial concentrations of metals, 0.5×10^{-4} mol of each metal in 100 mL. (Based on data from Reference 10.)

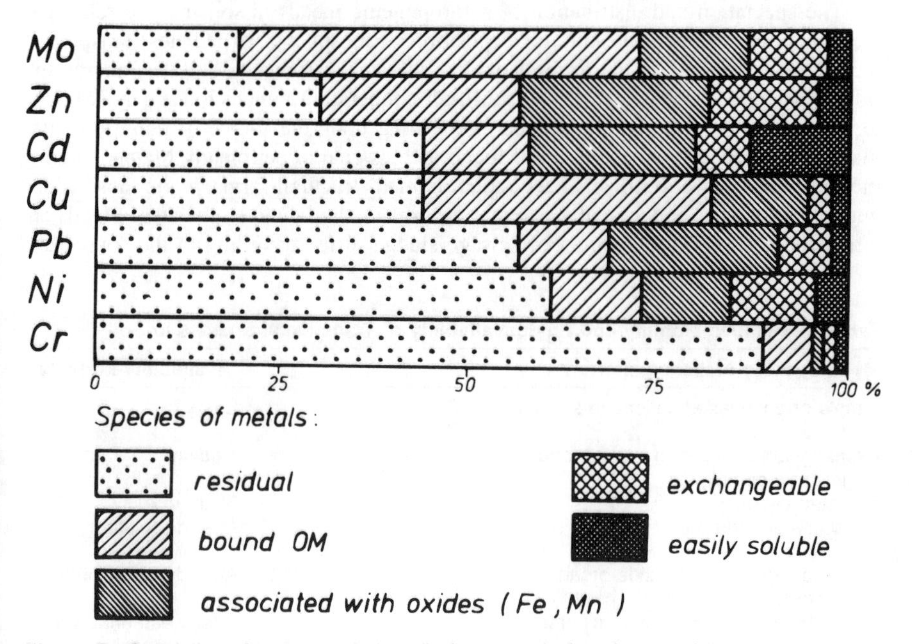

Figure 5 Speciation of trace metals in soils (in percent of total content).[11]

ganically complexed, exchangeable, and water-soluble species, while less-mobile metals (e.g., Cr and Pb) may mainly be bound onto silicate fraction; Cu and Mo may predominate in organically bound and exchangeable forms. Soluble, exchangeable, and chelated metal species are the most mobile fraction of metals in soils (Table 4). However, root exudates (i.e., consisting of organic acids) are known to facilitate, under certain conditions, the root absorption of trace metals existing in even sparingly soluble form.

The concentration of trace metals in soil solution is a good indicator of the mobile pool of metals in soil. Commonly reported ranges of trace metals in soil solution indicate that relatively mobile metals (e.g., Cd, Ni, Zn) occur in the solution phase in a relatively larger proportion than the hardly mobile metals (e.g., Co, Cr, Pb) (Table 5). Evaluation of the relative mobility and availability of trace metals should be related to their contents in both solid and aqueous phases, as well as their dissolution rate.[16]

C. ANTHROPOGENIC METALS

Anthropogenic metals enter the soils by a variety of pathways:[11,17,18] (1) aerial deposition, (2) fertilizer and pesticide application, (3) waste utilization, (4) dredged sediment disposal, and (5) river and irrigation waters. The common sources of trace metals in agriculture, namely, municipal sewage sludge and phosphatic fertilizers, will be discussed in more detail in a later section. The mobilization of metals from smelters and mine spoil by transport with seepage water, runoff, or by windblown dust may also be an important source of soil metals. Anthropogenic metals could also separate in soils, depending upon the reactant surface and binding sites with differing bonding energy. An example of differing forms of zinc as affected by its form added to soil is manifested in Figure 6.

The speciation and distribution of anthropogenic metals in soil may be related to their chemical forms at the time of deposition. Atmospheric particulates transporting metals are commonly in the form of oxides, silicates, carbonates, sulfates, and/or sulfides. Particulates of glassy structure from high-temperature combustion and those bound to polycyclic hydrocarbons may occur, depending upon the source of pollution; such may be the case in combusting fossil fuels, such as coal or petroleum. Metals entering the soil with mineral fertilizers have rather defined structural forms, while those added with plant residues and animal waste may be organically bound or chelated. Metals entering the soil with the addition of sewage sludge differ according to the source and

Table 4 Relative Mobility and Phytoavailability of Trace Metal Species in Soil

Metal Species or Association	Availability-Mobility
Simple or complexed cations in solution phase	Easy
Exchangeable cations in organic and inorganic complexes	Medium
Chelated cations	Slight to easy
Metal compounds precipitated on soil particles	After dissolution
Metals bound or fixed inside organic substances	After decomposition
Metals bound or fixed inside mineral particles	After weathering and/or decomposition

Table 5 Trace Metals in Natural Soil Solutions Obtained by Centrifugation from Different Soils ($\mu g\ L^{-1}$)

Metal	Range
Cd	3–5
Co	0.3–5
Cr	0.4–0.7
Cu	25–140
Mn	30–270
Mo	2–30
Ni	15–150
Pb	2–8
Zn	20–350

From Kabata-Pendias, A. and Pendias, H., *Trace Elements in Soils and Plants*, 2nd ed., CRC Press, Boca Raton, FL, 1992, 365. With permission.

treatment of the wastes. In dredged sediments, metals are likely to be fixed by organic substances, clay minerals, and Fe-Mn-Al hydrous oxides. Thus, anthropogenic metals may form different species in soils, depending upon the reactant surface, pH, redox potential, and other factors.

Recently, agricultural soils have been subjected to both input and output of micronutrients at much higher rates, and on a more extensive scale. Soils of Europe have been exposed to pollution for a much longer time than soils of other regions and to higher doses of

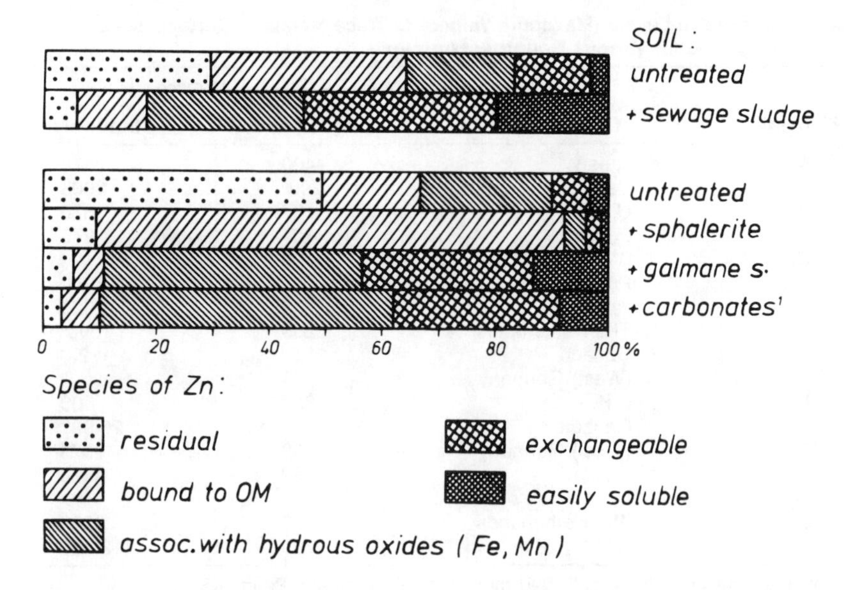

Figure 6 Impact of Zn from various sources (added at level 5000 mg kg^{-1}) on its speciation in the soils determined by sequential extraction method. (Based on data from References 19 and 20.)

Table 6 Total Atmospheric Deposition of Trace Metals on Land in Europe ($g\ ha^{-1}\ year^{-1}$)

Metal	Europe[21]	Poland[2]	(West) Germany[22]	Sweden[23]
Be	0.05	—	—	—
Cd	3	5	4.5	3.2
Cr	20	—	—	3
Cu	17	39	18	9
Mn	—	181	70	90
Ni	17	—	34	6
Pb	156	200	110	77
V	38	—	—	
Zn	88	540	210	331

Table 7 Metal Budgets of Surface Soils in Europe ($g\ ha^{-1}\ year^{-1}$)[11,22–24]

Ecosystem (Method)	Country	Cd	Cu	Mn	Pb	Zn
Pine forest (seepage water)	(West) Germany	3	10	–360	104	134
Spruce forest (lysimeter)	Sweden	–1	5	–600	75	–130
Farmland (drainage water)	Denmark	3	—	—	260	130
Farmland (seepage water)	Poland	2	14	90	160	360

Table 8 Excessive Levels (Maximum Values) of Trace Metals in Surface Soils Reported for Various Countries[a] ($mg\ kg^{-1}$)

		Source	
Element	Country	Agricultural[b]	Industrial[c]
As	Japan	400	2470
	Canada	290[d]	2000
Cd	U.K.	167	336
	Poland	107	270
Cr	(West) Germany	400	
	Poland		10,000
Cu	Japan	300	2020
	The Netherlands	265	1090[e]
Hg	Canada	11.5[d]	2.0
	(West) Germany	24	
Ni	U.K.	850	600
	Canada		26,000
Pb	(West) Germany	800	3075
	U.K.	390	4560
V	India		840[f]
Zn	The Netherlands	760	3625
	U.S.	765	12,400

[a] Compilation based on data collected by Kabata-Pendias and Pendias.[2]
[b] Mainly sludged farmland.
[c] Mining and metal-processing industry.
[d] After application of pesticides.
[e] Belgium.
[f] Vicinity of graphite industry.

fallout trace metals (Table 6). In a temperate climate, metals may migrate with the percolating water through the soil profile. However, the input-output budget of most metals in the surface layer of soils in Europe is generally positive, i.e., net input. However, accelerated leaching loss over the deposition of metals has been reported for acid forest soils in some Scandinavian countries and in Germany (Table 7).

III. EFFECTS OF INDUSTRY AND AGRICULTURE ON METALS IN SOILS AND PLANTS

Soils affected by industrial pollution are widely investigated and are known to be common in industrial regions. Metals from agricultural sources can also be enriched in soils (Table 8). Some contrasting features of soil contamination from the two main sources, fallout and soil-borne agricultural pollution, are presented in Table 9. In many cases, however, the input of metals into soils is from both sources and is rather difficult to distinguish. Often, soils become contaminated by several inorganic pollutants in correspondence with acid rain (mainly SO_2 and NO_x) and organic pollutants (e.g., polycyclic hydrocarbons). Long-term effects of such pollution on soil quality are not yet well understood, especially with respect to their phytoavailability and residence time in soils.

In the tropics, metal cycling is rapid and residence time in soil is expected for much shorter periods than in soils of other climatic zones. In arid and semi-arid regions, trace metals are likely to concentrate at the surface horizon, together with salts. Their cycling in an arid climate is controlled largely by wind erosion (in dry land) or by irrigation water. In sandy soils of temperate climate, most anthropogenic metals are likely to be mobile and transported to groundwater, while in clayey or organic-rich soils metals are accumulated at the surface layer.

The continuing increase of metallic pollutants in soils in parts of Europe has already been of real concern. De Bruijn and De Walle[25] calculated that, to attain a 50% increase in the metal levels in the soils of the Netherlands, it would take Cu, 25 to 30 years; Hg, 45 to 50 years; Zn, 50 to 60 years; Pb, 120 to 160 years; Cd, 120 to 300 years; Cr, 190 years; and Ni, 120 to 300 years. These rather short durations are alarming in view of what effect this might have on environmental health.

Table 9 Some Features of Soil Contamination with Metals from Aerial and Soil-Borne Sources

Aerial Pollution	Soil-Borne Pollution
Widespread	Localized
Very recent	Begining with fertilizer and waste application
Continuous deposition	Annual rate doses
Slow cumulative rise in concentrations	Variable loads
Accumulation in thin top layer	Distribution in organic and/or plowed horizons
Increased pollution of top plants rather than roots	Increased concentration in roots rather than top plants

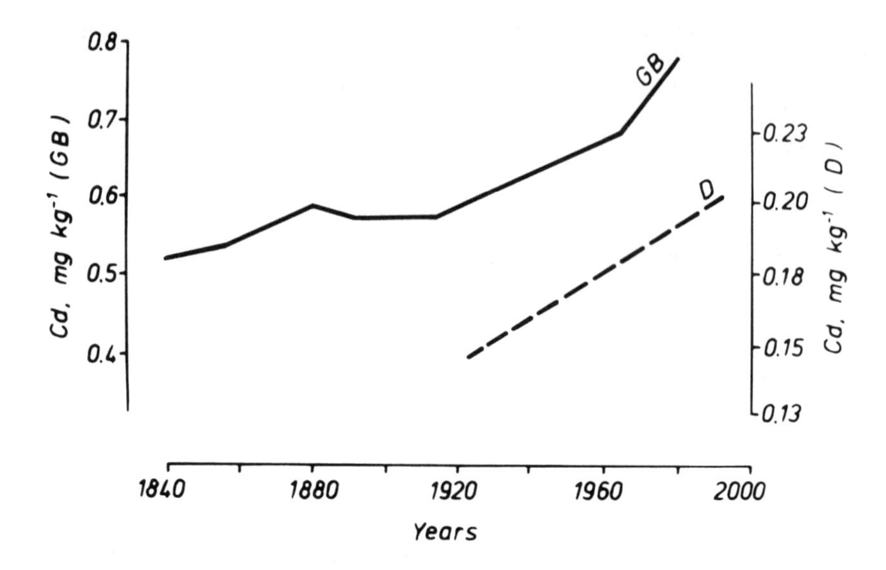

Figure 7 Changes in soil Cd levels in the unfertilized plots in Great Britain and Denmark. (Based on data from References 26 and 27.)

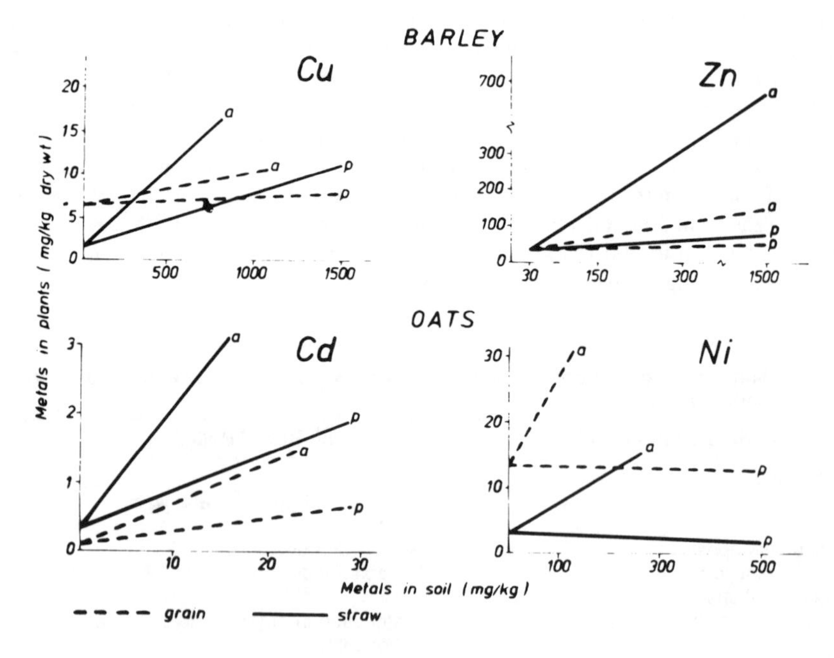

Figure 8 Metal uptake by barley and oats as influenced by their origin.[20,28,29] a = anthropogenic metals; p = pedogenic metals.

The accumulative impact of metallic pollution in certain soils of Europe from aerial sources is manifested by changes in the soil Cd level in unfertilized plots (Figure 7). However, the rate of increase of soil Cd from phosphate fertilizers is much more pronounced and was reported to have affected the level of this metal in crops.[27]

Among the soil factors, the origin of trace metals may also substantially influence their availability (Table 4). Anthropogenic metals in soils are, in most cases, more mobile than metals of indigenous origin, and are therefore more available to plants. Practically speaking, any metal added to soil could be taken up by plants at a higher rate than metals of lithogenic and pedogenic origin. These metals tend also to be more easily translocated to the grain of cereal crops (Figure 8).

Research usually indicates significant correlations between the metal contents of plants and their mobile-form concentrations in soils. This relationship may differ among metals and among plant parts. Data from Kloke[30] indicate that metals, including those previously reported to be fairly unavailable in typical agricultural soils, such as Hg and Pb, could be taken up by plants in acidic soil environment. These metals were translocated, resulting in a trend of higher concentrations in leaves than in fruits, tubers, or grains (Table 10). However, trace metals assimilated in above-ground plant tissues of plants could often be obscured by metals from atmospheric deposition. Dalenberg and van Driel[31] calculated fallout input to account for as much as 73 to 95% of the lead in leaves. The retranslocation from polluted leaves to roots has also been reported for some trace metals.[11]

Excessive bioavailable levels of trace metals may alter plant growth and are known to cause various changes in plant metabolism.[32] Tolerant plants, by virtue of their ability to be able to grow in polluted soils, may pose a great risk to the food chain by their accumulation of metals without exhibiting phytotoxicity. This is particularly of concern

Table 10 Cd, Hg, and Pb in Food Plants as a Function of Soil Cd, and Soil Hg, and Soil Pb at pH ca 6.0 (mg kg^{-1})[30]

	Soil Cd			Soil Hg			Soil Pb		
Plant	**7.3**	**54**	**209**	**0.16**	**54**	**150**	**36**	**1055**	**3800**
Potato									
Leaves	16.2	45.4	20.8	0.23	10.6	51.9	6.6	29.4	172.8
Tubers	0.8	1.6	1.8	<0.005	0.005	1.5	<0.05	0.3	1.1
Tomato									
Leaves	11.5	52.4	172.3	0.22	12.63	59.67	3.6	68.5	200.8
Fruits	1.6	2.3	5.6	<0.005	0.06	0.35	0.1	0.2	0.8
Cabbage, leaves									
Outer	1.6	11.1	67.0	—	—	—	3.8	4.5	4.8
Inner	0.7	4.6	27.7	—	—	—	2.4	2.6	2.5
Lettuce, leaves									
Outer	27.4	71.4	262.5	—	—	—	6.4	30.2	70.7
Inner	4.2	14.3	34.0	—	—	—	0.8	3.2	5.0
Oats									
Straw	—	—	—	0.02	0.09	0.41	—	—	—
Grains	—	—	—	0.01	0.01	0.04	—	—	—

for the metals Cd, Pb, and Mo that may be passed on to aerial plant tissues and then on to consumers.

A. MUNICIPAL SEWAGE SLUDGE AS A SPECIAL METAL SOURCE

With more stringent environmental regulations to enhance water quality, many countries now require prior treatment of domestic waste effluents before they are disposed of. This treatment results in municipal sewage sludge, a by-product of treating combined industrial waste effluents and residential waste waters. Because direct disposal of these wastes into oceans or open bodies of water is now prohibited, land disposal has become a more and more popular option. Recently, municipal sewage sludge has become more attractive for agricultural utilization because this material is characteristically high in the major plant nutrients nitrogen and phosphorus.[33-35] However, this material may also contain high concentrations of Cd, Zn, Cu, and other metals. More recently, a composting technology whereby sewage sludge is enriched in alkaline material has made this material more marketable for uses other than agriculture. They are now commercially available in the "bag" form for general residential and landscape uses.

There are a few potentially toxic metals most likely to occur in sewage sludge and, therefore, would be of interest in protecting the food chain in agricultural systems: in particular Cd, Cu, Ni, Pb, and Zn, and to a lesser extent As, Co, Mn, and Mo. Mercury also could be of concern especially if the sludge is to be used in mushroom culture.

Data in Table 11 indicate that there are wide ranges of metals in sewage sludge. It should be pointed out that sludges, even from rural areas where there was no industrial input, could also exhibit enriched metal levels. Some of these metals could have originated from the water delivery system, especially in old houses fitted with metal pipes.

Table 11 Ranges and Median Concentrations of Trace Elements in Dry Digested Sludges (mg/kg dry sludge)

	Reported Range		
Element	Minimum	Maximum	Median
As	1.1	230	10
Cd	1	3,410	10
Co	11.3	2,490	30
Cu	84	17,000	800
Cr	10	99,000	500
F	80	33,500	260
Fe	1,000	154,000	17,000
Hg	0.6	56	6
Mn	32	9,870	260
Mo	0.1	214	4
Ni	2	5,300	80
Pb	13	26,000	500
Sn	2.6	329	14
Se	1.7	17.2	5
Zn	101	49,000	1,700

From Logan, T. J. and Chaney, R. L., *Utilization of Municipal Wastewaters and Sludge on Land*, Page, A. L. et al., Eds., University of California, Riverside, 1983, 235. With permission.

Chang and co-workers[37] enumerated 12 exposure pathways and proposed standards to regulate 25 potential pollutants with regard to agricultural land application of sewage sludge in the U.S. The pollutants include various organics including the pesticides Aldrin, Chlordane, DDT, Heptachlor, Lindane, and the metals As, Cd, Cr, Cu, Pb, Cu, Mo, Ni, and Zn, as well as nonmetallic Se.

Plant uptake is one of the major pathways by which land-applied sewage sludge pollutants enter the food chain. In the food chain models, there are five different pathway scenarios on how sludge-borne metals could enrich the food chain. The most complete pathway is represented by sludge → soil → plant → animal → human. Among the various metals that are of potential concern, Cd ranks first as it could be readily taken up by plants via the roots, while at the same time may not be phytotoxic at concentrations that may substantially increase human exposure. The various factors that may influence the transfer of sludge-borne metals from the soil to the food chain will be discussed in more detail in a later section.

While certain metals could become fairly bioavailable for plant uptake, research has shown that most of these metals do not migrate substantially beyond the application zone (i.e., plow layer).[38–41] As discussed earlier, there are a few major factors that may induce redistribution of sludge-borne metals applied to land. They are tillage, leaching, erosion and/or runoff, and harvesting. Even in long-term sludge application studies no greater than 20% of the total applied was unaccounted for. Chang et al.[38] showed that, after 6 years of continually applying sludge at a cropland disposal site, over 90% of the applied heavy metals was found in the 0- to 15-cm soil depth. This depth is practically within the plow layer. Where sludges were incorporated in a much longer field study, McGrath and Lane[42] showed that, even 43 years after the initial application of metals (applications were continued from 1942 to 1961), about 80% of the metals was accounted for in the 0- to 20-cm plowed layer. Both long-term studies by Chang et al.[38] and McGrath and Lane[42] imply that metal loss due to crop harvest was minimal, which is usually in the neighborhood of less than 1% of the total amounts applied. In addition, these studies indicate that metal loss due to vertical migration of the metals via leaching is also minimal, which is also in the neighborhood of 1% of the total applied.[41]

In a redistribution study,[41] it was found that there was no significant enrichment with metals from sludge application below the 30-cm application layer. They found that vertical transport of metals in medium- to fine-textured soil was very small below the tillage zone. They hypothesize that vertical movement might have occurred due to particle transport facilitated by funnel action and gravity or water transport through cracks and root channels. A mass balance calculation for metals yielded an average recovery of about 99% of the total metals added. These investigations indicate that the amounts of metals applied that may be unaccounted for are possibly a result of leaching and erosion losses. Dowdy and Volk[40] reported that Zn is the element that most consistently has potential to move in soils below the depth of incorporation. We consider Cd to also have the same potential. In general, metal movement will most likely occur with large applications to a sandy, acid, low-organic matter soil that receives high rainfall or irrigation.

Recently, a methodology for establishing phytotoxicity criteria for selected metals (i.e., Zn, Ni, Cu, Cr) in agricultural land application of municipal sewage sludge was developed. However, it should be cautioned that the applicability of this model depends on the availability of good quality data, including the various factors that may affect plant uptake of metals.

IV. FACTORS AFFECTING METAL BEHAVIOR AND BIOAVAILABILITY IN SOIL

Once in soil, trace metals' behavior, bioavailability, and fate are subject to various factors. These factors can be classified into the following categories:

- Soil capacity
  - pH
  - Cation exchange capacity
  - Organic matter
  - Amount and type of clay
  - Oxides of Fe and Mn
  - Redox potential
- Plant capacity
  - Plant species
  - Plant cultivars
  - Plant parts and age
  - Ion interactions
- Environmental and miscellaneous factors
  - Climatic conditions (i.e., season)
  - Management practices
  - Irrigation water/salinity
  - Topography

The influence of these various factors on the biogeochemical cycling of metals has recently been nicely summarized.[1,2,11,43–47]

In general, various studies[39,44,48–51] have indicated that there is a concomitant decrease in the plant uptake of the various metals with the application of agricultural limestone or increasing soil pH. Because this practice is part of any agricultural management scheme and since it has been shown to be very effective in reducing metal uptake by crops, agricultural soils are required to be limed at about pH 6.5 when municipal sewage sludge is applied.[39,48–50] In addition, the higher the amount of soil organic matter and the higher the percentage of clay in the soil (i.e., expressed as higher cation exchange capacity), the higher the soil's sorption potential for the metal cations. Also, soils that are high in the hydrous oxides of Fe and Mn characteristically exhibit high sorption potential for metal cations.[1,36]

Crop species exhibit varying responses to sludge-borne metals. Generally, vegetable plants grown on soils high in Cd are especially responsive, and since many vegetables may accumulate large amounts of this metal without showing any phytotoxic symptom this may compromise the quality of the food chain.[36] Among plant species, Sauerbeck[46] has shown that dicotyledonous plants absorb more metals than the monocots. It has always been the general observation that the plant vegetative organs usually contain more metals than the generative plant parts. Cadmium has been demonstrated as the element of most concern in the food chain as a result of sewage sludge application. Factors that have been shown to be most significant in determining plant uptake of Cd are Cd application rate, soil pH, and plant species.[36] This information was used to develop regulations in the U.S. for land application of municipal sewage sludge that regulates soil pH and annual and cumulative Cd input on land.

Table 12 Recommended Upper Limits of Potentially Toxic Metals in Soil to Plow Depth after Application of Sewage Sludge and Maximum Rate of Metal Applications[35]

Element	Typical Total Metal Content of Uncontaminated Soil (mg/kg)	Recommended Upper Limit for Soil Concentrations after Sludge Additions (mg/kg)	Maximum Annual Rate of Metal Addition Based on a 10-Year Average (kg/ha/year)
Zinc (Zn)	80	300	15
Copper (Cu)	20	135	7.5
Nickel (Ni)	25	75	3
Cadmium (Cd)	0.5	3	0.15
Lead (Pb)	50	250	15
Mercury (Hg)	0.1	1	0.1
Molybdenum (Mo)	1	4	0.2
Chromium (Cr)	50	600	40
Selenium (Se)	0.5	3	0.15
Arsenic (As)	10	20	0.7

The maximum annual application rate of 0.15 kg Cd/ha/year in some European countries is more prohibitive than the 1.9 kg Cd/ha/year for land application limits for sewage sludge in the U.S. The European Community (EC) proposed interim guidelines for the upper limits of certain metals in soils to plow depth and maximum rates of metal application (Table 12). As earlier indicated by Chang et al.,[37] these values were set to ensure acceptable quality for the food chain in the absence of more-reliable predictive models of metal uptake by crops. It should be pointed out that the divergent recommended limits for metal annual input as well as metal levels in soils among countries, and even among the countries of the EC, may reflect the political and nonscientific community input in formulating these environmental regulations, or it may also indicate our lack of confidence due to gaps of data in our capacity to reliably predict the bioavailability of metals in the soil and their subsequent transfer in the food chain.

V. REMEDIATION AND USE OF SOILS CONTAMINATED WITH METALS

The assessment and classification of soils contaminated with trace metals and appropriate land use is of vital importance. In spite of elaborate monitoring programs, the net impact on the soil environment by large amounts of heavy metals is as yet unpredictable. Besides the consequential impact of metals from point sources, substantial loading of metals from diffuse sources, such as agricultural activities and long-range transport of atmospheric particulates, may also occur. Waste products are likely to accumulate because of the enactment and implementation of environmental regulations to enhance air and water qualities. Such products include municipal sewage sludge, coal fly ash, and others.

Surveys of background levels and soil contamination with metals have shown that soil metal pollution is a serious problem in some countries. Varying standards have been proposed in different countries for classifying soils contaminated with metals and the corresponding appropriate land use. The target values are usually established after the

Table 13 Classes of Soil (0- to 20-cm Layer) Pollution with Heavy Metals Proposed in Poland (mg kg^{-1})[52]

Metal	Soil Group[a]	Classes of Pollution[b] 0	I	II	III	IV	V
Cd	a	0.3	1	2	3	5	>5
	b	0.5	1.5	3	5	10	>10
	c	1	3	5	10	20	>20
Cu	a	15	30	50	80	300	>300
	b	25	50	80	100	500	>500
	c	40	70	100	150	750	>750
Ni	a	10	30	50	100	400	>400
	b	25	50	75	150	600	>600
	c	50	75	100	300	1000	>1000
Pb	a	30	70	100	500	2500	>2500
	b	50	100	250	1000	5000	>5000
	c	70	200	500	2000	7000	>7000
Zn	a	50	100	300	700	3000	>3000
	b	70	200	500	1500	5000	>5000
	c	100	300	1000	3000	8000	>8000

[a] Soil group: (a) sandy and medium, pH <5.5; (b) medium and heavy, pH <5.5; (c) sandy and rich in SOM, pH 5.5 to 6.5.
[b] Classes of pollution: (0) unpolluted, (I) slightly, (II) moderately, (III) considerably, (IV) heavily, (V) extremely.

ecological risk assessment and safety factor for metal exposure of man due to consumption of plant food. The greatest risk is related most often to intake of metals with home-grown vegetables. Therefore, the maximum acceptable concentration (MAC) values for metals in garden soil are the lowest than for other agricultural land uses.

An example of the classification of soil pollution with metals and corresponding recommended land use is given in Table 13. For any degree of pollution, three groups of soils are distinguished (e.g., sandy and acid, medium and acid, clayey or organic-rich and neutral).[52] The tentative recommended land use for a given class of pollution follows:

I. Soils slightly polluted can be used for all field crops, with the exception of growing vegetables for infants and children.
II. Soils moderately polluted are preferable for cereals, potatoes, sugar beet, and fodder plants. Growing some vegetables, e.g., lettuce and spinach, should be forbidden.
III. Soils moderately polluted present a risk of contamination of any crop. Agricultural practices to limit the metal uptake, and frequent control of plant food quality are recommended. Preferable are commercial plants and seed grasses.
IV. Soils heavily polluted should not be used for food plant production, especially when soils are acid and sandy. Recommended are commercial plants, and in particular, crops used for alcohol, technical oil, and energy production.
V. Soils extremely polluted should be excluded from agricultural use and, if possible, remediated from excess trace metals. On suitable sites, some commercial plants (see item IV) can be cultivated.

Methods for remediating metal-contaminated soils have been recently broadly discussed by several authors.[13,53–55] However, reclamation of degraded soil requires a full

Table 14 Trace Metal Removal by Crops from Soil

Metal	Content of Soil (kg ha^{-1})	Output with Plant: Reference g ha^{-1}	Reference %[a]	Accumulator g ha^{-1}	Accumulator %[a]
Mn	810	1000	0.1	5000	0.6
Cr	150	50	0.03	500	0.3
Zn	135	400	0.3	1500	1.0
Pb	75	100	0.1	500	0.6
Cu	45	100	0.2	500	1.0
Ni	39	50	0.1	100	0.3
Mo	6	30	0.5	250	4.0
Cd	1.5	1	0.06	100	10.0

[a] As percent of the total content of soil.

From Reference 57.

understanding of soil properties, processes, and the geochemical behavior of metals before and after enrichment. For soils not heavily polluted with trace metals, the advisable methods to prevent plant tissue contamination are based on two processes — immobilization of mobile fractions of metals and leaching of easily soluble fractions. Severely contaminated soils may need some drastic treatments, as, for example, extraction of soil with EDTA solution. Slightly polluted soils can be partly decontaminated by growing metal accumulator species or cultivars. This new clean-up technology known as "Green Remediation"[56] is becoming popular. However, the potential to remove excess trace metals by growing plants is not yet promising. Relative proportions of the metals removed by harvesting such crops, even at very high yields and plant tissue concentrations, rarely exceed 1% of their total contents in soil (Table 14). This estimation is related to the background concentrations of metals in soils. When a soil becomes polluted, the proportion of trace metals that may be removed by accumulating vegetation is even more negligible.[57]

It is necessary to emphasize that restoring the soil's original physical, chemical, and biotic properties may never be achieved. However, some restoration can be achieved by using some of the best available technology for a given kind of soil and type of soil pollution.

Proposed methods for cleaning soils degraded with metals are reviewed by Rulkens et al.:[17]

1. Treatment of excavated soil by extraction
2. *In situ* extractive treatment of soil
3. Treatment of soil by electroreclamation
4. *In situ* soil vapor extraction
5. Treatment of soil by biodegradation of substances containing trace metals
6. Treatment of soil by microbiological mobilization
7. Vegetative *in situ* extraction of trace metals

In most agricultural soil pollution, a logical approach is to control the transfer of metals from the soil to the food chain. This can be achieved by manipulating pH, redox, soil organic matter, as well as by cultivating species of plants that tend to exclude metals (Section IV). Also, deep plowing, mixing polluted topsoil with imported unpolluted soil, covering polluted soil with clean soil, or replacement of the polluted top layer have been demonstrated as promising options for arable soils.

VI. METALS IN SOIL AND SUSTAINABLE LAND USE

In agriculture, remediation of soils contaminated by trace metals is usually accomplished by applying lime, phosphate fertilizers, and organic matter. Although the addition of lime does not always immobilize trace metals, in most cases this treatment is quite effective in lowering metal bioavailability to plants, as well as impairing their mobility to groundwater. Exceptions are metals in their anionic form, such as molybdate, arsenate, chromate, etc., which may become more so with increasing pH. The sustainable and acceptable land use at different stages of pollution need to be patterned for a specific plant-soil system.

Standard values for trace metal levels in soil, as well as any remediation and land-use scenario should consider the risk of water pollution. Metal contamination of the water resources (e.g., groundwater, lakes, rivers) due to metal migration from soil has been reported.

Soils will continue to be subjected to metal enrichment from fertilizer application, pesticide application, waste disposal (especially sewage sludge), and industrial pollution. These activities will affect biological, chemical, and physical soil properties, which undoubtedly could alter the behavior of metals in soil. The mobility and phytoavailability of trace metals, in particular to crop plants, in soils that are acidic, alkaline, and saline and histosols should be further investigated in formulating general guidelines for sustainable land use. These are necessary for protecting or restoring the health of the soil.

VII. ASSESSMENT CRITERIA AND APPLICATION GUIDELINES FOR METALS

To ensure safety in the food chain as well as preventing movement of metals into the groundwater, several countries have established guidelines for certain potentially critical metals. Within the EC, for example, data in Table 12 indicate the recommended upper limits of potentially toxic elements in the plow layer after sewage sludge application and its corresponding maximum annual rate of metal applications.[58]

Soil plays a crucial role as a biotic compartment of the biosphere because it not only serves as a geochemical sink for contaminants, but also acts as a natural buffer controlling the transport of chemical constituents to the atmosphere, hydrosphere, and biota. Soil also functions as the most important substrate for the production of food and fiber. The sustainability of the various physical, biological, and chemical processes in soil may depend largely on the trace element balance, especially with the micronutrient metals.

As a sink, trace metals originating from various sources may accumulate in the surface of the soil. Their eventual fate depends on biogeochemical properties of soil and also on their physicochemical behavior. The persistence of trace metals in soil is reportedly much longer than in other compartments of the biosphere. Once deposited onto the soil, certain metals such as Pb and Cr may be virtually permanent.

The input-output balance of Cd, as an analog for trace metals, indicates that trace metal concentrations in surface soils are likely to increase (Table 15). More extensive soil

Table 15 Budgets of Cd in Agricultural Soils of Poland and Germany (g ha^{-1} yr^{-1})

	Poland	Germany
Input		
Fertilizers	1–2.5	1–6
Slurry	2.5	—
Sludges	1.5	<1–25
Harvest residues	3	0.3–8
Atmospheric	2.5–4	3–8
Total input	10.5–13.5	4.3–47
Output		
With crops	3	1–5
With seepage water	3	1–2
Total output	6	2–7
Balance (net gain)	4.5–7.5	1.3–40

From Kabata-Pendias, A. and Bruemmer, G. H., *Global Perspectives on Lead, Mercury and Cadmium Cycling in the Environment*, Hutchinson, T. C., Gordon, C. A., and Meema, K. M., Eds., SCOPE, Wiley Eastern, New Delhi, 1992, 107. With permission.

pollution occurs mainly in industrialized regions, and within the center of heavily populated areas where industrial emission, motor vehicles, and municipal wastes are the most important sources of metals. However, due to the long-distance aerial transport of trace pollutants, especially those that form volatile compounds (e.g., As, Sb, Se, and Hg), it is rather cumbersome to estimate the natural background values for certain elements in soil.

Consequences from excessive levels of metals in soil depend upon a myriad of complex reactions between the trace ions and components of soil, i.e., solid, liquid, and gaseous phases. Permissible levels of trace metals in soil can be calculated based on several factors including nonscientific ones. Thus, permissible contents of trace metals in surface soil may differ upon the local condition, land use, and country. Nevertheless, the maximum acceptable concentrations (MAC) of metals in soil, as estimated by various authors, are fairly comparable (Table 16).

Table 16 Proposals for MAC of Trace Metals Considered as Phytotoxic in Agricultural Soils (mg kg^{-1})

Metal	Austria[59]	Canada[60]	Poland[61]	Japan[62]	U.K.[63]	Germany[64]
As	50	25	30	15	20[a]	40 (50)[b]
Be	10		10			10 (20)
Cd	5	8	3		1 (3)	2 (5)
Co	50	25	50	50		—
Cr	100	75	100		50	200 (500)
Cu	100	100	100	125	50 (100)	50 (200)
Hg	5	0.3	5		2	10 (50)
Mo	10	2	10			—
Ni	100	100	100	100	30 (50)	100 (200)
Pb	100	200	100	400	50 (100)	500 (1000)
Tl						2 (20)
Zn	300	400	300	250	150 (300)	300 (600)

[a] Values are proposals of EEC for MAC in soils treated with sewage sludges. Values in parentheses are mandatory concentrations.
[b] Tolerable and toxic (in parentheses) contents.

Soil type, plant species, and environmental conditions influence the divergent impact of soil metals on plants. There are still not enough data, however, to finalize values for criteria needed to protect the soil and plants against trace metal pollution. Several ecological standards are used to assess the significance of trace metal level and their permissible loading into soils:

NOEC — no observed ecological consequences
LKE — lowest known effect
MAC — maximum acceptable concentration
MCA — maximum cumulative amount
RMCL — recommended maximum contaminant level
TTL — threshold trigger level
ATL — action trigger level
PAA — permissible annual application
MAL — maximum allowable loading
PEC — predicted environmental concentration
DWE — dose without effect
ELD — ecosystem lethal dose

In addition to ecological effects, criteria should also take into account adverse effects on crops, risk to livestock (particularly the hazard associated with the ingestion of soil particles during grazing), and potential bioaccumulation that may occur in the food chain. The potential contamination of human food is of special concern. Values for any critical concentration or soil loading of trace metals should be regarded as tentative until a more definitive evaluation of data and validation of models and standards are in place. It is relevant to evaluate permissible levels of application and/or loading of trace metals in relation to

1. Initial trace element content of soil
2. Total amount added of a particular metal and of all elements
3. Annual or cumulative total load of trace metals
4. Threshold values of trace metal concentration in soil
5. Input-output balance
6. Relative ratio between interacting elements
7. Soil characteristics (e.g., pH, redox, free carbonate, organic matter, clay content, etc.)
8. Equivalency of trace metal toxicity in plants
9. Plant sensitivity and adaptability

There is a real need for critical evaluation of information that should be employed to establish the criteria for "safe" limits for trace metal content of soil. Also, guidelines for allowable loading of these metals is of a basic importance, especially when sewage sludge is used in the agriculture. Values for MAL proposed in recent directives for practical regulation are relatively similar (Table 17).

A compendium of data presented in Table 17 shows the total contents of trace metals for MAC and MAL. Criteria based on the measurement of the amounts of "available" metals may present some complication because of the difficulty in deciding if "availability" should be measured in biological or chemical terms and the exact manner in which they can be determined.

Criteria for regulating the application of trace metals with sewage sludge and compost in agriculture should take into account, in addition to general ecological effects, the

Table 17 MAC and MAL of Trace Metal to Arable Soil

Metal	MAC[65] (mg kg^{-1})	MAL (kg ha^{-1} year^{-1}) U.K.[66]	European Communities[67]	Poland[65]
Zn	2000	15	30	10
Cr	1000	—	—	15
Pb	1000	15	15	10
Cu	1000	7.5	5	5
Ni	150	3	3	3
Cd	20	0.15	0.15	0.2
Hg	10	0.1	0.1	0.2

effects on the quality of crops and groundwater. Thus, the tolerable input of metals onto soil is often set to approximate the average uptake by crops. However, no equilibrium relation between input, concentration in the soil, and output has ever been achieved (Table 15). Because there is inadequacy of models for the trace ion uptake by plants and cation migration within the soil profile, application of waste materials should be prohibited if reference values ("target values") are exceeded (Table 17). However, recurrent monitoring of metal accumulation in agricultural soils treated with sewage sludge has shown that it will be unreasonable to assume that these metals could become immobilized in soil with time.[11]

Long-term studies on metal mobility in sludged soils show a slight but trendy increase of metal concentration in the 15- to 30-cm depth of the treated soil compared to the control. However, as discussed earlier, virtually no change in the metal concentration in the treated soil occurred below the 30-cm depth, in spite of the increased soil acidity and greater proportion of the mobile species.[68,69] However, the mobility of trace metals in soil depends largely upon the properties of the soils, which are known to change under variable climatic and hydrogeological conditions and land use. Therefore, the long-term potential capacity of soil to store large amounts of metals cannot be extrapolated from short-term effects of soil pollution and present land use. The concept of "sustainable land use" should also be fundamental in formulating environmental policy with respect to trace metal pollution.[57] Soil protection against excessive trace metal input is therefore a prerequisite for the evaluation of soil quality criteria.

ACKNOWLEDGMENTS

Financial support for D.C. Adriano was obtained from contract DE-AC09-76SR00-819 between the U.S. Department of Energy and the University of Georgia.

REFERENCES

1. Adriano, D.C., *Trace Elements in the Terrestrial Environment*, Springer-Verlag, New York, 1986, 533.
2. Kabata-Pendias, A. and Pendias, H., *Biogeochemistry of Trace Elements*, PWN, Warsaw, 1993, 364 (published in Polish).

3. Chen, J., Wei, F., Zheng, C., Wu, Y., and Adriano, D., Background concentrations of elements in soils of China, *Water, Air, Soil Pollut.*, 57, 699, 1991.
4. FAO/UNESCO, *Soil Map of the World*, Vol. 1, UNESCO, Paris, 1974.
5. Sposito, G. and Page, A.L., Cycling of metal ions in the soil environment, in *Metal Ions in Biological Systems*, Siegel, H., Ed., Marcel Dekker, New York, 1984, 287.
6. Harrison, P.R. and Rahn, K.A., Atmospheric particulate "pollutant", in *Review of Research on Modern Problems in Chemistry*, Siegel, F.R., Ed., UNESCO, Paris, 1979, 177.
7. Chesworth, W., Geochemistry of micronutrients, in *Micronutrients in Agriculture*, Mortvedt, J.T., Cox, E.R., Shuman, L.M., and Welch, R.M., Eds., Soil Science Society of America, Madison, WI, 1991, 1.
8. Thornton, I., Geochemical aspects of the distribution and forms of heavy metals in soils, in *Effect of Heavy Metal Pollution on Plants*, Vol. 2, Lepp, N.W., Ed., Applied Science, London, 1981, 1.
9. Kabata-Pendias, A. and Bruemmer, G.H., Ecological consequences of As, Cd, Hg, and Pb enrichment in European soils, in *Global Perspectives on Lead, Mercury and Cadmium Cycling in the Environment*, Hutchinson, T.C., Gordon, C.A., and Meema, K.M., Eds., SCOPE, Wiley Eastern, New Delhi, 1992, 107.
10. Gamble, D.C., Interaction between natural organic polymers and metals in soils and fresh water systems: equilibria, in *The Importance Speciation in Environmental Processes*, Brinckman, M. and Sadler, P.J., Eds., Springer-Verlag, Berlin, 1986, 217.
11. Kabata-Pendias, A. and Pendias, H., *Trace Elements in Soils and Plants*, 2nd ed., CRC Press, Boca Raton, FL, 1992, 365.
12. Alloway, B.J., The origin of heavy metals, in *Heavy Metals in Soils*, Alloway, B.J., Ed., Blacke & Wiley, New York, 1990, 29.
13. Schmitt, H.W. and Sticher, H., Heavy metal compounds in the soil, in *Metals and their Components in the Environment*, Merian, E., Ed., VCH, Weinheim, 1991, 311.
14. Hiller, D.A., Bruemmer, G.W., and Ackerman, D., Gehalte an Haut- und Nebenelementes in Konkretionen aus Oberboeden von Marsche- Untersuchungen mit eine Microsonde, *Z. Pflanzenernaerhr. Bodenkunde*, 1988.
15. Otte, M.L., Buijs, E.P., Riemer, J., and Broekman, R.A., The iron-plaque on the roots of saltmarsh plants: a barrier to heavy metal uptake?, in *Heavy Metals in the Environment*, Vol. 1, Lindberg, S.E. and Hutchinson, T.C., Eds., CEP Consl., Edinburgh, U.K., 1987, 407.
16. Bruemmer, G.W., Gerth, J., and Herms, U., Heavy metal species, mobility and availability in soil, *Z. Pflanzenernaehr. Bodenkunde*, 149, 382, 1986.
17. Rulkens, W.H., Grotenhuis, J.T.C., and Tichy, R., Methods for cleaning contaminated soils and sediments, presented at workshop Heavy Metal Pollution in Eastern and Central Europe: An Assessment of Current Problems and Solution, Prague, October 11–16, 1992.
18. Freedman, B. and Hutchinson, T.C., Sources of metal and elemental contamination of terrestrial environments, in *Effect of Heavy Metal Pollution on Plants*, Vol. 2, Lepp, N.W., Ed., Applied Science, London, 1981, 35.
19. Dudka, S., Piotrowska, M., Chlopecka, A., and Lopatek, J., Speciation, Mobility and Phytoavailabillity of Trace Metals in Sewage Sludge Amended Soils, Rep. 270, IUNG, Pulawy, Poland, 1990, 42.
20. Chlopecka, A., Forms of trace metals from inorganic sources in soils and amounts found in spring barley, *Water, Air Soil Pollut.*, 69, 127, 1993.

21. Pacyna, J.M., Atmospheric trace elements from natural and athropogenic sources, in *Toxic Metals in the Atmosphere*, Nriagu, J.O. and Davidson, C.I., Eds., John Wiley & Sons, London, 1986, 33.
22. Zoettl, H.W., Stahr, K., and Haedrich, F., Umsatz von Spurenelementen in der Baerhalde und ihren Oekosystem, *Mit. Dtsch. Bodenkundl. Ges.*, 29, 569, 1979.
23. Berkvist, B., Folkenson, L., and Berggren, D., Fluxes of Cu, Zn, Pb, Cr and Ni in temperate forest ecosystems, *Water Air Soil Pollut.*, 49, 217, 1989.
24. Mayer, R. and Schultz, R., Effects of soil acidification on heavy metal cycling in forest ecosystems, in *Heavy Metals in the Environment*, Vol. 2, Lindberg, S.E. and Hutchinson, T.C., Eds., CEP Consl., Edinburgh, U.K., 1987, 402.
25. De Bruijn, A.J. and De Walle, V.P., Standards for soil protection and remedial action in the Netherlands, in *Contaminated Soils '88*, Wolf, K., van den Brink, W.J., and Colon, F.J., Eds., Kluwer Academic, Dordrecht, The Netherlands, 1989, 339.
26. Christensen, T.N. and Tjell, J. Chr., Sustainable management of heavy metals in agriculture, in *Heavy Metals in the Environment*, Vol. 1, Lindberg, S.E. and Hutchinson, T.C., Eds., CEP Consl., Edinburgh, U.K., 1991, 40.
27. Jones, K.C., Symon, C.J., and Johston, A.E., Long-term changes in soil and cereal grain cadmium: studies at Rothamsted Experimental Station, in *Trace Substances in Environmental Health*, 21, Hemphill, D.D., Ed., University of Missouri, Columbia, 1987, 450.
28. Grupe, M. and Kuntze, H., Zur Ni-Mobilitaet einer geogen belasteten Braunerde, *Mitt. Dtsch. Bodenkudl. Gesellsch.*, 55, 333, 1987.
29. Grupe, M. and Kuntze, H., Zur Ermittlung der Schwermetallverfuegbarkeit lithogen und anthropogenbelasteter Standorte, *Z. Pflanzerernaehr. Bodenkd.*, 151, 319, 1988.
30. Kloke, A., Schadgas- und Schwermetallbelastungen von Boeden und Pflanzen, *Garden Umwelt*, 37, 1, 1985.
31. Dalenberg, J.W. and van Driel, W., Contribution of atmospheric deposition to heavy-metal concentration in field crops, *Netherlands J. Agric. Sci.*, 38, 367, 1990.
32. Ernst, W.H.O., Verkleij, J.A.C., and Schat, W., Metal tolerance in plants, *Acta Bot. Neerl.*, 41(3), 229, 1992.
33. Page, A.L., Logan, T.G., and Ryan, J.A., Eds., *Land Application of Sludge*, Lewis Publishers, Chelsea, MI, 1987, 168.
34. Tiller, K.G., Heavy metals in soils and their environmental significance, in *Advances in Soil Science*, Vol. 9, Stewart, B.A., Ed., Springer-Verlag, New York, 1989, 113.
35. Williams, J.H., Guidelines, Recommendation, Rules, and Regulation for Spreading Manures, Slurries, and Sludge in Arable and Grassland, Commission of the European Communities SL/124/88 XII/Env./14/18, 1988.
36. Logan, T.J. and Chaney, R.L., Utilization of municipal wastewater and sludge on land — metals, in *Utilization of Municipal Wastewaters and Sludge on Land*, Page, A.L. et al. Eds., University of California, Riverside, 1983, 235.
37. Chang, A.C., Granato, T.C., and Page, A.L., A methodology for establishing phytotoxicity criteria for chromium, copper, nickel and zinc in agricultural land application of municipal sewage sludges, *Environ. Qual.*, 21, 521, 1992.
38. Chang, A.C., Warneke, J.E., Page, A.L., and Lund, L.J., Accumulation of heavy metals in sewage sludge-treated soils, *Environ. Qual.*, 13, 87, 1984.
39. Dowdy, R.H. and Larson, W.E., Metal uptake by barley seedlings grown on soils amended with sewage sludge, *Environ. Qual.*, 4, 229, 1975.

40. Dowdy, R.H. and Volk, V.V., Movement of heavy metals in soils, in *Chemical Mobility and Reactivity in Soil Systems*, American Society of Agronomists, Madison, WI, 1983.
41. Yingming, L. and Corey, R.B., Redistribution of sludge-borne cadmium, copper, and zinc in a cultivated plot, *Environ. Qual.*, 22, 1, 1993.
42. McGrath, S.P. and Lane, P.W., An explanation for the apparent losses of metals in a long-term field experiment with sewage sludge, *Environ. Pollut.*, 60, 235, 1989.
43. Berrow, M.L. and Burridge, J.C., Uptake distribution and effects of metal compounds on plants, in *Metals and their Compounds in the Environment*, Merian, E., Ed., VCH Weinheim, 1991, 399.
44. Coker, E.G. and Matthews, P.J., Metals in sewage sludge and their potential effects in agriculture, *Water Sci. Technol.*, 15, 209, 1983.
45. Moraghan, J.T. and Mascagni, H.J., Environmental and soil factors affecting micronutrient deficiencies and toxicities, in *Micronutrients in Agriculture*, Mortvedt, J.T., Cox, E.R., Shuman, L.M., and Welch, R.M., Eds., Soil Science Society of America, Madison, WI, 1991, 371.
46. Sauerbeck, D.R., Plant, element and soil properties governing uptake and availability of heavy metals derived from sewage sludge, *Water, Air, Soil Pollut.*, 1991.
47. Webber, J., Trace metals in agriculture, in *Effect of Heavy Metal Pollution on Plants*, Vol. 2 Lepp, N.W., Ed., Applied Science, London, 1981, 159.
48. Heckman, J.R., Angle, J.S., and Chaney, R.L., Residual effects of sewage sludge on soybean. I. Accumulation of heavy metals, *Environ. Qual.*, 16, 113, 1987.
49. Kuo, S., Jellum, E.J., and Baker, A.S., Effects of soil type, liming, and sludge application on zinc and cadmium availability to swiss chard, *Soil Sci.*, 139, 122, 1985.
50. Pepper, I.L., Bezdicek, D.F., Baker, A.S., and Sims, J.M., Silage corn uptake of sludge-applied zinc and cadmium as affected by soil pH, *Environ. Qual.*, 12, 270, 1983.
51. Sanders, J.R., McGrath, S.P., and Adams, T. McM., Zinc, copper and nickel concentrations in rye grass grown on sewage sludge-contaminated soils of different pH, *Sci. Food Agric.*, 37, 961, 1986.
52. Kabata-Pendias, A., Piotrowska, M., and Witek, T., Standards for the assessment of soil and crop pollution with heavy metals as a guideline for agricultural land use, P-53, IUNG, Pulawy, Poland, 1993 (in Polish).
53. Ernst, W.H.O., Decontamination or consolidation of metal-contaminated soils by biological means, presented at workshop Heavy Metal Pollution in Eastern and Central Europe: An Assessment of Current Problems and Solution, Prague, October 11–16, 1992.
54. Vegter, J.J., Soil protection in the Netherlands, presented at workshop Heavy Metal Pollution in Eastern and Central Europe: An Assessment of Current Problems and Solution, Prague, October 11–16, 1992.
55. Logan, T.J., Reclamation of chemically degraded soils, in *Advances in Soil Science*, Lal, R. and Stewart, B.A., Eds., Springer-Verlag, New York, 1992, 13.
56. Baker, A.J.M. and Walker, P.L., Ecophysiology of metal uptake by tolerant plants, in *Heavy Metal Tolerance in Plants: Evolutionary Aspects*, Shaw, A.J., Ed., CRC Press, Boca Raton, FL, 1990, 155.
57. Kabata-Pendias, A., Maintaining of soil micronutrient status, presented at int. symp. Soil Resilience and Sustainable Land Use, Budapest, September 27 to October 3, 1992.

58. Ewers, U., Standards, guidelines, and legislative regulations concerning metals and their compounds, in *Metals and their Compounds in the Environment*, Merian, E., Ed., VCH, Weinheim, 1991, 687.
59. El-Bassam, N. and Tietjen, C., Municipal sludge as organic fertilizer with special reference to the heavy metals constituens, in *Soil Organic Matter Studies*, Vol. 2, International Atomic Energy Agency, Vienna, 1977, 253.
60. Linzon, S.N., Phytotoxicology Excessive Levels for Contaminants in Soils and Vegetation, Report of Ministry of the Environment, Ontario, 1978.
61. Kabata-Pendias, A., Effects of inorganic air pollutants on the chemical balance of agricultural eosystems, ecosystems, in *Effect of Air-Borne Pollution on Vegetation*, ECE, Warsaw, 1979, 134.
62. Kitagishi, K. and Yamane, I., Eds., *Heavy Metal Pollution in Soils of Japan*, Japan Science Society Press, Tokyo, 1981, 302.
63. Finnecy, E.E. and Pearce, K.K., Land contamination and reclamation, in *Understanding Our Environment*, Hester, R.E., Ed., Royal Society of Chemistry, London, 1986, 172.
64. Kloke, A. and Eikmann, Th., Nutzung- und Schutzbezogene Orientierungsdaten fuer (Schad) Stoffe in Boeden. Mitteilungen VDLUFA, Sonderdruck aus Heft 1, 8, 1991.
65. Kabata-Pendias, A. and Piotrowska, M., *Trace Elements as Criteria for Waste Use in Agriculture*, P33, IUNG, Pulawy, Poland, 1987, 46.
66. Code for Practice for Agricultural Use of Sewage Sludge, Department of the Environment, Her Majesty's Stationary Office, London, 1989.
67. Council Directive on the Protection of the Environment and in particular of the Soil, when Sewage Sludge is Used in the Agriculture, Official Journal of the European Communities, No. L. 181/6, 1986.
68. Williams, D.E., Vlamis, J., Pukite, A.H., and Corey, J.E., Metal movement in sludge amended soils: a nine-year study, *Soil Sci.*, 143, 124, 1987.
69. McGrath, S.P., Long-term studies of metal transfer following applications of sewage sludge, in *Pollutant Transport and Fate in Ecosystems*, Coughthray, P.J., Martin, M.H., and Unsworth, M.H., Eds., Blackwell Scientific, Oxford, U.K., 1987, 301.

CHAPTER 5

Manure

H. V. Eck and B. A. Stewart

I. INTRODUCTION

Livestock manure has been spread on the land for many centuries, at first, perhaps, as a method of disposal but later as fertilizer for crop production. Before the advent of inorganic fertilizers, it was the principal source of plant nutrients added to the soil.

0-87371-859-3/95/$0.00+$.50

Livestock feeding operations were farm sized and manure was spread on the farms. In the period since World War II, two things have happened to change attitudes toward manure. First, livestock and poultry production has become concentrated in large-scale, confinement-type enterprises. These include multi-hundred dairy cow operations, multi-thousand head beef and hog feedlots, and poultry enterprises with many hundreds of thousands of birds. Such large concentrations of animals have greatly magnified the problems of handling wastes, including health hazards and aesthetic nuisances. Second, relatively cheap sources of inorganic plant nutrients have become available, have been shown to be equally as efficient as manures, and have become accepted by farmers. Farmers find it more convenient and often more economical to spread small quantities of commercial fertilizers than to spread the rather large quantities of manure that would be required to supply plant nutrients equal to those in the concentrated inorganic fertilizers. Thus manure, once prized as a fertilizer material, has become a surplus by-product of the livestock feeding industry in the U.S.

It is interesting to note that the perception of animal manures in western Europe has paralleled that in the U.S. Wadman et al.[1] reviewed the development in the production and appreciation of animal manure in western Europe during the last 100 years. They concluded that, depending on the degree of intensification of livestock feeding, animal manures in western Europe have turned from a precious resource into a waste product.

A. CURRENT SOURCES AND SUPPLIES

In 1975, the Council for Agricultural Sciences and Technology (CAST)[2] estimated the production of domestic livestock and poultry manures at 340 million tons/year on a dry-weight basis. Assuming 20% solids, manure production was about 1.7 billion tons/year. Since 1975, numbers of cattle (*Bos taurus*) and sheep (*Ovis aries*) in the U.S. have declined (cattle, 25%; sheep, 22%) while numbers of hogs (*Sus scrofa*), chickens (*Gallus gallus*), and turkeys (*Meleagus gallopavo*) have increased (hogs, 11%; chickens, 70%; turkeys, 128%).[3] Assuming that the decline in livestock numbers is responsible for a 20% reduction in manure, current manure production is about 1.4 billion tons (wet weight) or 272 million tons (dry weight).

In 1975, CAST estimated that about one half of the manure produced was recoverable in a form that could be applied to cropland while the remainder was deposited in a diffuse manner on pastures and rangelands where it does not constitute a water-quality problem. Fedkiw[4] considered this estimate and the changes in livestock numbers and estimated that about two thirds (180 million tons dry weight) of the total production is available for use on cropland or must be disposed of in a manner that does not cause pollution.

The manure is extremely variable in composition, thus, in fertilizer value. The nutrient content depends on the type of animal, the ration fed, the amount and type of bedding material, the collection system, and the management between production and use.

Chicken manures containing very little bedding have relatively high nutrient contents and nitrogen (N) in a form that is almost all available during the year of application. However, bovine manures that have been aged by cycles of wetting and drying and have been leached with rainwater may have lost most of their value as fertilizers. The N content and the availability of the remaining N decrease with losses of ammonia and nitrate through volatilization and leaching. Phosphorus (P) and potassium (K) do not volatilize, are less subject to leaching than N, and are as readily absorbed by plants as are most sources of these elements.

Table 1 Chemical Composition of Various Animal Manures

Reference Source and Type of Animal Manures	Percent, Dry-Weight Basis									mg kg^{-1}										
	Ash	N	P	K	Ca	Mg	S	Na	"Salt"	Mn	Fe	B	Cu	Zn	Mo	Co	As	Al	Ba	Sr
Perkins and Parker[6]																				
Broiler litter (82 samples)	31.3	2.27	1.07	1.70	1.97	0.37	0.35			272	1244	33	29	128	12.6					
Hen litter (31 samples)	42.7	2.00	1.91	1.88	3.42	0.52	0.49			333	1347	28	31	120	13.5					
Atkinson et al.[7]																				
Various species (44 samples, fresh and composted)										201.1		20.2	15.6	96.2	2.05	1.04				
Abbott[8]																				
Dairy and feedlot corral scrapings (6 samples)		1.65	0.50	2.3				0.85	8.1											
El-Sabben et al.[9]																				
Broiler litter (33 samples)	12.5	4.55	1.6	1.9	1.9	0.47		0.58		242	490	41	105	253			12	305		
Hen litter (22 samples)	26.3	3.28	2.3	2.3	6.1	0.54		0.48		251	340	40	71	330			31	730		
Hileman[10]																				
Broiler litter (197 samples)		4.11	1.45	2.18	1.45	0.42		0.07		225	1000	44	32	125	4	1		860	26	64
Benne et al.[11]																				
Chicken from dropping boards on litter	36.1	3.4	0.9	0.8	8.1	0.63	0.68			196	1014	131	33	196	12					
	17.2	4.3	1.6	1.6	3.6	0.31	0.42			102	397	38	26	128	6					
Dairy cow	10.6	2.7	0.5	2.4	1.6	0.61	0.28			56	222	83	28	83	6					
Fattening cattle	10.2	3.5	1.0	2.3	0.55	0.46	0.39			23	182	91	23	68	2					
Hog	28.8	2.0	0.6	1.5	2.0	0.29	0.48			72	1002	143	18	215	4					
Horse	27.9	1.7	0.3	1.5	2.9	0.52	0.26			37	500	56	19	56	4					

Table 1 (*Continued*)

Reference Source and Type of Animal Manures	Percent, Dry-Weight Basis									mg kg^{-1}										
	Ash	N	P	K	Ca	Mg	S	Na	"Salt"	Mn	Fe	B	Cu	Zn	Mo	Co	As	Al	Ba	Sr
Sheep	10.0	4.0	0.6	2.9	1.9	0.60	0.29			32	518	32	16	81	3					
Taiganides and Hazen[12a]																				
Hen	24	5.6	1.9	1.7																
Swine	15	4.5	1.2	3.6																
Cattle	20	3.7	0.5	2.5																
Mathers et al.[13]																				
Cattle (23 feedlots)		2.04	0.81	2.28	1.98	0.76		1.13			3200			140						
Taiganides and Stroshine[14b]																				
Swine (calculated for fresh manure)	17.6	5.6	1.1	1.4																
Hen (calculated for fresh manure)	27.2	5.9	2.0	2.1																
Beef (calculated for fresh manure)	17.2	7.8	0.5	1.8																
Sheep (calculated for fresh manure)	15.3	4.0	0.6	2.9																
Dairy (calculated for fresh manure)	19.7	4.0	0.5	1.7																
D. Bell[15]																				
Chicken (40 samples)		3.68	2.04	1.54					6.15											
Salter and Schollenberger[16]																				
Horse		3.0	0.52	1.8	0.7															
Cattle		3.5	0.52	2.3	1.4															
Sheep		4.5	0.43	2.8	0.8															
Hogs		4.2	1.20	4.8	0.3															
Hen		2.5	0.87	0.8																

Note: Adapted and expanded from a similar table by Azevedo and Stout.[5]

[a] All values except ash and nitrogen were estimated from "air-dry" values reported (assumed to be 7% moisture).
[b] Figures reported by these authors are averages of values given by other authors.

Chemical compositions of some animal manures, as reported in the literature and compiled by Azevedo and Stout,[5] are reported in Table 1. Since age, degree of decomposition, inclusions of bedding and soil, and amount of urine caught with the feces modify the chemical compositions of manures and obscure the characteristics ascribable to fresh excreta from different species, the values in the table can only represent the likely concentrations of nutrient elements in manure.

If, on the average, manure contains 2% N, 0.5% P, and 1.5% K (dry-weight basis) and if one half of the N is available for uptake the 1st year, 180 million tons of manure would supply 1.8 million tons of available N, 900,000 tons of P, and 2.7 million tons of K. These quantities are equivalent to about 17, 24, and 64% of the quantities of N, P, and K used annually in chemical fertilizers in the U.S.

The large-scale feeding operations, especially those of beef cattle and poultry, have resulted in imbalances between livestock numbers and cropland base. Feed and nutrients are imported into feeding sites. Only about 28% of the grain produced on farms today, for example, is fed on farms where it is grown.[17] In 1989, 83.6% of the feeder cattle were fed in 1659 feedlots with one-time capacities of 1000 or more animal units. Only 16.4% of the feeder cattle were fed in the 45,244 feedlots with less than a one-time capacity of 100 units.[18] The increase in the proportion of feeder cattle marketed through feedlots with one-time capacities greater than 16,000 (from 27.6% in 1970 to 50.4% in 1989) is indicative of the trend toward larger feedlots.[18] Some 4.7% of the broiler facilities and 6.5% of the egg producers provide over 50% of the U.S. broiler and egg production.[19]

Problems of imbalance between livestock and poultry populations and the land base for recycling animal waste have not been widespread but can be major problems for localized areas. Where imbalances exist, manure must be hauled longer distances, resulting in increased costs for environmentally acceptable use or disposal. The Second RCA Appraisal of Soil, Water, and Related Resources reported, on a county-by-county basis, that the ratios of animal waste production to cropland and grassland base for recycling were generally low to medium throughout the country.[20] High ratios occurred only in a few areas: the New England dairy region; south central and southeast Pennsylvania; northern Virginia; western Maryland; the Tennessee, North Carolina, and northern Georgia Appalachian area; southern Mississippi; an area in east central California adjacent to the Nevada border; and several northwest Washington counties adjacent to Puget Sound. Other areas with high ratios include southern Delaware and eastern Maryland, Sonoma County, Marin County, and the Chino Basin in California; Maricopa County in Arizona; and the Tillamook area in Oregon. The EPA estimates that such imbalances occur in only 28 counties out of some 1500 or more engaged in livestock and poultry production.[21]

In some locales, the problem of imbalances between animal populations and the cropland base is being intensified by rapid growth of confined livestock and poultry production. Examples are in northwest Arkansas, where a high concentration of poultry operations has developed and some places in central and east Texas where dairy operations have expanded rapidly. Such expansion is expected to continue, especially for poultry, swine, and some dairy areas, through the decade of the 1990s.

Even though there may be sufficient land near feeding operations for economical use of the manure on crops or grazing land, some farmers hesitate to apply manure on their land. Some commercial feeding operations purchase feed from nearby farms and as a condition of their purchases require that the farmers use manure from the feedyard on their land. This type of arrangement can be very satisfactory to all parties because nutrients are recycled through the total production system. However, many farmers refuse to use manure on their land for various reasons. Some reasons most often stated are that manure from feedyards (1) often contains concrete or other foreign items; (2) is in poor physical condition (may contain large lumps) and cannot be spread uniformly; (3) contains weed seed (including noxious weeds); and (4) is highly variable and sometimes causes burning of crops. Also, trucks loaded with manure may cause serious compaction of the soil. Although some of the problems stated may not be as serious as perceived, they are valid concerns. However, most of them can be overcome by composting the manure.

Manure can be composted at the feedyard by the following process. It is removed from the pens and piled in windrows about 2 m wide and 1.3 to 1.6 m high. Water is added and the material is mixed thoroughly. It is allowed to ferment until the temperature reaches about 65°C and begins to decrease (about 1 week) when it is stirred again. The process is repeated six to eight times until composting is complete. About 30 to 50% of the manure mass is lost during the composting process so there is significantly less material to spread, and the compost is a more uniform product in excellent physical condition without offensive odors. It can be spread with fertilizer trucks equipped with high flotation tires that do not result in serious soil compaction. Also, composting kills much of the weed seed.[22,23] Lavake and Wiese[22] found that seed of Johnson grass (*Sorghum halepense*), pigweed (*Amaranthus retroflexus*), kochia (*Kochia scoparia*), barnyard grass (*Echinochloa crusgalli*), and grain sorghum (*Sorghum bicolor*) were killed after exposure for 1 day at 71°C in a feedlot manure compost pile and that of field bindweed (*Convolvulus arvensis*) was killed after 6 days at 71°C in the pile. Cudney et al.[23] sampled compost from five dairies and found that the effectiveness of composting varied among locations. Composting reduced numbers of viable weed seed but in no location was manure found to be completely free of viable weed seeds.

Much N and carbon (C) are lost in composting so the product contains about as much N but more P and K than the uncomposted manure. One composter of feedlot manure reports that his product contains 1.5% N, 0.66% P, and 1.7% K. The N remaining after composting may be less readily available to plants than that present before composting. Compost is more acceptable to farmers than manure except that the cost is sometimes higher than can be justified from the fertilizer value. There is a growing interest in composting among some feedyard owners because they realize that they must make their wastes acceptable to farmers even if they must bear part or all of the cost of composting. Much of this interest is being brought about by impending governmental regulations.

B. USE, FORMS, AND MODES OF APPLICATION

The principal use or method of disposal of manure is land application, either as fertilizer for crops or in disposal quantities. Some alternatives have been suggested and tried but none have been developed to the extent of utilizing significant quantities of waste. These include feeding to livestock, as a substrate for anaerobic digestion to produce biogas, as a fuel for combustion/gasification for electric power generation, and composting for use as fertilizer for crops or for use in landscaping, nurseries, and various urban applications. As regulations regarding waste disposal become more stringent and feedlot operators find it necessary to bear more of the cost of disposal, they may find composting to be an acceptable solution to waste disposal.

Depending on the livestock involved, the type of operation, and the chosen method of handling waste, it may be harvested and applied either in solid or liquid form. Wastes from dairies where milking barn floors are washed down may be handled as liquids or slurries as are wastes from swine operations where the animals are kept on slotted floors. Wastes from most beef cattle and poultry enterprises are harvested and applied in solid form. Slurries may be sprayed on the soil surface or injected into the soil while solid wastes are spread on the soil surface. The waste must be incorporated into the soil as soon after application as possible to prevent loss of N through volatilization.

Commercial equipment is available for spreading and injecting slurries and liquids as well as for spreading dry materials. Whether wastes are in solid or liquid form, even distribution of plant nutrients may be a problem. Application can only be as even as the waste is uniform in content and application equipment, though calibrated, may not be reliable in applying desired quantities. Prins and Snijders[24] presented data showing the necessity of having slurry well mixed and the variability in delivery of slurry by various spreaders. Seventeen types of slurry spreaders were each set to apply 25 m^3 ha^{-1}. Rates actually applied varied from 17 to 55 m^3 ha^{-1}. Solid waste, such as that from cattle feedlots, may vary widely in chemical content, moisture content, and physical condition, making uniform application of the material and distribution of plant nutrients difficult.

II. MOVEMENT AND REACTIONS IN SOIL

Most of the components of manure are immobile and unreactive in the soil while they remain in the organic form but become mobile and reactive when mineralized by microorganisms. The time required for mineralization depends on the nutrient content and organic composition of the manure, the soil to which it is applied, and the temperature and moisture conditions in the soil. Mineralization of N in various manures has been studied extensively and conditions for mineralization have been described.[5] Principal manure components and decomposition products are N, P, organic matter, pathogens, weed seeds, and salts.

A. NITROGEN

Organic N is mineralized to ammonia or to the end product of mineralization, the nitrate form. The negatively charged nitrate ion does not attach to clay particles, thus, it remains mobile in the soil solution. It is subject to uptake by plants and movement with water in the soil, thus, it is subject to leaching from the soil and entering groundwater. Ammoniacal N occurs in two forms: NH_3 (ammonia) and NH^+_4 (ammonium). Both are soluble in water. They exist in equilibrium in solution with the relative concentration of each depending on the pH and temperature of the solution. In the pH range of arable soils, most of the ammoniacal nitrogen that enters the soil is in the NH^+_4 form or is converted to that form. It is then immobilized through attachment to the negatively charged soil clay until it is taken up by plants or converted to nitrates.

B. PHOSPHORUS

When water-soluble inorganic P enters the soil, it is rapidly converted into water insoluble forms. Two mechanisms that have been shown to hold P in the soil are those involving calcium (Ca) ions and those involving iron (Fe) and aluminum (Al) ions. The Ca ions that hold P in a soil may be Ca ions in solution, exchangeable Ca ions forming calcium phosphates on the surface of the clay particles, or Ca ions anchored on the surface of calcium carbonate crystals. This process is of primary importance in inherently neutral or calcareous soils. The iron (Fe) and Al ions that hold P may either be present in films of hydrated oxides or the Al may be present as exchangeable cations if the soil is acid or a film of aluminum hydroxide if the soil had been acid and has had its pH raised by liming.[25]

Inorganic P in the soil occurs in two forms: "fixed" P and "labile" P. Fixed P is tightly adsorbed on or within soil particles while labile P is loosely bound to soil particles. The labile P, a small fraction of the total attached P, remains in equilibrium with soluble P. When the soluble P concentration is reduced by plant uptake, some labile P is transformed to soluble form to restore the equilibrium.[26] In a given soil, the larger the labile P pool, the higher the concentration of P in the soil solution.

Soils vary in ability to fix P. Although several characteristics such as mineral composition, pH, buffering capacity, and clay content influence P fixation capacity, soil texture is the most obvious indicator of fixation capacity. Usually the finer the texture, the more P a soil will hold. When P is applied to the soil surface, there is little downward movement of P until the fixation capacity of the surface layer is satisfied. Then, P in soil solution may move deeper into soil that is not P saturated. There is little movement of P through the soil into the groundwater until the fixation capacity of the soil is satisfied. Except on very sandy soils with little P fixation capacity, P pollution is more likely to occur in surface runoff, either in solution or as particulate material, than through movement into groundwater.

C. ORGANIC MATTER

Although small amounts of soluble organic matter may penetrate the soil through macropores, most of the added organic matter stays where it is deposited and serves as food for microorganisms. Organic residue remaining in the soil after microbial digestion becomes soil organic matter, is not mobile, and does not cause pollution. Adding manure regularly to an arable soil either increases its organic matter, if it had reached an equilibrium value, or reduces its rate of loss, if that is occurring.[25]

D. PATHOGENS

Livestock and poultry wastes contain microorganisms that are infectious to warm-blooded animals. These include bacteria, viruses, protozoa, parasites, and fungi. Bacterial diseases that can be spread through manure include salmonella, leptospirosis, anthrax, tularemia, brucellosis, erysipelas, tuberculosis, tetanus, colibacillosis, and others. Viral diseases include Newcastle, hog cholera, foot and mouth, and psittacosis. Protozoal, parasitic, and fungal diseases are less common than bacterial and viral diseases. Hull[27] compiled the list of diseases potentially transmitted by animal manures reported in Table 2.

Table 2 Diseases Potentially Transmitted by Animal Manures

Type of Disease	Etiologic Agent	Common Domestic Reservoirs	Method of Entry	Ref.
Bacterial				
Salmonellosis	*Salmonella* sp.	Most domestic animals and man	Ingestion	28
Leptospirosis	*Leptospira pomona*	Most domestic animals and man (urine)	Cutaneous and facial membranes	29
Anthrax	*Bacillus anthracis*	Most domestic animals and man (urine)	Cutaneous, inhalation, ingestion	30
Tuberculosis	*Mycobacterium tuberculosis*	Man, swine	Inhalation, ingestion	32
	M. avium	Poultry, swine		33
	M. bovis	Cattle, swine, man		
Johnes Disease (paratuberculosis)	*M. paratuberculosis*	Cattle, sheep, goats	Ingestion	34
Brucellosis	*Brucella abortus*	Cattle, swine, goats, man	Ingestion, inhalation, body openings	35
	B. melitensis	Goats, swine, cattle, man		
	B. suis	Swine, goats, cattle, man		
Listeriosis	*Listeria monocytogenes*	Cattle, sheep, swine, chickens, man, horse	Ingestion (?)	36
Tetanus	*Clostridium tetani*	Horse, sheep, man	Deep cutaneous	37
Tularemia	*Pasteurella tularensis*	Sheep, rabbits, man	Ingestion, inhalation, biting anthropods	
Erysipelas	*Erysipelothrix rhusiopathiae*	Swine, turkeys, man, decaying organic matter	Ingestion, cutaneous	38
Colibacilosis	*Escherichia coli* (some serotypes)	Most domestic animals and man	Ingestion, especially in newborn	39
Coliform mastitis-metritis	*E. coli* (some serotypes)	Cattle	Body openings	
Rickettsial				
Q fever	*Coxiella burneti*	Cattle, sheep, goats, man (vaginal discharges)	Dust inhalation, ingestion	39

Table 2 (*Continued*)

Type of Disease	Etiologic Agent	Common Domestic Reservoirs	Method of Entry	Ref.
Viral				
Newcastle	Virus	Avian species, man	Contact, dust	40
Hog cholera	Virus	Swine	Contact, ingestion	41
Foot and mouth	Virus	Cloven-foot animals, rarely man	Contact, ingestion	42
Psittacosis (ornothosis)	Virus	Avian species and man	Contact, dust inhalation	
Fungal				
Coccidioidomycosis	*Coccidioides immitus*	Cow, hog, horse, sheep, man, free-living	Dust inhalation, cutaneous	43
Histoplasmosis	*Histoplasma capsulatum*	Cow, horse, hen, man, free-living	Dust inhalation, some ingestion	43
Ringworm	Various *Microsporum* and *Trichophyton*	Most domestic animals and man	Contact	
Protozoal				
Coccidiosis	*Eimeria* sp.	Most domestic animals	Ingestion of oocytes	
Balantidiasis	*Balantidium coli*	Swine (normal), man	Ingestion of oocytes	44
Toxoplasmosis	*Toxoplasma* sp.	Many domestic animals and man	Uncertain	
Parasitic				
Ascariasis	*Ascaris lumbricoides*	Swine, man	Ingestion	
Sarcocystiasis	*Sarcocystis* sp.	Most domestic animals and man	Ingestion	

Note: Adapted and expanded by Azevedo and Stout[5] from a similar table of Hull.[27]

Some pathogens have relatively short survival time outside their host while others, such as anthrax, can survive for several decades. Salmonella can survive up to 1 year in liquid manure. Infections from field disposal of manure are uncommon. The potential for infection is greater from contaminated water supplies.[26,45]

E. WEED SEEDS

Manure from cattle and swine contains viable weed seeds while that from poultry does not unless it has bedding or other weed seed-bearing material in it. Weed seeds are broken down by the grinding action of sand in the gizzard of poultry but in cattle and swine many seeds pass through the digestive system and remain viable. In fact, the digestive system may loosen the hard seed coats of weeds, thus enhancing germination. In addition, seeds from weeds growing on and adjacent to sites of feeding operations may enter the manure. The use of manure on farms has sometimes been correlated with weedier fields compared with farms not using it.[23] This gives manure a poor reputation among farmers. Composting of manure will reduce but will not eliminate viable weed seeds in manure.

F. SALTS

In addition to N and P, manure contains substantial quantities of K, Ca, magnesium (Mg), sodium (Na), sulfur (S), and other elements. Manure, from which some N has been lost, may contain as much K and one half as much Na as it does N. When manure is used as a source of N for crop production, levels of these elements, like those of P, may accumulate in the soil. Potassium and Na may be retained on soil clay particles in exchangeable form and K may be fixed in the lattice structure of certain clays in some soils; however, both elements are more mobile than P and may be leached from the soil. Unless manure is applied at very high rates, the salts will not affect plant growth; however, salts leached from the soil may become contaminants of ground and surface water.

III. BENEFICIAL EFFECTS

A. SOIL FERTILITY

Manure is a valuable resource in increasing and maintaining soil fertility. It supplies nutrients for plant growth and organic matter for improving and maintaining soil physical properties. Plant nutrients from manure, when in inorganic form, are no different from those from other sources. However, in fresh manure, 60 to 80% of N and P occurs in organic form and must be mineralized before they can be absorbed by plants. Azevedo and Stout[5] stated that, "Considering the complexity of factors affecting the availability of N from manure in different soils, estimates of average N availability from manure can only be expressed in terms as a range of recoverable amounts." They cited several authors who made generalized statements regarding the availability of N in manures. Turk and Weidemann[46] suggested that the N in animal manures was about 30% as available to plants as was N in mineral fertilizers. Kesler[47] quotes University of Illinois scientists as saying that, as a rule of thumb, 40% of the N is available the 1st year, 30%

Table 3 Total Nitrogen Input from Various Types of Manure Required to Maintain a Yearly Mineralization Rate of 200 kg/ha during a 20-Year Period

	Number of Years Applied (nitrogen input, kg ha^{-1} year^{-1})							
Manure Type and Decay Series[a]	**1**	**2**	**3**	**4**	**5**	**10**	**15**	**20**
Chicken manure								
0.90, 0.10, 0.075, 0.05,	222	220	218	217	216	214	212	210
0.04, 0.03, 0.90, 0.10, 0.05	222	220	219	218	217	213	209	207
Fresh bovine waste, 3.5% N								
0.75, 0.15, 0.10, 0.075, 0.05,	267	253	246	242	240	231	223	218
0.04, 0.03, 0.75, 0.15, 0.10, 0.05	267	253	246	244	241	230	221	215
Dry corral manure, 2.5% N								
0.40, 0.25, 0.06, 0.03	500	312	349	332	326	295	272	255
0.40, 0.25, 0.06	500	312	349	316	308	258	232	218
Dry corral manure, 1.5% N								
0.35, 0.15, 0.10, 0.075, 0.05,	571	412	367	343	336	291	270	240
0.04, 0.35, 0.15, 0.10, 0.05	571	412	367	364	344	281	245	225
Dry corral manure, 1.0% N								
0.20, 0.10, 0.075, 0.05, 0.04,	1000	600	490	475	451	361	300	261
0.03, 0.20, 0.10, 0.05	1000	600	580	489	437	277	225	208

[a] The first of each pair of decay series presented is meant to represent a slower rate of mineralization of residual N, as would be the case for a colder climate.

From Pratt, P.F., Broadbent, F.E., and Martin, J.P., *Calif. Agric.*, 27(6), 10, 1973. With permission.

the 2nd, 20% the 3rd, and 10% the 4th. For poultry manure, Robertson and Wolford[48] estimate that 50% of the N can be utilized by crops the 1st year. Eno,[49] on the other hand, feels that 30 to 60% of the N in poultry manure can be made available within 6 weeks of application to soil. Martin[50] generalized that 50% of the N in steer, dairy, and poultry manures can become available to plants within 6 months of application.

Pratt et al.[51] estimated rates of annual release of N from various manures applied to irrigated lands. They expressed the availability of N in terms of "decay series" (Table 3). Azevedo and Stout[5] give the following example of the use of one decay series. "With dry corral manure the decay series 0.40, 0.25, 0.06 means that the first year's release of mineralized N is 40% of the amount applied. In the second year, 25% of the remaining organic N is released (15% of the original organic N) and during the third year 0.06 of the second year's residual organic N (2.7% of the original N applied) is mineralized. To achieve a constant release of N to growing crops each year, enough organic N must be added to supply sufficient mineralized N for the first year's crop needs (e.g., 250 kg of organic N from dry corral manure will supply 100 kg of mineralized N during the first year)." In successive years, decreasing amounts of manure are required to maintain the desired supply of mineralized N. The decay series concept can be used to match crop needs with organic N inputs and thus increase the efficiency of manure use.

The P in manure is rapidly hydrolyzed and chemically precipitated or adsorbed by soil minerals.[5] Potassium may be held on soil clays in exchangeable form or fixed in the lattice structure of certain clays in some soils. Goss and Stewart[52] reviewed the literature regarding use of animal manures as sources of P for plant growth. Although many reports indicated that P from manure increased both available P in the soil and crop yields, only a few compared manure with commercial forms of P. May and Martin[53] concluded that the yield response of P in steer, chicken, or dairy manure was equal to or slightly better than the response to commercial P. McAuliffe et al.[54] found manure an effective P source for plant growth, but plants fertilized with commercial sources of P removed a higher percentage of the added P than those fertilized with manure. Goss and Stewart[52] evaluated beef feedlot manure as a P source and compared the efficiency of P use from manure sources with that from superphosphate. They found that alfalfa (*Medicago sativa*) grown in soil treated with superphosphate removed a higher percentage of the added P than alfalfa grown in soil treated with manure. However, alfalfa grown in manured soil had the higher P utilization efficiency, as measured by yield increase per unit increase in P removed. They found that the higher utilization efficiency was due in part to more luxury consumption of P by the early cuttings on superphosphate treatments. The P in feedlot manure was released more slowly for plant uptake and the P level remained adequate for a longer period. Goss and Eck[55] compared beef feedlot manure and concentrated superphosphate (0-46-0) as sources of P for irrigated alfalfa and found the two sources to be equally efficient in producing hay yields. Other investigators have suggested that P and K in animal manures are at least as available as those supplied by mineral fertilizers.[16,46,47,56,57]

Manure has been shown to be a valuable source of Fe and zinc (Zn) in soils deficient in those elements. Thomas and Mathers[58] and Mathers et al.[59] conducted greenhouse and field studies showing that manure was more efficient than iron sulfate in supplying Fe to grain sorghum. Miller et al.[60] found that poultry manure corrected Fe and Zn deficiencies. They suggested that manure may supply chelating agents that aid

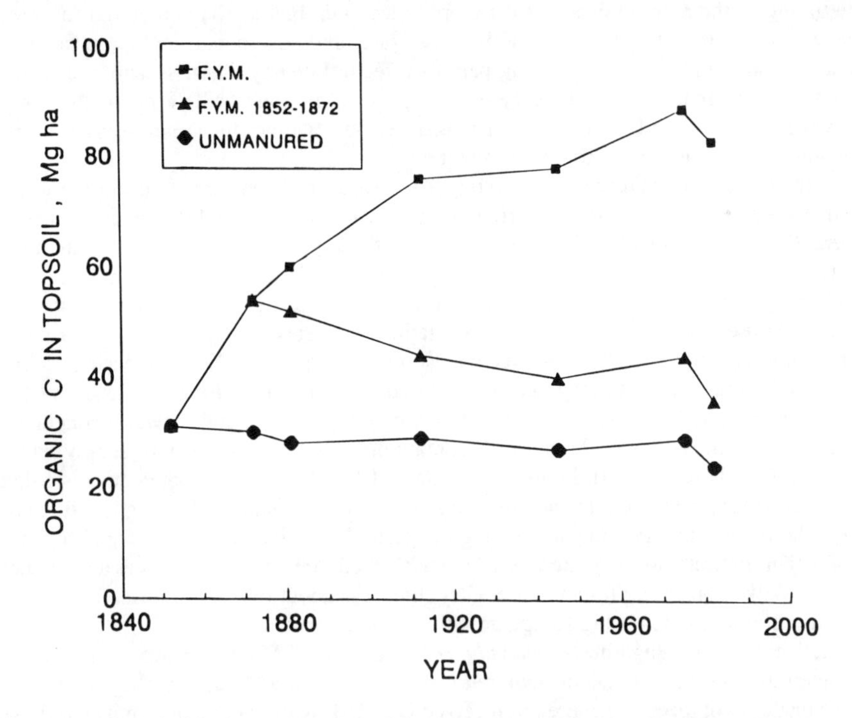

Figure 1 Organic C in the top 23 cm of soil from three plots of the Hoosfield Barley Experiment, Rothamsted, 1852 to 1982. F.Y.M., farmyard manure (35 Mg ha^{-1}) annually; F.Y.M., 1852 to 1872 farmyard manure (35 Mg ha^{-1}) 1852 to 1871, nothing thereafter; and unmanured. Data from Jenkinson, 1991.

in keeping micronutrients soluble, thus more available to plants. Tan et al.[61] found chelating liquids in water extracts of poultry litter that would chelate Fe. Chelated Fe may remain available to plants in calcareous soils.

Manure benefits soil fertility in other ways than supplying plant nutrients. It increases soil organic matter and improves soil physical properties. Increases in organic matter are long lasting as illustrated in Figure 1. Data from the Hoosfield Barley Experiment at Rothamsted[62] show that the effects of farmyard manure applications (35 Mg ha^{-1} $year^{-1}$ for 20 years) on soil organic matter were still measurable more than 100 years after manure applications were discontinued. Manure has been shown to increase soil aggregate stability,[63,64] increase water-holding capacity and decrease evaporation rate,[65] increase water infiltration,[66–68] decrease bulk density, and increase hydraulic conductivity.[60] These effects are less obvious than those from addition of plant nutrients and may become measurable only after sustained manure use. They contribute to improved soil tilth and are beneficial to plant growth.

B. CROP PRODUCTION

Manure supplies plant nutrients and improves soil physical conditions, thus benefiting plant growth and crop production. The efficiency of manure in supplying nutrients depends on the amount and timing of mineralization of the nutrients in the soil and the timing of the nutrient needs of the crop. Since microbial action is required for mineralization, timing of application, soil temperature, and soil moisture affect efficiency of use. Also, length of crop-growing periods affect efficiency. Short-season crops usually require abundant nutrients early in their growing periods while crops with longer growing periods absorb nutrients over longer periods of time; thus, early-season nutrient supplies may not be as important for them.

Effectiveness of manure in supplying N to crops has been measured in proportion to that supplied by inorganic N carriers. Herron and Erhart[70] found that manure N produced first crop yields of grain sorghum 35% of those produced by equivalent amounts of N as ammonium nitrate. However, residual effects of the manure over a 4-year period brought the efficiency of the manure to about 71% of that of the inorganic source. In greenhouse trials with grain sorghum, Mathers and Stewart[71] found that cattle feedlot manure N was about 42% as effective as ammonium sulfate N in increasing yields.

In a 14-year study utilizing beef feedlot manure as a source for irrigated corn (*Zea mays*), wheat (*Triticum aestivum*), and grain sorghum, Mathers and Stewart[72] concluded that annual applications of 22 Mg ha^{-1}, containing 35 to 50% water, will supply fertilizer needs of those crops. If the manure contained 1.34% N, the average content of that collected from 23 feedlots in the area,[13] about 300 kg of manure N was applied annually. The comparable treatment receiving inorganic N received about 150 kg N ha^{-1} annually. This indicates that manure was at least 50% as efficient in supplying N as ammonium nitrate. Wilkinson et al.[73] found that poultry litter was about one half as effective as ammonium nitrate in increasing forage growth of fescue pastures.

Eck et al.[74] grew sugarbeet (*Beta vulgaris*) on the site of Mathers' and Stewarts' study to determine residual effects of manure on sugarbeet yield and quality. They found that plots that did not receive manure but received N or N-P-K did not yield as much as those that received manure, indicating that sugarbeet responded to effects of manure other than the supply of N, P, and K. However, other researchers have not reported any effects of manure on sugarbeet growth beyond supplying plant nutrients. Gardner and Robertson[75] found N and P fertilizers and manures to be equal in their effects on sugarbeet yields. Halverson and Hartman[76] found that application of 22.4 Mg ha^{-1} of barnyard manure in alternate years produced higher sucrose yield than any other treatment without any apparent accumulation of soil nitrate-N and concluded that barnyard manure could be utilized to produce quality sugarbeets and dispose of a waste and potential pollution product.

The effects of plant nutrients supplied in manure on plant growth and crop production are easily measured but the effects of the organic matter are less obvious. In long-term field experiments in the Netherlands, Wadman[77] studied the effects of the organic component of manure on yields of sugarbeet and potatoes (*Solanum tuberosum*). In two of three experiments with sugarbeet, maximum yields of sugarbeet were not increased by organic N. In the third experiment, manure gave a considerable increase in sugar yield. Wadman suggested that it resulted from fertilizer N applied at the start of the growing season being leached beyond the root zone and that slowly released from the manure becoming the essential supply. However, this suggestion was not verified in the experimental field. Maximum yields of potatoes were possibly increased in one of two studies. Wadman suggested that the increase may have been due to K rather than to the slow release of organic N.

IV. ADVERSE EFFECTS

A. EFFECTS ON SOIL ENVIRONMENT

When applied in quantities that supply adequate but not excessive nutrients for crop requirements, manure is generally beneficial to the soil environment; however, excessive quantities of manure may be at least temporarily detrimental to the soil. In the 1960s when livestock feeding operations became concentrated and manure became a waste product to be disposed of, studies were conducted in which manure was applied at disposal rates. Mathers and Stewart[78] applied cattle feedlot waste on a clay loam soil at rates of 0, 22, 67, 134, 269, and 538 Mg ha^{-1}. Irrigated grain sorghum was grown. Ammonium and salt accumulations depressed yields on plots that received 269 and 538 Mg ha^{-1} of manure. Conductivity measurements, made at the end of the growing season, showed salt levels on those plots to be very high. All manure applications caused increased nitrate accumulation in the soil profile. Increases from up to 134 Mg ha^{-1} of manure were proportional to amounts of manure applied. Those from the two higher rates were smaller, indicating that decomposition and nitrification rates were slower where manure application rates were very high. Consequently, only part of the N in these large applications was nitrified. After 14 years of studies on the plots, Mathers and Stewart[72] found that the manure applications had positive effects on soil conditions but cautioned that manure applications of 67 Mg ha^{-1} or more may cause salt or high ammonia damage to emerging seedlings and N losses by nitrate leaching below the root zone.

Hileman[79] reported that soil chemical imbalance had been observed on fields that had been fertilized with poultry litter for more than 30 years. He initiated studies to determine the relationship of application of broiler litter to soil chemical properties. He studied rates of 11.2, 22.4, 33.6, and 44.8 Mg ha^{-1} on three soils: Ruston sandy loam, Sharkey clay loam, and Captina silt loam. He measured a rapid rise in soil pH on all soils which he attributed to the rapid release of ammonia from the raw litter. After 7 months, the soil pH was decreasing and a further decline with time was expected. The increase in pH was much greater on the sandier Ruston soil than on the two finer-textured soils.

Increases in K in the soil solution were even greater than amounts applied in the litter. Hileman indicated that K was released from clay minerals in the soil through the exchange for ammonia. He attributed the failure to obtain germination and growth in two successive plantings of fescue to high levels of K and ammonia in the soil. The litter applications had little effect on Ca levels in the soil solution. Magnesium levels, however, increased with increasing application rates on Ruston and Captina soils and showed a marked decrease on the Sharkey soil. Grass tetany in cattle is caused by low levels of Mg in forage. Since the Sharkey soil has a rather high level of native Mg, it is not regarded as tetany prone. However, results obtained in this study indicated that this is not necessarily true. Plants growing on this soil at the higher rates of broiler litter showed Mg deficiency. Hileman stated that soil type is crucial in determining potential tetany situations where high rates of broiler litter or fertilizers are applied. Broiler litter applications at the 11.2 Mg ha^{-1} rate caused severe salt toxicity on all three soils 3 months after the litter incorporation. This occurred under near-ideal greenhouse conditions.

Application of manure at disposal rates or overuse of manure as fertilizer can result in accumulation of excess salts or an imbalance of salts in the soil profile. Since N is usually the most limiting element to plant growth in most soils, manure is applied at rates required to supply the needed N. The manure, however, contains P, K, Ca, Mg, Na, S, micronutrients, and other essential and nonessential minerals. These elements may be in excess to plant needs and accumulate in the soil profile. If there is a high proportion of monovalent ions (Na, K, and sometimes ammonium), soil dispersion can increase and water penetration rates may be decreased.[80–81] However, Ca and Mg can cancel some of the deleterious effects of the monovalent ions. There has been some concern about influences that sodium chloride as a feed additive might have on soil salinity when animal manures are returned to the soil. Azevedo and Stout[5] state that amounts of Na salts in feed additives are much smaller than amounts of Na and K in feeds, thus adding little additional Na to the manure. However, they indicate that Na in animal manures is certainly of no benefit in terms of soil fertility or soil structural improvement and suggest that "sodium chloride additions to animal feeds should be minimized within the framework of modern feeding programs."

B. EFFECTS ON CROPS AND LIVESTOCK

Manure contains the elements essential for plant growth; thus, when manure is mixed with the soil at the correct rate and water is available, good plant growth results. However, failure to allow sufficient time for manure to react with the soil may prevent seedling germination and emergence, application on established grasslands may temporarily decrease yields, and excessive rates of application may affect crop quality.

The most immediate and visible deleterious effect of soil applied manure on crops is prevention of germination and emergences of seedlings. This effect results from accumulation of ammonia in the soil atmosphere, either from application of excessive rates of manure or from planting too soon after manure has been incorporated into the soil. Azevedo and Stout[5] state that urea in animal manure can be enzymatically converted to ammonia in 2 to 5 days, the conversion rates being governed by environmental factors affecting microbial activity and the amounts of urease enzyme present. They further state that little harm to the crop from ammonia toxicity should result from animals that excrete urea unless the manures are handled to preserve the urine-N or are applied to soil at high rates, because up to 90% of the urine-N can be lost by volatilization on a feedlot surface.[82] Poultry manure, however, contains most of its N in forms of uric acid from which ammonia is released in a series of enzymatic reactions. The ammonia is released slowly and moderate-to-heavy rates of poultry manure may severely damage plants as noted by Hileman.[79] The damage may be prevented by composting the manure or delaying planting until uric acid in the applied manure has been decomposed. Eno[49] recommends a delay of 4 weeks between the date of manure application and the date of planting. With any manure, it is advisable to time application so that the manure has time to pass the initial stages of decomposition before crops are planted.

Weeds compete with crops for water, plant nutrients, and light, thus reducing yields. Manure is a source of weed seeds, so it may increase weed populations in crops and grasslands. As previously mentioned, composting will reduce but will not eliminate viable weed seeds. Users of manure should watch their fields closely and eradicate any noxious weeds such as field bindweed and Johnson grass that may be introduced in manure. Enhanced weed control measures may be required where manure is spread.

Prins and Snijders[24] reviewed the negative effects of farmyard manure on grassland in the Netherlands. They listed the following four effects: (1) variability in nutrient content; (2) uneven distribution of nutrients from surface spreading and injection; (3) sward damage caused by surface spreading; and (4) sward damage caused by injection. While inorganic fertilizer application can be tailored to suit specific crops and timed exactly to suit plant requirements, manure supplies a variable range of nutrients. The variability is related to kind of animal, kind of feed, moisture content, and dilution and loss during storage. Also, when manure is applied as a slurry, uneven application of nutrients may result from incomplete mixing of the slurry in the storage tank. The problem for the farmer is that he does not know the exact nutrient concentration of the manure. The authors presented data showing that slurry spreaders were inconsistent in amounts of slurry applied and that P content of grass decreased with increasing distance from the tines when slurry was injected through tines spaced 50 cm apart. Tines spaced closer than 50 cm carry the risk of "rolling up" the sward. Sward damage from surface spreading results from smothering and scorching and is most severe at the first cut after application. The damage increased with increasing amounts of manure applied. Sward damage from injection equipment also occurs mostly in the first cuts after application.

Even though manure has been shown to be a good fertilizer source for sugarbeet, application rates must be limited to amounts that supply adequate but not excessive plant nutrients since excessive N decreases sugar content and excessive K and Na interfere with extraction of sugar from the beet extract.

Excessive soil N levels such as those that arise from disposal rates of manure affect the quality of forages much more than that of grain. The chemical content of grain remains comparatively constant while excessive N rates increase nitrate levels and cause cation imbalances in forage which result in health problems in grazing ruminants. Nitrate poisoning may result when the nitrate-N content of the total ration exceeds 0.21% (dry-weight basis). Grass tetany is caused by a hypomagnesia condition when the ratio of K to (Ca + Mg) in the ration exceeds 2.2. High rates of N fertilization have been shown to increase the ratio,[83] and the K in the manure would also be expected to increase the ratio. Bovine fat necrosis occurs in beef cows consuming tall fescue (*Fescuta arundenacea* Schreb) infected with the endophyte (*Acremonium coenophialum*). The condition is most likely to occur in cows grazing fescue that has received high rates of N fertilizer.[84] Stuedemann et al.[85] summarized the results of a 4-year study of broiler litter fertilization of tall fescue pastures on health and performance of beef cows. Three Kentucky 31 tall fescue pastures were fertilized at three different N levels: high N (703, 794, 483, 0 kg/ha from broiler litter in 1972, 1973, 1974, and 1975, respectively); moderate-N (224 kg N/ha/year from NH_4NO_3); and low-N (84 kg N in 1972 and 74 kg N/ha in 1973, 1974, and 1975, respectively, from NH_4NO_3). The pastures were stocked at the rate of 0.4 ha/Aberdeen Angus brood cow.

On the high-N fescue pasture from 1972 through 1975, there were 13 cases of grass tetany that resulted in 7 deaths among 45 cows. On the moderate-N fescue pasture, there were 7 cases that resulted in 5 deaths among 48 cows, and there were no grass tetany cases on the low-N fescue pastures.

The incidence of fat necroses was measured on the high- and medium-N pastures from 1969 through 1976 and on the low N pasture from 1972 through 1976. Of the cows tested from the high-N pasture, 52% had necrotic fat lesions. Percentages of cows with lesions on the medium-N and low-N pastures were 10 and 3, respectively. The relative incidence of fat necrosis remained approximately the same for 2 years after litter application had stopped. Stuedemann indicated that, because fat necrosis occurred on all three pastures, it should not be associated with any broiler litter effect other than its plant nutrient input.

Although plant levels of nitrate-N reached concentrations of 3300 and 2000 μg/g on the high- and moderate-N fertilized pastures, no direct evidence of nitrate poisoning was observed in cattle.

C. EFFECTS ON WATER QUALITY AND AQUATIC SYSTEMS

Nitrogen, P, organic matter, pathogens, and salts are components of manure that degrade water quality. Nitrogen and P are the most prevalent pollutants. In fresh manure, they are mostly in organic forms (60 to 80%) which are not mobile when applied to the soil unless runoff occurs and carries them away. They are, however, subject to mineralization by microorganisms.

1. Nitrogen

Nitrate N, the end product of N mineralization, remains mobile in the soil solution and is subject to leaching from the soil and entering groundwater. It may reach surface waters through the drains or through aquifers that discharge to surface waters.

The EPA has not yet proposed maximum levels for nitrates in freshwater aquatic systems. Nitrates are not toxic to fish except at very high concentrations. Concentrations in excess of 400 to 1000 mg/L may be toxic to freshwater salmonoids. Such concentrations are highly unlikely to occur from normal farming activities and would not result from natural causes.[26,86] The EPA makes the following statement in its report, "Quality Criteria for Water": "It is concluded that (1) levels of nitrate-nitrogen at or below 90 mg/L would not have adverse effects on warmwater fish . . . (2) nitrite-nitrogen at or below 5 mg/L should be protective of most warmwater fish . . . and (3) nitrite-nitrogen at or below 0.06 mg/L should be protective of salmonoid fishes. . . . These levels either are not known to occur or would be unlikely to occur in natural surface waters."[86]

Although nitrate levels in freshwaters may be comparatively high without causing deleterious effects on aquatic systems, tolerance levels in estuarine waters are much lower. This is not due to direct toxic effects of nitrate on aquatic organisms but to the indirect effect of reduced oxygen supply in bottom waters. Nitrate-N is regarded as a factor in eutrophication of brackish and saline waters of estuaries. Excess nitrate-N may cause excessive growth of algae. The algae die and as they decay the decay organisms deplete the supply of dissolved oxygen. Mass kills of fin fish and shellfish may occur.[4] In Chesapeake Bay, for example, the tolerance level associated with a healthy, high-quality ecological condition has been estimated at 0.6 mg nitrate-N/L. Ecological conditions are poor for levels above 1.8 mg/L.[4]

Because ammoniacal N is immobilized when it enters the soil and remains immobile until it is taken up by plants or converted to nitrates, little ammonia that enters the soil remains in that form if it enters groundwater or surface waters. However, ammonia from manure may enter aquatic systems through other channels. Ammonia volatilized from urine or feces on feedlot surfaces may be absorbed in water bodies and runoff from feedlots, manure piles, overflowing storage lagoons, springtime flushing from manure stored over winter on frozen soils, or other concentrations of manure may enter aquatic systems.

The EPA has found that levels of NH_3, in the range of 0.2 to 2.0 mg/L, have been toxic to freshwater fish. Concentrations below 0.2 mg/L may not kill a significant part of a fish population but may have adverse physiological or histo-pathological effects. The EPA has established a water-quality criterion of 0.02 mg/L for NH_3 (un-ionized ammonia) to assure safety for freshwater life forms that have not been investigated.[86] Natural waters generally contain concentrations of less than 0.1 mg/L of total free ammonia (NH_3 + NH_4).[26]

2. Phosphorus

The adverse effect of P in an aquatic system is that it, in combination with other plant nutrients, allows algae growth in freshwater environments. High concentrations of P promote growth of algae and aquatic weeds and accelerate eutrophication of lakes and reservoirs. As algae proliferate in response to elevated nutrient levels, sunlight cannot penetrate surface waters; thus, the growth of submerged vegetation is inhibited. Also, as the algae die and decay, the water is depleted of dissolved oxygen and fish kills may occur. Conditions favorable to algae growth vary in different lakes and in different areas making setting national criteria for P in lake waters or reservoirs difficult.[86]

Continual use of manure to supply the N needs of crops results in increased P content of the soil. The ratio of N to P in fresh manure is generally 3 or 4 to 1. Since a significant amount of N is lost by volatilization,[82] the N/P ratio of manure applied to the land is often less than 3. Therefore, when manure is applied at rates sufficient to supply adequate N for most cropping conditions, excess amounts of P and K are added. For example, Sharpley et al.[87] found that 8 years of continuous manure usage resulted in large accumulations of available P. Christie[88] also found very significant increases in extractable P in soil that had been treated with cow or pig slurries.

Increased P contents of surface soil increase the potential for soluble and sediment-bound P to be transported in runoff and therefore has water quality implications. Van Reimsdijk et al.[89] warned that land spreading of animal manure in quantities exceeding plant uptake of P results in an accumulation. They stated that, when the cumulative excess becomes large compared to the buffering capacity of the soil, P could leach to ground and surface water causing eutrophication. They concluded, however, that these negative effects develop only after a relatively long period of large manure applications, and that soils differ widely in their buffering capacities. Sharpley and Smith[90] recently developed an equation that predicts the soluble P concentration of runoff based on the available P content of the surface soil as determined by a soil test.

The potential P buildup from the use of manure poses a tremendous challenge for managing animal wastes. Historically, manure application rates have been mainly based on nitrogen loading rates with little attention being paid to P accumulations. However, with the growing environmental concern associated with P in surface water supplies, there are pressures mounting for limiting or even banning P additions to soils that exceed a certain level of plant-available P based on a soil test. Such criteria could result in manure application rates being significantly reduced or even eliminated. Michigan recently adopted guidelines that manure additions should be restricted to rates adequate to replace the P removed by crops once the soil test available P level reaches a value of 160. The EPA has proposed rules stating that, when local water quality is threatened by P, the application rate of manure should be limited to the crop uptake rate of P.[91] If P levels are used as the criteria for determining manure loading rates, supplementary N will often be required to supply N needs of crops.

The use of P as the criteria for determining manure loading rates may be appropriate, particularly in regions containing surface water where accelerated eutrophication can occur. In other areas, however, the benefits derived by using manures to enhance overall soil quality and as primary sources of N for supplementing plant growth may outweigh any potential negative effects associated with increased P levels.

There are no national environmental criteria for nonelemental forms of P in freshwater or marine ecosystems. However, to prevent biological nuisances and control accelerated or cultural eutrophication in standing waters, the EPA recommends that total P concentrations not exceed 50 μg/L (as P) in any stream at the point it enters a lake or reservoir and 25 μg/L within the lake or reservoir. P levels in the surface waters of most relatively uncontaminated lakes average from 10 to 30 μg/L. A desirable upper level to prevent plant nuisances in streams or other flowing water not discharging into lakes or reservoirs is 100 μg/L of total P.[86] Nonelemental forms of P are not toxic to fish and other aquatic organisms.

3. Organic Matter

The primary deleterious effect of manure organic matter on an aquatic system is depletion of dissolved oxygen in the water. When the organic matter enters the water, it is subject to decomposition by microorganisms. The microorganisms, principally bacteria, consume oxygen from the water as they consume the organic matter. Depletion of dissolved oxygen can cause fish kills in the same manner as those resulting from growth of algae. Generally, bacterial growth in the water is proportional to its food supply, the concentration of manure, so as growth increases the consumption of free oxygen is increased. The level of depletion will vary with the amount of waste added; the size, velocity and turbulence of the stream; the initial dissolved oxygen levels in the manure and in the stream; and the water temperature. Turbulence helps replenish dissolved oxygen; thus, turbulent streams can assimilate more waste than placid streams. Also, cold waters can hold more dissolved oxygen than warm waters.[4]

The secondary effect of manure organic matter is the release of its decomposition products into the aquatic system. As the microorganisms consume it and convert it to simpler organic forms or to nonorganic compounds, nitrate, P, carbon dioxide (CO_2), N, ammonia, hydrogen sulfide (H_2S), or other compounds may be released.

4. Pathogens

Pathogens that may enter water and aquatic systems from manure are listed in Table 2. As indicated earlier, some have relatively short survival times outside their host while others may live for years. It is impractical to conduct a separate test for each pathogen. The alternative is to substitute an indicator organism whose level of occurrence can be used as a measure of the likely presence of pathogens. Fecal coliform bacteria occur exclusively in warm-blooded animals in a greater number than the frequency of pathogens and their die off or growth rate is comparable to that of the pathogens; thus, they are a good indicator of the possible presence of pathogens and are used for that purpose.

D. EFFECTS ON ATMOSPHERE

Flies, odors, and dust are the principal atmospheric contaminants arising from manures. They are also referred to as "nuisance factors". Populations of cosmopolitan "filth flies" have increased dramatically with the advent of large, commercial, confined feeding facilities for animal production.[92,93] Anderson[92] states that more than 50 species of flies can be attracted to and breed in the accumulated mass of manure under confined poultry, although only one or two species are attracted to range poultry droppings. Azevedo and Stout[5] state that manure in California feedlots attracts, at most, seven species of flies but most of these are obnoxious pests to cattle and man. Individual cowpats in pastures may contain more than 40 species of flies although only 2 (horn fly, *Haematobia irritans*, and face fly, *Musca autumnalis*) can be considered serious pests.

The species of troublesome flies common to California's confined animal manures as summarized by Loomis[94] are reported in Table 4. Although the table was prepared for California, the flies are common to other areas. Each species has habitat preferences and will build up in great numbers only when their habitat requirements are best met. The house fly (*M. domestica*) and the stable fly (*Stomoxys calcitrans*) are usually the most abundant species found at cattle feedlots.

Of the atmospheric contaminants arising from manure, odor is probably the most noticeable but least definable and most difficult to control. The most offensive odors arising from manure are produced by anaerobic bacterial activity during fermentation of wet manure. White et al.[95] found hydrogen sulfide, methanethiol, dimethyl sulfide, diethyl sulfide, propyl acetate, *n*-butyl acetate, trimethylamine, and ethylamine to be fermentative decomposition products from dairy manure. Amines were found to be the major odorous compounds around cattle feedlots.[96] Merkel et al.[97] found the major constituents of odor in a confined swine building atmosphere to be of the amino and sulfide groups. Deibel[98] identified butyric acid, ethanol, and acetoin as the chief volatile compounds in accumulated poultry manure.

Dust arising from confined feeding facilities can pose problems for the general public as well as for the owners of the facilities. Dust concentrations may build up within facilities, causing discomfort and health problems for the animals and their caretakers. Dust arising from the facilities may be injurious to neighbors or to the general public. Azevedo and Stout[5] cite a case in which a feedlot owner was sued by a grapeyardist for dust damage in a nearby grape field and instances in which feedlot operators have been

Table 4 Domestic Flies Commonly Associated with Manures of Confined Animals in California

Type of Fly	Number Days for Life Cycle (egg to adult)[a]	Type of Manure[b] (record of fly in manure)				
		Cow	Steer	Hog	Horse	Poultry
Black blow fly, *Phormia regina*	10	—	X	—	—	X
Blue blow fly, *Calliphorinae* species	15	X	—	X	—	X
Green flow fly, *Phaenicia* species	8	X	—	X	—	X
Black garbage fly, *Ophyra* species	10	X	—	X	—	X
Coastal fly, *Fannia femoralis*	14	—	—	—	—	X
Flesh fly, *Sarcophagidae* species	8	X	X	X	—	X
House fly, *Musca domestica*	8	X	X	X	X	X
Little house fly, *Fannia canicularis*	24	X	—	X	—	X
Stable fly, *Stomoxys calcitrans*	21	X	X	X	X	X
False stable fly, *Muscina* species	14	X	X	X	X	X
Drone fly, *Tubifera tenax*	21	X	X	X	—	X

[a]Relative time during warm summer temperatures. Cooler or very hot temperatures will, respectively, retard or accelerate this developmental period.

[b]The symbol x denotes a recorded occurrence of that species of fly in that type of manure; the symbol — indicates that species of fly is not commonly associated with that particular type of manure.

From Reference 94.

cited by the California Highway Patrol for the traffic hazard created by dust blowing across highways. George et al.[99] also cite court cases in the Midwest in which flies, odors, and dust have been alleged to be injurious to neighbors.

The severity of dust nuisances around confined feeding facilities is highly dependent on the amount of moisture in the manures. In Texas Panhandle feedlots, at least 25% moisture was necessary to prevent blowing manure dust.[100] According to Azevedo and Stout,[5] the three major nuisances of animal manures, flies, odors, and dust, are intimately related by their dependence on moisture conditions. Problems can be minimized if manure moisture is kept low enough to preclude serious odors and fly proliferation and high enough to prevent dusty conditions. Skill and careful attention are required for successful management of manure moisture, but use of moisture-control techniques can be highly effective.

E. EFFECTS ON PEOPLE

Animal manures can be sources of diseases that affect people. Some of the more common diseases potentially transmitted by animal manures are listed in Table 2. With the exception of salmonellosis, the incidence of disease transmission from animals to man via contact with manure and other routes is on the decline.[5] The frequency of infections from field disposal of manures is low. The potential for infection is greater through contaminated water supplies.[26,45]

The nutrients N and P are the most prevalent water pollutants associated with animal manures. Nitrate N in drinking water has been determined to be the cause of methemoglobinemia in infants. Bacteria present in the digestive tracts, because of high stomach pH and other factors, can change nitrate into nitrite and transform hemoglobin to methemoglobin which does not carry oxygen. Thus, the oxygen carried by an infant's blood can be decreased and its body gradually suffocated. The EPA has established a maximum contaminant level (MCL) of 10 mg/L for N, measured as nitrate-N, in domestic drinking water supplies. It is known that many infants have drunk water with nitrate-N levels greater than 10 mg/L without developing methemoglobinemia, but differences in infant susceptibility are not yet understood.[86] There are questions about the indirect carcinogenicity of nitrate and about neurosystem, reproductive, and developmental effects of nitrates in drinking water but there is no conclusive evidence that adverse effects exist.[101,102] There are no defined human health risks for P in surface or groundwaters.[26]

V. WAYS OF MINIMIZING ADVERSE EFFECTS AND INCREASING BENEFITS

Proper management will both minimize adverse effects of manure and increase benefits from it. If land application of manure is planned and executed properly, it results in improved soil fertility and crop yields. Careless handling, however, can result in impaired soil productivity, degraded quality of surface and groundwater, and cause nuisance complaints by neighbors. Use of the soil for disposal of animal wastes and for crop production requires total nutrient management. To attain such management requires knowledge of inputs of plant nutrients, principally N and P from manure and other sources, and nutrient requirements of crops for expected yield levels. Then, manure and/or other fertilizers can be applied to supply nutrients at adequate but not excessive levels. The following actions or restrictions are required for proper use of animal manures on cropland and pastureland.

1. Testing of soils and manure for available amounts of nutrients to determine proper rates and schedules of manure application to achieve yield goals without overfertilization and nutrient loss to the environment. The amount of residual N or P in the soil can vary considerably from field to field, and season to season; thus, comprehensive soil testing is required.
2. Avoiding excessive applications of manure or commercial fertilizers on land where nutrients can leach to groundwater or runoff to surface waters.
3. Scheduling manure and commercial fertilizer applications in ways that assure maximum crop uptake during the growing season and protect water quality. When manure is applied for crops on arable soils, it should be applied and incorporated in the soil long enough before planting to allow mineralization and release of nitrogen before it is needed by the crop, yet close enough to planting so it will not be leached from the root zone before it is taken up by plants.
4. Incorporating manure into the soil when practical and appropriate and avoiding applications on steep slopes, snow, ice, and frozen ground to preclude flushing, runoff, or leaching due to heavy rainfall or the spring thaw.

Two excellent sources of information are available for use as aids in planning for land utilization of animal manures. Both sources contain sufficient information for designing plans for ecologically sound land application of manure. *The Agricultural Waste Management Field Handbook*[26] is a comprehensive compilation of information and data used by NRCS personnel. It provides specific guidance for planning, designing, and managing systems where agricultural wastes are involved. Of special interest to those interested in environmental aspects of manure are Chapter 3, "Agricultural Wastes and Water, Air, and Animal Resources"; Chapter 4, "Agricultural Waste Characteristics"; Chapter 5, "Role of Soils in Waste Management"; Chapter 6, "Role of Plants in Waste Management"; Chapter 7, "Geologic and Ground Water Considerations"; and Chapter 11, "Waste Utilization." The other publication is entitled *Animal Waste Utilization on Cropland and Pastureland, A Manual for Evaluating Agronomic and Environmental Effects.*[103] Stated objectives of the manual are to (1) provide information for applying animal wastes to land in terms of agronomic benefit and/or pollution potential; (2) provide basic information to enable planners to reduce or control non-point pollution from animal wastes applied to land; and (3) provide sufficient information to enable planners to integrate the many variables into beneficial land-application systems. Topics specially pertinent to animal waste utilization are (1) quantity and characteristics of animal wastes, (2) land application planning, (3) water quality, and (4) economic considerations. Illustrative problems and worksheets are given to estimate quantities of manure available for land application from specific livestock or poultry operations, determine the land-application rate of manures and feedlot runoff that can be used to supply N for crops without creating salinity problems, and to determine environmental effects of manure on an application site.

REFERENCES

1. Wadman, W.P., Sluijsmans, C.M.J., and de la Lande Cremer, L.C.N., Value of animal manures: changes in perception, in *Animal Manure on Grassland and Fodder Crops*, van der Meer, H.G., Unwin, R.J., van Dijk, T.A., and Ennik, G.C., Eds., Martinus Nijhoff, Dordrecht, The Netherlands, 1987, 1.
2. Council for Agricultural Science and Technology (CAST), Utilization of Animal Manure and Sewage Sludges in Food and Fiber Production, Report No. 41, Ames, IA, February 19, 1975, 1.
3. U.S. Department of Agriculture (USDA) Agricultural Statistics 1991, U.S. Government Printing Office, Washington, D.C., 1991, 246.
4. Fedkiw, J., Progress and Status of Livestock and Poultry Waste Management to Protect the Nation's Waters 1970–1992, A working paper of the USDA working group on water quality, U. S. Department of Agriculture, Washington, D.C., October 1992, 1.
5. Azevedo, J. and Stout, P.R., Farm animal manures: an overview of their role in the agricultural environment, Calif. Agric. Exp. Sta. Man., 44, 69, 71, 74, 1974.
6. Perkins, H.F. and Parker, M.B., Chemical composition of broiler and hen manures, University of Georgia Agric. Exp. Sta. Res. Bull. 90, 1, 1971.
7. Atkinson, H.J., Giles, G.R., and Desjardins, J.G., Trace element content of farmyard manure, *Can. J. Agric. Sci.*, 34, 76, 1954.
8. Abbott, J.L., Use animal manure effectively, University of Arizona Agric. Exp. Sta. Bull. A-55, 1968, 1.
9. El-Sabben, F.F., Long, T.A., Gentry, R.F., and Frear, D.E.H., The Influence of Various Factors on Poultry Litter Composition, Animal Waste Management, Cornell University Conference on Agricultural Waste Management, Ithaca, NY, 1969, 340.
10. Hileman, L.H., The Fertilizer Value of Broiler Litter, University of Arkansas Agr. Exp. Sta. Rep. Series 158, 1, 1967.
11. Benne, E.J., Hoglund, C.R., Longnecker, E.D., and Cook, R.L., Animal Manures—What Are They Worth Today?, Michigan State University Agr. Exp. Sta. Circ. Bull. 231, 1, 1961.
12. Taiganides, E.P. and Hazen, T.E., Properties of farm animals excreta, *Trans. Am. Soc. Agr. Eng.*, 9, 374, 1966.
13. Mathers, A.C., Stewart, B.A., Thomas, J.D., and Blair, B.J., Effects of Cattle Feedlot Manure on Crop Yields and Soil Conditions, Technical Report No. 11, USDA Southwestern Great Plains Research Center, Bushland, TX, 1973, 1.
14. Taiganides, E.P. and Stroshine, R.L., Impact of farm animal production and processing on the total environment, Livestock Waste Management and Pollution Abatement, Proc. Intl. Symp. on Livestock Wastes, Am. Soc. Agr. Engr. Publ. PROC-271, 1971, 95.
15. Bell, D., Chicken manure as a fertilizer, University of California Agr. Ext. Serv., 1971, 1.
16. Salter, R.M. and Schollenberger, C.J., Farm manure, Ohio Agr. Exp. Sta. Bull. 605, 1939, 1.
17. Department of Agriculture (USDA), 1989 Fact Book of Agriculture, Misc. Publ. No. 1063, Office of Public Affairs, Washington, D.C., 1989, 17.
18. Krause, K.R., Cattle Feeding, 1962–1989, Location and feedlot size AER 642, U.S. Department of Agriculture, Economic Research Service, Washington, D.C., 1991, 1.

19. U.S. Department of Commerce, Bureau of Census (USDOC/Census) Census of Agriculture, Washington, D.C., 1987.
20. U.S. Department of Agriculture, The Second RCA Appraisal, Soil, Water, and Related Resources on Nonfederal Land in the United States, Analysis of Condition and Trends, Washington, D.C., 1989, 108.
21. Long, C.M. and Painter, W., The impact of livestock waste on water resources in the United States, Proceedings of the National Workshop: National Livestock, Poultry, and Aquaculture Waste Management, July 29–31, 1991, American Society of Agricultural Engineers, St. Joseph, MI, 1992, 48.
22. Lavake, D.E. and Wiese, A.F., Effect of composting on weed seed germination, *Proc. South. Weed Sci. Soc.*, 30, 167, 1977.
23. Cudney, D.W., Wright, S.D., Shultz, T.A., and Reints, J.S., Weed seed in dairy manure depends on collection site, *Calif. Agric.*, 46(3), 31, 1992.
24. Prins, W.H. and Snijders, P.J.M., Negative effects of animal manure on grassland due to surface spreading and injection, in *Animal Manure on Grassland and Fodder Crops*, van der Meer, H.G., Unwin, R.J., van Dijk, T.A., and Ennik, G.C., Eds., Martinus Nijhoff, Dordrecht, The Netherlands, 1987, 119.
25. Russell, E.W., *Soil Conditions and Plant Growth*, 9th ed., Jarrold and Sons, Norwich, 1961, 480, 293.
26. U.S. Department of Agriculture, Soil Conservation Service (USDA/SCS) Agricultural Waste Management Field Handbook, National Engineering Handbook, Washington, D.C., 1992.
27. Hull, T.G., The role of different animals and birds in diseases transmitted to main, in *Diseases Transmitted from Animals to Man*, 5th ed., Hull, T.G., Ed., Charles C Thomas, Springfield, IL, 1963, 896.
28. Dack, G.M., Salmonella food infections, in *Diseases Transmitted from Animals to Man*, 5th ed., Hull, T.G., Ed., Charles C Thomas, Springfield, IL, 1963, 210.
29. Van Ness, G.B. and Manthei, C.A., Leptospirosis, *Animal Diseases*, U.S. Department of Agriculture Yearbook, 1956, 226.
30. Stein, C.D. and Van Ness, G.B., Anthrax, *Animal Diseases*, U.S. Department of Agriculture Yearbook, 1956, 229.
31. Stein, C.D., Anthrax, in *Diseases Transmitted from Animals to Man*, 5th ed., Hull, T.G., Ed., Charles C Thomas, Springfield, IL, 1963, 82.
32. Johnson, H.W. and Ranney, A.F., Tuberculosis and its eradication, Animal Diseases, U.S. Department of Agriculture Yearbook, 1956, 213.
33. Feldman, W.H., Tuberculosis, in *Diseases Transmitted from Animals to Man*, 5th ed., Hull, T.G., Ed., Charles C Thomas, Springfield, IL, 1963, 5.
34. Larsen, A.B. and Johnson, H.W., Paratuberculosis (Johnes Disease), *Animal Diseases*, U.S. Department of Agriculture Yearbook, 1956, 221.
35. Manthei, C.A., Kuttler, A.K., and Goode, E.R., Jr., Brucellosis, in *Animal Diseases*, U.S. Department of Agriculture Yearbook, 1956, 202.
36. Biester, H.E. and Schwarte, Listeriosis, in *Animal Diseases*, U.S. Department of Agriculture Yearbook, 1956, 239.
37. Stein, C.D., Tetanus, in *Animal Diseases*, U.S. Dept. Agric. Yearbook, 1956, 239.
38. Schuman, R.D., and Osteen, O.L., Turkey Erysipelas, *Animal Diseases*, U.S. Department of Agriculture Yearbook, 1956, 474.

39. Diesch, S.L., Disease transmission of water-borne organisms of animal origin, in *Agricultural Practices and Water Quality*, Wellrich, T.L. and Smith, G.E., Eds., Iowa State University Press, Ames, 1970, 265.
40. Osteen, O.L., Newcastle Disease, in *Animal Diseases*, U.S. Department of Agriculture Yearbook, 1956, 455.
41. Torrey, J.P., Hog cholera, in *Animal Diseases*, U.S. Department of Agriculture Yearbook, 1956, 354.
42. Shahan, M.S. and Traum, J., Foot-and-mouth disease, in *Animal Diseases*, U.S. Department of AgricultureYearbook, 1956, 186.
43. Bridges, C.H., Fungous diseases, in *Diseases Transmitted from Animals to Man*, 5th ed., Hull, T.G., Ed., Charles C Thomas, Springfield, IL, 1963, 453.
44. Faust, E.C., Infections produced by animal parasites, in *Diseases Transmitted from Animals to Man*, 5th ed., Hull, T.G., Ed., Charles C Thomas, Springfield, IL, 1963, 433.
45. Switzer-House, K.D., Livestock Manure, in Agricultural Management Practices for Improved Water Quality in the Canadian Great Lakes Basin, Land Resource Institute, contribution number 82010, Research Branch, Agriculture Canada, Ottawa, 1982, 69.
46. Turk, L.M. and Weidemann, A.G., Farm manure, Michigan State University Agr. Exp. Sta. Bull., 196, 1945, 1.
47. Kesler, R.P., Economic evaluation of liquid-manure disposal from confinement finishing of hogs, in *Management of Farm Animal Wastes*, Proc. Natl. Symp., Am. Soc. Agr. Eng. Publ. SP-0366, 1966, 122.
48. Robertson, L.S. and Wolford, J., The effect of application rate of chicken manure on the yield of corn, in Poultry Pollution: Problems and Solutions, Sheppard, C.C., Ed., Michigan State University Agr. Exp. Sta. Res. Rep. 177, 1970, 10.
49. Eno, C.F., Chicken manure — its production, value, preservation, and disposition, University of Florida Agr. Exp. Sta. Circ. S-140, 1966, 1.
50. Martin, W.E., Use of manure in agricultural production, Talk presented at 11th Annu. Dairy Cattle Day, Department of Animal Science, University of California, Davis, March 20, 1972.
51. Pratt, P.F., Broadbent, F.E., and Martin, J.P., Using organic wastes as nitrogen fertilizers, *Calif. Agric.* 27 (6), 10, 1973.
52. Goss, D.W. and Stewart, B.A., Efficiency of phosphorus utilization by alfalfa from manure and superphosphate, *Soil Sci. Soc. Am. J.*, 43, 523, 1979.
53. May, D.M. and Martin, W.E., Manures are good sources of phosphorus, *Calif. Agric.*, 20(7), 11, 1966.
54. McAuliffe, C., Peech, M., and Bradfield, R., The utilization by plants of phosphorus in farm manure. II. Availability to plants of organic and inorganic forms of phosphorus in sheep manure, *Soil Sci.*, 68, 185, 1949.
55. Goss, D.W. and Eck, H.V., P fertilizer for Alfalfa — Concentrated Superphosphate or Feedlot Manure?, Texas Agr. Exp. Sta. MP-1539, 1983, 1.
56. Bartholomew, R.P., The availability of potassium to plants as affected by barnyard manure, *Agron. J.*, 20, 55, 1968.
57. Smith, K.A. and van Dijk, T.A., Utilization of phosphorus and potassium from animal manures on grassland and forage crops, in *Animal Manure on Grassland and Fodder Crops*, van der Meer, H.G., Unwin, R.J., van Dijk, T.A., and Ennick, G.C., Eds., Martinus Nijhoff, Dordrecht, The Netherlands, 1987, 87.

58. Thomas, J.D. and Mathers, A.C., Manure and iron effects on sorghum growth on iron-deficient soil, *Agron. J.*, 71, 792, 1979.
59. Mathers, A.C., Thomas, J.D., Stewart, B.A., and Herring, J.E., Manure and inorganic fertilizer effects on sorghum and sunflower growth on iron-deficient soil, *Agron. J.*, 72, 1025, 1980.
60. Miller, B.F., Lindsay, W.L., and Parsa, A.A., Use of poultry manure for correction of Zn and Fe deficiencies in plants, in Proc. Animal Waste Manage. Conf., Jan. 1969, Cornell University, Ithaca, NY, 1969, 120.
61. Tan, K.H., Leonard, R.A., Bertrand, A.R., and Wilkinson, S., The metal complexing capacity and the nature of chelating ligands of water extract of poultry litter, *Soil Sci. Soc. Am. Proc.*, 35, 265, 1971.
62. Jenkinson, D.S., The Rothamsted long-term experiments: are they still of use?, *Agron. J.*, 83, 2, 1991.
63. Elson, J., A comparison of the effect of fertilizer and manure, organic matter, and carbon-nitrogen ratio on water-stable soil aggregates, *Soil Sci. Soc. Am. Proc.*, 6, 86, 1941.
64. Elson, J., A 4-yr study of the effects of crop, lime, manure, and fertilizer on macroaggregation of Dunnmore silt loam, *Soil Sci. Soc. Am. Proc.*, 8, 87, 1943.
65. Unger, P.W. and Stewart, B.A., Feedlot waste effects on soil conditions and water evaporation, *Soil Sci. Soc. Am. Proc.*, 38, 954, 1974.
66. Mathers, A.C., Stewart, B.A., and Thomas, J.D., Manure effects on water intake and runoff quality from irrigated grain sorghum plots, *Soil Sci. Soc. Am. J.*, 41, 782, 1977.
67. Mazurak, A.P., Cosper, H.R., and Rhoades, H.F., Rate of water entry into an irrigated chestnut soil as affected by 39-year of cropping and manurial practices, *Agron. J.*, 47, 490, 1955.
68. Swayder, F.N. and Stewart, B.A., The Effect of Feedlot Wastes on Water Relations of Pullman Clay Loam, ASAE Paper 72-959, ASAE, St. Joseph, MO, 1972.
69. Mathers, A.C. and Stewart, B.A., The effect of feedlot manure on soil physical and chemical properties, in Livestock Waste: A Renewable Resource. Proc. 4th Int. Symp. on Livestock Wastes, ASAE, St. Joseph, MO, 1980.
70. Herron, G.M. and Erhart, A.B., Value of manure on an irrigated calcareous soil, *Soil Sci. Soc. Am. Proc.*, 29, 278, 1965.
71. Mathers, A.C. and Stewart, B.A., Nitrogen transformations and plant growth as affected by applying large amounts of cattle feedlot wastes to soil, in Relationship of Agriculture to Soil and Water Pollution, Conf. on Agr. Waste Mgmt., Cornell University, Ithaca, NY, 1970, 207.
72. Mathers, A.C. and Stewart, B.A., Manure effects on crop yields and soil properties, *Trans. ASAE*, 27, 1022, 1984.
73. Wilkinson, S.R., Stuedemann, J.A., Williams, D.J., Jones, J.B., Jr., Dawson, R.N., and Jackson, W.A., Recycling broiler house litter on tall fescue pastures at disposal rates and evidence of beef cow health problems, in Livestock Waste Management and Pollution Abatement, Proc. Int. Symp. on Livestock Wastes, Am. Soc. Agr. Eng. Publ. PROC-271, 1971, 321.
74. Eck, H.V., Winter, S.R., and Smith, S.J., Sugarbeet yield and quality in relation to residual beef feedlot waste, *Agron. J.*, 82, 250, 1990.
75. Gardner, R. and Robertson, D.W., Comparison of the effects of manures and commercial fertilizers on the yield of sugar beets, in Proc. of 4th General Meeting, Am. Soc. Sugarbeet Technol., Denver, Feb. 12–14, 1946, Am. Soc. Sugarbeet Technol., Fort Collins, CO, 1947, 27.

76. Halverson, A.D. and Hartman, G.P., Long-term nitrogen rates and sources influence sugarbeet yields and quality, *Agron. J.*, 67, 389, 1975.
77. Wadman, W.P., Effect of organic manure on crop yield in long-term field experiments, *INTECOL Bull.*, 15, 13, 1987.
78. Mathers, A.C. and Stewart, B.A., Crop production and soil analyses as affected by applications of cattle feedlot waste, in Livestock Waste Management and Pollution Abatement, Proc. In. Symp. on Livestock Wastes, Columbus, OH, April 19–22, 1971, Am. Soc. Agr. Eng. Publ. PROC-271, 1971, 229.
79. Hileman, L.H., Effect of rate of poultry manure application on selected soil chemical properties, in Livestock Waste Management and Pollution Abatement, Proc. Int. Symposium on Livestock Wastes, Columbus, OH, April 19–22, 1971, Am. Soc. Agr. Engl. Publ. PROC-271, 1971, 247.
80. Manges, H.L., Schmid, L.A., and Murphy, L.S., Land disposal of cattle feedlot wastes, in Livestock Waste Management and Pollution Abatement, Proc. Int. Symp. on Livestock Wastes, Columbus, OH, April 19–22, 1971, Am. Soc. Agr. Eng. Publ. PROC-271, 1971, 62.
81. Travis, D.O., Powers, W.L., Murphy, L.S., and Lipper, R.I., Effect of feedlot lagoon water on some physical and chemical properties of soils, *Soil Sci. Soc. Am. Proc.*, 35, 122, 1971.
82. Stewart, B.A., Volatilization and nitrification of nitrogen from urine under simulated feedlot conditions, *Environ. Sci. Technol.* 4, 579, 1970.
83. Azevedo, J. and Rendig, V.V., Chemical composition and fertilizer response of two range plants in relation to grass tetany, *J. Range Manage.*, 25, 24, 1972.
84. Stuedemann, J.A., Fescue Toxicosis: Effects on Stocker and Cow-Calf Operations, Proc. Fescue Toxicosis Symposium, Plains Nutrition Council Meeting, April 15, Amarillo, TX, 1987, 1B.
85. Stuedemann, J.A., Wilkinson, S.R., Williams, D.J., Ciordia, H., Ernst, J.V., Jackson, W.A., and Jones, J.B., Jr., Long-term broiler litter fertilization of tall fescue pastures and health and performance of beef cows, in Managing Livestock Wastes, Proc., 3rd Int. Symp. on Livestock Wastes, ASAE PROC-275, ASAE, St. Joseph, MO, 1975, 264.
86. U.S. Environmental Protection Agency (USEPA), Quality Criteria for Water, EPA 440/5-86-001, Washington, D.C., 1986.
87. Sharpley, A.N., Smith, S.J., Stewart, B.A., and Mathers, A.C., Forms of P in soil receiving cattle feedlot waste, *J. Environ. Qual.*, 18, 313, 1984.
88. Christie, P., Long term effects of slurry on grassland, in *Animal Manure on Grassland and Fodder Crops*, van der Meer, H.G., Unwin, R.J., Van Dijk, T.A., and Ennik, G.C., Eds., Martinus Nijhoff, Dordrecht, The Netherlands, 1987, 301.
89. Van Reimsdijk, W.H., Lexmond, Th.M., Enfield, C.G., and van der Zee, S.E.A.T.M., Phosphorus and heavy metals: accumulation and consequences, in *Animal Manure on Grassland and Fodder Crops*, van der Meer, H.G., Unwin, R.J., Van Dijk, T.A., and Ennik, G.C., Eds., Martinus Nijhoff, Dordrecht, The Netherlands, 1987, 213.
90. Sharpley, A.N. and Smith, S.J., Prediction of soluble P transport in agricultural runoff, *J. Environ. Qual.*, 18, 313, 1989.

91. U.S. Environmental Protection Agency (USEPA), National Pollutant Discharge System General Permit and Reporting Requirements for Discharge from Concentrated Feeding Operation, Proposed Rules, Federal Register 57 No. 141, July 22, 1992, 32491.
92. Anderson, J.R., Biological interrelationships between feces and flies, in Management of Farm Animal Wastes, Proc. Natl. Symp., Am. Soc. Agr. Eng. Publ. SP-0366, 1966, 20.
93. Anderson, J.R., Recent developments in the control of some arthropods of public health and veterinary importance: muscoid flies, *Bull. Ent. Soc. Am.*, 12, 342, 1966.
94. Loomis, E.C., Anderson, J.R., and Deal, A.S., Identification of common flies associated with livestock and poultry, University of California Agric. Ext. Serv. Publ. AxT-236, 1967, 1.
95. White, R.K., Taiganides, E.P, and Cole, G.D., Chromatographic identification of malodors from dairy animal waste, in Livestock Waste Management and Pollution Abatement, Proc. Int. Symp. on Livestock Wastes, Am. Soc. Agr. Eng., Publ. PROC-271, 1971, 110.
96. Stephens, E.R., Identification of odors from cattle feedlots, *Calif. Agric.*, 25(1), 1, 1971.
97. Merkel, J.A., Hazen, T.E., and Mines, J.R., Identification of gases in a confinement swine building atmosphere, *Trans. Am. Soc. Agr. Eng.*, 12, 31, 1969.
98. Deibel, R.H., Biological aspects of the animal waste disposal problem, in *Agriculture and the Quality of our Environment*, Brady, N.C., Ed., American Association for the Advancement of Science, Washington, D.C., 1967, 395.
99. George, J.A., Fulhage, C.D., and Melvin, S.W., A summary of midwest livestock odor court actions, in Agricultural Waste Utilization and Management, Proc. 5th Int. Symp. on Agricultural Wastes, December 16–17, 1985, Chicago, IL, Am. Soc. of Agr. Eng., St. Joseph, MO, 1985, 431.
100. Anon. Dust, fly and odor control methods practiced by western feeders, Prep. by Texas Cattle Feeders Assoc., Texas Tech University Agr. Ext. Serv., No date, 1.
101. Keeney, D., Sources of nitrate to groundwater, *Crit. Rev. Environ. Control*, 16(3), 257, 1986.
102. Weisenberger, D.P., Potential health consequences of ground water contamination by nitrates in Nebraska, in *Nitrate Contamination*, NATO ASI Series Vol. G-30, Bogardis, I. and Kuzelka, R.D., Eds., Springer-Verlang, Berlin, 1991, 327.
103. Gilbertson, C.B., Nordstadt, F.A., Mathers, A.C., Holt, R.F., Barnett, A.P., McCalla, T.M., Onstad, C.A., and Young, R.A., Animal Waste Utilization on Cropland and Pastureland: A Manual for Evaluating Agronomic and Environmental Effects, USDA Report No. URR-6, U.S. Department of Agriculture, Science and Education Administration, Hyattsville, MD, 1979, 1.

CHAPTER 6

Sewage Sludge

N. V. Hue

0-87371-859-3/95/$0.00+$.50

I. INTRODUCTION

In the U.S., the Federal Water Pollution Control Act (better known as the Clean Water Act), as amended in October 1972, required both municipalities and industries to implement rigorous wastewater treatments to abate water pollution. As a result, the quantity of wastewater residues, i.e. sewage sludge, has increased manyfold. In 1983, the U.S. generated approximately 6.2×10^6 dry metric tons (Mg) of municipal sewage sludge per year, corresponding roughly to 26 kg per person annually.[1] Sludge production increased to 8.5×10^6 Mg/year in 1990, and is expected to reach approximately 12×10^6 Mg/year by the year 2000 as the U.S. population grows and as more advanced waste treatment processes are instituted.[2]

A similar problem also faces the European community. A 1984 report indicated that West Germany was producing 2.2×10^6 Mg/year of dry sewage sludge; the U.K., 1.5×10^6 Mg/year; and Italy, 1.2×10^6 Mg/year.[3]

Domestic and industrial discharges are probably the two major sources of sewage sludge production. Runoff can also be a significant contributing source, depending on whether the sewage system is separate or combined with storm drainage. To illustrate the magnitude of domestic influents, Hue and Ranjith[4] reported that over 0.4×10^6 m^3 of wastewater was treated daily from a population of 0.8×10^6 people of Oahu (Honolulu metropolitan area), Hawaii. Nearly 33,500 Mg sludge per year was produced as a result. In the U.S., industrial influents come mainly from iron and steel, chemicals, food processing, and pulp and paper manufacturing industries, and often produce sludges containing higher metal concentrations than the domestic counterpart.[5] However, recent efforts to improve source control and industrial pretreatment have significantly decreased the metal content of industrially originated sludges.[6] In this chapter, no distinction will be attempted between domestic and industrial sewage sludges unless deemed necessary, and the term "municipal" will be used to represent both sources.

A. WASTEWATER AND SLUDGE TREATMENT

Once pumped into a wastewater treatment plant, the municipal wastewater is first held in a large clarifying tank and undergoes primary treatment that removes readily settleable solids. The primary sludge produced contains 3 to 7% solids, and can be easily thickened or dewatered. Primary treatment typically produces 2.5×10^3 to 3×10^3 L of sludges per 10^6 L of wastewater. The clarified wastewater further undergoes secondary treatment, which often involves such biological processes as activated sludge system (seeding sludge into the wastewater stream) or trickling filter system having bacterial growth attached. Secondary treatment removes fine suspended solids and some dissolved solids, and produces secondary sludge.[7] Biological secondary treatment produces approximately 1.5 to 2×10^4 L of secondary sludge for each 10^6 L of sludge treated. Secondary sludge generally has 0.5 to 2.0% solids, and is more difficult to thicken and dewater than the primary sludge.[1] Tertiary treatment is applied to remove certain dissolved solids such as phosphorus (P). The treatment often involves the additions of chemicals such as lime, Al and Fe salts (alum, ferric chloride, ferrous sulfate), or organic polymers to wastewater.[8] Tertiary treatment produces approximately 1×10^4 L of tertiary sludge per 10^6 L of wastewater treated.

Raw sewage sludges, both primary and secondary, are odorous and pathogen rich. To reduce odor potential and pathogen levels, sludges must be treated and stabilized. Low pathogen levels and low odor potential are required for land application of sludges.[1]

Major stabilization processes include anaerobic digestion, aerobic digestion, lime addition, and composting. Anaerobic digestion is the most common process, by which sludge is retained in the absence of air for at least 60 days at 20°C or 15 days at 35°C to 55°C.[1] This process can biodegrade 40 to 50% of the volatile solids in a sludge by converting organic carbon to methane, which can be used as an energy source. The process also transforms part of the organic nitrogen (N) into ammoniacal N. The proportion of ammoniacal N to total N may increase from about 5% up to as high as 70% after anaerobic digestion.[9] Aerobic digestion is conducted by agitating sludge with air or oxygen to maintain aerobic conditions for 60 days at 15°C, 40 days at 20°C, or 10 days at 55 to 60°C. Volatile solids are also reduced significantly by this process, although not as much as in anaerobic digestion.[10] Because much of the volatile organic matter has been given off as CH_4 or CO_2, stabilized sludges tend to have less objectionable odor and higher N content than raw sludges. Lime addition is a stabilization process where sufficient lime is added to sludge to obtain pH 12 after 2 hours of contact. Other stabilization methods include heat treatment, irradiation, and pasteurization. In heat treatment, sludge is passed through a heated reactor/heat exchanger at 150 to 250°C with a minimum residence time of 30 min.[11] Pasteurization involves maintaining sludge at a minimum temperature of 70°C for at least 30 min. Treatment processes using lime or heat would effectively kill pathogens, but may not adequately stabilize the biodegradable organic matter. On the other hand, mesophilic anaerobic and aerobic digestions or sand bed drying stabilize the sludge, but may not adequately reduce pathogens. To ensure a complete destruction of pathogenic bacteria, viruses, protozoa, as well as parasites (e.g., helminth eggs), thermophilic digestions, composting, or a combination of the above processes are often required.

To save transportation and storage costs, sludge volume must be reduced by partially eliminating its water. Treatment processes such as thickening, conditioning, dewatering, composting, and drying can lower sludge water content and raise the solids percentage.

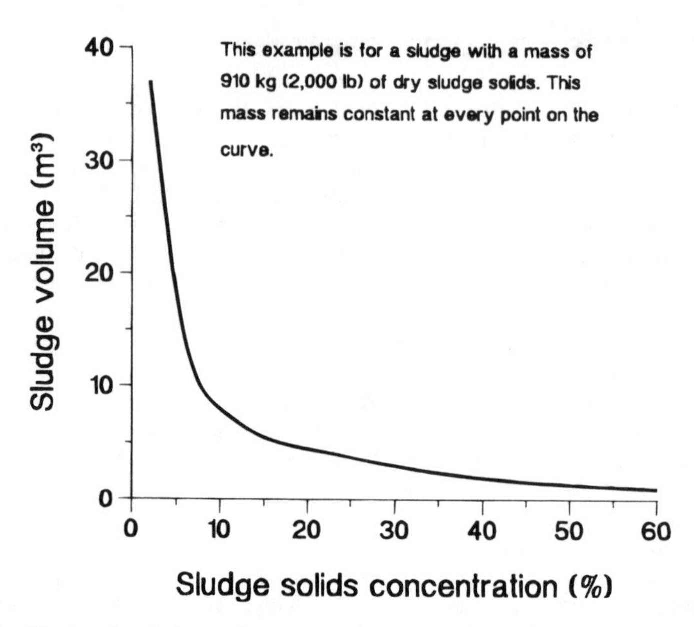

Figure 1 Change in sludge volume with sludge solids concentration. (Adapted from Reference 1.)

Raising the solids content from 5 to 10% halves the volume of the sludge (Figure 1). Thickening involves low-force separation of water and solids through gravity or flotation. Conditioning alters sludge properties to facilitate the separation of water from sludge. Lime, $FeCl_3$, or organic polymers are common sludge conditioners.[12] Dewatering can be accomplished through centrifugation, vacuum filter, filter press, sand bed drying, or heat drying. Dewatering processes produce a sludge cake with 15 to 25% solids for centrifugation or vacuum filter, 35 to 50% solids for filter press, 40 to 70% solids for composting, 50 to 80% solids for air drying bed, and 94 to 99% solids for heat drying.[12] Figure 2 summarizes the generation, treatment, and use/disposal of municipal sewage sludge.

B. SLUDGE CHARACTERISTICS

Municipal sewage contains various amounts of industrial wastes and street runoff, in addition to human excreta and residues from household activities such as cooking and laundering. Thus, important determinants of sewage sludge composition are wastewater treatment methods, industry profiles, and seasonal factors.[13,14] Nutrients (e.g., N, P), heavy metals (e.g., Cd, Pb), toxic organic chemicals [e.g., polychlorinated biphenyls (PCBs), polynuclear aromatic hydrocarbons (PAHs)], and pathogens are the main groups of sludge constituents controlling or limiting the use/disposal options of sludges.

1. Nutrients

Concentrations of plant macronutrients in sewage sludges vary widely (Table 1). For example, a survey of more than 190 sewage sludge samples from 8 states in the U.S.

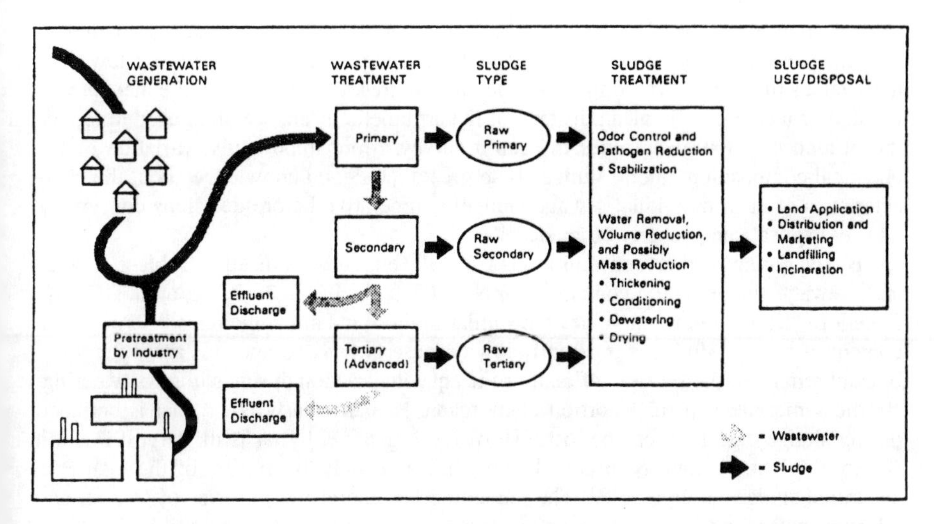

Figure 2 Generation, treatment, and disposal of municipal sewage sludge. (Adapted from Reference 1.)

Table 1 Total Concentration of Selected Plant Nutrients in U.S. Sewage Sludge

Variable	Total Nutrient, % Dry Weight N	P	K	Ca	Mg
	Sommers' Survey in 1976 (≅190 samples)[a]				
Range	0.1–17.6	0.1–14.3	0.02–2.64	0.1–25.0	0.03–1.97
Mean	3.9	2.5	0.4	4.9	0.54
Median	3.3	2.3	0.3	3.9	0.45
	Michigan's Survey in 1980 (>200 samples)[b]				
Range	0.2–21.0	0.1–15.0	0.1–6.5	0.5–17.0	0.1–2.5
Mean	3.5	2.2	0.5	4.0	0.7
Median	3.0	1.9	0.42	3.3	0.6
	New York's Survey in 1988 (15 samples)[c]				
Range	1.19–4.93	0.22–3.13	0.03–0.46	0.32–15.9	0.04–0.81
Mean	2.9	1.2	0.19	3.92	0.35
Median	2.78	0.87	0.15	2.17	0.34
	Hawaii's Survey in 1991–1992 (22 samples)[d]				
Range	1.32–6.25	0.26–0.85	0.01–0.13	0.5–6.81	0.08–0.80
Mean	3.8	0.6	0.06	1.78	0.29
Median	3.74	0.66	0.07	1.41	0.30
Average Median (4 surveys)	3.2	1.4	0.23	2.7	0.4

[a]Adapted with permission from Reference 10.
[b]Adapted with permission from Reference 14.
[c]Adapted with permission from Reference 15.
[d]From Hue (unpublished data).

showed total N ranging from 0.1 to 17.6% dry weight.[10] The statistical distribution of nutrient concentrations in sludges often follows a log-normal pattern with a few very high values that can be traced to a specific point source. For this reason, the median values and/or the geometric mean, instead of the arithmetic mean, are often used to reduce the influence of extreme values in such a survey. More importantly, variation of the chemical composition among sludges is so great that a good knowledge about the characteristics of each individual sewage sludge is imperative before decisions concerning its use/disposal can properly be made.[10,14]

By averaging median concentrations from all four surveys listed in Table 1, a "typical" sewage sludge would contain 3.2% N, 1.4% P, 0.23% K, 2.7% Ca, and 0.4% Mg. Except for K, these nutrient values are similar to those in animal manures.[16,17] Potassium content of sewage sludge is inherently low because most K compounds are water soluble and remain in the sewage effluent or the aqueous fraction during sludge dewatering. By the same mechanism, a portion of inorganic N, particularly NH_4^+, that is enriched during sludge digestion, can be lost.[10] However, organic N by far is the largest fraction (50 to 90%) of the total N in any sludge. Unlike N, only 10 to 30% of the total P in anaerobic sludges is organic P.[13] The remaining P is inorganic, in forms of Ca, Fe, and Al phosphates, and P sorbed on amorphous materials of Fe, Al, and Mn hydrous oxides.[10]

2. Heavy Metals

Although trace element is currently a preferred term, metals, heavy metals, and trace elements are used interchangeably in this chapter; strictly speaking As, Mo, and Se are not really metals. Some of these elements (e.g., Cu and Zn) are essential to plants and animals in small amounts, but become toxic in excess. In fact, concentrations of heavy metals in sludge are among the deciding factors for sludge utilization on lands because of their potential to damage crops and/or to enter the human food chain. A recent national survey of 209 wastewater treatment plants by the EPA shows that concentrations of sludge metals span many orders of magnitude (Table 2). For example, Cd levels range

Table 2 Total Concentration of Selected Heavy Metals in U.S. Sewage Sludge (mg/kg dry weight)

Element	National Sewage Sludge Survey (1990)[a]		Maximum in Proposed "Clean Sludge"[b]
	Range	Median	
As	0.3–315.6	6	100
Cd	0.7–8220	7	18
Cr	2.0–3750	40	2000
Cu	6.8–3120	463	1200
Hg	0.2–47.0	4	15
Mo	2.0–67.9	11	35
Ni	2.0–976	29	500
Pb	9.4–1670	106	300
Se	0.5–70.0	5	32
Zn	37.8–68,000	725	2700

[a] Adapted with permission from Reference 22.
[b] Adapted with permission from Reference 19. "Clean" means no observed adverse effects level (NOAEL) in Chaney's article.

from 0.7 to 8220 mg/kg. The high values usually come from industrial sources. Despite these large variations, median concentrations of the metals are relatively low and have been decreasing in time, suggesting an improvement in U.S. pollution control programs. As an illustration, median concentrations of Cd, Cu, Pb, and Zn were 7, 463, 106, and 725 mg/kg, respectively, in the 1990s survey (Table 2) as compared to 10, 800, 500, and 1700 mg/kg obtained in the early 1980s.[14] Based on the results of sewage sludge research during the past 2 decades,[18,19] a "clean sludge" category has been proposed (Table 2). "Clean" sludge would have no limit on its application rate to land.[20]

3. Toxic Organic Chemicals

Industrial wastes, household chemicals, and pesticides are main contributors to organic chemicals in sludges. These chemicals are of concern because of their known and unknown hazards to the public health and the environment. However, caution must be exercised when evaluating these organics because not only do they differ greatly in toxicity potential but also in frequency of occurrence and in concentrations found in sludges. As illustrated in Table 3, Aroclor 1254 (a PCB congener) was below the detection limit in all 431 sludges tested in one survey, but was positive in 40 out of 107 sludges tested in another survey.[21] Its concentration, when detected, ranged from 0.06 to 1960 mg/kg dry weight with a median of 5.35 mg/kg. Fortunately, the concentration of these toxic organics is sufficiently low in sludges from modern treatment plants that they would not pose a significant human health or environmental threat.[1,21] Nevertheless, given the inconsistency in terms of occurrence and concentration of these sludge-borne organics, individual sludge must be reasonably characterized before use or disposal.

Table 3 Frequency of Occurrence and Concentrations of Selected Toxic Organic Chemicals in Sewage Sludges

Chemical	No. of Sludges Tested	% Occurrence	Concentrations for Samples Testing Positive, i.e., >Detection Limits (mg/kg dry wt)	
			Range	Median
		Monocyclic Aromatics		
Chlorobenzene	158	6	2.06–846	10.2
p-Chlorotoluene	158	11	1.13–324	14.7
Hexacholrobenzene	237	43	0.000188–26.2	0.018
		Polynuclear Aromatics (PAH)		
Naphthaline	236	50	0.0554–6,610	30.3
Benzo (a) pyrene	12	100	0.12–9.14	0.88
		Polychlorinated Biphenyls (PCB)		
Aroclor[a] 1248	431	0	All samples < detection limit	
Aroclor 1254	107	39	0.0667–1,960	5.35
	431	0	All samples < detection limit	
	31	100	0.15–3.6	0.84
Aroclor 1260	111	58	0.0468–433	4.18
	431	0	All samples < detection limit	
	40	100	0.2–0.46	0.15

[a] Trade name of PCBs from Monsanto Chemical Co.

Adapted from Jacobs et al., *Land Application of Sludge*, Lewis Publishers, Chelsea, MI, 1987,101. With permission.

4. Pathogens

Bacteria, viruses, protozoa, and parasitic worms — some of them are pathogenic — are inevitably present in sewage and raw sludges. However, the number and types of organisms present vary, depending on such factors as population density, sanitary habits, and sludge treatments. Some sludge treatments, including anaerobic digestion, mesophilic aerobic digestion, and air drying, significantly reduce but do not completely eliminate pathogens. For this reason, they are called the processes to significantly reduce pathogens (PSRP) in regulatory terms. To virtually destroy these disease-causing organisms, thermophilic treatments of sewage sludges are often required. Thus, the latter treatments are called the processes to further reduce pathogens (PFRP).

The most common bacterial pathogens in sewage sludge are *Salmonella, Shigella*, and *Campylobacter*.[23] *Salmonella* can cause salmonellosis; *Shigella*, dysentery; and *Campylobacter*, gastroenteritis. Although *Escherichia coli* belongs to the *Shigella* spp., it is not considered pathogenic. It is often used to indicate the adequacy (or inadequacy) of a treatment process in reducing pathogens because *E. coli* is abundant in sludge.[1]

More than 110 different viruses may be present in raw sewage and sludge and the number is increasing.[23] Enteroviruses, which include Poliovirus, Echovirus, Coxsackievirus, and Hepatitis virus, can cause diseases from meningitis to infectious hepatitis. Reovirus and Adenovirus may cause respiratory infection. Viruses tend to sorb strongly onto sludge, often causing their number to be undercounted.[24] Over 90% of total waste-borne viruses are either inactivated or adsorbed onto sludge during a treatment process.[25] Sorbed viruses remain active and may survive longer than virus free in wastewater. However, infection is only possible when the virus is separated from sludge particles.[26]

Of the common protozoa that may be found in wastewater and sewage sludge, only three species are of major significance for disease transmission to humans: *Entamoeba histolytica, Giarda lambia*, and *Balantidium coli*.[23] All three can cause mild to severe diarrhea.

Eggs of Helminth parasites (intestinal worms), including *Ascaris lumbricoides* (round worm), *Ancyclostoma duodenale* (hookworm), *Trichuris trichiura* (whipworm), and *Taenia saginata* (tapeworm), tend to settle out with sludge solids during primary wastewater treatment.[24] These organisms are of particular concern because they can survive many forms of sludge treatment, and they can infect humans and animals even at small numbers.

Based on available research data, Sorber and Moore[27] concluded that (1) the number of *Salmonella* were reduced by 90% within 3 weeks in sludge-amended soils, (2) a median 90% of viruses were deactivated in 3 days when sludge was surface-applied in warm climates (vs. approximately 30 days in cold temperatures), and (3) maximal parasite survival, indicated by *Ascaris*, was relatively long with a median 90% reduction of 77 days.

C. CURRENT USAGE OR DISPOSAL

The large quantity (nearly 9×10^6 Mg/year of dry sewage sludge) that is currently produced in the U.S. must be moved out of wastewater treatment plants for disposal or preferably for beneficial use. Although all options have potential problems, some may be more acceptable than others for specific sludges under certain situations. Common modes of usage/disposal of sewage sludges include land application, marketing and distribution (usually in the form of compost), land filling, and incineration. Ocean dumping has been illegal in the U.S. since 1991.

1. Land Application

The practice of spreading sludge on or just below the land surface is probably the most widely used form. Currently over 25% of U.S. sludge is land applied.[2] In some states, such as Colorado, Florida, Oregon, and Washington, this percentage has passed 70%.[28] Sludge can be applied to agricultural land, forest land, or disturbed land (land reclamation) to improve soil physical (e.g., water retention and infiltration, aggregate stability) and chemical (e.g., plant nutrients, cation exchange capacity) characteristics. Land application can also serve as a sludge treatment. Sunlight, soil microorganisms, and desiccation help to destroy pathogens and many toxic organics in sludge. Reactions among soil constituents and sludge heavy metals and nutrients help immobilize the metals/nutrients, thereby reducing the potential for water pollution. However, a successful land application program must consider several factors, including (1) site characteristics, such as depth to groundwater, distance to surface water, slope of the site, soil permeability, mineralogy, pH, and public access;[1,29] (2) sludge application rates which are determined mainly by concentrations of nutrients, heavy metals, or toxic organics in the sludge;[30] and (3) method of application by which liquid or dewatered sludges are applied. Increasing solids content of the sludge requires additional equipment and facilities, but may be beneficial in the long term because it reduces sludge volume and mass, thereby lowering costs of handling such as storage and, particularly, transportation. Dewatered sludge is often surface-applied, then plowed or disked in when possible. Liquid sludge may be injected into the soil, sprayed, or spread over the soil surface. Subsurface application — injection of liquid sludge or incorporation of dewatered sludge — requires more equipment and energy but reduces odor and the potential for contact between the sludge and crops, animals, or humans.

Agricultural Application — Over 20 years of research have been devoted to the use of sewage sludge in agriculture.[30–37] Usually, application rates are limited by either N needs of the crop or by the annual or cumulative metals addition to the soil. As an illustration, assume that 180 kg N/ha must be applied to a corn (*Zea mays* L.) crop for a given target yield; the quantity of a hypothetical "clean" sludge containing 3.2% total N (1% inorganic N + 2.2% organic N with 20% mineralization the first year) that would be needed is $180/(0.01 + 0.2 \times 0.022) = 12{,}500$ kg/ha or 12.5 Mg/ha. As another example, the following scenario is used to show the calculation of sludge application rate based on heavy metal limitations. Assume that the maximum cumulative amount of Cd that can be added to a soil is 18 kg Cd/ha. This value was proposed by the EPA in the first draft of new regulations on sludge use/disposal (USEPA proposed Rule 40 CFR Part-503), and was judged reasonable by the peer review W-170 group;[20] further, assume that the application life of the site is 20 years. The annual Cd limit is $18/20 = 0.9$ kg Cd/year. If the sludge used contains 18 mg Cd/kg or 0.018 kg Cd/Mg sludge (the maximum Cd level for a proposed "clean" sludge, Table 2), then the maximum annual application rate is $0.9/0.018 = 50$ Mg sludge/year.

Generally, the annual application rates of sludge to agricultural land range from 2 to 70 Mg/ha, with 15 Mg/ha/year being typical.[1] Sometimes, application rates are based on P requirements of the crop. In this case, the rates are often lower than metal or N-based rates because (1) most crops require four to ten times less P than N,[38] (2) a major portion of sludge P is in inorganic (mostly bioavailable) forms,[13] while total P is about half that of total N in a sludge (Table 1). Sludge, however, is an imbalanced fertilizer because its K content is quite low (about 0.2%, Table 1). Thus, K must be supplemented either through an inorganic source such as KCl or an organic source such as wood ash or K-rich crop residues such as pineapple (*Ananus comosus*) leaves.[4]

Federal regulations (40 CFR-257) also require that, prior to land application, sludges must be treated by a PSRP. Public access to the sludge-applied land must be controlled for at least 12 months, grazing by animals whose products are consumed by humans must be prevented for at least 1 month, and growing edible crops must wait for at least 18 months. Otherwise, PFRP must be applied to the sludge.[1]

Forest Application — Sludge applications to forest lands have been in practice for more than 2 decades in the U.S., mainly by states on the West Coast and the Southeast.[39] Forest application of sludges provides many advantages: (1) its potential health risk is much lower than in agricultural application because forest products — from wild berries and mushrooms to game animals — contribute insignificantly to the human diet and (2) forest soils and vegetations are often more amenable to sludge application. Because of perennial stands with extensive roots, year-round application is possible if weather permits.[40] Regarding application rates, research at the University of Washington[40] has shown that forest growth did not slow down even 8 years after a one-time application of 94 dry Mg/ha (5-cm thick of a 18% solids sludge). Higher application rates up to 470 Mg/ha were also studied, but nitrate (NO_3^-) contamination of groundwater became apparent in addition to high seedling mortality caused by deer and small mammals (e.g., voles) damage. Thus, the researchers recommended that a sludge rate of 47 Mg/ha be applied every 5 years to maintain good tree growth, yet minimize NO_3^- leaching. For forest applications, special vehicles and equipment are required because of rough terrain and tree height. Four-wheel driven tankers equipped with a spraying cannon capable of shooting sludge at 18% solids 45 m into or over tree stands have been designed and operated by the municipality of Metropolitan Seattle (Metro) and the University of Washington to spray sludge.[40] An improved version having a spray nozzle that can be raised 10.4 m into the air has been perfected; with this device, sludge fertilization over 15-m tall trees is possible.[39]

Land Reclamation — Notable disturbed lands are surface mine spoils, mine tailings, quarries, clear forests, and completed landfills. They all have a common feature: the surface layer is extremely poor in plant nutrients and organic matter, and unable to sustain plant growth if left unamended.[41] Thus, using sewage sludges, which are rich in organic matter and reasonably high in plant nutrients, to restore these disturbed lands has met with remarkable success.[1,32,42] For example, Sopper[42] reported that a strip mine site in Pennsylvania, which was extremely acidic (pH 3.8), highly compacted, and devoid of vegetation, had a complete cover with orchardgrass (*Dactylis glomerata* L.) and Kentucky-31 tall fescue (*Festuca arundinacea* schreb.) 3 months after an application of 184 Mg/ha of a stabilized, dewatered sludge. (Before the sludge application, 8.4 Mg/ha of an agricultural lime was incorporated into the site to raise the spoil pH to >6 as required by the Pennsylvania's law). It was worth noting that lime, commercial fertilizers, and seeds had failed to revegetate this same site on several attempts a decade earlier. Sopper[42] further noted that (1) trace metal concentrations in the surface spoil (0 to 15 cm) amended with 184 Mg/ha sludge were increased slightly and (2) NO_3-N concentration in soil percolate at the 90-cm depth increased from 1.8 to 7.3 mg/L during the first year after sludge application, but returned to background levels (<1 mg/L) in subsequent years. Thus, it seems inevitable that some contamination of ground and surface waters might occur shortly after sludge amendment, especially when a typical one-time, relatively high application rate (100 to 400 Mg/ha) is used. Nevertheless, such adverse effects seem minimal when compared with environmental problems without reclamation.

2. Composting and Marketing

Sludge composting is a process in which dewatered sludge is commonly mixed with a bulking agent, such as wood chips, bark, shredded tires, cereal hulls, straw, or previously composted sludge and allowed to decompose aerobically for several weeks.[1] Although less common, liquid untreated sludges have also been used in composting.[43] Recently, the co-composting process of sewage sludge and yard waste (e.g., leaves, tree trimming, grass) has attracted strong interest in the U.S.[44]

The objective of composting is to aerobically decompose sludge into a humus-like product that can be marketed as a soil conditioner and/or a slow-release, low grade fertilizer. Some commercial products, such as Milorganite, which is derived from sewage sludge from Milwaukee, and Nitrohumus which uses sewage sludges from Los Angeles as its raw material, have been on the market throughout the U.S. for several years.[45]

Composting can be done by one of the following methods. (1) Windrow composting: long open-air piles of sludge-bulking agent mixture are turned periodically to provide adequate O_2 for the decomposition. If operated properly, the temperature in the pile can rise to 55 to 60°C or even higher.[46] The high temperatures destroy virtually all pathogens and parasites. (2) Aerated piles also called static piles: air is forcedly blown into the piles through perforated pipes embedded under the piles with no mechanical turning or agitation. (3) In-vessel composting: compostable mixture is enclosed in drums or containers where controlled aeration and mechanical agitation are provided. The latter process is most suitable for situations where land is limited or climates are cold, but it requires more sophisticated equipment and involves higher capital cost.

In the U.S. about 6% of the nation's sewage sludge is being composted;[2] the number is even smaller in the European community.[47] The main reason is high costs of composting relative to other alternatives for sludge use/disposal.

3. Landfilling and Incineration

Unlike composting and land application, landfilling and incineration — even when energy is recovered — are considered sludge "disposal" methods. Currently, about 25% of the U.S. sludges are landfilled, and 14% incinerated.[2] However, with more stringent regulations on air pollution control and scarcer landfill space, these two options may not be viable long-term solutions for most communities. A 1990 survey on sludge management in 21 states in the U.S. showed that New Jersey has completely banned landfilling of sludge except in emergency, North Carolina no longer allowed disposing of sludge on active landfills, and Missouri, Ohio, Oregon, and Washington had <10% sludge landfilled.[28]

Landfilling is a method in which sludge is deposited in a dedicated area, alone (monofill) or with other solid wastes (codisposal), and finally covered up with a soil layer. In monofill, sludge is often buried in trenches, which may be narrow (1 to 3 m wide) or wide (3 to 15 m). Wide trenches require sludges containing >30% solids so that haul vehicles can work within the trench.[1] In codisposal, sludge and municipal solid waste are deposited together in a landfill. The high moisture-absorbing capacity of the solid waste and the soil-conditioning characteristic of the sludge complement each other. The ratio of refuse to sludge is about 4 to 1 by weight for sludge with >20% solids.[48] In sludge land filling, two important parameters must be considered: leachate and biogas.[48,49] Leachate, which is generated from the excess moisture in the sludge and from rainwater, may contain significant amounts of heavy metals and toxic organics that could contaminate groundwater or even surface waters downslope. To minimize this potential problem, clay-based liners, synthetic liners, or both should be installed in a sludge landfill.[1] Biogas (mostly CH_4, some CO_2, and small amounts of H_2S) are generated during slow anaerobic decompostion of organic matter in sludge and solid waste. Gas venting or a collection system should be installed nearby the landfill to prevent potential gas accumulation and explosion.

Incineration is the burning of sludge solids in the presence of O_2. This process reduces the sludge to about 10 to 20% of its original size, destroys all pathogens, and degrades many toxic organics. Thus, incineration is a poplular option for many communities, especially where land is scarce or if the sludge contains high levels of pathogens or organics that are unsuitable for other use/disposal. The two common types of incinerators for burning sludge are multiple-hearth and fluidized-bed furnaces.[50] Multiple-hearth incinerators have been in use for many years, and are the most common (about 76%).[2] They are durable, simpler to operate, and more tolerant of variations in sludge quality and loading rates.[1] The incineration takes place on the middle hearths, where temperatures can reach 760 to 927°C (1400 to 1700°F).[50] However, multiple-hearth furnaces are not suitable for frequent stop-and-start situations, and many older units require costly upgrades to meet air quality requirements. Newer fluidized-bed furnaces, with more efficient combustion designs (even though the sludge incinerating temperatures only range between 760 and 816°C), can achieve better control on organic emissions. Currently, they constitute approximately 18% of sludge incinerators in use.[2] Regardless of incinerator designs, metals are not destroyed. Instead, they are concentrated in the ash and in particulate matter entrained in the exhaust gases. Thus, appropriate means of ash disposal must be instituted.

II. REACTIONS IN SOIL

Once applied to land, sludge undergoes numerous reactions; most are biologically mediated, but some are purely chemical. The reactions of sludge N, heavy metals, and toxic organics with soil constituents are of special interest to agronomists, environmental scientists, and epidemiologists.

A. TRANSFORMATION AND MOVEMENT OF SLUDGE N IN SOIL

Because sludge N can be a valuable nutrient for plants and a potential groundwater contaminant, its fate in soils has received considerable interest and research.[51-56] Figure 3 illustrates major transformation processes of sludge N after soil application.

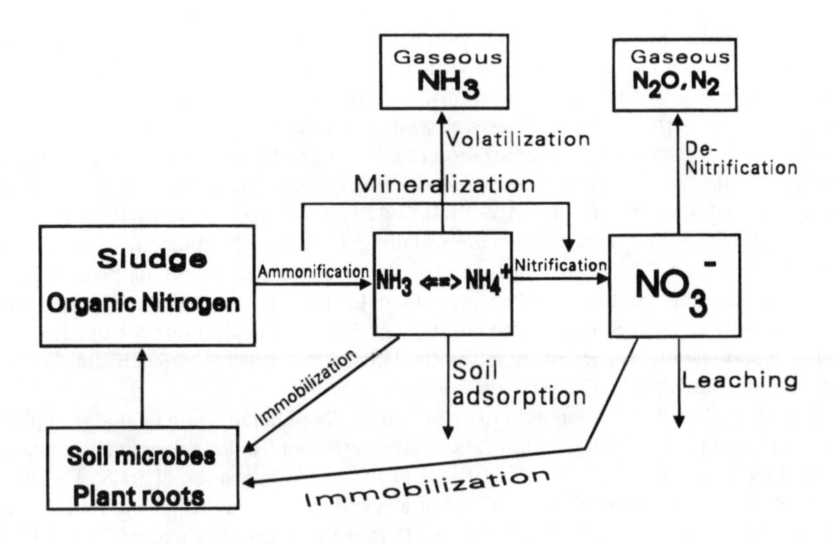

Figure 3 Major transformation processes of sludge N in soil.

1. Ammonification and Mineralization

Ammonification is a biological process by which organic N is transformed to NH_3 and NH_4^+ (aqueous NH_3), while mineralization is the conversion of organic N to inorganic N ($NH_4^+ + NO_3^-$). The formation of NH_3, and NH_4^+ in the presence of H_2O, results from the failure of microbes to use all of the N released by their metabolism of nitrogenous compounds (mostly proteins) in sludge for their cellular synthesis. Generally, microbes break down proteins into amino acids with the aid of appropriate enzymes. These amino acids, in turn, are incorporated into microbial cells, using organic carbon as the energy/substrate source. However, if the readily oxidizable carbon source is low relative to the amino acids, then the excess amino acids will be deaminated to yield NH_3 as illustrated below. (For a detailed discussion of the biochemistry of ammonification, see Reference 57.)

$$CH_3CHNH_2COOH + \tfrac{1}{2}O_2 \xrightarrow[\text{oxidase}]{\text{amino acid}} CH_3COCOOH + NH_3$$

Thus, the C/N ratio of sludge and sludge-amended soil is important to the ammonification: low C/N ratios (<15) strongly favor it.[56,58]

Since both NH_4^+ and NO_3^- are readily available for plant uptake, N mineralization has been traditionally studied in evaluating the agronomic value of sludge N.[52] It should be noted that sludge decomposition in soils (as measured by CO_2 production) does not necessarily coincide with N mineralization in the short term (the first few months for freshly added sludge). For example, Clark[56] reported that after 31 days of laboratory incubation, an anaerobically digested, liquid sludge mixed with a soil (pH 7.6) from Kansas City yielded 21.3% decomposition and 4.8% N mineralization, as compared to 24.8% decompostion and 13.2% N mineralization after 64 days. This distinction may be important for sludge management in providing N to annual crops. In the long term (≥1 year), however, the two processes are virtually parallel. Mineralization of soil-applied sludge N, like many other biochemical processes in soils, is dependent on the chemical and physical properties of the soil, as well as those of the sludge in addition to climatic factors.[55] This explains why mineralization rates vary greatly among raw, digested, and composted sludges.[58–60] Incubation studies with digested sludges have reported mineralization rates of 4 to 48% in 16 weeks,[61] 40 to 42% in 15 weeks,[59] and 1 to 58% in 26 weeks.[53] In Wisconsin where winters are usually cold, sludge N mineralization rates of 15 to 20, 6, 4, and 2% for the 1st, 2nd, 3rd, and 4th years after soil application have been suggested.[62] By contrast, workers in southern California proposed much higher rates of 35, 10, 6, and 5%.[63] Similarly, the constant 20% mineralization rate for the 1st year of sludge application, as recommended by the USEPA,[64] consistently underestimates actual mineralization, particularly for the southern U.S.[55,65]

First-order kinetics is often used to model/predict N mineralization of sludge applied to soils, implying that the N mineralization rate is affected by sludge application rate as reported by Terry et al.[66] The rate constant k of the general equation $N_m = N_o [1 - e^{-kt}]$, where N_m is the amount of N mineralized at a specific time t and N_o is the mineralization potential at time zero, has been reported to range between 0.02 and 0.07/week.[52,53,55] In contrast, the independence of N mineralization rate from sludge application rate has also been reported.[59,67]

2. NH_3 Volatilization and NH_4 Adsorption

Because NH_3 is a gas under normal temperatures and pressures, its potential loss to the atmosphere is substantial, especially if sludge is surface applied.[68] Environmental conditions that are conducive to NH_3 volatilization include (1) high soil pH: under alkaline conditions (pH >8.0), the reaction $NH_3 + H_2O \leftrightharpoons NH_4^+ + OH^-$ favors NH_3 over NH_4^+; (2) warm temperatures; (3) high wind; and (4) sandy soils with low cation exchange capacity (CEC).[69] King[70] reported gaseous N losses from 16 to 22% for incorporated sludge and 21 to 36% for surface-applied sludge. Ammonium ions, on the other hand, are predominant under acid and neutral conditions. As a cation, NH_4^+ can be attracted and retained by the negatively charged sites on clay surfaces, making it less susceptible to leaching, yet still available for plant uptake. Thus, soils with high clay or organic matter content and high CEC favor NH_4^+ adsorption. Furthermore, some clay minerals such as vermiculites and micas, because of their strongly negative charge and structural arrangement, can fix NH_4^+ irreversibly, making it unavailable to plants in the short term.

3. Nitrification and Denitrification

Nitrification is the transformation of NH_4^+ to NO_3^- by a two-step biochemical process.[71] In the first step, NH_4^+ is oxidized to NO_2^- usually by the bacterial genus *Nitrosomonas*: $NH_4^+ + 1\frac{1}{2} O_2 \rightarrow NO_2^- + H_2O + 2H^+$. In the second step, NO_2^- is oxidized to NO_3^- usually by *Nitrobacter*: $NO_2^- + \frac{1}{2} O_2 \rightarrow NO_3^-$. Two protons are released by the NH_4^+ oxidation, implying that (1) nitrification acidifies the soil and (2) *Nitrosomonas* is more sensitive to soil pH than *Nitrobacter*.[72,73] Thus, by raising pH, at least initially, through sludge addition, Terry et al.[66] observed a faster nitrification rate in sludge-amended soils relative to the control. In addtion to soil pH, nitrification is slowed down by moisture stress.[74] Unlike ammonification, nitrification rate is relatively independent of sludge application rate.[66]

Denitrification, the biological reversal of NO_3^- back to NO_2^- and finally to gaseous N_2O and N_2, is favored by (1) anaerobic conditions, (2) the abundance of decomposable organic materials, and (3) the presence of denitrifying bacteria, which are more diverse and numerous than nitrifying bacteria.[75] Under certain conditions, sludge applied to land is conducive to denitrification. For example, King[70] reported a 17% loss by denitrification during 5 weeks and a 20% loss during 22 weeks following surface application (6-cm depth) of a digested liquid sludge. Similarly, approximately 50% of sludge N applied to a forest in northwestern Lower Michigan could not be accounted for and was attributed mainly to denitrification loss.[60] From an agronomic viewpoint, denitrification has often been considered undesirable. However, recently its potential beneficial role in reducing excess N in soil and water below the root zone has been increasingly appreciated.[60,76,77]

4. *Nitrate Leaching*

Unlike phosphate or sulfate, NO_3^- has no known specific adsorption or precipitation in soil. Thus, if not taken up by plants, immobilized, or denitrified by soil microorganisms, NO_3^- will move rather freely with the soil water past the root zone and eventually entering groundwater or surface water downslope. Nitrate movement is controlled largely by two processes: convection and diffusion, collectively called dispersion.[78] Quantitatively, the downward movement (one-dimensional flow) of NO_3^- can be expressed by the following differential equation:

$$\frac{\partial C}{\partial t} = k\frac{(\partial^2 C)}{\partial Z^2} \qquad (1)$$

Where C is NO_3^- concentration; t is time; Z is soil depth; and k is the dispersion coefficient, characteristic of the soil pore space and flow process. Equation 1 can be used to predict NO_3^- concentration (thus, NO_3^- movement) for a given depth at a given time. For example, Balasubramanian et al.[79] used the integrated form of Equation 1 to successfully predict the movement of added NO_3^- in a clayey Oxisol in Hawaii. Since their soil was highly aggregated, resulting in a high proportion of micropores, NO_3 moved only about one third the distance of average pore water movement. Thus, the percolating water seemed to bypass most of the NO_3^- in the micropores. Furthermore, NO_3^- in micropores might undergo denitrification because O_2 is often depleted in such sites.[60]

Nitrate pollution of groundwater is a serious concern in sludge utilization on land (EPA, 1984). With this in mind, Brockway and Urie[60] used regression analysis to estimate sludge application rates that would not elevate NO_3 concentrations in a water-table aquifer (2.5 to 3.5-m depth) above the 10 mg NO_3-N/L limit set by the USEPA for drinking water. Allowable rates of sludge application were 9.5 Mg/ha (670 kg total N/ha) of a raw papermill sludge for a red pine (*Pinus resinosa* Ait.) plantation, 16.5 Mg/ha (990 kg N/ha) of an anaerobically digested municipal sludge for red pine and white pine (*Pinus strobus* L.) plantations, and 19.0 Mg/ha (1140 kg N/ha) of the same municipal sludge for aspen (*Populus grandidentata* Michx.) seedlings. These different sludge rates were necessary because (1) the undigested sludge mineralized faster and (2) the N requirement of young aspen stands was greatest. Varied N requirements with tree age were further demonstrated by Wells et al.[80] These authors found that NO_3^- leaching 4 months following applications of a liquid anaerobic sludge (at 800 kg N/ha) was greatest (>60 mg N/L in soil solution sampled at 1-m depth) under pine stands of less than 1 year old followed by stands of 28, 9, and 3 years old.

5. *Nitrogen Immobilization*

This process is loosely defined as the assimilation of inorganic N into cellular materials either by soil microorganisms or plant roots (below-ground biomass). Microbial immobilization of N has been suggested as an important pathway for NO_3^- retention in forest ecosystems.[81] This process generally occurs when the C/N ratio of a medium, including sludge, is greater than 30:1.[46] Composted sludges and those originating from paper industries often show N immobilization when applied to soils.[82] Mixing wood or bark with digested sludge at various ratios has been used to regulate N immobilization.[83] Below-ground biomass can be a significant N sink in forests. For example, McClaugherty et al.[84] reported that fine roots of hardwood and red pine stands can retain more than 73 and 44 kg N/ha/year, respectively.

B. FORMS AND MOVEMENT OF HEAVY METALS IN SLUDGE-AMENDED SOILS

An excellent review on heavy metals in sludges, soils, and sludge-amended soils has been presented by Lake.[85] Also a comprehensive book on the same subject has been edited and published by Alloway.[86] Readers are encouraged to consult these references for detailed information.

Generally, heavy metals in soils can occur in the following forms: (1) as ionic or complexed ions/molecules in the soil solution; (2) as exchangeable ions on organic and inorganic exchange sites of the soil/sludge matrix; (3) incorporated into or on the surface of crystalline or amorphous mineral precipitates such as Fe, Mn, and Al oxides, and (4) incorporated into or on microorganisms, biological residues, and solid organomineral complexes.[87] Distribution of a metal among these various forms is dependent on properties of the soil, the sludge, and the metal itself.[88–90] However, distinction among these metal forms in soils is not clear-cut, and remains largely a matter of definition based on experimental procedures.

Sequential chemical extractions have been preferred over single extractions for fractionating heavy metals in soils. For example, Kuo et al.[91] used $MgCl_2$, oxalate, citrate-dithionite-bicarbonate, and NaOCl to successively extract exchangeable, amorphous Fe oxides-associated, crystalline Fe oxides-occluded, and organic matter-bound forms of Cd, Cu, and Zn from Cu smelter-polluted soils of Washington state. Similarly, Emmerich et al.[92] used KNO_3, "ion-exchange water," NaOH, Na_2-EDTA, and hot HNO_3 to sequentially extract exchangeable, adsorbed, organically bound, carbonate, and residual forms of heavy metals in sludge-treated soils of California. Although such procedures have a certain logic, the diversity of extractants and extraction procedures used makes comparison among different studies difficult. Also, many of the reagents used are non-specific for a physical or chemical metal form.[93] Nevertheless, a general consensus among many studies[91,92,94] is that Cu is predominantly associated with organic matter, which is consistent with the known affinity of Cu for soil organic ligands.[95] Sesquioxides (R_2O_3), particularly amorphous Fe oxides, can adsorb a great proportion of soil Cu,[90,91] suggesting that Fe, Al, and Mn oxides may play an important role in immobilizing Cu; thus the content of R_2O_3 should be taken into account when setting the loading rates of metals on soils.[20]

The dominant form of Pb in soils is less clear-cut. Lead seems to precipitate as $PbCO_3$ in alkaline soils,[92,96] and to sorb onto iron oxides in acid soils.[90] Lead phosphates have also been proposed as the probable minerals controlling soluble Pb in noncalcareous soils.[96] Organically bound lead is thought to control the retention and mobility of the metal at least in sandy soils.[97] On the other hand, Harter[98] reported a good fit of Pb sorption data by the Langmuir isotherm and that soil pH and CEC were important in Pb immobilization. Soil pH and CEC also seem to control Cd retention in soils,[99] indicating that most of soil Cd is exchangeable.[100] Kuo et al.[91] reported that 30 to 60% of total Cd of metal-polluted soils is exchangeable in $MgCl_2$. This is consistent with the fact that most soil minerals from illite to kaolinite to ferrihydrite as well as peat show much higher selectivity for Cu and Pb than for Cd.[101] Along with Cd, exchangeable Zn appears to be the dominant form of Zn in soil.[102] However, oxide-bound forms of Zn may also be important.[91] For example, Hickey and Kittrick[94] reported that, on average, 39% of total Zn in three soils and a sediment was associated with Fe and Mn oxides. King[90] ranked the relative retention of heavy metals by the Ap and some B horizons of 13 soils from the Southeastern United States as follows: Pb > Sb > Cu > Cr > Zn > Ni > Co > Cd. He concluded that Fe oxides and clay content would be better parameters than soil CEC for setting limits on metal loading on land. Except for Zn, this order of metal retention is nearly identical to that proposed by Schnitzer and Khan[103] for metals on soil humus: Pb > Cu > Ni > Co > Zn > Cd.

Since heavy metals are strongly retained by soils in the solid forms, their soil-solution concentrations are usually low. For example, McGrath[104] reported that the soluble concentrations of Cd, Cu, Ni, Pb, and Zn were 0.015, 1.19, 0.38, 0.0015, and 0.58 mg/L, respectively, as measured in the drainage water of a long-term field plot that had received a cumulative amount of approximately 700 Mg/ha of sewage sludge for 20 years. These levels declined further to 0.004, 0.36, 0.10, 0.0014, and 0.27 mg/L 22 years after sludge applications were terminated. The very low concentrations of most heavy metals in solution coupled with the large number of possible chemical forms in which metals can exist make the determinations of metals difficult and help explain the limited literature on this subject.[85] Dudley et al.[105] passed saturation paste extracts of sludge-amended soils through a Sephadex G-15 gel filtration column to study the forms of soluble Cu, Ni, and Zn. Their data showed (1) virtually all soluble Cu was organically complexed during the 30-week incubation (mainly associated with a soluble organic C fraction having small molecular size containing high amide and, to a lesser extent, carboxylic functional groups). (2) Soluble Zn was mostly inorganic. (3) Souble Ni was organically complexed during the first 4 weeks of incubation, but partitioned almost evenly between organic and inorganic complexes as the incubation progressed. As discussed by Lake,[85] during the early phase of sludge-soil mixing, heavy metals appear to be most soluble, probably because of complexes with small organic ligands. As time progresses, microbial decomposition of such complexes, resulting in a loss of functional groups that bind metals, gradually shifts the metals toward more stable and insoluble forms. Therefore, while the speciation of heavy metals in sewage sludge applied at agronomic rates to soils is of significance during the initial stages of sludge-soil mixing, it is likely that such metals will eventually reach the long-term equilibrium of the soil irrespectively of the original nature of the sludge. This equilibrium would be controlled by the chemical properties of the soil. Emmerich et al.[106] found that total soluble concentrations of Cd, Cu, Ni, and Zn in soil saturation paste extracts increased by sewage sludge incorporation into the top 15 cm of three California soils that had been subjected to leaching for 25 months with approximately 5 m of Colorado River water. However, the authors concluded that (1) the concentration of the metals was low and in most cases within ranges of untreated soils; (2) between 50 and 60% for Cd, 60 and 70% for Ni, and 60 and 70% for Zn of the total metal in solution was in the free ionic form; and (3) copper was almost exclusively in organically complexed forms in sludge-treated layers, but below these layers Cu shifted to free ionic forms as soil pH decreased. Emmerich et al.[106] concluded that leaching of heavy metals through the soil profile was insignificant because of low soil-solution concentrations. McGrath[104] reached the same conclusion.

Movement of heavy metals in soils generally requires that the metal be in the solution phase or associated with mobile particulates. In the case of heavy metals, however, which have very low water solubility, physical mixing of soil during repeated cultivation may be the principal contributing factor to increased concentrations of many heavy metals in soils outside or below sludge application zones.[104] Using 4 *M* HNO_3-extractable metal as an indicator of sludge-derived metal, Dowdy and Volk[87] leached soil columns for 38 days with a total of 75 cm water and showed that Cd, Cr, Cu, Ni, Pb, and Zn did not move into soil horizons below the zone of sludge incorporation. This result is essentially identical to that of Emmerich et al.[92] The latter authors further suggested that the occurrence of metals in the stable organically bound carbonate and residual forms in the sludge, coupled with a shift toward the more stable form (residual) after soil incorporation, contributed to the lack of metal movement in soils. Under field conditions, Dowdy and Volk[87] reported that 4 *M* HNO_3-extractable Cd, Cr, Cu, Ni, Pb, and Zn concentrations of the 0- to 15-cm layer increased with a one-time sludge application of 100 Mg/ha to a Minnesota soil (Bold silt loam, pH 6.2) 4 years earlier. However, below a 23-cm depth there was no difference in heavy metal levels between the sludge-treated and the control soil. Kelling et al.[107] used DTPA-extractable metals to monitor up to 25 months the movement of heavy metals in two Wisconsin soils previously treated with 60 Mg/ha of sewage sludge. They found a slight increase of Cd, Cu, and Ni at the 15- to 30-cm depth, but felt the increase was related to tillage incorporation. However, Zn concentration did increase at depth greater than 30 cm, indicating some Zn movement. Movement of heavy metals in coarse-textured soils beneath sludge disposal ponds was studied by Lund et al.[108] HNO_3-extractable Cd, Cr, Cu, Ni, and Zn were greater in soils under disposal ponds than at adjacent control sites. Metal enrichment was evident to depths as great as 3 m under some ponds. Since the distributions of metals with depth were closely related to changes in chemical oxygen demand, the authors suggested that the metals moved as soluble metal-organic complexes.

Indications are that heavy metal movement, as a result of sludge applications to soil, is very limited, and is only likely in sandy, acid, low-organic matter soils that receive large applications of sludge coupled with high rainfall or irrigation. Among the metals, Zn has the highest potential for leaching. Physical movement through earthworm holes, root channels, or large cracks in soil or physical mixing through tillage may contribute more to metal enrichment of some subsoils than had previously been thought.

Table 4 Octanol-Water Partition Coefficient (K_{ow}), Henry's Constant (Hc′), Water Solubility, Concentration and Half-Life ($T_{1/2}$) of PAHs and PCBs Found in Sewage Sludge

Organics	log K_{ow}	Hc′	Water Solubility	In Sludge (mg/kg) Range	Median	$T_{1/2}$
		Polynuclear aromatic hydrocarbons (PAHs)				
Naphthalene	3.59	$4.79 \cdot 10^{-2}$	31.7	nd–5.8	1.0	<125 days
Phenanthrene	5.61	$1.63 \cdot 10^{-3}$	1.08	2.1–8.3	4.3	25 d–5.7 years
Fluoranthene	5.33	$2.69 \cdot 10^{-4}$	0.26	2.2–28.5	9.1	44–322 days
Pyrene	5.32	$2.10 \cdot 10^{-4}$	0.13	1.2–36.8	4.9	229 days
Benzo(b) fluoranthene	6.57	$4.96 \cdot 10^{-4}$	0.001	2.1–14.8	7.5	67–252 days
Benzo(a)pyrene	6.30	$6.46 \cdot 10^{-5}$	0.0016	0.1–7.5	2.6	269–420 days
Benzo(ghi) perylene	7.23	$2.23 \cdot 10^{-6}$	$2.65 \cdot 10^{-4}$	nd–0.3	0.2	<9.5 years
		Polychlorinated biphenyls (PCBs)				
Aroclor 1016	4.38	$1.32 \cdot 10^{-2}$	—	0.2–75	1.2	4 years
Aroclor 1232	4.54	$3.50 \cdot 10^{-2}$	—	—	—	4 years
Aroclor 1248	5.60	$1.60 \cdot 10^{-1}$	0.054	<Detection limits		4 years
Aroclor 1260	6.11	$3.40 \cdot 10^{-1}$	—	0.02–0.46	0.15	730 days–4 years

Adapted from Wild, S.R. and Jones, K.C., *Sci. Tot. Environ.*, 119, 85, 1992. With permission.

C. TRANSFORMATION AND MOVEMENT OF TRACE ORGANICS IN SLUDGE-AMENDED SOILS

Sewage sludge contains numerous synthetic organics in trace amounts as a result of household activities and industrial discharge (Table 3). These chemicals are large in numbers and differ widely in physiochemical as well as toxicological characteristics. Thus, it is not practical nor advisable to discuss their behavior in soils individually; instead only the basic principles governing their adsorption, leaching, volatilization, and degradation in soils will be reviewed in this chapter. Some focus will be given to PCBs and PAHs, the two groups of trace organics that cause most environmental concern in sewage sludge.

The following definitions are presented to aid in the ensuing discussion. (1) Octanol-water partition coefficient (K_{ow}) is the ratio of a chemical's concentration in *n*-octanol to that in pure water. As K_{ow} increases, the concentration of a chemical occurring in the aqueous phase decreases. Log K_{ow} is often preferred because K_{ow} values vary over several orders of magnitude. (2) Henry's constant (designated Hc′ in its dimensionless form) is the ratio of a chemical's concentration in the gas phase to its concentration in the aqueous phase. The Hc′ expresses the tendency for a chemical to move from the aqueous phase to the gas phase. (3) Half-life ($T_{1/2}$) is the time required for 50% of a chemical to be degraded. Wild and Jones[37] have compiled a comprehensive list of physio-chemical properties and concentrations of many organic chemicals in sewage sludge. A subset of their data concerning PAHs and PCBs is presented in Table 4.

1. *Adsorption and Leaching*

Sorption isotherms are frequently employed to describe the concentration-dependent partitioning of organic chemicals to soil particles. For nonionic, sparingly soluble chemicals occurring at trace concentrations in the soil environment, sorption isotherms are usually linear in the region of interest. Thus, sorption is readily expressed as a partitioning coefficient, K_d.[109] The K_d of many trace organics has been found to vary less with such soil properties as pH, sesquioxide content, type, and amount of clay than with organic carbon content of the soil and the K_{ow} of the chemical.[110] The ratio of K_d to organic C content of a soil is referred to as K_{oc}, which can be estimated from K_{ow}. For example,

$$\log K_{oc} = 0.909 \log K_{ow} + 0.088$$ [110]

or

$$\log K_{oc} = 1.00 \log K_{ow} - 0.317$$ [111]

Octanol:water partition coefficients have been used successfully to model/predict the adsorption of PAHs onto soils and sediments,[111] the sorption of PCBs on biomass,[112] and the half-life of 13 PAHs in long-term sludge-amended soils.[113] Wild et al.[113] further noted that the persistence of individual PAHs is related to the chemical's structure: benzo (ghi) perylene and coronene having six and seven fused benzene rings, respectively, were most persistent, while naphthalene with only two rings was least persistent. Based on log K_{ow}, Wild and Jones[37] propose an "adsorption potential" for screening a chemical's susceptibility to such processes as leaching and biodegradation. There are three classes: low adsorption potential, $\log K_{ow} < 2.5$; moderate, $2.5 < \log K_{ow} < 4.0$; and high, $\log K_{ow} > 4.0$.

Obviously, adsorption retards leaching, so the trace organics that have high log K_{ow} and low water solubility will be less readily leached. Thus, the movement of PCBs in soil seems minimal.[114] For example, no PCB movement beyond the incorporation zone was detected 3 weeks after sludge application at 1.6 to 15.0 Mg/ha to a Wisconsin silt loam. The sludges contained 25 to 75 mg/kg total PCBs, and the incorporation depth was 30 cm, resulting in 0.02 to 0.24 mg/kg PCB in soil.[115] Soil has been used to remove PCBs from wastewater in an overland flow experiment.[116] In this experiment, simulated wastewater containing 100 mg/L Aroclor 1242 was flowed at 0.75 cm/day from the upper end of a trough (2% slope) packed with 8 cm silt loam soil from Louisiana. After 3-day residence time 99.9% of the PCB had been removed. Soil analysis showed that 96% of the recoverable (after degradation) PCB was retained by a soil fraction making up 20% upslope of which the top 2 cm retained 82% PCB.

2. *Degradation*

The degradation rates of trace organics in sludge vary, but are generally slow. Otherwise, they would not have persisted through various wastewater and sludge treatments (Table 4). Degradation of organics can be abiotic (photolysis, hydrolysis, and oxidation) or biological. Some chemicals are more susceptible to abiotic degradation than others.[117] Phenolics and some PAHs are reportedly susceptible to photolysis,[118] and PAHs, with less than four benzene rings, in soils freshly amended with sludge can be degraded significantly by abiotic processes.[119] Biological degradation is probably the most important loss mechanism for many trace organics in soils. The rate of biodegradation is dependent not only on the chemical's properties, and the responsible microorganisms, but also on many environmental factors such as temperature, moisture, and soil pH. First-order kinetics is often used to describe this process, but other models such as logistic, logarithmic, or Monod with or without microbial growth have also been used.[120] In practice, degradation potential based on half-time data ($t_{1/2}$) may be useful, especially in screening of potentially toxic organics. For example, Wild and Jones[37] have proposed three degradability classes: fast if $t_{1/2} < 10$ days; moderate, $t_{1/2} > 10$ but < 50 days; and slow, $t_{1/2} > 50$ days. For PAHs and PCBs, degradation is usually slower as the number of fused benzene rings or substituted chlorines becomes higher. Tucker et al.[121] demonstrated that less-chlorinated PCBs were readily degraded while highly chlorinated PCBs were persisted in aerobic, activated sludge suspensions. Even after 5-month acclimation, degradation of tri- and tetra-chlorinated PCB congeners was very slow as compared to that of mono- and dichlorinated congeners. As degree of chlorination increased from about 20% (in Aroclor 1221) to about 55% (in Aroclor 1254) degradation decreased from approximately 80% to less than 20%. Strand et al.[122] have compiled an informative table on PCB degradation by various microorganisms under different environmental conditions.

3. *Volatilization*

Vapor phase partitioning of a chemical greatly influences the spread of that chemical in soil. A chemical with high vapor pressure may dissipate quickly into soil air, where it can move throughout the soil and across the soil surface.[123] Henry's constant (H_c) is often used to represent a chemical's volatility. Wild and Jones[37] proposed that chemicals with $H_c > 10^{-4}$ are susceptible to volatilization. Alternatively, they proposed the volatilization potential, based on H_c and K_{ow} as follows:

- High volatilization potential: $H_c > 10^{-4}$ and $K_{aw}/K_{ow} > 10^{-9}$, where $K_{aw} = C_{air}/C_{water}$, which is essentially H_c.
- Possible volatilization potential: $H_c > 10^{-4}$ or $K_{aw}/K_{ow} > 10^{-9}$.
- No volatilization potential: $H_c < 10^{-4}$ and $K_{aw}/K_{ow} < 10^{-9}$.

Since the K_{aw}/K_{ow} of PCBs is about $10^{-2}/10^{6} = 10^{-8}$ (Table 4), these compounds are moderately volatile. Similarly, the concentrations of PAHs in the above-ground tissues of carrot (*Daucus carota*) were found to be independent of soil PAH concentrations,[124] suggesting the insignificance of root uptake and the dominant role of the atmosphere in supplying these chemicals. In fact, PAH levels in olive fruits has been proposed as a measure of air pollution in Italy.[125]

III. BENEFICIAL EFFECTS

A. SOIL FERTILITY

A "typical" sewage sludge contains 3.2% N, 1.4% P, 0.2% K, 2.7% Ca, and 0.4% Mg (Table 1). Thus, it can be used as a low-grade N and P fertilizer, and a source of Ca and Mg as well. Caution, however, should be exercised when using sludge as a plant nutrient source: (1) supplemental K may be needed, because sludge K is inherently low and (2) site-specific and sludge-specific N release information should be used to estimate the appropriate sludge application rates for specific crop N requirements. Since direct effects on soil fertility have been discussed, this section will focus on the amendment effects of the sludge organic matter.

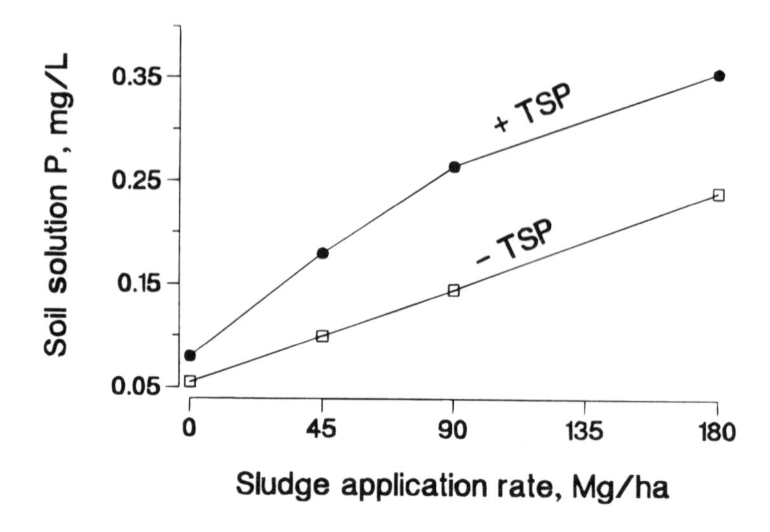

Figure 4 Changes in soil-solution P with sludge rate applied 5 years earlier, and with or without 120 kg P/ha as treble superphosphate (TSP). (Adapted from Hue, N.V., *Commun. Soil Sci. Plant. Anal.*, 21, 61, 1990. With permission.)

1. Phosphate Sorption

Given an average organic carbon content of 30%, sewage sludge is a rich source of organic matter.[10] Microbial activities gradually break down this organic matter, producing various organic compounds, particularly low-molecular-weight organic acids,[126] which can interact strongly with soil minerals and often reduce P sorption.[127] Recently, Hue[36] demonstrated that (1) 45, 90, and 180 Mg/ha of an anaerobically digested sludge (1.2% total P) applied to a high P-sorbing Oxisol 5 years earlier were equivalent to 262, 416, and 668 kg P/ha as treble super-phosphate based on the soil's P sorption isotherm and the change in soil-solution P concentration; (2) efficiency of P fertilizer was increased over 200% when freshly added to sludge-treated soil as compared to a no-sludge control (Figure 4); (3) decreased P sorption could be achieved with ≤45 Mg/ha, which was the lowest application rate used in the experiment.

Possible mechanisms responsible for the increased P efficiency include (1) direct contribution of sludge P; (2) competition of sludge-derived organic anions with phosphate for adsorption sites on the soil surface;[127]

$$\mathrm{Fe{-}O{-}P(=O)(OH){-}O{-}Fe} + \text{Oxalate} \longleftrightarrow \mathrm{Fe{-}O{-}C(=O){-}C(=O){-}O{-}Fe} + HPO_4^{2-}$$

and/or (3) complexation of organic anions with Al^{3+}, Fe^{3+}, or Ca^{2+}, thereby lowering the activities of these free cations in the soil solution.[16] This, in turn, increases the activity of soluble P in order to satisfy the thermodynamic principle of constant solubility product of soil P minerals.

2. Soil Acidity and Aluminum Detoxification

Aluminum complexation by organic matter has been proposed as a means to correct one aspect of soil acidity, more specifically, Al toxicity.[16,128] For agricultural lands, this detoxifying effect may not be evident or appreciated because, as required by law, the soil pH must be maintained above 6.0 for at least 3 years following sludge applications.[1,42] However, for forest lands where soils are often acidic and lime applications may not always be practical, this effect can contribute significantly to growth increases of the forest. In a greenhouse study, using an Al-sensitive legume as an indicator crop, Hue[16] showed that an application of 20 g/kg (approximately 40 Mg/ha) of an anaerobically digested sludge to a strongly acidic, Al-toxic Ultisol of Hawaii increased the shoot yield from 0.16 g/pot to 6.76 g/pot. [It is worth noting that a companion set of treatments receiving 2 to 8 cmol (OH) kg^{-1} as $Ca(OH)_2$ produced a maximum yield of only 2.86 g/pot.] Soil solution pH was increased slightly from 4.1 in the control to 4.7 in the sludge-treated soil, and the total soluble Al was decreased slightly to 0.47 mg/L, which would have been phytotoxic if present as Al^{3+}. Thus, the author concluded that the Al detoxification of the sludge was effected by two principal processes: (1) precipitation of Al by OH^- which was probably released by ligand exchange reactions between organic anions and terminal hydroxyls of Al or Fe oxides and (2) complexation of soluble Al by organic molecules, particularly low-molecular-weight organic acids.[129]

3. Cation Exchange Capacity (CEC)

The pk_a of most functional groups of organic matter is between 2 and 7.[130,131] Thus, under normal soil pH of 4 to 9, negatively charged sites are created by adding sludge to soil, and soil CEC increases as a result.[132] This CEC increase is particularly important for sandy or Al and Fe oxide-dominated soils where the initial CEC is low. For example, Epstein et al.[133] reported a CEC increase of nearly three-fold from 5.5 to 15.4 meq/100 g measured 1 month after 240 Mg/ha of an anaerobically digested sludge was applied to an Ultisol of Maryland. Similarly, Kladivko and Nelson[134] reported an average two-fold increase in CEC of three Alfisols from Indiana when 56 Mg/ha of sludge was rototilled into the soil. This sludge-induced increase in soil CEC, however, becomes less pronounced as time progresses because of the degradation of organic matter. For example, Kladivko and Nelson[134] applied 56 Mg/ha of sludge to an Alfisol which increased CEC from 9.7 to 24.0 meq/100g and from 8.2 to 17.9 meq/100g 2 months and 12 months, respectively, after the sludge application. The corresponding organic carbon contents were 0.95% in the control and 2.33% in the sludge-applied treatment after 2 months, and 1.00% (control) and 2.07% (sludged) after 12 months.

4. Soil Physical Properties

Soil organic matter, particularly polysaccharides, can bind soil particles into stable aggregates,[135] producing a loose, open, granular structure conducive to water and air permeability. In contrast, if organic matter or clay content is low, the soil aggregates are less stable in water and easily disperse into single-grain particles. Such soils become more closely packed and the bulk density increases. This explains why an application of 56 Mg/ha sludge to an Alfisol (Celina Series) decreased the soil bulk density from 1.28 to 1.19 g/cm^3, and increased mean weight diameter of soil aggregates from 0.59 to

1.49 mm and large pore space on a volume basis from 7.5 to 14.0%.[134] Similar results were obtained by Hall and Coker[136] when an undigested sludge was applied at rates >27 Mg/ha. In addition to these structural properties, soil hydraulic properties, such as water-holding capacity and hydraulic conductivity, are also increased by sludge additions.[136,137]

B. CROP PRODUCTION AND FOREST RESPONSE

Given the direct contribution of sludge to soil fertility, especially to N and P nutrition, and its indirect beneficial effects on soil physical and chemical properties, it is logical to expect positive growth responses of most plants to sludge when applied at agronomic rates.

For vegetables, Hemphill et al.[138] reported significant yield increases of lettuce (*Lactuca sativa*) and broccoli (*Brassica oleracea botrytis*) when grown on a Mollisol amended with a tannery waste at rates ranging from 32 to 192 Mg/ha. They estimated that the growth increase was approximately equal in effect to at least 112 kg N/ha as NH_4NO_3, and that plants grown on the waste-treated soil did not significantly contain higher concentrations of trace elements than the control, including Cr which was inherently high in the tannery waste.

Keefer et al.[139] measured heavy metals in edible and nonconsumable parts of radish (*Raphanus sativus*), carrots, cabbage (*B. olecacea capitata*), green beans (*Phaseolus vulgaris*), sweet corn (*Zea mays*), and tomatoes (*Lycopersicon esculentum*) field grown on a sandy loam Alfisol to which four sewage sludges were applied at 90 and 180 Mg/ha. Although the application rates were much higher than usual agronomic rates, Cd, Cr, and Pb concentrations in the edible parts of the vegetables from the sludge-treated plots were no more than 1.0 mg kg^{-1} above those from the untreated control plots. Nickel uptake, however, varied markedly with crops and sludges. Copper and Zn levels in vegetables grown on some sludge-treated soils were elevated, but not enough to cause alarm, according to the authors.

Quality of potatoes (*Solanum tuberosum* L.) grown on an acid (pH 5.2) soil amended with 15 cm of an anaerobically digested liquid sludge (4% solids) was compared with no sludge controls.[140] In 2 years of the study, sludge-grown potatoes consistently had higher total and protein N in the cortex tissue, but lower in the pith, than the controls. They also contained higher ascorbic acid than the controls in the first year, but lower in the second year. Phenolic content and enzymatic darkening in potatoes was not affected by the sludge amendment.

Ear yields of sweet corn grown on a Mollisol amended with three different sewage sludges were nearly double those of the control.[141] This study showed that (1) trace metal concentrations of kernels were much lower and less variable with sludge source than those of leaves, (2) liming the soil from pH 5.7 to 6.5 prior to sludge application had no consistent effect on ear or stover yield nor on metal contents of kernels, and (3) metal contents of kernels did not increase with second or third sludge applications. These results were supported by the work of Rappaport et al.,[142] who applied a high metal, aerobically digested sludge to three Ultisols of Virginia. Application rates ranged in 42 Mg/ha increments from 0 to 210 Mg/ha. (The highest sludge rate supplied 4.5 kg Cd, 760 kg Cu, 43 kg Ni, and 620 kg Zn ha^{-1}.) Corn grain and stover yields increased linearly with increasing sludge application rate, and metal concentrations in corn grain and ear leaves were within normal levels.

For small grains grown on a pH 5.5 soil of California, annual grain yields of barley (*Hordeum vulgare*) peaked when 90 Mg/ha of composted sludge and when 45 Mg/ha of "urban" sludge were annually applied for 7 years.[143] No phytotoxic effects were observed on any treatment. During the first 2 years, Zn and Cd concentrations in grain remained virtually unchanged with either sludge. During the last 4 years, Zn and Cd concentrations in grain and straw increased where the "urban" sludge was used. Straw consistently accumulated more Zn and Cd than grain. More importantly, over the entire 7-year period, Cd and Zn concentrations in grain produced by a chemical (NPKS) fertilizer were comparable to those in grain produced by the composted sludge at 45 Mg/ha. Grain yield of wheat (*Triticum aestivum*) on a sandy, pH 7.3 soil of Italy increased from 3.92 to 4.50 and 4.50 Mg/ha for the first year and from 2.85 to 3.01 and 4.34 Mg/ha for the second year as sludge rate increased from 0 to 50 and 100 Mg/ha, respectively.[144]

Forage yield of sudangrass (*Sorghum bicolor* L. Moench) also increased nearly two-fold compared to an NPK-fertilized treatment, when grown on an acid Oxisol that had received 45, 90, or 180 Mg/ha of a sludge 5 years earlier.[35] Sludge applications at 33.6, 67.2, and 89.6 Mg/ha to a calcareous Fe-deficient soil of New Mexico increased sorghum (*S. bicolor* L. Moench) dry matter from <1000 to >3000 kg/ha and grain yield from <400 to >1600 kg/ha.[145] Increased availabilities of micronutrients (Fe and Zn) and P, even 5 years after sludge application, were noted. In an experiment with tannery wastes applied to a Mollisol at 43, 86, and 171 Mg/ha the first year and at 23, 46, and 57 Mg/ha the second year, Hemphill et al.[146] showed that (1) tall fescue had the best growth on soil treated with the highest rates of the waste, (2) yields increased up to 3 years after one application of the waste, (3) no detrimental effects were noted on tall fescue during the 3-year study, and (4) although total Cr in soil increased as much as seven-fold with waste treatment, plant Cr increased only slightly or not at all.

Forests generally respond favorably to sludge applications.[147,148] Increases in fascicle dry weight and needle length of red pine were noted when fertilized with 4.8, 9.7, and 19.3 Mg/ha of an anaerobically digested sludge.[147] Understory biomass increased up to 132% in sludge-treated plots over the controls, and no metal toxicity symptoms were observed and sludge-treated understory vegetation remained green later into the growing season well after vegetation on untreated plots had begun to discolor and approach dormancy.

C. LIVESTOCK PRODUCTION

Sewage sludge can have beneficial effects — or more appropriately, no observed adverse effects — on animals through two main pathways: forage or feed grown on sludge-amended soils and direct ingestion of sludge.

Beef steers (*Bos taurus*) fed for 141 days a diet of corn grain, which was produced from soil treated with 7.6 cm/ha (20.2 Mg/ha) of a liquid-digested sludge, showed no reduction in performance (average daily gain increased from 0.98 to 1.04 kg and feed to gain ratio decreased from 11.1 to 10.2) or carcass quality.[149] Also, there were no significant differences in Cd, Co, Cr, Cu, Fe, Hg, Ni, Pb, and Zn concentrations in livers, kidneys, blood, or feces between steers fed the control corn diet and the sludge-fertilized corn diet. Similarly, there were no observable differences in average daily gain, feed-to-gain ratio, and daily feed intake between control swine and those fed corn grain produced from a plot that had received approximately 336 Mg/ha of a liquid sludge over a 3-year period before planting.[150] The trial began when pigs weighed approximately 17.6

kg and ended when pigs reached 90 kg. Since Cd, Ni, and Zn were higher in sludge-fertilized corn than the control corn, concentrations of Cd in kidney and liver and Ni in kidney of pigs fed sludge-fertilized corn were higher than those in the control pigs. However, no significant differences were observed in Cd or Ni concentrations in muscle or in hepatic microsomal mixed-function oxidase activity or in liver-to-body weight ratio. This indicates the absence or very low levels of organic toxicants in sludge-grown corn. These findings along with histopathologic analysis of various tissues for lesions led Lisk et al.[150] to conclude that pigs fed sludge-fertilized corn were not adversely affected. The effects of 3-year feeding of corn silage grown on soil amended with 0, 15, 30, and 45 Mg/ha/year of a sewage sludge containing 105 to 186 mg Cd/kg on the performance of dairy goats (*Capra hircus*) and market lambs (*Ovis aries*) were evaluated by Dowdy et al.[151] The authors reported that (1) dry matter intake, daily milk production, and feed efficiency of goats were not affected by 3 years of continuous consumption of high Cd silage and (2) average daily gains of lambs seemed to be higher than those of control lambs, whereas feed efficiency was not affected by sludge fertilization. Although the corn silage contained 5.3 mg Cd/kg and fairly low Zn (113 mg/kg) in year 3, Cd concentrations in goat milk,[152] heart, and muscle[153] were not increased. On the other hand, Cd concentrations in goat livers and kidneys were increased significantly, but still an order of magnitude less than the critical level for renal dysfunction.[153] These findings were in agreement with those of Telford et al.,[154] who conducted toxicologic studies with pregnant goats fed for 135 days a grass-legume silage grown on soil amended with 112 Mg/ha sludge. Differences in Cd concentration in goat milk or body tissues of the newborns were inconsistent. Livers of adult animals fed the sludge-treated diet contained more Cd. No changes in the tissue ultrastructure were observed by electron microscopy. Hinesly et al.[155] used Cd accumulating corn and soybeans grown on strip-mine spoil amended with 200 Mg/ha of sludge to formulate three diets containing 0.09, 0.57, and 0.97 mg Cd/kg for chickens. After 80 weeks of feeding, from chicks until laying hens, the highest dietary Cd level did not alter the Cd concentration in the brain, breast muscle, leg muscle, or eggs; feed consumption, body weight gains, rate of mortality, egg production, and egg quality remained unchanged.[155]

Direct feeding of sewage sludge to animals has also received considerable research. Baxter et al.[156] fed cows and young steers diets containing either 0 or 12% sludge for 9 months, followed by a 4-month withdrawal with no sludge. The large quantity of sludge consumed increased Cd, Cu, and Pb concentrations in kidneys and livers, yet the metal increases produced no ill effects on animal health or performance. Similarly, a feedlot diet with 7% sewage sludge did not appear to adversely affect growth or carcass characteristics of lambs in a 90-day feeding trial.[157] Elemental composition of blood, milk, and tissues (liver, kidney, spleen, bone, and muscle) of breeding ewes fed a diet containing 7% sludge for 2 years was little influenced by the diet. Cadmium in kidneys was 1.2- to 1.5-fold higher, but all values were <1 μg/g; Pb and Cd in livers were about 1.5 times higher; metal concentrations in bone, blood, and milk remained unchanged.[158] The study was continued for a total of 4 years, during which cytochrome P-450 (P-450) content and enzyme activities for xenobiotics biotransformations were assayed in livers after 2 years and in livers, kidneys, and ileal tissue after 4 years.[159] The results showed that dietary sludge did not increase P-450 and activities of biotransformational enzymes, that dietary sludge increased Cd levels in livers but not in kidneys or spleens, yet all Cd levels were within ranges for livestock fed conventional feed, and that no hazardous accumulation of toxic elements and little, if any, evidence of toxicosis existed.

Table 5 Total Metal Concentration and pH of the Topsoil (0- to 25-cm depth) from Bordeaux, France, as a Result of Sewage Sludge Applications

Cumulative Sludge Inputs (Mg/ha)	Metal (mg/kg)							
	Cd	Cu	Cr	Mn	Ni	Pb	Zn	pH
Experiment Bx 1: Started 1974, Measured 1989								
0	0.33	14.2	6.8	33.0	2.4	17.9	19.2	5.4
160	1.00	18.9	8.2	312.0	6.4	44.7	199.0	6.2
800	5.71	66.7	23.1	1789.0	30.8	189.0	1074.0	5.8
Experiment Bx 2: Started 1976, Ended 1984; Measured 1989								
0	1.30	4.5	3.7	23.0	3.6	10.6	8.1	5.8
50	27.9	15.6	7.2	33.0	73.6	22.1	45.7	6.8
300	96.0	45.5	21.0	69.0	247.0	44.3	155.3	7.1

Adapted from Juste, C. and Mench, M., *Biogeochemistry of Trace Metals*, Lewis Publishers, Chelsed. MI, 1992, 159. With permission.)

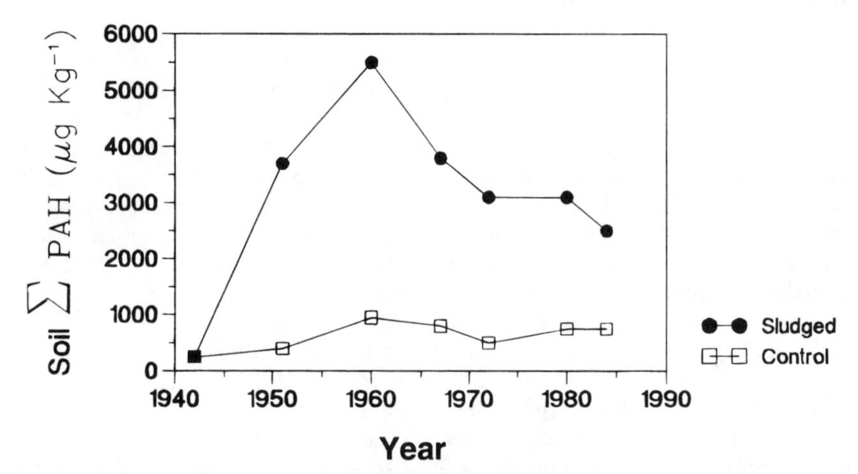

Figure 5 Total concentrations of polynuclear aromatic hydrocarbon (ΣPAH) in sludge-amended and control soils. Sludge applications were terminated in 1961. (Adapted from Wild et al., *Environ. Sci. Technol.*, 24, 1706, 1990. With permission.)

D. SOIL MICROORGANISMS AND MESOFAUNA

Heterotrophic soil microorganisms seem to be stimulated by sewage sludge application at agronomic rates probably because of improvement in soil fertility and soil structure by the sludge. Stadelmann and Furrer[160] reported a significant increase in yeasts and hyphal fungi, a 164 to 204% increase in aerobic bacteria and a 142 to 251% increase in actinomycetes relative to a mineral fertilizer control when 13.3 Mg/ha (approximately 400 kg/ha of total N or 200 kg/ha of available N) of a sewage sludge was applied annually to a sandy loam soil of Switzerland from 1977 to 1981. Numbers of autotrophic microorganisms such as algae, however, were decreased by the sludge treatment. Even at high rates (approximately 1500 kg/ha of total N, from two liquid sludges 1.6 and 3.5% total solids), there were no differences in numbers of total soil bacteria, actinomycetes, or fungi among sludge, pig manure, or mineral fertilizer treatments.[161] No adverse effects were found, as indicated by a diversity index, in bacterial communities of grassland soils receiving sewage sludge yearly from 1920 to the late 1960s.[162] A single application of sludge, 56 or 112 Mg/ha to a silt loam soil, increased the number of soybean rhizobia (*Bradyrhizobium japonium*) 11 years later, suggesting that sludge application did not have a long-term detrimental effect on soil rhizobial numbers, nor did it result in a shift in nodule serogroup distribution.[163] In forest soils, the applications of a well-decomposed liquid sludge (2% solids) reduced the number of mites (presumably because of interstitial clogging), but increased the number of more mobile Collembola (springtails), whereas the application of a solid sludge (30% solids) increased both mites and Collembola but did not change the community structure of the soil mesofauna.[164]

IV. POTENTIAL ADVERSE EFFECTS

A. EFFECTS ON SOIL ENVIRONMENT

Given the facts that (1) sludge contains a spectrum of trace elements and synthetic organics, many of which exist at very low concentrations or not at all in the unamended soils (Tables 2 and 3), (2) many trace elements hardly move beyond the plow layer, and (3) crop uptake usually accounts for less than 1% of inputs,[165] it is inevitable that sludge applications to soils, especially at high rates or with repeated frequency, will result in elevated concentrations of many potentially toxic elements. Long-term experiments with sludge in France[166] illustrate this point: in one experiment (Bx 1), sludge has been added continually since 1974 either annually at 10 Mg/ha or every other year at 100 Mg/ha; in another experiment (Bx 2), sludge was similarly applied but only from 1976 to 1980; soil data taken in 1989 are listed in Table 5.

Similarly, repeated applications of sludge from 1942 to 1961 to a soil in England[167] resulted in elevated levels of total PAHs in the soil measured 25 years (in 1984) after sludge application had been terminated (Figure 5).

It should be emphasized that metal bioavailability or toxicity does not necessarily parallel total concentrations of the metals. As discussed earlier, the availability of a metal

Table 6 Cadmium Concentration in Edible Tissues of Common Crops Grown on Soils with or without Sludge Amendments (mg/kg dry wt)

Crop	Cultivar	Crop Cd	Background Cd (estimate)	Increased Crop Cd
Lettuce	Tom Thumb	8.1	0.7	7.4
	Paris White Romaine	6.2	0.7	5.5
Cabbage	Greyhound	0.97	0.27	0.6
Turnip	Bruce	0.74	0.21	0.53
	Snowball	0.58	0.21	0.37
Barley	Julia	0.31	0.08	0.23
Radish	French Breakfast	0.33	0.21	0.12
Oat	Leander	0.18	0.08	0.10
Tomato	Moneymaker	0.40	0.32	0.08
Potato	Desiree	0.20	0.13	0.07
Sweet corn	Golden Earley	0.16	0.11	0.05
Maize	Caldera	0.13	0.08	0.05

Adapted with permission from Reference 174.

(as approximated by its activity in the soil solution) is controlled by many factors, including sludge properties, soil properties (e.g., pH, clay content, organic matter, Fe and Mn oxides, and interactions with other cations present), as well as the sludge × soil interactions.[4,126] By studying the activity of trace elements in soil solution, Corey et al.[89] proposed that it is possible to differentiate sludges that require loading limits, to prevent harmful metal accumulations in plants, from those that do not. In the long term, however, increased quantities of heavy metals in soils as a result of sludge applications are a justified concern because these elements, if not properly controlled, may threaten the health of living organisms.[168] For example, soil acidification, which readily occurs via sludge oxidation and ensuing nitrification, would make most trace elements more soluble and bioavailable. Christensen[169] showed a three-fold increase in water-soluble Cd as soil pH decreased one unit. Furthermore, these added metals, whether soluble or not, are effectively absorbed by such soil invertebrates as earthworms. Beyer et al.[170] reported that earthworms (Lumbricidae) in four sludge-amended soils contained more Cd (12 times), Cu (2.4 times), Zn (2.0 times), and Pb (1.2 times) than those from adjacent control sites. Such bioaccumulation has serious implications for the food chain. Significant decreases in the number of *Rhizobium japonicum* in soils incubated for 42 days with different rates of sludge (soil:sludge ratios = 13, 9, and 5) have also been attributed to metal toxicity from the sludge.[171] Recently, Giller et al.[172] attributed the absence of N fixation in white clover (*Trifolium repens* L.) grown on an English soil amended with sewage sludge for 20 years to metal toxicity, which selectively favored the survival of ineffective *Rhizobium* strains.

B. EFFECTS ON CROPS AND LIVESTOCK

Although many metals in sludge (e.g., Cd, Ni, Pb) are not essential for plant growth, they are taken up and accumulated by crops to various degrees.[173] Table 6, extracted from a comprehensive work of Chaney et al.,[174] shows Cd concentration in edible tissues of crops grown on long-term sludge-amended soils compared to the corresponding background Cd levels. From this table it is clear that the ability to accumulate metals varies widely with crops species and even with cultivars within a species. In fact, it has long been recognized that lettuce and tobacco (*Nicotiana tabacum* L.) are Cd accumulators,[173,174] whereas small grains (wheat, rice, barley) are not.[175] Within a plant, metal accumulation often follows the order grains and tubers < stems and leaves < roots.

Metal accumulation by crops, however, can go undetected, especially by the general public, because it occurs long before yield reduction or toxic symptom becomes apparent. At the extreme, if metal concentrations in soil (more correctly, in soil solution) are high enough, phytotoxicities occur, yields are reduced, and crops eventually die. Such observations led Chaney[176] to develop the soil/plant barrier concept which briefly states that higher animals in the food chain are partially protected from certain metal toxicity because phytotoxicities prevent crops from accumulating metals (e.g., Zn, Cu, Ni, Mn, As, B) to levels harmful to animals.

Lutrick et al.[177] reported severe phytotoxicity to soybeans grown on three acid Ultisols in Florida, which received cumulative amounts of 288 and 335 Mg/ha of a liquid sludge (2.6% solids, 53 mg Mn/kg and 2440 mg Zn/kg) during a 6-year period before planting. The seedlings became chlorotic and eventually died; their tissues contained 963 and 1188 mg Zn/kg, and 484 and 725 mg Mn/kg, respectively, suggesting that Zn and/or Mn toxicities were the cause of soybean injury. Interestingly, no reduction in yields was observed for maize and grain sorghum grown in the same treatments. Juste and Mench[166] reported a sharp decrease in grain yield of maize (about 50% of the control) grown on a soil amended with a high Cd and Ni sludge (experiment Bx 2). Phytotoxic symptoms (e.g., intervenal chlorosis and purple on stems) were visible in early growth stage of maize. Leaf composition (approximately 20 mg Ni/kg and 40 to 80 mg Cd/kg) indicated that Ni and Cd toxicities were likely, with Ni toxicity being the most probable cause.

Trace organics can enter crops by (1) uptake from the soil solution, then translocation from roots to shoots and/or (2) adsorption by roots or shoots of organics volatilized from the soil (vapor transport). Root crops are most susceptible to contamination by the vapor phase. Carrot peel (epidermis) is lipid rich, which serves as a sink for volatile lipophilic organics, including sludge-borne PCBs and PAHs.[124,178] Other root crops (e.g., sugar beet, onion, turnip) are much less effective in accumulating lipophilic organics in their edible roots, probably because their root surface is lower in lipids.[14]

Animal species, age, and diet affect the animal's tolerance to heavy metals: young animals are generally most sensitive, and interactions among elements in the diet strongly modify metal effects.[14] Cadmium often accumulates in kidney, and to a lesser extent liver, of sludge-fed animals as reported for swine fed a diet containing corn grain high in Cd and Cd/Zn ratio.[150] This corn was grown on a soil that had received 336 Mg/ha of sludge. It should be noted, however, that Cd concentrations in kidneys and liver of steers were not increased over the control when the steers were fed corn grown on a low Cd and low Cd/Zn sludge-amended soil.[149] Although a nonessential element, Co can accumulate in crops to levels potentially harmful to ruminants.[173] Sheep are sensitive to Cu toxicity, which may occur when they are fed plants containing >25 mg Cu/kg. Conversely, excessive Mo in forage as a result of sludge applications can poison grazing ruminant animals. Ingested Mo is transformed in the rumen to tetrathiomolybdate, which binds Cu, prevents Cu absorption from the gut, and minimizes Cu bioavailability. If forages contain 5 to 10 mg Mo/kg such that the Cu/Mo ratio <2, induced Cu deficiency may occur.[14] Generally, the availability of Mo for plant uptake increases and that of Cu decreases as soil pH rises. Thus, Mo toxicity to animals is most frequently associated with forage grown on alkaline soils. Copper deficiency in ruminants can also be induced by excessive dietary Zn (300 to 1000 mg Zn/kg). Although the soil/plant barrier may protect animals from excess Zn in crops, it is conceivable that direct ingestion of high Zn, low Cu sludges could lead to Zn-induced Cu deficiency in ruminants. Most grazing animals do consume modest amounts of soil. An extreme case of 14% dry matter intake as soil has been reported by Fries.[179] He reported that PCBs accumulation in milk and red meat occurred mainly through direct ingestion of sludge-amended soils. Concentrations of PCBs in fatty tissues and of Cd in kidney and liver of swine increased significantly when these animals were confined over two winters in small areas of a sludge-amended soil: PCB increased from 36 μg/kg fat in the control to 389 μg/kg in the highest sludge plot, kidney Cd from 0.35 to 4.71 mg/kg fresh weight, and liver Cd from 0.03 to 0.83 mg/kg.[180]

C. EFFECTS ON FORESTRY AND WILDLIFE

Sludge applications to forest lands often result in increased quantity and quality of vegetation, especially crude protein and P content.[181,182] While this represents an improvement in habitat quality for herbivores, it may lead to changes in the plant community via introduction of new species from seeds in the sludge, competition of plant species favored by nutrient enrichment, and browsing pressure on the treated area. Campa et al.[181] reported that browsing by white-tailed deer (*Odocoileus virginianus*) and elk (*Cervus elaphus canadensis*) decreased vertical and horizontal cover and the densities of key browse species in a 10-year-old aspen (*Populus* spp.) clear-cut area amended with approximately 10 Mg/ha of an anaerobically digested sludge. Regeneration of woody species was severely reduced by sludge application to recent clear-cuts in western Washington, owing to plant competition from rapid growth of herbaceous species and animal damage caused primarily by increased numbers of voles (*Microtus townsendii*) that girdled seedlings.[182] Conversely, wildlife species with habitat requirements for more open understories may decline in numbers.

Because sludge applications increased concentrations of many heavy metals in forest soils, accumulation of metals in animal tissues can occur.[183] Insectivores seem to be the most affected while granivores the least. Following an application of 51 Mg/ha sludge containing 50 mg Cd/kg, Cd levels in insectivorous Trowbridge's shrews (*Sorex trowbridgii*) exceeded those in granivorous deer mice (*Peromyscus maniculatus*) by an order of magnitude in kidneys and by two orders of magnitude in liver.[184] These metal concentrations were at least three times more than those of corresponding controls. Similarly, Cd levels in kidneys and liver of meadow voles (*M. pennsylvanicus*) enclosed in a field treated with 9 Mg/ha of Milorganite containing 60 mg Cd/kg increased significantly over a control.[185]

D. EFFECTS ON WATER QUALITY AND AQUATIC SYSTEMS

Sewage sludge applied to lands, if not properly managed, can cause (1) eutrophication of lakes and streams through surface run off and soil erosion[186] and (2) nitrate pollution of ground-water through leaching when N released by the sludge is in excess of or off-phase with plant needs.[54]

Eutrophication occurs when water is enriched with nutrients, especially N and P, which stimulate algal growth. The most visible effects of eutrophication are massive blooms of phytoplankton (e.g., diatoms, green algae, and blue-green algae). The decay of excessive algal blooms, as they die off and sink, cause depletion of dissolved oxygen or hypoxia (O_2 concentration <2 mg/L). This condition can lead to mass mortality of fish. Generally, eutrophication is favored when concentration of NO_3-N in water is >1.0 mg N/L.[77] Nitrogen-to-P ratio may be more important than absolute levels of either nutrient because algal cells require approximately 16 molecules of N to 1 molecule of P, 7 mg N to 1 mg P, for their synthesis.[187]

Groundwater is the source of drinking water for nearly 100 million people in the U.S.[186] Although the USEPA standard for NO_3 in drinking water is set at 10 mg N/L, a level above 3 mg N/L is considered to have been caused by human activities, primarily agricultural uses of N.[186] Applications of sludge to land, particularly forest land, often produce NO_3 levels much higher than the 10 mg N/L limit.[188–190] For example, Riekerk[188] reported that 247 Mg/ha dried digested sludge applied to a forest soil of western Washington increased NO_3 concentrations in a monitoring well (5- to 6-m depth) and a spring to as high as 30 mg N/L 1 year later. The subsoil solution, sampled at 150- to 180-cm depth beneath the site a few months after sludge application, contained 150 to 250 mg NO_3-N/L.

Although the tolerance level of livestock to NO_3-N in water is higher than the health standard for humans (discussed later), NO_3 can become toxic to ruminants, horses, and baby pigs. Levels of 40 to 100 mg NO_3-N/L are risky unless the feed is low in NO_3 or fortified with vitamin A. Water with ≥100 mg NO_3-N/L should not be used in animal production.[77]

Heavy metals with their strong reactions with soil constituents do not leach significantly into groundwater. Contaminations of water, if any, would be minimal and mainly through soil erosion. Nevertheless, in an aquarium model ecosystem with sludge-amended soils serving as the sediment, Lu et al.[191] demonstrated increased mortality and metal accumulation in alga (*Oedogonium cardiacum*), daphnia (*Daphnia magna*), mosquito larva (*Culex pipiens quinquefasciatus*), and mosquito fish (*Gambusia affinis*). The sludge-treated sediment contained 3.1, 8.4, 7.6, and 38 mg/kg of total Cd, Cu, Pb, and Zn, respectively. It is worth noting that chronic toxicity of Pb to fish (increased mortality, reduced hatching, and increased spinal curvature and blackening of tails) can occur at Pb concentrations below concentrations used to be considered safe for human drinking (0.05 mg Pb/L). For example, chronic Pb exposure to rainbow trout (*Salmo gairdneri*) as fingerlings caused blackening of the tails and spinal curvature (symptoms of neurotoxicity) at 0.013 to 0.020 mg Pb/L in a soft water with 27 mg/L hardness as $CaCO_3$.[192]

E. EFFECTS ON THE ATMOSPHERE

Aerosol may be generated during spraying of liquid sludge on land, and volatilization of some trace organics would occur during surface application of sludge.[193] However, these effects on the atmosphere, if any, are localized and are minimal compared with those from sludge incineration. Therefore, the following discussion will focus on potential adverse atmospheric effects when sewage sludge is incinerated.

Emissions from sludge (and municipal solid waste) incinerators include particulates, water vapor, CO, CO_2, SO_2, NO_x, hydrocarbons (organics), and metals.[194] Composition of the emissions depends on combustion temperature: high temperatures burn organic compounds more completely (thus reducing toxic organics), but increase metal emission.[195] Combustion temperatures commonly range between 700 and 1000°C, and are limited on the low end by the need to eliminate odorous matter in the flue gas and on the high end by the melting of the ash.[196] Atmospheric metal emissions from sludge incineration are a function of (1) metal concentration in the feed sludge, (2) amount of metal that is carried out in the flue gas, either adsorbed on particulates or as gas, and (3) type of air pollution controls in use. Some metals are concentrated on fine particulates during emission, perhaps because of greater specific surface of the small particles that are conducive to adsorption or condensation.[197] Particulates <20 μm in diameter are not easily captured in pollution control equipment. Thus, they readily enter the atmosphere from the stack. Once in the air, these particles readily disperse due to their small size, and can be inhaled, if <10 μm, and ingested.[198] Among the heavy metals commonly found in sewage sludge, Cd, Hg, and Pb are most volatile during incineration and highly toxic to living organisms.[195,198] Thus, further discussion on the three metals is warranted.

1. *Cadmium*

Cadmium is a volatile metal as its vapor pressure is 710 mm Hg at 760°C. Takeda and Hiraoka[196] reported a Cd residual ratio (the amount remaining in the ash to that in sludge feed) of 80% at 450°C and 25% at 800°C, and that most Cd apparently was lost to the atmosphere because the usual dust collectors (in the early 1970s) failed to trap it.

Similarly, Bennett and Knapp[199] reported that Cd enrichment in emissions was highest from an incinerator with a wet single-pass scrubber system which was inadequate for particulates <1 μm with which most emitted Cd was associated.

2. *Mercury*

Given its boiling temperature of 357.6°C at 0.1 MPa, and its high vapor pressure, 0.16 Pa at 20°C (or 14 mg Hg/m^3 air), Hg is probably the most volatile heavy metal in sludge incineration.[200] Most Hg in sludge feed is emitted as vapor or volatilized Hg compounds.[195] In a New Jersey fluidized bed incinerator, combustion temperatures averaged 788°C and nearly 98% of Hg in the feed sludge was emitted to the atmosphere; only 0.4% was retained in the ash and about 2% in the scrubber water.[197] Because Hg is difficult to eliminate from the gas stream by condensation, scrubbing, or filtration, and it is accutely neurotoxic, Kistler and Widmer[200] recommended strict controlling of its input into the sewering system.

3. *Lead*

Based on a vapor pressure of 3.5×10^{-2} mm Hg at 760°C and 1.3 mm Hg at 980°C, Pb is considered a nonvolatile metal. Lead chloride, however, is about two orders of magnitude more volatile than elemental Pb or lead oxide.[195] Furthermore, lead oxide may quickly condense on particulate matter as it cools. Lead emissions in vapor form are very sensitive to temperature: a 25-fold increase occurred when incinerator temperature was raised from 650 to 950°C; also, Pb was associated mainly with particles <3 μm.[197]

In addition to metals, incineration emissions may contain many toxic organics. Of most concerns are polychlorinated dibenzo-*p*-dioxins, polychlorinated dibenzofurans, and PAHs. Concentrations of these organics depend on combustion conditions: more complete combustion forms less organics. For this reason, air pollution from organics is much less from continuously operating incinerators with complete combustion than from batch-type incinerators with fluctuating temperatures during start-up and shut-down.[201]

F. EFFECTS ON PEOPLE

Sludge affects people indirectly, through eating crops or animals that were contaminated by sludge-amended soils or drinking water high in NO_3^-. Rare cases of children eating high-metal soils (the Pica child) that resulted perhaps from sludge applications have also been mentioned.[20,173] These contamination pathways imply chronic exposures: low doses over long periods of time. To illustrate the potential risk of sewage sludge to humans, the following health effects are discussed.

1. *Nitrate Toxicity*

Consumption of food or water high in NO_3^- may cause methemoglobinemia in infants, particularly those less than 3 months old.[77] Methemoglobinemia occurs when an oxidizing agent, such as nitrite, oxidizes the Fe in hemoglobin to form methemoglobin (metHb), thus reducing the oxygen-carrying capacity of blood. Young babies are highly susceptible to NO_2^- poisoning because certain bacteria in their digestive tract (with alkaline pH) are capable of reducing NO_3^- to NO_2^-, thus transforming hemoglobin to metHb. As the O_2 carried by the blood decreases, the body is suffocated and turns blue. Thus, the common name for methemoglobinemia is the "blue-baby syndrome". The symptoms of MetHb are easily recognizable and can be treated by injection with methylene blue.[54] The recovery is rapid and complete with no known long-term effect.[77] While most infants can tolerate NO_3^- in water at levels much higher than 10 mg N/L, some infants begin to exhibit metHb symptoms at levels only slightly higher than 10 mg N/L after a few days of consumption. To prevent toxicity, the USEPA recommends a maximum of 10 mg NO_3-N/L in drinking water.[77] Adults and older children are much less susceptible to MetHb because stomach HCl levels increase with age and kill most of the bacteria that can reduce NO_3^- to NO_2^-. For adults 10 mg NO_3-N/L is a lifetime health advisory level. The rationale for this advice is that, although NO_3^- itself is not carcinogenic, a small fraction of it could be reduced to NO_2^- and react with secondary and tertiary amines in the human body to form nitrosamines.[54] Nitrosamines have been identified as a potent carcinogen in animals, and may be an important carcinogen in humans. However, there is no conclusive evidence linking any type of cancer to NO_3^- in drinking water.[77]

2. Cadmium Toxicity

The potential risk of high Cd in food to human health was first recognized in the early 1970s when occurrence of Cd-related Itai-itai disease in Japan was reported.[173,202] In the 1960s a large number of Japanese farmers suffered Cd health effects after long-term consumption of Cd-enriched rice grown in paddies irrigated with Cd-polluted water from the Jinzu river.[202,203] A literature review by Nogawa[202] made the following points. (1) Average Cd levels of rice samples harvested in 1963 from the endemic area was 0.9 mg Cd/kg (wet weight), while levels in samples from the slightly endemic area and nonendemic areas were 0.3 and 0.15 mg Cd/kg, repectively. (2) Cadmium-related health problems (proteinuria and glucosuria) increased with increasing age and Cd concentration in rice. Among inhabitants older than 50 years, significant increases in proteinuria and glucosuria (the presence of proteins and sugars in urine, repectively) were observed if their rice staple contained 0.30 to 0.49 mg Cd/kg rice or more. The name "Itai-itai" (meaning "ouch ouch" in English) came from expressions of pain by women suffering bone fractures due to Cd-induced osteomalacia. This disease, however, often occurs only when individual eat high-Cd *and* low-Ca diets.[173] The early health effect of excessive chronic Cd exposure is the renal dysfunction (kidney disease), which expresses as proteinuria and glucosuria because Cd accumulates most in kidneys, and to a lesser extent in livers, with a mean residence time of 30 or more years.[175] The presence of elevated Cd triggers increased levels of metallothionein in the kidney and liver. The metallothionein is a group of low-molecular-weight (≅6 to 10 kDa), cytoplasmic proteins rich in cysteine (≅30% of its amino acids) but contains no aromatic amino acids or histidine. It has a high affinity for Cd and several other heavy metals (e.g., Zn, Cu, Hg). Seven atoms of Cd and/or other metals are bound to each molecule of metallothionein.[204] Binding of Cd by metallothionein may decrease Cd distribution to other sensitive organelles, and thus reduce Cd toxicity. On the other hand, this binding may also decrease Cd excretion, and thus prolong Cd biological life. Since Cd is highly persistent, a maximum Cd exposure limit has been proposed by the World Health Organization (WHO). WHO recommends a provisional tolerable weekly Cd intake (PTWI) of 400 to 500 μg for an adult.[203] This limit was based on the following assumptions: (1) all Cd comes from the diet (however, a sizeable portion may come from the air, especially from tobacco smoke), (2) 200 μg Cd/g (fresh weight) in the renal cortex is the critical concentration for kidney failure, (3) absorption by the gastrointestinal tract is 5% efficient, and (4) daily excretion is 0.005% of the total body burden.[175] Caution must be exercised when interpreting this PTWI because it was based on an average body weight of 70 kg. The PTWI should be lowered for populations with a smaller mean body size. For example, the PTWI for the Japanese population is 325 μg, corresponding to a maximum exposure of 1 μg Cd/kg body weight per day.[175] Also, children absorb Cd more efficiently than adults.[205]

Table 7 Exposure Assessment for Agricultural Land Applications of Sewage Sludge

Pathway		Most Exposed Individual	Most Limiting Pollutant
1	Sludge→soil→plant→human	General food chain	None
1F	Sludge→soil→plant→human	Home garden after 5 years	Cd, HCB, Pb
2F	Sludge→soil→human	Residential soil, 5 years	None
3	Sludge→soil→plant→animal→human	Farms; 40% of meat	HCB, DDT, PCB, Tox
4	Sludge→soil→animal→human	Farms; 40% of meat	None
5	Sludge→soil→plant→animal	Livestock feeds	Se, Mo, Cd, Zn
6	Sludge→soil→animal	Grazing livestock	None
7	Sludge→soil→plant	"Crops"	Cu, Zn, Ni
8	Sludge→soil→soil biota	Earthworms, slugs	Cu
9	Sludge→soil→soil biota→predator	Birds	Pb, Zn, aldrin
10	Sludge→soil→airborne dust→human	Tractor operator	None
11	Sludge→soil→surface water→human	Water quality criteria	None
12	Sludge→soil→air→human	Farm households	Possibly BaP, BEHP, DMNS
12W	Sludge→soil→groundwater→human	Well water on farms	Cd, TCE, Tox

Adapted with permission from Reference 210.

Table 8 Mean Values with Uncertainty for Some Commonplace Risks of Death

Action	Annual Risk	Uncertainty
Drinking water with EPA limit of chloroform[a]	6×10^{-7}	Factor of 10
Eating peanut butter[b] (four tablespoons/day)	8×10^{-6}	Factor of 3
Alcohol, light drinker	2×10^{-5}	Factor of 10
Frequent flying professor	5×10^{-5}	50%
Home accidents	1.1×10^{-4}	5%
Police killed in line of duty	2.2×10^{-4}	20%
Auto accidents	2.4×10^{-4}	10%
Cigarette smoking (one pack/day)	3.6×10^{-3}	Factor of 3
All cancers	2.8×10^{-3}	10%

[a]Chloroform is produced by reaction of chlorine with organic matter during the chlorination of surface waters to kill bacteria.
[b]Peanut butter may contain aflatoxin, which is a naturally occurring, cancer-causing chemical.

Adapted with permission from Reference 212.

3. Toxicities of PAHs and PCBs

Not all PAH compounds are toxic. However, some PAH members, such as benzo(a)pyrene, have been proven carcinogenic.[206] In fact, benzo(a)pyrene itself is not a carcinogen, but upon ingestion it is transformed by several enzyme systems in the body into 7,8-diol-9,10-epoxide of benzo(a)pyrene which is a potent mutagen.[207]

Toxic effects of PCBs were first recognized in 1968 when over 1600 people in Southwestern Japan were poisoned after consuming rice oil contaminated with a commercial PCB fluid; a similar poisoning incident occurred in Taiwan in 1979.[208] The PCB levels in the contaminated rice oil consumed by the Taiwanese averaged 430 mg/kg, which resulted in a blood serum PCB level of 42 μg/L as compared to 12 μg/L in a control group.[208] A common initial symptom of PCB toxicity was chloracne and related dermal problems. Other effects, such as teratogenicity, thymic and splenic atrophy, and impaired immune functions, have also been reported.[208] The most toxic PCB congeners are 3,3′,4,4′-tetra-, 3,3′,4,4′,5-penta-, and 3,3′,4,4′,5,5′-hexachloro-biphenyl which can assume coplanar conformations and are approximate isostereomers of the highly toxic hydrocarbon 2,3,7,8-tetrachlorodibenzo-*p*-dioxin (TCDD). TCDD exerts its toxicity by binding to the cytosolic aryl hydrocarbon receptor protein of the target tissues.[209]

V. MANAGEMENT OPTIONS

The preceding discussion has briefly addressed sludge properties, the transformation and movement of sludge constituents in soils, and beneficial effects as well as potentially adverse impacts of sludge use/disposal on man and the environment. To keep things in perspective, the following points should be noted. On the positive side: (1) sludge N, which is mostly organic, is less likely to cause groundwater pollution than chemical N fertilizers. Of course, sludge application rates must be based on the N need of the crops to be grown so that excessive supplies of N can be avoided to minimize leaching and surface water contamination. (2) Other nutrients, particularly P, are also valuable. In some soils, such as Fe-deficient calcareous soils, sludge can provide many essential micronutrients (e.g., Fe, Zn, Cu) even more effectively than commercial fertilizers. (3) The organic matter in sludge can adsorb and deactivate heavy metals and toxic organics to such an extent that a "clean sludge" category has been proposed,[18,210] and it can also markedly improve soil physical properties which are badly needed by disturbed or degraded lands. On the negative side: sewage sludge is contaminated with a "soup" of metals, organics, and pathogens. Most of these contaminants have no known adverse effects, but some show inconclusive (thus suspectable) effects, while a few have been proven hazardous to human health or the environment. Obviously, risks are involved in using sludges, and risk assessments should be made. We need to know how frequent (probability of occurrence) and how serious (severity of occurrence) are the adverse effects on human health and the environment when sludge is applied to lands relative to other options such as incineration or landfilling. Such information is essential for risk assessment because some events, such as the release of cyanide gas in Bhopal, India, are rare but with catastrophic consequences; others, such as auto accidents, occur rather regularly but with minor impacts.

It would be ideal if risk assessments could be derived from well-controlled epidemiological studies on the population of interest with realistic doses and exposures.

Unfortunately, there are no such studies because epidemiological research always has problems, including (1) too few subjects for confident conclusions, (2) interference from uncontrollable but unavoidable factors, (3) weak cause/effect relationships because of the chronic nature of the exposure, and (4) exposure levels many times greater than the standards being considered (e.g., soils contaminated with metals from smelters vs. those from sludge applications). Thus, in practice, risk assessment must be based on less-than-desired data involving extrapolation from high to low doses. Questions then arise in choosing models for dose/response relationship: is it linear as preferred by the EPA[211] or plateau (exponential) as advocated by Chaney[18] and Chang et al?[34] Furthermore, exposure assessment, including pathway and individual or population affected, should be performed as illustrated in Table 7 for agricultural land applications of sludge.

All risks are relative and people should be made aware of that because public acceptability of sludge utilization on lands not only depends on how "clean" the sludge is but also on public perception of risk. To broaden people's perspective, quantitative risk estimates for a variety of hazards should be expressed in some unidimensional index, such as annual probability of death or reduction of life expectancy, and compared with other risks as illustrated in Table 8.

Such risk comparisons, however, have no scientific value for acceptability of risk; some "safe" levels must be established. This approach applies the *de minimus* principle which advocates that below certain levels risks are too trivial to be regulated.[213] The "clean sludge" standards (Table 2) proposed by Chaney[210] and apparently adapted by the USEPA after some modifications[214,215] comply with this principle. Given the extensive public consultation and critical but constructive peer review of the regulations on sludge management (EPA regulations Part 503), hopefully the public health and environmental quality will be adequately protected with minimal costs to the society from the use/disposal of our own sewage sludge.

REFERENCES

1. EPA, Environmental Regulations and Technology: Use and Disposal of Municipal Wastewater Sludge, EPA-625/10-84-003, *EPA*, Washington, D.C., 1984.
2. Hasbach, A.C., Putting sludge to work, *Pollut. Eng.*, December, 62, 1991.
3. Calcutt, T. and Moss, J., Sewage sludge treatment and disposal. The way ahead, *Water Pollut. Control*, 83, 163, 1984.
4. Hue, N.V. and Ranjith, S.A., Sewage sludge in Hawaii: chemical composition and reactions with soils and plants, *J. Water, Air Soil Pollut.*, 72, 265, 1994.
5. Stephenson, T., Sources of heavy metals in wastewater, in *Heavy Metals in Wastewater and Sludge Treatment Processes*, Vol. 1, Lester, J.N., Ed., CRC, Boca Raton, FL, 1987, 31.
6. Bastian, R., Overview in sludge utilization, in *The Forest Alternative for Treatment and Utilization of Municipal and Industrial Wastes*, Cole, D.W., Henry, C.L., and Nutter, W.L., Eds., University of Washington, Seattle, 1986, 7.
7. Lester, J.N., Biological treatment, in *Heavy Metals in Wastewater and Sludge Treatment Processes*, Vol. 2, Lester, J.N., Ed., CRC, Boca Raton, FL, 1987, 15.
8. Bowker, R.P.G. and Stensel, H.D., *Phosphorus Removal from Wastewater*, Noyes Data Corporation, Park Ridge, NJ, 1990, 115.

9. Demuyncle, M., Nyns, E.J., and Naveau, H., Use of digested effluents in agriculture, in *Long-Term Effects of Sewage Sludge and Farm Slurries Applications*, Williams, J.H., Guidid, G., and L'Hermite, P., Eds., Elsevier, New York, 1985, 2.
10. Sommers, L.E., Chemical composition of sewage sludges and analysis of their potential use as fertilizers, *J. Environ. Qual.*, 6, 225, 1977.
11. LeBrun, T.J. and Tortorici, L.D., Thermal Treatment of Municipal Sewage Sludges, EPA project summary, EPA-600/52-84-104, EPA, Washington, D.C., 1984.
12. Process Design Manual for Dewatering Municipal Wastewater Sludges, EPA-625/1-82-014, EPA, Washington, D.C., 1982.
13. Sommers, L.E., Nelson, D.W., and Yost, K.J., Variable nature of chemical composition of sewage sludges, *J. Environ. Qual.*, 5, 303, 1976.
14. Hansen, L.G. and Chaney, R.L., Environmental and food chain effects of the agricultural use of sewage sludges, in *Reviews in Environmental Toxicology*, Vol. 1, Hodgson, E., Ed., Elsevier, Amsterdam, 1984, 103.
15. Mumma, R.D., Rashid, K.A., Raupach, D.C., Shane, B.S., Scarlet-Kranz, J.M., Bache, C.A., Gutenmann, W.H., and Lisk, D.J., Mutagens, toxicants, and other constituents in small city sludges in New York state, *Arch. Environ. Contam. Toxicol.*, 17, 657, 1988.
16. Hue, N.V., Correcting soil acidity of a highly weathered Ultisol with chicken manure and sewage sludge, *Commun. Soil Sci. Plant Anal.*, 23, 241, 1992.
17. Chaussod, R., Catroux, G., and Juste, C., Effects of anaerobic digestion of organic wastes on carbon and nitrogen mineralization rates: laboratory and field experiments, in *Efficient Land Use of Sludge and Manure*, Kofoed, A., Williams, J.H., and L'Hermite, P., Eds., Elsevier, New York, 1986, 24.
18. Chaney, R.L., Twenty years of land application research, *Biocycle*, 31, 54, 1990.
19. Chaney, R.L., Food chain impact, *Biocycle*, 31, 68, 1990.
20. Peer Review of Standards for the Disposal of Sewage Sludge, U.S. EPA proposed rule 40 CFR Parts-257 and 503, CSRS, University of California, Riverside, 1989.
21. Jacobs, W.L., O'Connor, G.A., Overcash, M.A., Zabik, M.J., and Rygiewicz, P., Effects of trace organics in sewage sludges on soil-plant systems and assessing their risk to humans, in *Land Application of Sludge*, Page, A.L., Logan, T.J., and Ryan, J.A., Eds., Lewis Publishers, Chelsea, MI, 1987, 101.
22. Kuchenrither, R.D. and McMillan, S.I., Preview analysis of national sludge survey, *Biocycle*, 32, 60, 1991.
23. Gerba, C.P., Pathogens, in *Utilization of Municipal Wastewater and Sludge on Land*, Page, A.L., Gleason, T.L., Smith, J.E., Iskandar, I.K., and Sommers, L.E., Eds., University of California, Riverside, 1983, 147.
24. Gaus, J., Brallier, S., Harrison, R., Coveny S., and Dempsey, J., Pathogen survival and transport in sludge amended soils, in *Literature Reviews on Environmental Effects of Sludge Management*, Henry, C.L. and Harrison, R.B., Eds., University of Washington, Seattle, 1991.
25. Metro, Draft Sludge Management Plan. Appendix A. Metro Sludge Quality: Monitoring Report and Literature Review, Municipality of metropolitan Seattle, 1983.
26. Gerba, C.P. and Bitton, G., Microbial pollutions: their survival and transport pattern to groundwater, in *Groundwater Pollution Microbiology*, Bitton, G. and Gerba, C. P., Eds., John Wiley & Sons, New York, 1984, 65.
27. Sorber, C.A. and Moore, B.E., Survival and transport of pathogens in sludge-amended soils. A critical literature review, NTIS, #PB87-180337, Washington, D.C., 1987.

28. Goldstein, N., Sludge management practices in the U.S., *Biocycle*, 32, 46, 1991.
29. Johannsen, C.J., *Site Selection and Land-Use Considerations*, North Central Regional Extension publication, No. 52, Michigan State University, East Lansing, 1977.
30. Galloway, H.M. and Jacobs, L.W., *Sewage sludges. Characteristics and management*, North Central Regional Extension publication, Michigan State University, East Lansing, No. 52, 1977.
31. *Recycling Municipal Sludges and Effluents on Land*, EPA, Washington, D.C., 1973.
32. Hinesly, T.D., Hansen, L.G., Bray, D.H., and Redborg, K.E., *Long-Term Use of Sewage Sludge on Agricultural and Disturbed Lands*, Project summary, EPA-600/S2-84-128, EPA, Washington, D.C., 1984.
33. Koskela, I., Long-term field experiments on the fertilizer value and soils ameliorating properties of dewatered sludges, in *Long-Term Effects of Sewage Sludge and Farm Slurries Applications*, Williams, J.H., Guidi, G., and L'Hermite, P., Eds., Elsevier, New York, 1985, 98.
34. Chang, A.C., Page, A.L., and Warneke, J.E., Long-term sludge applications on cadmium and zinc accumulation in Swiss chard and radish, *J. Environ. Qual.*, 16, 217, 1987.
35. Hue, N.V., Residual effects of sewage-sludge application on plant and soil-profile chemical composition, *Commun. Soil Sci. Plant Anal.*, 19, 1633, 1988.
36. Hue, N.V., Interaction of $Ca(H_2PO_4)_2$ applied to an Oxisol and previous sludge amendment: soil and crop response, *Commun. Soil Sci. Plant Anal.*, 21, 61, 1990.
37. Wild, S.R. and Jones, K.C., Organic chemicals entering agricultural soils in sewage sludges: screening for their potential to transfer to crop plants and livestock, *Sci. Tot. Environ.*, 119, 85, 1992.
38. Mengel, K. and Kirkby, E.A., *Principles of Plant Nutrition*, International Potash Institute, Berne, 1979, 593.
39. Nichols, C.G., U.S. forestry uses of minicipal sewage sludge, in *Alternative Uses for Sewage Sludge*, Hall, J.E., Ed., Pergamon, Press, Elmsford, NY, 1991.
40. Henry, C.L. and Cole, D.W., Pack forest sludge demonstration program: history and current activities, in *The Forest Alternative for Treatment and Utilization of Municipal and Industrial Wastes*, Cole, D.W., Henry, C.L., and Nutter, W.L., Eds., University of Washington, Seattle, 1986.
41. Byron, K.L. and Bradshaw, A.D., The potential value of sewage sludge in land reclamation, in *Alternative Uses for Sewage Sludge*, Hall, J.E., Ed., Pergamon, Press, Elmsford, NY, 1991, 11.
42. Sopper, W.E., Utilization of sewage sludge in the United States for mine land reclamation, in *Alternative Uses for Sewage Sludge*, Hall, J.E., Ed., Pergamon, Press, Elmsford, NY, 1991, 21.
43. Matthews, P.J. and Border, D.J., Compost — a sewage sludge resource for the future, in *Alternative Uses for Sewage Sludge*, Hall, J.E., Ed., Pergamon, Press, Elmsford, NY, 1991. 303.
44. Goldstein, N. and Riggle, D., Sludge composting maintains momentum, *Biocycle*, December, 26, 1990.
45. Logsdon, G., Selling sludge compost, *Biocycle*, May, 75, 1990.
46. Understanding the process. The art and science of composting, *Biocycle*, Press PA, 1991, 14.
47. Martel, J.L., Economics and marketing of urban sludge composts in the EEC, in *Alternative Uses for Sewage Sludge*, Hall, J.E., Ed., Pergamon Press, Elmsford, NY, 1991, 291.

48. De Bekker, P.H.A.M.J. and van den Berg, J.J., Landfilling with sewage sludge, in *Treatment of Sewage Sludge*, Bruce, A.M., Colin, F., and Newman, P.J., Eds., Elsevier, New York, 1989, 72.
49. Boari, G., Mancini, I.M., and Spinosa, L., Landfill leachate: operating modalities for its optimal treatment, in *Treatment of Sewage Sludge*, Bruce, Colin, and Newman, Eds., Elsevier, New York, 1989, 14.
50. Municipal wastewater sludge combustion technology, EPA/625/4-85-015, EPA, Washington, D.C., 1985.
51. Parker, C.F. and Sommers, L.E., Mineralization of nitrogen in sewage sludges, *J. Environ. Qual.*, 12, 150, 1983.
52. Lindemann, W.C. and Cardenas, M., Nitrogen mineralization potential and nitrogen transformations of sludge-amended soil, *Soil Sci. Soc. Am. J.*, 48, 1072, 1984.
53. Chae, Y.M. and Tabatabai, M.A., Mineralization of nitrogen in soils amended with organic wastes, *J. Environ. Qual.*, 15, 193, 1986.
54. Keeney, D., Sources of nitrate to ground water, *Crit. Rev. Environ. Control*, 16, 257, 1986.
55. Gilmour, J.T. and Clark, M.D., Nitrogen release from wastewater sludge: a site specific approach, *J. Water Pollut. Control Fed.*, 60, 494, 1988.
56. Clark, M.D., Annual Report to W-170 Regional Project, Las Vegas, 1991.
57. Ladd, J.N. and Jackson, R.B., Biochemistry of ammonification, in *Nitrogen in Agricultural Soils*, (*Agronomy* #22), Stevenson, F.J., Ed., American Society of Agronomists, Madison, WI, 1982, 173.
58. Hall, J.E., Predicting the nitrogen values of sewage sludges, in *Processing and Use of Sewage Sludge*, L'Hermite, P. and Ott, H., Eds., D. Reidel, Dordrecht, The Netherlands, 1984, 268.
59. Epstein, E., Keane, D.B., Meisinger, J.J., and Legg, J.O., Mineralization of nitrogen from sewage sludge and sludge compost, *J. Environ. Qual.*, 7, 217, 1978.
60. Brockway, D.G. and Urie, D.H., Determining sludge fertilization rates for forests from nitrate-N in leachate and groundwater, *J. Environ. Qual.*, 12, 487, 1983.
61. Ryan, J.A., Keeney, D.R., and Walsh, L.M., Nitrogen transformations and availability of an anaerobically digested sewage sludge in soil, *J. Environ. Qual.*, 2, 489, 1973.
62. Keeney, D.R., Lee, K.W., and Walsh, L.M., Guidelines for the application of wastewater sludge to agricultural lands in Wisconsin, Wisconsin Department Natural Resources Tech. Bull., 1975, 88.
63. Pratt, P., Broadbent, F.E., and Martin, J.P., Using organic wastes as nitrogen fertilizers, *Calif. Agric.*, 27, 10, 1973.
64. EPA, Process design for agricultural utilization, in *Process Design Mannual for Land Application of Municipal Sludge*, EPA-625/1-83-016, EPA, Washington, D.C., 1983, 6.1.
65. Sims, J.T. and Boswell, F.C., The influence of organic wastes and inorganic nitrogen sources on soil nitrogen, yield, and elemental composition of corn, *J. Environ. Qual.*, 9, 512, 1980.
66. Terry, R.E., Nelson, D.W., and Sommers, L.E., Nitrogen transformations in sewage sludge-amended soils as affected by soil environmental factors, *Soil Sci. Soc. Am. J.*, 45, 506, 1981.
67. Magdoff, F.R. and Amadon, J.F., Nitrogen availability from sewage sludge, *J. Environ. Qual.*, 9, 451, 1980.

68. Kirkham, M.B., Disposal of sludge on land: effects on soils, plants, and ground water, *Compost Sci.*, March–April, 6, 1974.
69. Stevenson, F.J., Origin and distribution of nitrogen in soil, in *Nitrogen in Agricultural Soils*, (*Agronomy*, #22), Stevenson, F.J., Ed., American Society of Agronomists, Madison, WI, 1982, 1.
70. King, L.D., Mineralization and gaseous loss of nitrogen in soil-applied liquid sewage sludge, *J. Environ. Qual.*, 2, 356, 1973.
71. Schmidt, E.L., Nitrification in soil, in *Nitrogen in Agricultural Soils,* (*Agronomy* #22), Stevenson, F.E., Ed., American Society of Agronomists, WI, Madison, 1982, 253.
72. Hue, N.V., *Effects of Phosphorus Levels and Clays on the Nitrification Process*, Ph.D. dissertation, Auburn University, Auburn, AL, 1981.
73. Hue, N.V. and Adams F., Effect of phosphorus level on nitrification rates in three low-phosphorus Ultisols, *Soil Sci.*, 137, 324, 1984.
74. Keeney, D.R., Sahrawat, K.L., and Adams, S.S., Carbon dioxide concentration in soil: effects on nitrification, denitrification and associated nitrous oxide production, *Soil Biol. Biochem.*, 17, 571, 1985.
75. Firestone, M.K., Biological denitrification, in *Nitrogen in Agricultural Soils,* (*Agronomy* #22), Stevenson, F.E., Ed., American Society of Agronomists, Madison, WI, 1982, 289.
76. Raveh, A. and Avnimelech, Y., Minimizing nitrate seepage from the Hula Valley into lake Kinneret (Sea of Galilee). I. Enhancement of nitrate reduction by sprinkling and flooding, *J. Environ. Qual.*, 2, 455, 1973.
77. Fedkiw, J., Nitrate Occurrence in U.S. Waters, USDA, Washington, D.C., 1991, 35.
78. Davidson, J.M., Rao, P.S.C., and Nkedi-Kizza, P., Physical processes influencing water and solute transport in soils, in *Chemical Mobility and Reactivity in Soil Systems*, SSSA Special Publ. No. 11, Soil Science Society of America, Madison, WI, 1983, 35.
79. Balasubramanian, V., Kanehiro, Y., Rao, P.S.C., and Green, R.E., Field study of solute movement in a highly aggregated Oxisol with intermittent flooding. I. Nitrate, *J. Environ. Qual.*, 2, 359, 1973.
80. Wells, C.G., Murphy, C.E., Davis, C., Stone, D.M., and Hollod, G.J., Effect of sewage sludge from two sources on element flux in soil solution of loblolly pine plantations, in *The Forest Alternative for Treatment of Municipal and Industrial Wastes*, Cole, D.W., Henry, C.L., and Nutter, W.L., Eds., University of Washington, Seattle, 1986, 154.
81. Davidson, E.A., Hart, S.C., and Firestone, M.K., Internal cycling of nitrate in soils of nitrate in soils of a mature coniferous forest, *Ecology*, 73, 1148, 1992.
82. King, L.D., Availability of nitrogen in municipal, industrial and animal wastes, *J. Environ. Qual.*, 13, 609, 1984.
83. Sabey, B.R., Agbim, N.N., and Markstrom, D.C., Land application of sewage sludge. III. Nitrate accumulation and wheat growth resulting from addition of sewage sludge and wood wastes to soils, *J. Environ. Qual.*, 4, 388, 1975.
84. McClaugherty, S.A., Aber, J.D., and Melillo, J.M., The role of fine roots in organic matter and nitrogen budgets of two forested ecosystems, *Ecology*, 63, 1481, 1982.
85. Lake, D.L., Sludge disposal to land, in *Heavy Metals in Wastewater and Sludge Treatment Processes*, Vol. 2, Lester, J.N., Ed., CRC Press, Boca Raton, FL, 1987, 91.
86. Alloway, B.J., *Heavy Metals in Soils*, Blackie & Sons, Glasgow, 1990, 339.

87. Dowdy, R.H. and Volk, V.V., Movement of heavy metals in soils, in *Chemical Mobility and Reactivity in Soil Systems*, SSSA Special Publ. No. 11, Soil Science Society of America, Madison, WI, 1983, 229.
88. Sommers, L., Volk, V.V., Giordano, P.M., Sopper, W.E., and Bastian, R., Effects of soil properties on accumulation of trace elements by crops, in *Land Application of Sludge*, Page, A.L., Logan, T., and Ryan, J., Eds., Lewis Publishers, Chelsea, MI, 1987, 5.
89. Corey, R.B., King, L.D., Lue-Hing, C., Fanning, D.S., Street, J.J., and Walker, J.M., Effects of sludge properties on accumulation of trace elements by crops, in *Land Application of Sludge*, Page, A.L., Logan, T., and Ryan, J., Eds., Lewis Publishers, Chelsea, MI, 1987, 25.
90. King, L.D., Retention of metals by several soils of the Southeastern United States, *J. Environ. Qual.*, 17, 239, 1988.
91. Kuo, S., Heilman, P.E., and Baker, A.S., Distribution and forms of copper, zinc, cadmium, iron, and manganese in soils near a copper smelter, *Soil Sci.*, 135, 101, 1983.
92. Emmerich, W.E., Lund, L.J., Page, A.L., and Chang, A.C., Solid phase forms of heavy metals in sewage sludge-treated soils, *J. Environ. Qual.*, 11, 178, 1982.
93. Stover, R.C., Sommers, L.E., and Silviera, D.J., Evaluation of metals in wastewater sludge, *J. Water Pollut. Control Fed.*, 48, 2165, 1976.
94. Hickey, M.G. and Kittrick, J.A., Chemical partitioning of cadmium, copper, nickel and zinc in soils and sediments containing high levels of heavy metals, *J. Environ. Qual.*, 13, 372, 1984.
95. Stevenson, F.J. and Ardakani, M.S., Organic matter reactions involving micronutrients in soils, in *Micronutrients in Agriculture*, Soil Science Society of America, Madison, WI, 1972, 79.
96. Santillan-Medrano, J. and Jurinak, J.J., The chemistry of lead and cadmium in soil:solid phase formation, *Soil Sci. Soc. Am. Proc.*, 39, 851, 1975.
97. Miller, W.P., McFee, W.W., and Kelly, J.M., Mobility and retention of heavy metals in sandy soils, *J. Environ. Qual.*, 12, 579, 1983.
98. Harter, R.D., Adsorption of copper and lead by Ap and B2 horizons of several Northeastern United States soils, *Soil Sci. Soc. Am. J.*, 43, 679, 1979.
99. Alloway, B.J., Cadmium, in *Heavy Metals in Soils*, Alloway, B.J., Ed., Blackie & Sons, Glasgow, 1990, 100.
100. Soon, Y.K., Solubility and sorption of cadmium in soils amended with sewage sludge, *J. Soil Sci.*, 32, 85, 1981.
101. Alloway, B.J., Soil processes and the behavior of metals, in *Heavy Metals in Soils*, Alloway, B.J., Ed., Blackie & Sons, Glasgow, 1990, 7.
102. Kiekens, L., Zinc, in *Heavy Metals in Soils*, Alloway, B.J., Ed., Blackie & Sons, Glasgow, 1990, 261.
103. Schnitzer, M. and Khan, S.U., *Humic Substances in the Environment*, Marcel Dekker, New York, 1972, 327.
104. McGrath, S.P., Long-term studies of metal transfers following application of sewage sludge, in *Pollutant Transport and Fate in Ecosystems*, Bull. Ecological Society, Special Publ. No. 6, Coughtrey, P.S., et al., Eds., Blackwell Scientific, Oxford, 1987, 301.
105. Dudley, L.M., McNeal, B.L., Baham, J.E., Coray, C.S., and Cheng, H.H., Characterization of soluble organic compounds and complexation of copper, nickel, and zinc in extracts of sludge-amended soils, *J. Environ. Qual.*, 16, 341, 1987.

106. Emmerich, W.E., Lund, L.J., Page, A.L., and Chang, A.C., Predicted solution phase forms of heavy metals in sewage sludge-treated soils, *J. Environ. Qual.*, 11, 182, 1982.
107. Kelling, K.A., Keeney, D.R., Walsh, L.M., and Ryan, J.A., A field study of the agricultural use of sewage sludge. III. Effect on uptake and extractability of sludge-borne metals, *J. Environ. Qual.*, 6, 352, 1977.
108. Lund, L.J., Page, A.L., and Nelson, C.O., Movement of heavy metals below sewage disposal ponds, *J. Environ. Qual.*, 5, 330, 1976.
109. Karickhoff, S.W., Brown, D.S., and Scott, T.A., Sorption of hydrophobic pollutants on natural sediments, *Water Res.*, 13, 241, 1979.
110. Hassett, J.J. and Banwart, W.L., The sorption of nonpolar organics by soils and sediments, in *Reactions and Movements of Organic Chemicals in Soils*, SSSA Special Publ., No. 22, Sawhney, B.L. and Brown, K., Eds., Soil Science Society of America, Madison, WI, 1989, 31.
111. Means, J.C., Wood, G.S., Hassett, J.J., and Banwart, W.L., Sorption of polynuclear aromatic hydrocarbons by sediments and soils, *Environ. Sci. Technol.*, 14, 1524, 1980.
112. Bell, J.P. and Tsezos, M., Removal of hazardous organic pollutants by biomass adsorption, *J. Water Pollut. Control Fed.*, 59, 191, 1987.
113. Wild, S.R., Berrow, M.L., and Jones, K.C., The persistence of polynuclear aromatic hydrocarbons (PAHs) in sewage sludge amended agricultural soils, *Environ. Pollut.*, 72, 141, 1991.
114. Anderson, M.R., and Pankow, J.F., A case study of a chemical spill: polychlorinated biphenyls (PCBs). III. PCB sorption and retardation in soil underlying site, *Water Resour. Res.*, 22, 1051, 1986.
115. Madison Metropolitan Sewerage, Interim draft report on PCB field studies, Madison, WI, 1988.
116. Pardue, J.H., DeLaune, R.D., and Patrick, W.H., Jr., Removal of PCBs from wastewater in a simulated overland flow treatment system, *Water Res.*, 22, 1011, 1988.
117. Wolfe, N.L., Metwally, M.E., and Moftah, A.E., Hydrolytic transformations of organic chemicals in the environment, in *Reactions and Movements of Organic Chemicals in Soils*, SSSA Special Publ. No. 22, Sawhney, B.L. and Brown, K., Eds., Soil Science Society of America, Madison, WI, 1989, 229.
118. Overcash, M.R., Land treatment of municipal effluent and sludge: specific organic compounds, in *Proc. Workshop on Utilization of Municipal Wastewater and Sludge on Land*, Page, A.L., Gleason, T.L., Smith, J.E., Iskandar, I.K., and Sommers, L.E., Eds., University of California, Riverside, 1983, 199.
119. Wild, S.R. and Jones, K.C., Biological and abiotic losses of polynuclear aromatic hydrocarbons (PAHs) soils freshly amended with sewage sludge, *Environ. Toxicol. Chem.*, 12, unpublished data, 1993.
120. Alexander, M. and Scow K.M., Kinetics of biodegradation in soil, in *Reactions and Movement of Organic Chemicals in Soils*, SSSA Special Publ. No. 22, Sawhney, B.L. and Brown, K., Eds., Soil Science Society of America, Madison, WI, 1989, 243.
121. Tucker, E.S., Saeger, V.W., and Hicks, O., Activated sludge primary biodegradation of polychlorinated biphenyls, *Bull. Environ. Contam. Toxicol.*, 14, 705, 1975.
122. Strand, S.E., Brallier, S., and Harrison, R.B., The fate of trace synthetic organics in sludge applied to soil, in *Literature Reviews on Environmental Effects of Sludge Management*, Henry, C.L. and Harrison, R.B., Eds., University of Washington, Seattle, 1991.

123. Glotfelty, D.E. and Schomburg, C.J., Volatilization of pesticides from soil, in *Reactions and Movements of Organic Chemicals in Soils*, SSSA Special Publ., No. 22, Soil Science Society of America, Madison, WI, 1989, 181.
124. Wild, S.R. and Jones, K.C., Polynuclear aromatic hydrocarbon uptake by carrots grown in sludge-amended soil, *J. Environ. Qual.*, 21, 217, 1992.
125. Ignesti, G., Lodovici, M., Dolara, P., Lucia, P., and Grechi, D., Polycyclic aromatic hydrocarbons in olive fruits as a measure of air pollution in the valley of Florence (Italy), *Bull. Environ. Contam. Toxicol.*, 48, 809, 1992.
126. Hue, N.V., A possible mechanism for manganese phytotoxicity in Hawaii soils amended with a low-manganese sewage sludge, *J. Environ. Qual.*, 17, 473, 1988.
127. Hue, N.V., Effects of organic acids/anions on P sorption and phytoavailability in soils with different mineralogies, *Soil Sci.*, 152, 463, 1991.
128. Licudine, D.L. and Hue, N.V., Liming effects of organic manure, *Agron. Abst.*, 1992, 283.
129. Hue, N.V., Craddock, G., and Adams, F., Effect of organic acids on aluminum toxicity in subsoils, *Soil Sci. Soc. Am. J.*, 50, 28, 1986.
130. Sposito, G., Soil organic matter, in *The Chemistry of Soils*, Oxford University Press, Oxford, 1989, 42.
131. Thomas, G.W. and Hargrove, W.L., The chemistry of soil acidity, in *Soil Acidity and Liming*, Adams, F., Ed., Soil Science Society of America, Madison, WI, 1984, 3.
132. Guidi, G. and Hall, J.E., Effects of sewage sludge on the physical and chemical properties of soil, in *Processing and Use of Sewage Sludge*, L'Hermite, P. and Ott, H., Eds., D. Reidel, Dordrecht, The Netherlands, 1984, 295.
133. Epstein, E., Taylor, J.M., and Chaney, R.L., Effects of sewage sludge and sludge compost applied to soil on some soil physical and chemical properties, *J. Environ. Qual.*, 5, 422, 1976.
134. Kladivko, E.J. and Nelson, D.W., Changes in soil properties from application of anaerobic sludge, *J. Water Pollut. Control Fed.*, 51, 325, 1979.
135. Bohn, H., McNeal, B., and O'Connor, G., Soil organic matter, in *Soil Chemistry*, John Wiley & Sons, New York, 1985, 135.
136. Hall, J.E. and Coker, E.G., Some effects of sewage sludge on soil physical conditions and plant growth, in *The Influence of Sewage Sludge Application on Physical and Biological Properties of Soils*, Catroux, G., L'Hermite, P., and Suess, E., Eds., D. Reidel, Dordrecht, The Netherlands, 1983, 43.
137. Gupta, S.C., Dowdy, R.H., and Larson, W.E., Hydraulic and thermal properties of a sandy soil as influenced by incorporation of sewage sludge, *Soil Sci. Soc. Am. J.*, 41, 601, 1977.
138. Hemphill, D.D., Jr., Volk, V.V., Sheets, P.J., and Wickliff, C., Lettuce and broccoli response and soil properties resulting from tannery waste applications, *J. Environ. Qual.*, 14, 159, 1985.
139. Keefer, R.F., Singh, R.N., and Horvath, D.J., Chemical composition of vegetables grown on an agricultural soil amended with sewage sludges, *J. Environ. Qual.*, 15, 146, 1986.
140. Mondy, N.I., Naylor, L.M., and Phillips, J.C., Quality of potatoes grown in soils amended with sewage sludge, *J. Agric. Food Chem.*, 33, 229, 1985.
141. Hemphill, D.D., Jr., Jackson, T.L., Martin, L.W., Kiemnec, G.L., Hanson, D., and Volk, V.V., Sweet corn response to application of three sewage sludges, *J. Environ. Qual.*, 11, 191, 1982.

142. Rappaport, B.D., Martens, D.C., Reneau, R.B., Jr., and Simpson, T.W., Metal availability in sludge-amended soils with elevated metal levels, *J. Environ. Qual.*, 17, 42, 1988.
143. Vlamis, J., Willians, D.E., Corey, J.E., Page, A.L., and Ganje, T.J., Zinc and cadmium uptake by barley in field plots fertilized seven years with urban and suburban sludge, *Soil Sci.*, 139, 81, 1.
144. Consiglio, M., Barberis, R., Piccone, G., DeLuca, G., and Trombetta, A., Productivity and quality of cereal crops grown on sludge-treated soil, in *Processing and Use of Organic Sludge and Liquid Agricultural Wastes*, L'Hermite, P., Ed., D. Reidel, Dordrecht, The Netherlands, 1986, 436.
145. McCaslin, B.D., Davis, J.G., Cihacek, L., and Schluter, L.A., Sorghum yield and soil analysis from sludge-amended calcareous iron-deficient soil, *Agron. J.*, 79, 204, 1987.
146. Hemphill, D.D., Jr., Volk, V.V., Sheets, P.J., and Wickliff, C., Fescue response and soil properties following soil amendment with tannery wastes, *Agron. J.*, 76, 719, 1985.147.
147. Brockway, D.G., Forest floor, soil, and vegetation responses to sludge fertilization in red and white pine plantations, *Soil Sci. Am. J.*, 47, 776, 1983.
148. Henry, C.L., Growth response, mortality, and foliar nitrogen concentrations of four free species treated with pulp and paper and municipal sludges, in *The Forest Alternative for Treatment and Utilization of Municipal and Industrial Wastes*, Cole, D.W., Henry, C.L., and Nutter,W.L., Eds., University of Washington, Seattle, 1986, 258.
149. Bertrand, J.E., Lutrick, M.C., Breland, H.L., and West, R.L., Effects of dried digested sludge on performance, carcass quality and tissue residues in beef steers, *J. Animal Sci.*, 50, 35, 1980.
150. Lisk, D.J., Boyd, R.D., Telford, J.N., Babish, J.G., Stoewsand, G.S., Bache, C.A., and Gutenmann, W.H., Toxicologic studies with swine fed corn grown on municipal sewage sludge-amended soil, *J. Animal Sci.*, 55, 613, 1982.
151. Dowdy, R.H., Bray, B.J., Goodrich, R.D., Marten, G.C., Pamp, D.E., and Larson, W.E., Performance of goats and lambs fed corn silage produced on sludge-amended soil, *J. Environ. Qual.*, 12, 467, 1983.
152. Dowdy, R.H., Bray, B.H., and Goodrich, R.D., Trace metal and mineral composition of milk and blood from goats fed silage produced on sludge-amended soil, *J. Environ. Qual.*, 12, 473, 1983.
153. Bray, B.J., Dowdy, R.H., Goodrich, R.D., and Pamp, D.E., Trace metal accumulation in tissues of goats fed silage produced on sewage sludge-amended soil, *J. Environ. Qual.*, 14, 114, 1985.
154. Telford, J.N., Babish, J.G., Johnson, B.E., Thonney, M.L., Currie, W.B., Bache, C.A., Gutenmann, W.H., and Lisk, D.J., Toxicologic studies with pregnant goats fed grass-legume silage grown on municipal sludge-amended subsoil, *Arch. Environ. Contam. Toxicol.*, 13, 635, 1984.
155. Hinesly, T.D., Hansen, L.G., Bray, D.J., and Redborg, R.E., Transfer of sludge-borne cadmium through plants to chickens, *J. Agric. Food Chem.*, 33, 173, 1985.
156. Baxter, J.C., Barry, B., Johnson, D.E., and Kienholz, E.W., Heavy metal retention in cattle tissues from ingestion of sewage sludge, *J. Environ. Qual.*, 11, 616, 1982.
157. Sanson, D.W., Hallford, D.M., and Smith, G.S., Effects of dietary sewage solids on feedlot performance, carcass characteristics, serum constituents and tissue elements of growing lambs, *J. Anim. Sci.*, 59, 425, 1984.

158. Sanson, D.W., Hallford, D.M., and Smith, G.S., Effects of long-term consumption of sewage solids on blood, milk and tissue elemental composition of breeding ewes, *J. Anim. Sci.*, 59, 416, 1984.
159. Smith, G.S., Hallford, D.M., and Watkins, J.B., III, Toxicological effects of gamma-irradiated sewage solids fed as seven-percent of diet to sheep for four years, *J. Anim. Sci.*, 61, 931, 1985.
160. Stadelmann, X. and Furrer, O.J., Influence of sewage sludge application on organic matter content, micro-organisms and microbial activities of a sandy loam soil, in *The Influence of Sewage Sludge Application on Physical and Biological Properties of Soils*, Catroux, G., L'Hermite, P., and Suess, E., Eds., D. Reidel, Dordrecht, The Netherlands, 1983, 141.
161. Pera, A., Giovannetti, M., Vallini, G., and DeBertoldi, M., Land application of sludge: effects on soil microflora, in *The Influence of Sewage Sludge Application on Physical and Biological Properties of Soils*, Catroux, G., L'Hermite, P., and Suess, E., Eds., D. Reidel, Dordrecht, The Netherlands, 1983, 208.
162. Barkay, T., Tripp, S.C., and Olson, B.H., Effect of metal-rich sewage sludge application on the bacterial communities of grasslands, *Appl. Environ. Microbiol.*, 49, 333, 1985.
163. Kinkle, B.K., Angle, J.S., and Keyser, H.H., Long-term effects of metal-rich sewage sludge application on soil populations of *Bradyrhizobium japonicum*, *Appl. Environ. Microbiol.*, 53, 315, 1987.
164. MacConnell, G.S., Wells, C.G., and Metz, L.J., Influence of municipal sludge on forest soil mesofauna, in *The Forest Alternative for Treatment and Utilization of Municipal and Industrial Wastes*, Cole, D.W., Henry, C.L., and Nutter, W.L., Eds., University of Washington, Seattle, 1986, 177.
165. Chang, A.C., Warneke, J.E., Page, A.L., and Lund, L.J., Accumulation of heavy metals in sewage sludge-treated soils, *J. Environ. Qual.*, 13, 87, 1984.
166. Juste, C. and Mench, M., Long-term application of sewage sludge and its effects on metal uptake by crops, in *Biogeochemistry of Trace Metals*, Adriano, D.C., Ed., Lewis, Publishers, Chelsea, MI, 1992, 159.
167. Wild, S.R., Waterhouse, K.S., McGrath, S.P., and Jones, K.C., Organic contaminants in an agricultural soil with a known history of sewage sludge amendments: polynuclear aromatic hydrocarbons, *Environ. Sci. Technol.*, 24, 1706, 1990.
168. Purves, D., *Trace-Element Contamination of the Environment*, Elsevier, Amsterdam, 1985, 243.
169. Christensen, T.H., Cadmium soil sorption at low concentrations. I. Effect of time, cadmium load, pH, and calcium, *Water, Air, Soil Pollut.*, 21, 105, 1984.
170. Beyer, W.N., Chaney, R.L., and Mulhern, B.M., Heavy metal concentrations in earthworms from soil amended with sewage sludge, *J. Environ. Qual.*, 11, 81, 1982.
171. Reddy, G.B., Cheng, C.N., and Dunn, S.J., Survival of *Rhizobium japonicum* in soil-sludge environment, *Soil Biol. Biochem.*, 15, 343, 1983.
172. Giller, K.E., McGrath, S.P., and Hirsch, P.R., Absence of nitrogen fixation in clover grown on soil subject to long-term contamination with heavy metals is due to survival of only ineffective *Rhizobium*, *Soil Biol. Biochem.*, 21, 841, 1989.
173. Logan, T. and Chaney, R.L., Metals, in *Utilization of Municipal Wastewater and Sludge on Land*, Page, A.L., Gleason, T.L., Smith, J.E., Iskandar, I.K., and Sommers, L.E., Eds., University of Calif., Riverside, 1983, 235.

174. Chaney, R.L., Bruins, J.F., Baker, D.E., Korcak, R.F., Smith, J.E., and Cole, D., Transfer of sludge-applied trace elements to the food chain, in *Land Application of Sludge*, Page, A.L., Logan, T., and Ryan, J., Eds., Lewis Publishers, Chelsea, MI, 1987, 67.
175. Jackson, A.P., and Alloway, B.J., The transfer of cadmium from agricultural soils to the human food chain, in *Biogeochemistry of Trace Metals*, Adriano, D.C., Ed., Lewis, Publishers, Chelsea, MI, 1992, 229.
176. Chaney, R.L., Health risks assocated with toxic metals in municipal sludge, in *Sludge—Health Risks of Land Application*, Bitton, G., Damron, B.L., Edds, G.T., and Davidson, J.M., Eds., Ann Arbor Science, Ann Arbor, MI, 1980, 59.
177. Lutrick, M.C., Robertson, W.K., and Cornell, J.A., Heavy applications of liquid-digested sludge on three Ultisols. II. Effects on mineral uptake and crop yield, *J. Environ. Qual.*, 11, 283, 1982.
178. O'Connor, G.A., Kiehl, D., Eiceman, G.A., and Ryan, J.A., Plant uptake of sludge-borne PCBs, *J. Environ. Qual.*, 19, 113, 1990.
179. Fries, G.F., Potential polychlorinated biphenyl residues in animal products from application of contaminated sewage sludge to land, *J. Environ. Qual.*, 11, 14, 1982.
180. Hansen, L.G., Washko, P.W., Tuinstra, L.G.M.Th., Dorn, S.B., and Hinesly, T.D., Polychlorinated biphenyl, pesticide, and heavy metal residues in swine foraging on sewage sludge amended soils, *J. Agric. Food Chem.*, 29, 1012, 1981.
181. Campa, H., Woodyard, D.K., and Haufler, J.B., Deer and elk use of forages treated with municipal sewage sludge, in *The Forest Alternative for Treatment and Utilization of Municipal and Industrial Wastes*, Cole, D.W., Henry, C.L., and Nutter, W.L., Eds., University of Washington, Seattle, 1986, 188.
182. Haufler, J.B. and West, S.D., Wildlife responses to forest application of sewage sludge, in *The Forest Alternative for Treatment and Utilization of Municipal and Industrial Wastes*, Cole, D.W., Henry, C.L., and Nutter, W.L., Eds., University of Washington, Seattle, 1986, 110.
183. Sidle, R.C. and Kardos, L.T., Transport of heavy metal in a sludge-treated forest area, *J. Environ. Qual.*, 6, 431, 1977.
184. Hegstrom, L.J. and West, S.D., Heavy metal accumulation in small mammals following sewage sludge application to forests, *J. Environ. Qual.*, 18, 345, 1989.
185. Anderson, T.J., Barrett, G.W., Clark, C.S., Elia, V.J. and Majeti, V.A., Metal concentrations in tissues of meadow voles from sewage sludge-treated fields, *J. Environ. Qual.*, 11, 272, 1982.
186. National Research Council, *Alternative Agriculture*, National Academy of Sciences, Washington, D.C., 1989.
187. EPA, Technical guidance manual for performing waste load allocations, in *Streams and Rivers*, EPA-440/4-84-021, EPA, Washington, D.C., 1983, chap. 2.
188. Riekerk, H., The behavior of nutrient elements added to a forest soil with sewage sludge, *Soil Sci. Soc. Am. J.*, 42, 810, 1978.
189. Soon, Y.K., Bates, T.E., Beauchamp, E.G. and Moyer, J.R., Land application of chemically treated sewage sludge. I. Effects on crop yield and nitrogen availability, *J. Environ. Qual.*, 7, 264, 1978.
190. Sidle, R.C. and Kardos, L.T., Nitrate leaching in a sludge-treated forest soil, *Soil Sci. Soc. Am. J.*, 43, 278, 1979.
191. Lu, P.Y., Metcalf, R.L., Furman, R., Vogel, R.and Hasset, J., Model ecosystem studies of lead and cadmium and of urban sewage sludge containing these elements, *J. Environ. Qual.*, 4, 505, 1975.

192. Demayo, A., Taylor, M.C., Taylor, K.W. and Hodson, P.V., Toxic effects of lead and lead compounds on human health, aquatic life, wildlife plants, and livestock, *Crit. Rev. Environ. Control*, 12, 257, 1980.
193. Kowal, N.E., An overview of public health effects, in *Utilization of Municipal Wastewater and Sludge on Land*, Page, A.L., Gleason, T.L., III, Smith, J.E., Jr., Iskandar, I.K., and Sommers, L.E., Eds., University of California, Riverside, 1983, 329.
194. Dhargalkar, P.H. and Goldbach, K., Control of heavy metal emissions from waste incinerator, in *Control and Fate of Atmospheric Trace Metals*, Pacyna, J.M. and Ottar, B., Eds., Kluwer Academic, Dordrecht, The Netherlands, 1989, 33.
195. Gerstle, R.W. and Albrinck, D.N., Atmospheric emissions of metals from sewage sludge incineration, *J. Air Pollut. Control Assoc.*, 32, 1119, 1982.
196. Takeda, N. and Hiraoka, M., Combined process of pyrolysis and combustion for sludge disposal, *Environ. Sci. Technol.*, 10, 1147.
197. Newman, M.E., Smith, S.R., Henry, C.L. and Moll, T.L., Air emissions and ash resulting from incineration of municipal sewage sludge, in *Literature Reviews on Environmental Effects of Sludge Management*, Henry, C.L. and Harrison, R.B., Eds., University of Washington, Seattle, 1991, 16.
198. Denison, R.A. and Silbergeld, E.K., Risks of municipal solid waste incineration: an environmental perspective, *Risk Anal.*, 8, 343, 1988.
199. Bennett, R.L. and Knapp, K.T., Characterization of particulate emissions from municipal wastewater sludge incinerators, *Environ. Sci. Technol.*, 16, 831, 1982.
200. Kistler, R.C. and Widmer, F., Behavior of chromium, nickel, copper, zinc, cadmium, mercury, and lead during the pyrolysis of sewage sludge, *Environ. Sci. Technol.*, 21, 704, 1987.
201. Kamiya, A. and Ose, Y., Mutagenic activity and PAH analysis in municipal incinerators, *Sci. Total Environ.*, 61, 37, 1987.
202. Nogawa, K., Itai-itai disease and follow-up studies, in *Cadmium in the Environment. Part II: Health Effects*, Nriagu, J.O., Ed., John Wiley & Sons, New York, 1981, 1.
203. WNO, *Cadmium. Environmental Health Criteria*, World Health Organization (WHO), No. 134, Geneva, 1992.
204. Kotsonis, F.N. and Klaassen, C.D., Metallothionein and its interactions with cadmium, in *Cadmium in the Environment. Part II: Health Effects*, Nriagu, J. O., Ed., John Wiley & Sons, New York, 1981, 595.
205. Burger, J., Cooper, K. and Gochfeld, M., Exposure assessment for heavy metal ingestion from a sport fish in Puerto Rico: estimating risk for local fishermen, *J. Toxicol. Environ. Health*, 36, 355, 1992.
206. Boulos, B.M. and Smolinski, A.V., Assessing hazards to public health in exposure to polynuclear aromatic hydrocarbons (PAH), in *Polynuclear Aromatic Hydrocarbons: Chemistry, Characterization and Carcinogenesis*, Cooke, M. and Dennis, A.J., Eds., Battelle, Columbus, OH, 1986, 99.
207. Manahan, S.E., *Toxicological Chemistry: A Guide to Toxic Substances in Chemistry*, Lewis Publishers, Chelsea, MI, 1989, 186.
208. Safe, S., Polychlorinated biphenyls. Human health effects, in *Hazards, Decontamination, and Replacement of PCB. A Comprehensive Guide*, Crine, J. P., Ed., Plenum Press, New York, 1986, 51.
209. Safe, S., Polychorinated biphenyls (PCBs) and polybrominated biphenyls (PBBs): biochemistry, toxicology and mechanism of action, *Crit. Rev. Toxicol.*, 13, 319, 1984.

210. Chaney, R.L., Scientific analysis of proposed sludge rule, *Biocycle*, 30, 80, 1989.
211. Russell, M. and Gruber, M., Risk assessment in environmental policy-making, *Science*, 236, 286, 1987.
212. Wilson, R. and Crouch, E.A.C., Risk assessment and comparisons: an introduction, *Science*, 236, 267, 1987.
213. Cohrssen, J.J. and Covello, V.T., Risk Analysis: A Guide to Principles and Methods for Analyzing Health and Environmental Risks, Council on Environmental Quality, Executive Office of the President, Washington, D.C., 1989.
214. Donovan, J.F., Part 503 sludge management rule nears completion, *Biocycle*, 33, 82, 1992.
215. Chaney, R.L., Annual Report to W-170 Regional Project, Las Vegas, 1993.

CHAPTER 7

Compost

Margie Lynn Stratton, Allen V. Barker, and Jack E. Rechcigl

0-87371-859-3/95/$0.00+$.50

I. INTRODUCTION

Composting is defined in general terms as the practice of employing biological reduction of organic wastes to humus or humus-like substances. Traditionally, composting transforms biodegradable organic wastes into a soil amendment or fertilizer, sometimes referred to as artificial manures.[1] Composting provides an on-farm means of utilizing plant residues, such as straw, that if incorporated into soil directly without additional fertilizer, produce nutrient deficiencies in crops. Composting is used also as a means of converting objectionable wastes, such as *biosolids* (sewage sludge), garbage, organic trash, food processing wastes, and farm manures, into materials suitable for application to land. The benefits of additions of compost-generated humus to soils are the same as the benefits imparted by rich natural humus levels in soils. Compost enriches soils with plant nutrients and improves physical features of soil, such as tilth and water-holding capacity.

Composting is one phase of a continuous cycling of nutrients and carbon and can be described by arbitrarily beginning with *synthesis*. Chlorophyll-bearing plants, bacteria, and algae are the principal organisms that synthesize organic matter from mineral compounds (plant nutrients) derived from the air, the soil, the sea, or from other substrates on which these organisms grow. Animals and other nonchlorophyllous organisms (*heterotrophic* organisms) use these organic-matter synthesizers as sources of food. The heterotrophic organisms do not digest their food completely. They eliminate or leave behind materials that are food for other heterotrophic microorganisms, namely bacteria, fungi, and protozoa in soil, in sea, or in composts and other media. As a result of this sequence of processes, the nutrients originally synthesized into organic matter are recirculated in a process called *mineralization*.

The plant and animal residues, however, do not become mineralized completely in the short or intermediate term. A part of the residues is resistant to biological decay and remains for some time in a modified state that may accumulate. *Stabilization* is the term applied to the conversion of organic matter into this modified state or into intermediate states that are less reactive or less subject to biological decomposition than the original source of food for the heterotrophic organisms. In soil and in compost piles, the resistant material is a dark brown to black substance of complex chemical and physical nature. This material is *humus*.[2]

The most commonly composted materials include those which are familiar to most people, kitchen vegetable scraps, yard clippings, and wastes from landscaping and farming activities. Most of these materials, with the exception of the most recalcitrant woods and leaves, are easily compostable in the personal backyard compost pile. On an industrial level, many communities are investigating roadside pickup, chipping of materials, and large-scale composting of greens wastes in a municipal setting. Composting of biosolids or animal manures is a means of reduction or stabilization of wastes prior to utilization on land or disposal.

Satisfactory composting requires much effort to ensure that the process is environmentally safe and that the end product will be beneficial to the land to which it is applied. Composting of all wastes requires a type of engineered technology that is specific to each waste. Methods of municipal and industrial composting are complex and often require scientific investigation to achieve acceptable results.[3]

Two basic distinctions in composting methods can be made regarding oxygenation. *Aerobic* composting includes air or oxygen in the process whereas *anaerobic* composting involves a system in which oxygen is depleted. Other modifications to the process may include, among others, control of heat, chemical predigestion, presorting of materials, and inoculation with microorganisms. Adaptations in the methodology are driven by the need to dispose or recycle great masses and types of substances, as well as the need for the process to remain economically justifiable. The disposal of compostable materials in public landfills has been banned by several states in the U.S., providing further impetus for the development of improved composting technology and systems.

In 1994, the United States generated 188 million Mg of solid waste,[4] and by other estimates, in 1989, about 255 million Mg.[5] In 1985, about 83% of municipal solid wastes (MSW) was landfilled compared with 72% in 1994.[5,6] Eleven percent of the MSW was recycled in 1990, and 17% was recycled in 1994, of which 4.3 million Mg was composted, approximately 2.5% of the waste total.[5,6] MSW landfills numbered about 6500 in 1986.[6] Two-thirds of all landfills in the United States have closed since 1970 with 1000 closing in 1990 and 514 closing in 1991.[5] In 1994, a total of 5812 solid waste landfills remained open to serve the entire country, a third of which are expected to close during 1995, a rate which is expected to leave just 1200 landfills open by the year 2010.[5] No new landfills will open to replace the closed facilities.

Landfill disposal is at great cost ranging from $11 per Mg in Montana to $36 in Florida and Iowa to $50 in Massachusetts and to $75 in Minnesota.[7] Since 1989, Oyster Bay, Long Island, residents have composted autumn leaves, reducing trash volume by half for which $104 to $125 per Mg was paid for disposal.[5] About 10,000 Mg of compost remained for use in the community.[5] In Bowling Green, Kentucky, leaf compost is sold to landscapers for $4.60 per Mg, reducing annual leaf disposal costs by $200,000.[5] In Islip, New York, $5 million are saved each year by composting grass clippings.[5] The city of Seattle may lead the country in yard waste reduction with its Master Composter program through which 70,000 compost bins have been distributed; $18 per Mg are saved in disposal fees, and 40% of all residential garbage is expected to be home-composted.[5]

As of early 1992, 2200 yard waste composting programs existed in the United States, with an estimated 30 to 60% of household wastes composted.[5] The U.S. Environmental Protection Agency (EPA) estimated that of the 32 million Mg of yard trimmings generated in 1990, 3.8 million Mg (12%) were collected and composted.[8] Based on EPA estimates, Kashmanian[9] estimates 15.5 million Mg of yard clippings will be produced in 1996, which after composting should yield up to 7.7 million Mg of end product.

By 1990, 13 MSW composting facilities were operating, with 15 more preparing to begin operations by 1992.[10,11] By 1996, Washington D.C. and 23 states that have already written bans against landfilling yard clippings will see those bans take effect.[9] These 23 states contain 83% of the nation's facilities that compost yard wastes; therefore, a significant impact on the amount of yard trimmings composted municipally is expected in the near future.[9] In addition, many more composting facilities are expected to be operating before the end of this century.

Uses of compost are increasing as scientists concentrate efforts in this area. Compost has traditionally been used as a soil amendment for garden plants and as a substrate for mushroom culture (*Agaricus bisporus*).[12–15] Recently, compost use has been suggested for pasture and orchard improvement, commercial vegetable production, containerized nursery crop operations, and turfgrass production.[16–34] Other end uses of compost include incineration for energy, methane generation, heat generation, biomass production, and use as a biofilter for various cleansing operations.[35–51]

II. MATERIALS AND PROCESSES

All methods of composting involve decomposition and stabilization of organic substrates under biologically-produced conditions.[52] With careful attention to materials and processes, the final organic product is considered sufficiently stable (inactive) for storage or application to land without adverse effects on agriculture or on the environment. Although in simplified terms composting is the decaying of organic matter, the process is intricate involving many different kinds of microorganisms and substrates and producing several by-products enroute to the stabilized product. Methods of composting have been refined to speed decomposition, to reduce odors from some by-products, to increase benefits to the environment, to protect the health of workers and users, and to adapt the processes to various scales of operation. In general, methods are adapted around processes that involve aerobic or anaerobic decomposition.

A. AEROBIC AND ANAEROBIC COMPOSTING

Aerobic decomposition involves methods that infuse oxygen into the decaying mass, whereas anaerobic decomposition proceeds after oxygen has been exhausted by microbial action. Aerobic and anaerobic processes can occur in the same mass. Undisturbed piles of decaying material typically have aerobic zones near the exteriors of the piles and anaerobic zones in the interiors of the piles. Piles that are turned for aeration usually have periods of aerobic decomposition throughout the piles, but the interior can become anaerobic as the oxygen is consumed.

Although the finished composts from aerobic or anaerobic decomposition may be similar, intermediates and by-products differ between the processes. Aerobic processes are more active metabolically and, hence, are more rapid than anaerobic processes. Temperatures generated by aerobic digestion are usually in the range of 45 to 65°C, but may be 60 to 70°C or higher.[53] Microorganisms facilitating this decomposition are termed *thermophilic*. Temperatures in anaerobic systems range from 15 to 45°C, with some systems generating temperatures of 38 to 55°C.[53] Organisms carrying out the decomposition processes in anaerobic masses are called *mesophilic* microorganisms. Carbon dioxide is a major product of aerobic processes, whereas CH_4 and relatively small amounts of CO_2 are generated by anaerobic processes.[54] Intermediates such as mercaptans and gaseous sulfides, are released during anaerobic decomposition, giving noxious odors in the environment.[53,55]

B. CARBON:NITROGEN RATIO

Many raw materials for composting are carbonaceous materials, rich in C and low in N. Paper, twigs, wood chips, dead leaves, and residues of dead plants may have C:N ratios exceeding 200:1.[56] Before these materials can be added to agricultural land the C:N ratio must be narrowed to about 35:1[57] or by some reports to 20:1.[53,58] Addition of materials with wide C:N ratios to soil induces microbial consumption of soil-borne nutrients (*immobilization*).

Composting accomplishes a narrowing of C:N ratios.[53] Carbonaceous materials alone, such as wood chips, decay slowly in composting; therefore, N-rich materials such as grass clippings, biosolids, and farm manures which have C:N ratios below 35:1 may be mixed with carbonaceous materials to accelerate rates of composting.[54,56,59] Mixing coarse carbonaceous materials with nitrogenous materials also imparts porosity in the mass (*bulking*). Finished composts have C:N ratios of 15:1 to 30:1.[53,60]

The narrowing of the C:N ratio during composting is accomplished by microorganisms, which use the carbonaceous materials as their source of C, use the nitrogenous materials as their source of N, and consume other mineral nutrients, such as P, S, Ca, Mg, and K. During composting, C is lost to the atmosphere mainly as CO_2, and N is lost mainly as NH_3 gas; however, the loss of C as CO_2 exceeds the loss of N as NH_3. The end result is a narrowing of the C:N ratio. The C:N ratio of the bodies of microorganisms ranges from 5:1 to 15:1[57] or by some reports from 15:1 to 30:1.[58] During composting, a series of microorganisms grow and die contributing additional N-rich material to the mass, narrowing the C:N ratio.

The C:N ratio is not the only factor governing the rate of decomposition of composting materials. Materials such as paper, wood chips, and sawdust with high lignin contents are recalcitrant to decay.[61] Even though lignin may have a wide C:N ratio, it does not contribute to immobilization of N or other nutrients. Peat moss is a lignin-rich material which can be added to potting media with little or no possibility of nutrient immobilization; yet, the benefits of the organic matter in the media are realized. Compost is suggested as a substitute for peat moss in soil-based or soilless media.[32] Compost for this purpose must be at a stable C:N ratio so that the compost will have no effect on nutrient availability and will be long lasting as an amendment.

C. MICROORGANISMS

Constituents of compostable materials, whether of plant origin or from urban waste, are primarily of three chemical fractions: simple and polymeric carbohydrates (sugars, starch, cellulose, hemicellulose, pectin), lignin, and nitrogenous compounds (amino acids, proteins, inorganic N). The sugars, starch, pectin, amino acids, proteins, and inorganic N are easily decomposable substrates. Cellulose is more resistant to degradation than sugars, starch, and pectin. Hemicellulose is associated with woody tissues containing lignin. Decomposition of cellulose and hemicellulose is the second phase in stabilization. Lignin is the most recalcitrant to decay of the plant polymers.

The relative composition and status of composting organic matter governs the microbial populations in the system and the rate of decomposition of organic materials. Various microorganisms have substrate-specific enzyme complexes, and populations vary as the substrate changes as a result of microbial activity. Initial activity is by fermenting organisms (largely bacteria) that utilize the easily decomposable sugars, starch, amino acids, proteins, and inorganic N.[62] The heat generated by composting is high during the activity of these organisms. The mass of composting material during the activity of the fermenting organisms is usually quite acidic, being around pH 5.[53]

Cellulose, hemicellulose, and their complexes with lignin (*lignocellulose*) remain after the easily decomposable materials are depleted. Lignocelluloses are the principal substrates in the production of composts.[62] Cellulolytic fungi (e.g., *Trichoderma sp.*) attack these complexes, which are more resistant to microbial degradation than sugars and starches.[57] Lignin may be linked in secondary reactions with cellulose (or hemicellulose). These cross-linked compounds are recalcitrant to decay. Thermophilic actinomycetes (e.g., *Thermonosporo mesophila*) function in degradation of lignin, lignocellulose, and cellulose.[57] These thermophilic organisms increase in number after those which have degraded the easily decomposable materials have diminished. Decay of lignin occurs by mesophilic organisms in the phase following the thermophilic phase.[57]

End-product accumulation from the metabolism of a single species may result in enzyme inactivation thereby inhibiting the metabolic activity of that organism. A consortium of organisms metabolizes the end products. Although a species may dominate during certain phases of composting, a microbial consortium of bacteria, actinomycetes, and fungi is active in compost piles.[54] Protozoa and viruses, although of limited decomposing capability, are also reported to be present in compost.[52] The consortium has organisms that thrive on the products of intermediary metabolism of other organisms and aid in removal of inhibitory end products.[62] Organisms that die after the depletion of simple carbonaceous and nitrogenous substances are themselves substrate for other organisms in the consortium. Dead and living microorganisms can comprise as much as 25% of the total weight of compost.[54]

Golueke[63] estimates that 80 to 90% of the microbial activity in composting may be attributed to bacteria. The numerous bacteria important in composting use carbonaceous substrates of sugar, starch, cellulose, and lignin. Some bacteria which degrade carbon substrates are listed in Table 1.

Bacteria also utilize substrates of urea, protein, or inorganic N. *Nitrosomonas* and *Nitrobacter* oxidize NH_4^+, which is generated during composting, to NO_2^- and NO_3^- in a sequence of reactions.[57] Bacteria active in composting are aerobic or anaerobic. In most composting systems, aeration is considered essential for maximum rates of degradation, but aeration inhibits the N-fixing bacterium, *Clostridium butyricum*, which has been suggested as being beneficial to composting through enrichment of the substrate with N. Inoculation of piles with *Clostridium* has been suggested as a means of supplying N to composting systems.[62]

Fungi that utilize compostable substrates for carbon and energy are important in composting. Both aerobic and anaerobic fungi are active in composting. Fungi are reported to require less nitrogen and water than do bacteria or actinomycetes.[52] Fungi may proliferate at pHs which are too acidic or alkaline to support bacteria or actinomycetes.[52,57] The genera of fungi capable of decomposing cellulose, hemicellulose, lignin, starch, sugars, and pectin are numerous (Table 2).

Actinomycetes, along with fungi and aerobic bacteria comprise the three most important groups of microorganisms in organic matter degradation. High pH favors actinomycetes, while cool and wet conditions below pH 5 suppress actinomycetes.[57] Some genera of actinomycetes found in composts are listed in Table 3.

TABLE 1 Some Genera of Bacteria and Substrates in Composting[57]

Substrate				
Cellulose	**Hemicelluloses**	**Lignin**	**Starch and Sugars**	**Pectins**
Bacillus	*Bacillus*	*Arthrobacter*	*Bacillus*	*Arthrobacter*
Cellulomonas	*Ctyophaga*	*Flav obacterium*	*Chromobacterium*	*Bacillus*
Clostridium	*Erwinia*	*Micrococcus*	*Clostridium*	*Clostridium*
Corynebacterium	*Pseudomonas*	*Pseudomonas*	*Cytophaga*	*Corynebacterium*
Cytophaga		*Xanthomonas*	*Flavobacterium*	*Erwinia*
Polyangium			*Micrococcus*	*Flavobacterium*
Pseudomonas			*Pseudomonas*	*Micrococcus*
Sporocytophaga				*Pseudomonas*
Vibrio				*Xanthomonas*

Methane-generating bacteria are strict anaerobes;[57,64] therefore, for the specific purpose of generating methane for fuel, development of anaerobic composting conditions is necessary.[64] Waterlogging and use of enclosed chambers lead to rapid development of anaerobiosis. Methane-generating bacteria are from the family *Methanobacteriacea*, which includes genera such as *Methanobacterium, Methanosarcina, Methanococcus*, and *Methanospirillum*.[57] Other volatile compounds that may be produced by anaerobiosis are ethylene, ethane, acetylene, propane, propylene, and butane.[57] These compounds are important in industry and as fuels.

Paraffin, kerosene, gasoline, mineral and lubricating oils, asphalts, tars, natural and synthetic rubbers, methane, ethane, butane, pentane, hexane, and other noncyclic, straight-chained carbon compounds can be decomposed by microorganisms in composts.[57] Rate of decomposition of these compounds is affected by length of carbon chain, with short chains generally being more readily decomposed than long-chained molecules.[57] Biodegradation of these and other *aliphatic* compounds occurs most readily in media with oxygen, above pH 5, and in a range of temperatures from 0 to 55°C.[57] Bacteria

TABLE 2 Some Genera of Fungi and Substrates in Composting[57]

Substrate				
Cellulose	**Hemicelluloses**	**Lignin**	**Starch and Sugars**	**Pectin**
Alternaria	*Alternaria*	*Agaricus*	*Aspergillus*	*Alternaria*
Aspergillus	*Aspergillus*	*Armillaria*	*Fomes*	*Aspergillus*
Chaetomium	*Chaetomium*	*Clavaria*	*Fusarium*	*Botrytis*
Coprinus	*Fusarium*	*Clitocybe*	*Polyporus*	*Fusarium*
Fomes	*Glomerella*	*Collybia*	*Rhizopus*	*Geotrichum*
Fusarium	*Penicillium*	*Cortinellus*		*Monilia*
Myrothecium	*Trichoderma*	*Fomes*		*Penicillium*
Penicillium		*Ganoderma*		*Rhizoctonia*
Polyporus		*Lenzites*		*Rhizopus*
Rhizoctonia		*Marasmius*		
Rhizopus		*Mycena*		
Trametes		*Panus*		
Trichoderma		*Pholiota*		
Trichothecium		*Pleurotus*		
Verticillium		*Polyporus*		
Zygorhynchus		*Polystictus*		
		Poria		
		Schizophyllum		
		Stereum		
		Trametes		
		Ustulina		

TABLE 3 Some Genera of Actinomycetes and Substrates in Composts[57]

Substrate				
Cellulose	**Hemicelluloses**	**Lignin**	**Starch and Sugars**	**Pectins**
Micromonospora	*Streptomyces*	*Micromonospora*	*Actinoplanes*	*Thermonospora*
Nocardia		*Nocardia*	*Microbiospora*	
Streptomyces		*Streptomyces*	*Micromonospora*	
Streptosporangium			*Streptomyces*	
Thermonospora			*Streptosporangium*	

and fungal genera that utilize aliphatic compounds include *Mycobacterium, Pseudomonas, Corynebacterium, Acinetobacter, Bacillus, Nocardia, Streptomyces, Candida*, and *Rhodotorula*.[57]

Aromatic compounds are derivatives of the cyclic compound benzene[65] and include substances such as benzene, toluene, xylene, and napthalene. Many insecticides, fungicides, and herbicides are aromatic derivatives. Organisms that decompose aromatic compounds are expected to receive increased attention in the future as composting for bioremediation expands. The bacterial genera, *Pseudomonas, Mycobacterium, Arthrobacter, Bacillus, Nocardia*, and occasional fungi are the dominant microorganisms utilizing aromatic compounds.[57]

Thermophilic conditions may kill beneficial as well as pathogenic microorganisms; therefore, inoculation with beneficial organisms after composting may be recommended to ensure that the disease-suppressing properties of composts are present.[3] Other inoculations that have been suggested are the additions of N-fixing bacteria (e.g., *Clostridium* sp.), which would enrich the composting mass with N and possibly accelerate rates of degradation.[62] Beneficial effects of composting microorganisms are discussed in detail later in this chapter.

D. COMPOST MATURITY

Curing, stabilization, or maturation of compost is required if the product is malodorous, only partially decomposed, or is derived from an anaerobic composting process. The degree of stabilization required in composting is dictated by the desired use of the end product. Especially for horticultural use, the compost should not be immature. Complaints of phytotoxicities from immature MSW compost have greatly reduced the potential use of this compost as a fertilizer in Italy.[66] Root injury has occurred after the introduction of immature composts to the root zone.[67] Such injury has been attributed to the high C:N ratio and NH_4 content[63] and phytotoxins such as phenolic acids, acetic acid, and other volatile (short-chain) fatty acids present in immature composts.[68,69] Ethylene, an inhibitor of root elongation, may be present in immature composts from anaerobic composting.[34]

Haug suggests some guidelines for measuring the degree of stabilization: (a) decline in temperature at the end of batch composting, (b) a low level of self-heating in the final product, (c) analysis of organic content yielding a desirable C:N ratio, (d) O_2 uptake rate of end product 1/30 that of substrate, (e) presence of NO_3^- with concurrent absence of NH_3, and starch, (f) lack of insect attraction or insect larvae, (g) characteristic lessening of obnoxious odor during composting and absence of odor upon rewetting of the end product, (h) rise in redox potential, and (i) experience of the operator.[52] Poincelot[54] suggests that compost is stabilized when decomposition no longer uses N and the C:N ratio is 10 to 12. Zucconi and De Bertoldi[70] suggest using a germination bioassay to assess maturity of compost. Brinton and Droffner[71] define stability in composting as low CO_2 respiration and lack of continued self-heating. Plant nutrient availability, cation exchange capacity, electrical conductivity, C:N ratio, NO_3^--N content, and respiration have been used to assess compost maturity.[72–75] Some researchers report that total C:N ratio is not usually a good indicator of maturity for organic wastes.[73–75] Several methods of assessing compost maturity have been reviewed, with the authors suggesting that no method is adequate to assess the maturity of composts from all the different types of substrates.[76]

Vermicomposting, composting with earthworm culture, is one method of curing or maturing composts. The earthworm, *Eisenia foetida*, has been used successfully in composting and in stabilization of sludges.[77–80] In temperate climates, *Eisenia foetida, Dendrobaena veneta*, and *Lumbricus rubellus*, and in tropical climates, *Eudrilus eugeniae* and *Perionyx excavatus*, have been used in composting of manures and plant residues.[81] Interest in vermicomposting of sewage sludge, animal manures, and vegetable wastes is increasing. Vermicomposting of many organic wastes has been studied.[82–85] Aerobic sewage sludge and animal manure have been shown to be good substrates for earthworm growth.[86–88] Cattle (*Bos taurus*) biosolids, brewery waste, spent mushroom compost, or potato wastes may be used for earthworm culture without prior composting, but pig (*Sus scrofa.*) wastes need to be composted for 2 weeks prior to worm culture.[81] Poultry biosolids need to be leached of salts, and NH_3 must be allowed to volatilize until acceptable levels are reached.[81] Culture of earthworms in animal or vegetable wastes require temperatures of 4 to 30°C (15 to 20° optimum), moisture content 60 to 90% (80 to 90% optimum), aerobic conditions, NH_3 below 0.5 mg g^{-1}, salt content below 0.5%, and pH between 5 and 9.[81]

Vermicomposted cattle solids have been analyzed before and after vermicomposting with the results that K and NO_3^- increased, pH stayed about the same (7.4 to 8.6), and NH_4^+ decreased significantly.[81] Earthworm culture is reported to increase the overall rate of decomposition,[89–90] decrease the proportion of anaerobic to aerobic decomposition resulting in a decrease of CH_4 production and volatile S compounds,[89] and decrease the incidence of pathogenic bacteria.[91,92]

E. COMPOSTING TECHNOLOGIES

Some considerations in the design of a composting facility include technology to be used, capacity required, on-site process flow of materials, equipment required, and environmental safeguards.[93] Site selection, layout of storage areas, control of surface runoff, erosion, and sediment, containment and treatment of leachate, and safety, security, and public education add complexity to the facility design.[94] Technology refers to the type of composting process used, static pile, windrow, in-vessel, or other process.[95] Some researchers suggest that system designs for composting should approximate some general parameters: C:N ratio of near 30:1, internal oxygen supply of 5 to 15% (porosity of 30%), internal temperatures near 60°C, and moisture content of 50%.[96,97] Anaerobic rather than aerobic composting systems are sometimes used for municipal composting. Several high rate anaerobic digesters have been developed.[98] MSW may be composted anaerobically in these digesters.[99,100]

1. Municipal Solid Waste

To achieve effective composting of MSW many types of facilities have been employed. Since 1925, when the Bangalore or Indore system was developed by Sir Albert Howard, several adaptations have been made with varying degrees of success (Table 4).

MSW presents unique problems for waste materials handlers. As can be seen in Table 4, the great diversity in compostable materials and available equipment has led to many variations in facility design. Haug[52] cautions that small scale systems which appear simple become complex on a large scale. Steps in MSW composting are presented in a flow chart (Figure 1).

TABLE 4 Some Composting Processes for Municipal Solid Wastes

Type	Name	Description
Nonstatic Solids Bed	Bangalore (Indore)	Trench in ground; alternate layers; no grinding; turn often by hand; 120–180 days; developed in India.[52,53]
	Open Windrow	Elongate mounds; refuse in ground; aeration is either forced or by turning frequently; once used in Mobile, AL, Boulder, CO, Johnson City, TN, Mexico City.[52,53]
	Open Land	Refuse spread in furrows; turned 1–2 times per week for 5 weeks; product ground; 1 ha needed to process 49 Mg day^{-1}.[37] Highly mechanized operations require 1/2 as much land.[101]
	Van Maanen	Elongated mounds; raw refuse; 120–180 days; turned by grab crane; problems with rodents, files, odors; first used in Netherlands 1931.[52,53]
	Others	Many names for modified windrow systems; the modification usually consisting of the name and type of equipment used.
Static Solids Bed	Brikollari (Caspari Briquet System)	Material ground and compressed into blocks and stacked for 30–40 days; air forced through stacks; blocks later grounded; used in Germany and Switzerland.[52,53]
Vertical Flow Reactor	Earp-Thomas	Silo type with 8 decks stacked vertically; compost agitated and moved downward; air passes up through silo; 2–3 days followed by windrowing; used in Korea, Italy, and Switzerland.[52,53]
	Fraser-Eweson	Ground refuse in vertical silo; perforated platforms; forced air up through platforms; refuse down; 4–5 days; problems with solids going through platforms; plant once operated in Springfield, MA, 1954–1962, now closed.[52,53]
	Jersey (John Thompson)	Structure with 6 floors; from each floor refuse is dumped onto the lower floor; aeration occurs while dumping; 6 days, 6–8 weeks curing time; used in Thailand.[52,53]
	Naturizer (International)	Conveyor belts designed to pass material from belt to belt; air passes upward through digestion area; 6–8 days; once used in San Fernando, CA, and St. Petersburg, FL; problems with odors.[52,53]
	Riker	Four-story high bins with hinged floors; materials dropped to each lower floor; forced aeration; 20–38 days; problems maintaining aeration; once used in Williamston, MI.[52,53]
	T. A. Crane	Two cells, three horizontal decks each, horizontal ribbon screws to circulate refuse; air diffuses up from bottom; 3 days in composter; 7 days curing in bins; used in Japan.[52,53]
	Varro	Eight deck digester; ground refuse moved downward from deck to deck; forced aeration on each deck, 40 h; output is dried, ground, and used as wallboard, fertilizer, soil conditioner; facility was built in Brooklyn, NY, in 1971.[52]

TABLE 4 (*Continued*)

Type	Name	Description
Horizonal and Inclined Flow Reactor	Dano	Rotating drum; slightly inclined; kept half full of refuse; no grinding; 1–5 days in composter followed by curing in windrow; forced aeration in drum; very popular worldwide; 160 facilities in 1972.[52–54]
	Fermascreen	Rotating six-sided drums; three sides are screens; refuse is ground and loaded into drum with screens closed; screens later opened for aeration; 4 days.[52,53]
	Geochemical–Eweson	Refuse not ground; placed in rotating drums set in series on slight incline; refuse transferred down line from drum to drum every 1–2 days; 3–6 days total; output is screened.[52]
	Fairfield-Hardy	Circular tank with vertical screws; forced aeration; 5 days.[52,53]
	Snell	Rectangular tank 2.5 m-deep with porous floor, air ducts for forced aeration; inclined tank; 5–8 days.[52]
	Metro-Waste	Rectangular tanks, 6 m-wide, 3 m-deep, 60 to 120 m long; refuse is ground; 7 days; specialized agitator moves on rails in tanks.[52,53]
	Tollemache	Similar to Metro-Waste; built in Spain and Rhodesia.[52,53]

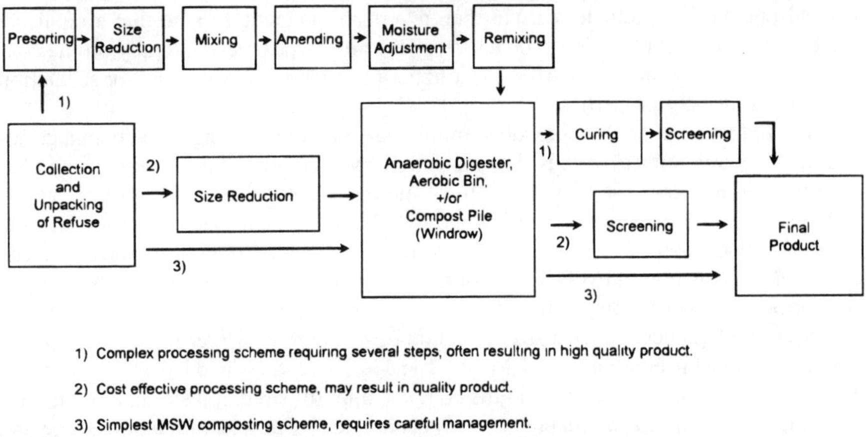

Figure 1. Processing schemes for composting municipal solid waste (MSW).

Many communities are requesting, or even requiring, separation of plastics, glass, and newspaper in the household prior to placing trash or garbage at the curb for collection. Plastic, clear glass, and paper are commonly recycled and taken out of the compostable waste stream. Many communities remove aluminum beverage cans from the waste stream through a deposit system or voluntary recycling.

In some communities, newspaper is removed from the MSW at curbside, prior to collection, and is incinerated or used as fuel in an energy recovery system.[40] Oven-dried newspaper can produce 2.9×10^6 J kg^{-1}.[37] If paper is subjected to heat, but not O_2, (*pyrolysis*), then CH_4 is produced, which can then be recovered and used as fuel.[37] Paper is also important as a recycled material for paper production, feedstock in various fermentations, and as a bulking agent for composting.[36] Despite these community efforts at curbside recycling, plastics, glass, newspaper, and aluminum often contaminate the compostable waste stream. Other materials such as food cans and other metal items, old appliances, tires, and other household goods should be removed from the compostable materials prior to processing at the MSW compost facility.

After the refuse is collected, it is brought to the MSW compost facility and unpacked prior to presorting. Presorting is considered an optional process; however, proponents of presorting argue that a more uniform end product without contaminants can be achieved by presorting. In the United States, labor-intensive manual presorting is considered cost prohibitive.[40,59] Mechanical presorting may be used.[102] Mechanical presorting utilizes size, density, or magnetic properties of the materials and often consists of air classification, ballistic separation, magnetic sorting, or a combination of the three.[59]

Air classification utilizing a forced upward flow of air reportedly has removed metal, glass, rocks, rubber, and wood[103] and is especially useful in removing combustibles from the waste stream.[36] Ballistic separation uses a high-speed conveyor belt with sectioned bins at the end to separate refuse by weight. Magnetic presorting or ferromagnetic separation utilizes permanent electromagnets, rotating drum, suspended, or pulley-type magnetic separators. A powerful magnet is passed over the refuse, separating the ferrous metal from the rest of the refuse. Aluminum and other nonferrous metals presented a sorting problem until polyphase alternating current electromagnetic field technology was developed.[104] This technology depends upon the conducting ability of nonferrous metals and operates by inducing current then reversing magnetic flux so that a repulsive force is created.[104] The repulsive force sweeps the nonferrous metal laterally off the conveyor belt.[104] Glass can be removed by utilizing methods such as flotation or separation that rely on density properties.[40]

Size reduction of MSW (shredding, milling, hammering, rasping, pulverizing, grinding, and comminution)[40] speeds decomposition by increasing initial aeration and increasing surface area, allowing refuse to be more susceptible to decomposition by microorganisms.[54] Golueke and Diaz[59] suggested shredding as a mechanism to raise the ratio of surface area to mass of particles to be composted. Grinding or shredding can double the production of evolved CO_2 during the thermophillic stage of composting, an indication of active decomposition.[54]

Neal and Schubel[37] suggest size reduction of refuse to no larger than 7.5-cm diameter for successful composting of MSW. The ideal size for shredded refuse is determined by (a) resultant pore space in the compost pile, (b) structural strength of the substrate, (c) suitability to automation, and (d) economics.[59] The size of the pore space and the structural strength of the material ultimately determine the degree of aerobic microbial activity that can be sustained. Therefore, Golueke and Diaz[59] suggest different size limits for different materials, roughly 1.3 to 5 cm for paper and vegetative matter and less than 1.3 cm for wood products. Powerful size reduction equipment will handle most objects found in unsorted municipal refuse, such as home appliances, solid wood, and even tires on wheels.[40] Some researchers state that size reduction results in a material that is easier to handle and moisten.[54] Mechanical mixing may occur after size reduction, or the process of size reduction and the later process of turning may be relied upon to mix the refuse. To some extent size reduction mixes the acidic and alkaline materials and distributes microorganisms.[105] Consequently, mixing as a separate process is often excluded due to cost.

Amendments such as inorganic fertilizer, sewage, bark chips, coal ash, or microorganisms may be added to MSW during grinding, mixing, or bulk processes. Other amendments are used with other substrates and are discussed under the individual sections discussing substrate types. Sewage is often used as an amendment to supply N in high C municipal wastes, such as paper and cellulosic household refuse.[106] To adjust the C:N ratio, slow-release inorganic N fertilizer also has been suggested as an additive.[107] Phosphate fertilizer has been suggested as an additive to speed decomposition of compost; however, adding an amount of calcium phosphate greater than 2% of the substrate may inhibit decomposition.[54] The action of the phosphate is reported to increase populations of cellulose-decomposing organisms.[54] Poincelot[54] cautions that the addition of phosphate is not usually needed in composting and any benefit may be outweighed by the additional cost of materials and handling.[54] Adding $CaCO_3$ or $Ca(OH)_2$ to acidic compost substrate to raise pH can accelerate the rate of decomposition.[54] The slight improvement from liming is usually outweighed by the additional cost and the loss of N as NH_3 due to volatilization that occurs at higher pHs.[54]

Bulking agents may be added to the compost mixture to improve porosity. Bark chips, straw, or rice hulls have been used as bulking agents to absorb moisture, serve as a carbon substrate, and decrease bulk density.[56] The size and quantity of the bulking agent must be controlled to maintain porosity and aerobic conditions in the compost medium.[52] Some decomposition of the bulking material will occur requiring periodic replacement.[52]

In general, additives do not seem justified except in certain circumstances.[54] If sawdust is to be composted, the fungus *Coprinus ephemerus*, a cellulose decomposer, may be added to accelerate decomposition of cellulose.[54] If *Coprinus* is added, Poincelot[54] suggests concurrent addition of N, P, and K fertilizer for the nutrition of the fungus, shortening the composting time of sawdust from 1 to 2 years to 3 months, and resulting in a compost that did not require additional N upon application to the soil.

After the high temperatures of composting have killed beneficial microorganisms, inoculation with beneficial organisms often is recommended to ensure that the disease-suppressing properties of composts are present.[3] Other inoculations that have been suggested are the additions of N-fixing bacteria (e.g., *Clostridium* sp.), which would enrich the composting mass with N and possibly accelerate rates of degradation.[62] Inoculation of anaerobic digestion with methane-generating microorganisms may be proposed, as well as inoculation with specific microorganisms to decompose specific industrial solvents or wastes.

The maximum recommended moisture content for effective MSW composting is between 55 and 65% of the total weight.[63] Less than 45% moisture creates an environment that is limiting to the growth of microorganisms, and below 12% moisture biological activity ceases.[56] The upper limit of moisture addition for mechanical operation within the composting facility is considered to be 65 to 70% as values higher than this result in sticking and clogging of the machinery.[56] Haug[52] suggests a turning schedule for MSW compost that is dependent upon moisture content. If the MSW to be composted is <40% moisture, water is added; at 40 to 60% moisture, compost should be turned at 3 day intervals for a total of four turns; at 60 to 70% moisture, compost should be turned at 2 day intervals for a total of five turns; at >70% moisture, the compost must be turned daily until 70% moisture is achieved.[52]

At this point the prepared substrate is ready for piling into windrows. Haug[52] states that windrowing as a final curing process is necessary for MSW as the organics in MSW are typically resistant to decomposition. An optimum height of 1.2 to 1.8 m is suggested for windrows as taller piles may become too compressed (anaerobic) or maintain temperatures that are too high, whereas piles lower than 1.2 m may result in temperatures which are too low.[54] Within the nonaerated windrow turning, is used to aerate the pile, and the timing of turning is often based on moisture and temperature levels as stated above. Poincelot[54] suggests O_2 concentration as a more accurate indicator of the need for turning; however, it is realized that this is a more expensive and time consuming measurement than moisture or temperature. Turning two to three times per week can prevent anaerobiosis; daily turning can reverse anaerobic conditions.[54]

The final step before marketing of the end product, use of the product in horticultural operations, or land application of the compost is often a screening procedure.[52,59]

2. Composted Sewage

Sewage can be composted on a large scale in a municipal setting,[108,109] or in an individual biological waste facility accompanying remote cabins in recreation areas[110,111] or in any number of variations somewhere between the two extremes. Composting of sewage (biosolids) and composting facilities in many countries, as well as compost applications, are discussed in some detail in a book by Shuval, Gunnerson, and Julius.[112] Composting of sewage can involve drying, amendment with bulking agents, pH adjustment, aerobic or anaerobic decomposition, microbial inoculation, curing, vermicomposting, or other methods designed to give more control in the process, lessen odors, or improve the end product. Bulking materials can affect both the system and the end product[62,113,114] and allow rapid initial decomposition.[115] In some systems both wood chips and leaves, or leaves alone, were added to sewage as bulking agents.[116] Other bulking agents such as bark, shredded tires, cereal hulls, and straw have been used.[117] Amendments such as fly ash have been used with sewage sludge in the composting process.[118]

The composting of sewage biosolids with various amendments (co-composting) is of recent increased interest to many scientists. Through decomposition composting can allow the sewage wastes to be stored, handled, or applied to soil.[63,119] The process is similar to all composting in that microorganisms are encouraged to decompose the substrates. Sewage is addressed in detail in Chapter 6.

3. Vegetative Wastes

Yard and lawn trimmings and kitchen wastes comprise much of the composted waste in a municipal setting. Mixes of leaves and grass have been composted and studied in the laboratory.[120] Composting of vegetable waste has proven successful with adjustments made to the process as described for MSW composting.[121,122] Vegetable wastes may also be composted by vermicomposting.[81]

It is considered especially important to separate at the source vegetable wastes from other refuse.[123–128] Once separated from the MSW mainstream, various amendments and processes may be used for composting of vegetative waste. Daily wastes from garden produce markets in Florence, Italy, with wood shavings as a bulking agent were successfully composted for 35 days in a specially designed open reactor and then matured.[121] Anaerobic digestion at 55°C for 42 days has been used to compost kitchen wastes with added waste paper.[122] If anaerobic digestion is not used, forced aeration and frequent turning are needed when kitchen wastes are composted.[129]

Wheat straw, corn stalks, maize cobs, brewer's grains, seed meals, cotton hulls, sugarcane bagasse, and other agricultural residues high in lignocellulose can be composted in a fermentation reactor.[130–134] The thermoregulated, static, aerated conditions for this type of composting were established by Laborde.[135,136] Under aerobic conditions optimal decomposition has been reported at 60°[137] or 55°C.[133] However, these temperatures are considered too high for the thermophilic fungi needed to decompose residues high in lignocellulose, therefore, some researchers propose reactors operate at 50 to 55°C.[138,139] Some researchers have reported cellulase activity at 35° and 45°C;[130] therefore, composting of residues high in lignocellulose may occur at lower temperatures than previously reported. Materials high in lignocellulose content may pose problems in composting as they may remain wet and compressed resulting in anaerobic areas within the reactor.[130]

Composting of tree bark has been studied in small reactors (26-L) with constant aeration and with a moisture content of 60%.[140] The recent development of these low-cost, highly-controlled reactors[141] has allowed the investigation of composting a wide variety of substrates.[142] The development of reactors allows a measure of control which should result in decreased composting time, lower energy costs, and a more consistent product. Another vegetative waste which has been composted is water hyacinth (*Eichornia crassipes*).[143,144] Water hyacinth composting was greatly improved if plant material was first dehydrated, a high C substrate was added, and gizzard shad (*Dorosoma cepedianum*), poultry or cattle manure, or scallop (*Pecten* sp.) offal was added.[143]

4. Farm Animal Wastes

Environmental aspects of manure are discussed in detail in Chapter 5. Cattle manure solids may be composted after separation of solids from slurry,[145] composted in turned windrows,[146] composted in bins, composted in agitated solids bed reactors,[52] anaerobically digested prior to vermicomposting,[147,148] and composted with crop and forest residues[149] or wood shavings, sawdust, or peat moss.[150] Poultry manure has been composted with woollypod vetch (*Vicia dasycarpa* var. *lana*),[151] as well as with sawdust or ground corncobs.[152] Separated swine (*Sus scrofa.*) wastes were composted with peat and sawdust as bulking agents,[153] with alkaline fly ash or coal ash, and cement kiln dust.[154,155] Feather meal has been composted in a mix with peat.[153] These are just a few examples of farm wastes composting which are currently studied.

5. Industrial Wastes

Various industrial wastes, especially cellulosic or food wastes, may be composted. For example, the anaerobic decomposition of paper mill sludge in a high-rate reactor with rumen microorganisms has been shown to efficiently digest cellulosic wastes.[156,157] The composting of waste paper with Lantana straw (*Lantana camara*) has also been investigated.[14] Instant coffee (*Coffea* sp.) production by-products may be digested in an anaerobic thermophilic process resulting in *biogas* production during the process and an end product compost considered by the authors to be suitable for horticultural use.[158] Biogas is the name given to the mixture of gases (mostly CH_4, CO_2, and some traces of H_2S, N_2, O_2) evolved from anaerobic decomposition of organic residues.[45,64] Coffee wastes have also been composted anaerobically using an Upflow Anaerobic Sludge Blanket (UASB) reactor.[159]

The composting of press-molded, wood-fiber pallets has been accomplished in a pilot scale evaluation experiment by Keener.[160] Pulverized pallets have also been composted with sewage sludge and wood ash using an in-vessel process.[161] Composting of logyard residues for a variety of end uses is being encouraged in the logging and lumber industry.[162]

A composting method for fish offal (fish scrap or waste) was developed by Mathur et al.[163] This method involved using peat as a bulking agent and adding seaweed and crab scrap to the fish scrap.[163] The compost extracts of fish offal and peat compost may be used as a substrate for fermentation and microbial biomass production.[164] Another compost mixture that has been investigated includes sawdust, peat, poultry manure, and crab processing wastes with fish offal.[114] Fish wastes have also been composted in static pile windrows with wood chips as a bulking agent.[165]

Slaughterhouse wastes have been anaerobically digested, cured, and used horticulturally,[166] and silkworm (*Bombyx mori*) litter has been composted for 60 days in an aerobic pit then used instead of fertilizer.[24] Distillery waste which has been anaerobically digested has been used to fertilize St. Augustine grass (*Stenotaphum secundatum*).[167] The industrial wastes presented here are just a few examples of wastes which are composted and studied.

6. Biogas Production

Biogas production for energy use may be one important objective of anaerobic thermophilic digestion of industrial wastes. Many types of organic residues have been composted anaerobically for the production of biogas. Biogas (mostly CH_4, CO_2, and some traces of H_2S, N_2, O_2) is evolved from anaerobic decomposition of organic residues.[45,64] Distillery wastes have been anaerobically digested for biogas production.[167] Waste paper products, along with MSW are among the many substrates used in biogas production.[45] Recently, disposable diapers with MSW have been studied for the feasibility of composting for biogas production.[45]

III. BENEFICIAL EFFECTS

Many beneficial effects are related to the process of composting and the uses of its end product.[168] Composting can have permanent and far-reaching effects on the quality of the environment by removing large amounts of refuse from the waste stream, thereby reducing landfilling and incineration of wastes. Compost has an agronomic and economic value that is based on the value of composts as immediate sources of plant nutrients.[169] Another beneficial effect to the environment is the production of biogas from anaerobic composting. The use of biogas as a fuel has a caloric value of 4450 to 6230 kcal m^{-3}.[35] Biogas contains approximately 65% CH_4 and as an alternative energy source may reduce the need for fossil fuels, which are nonrenewable resources.[35] Predicting the full extent of the benefits derived from CH_4 production and use is difficult, as this aspect of composting is still considered to be experimental. Important long-term benefits from the application of composts to agricultural land include the residual value of nutrients and the physical improvements in soils amended with composts. All, and more, of the compost produced in the United States from municipal wastes could be used in agriculture resulting in great potential benefits from composting such as increased yields and food production and improved soil and environmental conditions.[170]

A. EFFECTS ON WASTE STREAM

The contribution to environmental preservation afforded by composting is substantial in terms of amounts of materials removed from the traditional waste stream. In 1994, the United States generated 188 million Mg of garbage and trash.[4] Others estimated that 255 million Mg were produced in 1989.[5] About half, with estimates ranging from 30 to 60%, of the total MSW is considered compostable.[5] Composting of MSW has advantages over traditional disposal practices such as landfilling or incineration.[171] The advantages include potentially lower operational costs, decreased air and water pollution,[172] and production of beneficial end products.[173] If an estimate of 200 million Mg of waste annually is used as an average, a minimum of 60 million Mg of waste may be composted and removed from the traditional waste stream and cycled into agricultural uses. Potential usage in agriculture may be as much as 450 million Mg per year.

Sewage biosolids and animal feces can also be removed from the waste stream by composting. These materials often are composted with other municipal, agricultural, and forest wastes, resulting in reduction and recycling of wastes from several industries and domiciles. Industrial wastes such as pharmaceutical fermentation residues, cranberry wastes, and food flavoring wastes have been removed from the mainstream and composted.[174] Wastes from the coffee industry,[158] papermills,[156,157] lumber industry,[160,162] fishing industries,[114,163,164] silk industry,[22] slaughterhouses,[166] and distilleries[167] have all been removed from the waste stream and composted. Disposable diapers may also be composted instead of landfilled.[45]

B. EFFECTS ON WATER AND AIR QUALITY

1. Biofiltration

Biofiltration of water and air is a relatively new advancement in bioremediation. A trickle biofiltration apparatus for bioremediation of contaminated water is shown in Figure 2.

Sand, tree bark, wood chip, and leaf compost have been used as biofilters to treat NO_3^- contamination in agricultural runoff.[175] Composts from straw-manure mixes and spent mushroom composts used as biofilters were effective in treating acid coal mine drainage,[176] increasing the pH while decreasing soluble Fe, Mn, Al, and SO_4^{-2} concentrations to an even greater extent than sphagnum peat biofilters.[177] The treatment of mine drainage was least effective during winter than in other seasons, as microbial activity is responsive to seasonal changes.[177] Water with 3 to 6 mg NO_3 N L^{-1} was treated in individual reactors at the rate of 10 to 60 L day^{-1} resulting in a significant decrease in NO_3-N.[175] Use of biofiltration for contaminated water may have great potential for the treatment of many different pollutants. Inoculation of the compost biofilter with microorganisms specific to the pollutant may improve the efficiency of the biofilter.

Composting can remove unstable N compounds leaving more stable organic forms of N, which must be decomposed by microrganisms before being released into the soil solution thereby allowing less available NO_3^- to be leached.[27] In a series of field experiments NO_3^- concentrations under compost-amended plots was 10 mg N L^{-1}, but concentrations under chemically fertilized plots reached 14.7 mg N L^{-1}.[27] Potting media containing 25, 50, or 100% MSW leached less N than a conventional medium.[178]

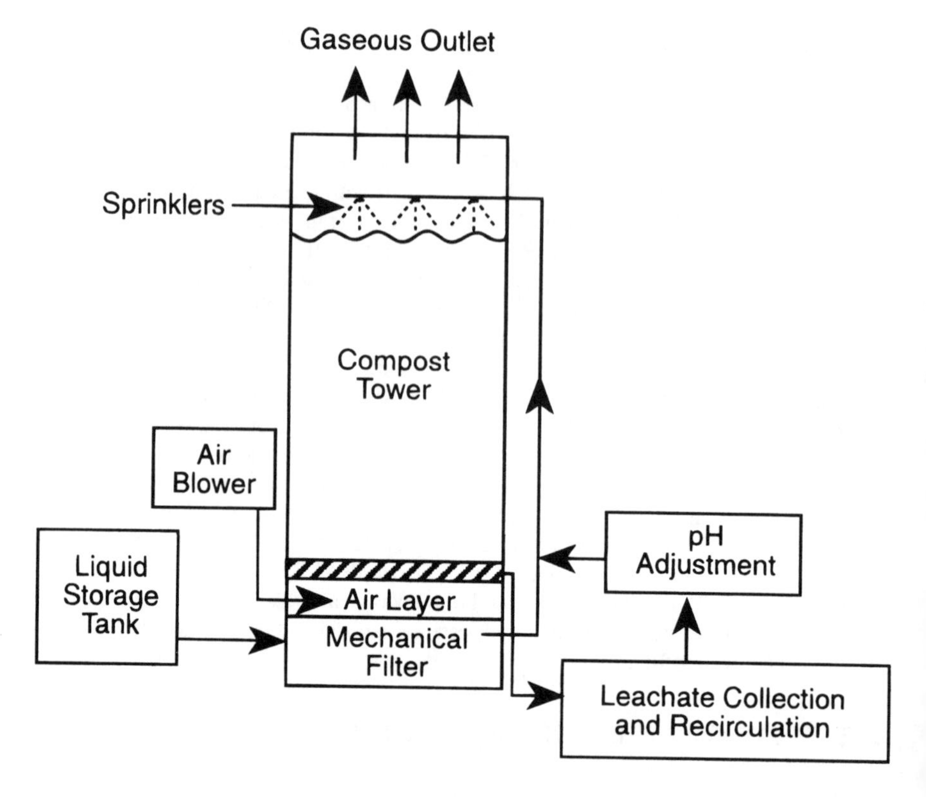

Figure 2 Use of compost for biofiltration of liquid.

Leaching of NO_3^- into groundwater and surface waters may be decreased on agricultural land where compost has been applied.

Control of leaching of organic substances is a growing area of research with advances expected in the future. Lower concentrations of insecticides and fungicides leached through MSW-amended sand than through sand alone.[19] Pentachlorophenol, a wood treatment and herbicide, was more strongly sorbed by increased organic matter in soil.[179] Similar effects of organic matter on sorption in soil were reported for para-dichlorobenzene and 2,4-dichlorophenol.[180] Increased organic matter in compost-amended soil can be expected to decrease the concentrations of many other organic contaminants in leachate from contaminated soil.

Biofiltration of air can be useful in many malodorous industries. The use of compost as a biofilter for contaminated air has been studied by several researchers.[47,48,181–188] An example of a system for using compost as a biofilter for contaminated air is presented in Figure 3. Biofilters have been used for removal of noxious odors, such as S compounds and esters,[51] volatile organics,[50,186] ethanol vapors,[48] aromatic compounds, such as styrene and toluene,[189,190] aliphatic compounds, such as propane and isobutane,[191] and esters and alcohols.[184,190] A compost biofilter removed >90% of toluene (at 10 to 20 mg kg $^{-1}$) with an efficiency 40 to 80 times greater than that of previously existing biofilters.[49] Hydrogen sulfide removal exceeded 99.9% in one air biofiltration system.[51]

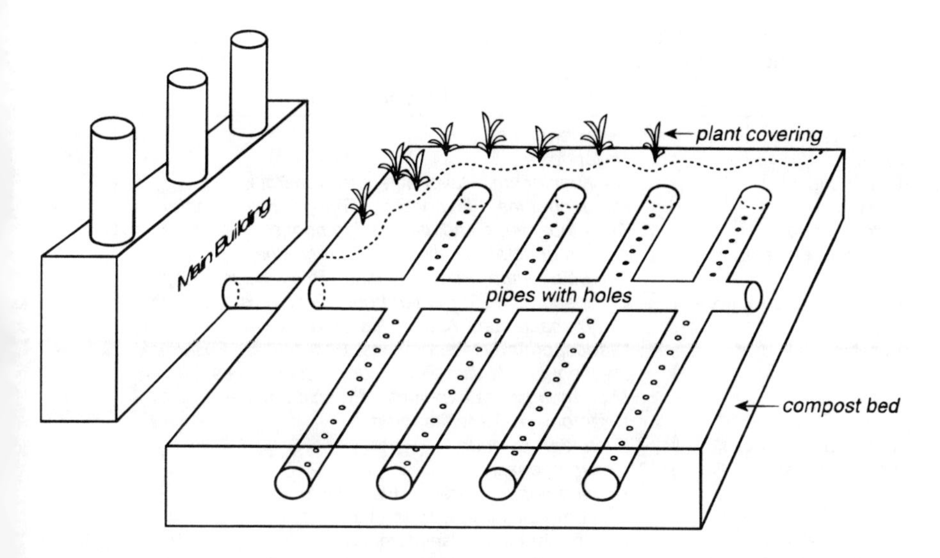

Figure 3 Cut-away view of air biofiltration bed.

C. EFFECTS ON SOIL CONTAMINANTS AND SOIL QUALITY

1. Bioremediation

Organic compounds which may be considered hazardous in soils often can be decomposed to less hazardous materials by micoorganisms active in composting (Table 5).

Composting is suggested for bioremediation of soils contaminated by some organic compounds.[193] Enhanced decomposition by composting has been reported for explosives,[194] pesticides,[195] industrial wastes,[196] and petroleum hydrocarbons.[197] Savage et al.[192] listed microorganisms active in composting soil contaminants (Table 5). This listing may be valuable to practitioners who have interests in inoculating compost piles with organisms with reported activities in metabolism of specific contaminants. Specific examples of composting of soil contaminants including or in addition to those mentioned in Table 5 are in the following discussions to indicate substrates and their rates of decomposition.

Soils containing chlorophenol[198] or low levels of aliphatic hydrocarbons[193] were decontaminated by composting. Polynuclear aromatic hydrocarbons (PAHs) some of the most recalcitrant and reportedly toxic and carcinogenic organic constituents of petroleum and chemical processing industries[199,200] may be decomposed microbially by composting.[201] The white rot fungus, *Phanerochaete chrysosporium*, was reported to metabolize benzo(a)pyrene and other polynuclear hydrocarbons.[192,202,203] Naphthalene is decomposed completely in 7 days by composting, whereas pyrene concentrations are decreased moderately in 30 days.[204] Benzoanthracene levels fell 25% in 7 days and 42% in 30 days as reported by Snell in 1982.[204] After 18 to 20 days of composting, 10% of phenathrene was decomposed, and 1.6 to 4.9% of carbaryl was decomposed.[205] Other researchers have reported that pentachlorophenol was decomposed 44 to 60% by com-

TABLE 5 Hazardous Wastes Substrates and the Associated Decomposing Microorganisms[192]

Waste	Microorganisms
Benzene	*Pseudomonas*
Biphenyl	*Beijerinckia*
Crude oils	*Brevibacterium, Flavobacterium Nocardia, Pseudomonas, Flavobacter, Vibrio, Achromobacter*
Hexadecane	*Acinetobacter, Candida, Pseudomonas*
Industrial effluents	*Achromobacter, Alcaligenes, Comomonas, Flavobacterium, Pseudomonas, Thiobacillus, Zoogloea*
Kerosene and jet fuels	*Cladosporium, Hormodendrum, Torulopsis, Candida, Corynebacterium, Aspergillus, Botrytis*
Paraffin	*Cladosporium, Debaromyces, Endomyces, Fusarium, Hansenula, Monilia, Penicillium, Actinomyces, Micromonospora, Nocardia, Proactinomyces, Streptomyces, Trichosporon*
Malathion	*Trichoderma viride*
Naphthalene	*Pseudomonas*
Pesticides	*Pseudomonas, Bacillus, Flavobacterium, Achromobacter, Nocardia, Aspergillus*
Phenol cresols	Mutant strains of *Pseudomonas*

posting in spent sawdust mushroom (*Lentinus edodes*) compost.[206] Certain pesticides, chlorinated hydrocarbons such as chlordane, heptachlor, and aldrin, are persistent in the environment (and may be present in soils for some time after use).[207,208] In general, under aerobic conditions, compounds of high molecular weight with four or more fused carbon rings may be mineralized at minimal rates.[209–213] Compounds with fewer rings are decomposed more readily. Savage et al.[192] listed some organisms that decompose chlorinated hydrocarbons (Table 6).

Soils contaminated by chlorinated hydrocarbons such as chlordane, heptachlor, or aldrin could be composted with the microorganisms listed in Table 6. Chlordane, heptachlor, and aldrin have been replaced by biodegradable products such as chlorpyrifos and isofenphos.[214] The decomposition of the newer compounds has been verified under laboratory conditions.[205] Decomposition of three organophosphorus pesticides, chlorpyrifos, isofenphos, and diazinon, was studied after their application to turfgrass and subsequent composting of the grass clippings.[214] The insecticides decomposed rapidly with <1 mg kg^{-1} of the pesticides present after 14 days and no detectable residues present after 21 days.[214] Sears and Chapman[208] reported similar results. Carbaryl, one of the most widely used pesticides, is metabolized by soil organisms in 8 to 9 days.[215–217] The half-life of carbaryl in spent mushroom compost was less than 3 h, a significant improvement in decomposition time over that in soil alone, which was attributed to the composting process.[218]

2. Soil Quality

Along with soil chemical properties, physical properties such as bulk density, cation exchange capacity (CEC), water-holding capacity, porosity, and aggregate stability are important aspects of soil quality[219] which may be influenced by compost applications.[171] Soil quality and environment and health aspects are discussed in detail in a book by Doran et al.[220] Several researchers have studied the effects of composts on soil qual-

TABLE 6 Some Organochlorine Pesticides Substrates and the Associated Decomposing Microorganisms[192]

Pesticides	Microorganisms
Endrin	*Arthrobacter, Bacillus*
Lindane	*Clostridium, Escherichia, Hydrogenomonas, Klebsiella*
Aldrin	*Micrococcus, Proteus, Pseudomonas*
Heptachlor	*Pseudomonas*
Dieldrin	*Pseudomonas, Nocardia*
p-Chloronitrobenzene (PCNB)	*Streptomyces, Aspergillus, Fusarium, Mucor, Trichoderma, Saccharomyces*

ity.[142,221–226] Reported changes in soil quality occurred with compost applications as low as 13.6 Mg ha^{-1}[223] but in other cases effects were not detected until the application was 149 Mg ha^{-1}.[224] The most dramatic soil quality responses to the application of compost to land have been on marginal soils with poor soil structure and low levels of soil organic matter and plant nutrients.[225,226]

Repeated applications of MSW composts to agricultural lands can result in significant improvements in physical, microbiological, and chemical properties of soils similar to the effects imparted on soils after sewage biosolids or other types of organic matter are applied.[227–229] The effects of compost applications to European soils were reviewed by Gallardo-Lara and Nogales[230] and Guidi and Petruzzelli,[231] who indicated that the positive effects well outweighed the negative effects. Changes in soil physical properties generally are attributed to an increase in organic matter from the addition of composts.

A significant beneficial effect of compost applications to soil is improvement of soil structure due to the increased integrity of aggregates stabilized by the interaction of microorganisms and the mineral fractions of the soil.[230,231] Capriel et al.[232] reported significant correlations between soil aggregate stability and soil microbial biomass. Little improvement in soil structure was detected if organic residues were applied to sterilized soil indicating the importance of microorganisms.[233] Adding organic matter increases the growth of microorganisms, with the growth also being influenced by the seasons.[234–236] With stabilization of aggregates, bulk density is decreased and porosity is increased.[210,224] Water-holding capacity is increased by applications of compost to soil.[223,237]

The high content of organic matter in compost and the resultant effects of the organic matter on the humic fractions and nutrients in soil increase microbial populations, activity, and enzyme production, which in turn increase aggregate stability.[238–241] Under laboratory conditions unamended or MSW compost-amended soil was incubated for 1 year.[241] The contents of C, N, and S were higher in microbial biomass from amended soil than in microbial biomass from unamended soil.[241] The activities of phosphatase, urease, protease, deaminase, and arylsulphatase in soil were increased by the addition of MSW compost.[241] Soil microbial biomass and enzyme activity are considered important indicators of soil quality improvement due to the addition of organic matter.[241]

Compaction generally refers to an increase in soil bulk density.[242] One beneficial effect of organic matter or compost additions to soil is the decrease in soil bulk density.[243,244] In a study in Washington, D.C., composted sewage biosolids or composted MSW were surface-applied in a highly visible parkland restoration project.[245] The parkland had received many visitors prior to the experiment and the soil had been heavily compacted by foot traffic. Improvements to the soil after compost application included an increase in water infiltration rate, a decrease in bulk density, and an increase in pore volume.[245] Recent research also indicates that soil treated with MSW compost is less compactable than untreated soil.[246]

Organic matter content of soil is significantly correlated with erosion prevention.[247] Several investigators stressed the importance of organic matter in water-stable aggregates through the formation of organo-mineral complexes.[238–240] The extent to which a soil erodes depends on the strength of soil aggregates to withstand raindrop impact and surface flow.[248] Humic and fulvic acids and humins are important as persistent binding agents in mineral-organic complexes, and 52 to 92% of soil organic matter may be involved in these complexes.[249,250] Fungal hyphae are associated with temporary binding agents which may impart a degree of short-term stability to aggregates.[238]

Compost applications to prevent erosion can be an important measure in soil conservation. Through increases in soil organic matter, compost applications to soil are effective in reducing soil erosion and water runoff.[251] At 320 Mg ha^{-1}, (about 2.5 cm thick), compost application virtually eliminated erosion on steeply sloping vineyards.[252] Digested papermill sludge amended with N and P reduced erosion on a steeply sloping strip-mine reclamation site.[253]

Negatively charged soil colloids attract positively charged ions (*cations*). CEC is dependent on the intensity of the negative charges of the colloids. CEC is important in plant nutrition and fertility management as it effectively constitutes temporary storage for cations and, therefore, is considered an indicator of the nutrient-holding capacity of a soil. A low CEC may allow for greater leaching of cations. Addition of organic matter to soil may increase CEC significantly.[254] For each percent of humus in the soil, CEC is increased 2 meq $100g^{-1}$.[254] CEC of soil is increased by compost applications.[237] This effect is primarily due to dissociation of H^+ from weak acids in organic matter. The cation exchange sites of organic matter are mainly the carboxylic and phenolic groups.[255]

D. EFFECTS ON CROP YIELDS AND PLANT NUTRITION

1. Field

Improvements in the physical, chemical, and microbiological characteristics of cultivated soil and increased crop production from additions of compost to soil have been reported.[256–259] Some of these soil improvements are long-term and can have significant effects on yield. Composted animal and plant residues continued to improve sunflower yields after 6 years of cropping.[260] Water hyacinth (*Eichhornia crassipes*) compost applied to a rice-based (*Oryza sativa*) cropping system has been reported to increase rice yield.[144] Yields of wheat (*Triticum aestivum*) and gram (*Cicer arietinum*) grown after the rice harvest were increased significantly without any further additions of inorganic fertilizer in the same study.[144] Land-applied, poultry manure-based compost has been reported to produce a gradual, long-term response in crop yield.[261,262]

Seedlings of onion (*Allium cepa*), lettuce (*Lactuca sativa*), or snapdragon (*Antirrhinum majus*) transplanted to plots amended with composted sewage biosolids had more vigorous growth and produced higher yields than those transplanted to unamended plots.[16] In these experiments, compost-amended plots also produced increased growth of turfgrass (*Festuca arundinacea*) compared to turfgrass grown without compost amendment.[16] Recently, Astier et al.[151] reported increased yields of broccoli (*Brassica oleracea*) after woollypod vetch (*Vicia dasycarpa* var. *lana*) was added to composted poultry manure prior to land application. Composted poultry manure or woolypod vetch also increased yield of broccoli in this study.[151] Amending calcareous soils with MSW compost increased the growth and yield of tomato (*Lycopersicon esculentum*) and squash (*Curcurbita maxima*).[29] Mays and Giordano[28] landspread MSW compost annually and grew corn (*Zea mays L.*) for 14 years with yield increases of 55 to 153% over unamended soils. Composted plant material added to soil increased the yield of sweet potato (*Ipomoea batatas*)[263–268] and corn.[269]

Much of the effects of application of compost to crop yield and production is derived from the plant nutrients (particularly N) in composts.[29,270,271] A 24-year study of the long-term effects of soil applications of compost of garbage and biosolids showed substantial increases in soil N.[272] Relatively high applications of composted manure (19 to 30 Mg ha^{-1}) are required if all of the necessary N is to be supplied solely from compost in the short-term.[273–275] Composted pig slurry applied to wheat in field experiments significantly increased leaf K.[276] Increased plant nutrition from composted animal manure applied to soil also increased the protein content of potato (*Solanum tuberosum*).[277]

Other researchers report an increase in plant nutrients from an increase in organic matter due to amendment with MSW compost.[171,278] In a laboratory experiment, a sandy loam or a clayey silt was incubated for 12 months with or without MSW compost.[279] Total N, soluble P, and exchangeable K increased in the MSW-amended soil compared to the unamended soil.[279] Koma Alimu and Janssen[280] reported that yield responses from MSW additions to soil in their experiments were due to increases in K.

Several studies suggest that organic sources of P are more effective for plant absorption than inorganic sources.[281–285] This result may arise from increases in soluble P,[286] increases in the microbial P pool,[287] or increased hydrolysis of organic P by microbial activity.[284] The increase in availability of soluble P from additions of compost is an effect that is described as resulting from phosphohumic complexes that minimize immobilization processes, anion replacement of phosphate by humate ions, and coating of sesquioxide particles by humus to form a cover which reduces the phosphate-fixing capacity of the soil.[281,283,288] In a field study with broccoli, P use efficiency was increased with compost amendments.[289]

Sulfur availability to plants may be enhanced by applications of composts to agricultural land.[290] Although more research is needed in this area and some reports are inconclusive or conflicting , a review by Gallardo-Lara and Nogales[230] has addressed some of the effects of MSW compost on N, P, and S in soils.

Earthworm casts are reported to be higher in available N, P, Ca, and Mg than the substrates upon which they feed.[291,292] Earthworm casts have high levels of labile P and phosphatases[293] due to the presence of alkaline phosphatases in the gut of the earthworm.[294] The activity and numbers of microorganisms producing acid phosphatases is increased near earthworm casts in the soil.[293–295] Combined together, these specific effects appear to raise P availability in soil amended with vermicompost.[289]

Composted MSW may have a high pH and contain large amounts of $CaCO_3$.[70] Therefore, applications of MSW compost to acid soils may be beneficial. Compost amendment increased pH in a soil with low buffering capacity.[289] Compost amendments to a field experiment with broccoli resulted in a significant increase in soil pH, and the pH increased with increasing applications of compost.[27] Leaf compost added to soil for 12 years increased the pH from 5.6 to 6.5 in unlimed plots.[27]

The pH and the organic matter content of soils affect ion absorption by plants.[270,296–299] An increase in pH can bring about strong adsorption or, in some cases, precipitation of Cd, Mn, Pb, and Zn, among other metals, which in turn allows for lower accumulation of these metals in plant tissue.[270,297] Decreases in accumulation of Cd by plants with increasing soil pH is well documented.[296,298–300] Lettuce, spinach (*Spinacia oleracea*), carrot (*Daucus carota*), and red beet (*Beta vulgaris*) grown in soil amended with sewage sludge accumulated less Pb than plants grown on unamended soil.[301] Organic matter complexes Cu and reduces its availability and potential for toxicity to plants.[270] Availability of Zn can be decreased by organic matter additions to soil by formation of insoluble Zn organic complexes with humic acids, thereby lessening risk of Zn toxicity of plants.[302] Therefore, some beneficial effects to heavy metal concentrations in soils can be achieved by applying compost to agricultural land. Adverse effects of heavy metal accumulations in soils is discussed in some detail in the next section of this chapter.

Minor essential elements may be supplied by compost amendments to deficient croplands. Increases in Zn concentrations in soil were noted after the addition of compost in one study.[279] Depending on soil conditions, compost may increase exchangeable and water-soluble Mn.[288] Although Zn and Mn can be toxic to plants, their introduction into aerobic soil from compost is not usually considered a problem[303] as high concentrations of Fe in composts inhibit Zn uptake by plants, and Mn availability to plants is reduced by the oxidizing conditions in most soils.[304] Copper is bound very tightly to inorganic exchange sites in soils and is not readily available to plants.[305] Iron, Cu, Mn and Zn are essential elements that may be supplied by biosolids compost additions.[288] For example, Fe deficiency in corn was corrected by application of Fe-rich biosolids.[306]

2. Potting Media

Horticultural uses of MSW composts have been reviewed by Rosen et al.[307] Other types of composts also have been shown to be beneficial for horticultural use. Digested coffee slurry effectively replaced peat in a potting substrate for *Gypsophila*, *Lysimachia*, and *Phlox* with root growth of *Lysimachia* and shoot growth of *Phlox* being enhanced by the coffee residues.[158] Mature compost prepared from vegetable market refuse and the stomach contents of slaughtered cattle was added to sandy potting soil and doubled the yield of sunflowers (*Helianthus* sp.).[26] Composted biosolids or composted refuse used as amendments for a sandy soil potting mix increased shoot dry weight and flower production of petunia (*Petunia grandiflora*).[32,33] In another series of experiments, growth was greater in compost-amended media than that of plants grown with peat amendment.[32]

Composts can be used to replace peat or used in addition to peat in potting media.[145,297,308,309] MSW compost has also been used as a component of growing media for potted plant culture.[145,308,310] Siminis and Manios[308] grew rooted cuttings of *Ficus benjamina* successfully in a medium with peat and 20% MSW compost. Sod may also be produced using MSW. Mixtures of Kentucky bluegrass (*Poa pratensis*), red fescue (*Festuca rubra*), and tall fescue (*F. arundinacea*) were sown in composted biosolids with a marketable sod produced significantly earlier than the 12 to 18 months for traditional sod production in soil.[311] Shortened production time also was reported by Neel et al.[312] who produced harvestable sod of bahiagrass (*Paspalum notatum*) and bermudagrass (*Cynodon dactylon*) on a medium of composted sewage biosolids alone or composted biosolids and other wastes. Heavy fractions of MSW compost produced a marketable Kentucky bluegrass sod in a shorter period of time than conventional sod production in soil.[20] St. Augustine sod (*Stenotaphrum secundatum*) was produced on 100% MSW compost in 5 months if additional nitrogen fertilizer was used.[19] Modern potting composts are discussed in detail in a chapter by Bunt.[313]

Growth, plant quality, yield, and nutrition of potted plants grown with composts in the media have generally been positively affected by the additions of composts. A mixture of composted cotton burrs (*Gossypium* sp.), spaghnum peat, and composted pine bark was used to grow rooted cuttings of poinsettia (*Euphorbia pulcherrima*) in a greenhouse.[33] The number of branches and bracts, days to bloom, and plant grade after 30 days was equal to that of plants grown in a peat-bark mix.[33]

Corn grown in pots in a greenhouse had sufficient P in tissue from addition of composted plant residue.[314] An unamended soil required 300 mg kg^{-1} P supplement, whereas 75% compost mix supplied P to maintain 0.20 mg L^{-1} in soil solution, which was sufficient for plant growth. Plant analyses have indicated that adequate N and K were supplied by refuse compost and that adequate N, P, and K were supplied by composted sewage sludge.[32]

Along with use on field crops,[315] fruits,[316] and vegetables,[27,317–319] spent mushroom compost has been used with positive results to amend potting soil used to grow greenhouse crops[320] and foliage plants.[321] Spent mushroom compost made of decomposed straw horse manure, peat, limestone chips, gypsum, cottonseed meal, urea, and residual fungal mycelia was used as a potting medium to grow lettuce, tomato, marigold (*Tagetes patula*) and broccoli.[23] All crops in this experiment grew better with 12.5% of the spent mushroom compost in the medium than plants grown without the compost added.[23] Dogwood (*Cornus alba*), forsythia (*Forsythia x intermedia*), and weigela (*Weigela florida*) grew significantly better in media with spent mushroom compost added than in unamended media.[18]

E. CONTROL OF PLANT DISEASES AND NEMATODES

Suppression of soil-borne plant pathogens after applications of organic and especially composted organic materials has been reported.[322–326] Applications of composted organic residues suppress *Phytophthora cinnamoni, Rhizoctonium solani, Fusarium oxysporum*, and *Pythium aphanidermatum* with many crops.[25,296,327–329]

Poplar bark compost inhibited fusarium wilt (*Fusarium oxysporum f.* sp. *dianthi*) in greenhouse experiments in potting media with carnation (*Dianthus caryophyllus*).[330] In another experiment, composted separated manure and composted grape (*Vitis vinifera*) marc suppressed *Pythium* damping-off disease.[25] Composted olive (*Olea europea*) pomace or a commercial composted pine bark incorporated into a potting medium instead of peat suppressed fusarium wilt of carnation.[331]

Among other disease-suppressing organisms are mesophilic strains of bacteria (*Bacillus* sp.).[332–334] A species of bacteria commonly found in compost, *Bacillus subtilis*, has been shown to produce antifungal volatiles,[335] which suppress plant pathogens through antibiosis.[336–343] Wood waste compost was at least as effective as fungicides in controlling *Phytophthora* root rots.[333] Topdressing with composts made with plant refuse or animal wastes suppressed dollar spot (*Sclerotinia homoeocarpa*) of turfgrass as effectively as a commercial fungicide.[344] Extract of spent mushroom compost inhibited *in vitro Venturia inaequalis*, the causal agent of apple scab.[334] Suppression of *Rhizoctonium* in compost media has been reported by several researchers.[327,345–349] Composted manure suppressed damping-off by *Rhizoctonium solani* in potting media experiments with radish seedlings (*Raphanus sativus*).[350] Composted grape (*Vitas vinifera*) marc and composted cattle manure suppressed diseases caused by *Rhizoctonium solani* and *Sclerotium rolfsii*.[351,352]

The suppressive action of beneficial organisms against pathogens is significant only in mature composts. Composted sewage biosolids cured for 4 months, mixed into potting media, and stored for an additional 4 weeks prior to planting cucumbers (*Cucumis sativus*) suppressed *Rhizoctonium* and *Pythium*.[353] Composted biosolids cured for 4 months without additional treatment prior to incorporation into media suppressed *Pythium* but not *Rhizoctonium solani*.[353] Uncured sewage compost suppressed neither plant disease.[353]

The majority of plant pathogens—nematodes, viruses, bacteria, and fungi—are sensitive to conditions in compost heaps. Some control of plant diseases is accomplished by heat and toxins generated during composting.[354] Although it has a major function in control of pathogens in compost-amended soil, microbial competition has a minor role in inactivation of plant diseases during composting. However, mechanisms of suppression have been proposed including microbial suppression of pathogen infection, sporulation, and survival.[355]

Other researchers have proposed two biological mechanisms for disease suppression of root pathogens in media with mature compost amendment.[333] *Pythium* and *Phytophthora* are inhibited by general suppression whereas *Rhizoctonium solani* is inhibited by a specific suppression.[333] The mechanism of suppression of *Fusarium* by composted bark is not clearly understood.[333] However, an antifungal compound, sitosterol, a secondary plant metabolite, was isolated from bark compost by Ueda et al.[356] Cholesterol, campesterol, and stigmasterol have also been isolated, and this group of sterols had strong antifungal activity against *Fusarium oxysporum f.* sp. *cucumerinum*.[356] Cholesterol was also reported as suppressive to *Cochliobolus miyabeanus, Gibberella zeae, Glomerella cingulata, Alternaria alternata*, and *Rhizoctonium solani*.[356] Extracts of bark compost were effective against *Fusarium oxysporum f.* sp. *cucumerinum, Gibberella zeae, Helminthosporium sigmoideum*, and *Glomerella cingulata*.[357]

Trichoderma, in addition to being a strong colonizer of compostable materials, has potential to control plant diseases.[329,333,358] Potting media containing high levels of *Trichoderma* have been reported to suppress *Pythium* and *Rhizoctonium*.[358] Some *Trichoderma* strains (TH1 and 8MF2) have biocontrol potential and have been reported to promote growth if added to compost after autoclaving.[359,360] Germination and growth of lettuce was generally increased with *Trichoderma*, but effects varied depending on the temperature to which *Trichoderma* was exposed, strain used, age of culture, and inoculum application rate.[361,362] Heat-treated *Trichoderma* increased growth of lettuce.[361] *Trichoderma* sp. also enhanced growth and development of vesicular-arbuscular mycorrhizal fungal (VAMF) mycelium.[363] Mycorrhizae can enhance nutrient uptake, promote plant uniformity, and reduce transplanting injury.[364] Other microorganisms, such as thermophilic fungi (*Scytalidium thermophilum*) promoted growth of the edible mushroom (*Agaricus bisporus*).[15] The response was attributed to increased CO_2 production.[15]

Soil amendments have been reviewed for their efficacy in suppressing nematode populations.[365] Organic amendments such as sawdust, green manure, poultry manure, and compost have reduced nematode populations.[365] Land applications of MSW compost decreased juvenile populations of *Meloidogyne incognita* in field squash in southern Florida but did not affect the plant-parasitic stages of nematode populations.[366] Products containing chitin, such as blue crab (*Callinectes sapidus*) scrap, have been shown to suppress nematode populations.[367–369] A compost of blue crab scrap and cypress (*Taxodium distichum*) sawdust suppressed populations of nematodes (*Meloidogyne javanica*) in container-grown tomato.[370] Thus, compost may have many beneficial effects on plant diseases and pests.

IV. ADVERSE EFFECTS

A. EFFECTS ON WATER AND AIR QUALITY

Historically, leaching of NO_3^- from fertilization practices has been a concern of farmers, scientists, and environmentalists.[371] Nitrate is highly mobile in the soil profile and can migrate downward into aquifers or laterally into surface water bodies.[372,373] Leaching of NO_3^- can be significant in areas of high precipitation and sandy soils. Nitrates are the most common contaminant in wells used for drinking water.[374] The U.S. EPA recommends a maximum of 10 mg L^{-1} NO_3-N in drinking water.[375] Ingestion of NO_3^- rich water or foods may cause methemoglobinemia in babies or other susceptible individuals and animals.[375-377] Nitrogen-enriched surface waters may result in eutrophication in ponds, lakes, and estuaries.[373,378] For these reasons, it is desirable to use caution in application of nitrogenous materials to land.

Animal manures and sewage sludge may be high in N compounds, which are mineralized to NO_3^-. If nitrogenous materials are applied to land in amounts excessive for crop uptake of N then leaching of NO_3^- is possible. Composting the nitrogenous materials allows a measure of control of the unstable N compounds.[27] Available N in raw manure or sewage may be immobilized by microorganisms during composting allowing a more stable, slow-release form of N for crop use. Therefore, composting of manure or sewage can give a measure of control in prevention of NO_3^- leaching to ground or surface waters.

Some researchers report that heavy metals such as Zn, Ni, and Cd may be readily leached in seepage water.[379,380] However, other studies indicated that heavy metals remained in the horizon where deposited[381] or in the top 30 cm of soil with MSW compost added annually for 14 years.[29] Field studies with repeated sludge biosolids applications over a 6-year period indicated that 90% of metals remained in the soil layer where applied.[382]

Organic dusts contain actinomycetes, bacteria, fungi, arthropods, and protozoa. Of major concern are the fungus *Aspergillus fumigatus*, cell walls of Gram-negative bacteria (endotoxins), and β-1, 3 glucans from the cell walls of fungi and mycotoxins.[383] These materials are found in aerosols generated from a wide variety of organic wastes including grass clippings, wood chips, food and household wastes, agricultural wastes, and sewage sludge.[383] Neighborhood and occupational exposure through inhalation of dusts may be concerns. *Aspergillus fumigatus* and *Thermoactinomyces vulgaris* are common where compost is disturbed.[383] Microorganisms at compost sites, in compost products, and raw materials are found in ambient air environments.[383] However, normal outdoor aerospora are dominated by *Alternaria, Cladosporium, Aspergillus*, and *Penicillium* even if no composting facility is nearby.

Organic dusts of some of the components of compost upon inhalation have been reported to cause some upper respiratory disorders in some people. Most of these organic dusts must be inhaled for several years at relatively high levels before symptoms such as those of chronic bronchitis appear.[383] Composts, grain dusts, moldy hay and straw, and other compostable materials may support populations of thermophilic actinomycetes, fungi, bacteria, and other microorganisms. Several weeks to years of exposure may inflame mucous tissues, precipitate bouts of asthma, or bring about responses in the lungs of sensitive individuals.[384-391] Many of the responsible microorganisms are commonly found in the home, in carpeting, in the air, on air conditioner filters, humidifiers, and many other locations throughout the home.[383] Millner et al.[392] reported that from 0 to 59 colony forming units (CFU) m^{-3} of thermotolerant and thermophilic actinomycetes were collected in air samples near a site unaffected by composting. Many of the actinomycetes found active in composting are components of natural air.[392]

Aspergillus fumigatus is considered ubiquitous.[393] It is associated with soils, animal and bird droppings, hay, corn, straws, grasses, and compost.[393] Airborne *A. fumigatus* has been found at many sites.[393] The concentrations in air of *A. fumigatus* have been compared to two other fungal allergens, *Cladosporium* and *Alternaria* which are greater in number, however, *A. fumigatus* is considered serious because of its multiple potential health effects in humans.[383] Airborne *A. fumigatus* was found at a yard waste composting site at 0 to 2648 CFU m^{-3} on-site with 0 to 11 CFU m^{-3} downwind a distance of 1.6 to 150 meters.[394] In other research, average viable airborne *A. fumigatus* spores were reported in concentrations ranging from 300 to 710 spores m^{-3} at 500 to 800 meters from the compost site, while background concentration was 65 spores m^{-3}.[383]

The effects of composting on bioaerosol production have been studied for more than 20 years.[395–403] The data indicated that at distances of 76 to 150 meters, the concentrations of *A. fumigatus* were at or below background concentrations. Some researchers report increased concentrations of *A. fumigatus* surrounding composting facilities.[383,400] However, the general population is not considered at risk to systemic or tissue infections from compost-associated bioaerosol emissions.[383] Worker populations at such facilities thus far have not shown any significant differences in overall body or respiratory fitness compared to unexposed persons.[383] Infective doses of human respiratory pathogens have not been identified as causing worker or environmental health problems.[404]

Control of odors is a necessary management concern for composting of organic wastes. Although composting can transform odorous wastes, such as farm manures and sewage, into nonodorous soil conditioners, composting also may generate odors. Bad odors from composts are considered an environmental affront, and concerns about bad odors limit site selection of composting plants. Malodorous emissions from organic decomposition, especially anaerobic decomposition, can have detrimental esthetic and economic effects on a community.[405] In health issues, odors have been reported anecdotally to incite allergic responses, poor appetite, lower water consumption, impaired respiration, nausea, vomiting, insomnia, and mental stress.[406] Many of the chemicals emitted from organic decomposition that are considered offensive are in very low concentrations in the air.[406] The compounds considered malorodorous include sulfurous and reduced N compounds, fatty acids (acetic, propionic, butyric), amines, ammonia, cadaverine, putrescine, terpines, and aromatic compounds.[406–412] Methods of managing offensive odors during composting are addressed in Section V (Minimizing Adverse Effects and Increasing Beneficial Effects).

B. EFFECTS ON SOILS AND CROPS

Compost with a high C:N ratio could cause plants and microorganisms to compete for N. Microorganisms often, compete effectively and the N is assimilated into the microorganisms cells (immobilization). Thus, composts with a high C:N ratio may leave little N available for plant development. Immature composts may have high C:N ratios (greater than 40:1), which initially immobilize N after application of compost to soil.[19] Additions of N fertilizer may overcome the problem, as was the case in research in which St. Augustine turfgrass sod was produced in 5 months on 100% MSW compost with additional N fertilizer.[19]

Adverse effects of MSW composts have been reported on plant yield, seed germination, and root growth.[413–416] Immaturity of compost is a factor in the suppressive effects on plant growth. Extracts of fresh MSW compost suppressed germination and inhibited root elongation of Chinese white cabbage (*Brassica chinensis*) and tomato, but extracts from 6-week-old compost and 12-month-old compost did not affect plants as much as the extracts of fresh compost.[34] Lettuce, tomato, marigold, and broccoli container-grown in aged compost (50% of medium) had significantly greater dry weights than plants grown in medium with 50% fresh compost by volume.[23] Very little growth of lettuce occurred in mixes with 38 or 50% fresh compost.[23,417,418] A significant negative relationship existed between root length and seed germination, and the concentration of ethylene oxide in the water extracts.[34] High concentrations of ethylene (1 to 10 μl L^{-1} of soil atmosphere) have been reported to inhibit root extension.[419] Low molecular fractions of fresh MSW compost also have been reported as being phytotoxic to several vegetable crops.[420] Another component of immature compost, NH_4^+, may also inhibit plant growth.[421] Acetic, phenylacetic, benzoic, propionic, isobutyric, butyric, and isovaleric acids have been isolated from fresh compost.[69,422–426] However, only a small amount of acetic acid was found in mature compost.[427]

Organic contaminants, such as polychlorinated biphenyls (PCBs) and polycyclic aromatic hydrocarbons (PAHs), may be present in composts.[230] PAHs and phenanthrene may be present in sludge biosolids.[201,428] Pesticides may be present on plant residues included in compost.[214] Most of these organic contaminants are decomposed during 6 months of composting.[171,214] Organic contaminants in MSW composts that are land applied are not expected to leach as vertical leaching becomes a problem only at extremely low soil organic content.[429]

Repeated applications of composts to soils has led to concerns of accumulation of metals in soils.[30,230] Composts from MSW and sewage biosolids may contain Zn, Al, Cd, Pb, Fe, Cu, Mn, and Mo. Concentrations of these heavy or trace metals in composts often are higher than most agricultural soils, particularly Cu, Pb, and Zn.[237,430–433] Heavy metals may be phytotoxic or accumulated in food and feed plants,[434,435] and historically Cd consumption by humans and animals has been a concern.[436,437] The occurrence and fate of heavy metals in soils from compost application may be highly dependent upon management of composting processes and applications, as discussed in Section V (Minimizing Adverse Effects and Maximizing Beneficial Effects).

Increased soluble salts in soils can occur from amendment with composted sewage sludge, MSW, and animal manures.[230,438] The potential for high salinity in mixes has limited the use of some composts as a soil amendment for potted crops.[439] However, leaching of salts was rapid in experiments growing woody ornamental plants in pots with trickle irrigation.[439,440]

C. EFFECTS ON HUMANS AND ANIMALS

Limitations to the widespread use of MSW and biosolids as fertilizers or composts have been tied to potential phytotoxicity and animal and human health hazards associated with plant and soil contents of heavy metals.[441,442] Humans, livestock, and wildlife can be exposed to heavy metals through ingestion of soil or plants that have accumulated heavy metals.[437,443] However, negligible increases in heavy metal concentration of tomato and squash occurred in an experiment where these crops were grown on calcareous soil with MSW compost amendment.[29] In a field experiment with tomato in soil amended with MSW compost, differences in food content included lowered concentrations of Be, Cd, and Mn.[444] Significant differences in pear (*Pyris communis*) peel Cu and Zn were noted if sewage biosolids compost was added to soil; however, the concentrations of Cu and Zn were still within normal limits.[445] An increase in the total content of any trace element in compost-treated soil does not necessarily lead to an increase in plant uptake of that element.[446,447] Chaney[448] asserted in a public health review that Cd uptake from MSW is too low to pose a serious health threat.

Because of concerns for public health, much attention has been given to controlling the release of metals into wastewater and solid waste, so that today heavy metals in biosolids and MSW are much less than in the past. The U.S. EPA regulates treatment of biosolids to eliminate pathogens and to limit heavy metals.[449] Conditions under which metals are released depend on pH of soil and compost.[258] Adverse effects are found only after highly contaminated composts of MSW or sewage biosolids are applied at high rates to strongly acid soils.[450] Presorting and other risk management strategies are discussed in the next section (Section V).

Incidence of some nuisance organisms near composting facilities has been a concern. At one composting facility in the central United States it was suggested that leaf compost attracted the Norway rat (*Rattus norvegicus*).[451] Rodents are prey to many bird species.[452,453] An increase in bird populations might be expected at compost sites; however, no increases in numbers of species or in overall numbers of birds or rodents were observed over background numbers from noncompost areas.[451] Upon further observation the authors noted that the Norway rat was present only in urban composting areas.[451]

Flies surrounding compost facilities or compost applications have long been considered a nuisance.[454] In one study in Florida, the green bottle blow fly (*Phaenicia cuprina*) comprised more than 90% of the fly population surrounding the composting facility.[454] Seasonal fluctuations in fly populations may be observed.[454] The major source of larvae was determined to be incoming refuse.[454] At facilities with poor management and control, fly infestations may indeed become a nuisance.

V. MINIMIZING ADVERSE EFFECTS AND INCREASING BENEFICIAL EFFECTS

A. PRESORTING, SITE CHOICE, AND FACILITY DESIGN

Most opposition to composting has risen from opinions regarding odor, presence of foreign or inert material (wood, plastics, glass) large pieces of metal contamination with trace metals, toxic organic chemicals, and pathogens.[430,455,456] Presorting can effectively reduce the presence of objectionable materials, such as heavy metals and toxic organic substances, large pieces of wood, metal, and glass[430,455] and slowly composable substrates which may reduce compost quality. Historically, heavy metals in land-applied compost have been a major concern for health professionals and environmentalists. It has been considered critical to prevent heavy metals such as Pb and Cd from entering the food chain.[434] The most effective modification in processing of composts, especially MSW composts, is the separation of raw materials prior to processing.[171] Presorting can reduce or remove the risk of contamination of composts with heavy metals.[455] Hence, compost manufacturers should be attentive to rejection of metal-rich components especially those components rich in Pb or Cd, contaminated sewage sludge, manures, and household articles, such as paints, batteries, appliances and similar contaminants.

Guidelines for limits of heavy metals in composts often are derived from the maximum pollutant concentrations in clean sewage sludge suggested by the U.S. EPA, commonly called the 503 regulations.[449] In mg kg^{-1} the U.S. EPA concentrations are As:41, Cd:39, Cr:1200, Cu:1500, Pb:300, Hg:17, Mo:18, Ni:420, Se:36, and Zn:2800.[449] Some states, such as Minnesota have published guidelines for Class 1 compost that are more stringent for most metals than the EPA guidelines for sewage.[459] Analysis of substrate intended for composting is an important measure of control in excluding undesirable components such as heavy metals, organic contaminants, or other unsuitable materials. Several researchers report analyses of sample composts and methods for evaluating risks after such analyses have been performed.[74,301,460–462] Legislation and regulation have also been instrumental in excluding pesticides and other organic contaminants from the waste stream for composting. Removal from the waste stream of such items as household or industrial pesticides, cleaners, and solvents impart an important measure of control in reducing contaminants in composting. Compost manufacturers must be diligent in excluding chemical contaminants and in complying with regulations to ensure no contamination of composts with these chemicals.

Summer composting of MSW compared to the same materials and processes in winter had different degrees of humification and different heavy metals-humic associations.[463] More research on the seasonal aspects of MSW composting and heavy metals is needed. There may be some potential for manipulation of processes to achieve maximum binding of heavy metals. Land application of immature compost can lead to a decrease in O_2 concentration in soil and an increase in reducing conditions,[464] which in turn could increase the mobility of some trace metals.[441] Care must be taken to ensure maturity of compost as described in the processing section discussing maturity.

Site choice and facility design can be highly effective in minimizing perceived or actual adverse effects of composting. The area of the site that is used for storing, windrowing, mixing, or screening the compost should be located as far downwind as possible from public areas. Enclosing the materials handling site can greatly reduce the size of the buffer area required between materials handling and the public.[383] In areas where winters are cold and the exterior of the windrow freezes, the application of materials over the frozen crust may allow runoff, rather than perolation through the piles.

Dry climates and areas with prevailing windy conditions pose some potential problems in the control of bioaerosols; therefore, enclosure may be considered.[465] A design goal to reduce exposure to bioaerosols could make use of topography and meteorological characteristics to maximize diffusion and distribute bioerosols over a wide area. Or conversely, the design parameters may be such that the desired outcome is more of a retainment of the bioaerosols. Site selection is of primary importance in the control of bioaerosols and odors.[383] Emissions generated during delivery of raw materials and composting can be avoided by using closed facilities, exhaust air collection systems, or biofiltration of exhaust.[412,466] The design for controlling potential pollution from composting facilities includes (a) control of wind movement, (b) control of snow distribution, (c) diversion of overland flows of water from offsite, (d) control of runoff, transport, and soil erosion from the site, and (e) maintenance of adequate drainage if high water table or surface inundation is possible.[465]

Trees and shrubs can be used for noise abatement[467,468] and windbreaks.[469] Barriers, snow fencing, and level bench terraces may be used to divert and control snowmelt.[470–472] Proper site preparation and compaction of soil in construction will help to control overland flows of water from offsite.[465] Runoff and erosion can be controlled by using terraces.[473] Level bench terraces can be used to impound potential runoff.[474,475] To remove standing water, the installation of open or covered drains[476] or tile drains[477] may be considered. If a site is well chosen to accommodate composting none of these preparatory measures may be necessary.

B. PROCESSING

Among the most effective controls for composting are regulating temperature, moisture content, degree of aeration, and duration of composting.[171] Optimization of composting parameters such as moisture content, temperature, aeration, and duration has significant effects on compost disinfection, pathogen destruction, and odor minimization.[455] Adequate aeration during processing is especially important in odor reduction.[478]

Maturation and curing is important to prevent N immobilization and NH_3 and or NH_4^+ toxicity to plants.[171,479] Immature compost can inhibit tomato seedling development,[480] and can decrease top growth of sweet william (*Dianthus barbatus*) and pansy (*Viola x Wittrockiana*).[31] Improper storage of composted sewage sludge can allow anaerobic conditions and accumulation of acetic acid and alcohol.[481] Several methods for assessing compost maturity have been reviewed by He et al.[171] and Inbar et al.[479]

Additives or bulking agents may give some control of some contaminants. With 40% (by volume) leaves added to sewage sludge compost heavy metal concentrations were lower than with sewage sludge compost alone.[116] A fish waste composting method was designed by Mathur et al.[163] which combined fish wastes with sphagnum peat. The author claims that nearly all of the NH_4^+ produced by the fish wastes is adsorbed by the moist, acidic peat.[163] Thus, choice of additives or bulking agents in composting procedures can allow an added measure of control. The risks from high NH_4^+ or soluble salt levels can be minimized by limiting the amount of compost to 33% of the mix by volume.[418] A computer program has been written to aid in decision making in co-composting ratios.[482]

Pathogenicity is one of the major concerns from use of MSW and sewage biosolids in agriculture. The objective of regulations from the U.S. EPA and other agencies for treatment of biosolids is to kill parasitic worm (helminth) eggs and pathogenic viruses and bacteria and to reduce vector (flies and rodents) attraction. Biosolids that meet the standards of pathogen reduction set by the U.S. EPA are called Class A. Composting can be used to eliminate pathogens from MSW and biosolids for land application. Composting should be carried out at relatively high temperatures (> 55°C) for 3 to 15 days, depending on the process of composting.[483] Pretreatment of biosolids by liming, anaerobic digestion, irradiation, or oxyozonation substantially reduces pathogens, relative to those present in raw biosolids. Considerable advantages in safety and destruction of pathogens can be obtained by use of pretreated sewage biosolids.

Salmonella bacteria are recalcitrant to control and are the most frequent pathogen in composted biosolids. Governmental regulations are directed toward ensuring that these bacteria are controlled. Bacterial pathogens have the capacity to regrow in composts if they are not sufficiently reduced by heating during composting. Only a small percentage of the pathogenic bacterial population may be exposed to lethal temperatures during a given period.[484] However, pathogenic viruses and parasitic helminth eggs do not present health hazards in composts.[483,484] Composting reduces these organisms below detectable limits, and they do not regrow in composts. Polio-type viruses are short-lived in the compost process.[484]

Temperature of composting must be sufficient (60 to 70°C) to destroy pathogenic organisms, weed seeds, and fly ova.[60,485] Phytotoxicity was reduced quicker in compost at 50°C than at 40°C.[486] *Aspergillus fumigatus* spore production less was at 50°C than at 40°C.[486] *Salmonella typhimurion, S. cairo, S. infantis, S. typhi*, and *S. paratyphis B* were killed at 50°C provided the organisms were in direct contact with compost.[110] After windrow composting of raw sewage and ground hardwood bark for 36 h at 60°C, no evidence of experimentally-added human pathogens *Escherichia coli, Candida albicans*, and *Salmonella heidleberg*, could be found.[487] Similar results were reported in compost to which *Salmonella newport, Ascaris lumbricoides, Candida albicans*, and poliovirus type I were added.[488]

Controlling temperature and moisture during composting, timing for minimization of disturbance during turning, site enclosure, biofiltration, and site topographic and landscape design are suggested as controls for bioaerosol production.[383] Bioaerosol monitoring is also suggested. Mechanical agitation of the compost pile can increase the concentration of bioaerosols surrounding a composting facility.[392] Agitation of composts by turning, screening, and transport increased the air levels of *Aspergillus fumigatus*.[400] Other factors that increase bioaerosols include mechanical agitation by wheels, physical handling of equipment, downdrafts onto dust-laded surfaces, movement of stored wood chips, and screening of compost.[392] Curing the compost for one month or more reduced the levels of *A. fumigatus*.[395] Mechanical agitation of the compost should be minimized wherever possible, and timing of agitation should coincide with the stage of composting when the microorganism populations are at a minimum, the wind is minimal, and the population of sensitive receptors is lowest.[383] Choices of materials composted can affect the need for handling and agitation. If leaves were added to sewage biosolid, composting of the material required 16% more handling than with sewage biosolids alone.[116]

Millner[383] stresses that the temperature and moisture conditions of the bulking agents used in sludge composts should be managed and controlled so that production of bioaerosols is minimized. Moisture for bioaerosol reduction should be 40 to 70%.[173,489,490] Spraying the compost and the production area with water can reduce production of dusts and bioaerosols.[491]

Use of a closed reactor system reduces human exposure to bronchial pathogens.[142] Site enclosure can reduce exposure of the public to bioaerosols. Workers may be subjected to higher concentrations of bioaerosols.[397,492] At one site, worker health over a period of years showed no evidence of adverse effects related to exposure to *A. fumigatus*.[493] A thorough discussion of bioaerosols associated with composting has been addressed by Millner et al.[383]

Site enclosure, especially for the unpacking process, and managed air streams are considered effective in reducing odors. An additional precaution is the refraining from turning of the windrow for the first two weeks.[412] Hydrogen sulfide is a particularly noxious odor. The odor threshold of H_2S is reported to be 4.7×10^{-4} µl L^{-1}.[159] Use of biofiltration to cleanse air of H_2S during the highly odorous process of anaerobically digesting coffee waste has been suggested.[159] A compost biofilter of 24 m^3 removed 2.01 g H_2S m^{-3} h^{-1}, and H_2S was not detected.[159] To reduce the risks of odor from biofiltration, the compost filtration medium must be kept moist, and the concentration of H_2S at the intake must not be so high as to overwhelm the sulfur-oxidizing microorganisms.[51] In general, finished composts do not have bad odors that affect consumers' reactions to use of composts.[494]

Fly populations at composting facilities may be reduced as much as 63% simply through good sanitation and procedural changes.[454] Most of the larvae in one study resulted from incoming refuse[454] so that inspection and exclusion may be one method of reducing fly populations. Pesticide-laden sugar-bait applications to the substrate and sticky tapes were effective in fly population reduction.[454] Breeding of flies was limited to the top 2.5 cm apparently due to cool temperatures in that zone indicating that high temperatures may reduce fly populations. Optimum moisture for fly breeding was 75%. At 45 to 55% moisture, hatching and development of pupae were limited to 0 to 14% of the eggs deposited in compost.[454] Since 75% moisture and low temperatures are limiting to composting microorganisms but optimum for fly breeding, good management practices and control of composting parameters can be expected to reduce fly populations associated with composting. In recent research, control of house flies (*Musca domestica*) and the fruit fly (*Drosophila melanogaster*) was proposed through encouragement of the growth of a predatory mite (*Hypoaspis solimani n.* sp.).[495]

C. COMPOST APPLICATIONS TO LAND

Careful management decisions are required in land application of composts. Mature composts should be used for land application. Compost should be analyzed and care taken to ensure N is not applied in excess of crop uptake. Choice of application site should be carefully made to limit runoff and leaching. Soils high in N are not appropriate for high N waste application.[496] Some authors suggest choice of soils with high ph buffering capacity.[289] High compost salinity can be detrimental to plant growth and can be avoided.[497] Increasing the contribution of biosolids, animal wastes, or disposable diapers to compost can increase soluble salts; therefore, it might be prudent to limit the amount of these items in the compost. Some sewage sludges may be relatively low in soluble salts as much has been removed through the removal of liquid effluent portion.[498] Chaney and Giordano[499] suggest judicious application of wastes on agricultural land including (a) excluding wastes contaminated by large amounts of plant micronutrients, (b) limiting maximum total applications to levels safe for crops and the food chain, and (c) managing waste-amended soils according to plant and soil tests.

VI. SUMMARY

Many beneficial effects are derived from composting, from the removal of materials in the waste stream to the increase in food supplies that can be achieved through crop yield increases with compost application to agricultural soils. Benefits such as improvements in water, air, and soil quality can be significant. Soil quality may be improved through the effects of compost application on bulk density, water-holding capacity, porosity, aggregate stability, erosion, and bioremediation of previously contaminated soils. Water quality may be improved through the removal of materials from the waste stream, binding of leachable compounds, the decomposition of potentially hazardous contaminants, and the biofiltration of previously contaminated water. Beneficial effects of composting on air quality may include removal of some materials from the waste stream and reduction in incineration, as well as reduced dependence on nonrenewable, air polluting fossil fuels if use of biogas as an energy source is employed. A significant and important contribution of composting to air quality is the removal of odors and contaminants in air exhaust from industrial processes (biofiltration). Bioremediation, and biofiltration through the use of composts as filters, are areas of growing interest to scientists, environmentalists, and governing regulators.

Beneficial effects of compost applications to crops are many and varied. Most are due to soil quality improvement and nutrient enhancement, and result in increases in crop quality and yield. Compost applications have significantly positive effects on plant disease suppression, especially soil-borne root pathogens. Compost has been used increasingly as a peat substitute in potting media, reducing reliance on this diminishing resource.

Adverse effects from composting processes are almost always attributable to poor management or control of the process, such as lack of presorting out contaminants, incorrect control of parameters, failure to mature compost, or poor facility siting or design. Adverse effects from composting applications are usually attributable to poor site choice, application of immature composts or overapplication of composts. Minimizing the adverse effects and maximizing the benefits of composting can be achieved through careful consideration to processing control, management, and application.

ACKNOWLEDGMENTS

The authors are grateful to Douglas G. Kelly for his research support and editorial assistance in the writing of this chapter, and to Hai Nguyen for his contributions.

REFERENCES

1. Allison, F.E., *Soil Organic Matter and its Role in Crop Production*, Elsevier, Amsterdam, 1973.
2. Waksman, S.A., *Humus*, Williams and Wilkins, Baltimore, MD, 1936.
3. Hoitink, H.A.J. and Keener, H.M., *Science and Engineering of Composting: Design, Environmental, Microbiological and Utilization Aspects*, Renaissance Publishing, Worthington, OH, 1993.
4. U.S. Environmental Protection Agency, *Characterization of Municipal Solid Waste in the United States, 1994 Update*, EPA 530-S-94-042, 1994.
5. Christopher, T. and Asher, M., *Compost This Book!*, Sierra Club Books, San Francisco, 1994.
6. Carra, J.S. and Cossu, R., *International Perspectives on Municipal Solid Wastes and Sanitary Landfilling*, Academic Press, San Diego, 1990.
7. Glenn, J., *The State of Garbage in America*, BioCycle, 33(4), 46, 1992.
8. U.S. Environmental Protection Agency, *Characterization of Municipal Solid Waste in the United States: 1992 Update*, United States Environmental Protection Agency, Office of Solid Waste and Emergency, EPA 530-R-92-019, 1992.
9. Kashmanian, R.M., Quantifying the amount of yard trimmings to be composted in the United States in 1996, *Compost Sci. Util.*, 1(3), 22, 1993.

10. Goldstein, N. and Spencer, R., Solid waste composting in the United States, *BioCycle* 31(11), 46, 1990.
11. Goldstein, N., Solid waste composting project update, *BioCycle*, 32(5), 38, 1991.
12. Kitto, D., *Composting the Organic Natural Way*, Thorsons Publishing Group, Wellingborough, Northamptonshire, England, 1988.
13. Minnich, J. and Hunt, M., *The Rodale Guide to Composting*, Rodale Press, Emmaus, PA, 1979.
14. Bisht, N.S. and Harsh, N.S.K., Biodegradation of *Lantana camara* and waste-paper to cultivate *Agaricus bisporus (Lange) singer, Agric. Wastes*, 12(3), 167, 1985.
15. Weigant, W.M., Wery, J., Buitenhuis, E.T., and de Bont, J.A.M., Growth-promoting effect of Thermophilic fungi on the myceloim of the edible mushroom *Agaricus bisporus, Appl. Environ. Microbiol.*, 58(8), 2654, 1992.
16. Bevacqua, R.F. and Mellano, V.J., Sewage sludge compost's cumulative effects on crop growth and soil properties, *Compost Sci. Util.*, 1(3), 34, 1993.
17. Bugbee, G.J., Frink, C.R., and Migneault, D., Growth of perennials and leaching of heavy metals in media amended with municipal leaf, sewage sludge and street sand compost, *J. Environ. Hort.*, 9(1), 47, 1991.
18. Chong, C., Cline, R.A., and Rinker, D.L., Bark- and peat-amended spent mushroom compost for containerized culture of shrubs, *HortScience*, 29(7), 781, 1994.
19. Cisar, J.L., Municipal solid waste compost offers a new soil amendment source for turf, *Grounds Maint.*, 29(3), 52, 1994.
20. Flanagan, M.S., Schmidt, R.E., and Reneau, R.B., Jr., Municipal solid waste heavy fraction for production of turfgrass sod, *HortScience*, 28(9), 914, 1993.
21. Hornick, S.B., Use of organic amendments to increase the productivity of sand and gravel spoils: effect on yield and composition of sweet corn, *Am. J. Alter. Agric.*, 3(4), 156, 1988.
22. Korcak, R.F., Renovation of a pear orchard site with sludge compost, *Commun. Soil Sci. Plant Anal.*, 17(11), 1159, 1986.
23. Lohr, V.I. and Coffey, D.L., Growth responses of seedlings to varying rates of fresh and aged spent mushroom compost, *HortScience*, 22(5), 913, 1987.
24. Madan, M. and Vasudevan, P., Silkworm litter: use as nitrogen replacement for vegetable crop cultivation and substrate for mushroom cultivation, *Biol. Wastes*, 27(3), 209, 1989.
25. Mandelbaum, R., Hadar, Y., and Chen, Y., Composting for agricultural wastes for their use as container media: effect of heat treatments on suppression of *Pythium aphanidermatum* and microbial activities in substrates containing compost, *Biol. Wastes*, 26, 261, 1988.
26. Marchesini, A., Allievi, L., Comotti, E., and Ferrari, A., Long-term effects of quality-compost treatment on soil, *Plant Soil*, 106, 253, 1988.
27. Maynard, A.A. and Hill, D.E., Impact of compost on vegetable yields, *BioCycle*, 35(3), 66, 1994.
28. Mays, D.A. and Giordano, P.M., Landspreading municipal waste compost, *BioCycle*, 30(3), 37, 1989.
29. Ozores-Hampton, M., Schaffer, B., Bryan, H.H., and Hanlon, E.A., Nutrient concentrations, growth and yield of tomato and squash in municipal solid-waste-amended soil, *HortScience*, 29(7), 785, 1994.
30. Petruzzelli, G., Lubrano, L., and Guidi, G., Uptake by corn and chemically extractability of heavy metals from a four year compost treated soil, *Plant Soil*, 116, 23, 1989.

31. Purman, J,R. and Gouin, F.R., Influence of compost aging and fertilizer regimes on the growth of bedding plants, transplants and poinsettia, *J. Environ. Hort.*, 10(1), 522, 1992.
32. Smith, S.R., Sewage sludge and refuse composts as peat alternatives for conditioning impoverished soils: effects on the growth response and mineral status of *Petunia grandiflora, J. Hort. Sci.*, 67(5), 703, 1992.
33. Wang, Y. and Blessington, T.M., Growth and interior performance of poinsettia in media containing composted cotton burrs, *HortScience*, 25(4), 407, 1990.
34. Wong, M.H. and Chu, L.M., The responses of edible crops treated with extracts of refuse compost of different ages, *Agric. Wastes*, 14(1), 63, 1985.
35. Polprasert, C., *Organic Waste Recycling*, John Wiley and Sons, New York, 1989.
36. Diaz, L.F., Savage, G.M., and Golueke, C.G., *Resource Recovery from Municipal Solid Wastes, Volume 1, Primary Processing*, CRC Press, Boca Raton, FL, 1982.
37. Neal, H.A. and Schubel, J.R., *Solid Waste Management and the Environment: The Mounting Garbage and Trash Crisis*, Prentice-Hall, Englewood Cliffs, NJ, 1987.
38. General Electric Company, *Solid Waste Management Technology Assessment*, Van Nostrand Reinhold, New York, 1975.
39. Rimberg, D., *Municipal Solid Waste Management*, Noyes Data Corp., Park Ridge, NJ, 1975.
40. Weinstein, J.J. and Toro, R.F., *Thermal Processing of Municipal Solid Waste for Resource and Energy Recovery*, Ann Arbor Science, Ann Arbor, MI, 1976.
41. Holdom, R.S. and Winstrom-Olsen, B., Fertilizers and energy from anaerobic fermentation of solid wastes, in *Handbook of Organic Waste Conversion*, Bewick, M. W. M., Ed., Reinhold Company, New York, 1980, 383.
42. Veroughstraete, A., Nyns, E.J., and Naveau, H.P., Unit of bioengineering, in *Composting of Agricultural and Other Wastes*, Gasser, J. K. R., Ed., Elsevier, London, 1985, 135.
43. Baines, S., Svoboda, I.F., and Evans, M.R., Heat from aerobic treatment of liquid animal wastes, in *Composting of Agricultural and Other Wastes*, Gasser, J. K. R., Ed., Elsevier, London, 1985, 147.
44. Thostrup, P., Heat recovery from composting solid manure, in *Composting of Agricultural and Other Wastes*, Gasser, J. K. R., Ed., Elsevier, London, 1985, 167.
45. Scherer, P., Lentz, R., and Carra, R, Cobiogasification of wastepaper products with separately collected municipal solid waste, *Compost Sci. Util.*, 1(4), 31, 1993.
46. Wong, M.H., Cultivation of microlagae in refuse compost and soy-bean waste extracts, *Agric. Wastes*, 12930, 225, 1985.
47. Bohn, H., Consider biofiltration for decontaminating gases, *Chem. Eng. Prog.*, 88(4), 34, 1992.
48. Hodge, D.S. and Devinny, J.S., Biofilter treatment of ethanol vapors, *Environ. Progress*, 13(3), 167, 1994.
49. Liu, P.K.T., Gregg, R.L., Saboll, H.K., and Barkley, N., Engineered biofilter for removing organic contaminants in air, *Air Waste*, 44, 299, 1994.
50. Togna, A.P. and Singh, M., Biological vapor-phase treatment using biofilter and biotrickling filter reactors: practical operating regimes, *Environ. Progress*, 13(2), 94, 1994.
51. Yang, Y. and Allen, E.R., Biofiltration control of hydrogen sulfide. II. Kinetics, biofilter performance, and maintenance, *J. Air Waste Manage. Assoc.*, 44, 1315, 1994.
52. Haug, T.H., *Compost Engineering, Principles and Practice*, Ann Arbor Science Publishers, Ann Arbor, MI, 1980.

53. Wiles, C.C., Composting of refuse, in *1977 National Conference on Composting of Municipal Residues and Sludges*, August 23–25, 1977, Information Transfer, Inc. 1978, 20.
54. Poincelot, R.P., The biochemistry of composting, in *1977 National Conference on Composting of Municipal Residues and Sludges*, August 23–25, 1977, Information Transfer, Inc., 1978, 33.
55. Cameron, R.D. and Koch, F.A., Toxicity of landfill leachate, *J. Water Poll. Con. Fed.*, 52(4), 760, 1980.
56. Golueke, C.G., Effect of management processes on the quality of compost materials, in *1977 National Conference on Composting of Municipal Residues and Sludges*, August 23–25, 1977, Information Transfer, Inc. 1978, 51.
57. Alexander, M., *Introduction to Soil Microbiology*, 2nd ed., Wiley Publishers, New York, 1977.
58. Hughes, E.G., The composting of municipal wastes, in *Handbook of Organic Waste Conversion*, Bewick, M. W. M., Ed., Van Nostrand Reinhold, New York, 1980, 108.
59. Golueke, C.G. and Diaz, L.F., Effect of management processes on the quality of compost materials, in *1977 National Conference on Composting of Municipal Residues and Sludges*, August 23–25, 1977, Information Transfer, Inc., 1978, 51.
60. Wiles, C.C. and Stone, G.E., Composting at Johnson City, Final report on joint USEPA-TVA Project, SW-31r.2, U.S. Environmental Protection Agency, 1975.
61. Gossett, J.M. and McCarty, P.L., *Heat Treatment of Refuse for Increasing Anaerobic Biodegradability*, Civil Engineering Technical Report 192, Stanford University, Stanford, CA, 1975.
62. Lynch, J.M., Substrate availability in the production of composts, in *Science and Engineering of Composting: Design, Environmental, Microbiological and Utilization Aspects*, Hoitink, H.A.J. and Keener, H.M., Eds., Renaissance Publications, Worthington, OH, 1993, 24.
63. Golueke, C.G., *Biological Reclamation of Solid Wastes*, Rodale Press, Emmaus, PA, 1977.
64. Knowles, R., Methane: process of production and consumption, in *Agricultural Ecosystem Effects on Trace Gases and Global Climate Change*, Peterson, G.A., Baenziger, P.S., and Luxmoore, R.J., Eds., ASA Special Publication, Madison, WI, 1993, 145.
65. Morrison, R.T. and Boyd, R.N., *Organic Chemistry: Fourth Edition*, Allyn and Bacon, Boston, 1983, 1009.
66. Zucconi, F., Forte, M., Monaco, A., and de Bertoldi, M., Biological evaluation of compost maturity, in *Composting Theory and Practice for City, Industry and Farm*, Staff of Compost Science/Land Utilization, Eds., JG Press, Emmaus, PA, 1982, 34.
67. Zucconi, F., Pera, A., Forte, M., and de Bertoldi, M., Evaluating toxicity of immature compost, in *Composting Theory and Practice for City, Industry and Farm*, Staff of Compost Science/Land Utilization, Eds., JG Press, Emmaus, PA, 1982, 61.
68. Harper, S.H.T. and Lynch, J.M., The role of water-soluble components in phytotoxicity from decomposing straw, *Plant Soil*, 65, 11, 1982.
69. Lynch, J.M., Production and phytotoxicity of acetic acid in anaerobic soils containing plant residues, *Soil Biol. Biochem.*, 10, 131, 1978.
70. Zucconi, F. and de Bertoldi, M., Compost specifications for the production and characterization of compost from municipal solid waste, in *Compost: Production, Quality and Use*, de Bertoldi, M., Ed., Elsevier, London, 1987, 30.

71. Brinton, W.F. and Droffner, M.W., Microbial approaches to characterization of composting processes, *Compost Sci. Util.*, 2(3), 12, 1994.
72. Chanyasak, V. and Kubota, H., Carbon/organic nitrogen ration in water extract as measure of composting degradation, *J. Ferment. Technol.*, 59(3), 215, 1981.
73. Garcia, C., Hernadez, T. and Costa, F., Changes in carbon fractions during composting and maturation of organic wastes, *Environ. Manag.*, 15, 433, 1991.
74. Garcia, C., Hernadez, T., Costa, F., and Ayuso, M., Evaluation of the maturity of municipal solid waste compost using simple chemical parameters, *Commun. Soil Sci. Plant Anal.*, 23, 1501, 1992.
75. Katayama, A., Application of Gel Chromatography for Monitoring Decomposition of Organic Wastes in Soil, Dissertation, the Graduate School of Toyko Institute of Technology, Department of Environmental Chemistry and Engineering, 1985, 113.
76. Jimenez, E.I. and Garcia, V.P., Evaluation of city refuse compost maturity: a review, *Biol. Wastes*, 27, 115, 1989.
77. Neuhauser, E.F., Hartenstein, R., and Kaplan, D.L., A second progress report on the potential use of earthworms in sludge management, in *Proc. of the Conference on Sludge Composting*, November 14–17, Sponsored by Information Transfer, Inc., Silver Spring, MD, 1979.
78. Camp, Dresser, and McKee, Inc., *Engineering Assessment of Vermicomposting Municipal Wastewater Sludges*, prepared for the Municipal Environmental Research Laboratory, Office of Research and Development, U.S. Environmental Protection Agency, Cincinnati, OH, EPA 600/2-81-075, 1981.
79. Dinges, R., *Natural Systems for Water Pollution Control*, Van Nostrand Reinhold Company, New York, 1981.
80. Loehr, R.C., Martin, J.H., Neuhauser, E.F., and Malecki, M.R., *Waste Management Using Earthworms—Engineering and Scientific Relationships*, PB84-193218, NTIS, Springfield, VA, 1984.
81. Edwards, C.A., Breakdown of animal, vegetable and industrial organic wastes by earthworms, in *Earthworms in Waste and Environmental Management*, Edwards, C.A. and Neuhauser, E.F., Eds., SPB Academic Publishing, The Hague, The Netherlands, 1988, 21.
82. Edwards, C.A., Earthworms, organic waste and food, *Span. Shell Chem. Co.*, 26(3), 106, 1983.
83. Huhta, V. and Haimi, J., Reproduction and biomass of *Eisenia foetida* in domestic waste, in *Earthworms in Waste and Environmental Management*, Edwards, C.A. and Neuhauser, E.F., Eds., SPB Academic Publishing, The Hague, The Netherlands, 1988.
84. Loehr, R.C., Martin, J.H. Jr., and Neuhauser, E.F., Stabilization of liquid municipal sludge using earthworms, in *Earthworms in Waste and Environmental Management*, Edwards, C.A. and Neuhauser, E.F., Eds., SPB Academic Publishing, The Hague, The Netherlands, 1988, 95.
85. Appelhof, M., Domestic vermicomposting systems, in *Earthworms in Waste and Environmental Management*, Edwards, C.A. and Neuhauser, E.F., Eds., SPB Academic Publishing, The Hague, The Netherlands, 1988, 157.
86. Hartenstein R., Neuhauser, E.F., and Kaplan, D.L., Reproductive potential of the earthworm *Eisenia foetida, Oecologia*, (Berlin), 43, 329, 1979.
87. Neuhauser, E.F., Hartenstein, R., and Kaplan, D.L., Growth of the earthworm *Eisenia foetida* in relation to population density and food rationing, *Oikos*, 35, 93, 1980.

88. Tomlin, A.D. and Miller, J.J., Development and fecundity of the manure worm, *Eisenia foetida (Annelida: Lumbricidae)* under laboratory conditions, in *Proc. of VII-International Colloquium of Soil Zoology*, Dindal, D., Ed., 1980, 673.
89. Mitchell, M.J., Hornor, S.G., and Abrams, B.L., Decomposition of sewage sludge in drying beds and the potential role of the earthworm, *Eisenia foetida, J. Environ. Qual.*, 9, 373, 1980.
90. Horner, S.G. and Mitchell, M.J., Effect of the earthworm *Eisenia foetida* (*Oligochaeta*), on fluxes of volatile carbon and sulfur compounds from sewage sludge, *Soil Biol. Biochem.*, 13, 367, 1981.
91. Brown, B.A. and Mitchell, M.J., Role of the earthworm, *Eisenia foetida*, in affecting survival of *Salmonella enteritidis ser. typhimurium, Pedobiologia*, 22, 434, 81.
92. Roch, P.H., Valembois, P., and Lassegues, M., Biochemical particularities of the antibacterial factor of the two subspecies *Eisenia fetida fetida* and *Eisenia fetida andrei, Am. Zool.*, 20, 794, 1980.
93. Diener, R.G., Collins, A.R., Martin, J.H., and Byran, W.B., Composting of source-separated municipal solid waste for agricultural utilization: a conceptual approach for closing the loop, *Appl. Eng. Agric.*, 9(5), 427, 1993.
94. Richard, T. and Chadsey, M., Environmental impact of yard waste composting, *BioCycle*, 31(4), 42, 1990.
95. Rynk, R., *On-Farm Composting Handbook*, Northeast Regional Agricultural Engineering Service, Ithaca, NY, 1992.
96. Kuter, G.A., Hoitink, H.A.J., and Rossman, L.A., Effects of aeration and temperature on composting of municipal sludge in a full-scale vessel system, *J. Water Poll. Control Fed.*, 57(4), 309, 1985.
97. Willson, G.B., Measuring compost stability, *Biocycle*, 27(7), 34, 1986.
98. van de Berg, L., Developments in methanogenesis from industrial waste water, *Can. J. Microbiol.*, 30, 975, 1984.
99. O'Keefe, D.M., Chynoweth, A.W., Barkdoll, A.W., Nordstadt, F.A., Owens, J.M., and Sifontes, J., Sequential batch anaerobic composting of municipal solid waste (MSW) and yard waste, *Water Sci. Tech.*, 27(2), 77, 1993.
100. Mata-Alverez, J., Cecchi, F., Pavan, P., and Bassetti, A., Semi-dry thermophilic anaerobic digestion of fresh and pre-composted organic fraction of municipal solid waste (MSW) digester performance, *Water Sci. Tech.*, 27(2), 87, 1993.
101. Tchobanoglous G., Theisen, H., and Eliassen, R., *Solid Wastes: Engineering Principles and Management Issues*, McGraw Hill, NY, 1977, 290.
102. Hill, S.A., Appropriate technology for resource recovery, *NCRR Bull.*, 11(1), 3, 1981.
103. Boettcher, R.A., *Air Classification of Solid Wastes*, U.S. Environmental Protection Agency, SW-30c, U.S. Government Printing Office, Washington, D.C., 1972.
104. Campbell, J.A., *Electromagnetic Separation of Aluminum and Nonferrous Metals*, Presented at 103rd American Institute of Mechanical Engineers Meeting, Combustion Power Company, Inc., Dallas, TX, February 24–28, 1974.
105. Snell, J.R., Proper grinding for efficient composting, *BioCycle*, 32(4), 54, 1991.
106. Diaz, L.F., Golueke, C.G., Lafrenz, D., and Chaser, B., Composting combined refuse and sewage sludge, in *Composting: Theory and Practice for City, Industry and Farm*, Staff of Compost Science/Land Utilization, Eds., JG Press, Emmaus, PA, 1982, 114.
107. Tillet, N.D., Miles, S.J., and Lane, A.G., Techniques for adding minor ingredients to containerized nursery stock compost, *J. Agric. Eng. Res.*, 25, 241, 1992.

108. Hay, J.C. and Kuchenrither, R.D., Fundamentals and application of windrow composting, *J. Environ. Eng.*, 116(4), 746, 1990.
109. Ishii, H., Tanaka, K., Aoki, M., Murakami, T., and Yamada, M., Sewage sludge composting process by static pile method, *Water Sci. Technol.*, 23(10–12), 1979, 1991.
110. Fay, S.C. and Leonard, R.E., Composting privy wastes at recreation sites, *Compost Sci. Util.*, 20(2), 36, 1979.
111. Fay, S.C., *The Composting Option for Human Waste Disposal in the Back Country*, Department of Agriculture, Forest Service, Northeastern Forest Experiment Station, 1977.
112. Shuval, H, I., Gunnerson, C.G., and DeAnne, S.J., Eds., *Appropriate Technology for Water Supply and Sanitation: Night Soil Composting*, The World Bank, 1981.
113. Bowen, P.T., Jackson, M.K., Corbitt, R.A., and Gonce, N., Sludge treatment, utilization, and disposal, *Water Environ. Res.*, 65(4), 360, 1993.
114. Martin, A.M., Evans, J., Porter, D., and Patel, T.R., Comparative effects of peat and compost employed as bulking agents in composting, *Bioresour. Technol.*, 44(1), 65, 1993.
115. Finstein, M.S. and Hogan, J.A., Integrating of composting process microbiology, facility structure and decision-making, in *Science and Engineering of Composting: Design, Environmental, Microbiological and Utilization Aspects*, Hoitink, H.A.J. and Keener, H.M., Eds., Renaissance Publications, Worthington, OH, 1993, 1.
116. Elwell, D.L., Keener, H.M., Hoitink, H.A.J., Hansen, R.C., and Hoff, J., Pilot and full scale evaluations of leaves as an amendment in sewage sludge composting, *Compost Sci. Util.*, 2(2), 55, 1994.
117. U.S. Environmental Protection Agency, *Environmental Regulations and Technology: Use and Disposal of Municipal Wastewater Sludge*, EPA-625/10-84-003, 1984.
118. Pichtel, J.R., Microbial respiration in fly ash/sewage sludge-amended soils, *Environ. Pollut.*, 63, 225, 1990.
119. Keener, H.M., Marugg, C., Hansen, R.C., and Hoitink, H.A.J., Optimizing the efficiency of the composting process, in *Science and Engineering of Composting: Design, Environmental, Microbiological and Utilization Aspects*, Hoitink, H.A.J. and Keener, H.M., Eds., Renaissance Publications, Worthington, OH, 1993, 59.
120. Mitchell, F.C., Jr., Reddy, C A. and Forney, L.J., Yard waste composting: studies using different mixes of leaves and grass in a laboratory scale system, *Compost Sci.Util.*, 1(3), 85, 1993.
121. Vallini, G., Pera, A., Valdrighi, M., and Cecchi, F., Process constraints in source-collected vegetable waste composting, *Water Sci. Technol.*, 28(2), 229, 1993.
122. Vermeulen, J., Huysmans, A., Crespo, M., Van Lierde, A., De Rycke, A., and Verstraete, W., Processing of biowaste by anaerobic composting to plant growth substrates, *Water Sci. Technol.* 27(2), 109, 1993.
123. Risch, B.W.K., Solid waste management concepts in Europe, in *Recycling International: Recovery of Energy and Material from Residues and Waste*, Thome-Kozmiensky K.J., Ed., E. Freitag-Verlag fur Umwelttechnik, Berlin, 1982, 47.
124. Gertman, R., Recycling in Davis, California, *Waste Manag. Res.*, 2, 293, 1984.
125. Shimizu, A., Toyohashi city's total waste utilization system, *Waste Manag. Res.*, 2, 91, 1984.
126. Vallini, G. and Pera, A., Green compost production from vegetable wastes separately collected in metropolitan garden-produce markets, *Biol. Wastes*, 29, 33, 1989.

127. Van Roosmalen, G.R. and Van De Langerijt, J.C., "Green Waste" composting in the Netherlands, *BioCycle*, 30(7), 32, 1989.
128. Vallini, G., Pera, A., Sorace, G., Cecchi, C., and Manetti, P., Green composting, *BioCycle*, 31(6), 46, 1990.
129. Lechner, P., Composting techniques, *Proc. of the ISWA Int. Congress on Energy and Materials Recovery from Wastes*, Perugia, Italy, VIII, 1988.
130. Bono, J.J., Chalaux, N., and Chabbert, B., Bench-scale composting of two agricultural wastes, *Bioresour. Technol.*, 40, 119, 1992.
131. Bagstam, G., Enebo, L., Lindell, T., and Swensson, H., Experiments made in bench-scale composters. I., *Apparatus Vatten*, 4, 358, 1974.
132. Kneebone, L.R. and Masson, E.C., Sugarcane bagasse as a bulk ingredient in mushroom compost, *Mushroom Sci.*, 8, 321, 1972.
133. Sikora, L.J. and Sowers, H.A., Effect of temperature control of the composting processes, *J. Environ. Qual.*, 14, 434, 1985.
134. Sayag, D. and Andre, L., Integrated waste management options, *BioCycle*, 28(6), 56, 1987.
135. Laborde, J., Olivier, J.M., Houdeau, G., and Delpech, P., Indoor static composting for mushroom (*Agaricus bisporus Lge. Sing*) cultivation, in *Cultivating Edible Fungi*, West, P.J., Royse, D.J., and Billman, R.P., Eds., Elsevier, Amsterdam, 1986, 91.
136. Laborde, J., Houdeau, G., Bes, B., Olivier, J.M., and Delpech, P., Indoor static composting: description of the process. Analysis. Main results, *Mushroom Sci.*, 12, 457, 1989.
137. Suler, D.J. and Finstein, M.S., Effect of temperature, aeration, and moisture on CO_2 formation in bench-scale, continuously thermophilic composting of solid waste, *Appl. Environ. Microbiol.*, 33(2), 345, 1977.
138. Laborde, J., Delmas, J., and D'Hardemare, G., Microbiological balance of composts, *Mushroom Sci.*, 7, 187, 1968.
139. Laborde, J., Delmas, J., Lamau, J.L., and Berthaud, J., Express preparation of substrates (P.E.S.), *Mushroom Sci.*, 8, 675, 1971.
140. Campbell, C.D., Darbyshire, J.F., and Anderson, J.G., The composting of tree bark in small reactors–self-heating experiments, *Biol. Wastes*, 31, 145, 1990.
141. Stentiford, E.I., Recent developments in composting in *Compost Production, Quality and Use*, de Bertoldi, M., Fernanti, M.P., L'Hermite, P., and Zucconi, F., Eds., Elsevier Applied Science, London, 1987, 52.
142. de Bertoldi, M., Ferranti, M.P., L'Hermite, P., and Zucconi, F., Eds., *Compost: Production, Quality and Use*, Elsevier Applied Science, London, 1987.
143. Hackett, W.C. and Thompson, D.V.M., *Biofermentation Pilot Study for Water Hyacinth and Gizzard Shad Composting*, Argon Corporation, Palatka, FL, 1991.
144. Sharma, A.R. and Mittra, B.N., Effect of different rates of application of organic and nitrogen fertilizers in a rice-based cropping system, *J. Agric. Sci.*, 117(3), 313, 1991.
145. Chen, Y., Inbar, Y., and Hadar, Y., Composted agricultural wastes as potting media for ornamental plants, *Soil Sci.*, 145(4), 298, 1988.
146. Bujang, K.B. and Lopez-Real, J.M., Composting for the treatment of cattle wastes, *Compost Sci. Util.*, 1(3), 38,1993.
147. James, P.J. and Campbell, R.J., The economics of anaerobic digestion on farms, in *Anaerobic Digestion of Livestock Wastes*, NIRD Technical Bulletin No. 2, 1985, 73.
148. West, R., *Matching Energy Supply and Demand in Anaerobic Digestion of Farm Waste*, NIRD Technical Bulletin No. 7, 1985, 55.

149. Hong, J.H., Matsuda, J., and Ikeuchi, Y., High rapid composting of dairy cattle manure with crop and forest residues, *Trans ASAE*, 533, 1983.
150. N'Dayegamiye, A. and Isgan, D., Chemical and biological changes in compost of wood shavings, sawdust and peat moss, *Can. J. Soil Sci.*, 71, 475, 1991.
151. Astier, M., Gersper, P.L., and Buchanan, M., Combining legumes and compost: a viable alternative for farmers in conversion to organic agriculture, *Compost Sci. Util.*, 2(1), 80, 1994.
152. Hansen, R.C., Keener, H.M., and Hoitink, H.A.J., Poultry manure composting: an exploratory study, *Trans. ASAE*, 32, 2151, 1989
153. Lo, K.V., Lau, A.K., and Liao, P.H., Composting of separated solid swine wastes, *J. Agric. Eng. Res.*, 54(4), 307, 1993.
154. Vincini, M., Carini, F., and Silva, S., Use of alkaline fly ash as an amendment for swine manure, *Biores. Technol.*, 49, 213, 1994.
155. Gagnon, B., Rioux, C, and Chagnon, J., Evolution of chemical composition and microbial activity during storage of compost-based mixes, *Compost Sci. Util.*, 1(3), 15, 1993.
156. Gijzen H.J., Schoenmakers, T.J.M., Caerteling C.G.M., and Vogels, G.D., Anaerobic degradation of papermill sludge in a two-phase digester containing rumen microorganisms and colonized polyurethane foam, *Biotechnol. Lett.*, 10(1), 61, 1988.
157. Gijzen, H.J., Lubberding, H.J., Verhagen, F.J., Zwart K.B., and Vogels, G.D., Application of rumen organisms for an enhanced anaerobic degradation of solid organic waste materials, *Biol. Wastes*, 22, 81, 1987.
158. Kostenberg, D. and Marchaim, C.U., Solid waste from the instant coffee industry as a substrate for anaerobic thermophilic digestion, *Water Sci. Technol.*, 27(2), 97, 1993.
159. Hajipakkos, The application of a full scale UASB plant for the treatment of coffee waste, *Water Sci. Technol.*, 25(1), 17, 1992.
160. Keener, H.M., Dick, W.A., Marugg, C., and Hansen, R.C., Composting spent press-molded, wood fiber pallets bonded with urea formaldehyde: a pilot scale evaluation, *Compost Sci. Util.*, 2(3), 73, 1994.
161. Riggle, D., Pulverized pallets become compost amendment, *BioCycle*, 35(1), 40, 1994.
162. Campbell A.G. and Tripepi, R.R., Logyard residues: products, markets, and research needs, *For. Prod. J.*, 42(9), 60, 1992.
163. Mathur, S.P., Daigle, J.-Y., Levesque, M., and Dinal, H., The feasibility of preparing high quality compost from fish scrap and peat with seaweeds or crab scrap, *Biol. Agric. Hort.*, 4, 27, 1986.
164. Martin, A.M. and Chintalapati, S.P., Fish offal-peat compost extracts as fermentation substrate, *Biol. Wastes*, 27(4), 281, 1989.
165. Frederick, L., Turning fishery wastes into saleable compost, *BioCycle*, 32(9), 70, 1991.
166. Marchaim, U., Levanon, D., Danai, O., Musaphy, S., Chen, Y., Inbar, Y., and Klinger, I.A., Suggested solution for slaughterhouse wastes: uses of the residual materials after anaerobic digestion, *Bioresour. Technol.*, 37(2), 127, 1991.
167. Sweeney, D.W. and Graetz, D.A., Application of distillery waste anaerobic digester effluent to St. Augustine grass, *Agric. Ecosystems Environ.*, 33, 341, 1991.
168. Golueke, C.G., Composting: a review of rationale, principles and public health, in *Composting: Theory and Practice for City, Industry and Farm*, staff of Compost Science/Land Utilization, Eds., JG Press, Emmaus, PA, 1981, 19.

169. Parr, J.F. and Hornick, S.B., Utilization of municipal wastes, in *Applications in Agricultural and Environmental Management*, Metting, F. B., Ed., Marcel Dekker, New York, 1993, 545.
170. Hyatt, G.W., Economic, scientific and infrastructural basis for using municipal composts in agriculture, in *Agricultural Utilization of Urban and Industrial By-products*, Karlen, D. L., Ed., ASA Spec. Publ. 58, ASA, CSSA, SSSA, Madison, WI, 1995, 19.
171. He, X.T., Traina, S.J., and Logan, T.J., Chemical properties of municipal solid waste composts, *J. Environ. Qual.*, 21, 318, 1992.
172. Airan, D.S. and Bell, J.H., Resource recovery through composting—a sleeping giant, in *Proc. 1980 National Waste Process Conf.*, Washington, D.C., May 11–14, 1980, Am. Soc. Mech. Eng., New York, 1980, 121.
173. Poincelot, R.P., A scientific examination of the principles and practice of composting, *Compost Sci. Util.*, 15, 24, 1974.
174. Bugbee, G.J. and Frink, C.R., Composted waste as a peat substitute in peat to lite media, *HortScience*, 24(4), 625, 1989.
175. Blowes, D.W., Robertson, W.D., Ptacek, C. J., and Merkely, C., Removal of agricultural nitrate from tile-drainage effluent water using in-line bioreactors, *J. Contam. Hydrol.*, 15(3), 207, 1994.
176. Stark, L.R., Benerick, W.R., Williams, F.M., Stevens, S.E., Jr., and Wuest, P.J., Restoring the capacity of spent mushroom compost to treat coal mine drainage by reducing the inflow rate: a microcosm experiment, *Water, Air, Soil Pollut.*, 75, 405, 1994.
177. Wieder, R.K., Ion input/output budgets for five wetlands constructed for acid coal mine drainage treatment, *Water, Air, Soil Pollut.*, 71, 231, 1993.
178. Bugbee, G.J., Growth of rudbeckia and leaching of nitrates in potting media amended with composted coffee processing residue, municipal solid waste and sewage sludge, *Compost Sci. Util.*, 2(1), 72, 1994.
179. Banerji, S.K., Piontek, K., and O'Connor, J.T., Pentachlorophenol adsorption on soils and its potential for migration into ground water, in *Hazardous and Industrial Solid Waste Testing and Disposal: Sixth Volume*, ASTM STP 933, Lorenzen, D., Conway, R.A. Jackson, L.P., Perket, C.L., Itamza, A., and Lacy, W.J., Eds., ASTM, Philadelphia, Pa, 1968, 120.
180. Uchrin, C.G. and Katz, J., Sorption kinetics of competing organic substances on New Jersey coastal plain aquifer soils, in *Hazardous and Industrial Solid Waste Testing and Disposal: Sixth Volume*, ASTM STP 933, Lorenzen, D., Conway, R. A., Jackson, L. P., Perket, C. L., Hamza, A., and Lacy, W. J., Eds., ASTM, Philadelphia, Pa, 1986, 140.
181. Bohn, H. and Rohn, R., Soil beds seek out air pollutants, *Chem. Eng.*, 95(6), 73, 1988.
182. Devinny, J.S., Medina, V.F., and Hodge, D.S., Bench testing of fuel vapor treatment by biofiltration, in *Proc. National Research and Development Conference on the Control of Hazardous Materials*, Hazardous Materials Control Institute, Anaheim, Ca, February 20–22, 1991.
183. Douglass, R.H., Armstrong, J.M., and Korreck, W.M., Design of a packed column bioreactor for on-site treatment of air stripper off gas, in *Proc. Battelle International Symposium on In Situ and On-Site Bioreclamation*, San Diego, March 19–21, 1991.
184. Hodge, D.S., Medina, V.F., Islander, R.L., and Devinny, J.S., Treatment of hydrocarbon fuel vapors in biofilters, *Environ. Technol.*, 122, 655, 1991.
185. Kosky, K.F. and Neff, C.R., *Innovation for Treating Gaseous Emissions Using Biofiltration Technology*, Report from Biofiltration, Inc., Gainesville, FL, 1990.

186. Leson, G. and Winer, A.M., Biofiltration: an innovative air pollution control technology for VOC emissions, *J. AWMA*, 41(8), 1045, 1991.
187. Leson, G., Winer, A.M., and Hodge, D.S., Application of biofiltration to the control of air toxins and other VOC emissions, in *Proc. 84th Annual Meeting of the Air Waste Management Association*, Vancouver, British Columbia, Canada, 1991.
188. Ottengraf, S.P.P., Exhause fase purification, in *Biotechnology*, 8, Rehm, H.J., and Reed, F., Eds., VCH Verlagsgesellsch., Weinheim, 1986.
189. Ottengraf, S.P.P., Meester, J.J.P., van den Oever, A.H.C., and Rozeman, H.R., Biological elimination of volatile zenobiotic compounds in biofilters, *Bioprocess Eng.*, 1, 61, 1986.
190. Ottengraf, S.P.P. and van den Oever, A.H.C., Kinetics of organic compound removal from waste gases with a biological filter, *Biotechnol. Bioeng.*, 25, 3089, 1983.
191. Kampbell, D.H., Wilson, J.T., Read, H.W., and Stockdalc, T.T., Removal of volatile aliphatic hydrocarbons in a soil bioreactor, *JAPCA*, 37(10), 1236, 1987.
192. Savage, G.M., Diaz, L.F., and Golueke, C.G., Disposing of organic hazardous wastes by composting, *BioCycle*, 26(1), 31, 1985.
193. Kamnikar, B., Bioremediation of contaminated soil, *Poll. Eng.*, 24(21), 50, 1992.
194. Williams, R.T. and Keehan, K.R., Hazardous and industrial waste composting, in *Science and Engineering of Composting: Design, Environmental, Microbiological, and Utilization Aspects*, Hoitink, H. A. J. and Keener, H. M., Eds., Renaissance Publications, Worthington, OH, 1993, 363.
195. Michel, F.C., Reddy, C.A., and Forney, L.J., Fate of certain lawn care pesticides during yard waste composting, *Proc. Composting Council's 4th National Conference*, November 17–19, The Composting Council, Washington, D.C., 1993, 5.
196. Mays, M.K., Sikora, L.J., Hatton, J.W., and Lucia, S.M., Composting as a method for hazardous waste treatment, in *Superfund '89: Proc. 10th National Conference*, November 27–29, U.S. Environmental Protection Agency, Washington, D.C., 1989, 298.
197. Fyock, O.L., Nordrum, S.B., Fogel, S., and Findley, M., Pilot scale composting of petroleum production sludges, in *3rd Annual Symposium on Environmental Protection in the Energy Industry: Treatment and Disposal of Petroleum Sludges* Sublette, K. L. and Harris, T. M., Eds., December 12, 1991, University of Tulsa, OK, 1991.
198. Valo, R. and Salkinoja-Salonene, Bioreclamation of chlorophenol-contaminated soil by composting, *Appl. Microbiol. Biotechnol.*, 25, 68, 1986.
199. McFarland, M.J., Qui, X.J., Sims, J.L., Randolph, M.E., and Sims, R.C., Remediation of petroleum impacted by soils in fungal compost bioreactors, *Water Sci. Technol. J. Int. Assoc. Water Pollut. Res. Control*, 25(3), 197, 1992.
200. Cerniglia, C.E., Aromatic hydrocarbons: metabolism by bacteria, fungi, and algae, in *Review in Biochemical Toxicology*, Vol. 3, Hodgson, E., Bond, J.R. and Philpot, R. M., Eds., Elsevier, New York, 1981, 321.
201. Crawford, S.L., Johnson, G.E., and Goetz, F.E., The potential for bioremediation of soils containing PAHs by composting, *Compost Sci. Util.*, 1(3), 41, 1993.
202. Haemmerli, S.D., Leisola, M.S.A., Sanglard, D., and Fiechter, A., Oxidation of benzo(a)pyrene by extracellular ligninases of *Phanerochaete chrysosporium, J. Biol. Chem.*, 261, 6900, 1986.
203. Sanglard, D., Leisola, M.S.A., and Fiechter, A., Role of extracellular ligninases in biodegradation of benzo(a)pyrene by *Phanerochaete chrysosporium, Enzyme Micro. Technol.*, 8, 209, 1986.
204. Snell, J., Rate of biodegradation of toxic compounds while in contact with organics which are actively composting, *Snell Environmental Group*, NTIS, 1982.

205. Racke, K.D. and Frink, C.R., Fate of organic contaminants during sewage sludge composting, *Bull. Environ. Contamin. Technol.*, 42, 526, 1989.
206. Oreke, B.C., Smith, J.E., Paterson, A., and Watson-Craik, I.A., Aerobic metabolism of pentachlorophenol by spent sawdust culture of "Shitake" mushroom (*Lentinus edodes*) in soil, *Biotechnol. Lett.*, 15(10), 1077, 1993.
207. Nash, R.G. and Woolson, E.A., Persistence of some hydrocarbons insecticides in soils, *Science*, 157, 924, 1967.
208. Sears, M.K. and Chapman, R.A., Persistence and movement of four insecticides applied to turfgrass, *J. Econ. Entomol.*, 72, 272, 1979.
209. Heitkamp, M.A., Franklin, W., and Cerniglia, C.E., Pyrene degradation by a *Mycobacterium* sp.: Identification of ring oxidation and ring fussion products, *Appl. Environ. Microbiol.*, 54(10), 2556, 1988.
210. Heitkamp, M.A. and Cerniglia, C.E., Polycyclic aromatic hydrocarbon degradation by a *Mycobacterium* sp. in microcosms containing sediment and water from a pristine ecosystem, *Appl. Environ. Microbiol.*, 55(8), 1968, 1989.
211. Keck, J., Sims, R.C., Coover, M., Park, K., and Symons, B., Evidence for cooxidation of polynuclear aromatic hydrocarbons in soil, *Water Res.*, 12(12), 1467, 1989.
212. McFarland, M.J., Qui, X.J., April, W.A., and Sims, R.C., Biological composting of petroleum waste organics using the white rot fungus *Phanerochaete chrysosporium, Proc. Institute of Gas Technology's Second International Symposium on Gas, Oil, Coal and Environmental Technology*, New Orleans, LA, 1989.
213. Sims, R.C., Doucette, W.J., McLean, J.E., Grenny, W.J., and Dupont, R.R., *Treatment Potential of 56 EPA Listed Hazardous Chemicals in Soil*, EPA/600/6-88/001, U.S. Environmental Protection Agency, Ada, OK, 1988.
214. Lemmon, C.R. and Pylypiw, H.M., Degradation of diazinon, chlorpyrifos, isofenphos, and pendimethalin in grass and compost, *Bull. Environ. Contam. Toxicol.*, 48, 409, 1992.
215. Johnson, D.J. and Stansbury, H.A., Adaptation of sevin insecticide (carbaryll) residue method to various crops, *J. Agric. Food Chem.*, 13, 35, 1965.
216. Larkin M.J. and Day, M.J., The metabolism of carbaryl by three bacterial isolates, *Pseudemonas* sp. (NCIB 12042 & 12043) and *Rhodococcus* sp. (NCIB 122038) from garden soil, *J. Appl. Bacteriol.*, 60, 233, 1986.
217. Rajagopal, B.S., Rao, V.R., Nagendrappa, G., and Sethunathan, M., Metabolism of carbaryl and carbofuran by soil-enrichment and bacterial cultures, *Can. J. Microbiol.*, 30, 1458, 1984.
218. Kuo, W.S. and Regan, R.W., Sr., Degradation of carbaryl and 1-naphthol by spent mushroom compost microorganisms, *Water Sci. Technol.*, 26(9–11), 2081, 1992.
219. Doran, J.W., and Parking, T.B., Defining and assessing soil quality, in *Defining Soil Quality for a Sustainable Environment*, Doran, J. W., Coleman, D. C., Bezdicek, D. F., and Steward, B. A., Eds., SSSA Special Publication Number 35, SSSA, Madison, WI, 1994, 3.
220. Doran, J.W., Coleman, D.C., Bezdicek, D.F., and Steward, B.A., *Defining Soil Quality for a Sustainable Environment*, SSSA Special Publication Number 35, SSSA, Madison, WI, 1994.
221. Cook, B.D., Halbach, T.R., Rosen, C.J., and Moncrief, J.R., Effect of a waste stream component on the agronomic properties of municipal solid waste compost, *Compost Sci. Util.*, 2(2), 75, 1994.
222. Elliott, L.F. and Stevenson, F.J., *Soils for Management of Organic Wastes and Waste Waters*, SSSA, Madison, WI, 1977.

223. Hernando, S., Lobo, M.C., and Polo, A., Effect of application of a municipal refuse compost on the physical and chemical properties of a soil, *Sci. Total Environ.*, 82, 589, 1989.
224. Mays, D.A., Terman, G.L., and Duggan, J.C., Municipal composts: effects on crop yields and soil properties, *J. Environ. Qual.*, 2, 89, 1973.
225. Hortenstein, C.C. and Rothwell, D.F., Use of municipal compost in reclamation of phosphate-mining sand tailings, *J. Environ. Qual.*, 1, 415, 1972.
226. Scanlon, D.H., Duggan, C., and Bean, S.D., Evaluation of municipal compost for strip mine reclamation, *Compost Sci. Util.*, 14, 4, 1973.
227. Epstein, E., Effect of sewage sludge on some soil physical properties, *J. Environ. Qual.*, 4(1), 139, 1975.
228. Khaleel, R., Reddy, K.R., and Overcash, M.R., Changes in soil physical properties due to organic waste applications: a review, *J. Environ. Qual.*, 10, 133, 1981.
229. Page, A.L., Logan, T.J., and Ryan, J.A., *Land Application of Sludge: Food Chain Implications*, Lewis Publishers, Chelsea, MI, 1987.
230. Gallardo-Lara, F. and Nogales, R., Effect of the application of town refuse compost on the soil-plant system: a review, *Biol. Wastes*, 19, 35, 1987.
231. Guidi, G. and Petruzzelli, Effect of compost on chemical and physical properties of soil, in *Compost Production and Use: Technology, Management, Application and Legislation*, Proc. Int. Symp. on Compost, S. Michele all'Adige, Italy, June 20–23, 1989, 53.
232. Capriel, P., Beck, R., Borchert, H., and Harter, P., Relationship between soil aliphatic fraction extracted with supercritical hexane, soil microbial biomass, and soil aggregate stability, *Soil Sci. Soc. Am. J.*, 54, 415, 1990.
233. Chesters, G., Attoe, O.J., and Allen, O.N., Soil aggregation in relation to various soil constituents, *Soil Sci. Soc. Am. Proc.*, 21, 272, 1957.
234. Guidi, G., Pera, A., Giovannetti, M., Poggio, G., and de Bertoldi, M., Variations of soil structure and microbial population in a compost amended soil, *Plant Soil*, 106(1) 113, 1988.
235. Lynch, J.M., Promotion and inhibition of soil aggregate stabilization by soil microorganisms, *J. Gen. Microbiol.*, 126, 371, 1981.
236. Tisdall, J.M., Cockroft, B., and Uren, N.C., The stability of soil aggregates as affected by organic materials, microbial activity and physical disruption, *Aust. J. Soil Res.*, 16, 9, 1978.
237. Bengston, G.W. and Cornette, J.J., Disposal of composted municipal waste in a plantation of young slash pine: effects on soil and trees, *J. Environ. Qual.*, 2, 441, 1973.
238. Tisdale, J.M. and Oades, J.M., Organic matter and water-stable aggregates in soil, *J. Soil Sci.*, 33, 141, 1982.
239. Dong, A., Chester, G. and Simsiman, G.V., Soil dispersibility, *J. Soil Sci.*, 136, 208, 1983.
240. Haynes, R.J. and Swift, R.S., Stability of soil aggregates in relation to organic constituents and soil water content, *J. Soil Sci.*, 41, 73, 1990.
241. Perucci, P., Effect of the addition of municipal solid-waste compost on microbial biomass and enzyme activities in soil, *Biol. Fertil. Soils*, 10(3), 221, 1990.
242. Chang, A.C., Page, A.L., and Warneke, J.E., Soil conditioning effects of municipal sludge compost, *J. Environ. Eng.*, 109(3), 574, 1983.
243. De Smet, J., Wontroba, J., De Bood, M., and Hartmann, R., Effect of application of pig slurry on soil penetration resistance and sugar beet emergence, *Soil Tillage Res.*, 19, 297, 1991.

244. Soane, B.D., The role of the organic matter in soil compactability: a review of some practical aspects, *Soil Tillage Res.*, 16, 179, 1990.
245. Cook, R.N., Patterson, J.C., and Short, J.R., Compost saves money in parkland restoration, *Compost Sci. Util.*, 20(2), 43, 1979.
246. Spugnoli, P., Partent, A., and Baldi, F., Compaction of soil treated with municipal solid waste compost using low-pressure and traditional tyres, *J. Agric. Eng. Res.*, 56, 189, 1993.
247. Young, R.A. and Onstad, C.A., Characterization of rill and interrill eroded soil, *Trans. ASAE*, 21, 1126, 1978.
248. Meyer, L.D., How rain intensity affects interrill erosion, *Trans. ASAE*, 24, 1472, 1981.
249. Edwards, A.P. and Bremner, J.M., Microaggregates in soils, *J. Soil Sci.*, 18, 64, 1967.
250. Hamblin, A.P., Structural features of affrefates in some east anglian silt soils, *J. Soil Sci.*, 28, 23, 1977.
251. Nearing, M.A., Deer-Ascough, L., and Laflen, J. M., Sensitivity analysis of the WEPP hillslope profile erosion model, *Trans. ASAE*, 33, 839, 1990.
252. Hart, S.A., *Solid Waste Management/Composting: European Activity and American Potential*, Report No. SW-2c, U.S. Environmental Protection Agency, Cincinnati, Ohio.
253. Watson, M.R. and Hoitnink, H.A.J., Long term effects of papermill sludge in stripmine reclamation, *Ohio Rep. Res. Dev. Agric. Home Econ. Nat. Res. Ohio Agric. Res. Dev. Cent.*, 70(2), 19, 1985.
254. Sopher C.D. and Baird, J.V., *Soils and Soil Management*, Reston Publishing, Restan, VA, 1978, 81.
255. Schnitzer, M. and Skinner, S.I.M., Organo-metallic interactions in soils. IV. Carboxyl and hydroxyl groups in organic matter and metal retention, *Soil Sci.*, 99, 278, 1965.
256. Selby, M., Carruth, J., and Golob, B., End use markets for MSW compost, *BioCycle*, 30(11), 56, 1989.
257. Steffen, R., The value of composted organic matter in building soil fertility, *Compost Sci. Util.*, 20, 34, 1979.
258. Darmody, R.G., Foss, J.E., McIntosh, M., and Wolf, D.C., Municipal sewage sludge compost-amended soils: some spatiotemporal treatment effects, *J. Environ. Qual.*, 12, 231, 1983.
259. Pera, A., Valini, G., Sireno, I., Bianchin, M.L., and de Bertoldi, M., Effect of organic matter on *rhizosphere* microorganisms and root development of Sorghum plants in two different soils, *Plant Soil*, 74, 3, 1983.
260. Allievi, L., Marchesinin, A., Salardi, C., Piano, V., and Ferrari, A., Plant quality and soil residual fertility six years after a compost treatment, *Bioresour Technol.*, 43, 85, 1993.
261. Altierii, M.A., Trujillo, J., Astier, M., Gersper, P.P., and Bakx, W., Low-input technology proves viable for limited-resource farmers in Salinas Valley, *Calif. Agric.*, 45(2), 20, 1991.
262. Astier, M., *Developing Low-input Energy Saving Vegetables Cropping Systems for Small Farmers in Salinas Valley*, Internal Report, Division of Biological Control and Department of Soil Science, University of California, Berkeley, ACBE, Washington, D.C., 1990.
263. Waddell, E., The mound builders: agricultural practices, environment, and society in the Central Highlands of New Guinea, *Am. Ethanol. Soc. Monogr.*, 53, 253, 1972.

264. Wohlt, P.B., Kandep: challenge for development, in *Technical Bulletin No. 2*, Division of Primary Industry, Department of Enga Province, PNG, 1986.
265. Wohlt, P.B., Subsistence systems of Enga Province, in *Technical Bulletin No. 3*, Division of Primary Industry, Department of Enga Province, PNG, 1986.
266. Floyd, C.N., Lefroy, R.D.B., and D'Souza, E.J., Composting and crop production of volcanic ash soils in the Southern Highlands of Papua New Giunea, *AFTSEMU Technical Report*, 12, 1985.
267. Floyd, C.N., D'Souza, E.J., and Lefroy, R.D.B., Soil fertility and sweet potato production on volcanic ash soils in the highlands of Papua New Guinea, *Field Crops Res.*, 19, 1, 1988.
268. Preston, S.R., Investigation of compost x fertilizer interactions in sweet potato grown on volcanic ash soils in the highlands of Papua New Guinea, *Trop. Agric.*, 67(3), 239, 1990.
269. Hue, N.V., Ikawa, H., and Silva, J.A., Increasing plant-available phosphorus in an ultisol with a yard-waste compost, *Commun. Soil Sci. Plant Anal.*, 25(19/20), 3291, 1994.
270. Woodbury, P.B., Trace elements in municipal solid waste composts: a review of potential detrimental effects on plants, soil biota, and water quality, *Biomass Bioenergy*, 3, 239, 1992.
271. Maynard, A., Evaluating the suitability of MSW compost as a soil amendment in field grown tomatoes. Part A: yield of tomatoes, *Compost, Sci. Util.*, 1(2), 34, 1993.
272. Werner, W., Scherer, H.W.S., and Olfs, H.W., Influence of long-term application of sewage sludge and compost from garbage with sewage sludge on soil fertility criteria, *Z. Acker. Pflanzenbau.*, 160(3), 173, 1988.
273. Maynard, A.A., Agricultural compost as amendments reduce nitrate leaching from soil, *Frontiers Plant Sci.*, Fall, 3, 1989.
274. Pratt, P.F. and Castellanos, J.Z., Available nitrogen from animal manures, *Calif. Agric.*, 35(7/8), 24, 1981.
275. Brinton, W.F., Nitrogen response of maize to fresh and composted manure, *Biol. Agric. Hort.*, 3, 55, 1981.
276. Gonzalez, J.L., Benitez, I.C., Perez, M.I., and Median, M., Pig-slurry composts as wheat fertilizers, *Bioresour. Technol.*, 40(2), 125, 1991.
277. Srikumar, T.S. and Ockerman, P.A., The effects of fertilization and manuring on the content of some nutrients in potato (*var. provita*), *Food Chem.*, 37, 47, 1990.
278. Shiralipour, A., McConnell, D.B., and Smith, W.H., Uses and benefits of MSW compost: a review and assessment, *Biomass Bioenergy*, 3, 297, 1992.
279. Giusquiani, P.L., Marucchini, C., and Businelli, M., Chemical properties of soils amended with compost of urban waste, *Plant Soil*, 109, 73, 1988.
280. Koma Alimu, F.X. and Janssen, B.H., Evaluation of municipal refuse from Dohomey (*Benin*) as an organic manure, *Proc. Soil Organic Matter Studies Symp.*, Braunschweig, 1976, 6.
281. Sample, E.C., Soper, R.J., and Racz, G.J., Reactions of phosphate fertilizers in soils, in *The Role of Phosphorus in Agriculture*, Khasawneh, F.E., Sample, E.C., and Kamprath, E.J., Eds., ASA-CSSA-SSSA, Madison, WI, 1980, 263.
282. Meek, B.D., Graham, L., and Donovan, T., Long-term effects of manure on soil nitrogen, phosphorus, potassium, sodium, organic matter, and water infiltration rates, *Soil Sci. Soc. Am. J.*, 46, 1014, 1982.

283. Swaider, J.M. and Morse, R.D., Influence of organic amendments on solution phosphorus requirements for vegetables in mine-spoil, *J. Am. Soc. Hort. Sci.*, 109, 150, 1984.
284. Mishra, M.M. and Bangar, K.C., Rock phosphate composting: transformation of phosphorus forms and mechanisms of solubilization, *Biol. Agric. Hort.*, 3, 331, 1986.
285. Singh, C.P., Singh, Y.P., and Singh, M., Effect of different carbonaceous compounds on the transformation of soil nutrients. II. Immobilization and mineralization of phosphorus, *Biol. Agric. Hort.*, 4, 301, 1987.
286. Azvedo, J. and Stout, P.R., *Farm Animal Manures: An Overview of Their Role in Agricultural Environment*, Calif. Agric. Expt. Sta. Serv. Manual #44, University of California, 1974, 108.
287. Coleman, D.C., Reid, C.P.P., and Cole, C.V., Biological strategies of nutrient cycling in soil systems, *Adv. Ecology*, 13, 1, 1983.
288. Tisdale, S.L., Nelson, W.L., and Beaton, J.D., *Soil Fertility and Fertilizers*, Macmillan Publishing, New York, 1985.
289. Buchanan, R.A. and Gliessman, S.R., The influence of conventional and compost fertilization on phosphorus use efficiency by broccoli in a phosphorus deficient soil, *Am. J. Alt. Agric.*, 5, 38, 1990.
290. Gallardo-Lara, F., Navarra, A., and Nogales, R., Extractable sulphate in two soils of contrasting pH affected by applied town refuse compost and agricultural wastes, *Biol. Wastes*, 33(1), 39, 1990.
291. Vleeschauwer, D.D. and Lal, R., Properties of work casts under tropical regrowth, *Soil Sci.*, 132(2), 175, 1981.
292. Satchell, J.E., Earthworm microbiology, in *Earthworm Ecology*, Satchell, J.E., Ed., Chapman and Hall, 1983, 495.
293. Satchell, J.E. and Martin, K., Phosphatase activity in earthworm species, *Soil Biol. Biochem.*, 16(2), 191, 1984.
294. Stewart, B.A. and Chaney, R.L., Wastes: use or discard, *Proc. Soil Conserv. Soc. Am.*, 30, 160, 1975.
295. Satchell, J.E., Martin, K., and Krishnamoorthy, R.V., Stimulation of microbial phosphatase production by earthworm activity, *Soil Biol. Biochem.*, 16, 195, 1984.
296. Heckman, J.R., Angle, J.S., and Chaney, R.L., Residual effects of sewage sludge on soybean. A. Accumulation of heavy metals, *J. Environ. Qual.*, 16, 113, 1987
297. Hue, N.V., A possible mechanism for manganese phytotoxity in Hawaii soils amended with a low-manganese sewage sludge, *J. Environ. Qual.*, 17, 473, 1988.
298. Street, J.J., Lindsay, W.L., and Sabey, B.R., Solubility and plant uptake of cadmium in soils amended with cadmium and sewage sludge, *J. Environ. Qual.*, 6, 72, 1977.
299. Tadesse, W., Shuford, J.W., Taylor, R.W., Adriano, D.C., and Sajwan, K.S., Comparative availability to wheat of metals from sewage sludge and inorganic salts, *Water, Air, Soil Pollut.*, 55, 397, 1991.
300. Williams, C.H. and David, D.J., The accumulation in soil of cadmium residues form phosphate fertilizers and their effect of the cadmium content of plants, *Soil Sci.*, 121, 86, 1976.
301. Biocycle, Sludge compost "binds" lead in contaminated soil, *Biocycle*, 32(10), 82, 1991.
302. Stevenson, F.J. and Ardakani, M.S., Organic matter reactions involving micronutrients in soils, in *Micronutrients in Agriculture*, Mortvedt, J.J., Ed., SSSA, Madison, WI, 1972, 79.
303. Leeper. G.W., *Managing the Heavy Metals on the Land*, Marcel Dekker, New York, 1978.

304. Mengel, K. and Kirkby, E.A., *Principles of Plant Nutrition*, 4th ed. International Potash Institute, Worblaufen-Bern, Switzerland, 1987, 513.
305. Grimme, H., Aluminum induced magnesium deficiency in oats, *Z. Pflanzenernahr. Bodenk.*, 146, 666, 1983.
306. Juste, C. and Mench, M., Long-term application of sewage sludge and its effects on metal uptake by crops, in *Biogeochemistry of Trace Metals*, Adriano, D.C., Ed., Lewis Publishers, Boca Raton, FL, 1992, 159.
307. Rosen, C.J., Halbach, T.R., and Swanson, B.T., Horticultural uses of municipal solid waste composts, *HortTechnology*, 3(2), 167, 1991.
308. Siminis, H.I. and Manios, V.I., Mixing peat with MSW compost, *BioCycle*, 31(11), 60, 1990.
309. Raviv, M., Chen, Y., and Inbar, R., Peat and peat substrates as growth media for container grown plants, in *The Role of Organic Matter in Modern Agriculture*, Chen, Y., and Avnimelech, Y., Eds., Martinus Nijhof, Dordrecht, The Netherlands, 1986.
310. Keeling, A.A., Mullett, J.A.J., Paton, I.K., Bragg, N., Chambers, B.J., Harvey, P.J., and Manasse, R.S., Refuse-derived humus: a plant growth medium, in *Advances in Soil Organic Matter Research and the Impact on Agriculture and the Environment*, Wilson, W., Ed., Royal Society of Chemistry, Cambridge, 1991, 365.
311. Murray, J.J., Patterson, J.C., and Wehner, D.J., Use of sewage sludge compost in turfgrass production, in *Proc. Natl. Conf. Ind. Municipal Sludge Utilization Disposal*, Washington, D.C., 1980, 28.
312. Neel, P.L., Burt, E.O., Busey, P., and Snyder, G.H., Sod production in shallow beds of waste material, *J. Am. Soc. Hort. Sci.*, 103, 549, 1978.
313. Bunt, A.C., Modern Potting Composts, a Manual on the Preparation and Use of Growing Media for Pot Plants, The Pennsylvania State University Press, Great Britain, 1976.
314. Alt, D., Peters, I., and Fokken, H., Estimation of phosphorus availability in composts and compost/peat mixtures by different extraction methods, *Commun. Soil Sci. Plant Anal.*, 25(11/12), 2063, 1994.
315. Wuest, P.P.J., *Development of Procedure for Using and Storing Spent Mushroom Compost to Reduce the Risk of Lowering Water Quality*, Final Research Project Report, Pennsylvania Department of Agriculture, University Park, 1991.
316. Robbins, S.H., Reghetti, T.L., Fallahi, R., Dixon, A.R., and Chapin, M.H., Influence of trenching, soil amendments, and mulching on the mineral content, growth, yield, and quality of "Italian" prunes, *Commun. Soil Sci. Plant Anal.*, 17, 457, 1986.
317. Maher, M.J., Spent mushroom compost (SMC) as a nutrient source in peat-based potting substrates, *Mushroom Sci.*, 12(1), 645, 1991.
318. Male, R.T., The use of spent mushroom compost in vegetable production, *Mushroom Sci.*, 11(1), 111, 1981.
319. Wang, H.S., Lohr, V.I., and Coffey, D.L., Spent mushroom compost as a soil amendment for vegetables, *J. Am. Soc. Hort. Sci.*, 109, 698, 1984.
320. White, J.W., Mushroom casing soil and sphagnum moss peat: growing media for Easter lillies, *Mushroom News*, 24(6), 17, 1976.
321. Henny, B.K., Production of six foliage crops in spent mushroom compost potting mixes, *Proc. Fl. State Hort. Soc.*, 92, 330, 1979.
322. Cook, R.J. and Baker, K.F., *The nature and practice of biological control of plant pathogens*, Am. Phyto. Soc., St. Paul, MN, 1983.

323. Thurston, H.D., *Sustainable Practices for Plant Disease Management in Traditional Farming Systems*, Westview Press, Boulder, CO, 1991.
324. Millner, P.D., Lumsden, R.D., and Lewis, J.A., Controlling plant disease with sludge compost, *BioCycle*, 23(4), 50, 1982.
325. Hoitink, H.A.J. and Kuter, J.A., Role of composts in suppression of soil-borne plant pathogens of ornamental plants, *BioCycle*, 25(4), 40, 1984.
326. Phae, C.G., Sasaki, Shoda, M., and Kubota, H., Characteristics of *Bacillus subrilis* isolated from composts suppressing phytopathogencie microorganisms, *Soil Sci. Plant Nutr.*, 36, 575, 1990.
327. Nelson, E.B. and Hoitink, H.A.J., Factors affecting suppression of *Rhizoctonium solani* in container media, *Phytopathology*, 72, 275, 1982.
328. Spring, D.E., Ellis, M.A., Spotts, R.A., Hoitink, H.A.J., and Schmitthenner, A.F., Suppression of the apple collar rot pathogen in composted hardwood bark, *Phytopathology*, 70, 1209, 1980.
329. Hoitink, H.A.J. and Fahy, P.C., Basis for the control of soil borne plant pathogens with composts, *Annu. Rev. Phytopath.*, 24, 93, 1986.
330. Pera, A. and Filippi, C., Controlling of fusarium wilt in carnation with bark compost, *Biol. Wastes*, 22, 218, 1987.
331. Pera, J. and Calvert, C., Suppression of fusarium wilt of carnation in a composted pine bark and composted olive pumice, *Plant Dis.*, 73, 699, 1989.
332. Hardy, G.E., St. J. and Sivasithamparam, K., Suppression of *Phytophthora* root rot by a composted Eucalyptus bark mix, *Aust. J. Bot.*, 39, 154, 1991.
333. Hoitink, H.A.J., Inbar, R., and Boehm, M.J., Compost can suppress soil-borne diseases in container media, *Am. Nurseryman*, 178(16), 91, 1993.
334. Yohalem, D.S., Harris, R.F., and Andres, J.H., Aqueous extracts of spent mushroom substrate for foliar disease control, *Compost Sci. Util.*, 2(4), 67, 1994.
335. Fiddaman, P.J. and Rossal, S., Effect of substrate on the production of antifungal volatiles from *Bacillus subtilis, J. Appl. Bacteriol*, 76(4), 395, 1993.
336. Cubeta, M.A., Hartman, G.L., and Sinclair, J.B., Interaction between *Bacillus subtilis, J. Phytopathol.*, 124, 207, 1985.
337. Tschen, J.S.M. and Kuo, W.L., Antibiotic inhibition and control of *Rhizoctonium solina* by *Bacillus subtillis*, *Plant Protect. Bull. (Taiwan R.O.C.)*, 27, 95, 1985.
338. Gupta, V.K. and Uthede, R.S., Factors affecting the production of antifungal compounds by *Enterbacter aerogenes* and *Bacillus subtilis*, antagonists of *Phytopthora cactorum*, *J. Phytopathol.*, 117, 9, 1986.
339. Loeffler, W., Tschen, J.S.M., Vanittanatkcon, M., Kugler, M., Knorpp, E., Hsieh, T.F., and Wu, T.G., Antifungal effects of *bacilysin* and *fengymycin* from *Bacillus subtilis* F-29-3. A comparison with activities of other *Bacillus* antibiotics, *J. Phytopathol.*, 76, 136, 1986.
340. Seifert, K.A., Hamilton, W.E., Breuil, C., and Best, M., Evaluation of *Bacillus subtilis* C186 as a potential biological control of sappstain and mold on unseasoned lumber, *Can. J. Microbiol.*, 33 1103, 1987.
341. Fravel, D.R., Role of antibiosis in the biocontrol of plant diseases, *Annu. Rev. Phytopathol.*, 26, 75, 1988.
342. Pusey, P.L., Hotchkiss, M.W., Dulmage, H.R., Baumgarder, R.A., Zehr, E.I., Reilly, C.C., and Wilson, C.L., Pilot tests for commercial production and application of *Bacillus subtilis* (B-3) for post-harvest control of peach brown rot, *Plant Dis.*, 72, 622, 1988.

343. Ferreira, J.H.S., Matther, F.N., and Thomas, A.C., Biological control of *Eutypa lata* on grapevine by an antagonistic strain of *Bacillus subtilis, J. Phytopathol.*, 81, 283, 1991.
344. Nelson, E.B. and Craft, C.M., Suppression of dollar spot on bentgrass and annual bluegrass turf with compost-amended topdressings, *Plant Dis.*, 76(9), 954, 1991.
345. Chen, W., Hoitink, H.A.J., and Schmitthenner, A.F., Factors affecting the suppression of *Pythium* damping-off in container media amended with composts, *Phytopathology*, 77, 755, 1988.
346. Chen, W., Hoitink, H.A.J., Schmitthenner, A.F., and Touvinen, O.H., The role of microbial activity in suppression of damping-off caused by *Pythium ultimum*, *Phytopathology*, 78, 314, 1988.
347. Kuter, G.A., Nelson, E.B., Hoitink, H.A.J., and Madden, L.V., Fungal populations in container media amended with composted hardwood bark suppressive and conducive to *Rhizoctonia* damping-off, *Phytopathology*, 73, 1450, 1983.
348. Kwok, O.C.H., Fahy, P.C., Hoitink, H.A.J., and Kuter, G.A., Interactions between bacteria and *Trichoderma hamatum* in suppression of *Rhizoctonia* damping-off in bark compost media, *Phytopathology* 77, 1206, 1987.
349. Tunlid, A., Hoitink, H.A.J., Low, C., and White, D.C., Characterization of bacteria that suppress *Rhizoctonia* damping-off in bark compost media by analysis of fatty acid biomarkers, *Appl. Environ. Microbiol.*, 55(6), 1368, 1989.
350. Voland, R.P. and Epstein, A.H., Development of suppressiveness to diseases caused by *Rhizoctonia solani* in soils amended with composted and noncomposted manure, *Plant Dis.*, 78(5), 461, 1994.
351. Mandelbaum, R., Gorodecki, B., and Hadar, Y., The use of composts for disease suppressive container media, *Phytoparasitica*, 13, 158, 1985.
352. Gorodecki, B. and Hadar, Y., Suppression of *Rhizoctonia solani* and *Sclerotium rolfsii* disease in container media containing composted separated cattle manure and composted grape marc, *Crop Protect.*, 9, 271, 1990.
353. Kuter, G.A., Hoitink, H.A.J. and Chen, W., Effects of municipal sludge compost curing time on suppression of *Pythium* and *Rhizoctonia* diseases of ornamental plants, *Plant Dis.*, 72(9), 751, 1988.
354. Bollen, G.J., Factors involved in inactivation of plant pathogens during composting of crop residues, in *Science and Engineering of Composting: Design, Environmental, Microbiological and Utilization Aspects*, Hoitink, H.A.J. and Keener, H.M., Eds., Renaissance Publications, Worthington, OH, 1993, 301.
355. Fokkema, N.J., Opportunities and problems of control of foliar pathogens with microorganisms, *Sp. estic, Science*, 37(4), 411, 1993.
356. Ueda, R., Kai, H. and Taniguchi, E., Growth inhibition of soil-borne pathogenic fungi by typical sterols, *Soil Biol. Biochem.*, 22(7), 987, 1990.
357. Kai, H., Tohru, U. and Sakaguchi, M., Antimicrobial activity of bark-compost extracts, *Soil Biol. Biochem.*, 22(7), 983, 1990.
358. Papavizas, G.C., *Trichoderma* and *gliocladium*: biology, ecology, and potential for biocontrol, *Annu. Rev. Phytopathol.*, 23, 23, 1985.
359. Jackson, A.M., Whipps, J.M., and Lynch, J.M., *In vitro* screening for the identification of potential biocontrol agents of Allium white rot, *Mycol. Res.*, 95(4), 430, 1991.
360. Lynch, J.M., *In vitro* identification of *Trichoderma harzianum* as potential antagonist of plant pathogens, *Curr. Microbiol.*, 16, 49, 1987.
361. Ousley, M.A., Lynch, J.M., and Whipps, J.M., Effect of *Trichoderma* on plant growth: a balance between inhibition and growth promotion, *Microbiol. Ecol.*, 26, 277, 1993.

362. Ousley, M.A., Lynch, J.M., and Whipps, J.M., Potential of *Trichoderman sp.* as consistent plant growth stimulators, *Biol. Fertil. Soils*, 17, 85, 1994.
363. Calvet, C., Barea, J.M. and Pera, J., *In vitro* interactions between the vescular-arbuscular mycorrhizal fungus *Glomus mosseae* and some saprophytic fungi isolated from organic substrates, *Soil Biol. Biochem.*, 24(8), 775, 1993.
364. Biermann, B.J. and Linderman, R.G., Increased geranium growth using pre-transplanted inoculation with mycorrhizal fungus, *J. Am. Soc. Hort. Sci.*, 108, 972, 1983.
365. Muller, R. and Gooch, P.S., Organic amendments in nematode control. An examination of the literature, *Nematropica*, 12, 319, 1982.
366. Mannion, C.M., Schaffer, B., Ozores-Hampton, M., Bryan, H.H., and McSorley, R., Nematode population dynamics in municipal solid waste-amended soil during tomato and squash cultivation, *Nematropica*, 224(1), 117, 1994.
367. Mankau, R. and Das, S., Effect of organic materials on nematode bionomics in citrus and root-knot nematode infested soil, *J. Nematol.*, 4, 138, 1974.
368. Rodrigues-Kabana, R., Morgan-Jones, G., and Ownley-Ginitis, B., Effects of chitin amendments to soil on *heterodera glycines*, nocrobial populations and colonization of cysts by fungi, *Nematropica*, 14, 10, 1984.
369. Rodrigues-Kabana, R., Boube, D. and Young, R.W., Chitinous materials from blue crab for control of root-knot nematode. I. Effect of urea and enzymatic studies, *Nematropica*, 19, 53, 1989.
370. Rich, J.R. and Hodge, C.H., Utilization of blue crab scrap compost to suppress *Meloidogyne javanica* on tomato, *Nematropica*, 23(1), 1, 1993.
371. Allaway, W.H., The effect of soils and fertilizers on human and animal nutrition, *Agricultural Information Bulletin 378*, U.S. Department of Agriculture, Washington, D.C., 1975.
372. Schepers, J., Frank, K., and Watts, D., *Influence of Irrigation and Nitrogen Fertilizer on Ground-Water*, Proc. Int. Union of Geodesy and Geophysics, Hamburg, 1984.
373. Frink, C.R., Estimating nutrient exports to estuaries, *J. Environ. Qual.*, 20(4), 717, 1991.
374. U.S. Environmental Protection Agency, *National Pesticide Survey. Summary of Results of EPA's National Survey of Pesticides in Drinking Water Wells*, U.S. EPA, Washington, D.C., 1993, 1.
375. Fedkiw, J. Nitrate occurrence in U.S. waters, *U.S. Department of Agriculture,* Washington D.C., 1991.
376. Lee, D.H.K., Nitrates, nitrites and methemoglobinemia, *Environ. Res.*, 3, 484, 1970.
377. Lee, D.H.K., Nitrates, nitrites and methemoglobinemia, *Environ. Res*, 2, 1, 1970.
378. Parr, J.F., Chemical and biochemical considerations for maximizing the efficiency of fertilizer nitrogen, *J. Environ. Qual.*, 2, 75, 1973.
379. Fiskell, J.G.A. and Pritchett, W.L., Profile distribution of phosphate and metals in forest soils amended with garbage compost, *Soil Crop Sci. Soc. Flordia Proc.*, 39, 23, 1980.
380. Schultz, R., Rates of accumulation and mobilization of heavy metals in forest soils, in *Proc. Int. Conf. Heavy Metals in the Environment*, Vernet, J. P., Ed., Geneva, CEP Consultant Ltd., Edinburgh, 2, 36, 1989.
381. Sommers, L.E., Nelson, D.W., and Silviera, D.J, Transformation of carbon, nitrogen, and metals in soils treated with waste materials, *J. Environ. Qual.*, 8, 287, 1979.
382. Chang, A.C., Warneke, J.E., Page, A.L. and Lund, L.J., Accumulation of heavy metals in sewage sludge-treated soils, *J. Environ. Qual.*, 13, 87, 1984.

383. Millner, P.D., Olenchock, S.A., Epstein, E., Rylander, R., M.D., Haines, J., Walker, J., Ooi, B.L., Horne, E. and Maritato. M., Bioaerosols associated with composting facilities, *Compost Sci. Util.*, 2(4), 6, 1994.
384. Richerson, H.B., Unifying concepts underlying the effects of organic dust exposure, *Am. J. Indus. Med.*, 17, 139, 1990.
385. Richerson, H.B., Hypersensitivity pneumonitis, in *Organic Dusts: Exposure, Effects and Prevention*, Rylander, R. and Jacobs, R.R., Eds., Lewis Publishers, Chicago, 1993.
386. Castellan, R.M., Olenchock, S.A., Kinsley, K.B. and Hankinson, J.L., Inhaled endotoxin and decreased sprirometric values: an exposure-response relation for cotton dust, *New Engl. J. Med.*, 317, 605, 1987.
387. Fink, J., Hypersensitivity pneumonitis, in *Occupational Respiratory Diseases*, Merchant, J., Ed., Department of Health and Human Services (NIOSH), Publication No. 86–102, U.S. Government Printing Office, Washington, D.C., 1986, 801.
388. Denning, D.W. and Stevens, D.A., The treatment of invasive *aspergillosis, Rev. Infect. Dis.*, 12, 1147, 1990.
389. Meeker, D.P., Gephardy, G.N., Cordasco, E.M., Jr. and Wiedemann, H.P., Hypersensitivity pneumonitis versus invasive pulmonary aspergillosis: two cases with unusual pathologic findings and review of the literature, *Am. Rev. Respir. Dis.*, 143, 431, 1991.
390. Zuk, J.A., King, D., Zakhour, H.D. and Delaney, J.C., Locallly invasive pulmonary *aspergillosis* occurring in a gardener: an occupational hazard, *Thorax.*, 44, 678, 1989.
391. Stevens, D.A., Aspergillosis, in *Cecil Textbook of Medicine*, 19th ed., Wyngaarden, J.B., Smith, L.H., Jr. and Bennett, J.C., Eds., W.B. Saunders, Philadelphia, 1992.
392. Millner, P.D., Bassett, D., and Marsh, P.B., Dispersal of *Aspergillus fumigatus* from sewage sludge compost piles subjected to mechanical agitation in open air, *Appl. Environ. Microbiol.*, 39, 1000, 1980.
393. Domsch, K.H., Gams, W. and Anderson, T.H., *Compendium of Soil Fungi*, Academic Press, London, 1980.
394. E&A Environmental Consultants, Inc., *Report to Earthgro*, Lebanon, CT, 1993.
395. Millner, P.D., Marsh, P.B., Snowden, R.B. and Parr, J.F., Occurrence of *Aspergillus fumigatus* during composting of sewage sludge, *Appl. Environ. Microbiol.*, 34, 764, 1977.
396. Hampton Roads Sanitation District, *Aspergillus fumigatus: A Background Report*, Hampton Roads, VA, 1981.
397. Lees, P.S.J. and Tockman, M.S., *Evaluation of Possible Public Health Impact of WSSC Site II Sewage Sludge Composting Operations*, Johns Hopkins University, School of Hygiene and Public Health, Report Prepared for Maryland Department of Health and Mental Hygiene, Baltimore, 1987.
398. Clayton Environmental Consultants, Ltd., *Air Sampling Program for Total Coliforms, Particulates and Fungal Spores at Selected Areas in the Windsor West Pollution Control Plant*, Windsor, Ontario, 1983.
399. ERCO, *Monitoring of Aspergillus fumigatus Associated with Municipal Sewage Sludge Composting Operations in the State of Maine*, Energy Resources Co., Inc., November, 1980.
400. Kothary, M.H., Chase, T., Jr. and Macmillan, J.D., Levels of *Aspergillus fumigatus* in air and in compost at sewage sludge composting site, *Environ. Pollut. Ser. A. Ecol. Biol.*, 34, 1, 1984.

401. Boutin, P., Torre, M. and Moline, J., Bacterial and fungal atmospheric contamination at refuse composting plants: a preliminary study, in *Compost: Production, Quality and Use*, de Bertoldi, M., Fernati, M.P. and L'Hermite, P., Eds., Elsevier Applied Science, London, 1987, 266.
402. Lundholm, M. and Rylander, R., Occupational symptoms among compost workers, *J. Occup. Med.*, 22, 256, 1980.
403. Clark, C.S., Rylander, R. and Larsson, L., Levels of gram-negative bacteria, *Aspergillus fumigatus*, dust and endotoxin at compost plants, *Appl. Environ. Microbiol.*, 5, 1501, 1983.
404. Epstein, E., Neighborhood and worker protection for composting facilities: issues and actions, in *Science and Engineering of Composting: Design, Environmental, Microbiological and Utilization Aspects*, Hoitink, H.R.J. and Keener, H.M., Eds., Renaissance Publications, Worthington, OH, 1993, 319.
405. Sullivan, R.J., *Air Pollution Aspects of Odorous Compounds*, Prepared for the National Air Pollution Control Administration Consumer Protection and Environmental Health Services Department of Health, Education, and Welfare, Contract No. PH-22-68-225, 1969.
406. Mosier, A.R., Morrison, S.M. and Elmund, G.K., Odors and emissions for organic wastes, in *Soils for Management of Organic Wastes and Waste Waters*, Madison, Wisconsin, 1977, 530.
407. Miller, F.C., Minimizing odor generation, in *Science and Engineering of Compost: Design, Environmental, Microbiological and Utilization Aspects*, Hoitink, H.A.J. and Kenner, H.M., Eds., Renaissance Publications, Worthington, OH, 1993, 219.
408. Walker, J.M., Control of composting odors, in *Science and Engineering of Compost: Design, Environmental, Microbiological and Utilization Aspects*, Hoitink, H.A.J. and Kenner, H.M., Eds. Renaissance Publications, Worthington, OH, 1993, 185.
409. Kanagawa, R. and Mikami, E., Removal of methanethiol, dimethyl sulfide, dimethyl disulfide, and hydrogen sulfide from contaminated air by *Thiobacillus thioparus tk-m*, *Appl. Environ. Microbiol.*, 55, 555, 1989.
410. Smith, N.A. and Kelly, D.P., Oxidation of carbon disulphide as the sole source of energy for the autotrophic growth of *Thiobacillus thioparus* strain tk-m., *J. Gen. Microbiol.*, 134, 3041, 1988.
411. Suylen, G.M. and Kuenen, J.G., Chemostat enrichment and isolation of *Hyphomicrobium ef*, a dimethyl sulphide oxidizing methylotroph and reevaluation of *Thiobacillus ms1*, *Antonie van Leeuwenhoek, J. Microbiol. Serol.*, 52, 281, 1986.
412. Derikx, P.J.L., Simons, F.H.M., Op den Camp, H.J.M., van der Drift, C., Van Griensven, L.J.L.D. and Vogels, G.D., Evolution of volative sulfur compounds during laboratory-scale incubations and indoor preparation of compost used as a substrate in mushroom cultivation, *Applied Environ. Micro.*, 57(2), 563, 1991.
413. Hinesly, T.D. and Sosewith, B., Digested sewage sludge disposal on crop land, *J. Water Pollut. Control Fed.*, 42, 822, 1969.
414. McCalla, R.M., Frederick, L.R. and Palmer, F.L., Manure decomposition and fate of breakdown products in soil, in *Agricultural Practices and Water Quality*, Willrich, R.L. and Smith, G.E., Eds., The Iowa State University Press,/Ames, 1970, 241.
415. Hunt, P.G., Hortenstine, C.C. and Eno, C.F., Direct and indirect effects of composted municipal refuse on plant seed germination, *Soil Crop Sci. Soc. Fl. Proc.*, 32, 92, 1973.
416. Sabey, B.R. and Hart, W.E., Land application of sewage sludge. I. Effect on growth and chemical composition of plants, *J. Environ. Qual.*, 4, 252, 1975.

417. Lohr, V.I., O'Brien, R.G. and Coffey, D.L., Spent mushroom compost in soilless media and its effects on the yield and quality of transplants, *J. Am. Soc. Hort. Sci.*, 19, 681, 1984.
418. Rathier, R.M., Spent mushroom compost for greenhouse crops, *Conn. Greenhouse Newslett.*, 109, 6, 1982.
419. Jackson, M.B., Drew, M.C. and Giffard, S.C., Effects of spraying ethylene to the root system of *Zea mays* on growth and nutrient concentration in relation to flooding tolerance, *Physiol. Plant.*, 52, 23, 1981.
420. Keeling, A.A., Paton, I.K. and Mullet, J.A.J., Germination and growth of plants in media containing unstable refuse-dervived compost, *Soil Biol. Biochem.*, 26(6), 767, 1994.
421. Lohr, V.I., Wang, S.H. and Wolt, J.D., Physical and chemical characteristics of fresh and aged spent mushroom compost, *HortScience*, 19, 681, 1984.
422. Lynch, J.M., Degradation of straw by soil microorganisms and its effect on seed germination, *Proc. of the Soc. for Gen. Microbiol.*, 3, 90, 1976.
423. Toussoun, T.A., Wienhold, A.R., Linderman, R.G. and Patrick, Z.A., Nature of phytotoxic substances produced during plant residues decomposition in soil, *Phytopathology*, 58, 41, 1968.
424. Chanyasak, V., Katayama, A., Hirai, M.F., Mori, S., Kubota, H., Effects of compost maturity on growth on komatsuna (*Brassica rapa Var. Pervidis*) in Neubauer's pot, *Soil Sci. Plant Nutr.*, 29, 239, 1983.
425. Still, S.M., Dirr, M.A. and Gartner, J.B., Phytotoxic effect of several bark extracts on mung bean and cucumber growth, *J. Am. Soc. Hort. Sci.*, 101, 34, 1976.
426. Solbraa, K., Composting of bark. IV. Potential growth reducing compounds and elements in bark, *Medd. NorskInst. Skogforsk.*, 34, 443, 1979.
427. DeVleeschauwer, D., Verdonck, O. and Van Assche, P., Phytotoxicity of refuse compost, *BioCycle*, 22(1), 44, 1981.
428. Wild, S.R. and Jones, K.C., Organic chemicals entering agricultural soils in sewage sludges: screening for their potential to transfer to crop plants and livestock, *Sci. Tot. Environ.*, 119, 85, 1992.
429. Hsu, S.M., Schnoor, J.L., Licht, L.A., St. Clair, M.A. and Fannin, S.A., Fate and transport of organic compounds in municipal solid waste compost, *Compost Sci. Util.*, 1(4), 36, 1993.
430. Terman, G.L. and Mays, D.A., Utilization of municipal solid waste compost: research results at Muscle Shoals, Alabama, *Compost Sci. Util.*, 14, 18, 1973.
431. Petruzzelli, G.G., Lurano, L. and Guidi, G., Heavy metal extractability, *BioCycle*, 26(11), 46, 1985
432. Gonzales-Vila, F.J., Saiz-Jimenez, C. and Martin, F., Identification of free organic chemicals found in composted municipal refuse, *J. Environ. Qual.*, 11, 251, 1982.
433. De Haan, S., Results of municipal waste compost research over more than fifty years at the Institute for Soil Fertility at Haren Groningen, the Netherlands, *J. Agric. Sci.*, 29, 49, 1981.
434. Chaney, R.L., Health risks associated with toxic metals in municipal sludge, in *Sludge—Health Risks of Land Application*, Bitton, G., Damron, B., Edds, G., and Davidson, J., Eds., Ann Arbor Science Ann Arbor, 1980, 59.
435. Yuran, G.T. and Harrison, H.C., Effects of genotype and sludge on cadmium concentration in lettuce leaf tissue, *J. Am. Soc. Hort. Sci.*, 111, 491, 1986.
436. Nogawa, K., Itai-itai disease and follow-up studies, in *Cadmium in the Environment, Part II: Health Effects*, Nriagu, J.O., Ed., John Wiley & Sons, New York, 1981, 1.

437. Logan, T. and Chaney, R.L., Metals, in *Utililization of Municipal Wastewater and Sludge on Land*, Page, Gleason, Smith, Iskandar and Sommers, Eds., Univ. of California, Riverside, 1983, 235.
438. Bevacqua, R.F. and Mellano, V.J., Cumulative effects of sludge compost on crop yields and soil properties, *Commun. Soil Sci. Plant Anal.*, 25(3&4), 395, 1994.
439. Chong, C., Cline, D.L., Rinker, D.L., and Allen, O.B., Growth and mineral nutrient status of containerized woody species in media amended with spent mushroom compost, *J. Am. Soc. Hort. Sci.*, 116, 242, 1991.
440. Chong, C., Cline, D.L., Rinker, D.L., and Hamersma, B., An overview of reutilization of spent mushroom compost in nursery container culture, *Landscape Trades*, 13(11), 14, 1991.
441. Cottenie, A., Sludge treatment and disposal in relation to heavy metals, in *Int. Conf. Heavy Metal Environ*, Amsterdam, September 1981, Commission of the European Communities, Amsterdam, the Netherlands, 1981, 167.
442. Chaney, L.R., *Food Chain Pathways for Toxic Metals and Toxic Organics in Wastes*, Francis & Anerbach, Eds., Butterworth Pub., Ann Arbor Science Pub., 1983, 179.
443. Beyer, W.N., Connor, E.E., and Gerould, S., Estimates of soil ingestion by wildlife. *J. Wildl. Manage*, 58, 375, 1994.
444. Stillwell, D.E., Evaluation of the suitability of MSW compost as a soil amendment in field grown tomatoes part b: elemental analysis, *Compost Sci. Util.*, 1(3), 66, 1993.
445. Shear, C.B. and Faust, M., Nutritional ranges in deciduous tree fruits and nuts, *Hort. Rev.*, 2, 142, 1980.
446. Andersson, A., Composted municipal refuse as fertilizer and soil conditioner. Effects on the contents of heavy metals in soil and plant, as compared to sewage sludge, manure and commercial fertilizers, in *Utilization of Sewage Sludge on Land: Rates of Application and Long-Term Effects of Metal*, Berglund, S., et al., Eds., D. Reidel Boston, 1983, 146.
447. Barbera, A., Extraction and dosage of heavy metals from compost-amended soils, in *Compost: Production, Quality and Use*, de Bertoldi, M., et al., Eds., Elsevier Applied Science, London, 1987, 598.
448. Chaney, R.L., Land application of composted municipal solid waste: public health, safety, and environmental issues, *Proc. Natl. Conf. Solid Waste Composting Council*, Falls Church, VA, 1991, 13.
449. U.S. Environmental Protection Agency, Standards for the disposal of sewage sludge: proposed rule 40, CFR parts 27 and 503, *Fed. Regist.*, 54(23), 5746, 1990.
450. Chaney, R.L. and Ryan, J.A., Heavy metals and toxic organic pollutants in MSW-composts: research results on phytoavailability, bioavailability, fate, *etc.*, in *Science and Engineering of Compost: Design, Environmental, Microbiological and Utilization Aspects*, Hoitink, H. A. J. and Kenner, H. M., Renaissance Publications, Worthington, OH, 1993, 451.
451. Gabrey, S.W., Belant, J.L., Dolbeer, R.A., and Bernhardt, G.E., Bird and rodent abundance at yard-waste compost facilities in northern Ohio, *Wildl. Soc. Bull.*, 22(2), 288, 1994.
452. Baker, J.A. and Brooks, R.J., Raptor and vole populations at an airport, *J. Wildl. Manage.* 45, 390, 1981.
453. Johnsgaard, P.A., *Hawks, Eagles, and Falcons of North America*, Smithsonian Press, Washington, D.C., 1990, 403.
454. Alverez, C.G., *Ecology and Control of the Principal Flies Associated with a Compost Plan*, Dissertation, University of Florida, Gainesville, 1971.

455. Richard, T., Clean compost production, *BioCycle*, 3(2), 46, 1990.
456. Epstein, E. and Epstein, J.I., Public health issues and composting, *BioCycle*, 30(8), 50, 1989.
457. De Haan, F.A.M., General aspects of compost uses in agriculture (with respect to soil quality), in *Compost Production and Use: Technology, Management, Application and Legislation*, Proc. Int. Symp. Compost, S. Michele all'Adige, Italy, 1989.
458. Walker, J.M. and O'Donnell, M.J., Comparative assessment of MSW compost characteristics, *BioCycle*, 32(8), 69, 1991.
459. Minnesota Pollution Control Agency, *Solid Waste Management Rules*, State Office Bldg., St. Paul, 1989, 151.
460. Golueke, C.G., Diaz, L.F., and Gurkewitz, S., Physio-chemical comparison: technical analysis of multi-compost products, *BioCycle*, 30(6), 55, 1989.
461. Yeager, J. and Hain, K.E., Risk assessment re: beneficial uses of composted sewage sludge, *Pro. National Conference on Municipal and Industrial Sludge Composting*. Operation Design, Marketing, Health Issues, Hazardous Materials Control Research Institute, !980. 173.
462. Dyer, J.M. and Razvi, A.S., Assessing risk of solid waste compost, *BioCycle*, 28(3), 31, 1987.
463. Ciavatta, C., Govi, M., Simoni, A., and Sequi, P., Evaluation of heavy metals during stabilization of organic matter in compost produced with municipal solid wastes, *Bioresour. Technol.*, 43(2), 147, 1993.
464. Ahrens, E. and Farkasdy, G., Investigation on nitrogen mineralization in soil after application of town refuse of different degrees of decomposition, *Sonderh. Landw. Forsch.*, 23, 77, 1969.
465. Norstadt, F.A., Swanson, N.P., and Sabey, B.R., Site design and management for utilization and disposal of organic wastes, in *Soils for Management of Organic Wastes and Waste Waters*, Elliot, L.F. and Stevenson, F.J., Eds., SSSA, Madison, WI, 1977, 348.
466. Bidlingmaier, W., Odour emissions from composting plants, *Compost Sci. Util.*, 1(4), 64, 1993.
467. Cook, D.L. and Van Haverbeke, D.F., Trees and shrubs for noise abatement, *Nebr. Agric. Exp. Stn. Res. Bull.*, 246, 77, 1974.
468. Van Haverbeke, D.F., and Cook, D.I., Green mufflers, *Am. For.*, 78, 28, 1972.
469. Woodruff, N.P., Read, R.A., and Chepil, W.S., Influence of a field windbreak on summer wind movement and air temperature, *Kans. Agric. Exp. Stn. Tech. Bull.*, 100, 24, 1959.
470. Woodruff, N.P., Shelterbelt and surface barrier effects on wind velocities, evaporation, house heating, snowdrifting, *Kans. State Coll. Tech. Bull.*, 77, 27, 1954.
471. Skidmore, E.L. and Hagen, L.J., Evapotranspiration in sheltered areas as influenced by windbreak porosity, *Agric. Meteorol.*, 7, 363, 1970.
472. Willis, W.O. and Haas, H.J., Snow and snowmelt management with level benches, small grain stubble and windbreaks, in *Proc. Snow and Ice in Relation to Wildlife Recreation Symp.*, Iowa State University, Ames, 1971, 89.
473. Jacobson, P., Mechanics of water erosion, in *Agricultural Engineers Handbook*, McGraw-Hill, New York, 1961, 401.
474. Black, A.A., Conservation bench terraces in Montana, *Trans. ASAE*, 11, 393, 1968.
475. Haas, J.H. and Willis, W.O., Conservation bench terraces in North Dakota, *Trans. ASAE*, 11, 396, 1968.
476. Donnan, W.E. and Schwab, G.O., Current drainage methods in the USA, *Agronomy*, 17, 93, 1974.

477. Sutton, J.G., Agricultural drainage, in *Agricultural Engineers Handbook*, McGraw-Hill, New York, 1961, 356.
478. Epstein, E., William, G.B., Burge, W.D., Mullen, D.C., and Enkiri, N.K., A forced aeration system for composting waste water sludge, *Water Pollut. Control Fed.* 48, 688, 1976.
479. Inbar, Y., Chen, Y., Hadar, Y., and Hoitink, H.A.J., New approaches to compost maturity, *BioCycle*, 31(12), 64, 1990.
480. Hadar, Y., Inbar, Y., and Chen, Y., Effect of compost maturity on tomato seedling growth, *Scientia Horticulturae*, 27(3/4), 199, 1985.
481. Gouin, F.R., Utilization of sewage sludge compost in horticulture, *HortTechnology*, 3(2), 161, 1993.
482. Fitzpatrick, G.E., A program for determination compost blending ratios, *Compost Sci. Util.*, vol. 1, 1993.
483. Farrell, J.B., Fecal pathogen control during composting, in *Science and Engineering of Compost: Design, Environmental, Microbiological and Utilization Aspects*, Hoitink, H.A.J. and Kenner, H.M., Eds., Renaissance Publications, Worthington, OH, 1993, 282.
484. Cooper, R.C. and Golueke, C.G., Survival of enteric bacteria and viruses in compost and its leachate, *Compost Sci.*, 20(2), 29, 1978.
485. Gaby, W.L., *Evaluation of Health Hazards Associated with Solid Waste/Sewage Sludge Mixtures*, EPA-670/2-75-023, U.S. Environmental Protection Agency, Washington, D.C., 1975.
486. Campbell, C.D. and Darbyshire, J.F., The composting of tree bark in small reactors: self-heating experiments, *Biol. Wastes*, 31(2), 145, 1990.
487. Walke, R.H., *The Preparation, Characterization and Agricultural Use of Bark-Sewage Compost*, Ph.D. Thesis, University of New Hampshire, Durham, 1975.
488. Wiley, B.B. and Westerberg, S.C., Survival of human pathogens in composted sewage, *Appl. Microbiol.*, 18(6), 994, 1969.
489. Nell, J.H. and Krige, P.R., The disposal of solid abattoir waste by composting, *Water Res.*, 5, 1177, 1971.
490. Gray, K.P., Sherman, K., and Biddlestone, A.J., Review of composting, the practical process, *Process Biochem.*, 2, 22, 1971.
491. U.S. Environmental Protection Agency, *Guidelines for Controlling Fugitive Emissions*, OAQPS RTP-NC, Washington, D.C., 1978.
492. General Physics Corp., Environmental Sciences Division, *Data Results: Revised Bioaerosol Monitoring Program for the Washington Suburban Sanitary Commission*, Montgomery County Regional Composting Facility, General Physics Corporation, Silver Spring, MD, 1991.
493. Chesapeake Occupational Health Service, *Health Surveillance Program for Compost Worker: An Epidemiologic Review*, WSSC Site II, Silver Spring, MD, 1991.
494. Miller, F.C., Minimizing odor generation, in *Science and Engineering of Composting: Design, Environmental, Microbiological and Utilization Aspects*, Hoitink, H.A.J. and Keener, H.M., Eds., Renaissance Publications, Worthington, OH, 1993, 219.
495. Nawar, M.S., Shereef, G.M., and Ahmed, M.A., Influence of food on development and reproduction of *Hypoaspis solimani N. SP. (Acari: Laelapidae)*, *Insect Sci. Applic.*, 14(3), 343, 1993.
496. Bauhus, J. and Meiwes, K.J., Potential use of plant residue wastes in forests of northwestern Germany, *For. Ecol. Manage.*, 66, 87, 1994.
497. Brito, L.M.C.M., *An Analysis of the Uses of Composted Sewage Sludge and Municipal*

Waste as Soil Amendments on the Growth and Mineral Content of Lettuce (Lactuca sativa L.), Thesis, University of Reading, U.K., 1989.

498. Stewart, B.A. and Meek, B.D., Soluble salt considerations with waste application, in *Soils for Management of Organic Wastes and Waste Waters*, Elliot, L.F. and Stevenson, F.J., Eds., SSSA, Madison, WI, 1977, 218.
499. Chaney, R.L. and Giordano, P.M., Microelements as related to plant deficiencies and toxicities, in *Soils for Management of Organic Wastes and Waste Waters*, Elliot, L.F. and Stevenson, F.J., Eds., SSSA, Madison, WI, 1977, 234.

CHAPTER 8

The Influence of Waste Amendments on Soil Properties

R. L. Hill and B. R. James

I. INTRODUCTION

With the increased environmental awareness in the U.S. since 1970, there has been a desire to reduce point source loadings of contaminants from waste materials to surface and groundwater and to seek alternate routes of waste disposal. The use of waste amendments on land has gained widespread attention as an alternative route of waste disposal. Much of this attention has focused on the application to the soil of animal or municipal waste amendments rich in plant nutrients. An understanding of how these amendments affect soil physical and chemical properties is germane to predicting long-term changes in soil and water quality.

0-87371-859-3/95/$0.00+$.50

The application of waste amendments to land may have beneficial or detrimental effects on soil physical and chemical properties, depending upon the characteristics of the amendment and the soil. Waste materials that are the most likely to influence soil physical properties, and that are commonly available in large amounts regardless of geographical location, are animal manures, sewage sludges, and compost-type solid waste materials. These types of waste amendments may make substantial organic carbon (C) contributions to the soil if the waste amendment is in a solid form or has been dewatered. Organic C is the constituent of waste amendments that is considered to be the most likely to influence soil physical and chemical properties.[1,2] The effects of waste and organic C amendments on soil properties are somewhat difficult to evaluate based upon the wide range of waste amendments that have been evaluated when applied to a broad range of soil types and textures. Previous studies have been characterized by different depths of incorporation, different loading rates, different C mineralization rates, and different incubation times prior to measurement of different soil properties.

There have been several excellent reviews conducted on the effects of organic amendments being added to soils on soil physical properties.[1-4] There is limited consistent information in the literature concerning the effects of organic amendments and resultant changes in soil chemical properties. The present review will attempt to summarize the results and analysis of previous studies and review articles on soil properties. The information has been presented in a manner easily understood by an individual with a general scientific background and does not necessarily require a strong background in soil science for interpretation.

II. CARBON AVAILABILITY IN WASTE AMENDMENTS

The availability of C in any waste amendment is dependent upon the rate of waste degradation and subsequent release of the CO_2. Table 1 illustrates the increases in soil C from the addition of several different waste amendments to soils with varying textures under differing climatic regimes. The degradation of the waste amendment is partially dependent upon climatic factors of moisture and temperature as they influence chemical reactions and microbial populations. Degradation is also partially dependent upon the chemical composition of the waste amendment with respect to the carbohydrate, protein, lignin, fat, and wax content of the material. The carbon-to-nitrogen ratio (C/N) is a characteristic of the waste amendment which is commonly considered as being integral to the rate of decomposition and release of NH_4^+. A C/N ratio of approximately 20:1 is a cutoff value above which heterotrophic microbial respiration results in no net release of NH_4^+ as a part of decomposing organic C to CO_2. Recalcitrant forms of C (e.g., lignin) also may remain in the soil or intermediate products of respiration may result in an accumulation of partially or fully humidified forms of organic matter (e.g., peat soils or wetlands high in organic matter content).

Management factors influencing waste degradation and C availability include quantity of the waste amendment applied per unit area, method of soil incorporation, the frequency and timing of multiple applications, and the crop grown after addition of the waste. The soil type, texture, slope, aspect, and soil physical and chemical properties could also have significant effects. Although adapted from a table summarizing the effects of green manures on soil properties, Table 2 summarizes some of the same factors influencing the degradation of organic amendments when added to soil. It should be re-

Table 1 Soil Carbon and Bulk Density Changes Resulting from Applications of Various Organic-Rich Amendments

Type of Amendment	Study Period (years)	Soil Type	Waste Application Rate (mt ha^{-1} year^{-1})	Carbon Application Rate (mt ha^{-1} year^{-1})	Net Increase Soil Carbon (%)	Bulk Density (g/cm^3)	Ref.
Municipal compost	2	Silt loam	23	6.2	0.21	1.31 (1.37)	9
			41	11.7	0.25	1.27	
			82	23.5	0.63	1.24	
			164	46.8	1.56	1.12	
Cattle manure	4	Clay loam	22	7.0	0.42	1.33 (1.37)	14
			67	21.4	0.69	1.28	
			134	42.9	0.80	1.20	
			268	85.8	0.68	1.12	
Cattle manure	2	Silt loam	90	28.8	0.50	1.00 (1.02)	15
			200	64.0	1.05	1.00	
			415	132.8	2.55	0.85	
Sewage sludge	2	Sand	112	15.7	0.44	1.37 (1.43)	12
			225	31.5	1.30	1.24	
			450	63.0	2.69	1.03	
Municipal waste	1	Loam	188	69.6	0.59	1.25 (1.38)	10
Sewage sludge	1	Silt loam	22.4	4.3	0.63	1.27 (1.35)	8
			56	10.9[a]	1.07	1.29	
			56	10.9[b]	1.27	1.09	
			89.6	17.4	1.14	1.21	
Poultry manure	5	Clay loam	27	9.9		0.97 (1.11)	11
			56	20.6		0.94	
			85	31.2		0.80	
			110	40.4		0.78	

Note: Values in parentheses are for control areas.

[a] Rototill method of incorporation.
[b] Disk method of incorporation.

Table 2 Factors Influencing the Degradation of Organic Amendments in Soil

Climate
Temperature, solar radiation
Precipitation, evaporation
Soil/site
Geographical location (elevation, slope, aspect)
Texture
Soil structure, compaction
Native organic matter and humus content
Soil temperature
Soil moisture, aeration
pH
Mineral ion content
Plant cover (species, density, distribution, history of site)
Microbial populations (species, density, distribution, history of site)
Faunal populations (species, density, distribution, history of site)
Use of fertilizers, lime, mulches, and pesticides
Tillage, cultivation, drainage, irrigation
Amendment
Composition (e.g., carbohydrates, proteins, lignins, fats, waxes)
Quantity added per unit area
Moisture content
C/N ratio
Mineral ion content
Timing, method, and frequency of incorporation

Adapted from Reference 4.

membered that complete decomposition of organic C to CO_2 will have no residual effect on soil physical properties, while incomplete oxidation of organic C may result in organic acid production, with concomitant effects on soil pH, mineral weathering, metal chelation, sorption processes, and oxidation-reduction status (pe).

III. SOIL PHYSICAL PROPERTIES INFLUENCED BY ORGANIC AMENDMENTS

A. BULK DENSITY

Bulk density (D_b) is defined as the mass of dry soil solids per unit volume of soil. The addition of organic amendments may cause reductions in D_b because (1) the particle density of organic particles is lower than the particle densities of mineral particles and mixing causes a dilution effect and (2) the organic substances facilitate increased soil aggregation, thereby, increasing soil porosity or void areas and reducing the D_b.[5,6]

Klute and Jacob reported over 40 years ago that the addition of organic-rich amendments results in corresponding reductions in D_b which are related to the magnitude and frequency of waste applications.[7] Since that time, there have been several reports of reduced D_b with increased rates of organic amendment applications.[8–11]

Bulk density has been shown to decrease linearly with increasing amounts of organic matter additions in sludge.[12] In an 11-year study, Mathers and Stewart reported D_b decreased following manure incorporation with a high correlation exhibited between soil organic matter and D_b.[13] When the organic matter levels decreased over time, the D_b

tended to increase. Unger and Stewart also reported that D_b decreased with increased manure applications.[14] Changes were credited to the dilution effect from the addition of organic matter, but were also partially attributed to increased aggregation. Reduced D_b values following manure applications has also been attributed to the decreased particle density of solids within a volume of soil.[15] Sommerfeldt and Chang reported D_b decreased with increased manure applications 5 years after addition and note that a similar change in D_b was also present 12 years after manure additions.[16,17]

Khaleel et al. formulated a linear regression equation to express the relationship between the percent change in bulk density (ΔD_b) and percent change in organic carbon (ΔC) relative to values observed in control areas:

$$\Delta D_b = 3.99 + 6.62(\Delta C) \qquad r^2 = 0.69$$

which was based on 12 studies that ranged in duration from 1 to 85 years.[1]

B. AGGREGATION AND AGGREGATE STABILITY

Soil aggregation may be characterized in terms of the size distribution of aggregates present in the soil and by the water stability of those aggregates. Organic matter is a primary factor responsible for the stability of surface aggregates.[4] The addition of organic matter to soils through amendments has been shown to enhance soil aggregation and stability.[2,8,10,11,14,15,18] Although the changes in aggregate properties that occur within a soil through organic matter amendments vary greatly, depending on the physical and chemical properties of the soil and the amendment, the process has been shown to be primarily microbiological in nature.[2,19] The persistence of these changes in aggregation or stability will also be dependent on the properties of the soil and the amendment as they interact with climatic conditions.

In a laboratory study, Epstein reported that a silt loam soil that had been amended with raw sludge had a greater percentage of water-stable aggregates during the first 118 days than the same soil that had been amended with digested sludge.[18] After 175 days, the percentage of stable aggregates for the two sludge treatments was the same, but averaged 34% compared to 17% for the original soil.

In a 5-year study utilizing poultry manure on a clay loam soil, Weil and Kroontje reported an increase in aggregates greater than 2-mm diameter which was accompanied by an increase in water-stable aggregates from 73% to nearly 94%.[11] In a field study using two silt loam soils and a sandy loam soil, the mean weight diameter of water-stable aggregates in the top 5 cm of soil increased as much as four times over that of the control areas that received no sludge. The gains in aggregate stability diminished over the winter months, and this was attributed to physical stresses such as freeze-thaw, but the sludge-treated areas continued to exhibit greater aggregation than their respective control areas.

A clay loam soil that had received annual applications of feedlot wastes for 4 years exhibited no changes in the percentage of water stable aggregates from the control when applied at 22 and 67 tons/ha.[14] Some changes in the water-stable aggregates occurred at the 134 tons/ha rate, but major changes occurred at the 268 tons/ha rate. There was a decrease in the 0.25-mm diameter-sized aggregates, but a significant increase in the percentage of water stable aggregates in the 1-, 2-, and 4-mm diameter-sized range. It was felt that the feedlot manure lowered the percentage of small water-stable aggregates and increased the percentage of large water-stable aggregates when applied at high rates.

Tiarks et al. determined that the geometric mean diameter of water-stable aggregates increased exponentially with the amount of manure applied.[15] This observation was contrary to previous results which suggested that aggregate stability increased logarithmically as organic matter increased or that a linear relationship existed between the percentage of water-stable aggregates and the soil organic carbon content.[20,21] Tiarks et al. came to the conclusion that the geometric mean weight diameter increased exponentially with increasing amounts of manure additions, but that the water-stable aggregates of a given size class increased linearly with manure additions when the soil organic C content was greater than 1.5%.[15]

C. SOIL WATER RETENTION AND POROSITY

Soil water retention refers to the quantity of water retained within a soil and the potential energy specifically associated with that quantity of water. The water is retained within soil pores and is, therefore, directly related to the volume and number of any one size of pore and to the soil's pore size distribution. Water is also directly adsorbed to particle surfaces and is directly related to the soil's specific surface. The water that is retained in soil pores is determined by the effective diameter of the pores and the forces of capillarity. This water is more easily removed than adsorbed water and is retained at higher levels of potential energy. The water within the interval after which water has freely drained in response to the force of gravity (i.e., field capacity) and water still retained when plants irreversibly wilt (i.e., wilting point) is labeled plant-available water. The potential energy status within the soil that corresponds to field capacity and wilting point are classically considered those water contents at –33 and –1500 kPa (–0.33 to –15 atm), respectively.

Bouyoucos presented the results of a laboratory study in 1937 showing that the addition of organic-rich amendments to soils resulted in increased values of water-holding capacity, wilting point, and available water.[22] The increases in available water were valid whether the results were computed on a volume or weight basis and were greater in light-textured soils than in clays. Several researchers have corroborated his findings in reporting that, when organic amendments are added to soil, there has been a corresponding increase in soil water retention.[4,8,10–12,14,23] Table 3 illustrates the effects occurring in water-holding capacity and available water from adding organic amendments to different soil textures.

In a laboratory study comparing amendments of raw and digested sewage sludge applied to a silt loam soil, Epstein reported that raw sludge retained twice the amount of water as the digested sludge.[18] The digested sludge retained approximately ten times the amount of water as the soil. When the raw sludge was mixed with the soil, the available water content progressively decreased and came to an "equilibrium" status more rapidly at higher temperatures indicating that microbial activity was involved in the process although there were no appreciable differences in the total carbon contents of the sludge-soil mixtures being evaluated at different temperatures.

Water retention increased when sludge was added to a coarse-textured sandy soil with most of the change occurring at potentials less than –1500 kPa.[12] There was not, however, an accompanying increase in the amount of plant-available water. Addition of

Table 3 Water-Holding Capacity (WHC) of Various Soils Resulting from Applications of Organic-Rich Amendments

Type of Waste	Study Period (years)	Soil Type	Waste Application Rate (mt ha^{-1} year^{-1})	WHC at Field Capacity (% by weight)	WHC at Wilting Point (% by weight)	Available Water Capacity (% by weight)	Ref.
Cattle manure	4	Clay loam	22	28.6[a] (28.0)	18.9 (18.2)	9.7 (9.8)	14
			67	29.2	18.7	10.5	
			134	30.3	19.5	10.8	
			268	32.3	19.3	13.0	
Sewage sludge	2	Sand	112	7.0 (5.3)	5.5 (3.8)	3.0 (2.6)	12
			225	9.7	9.2	2.3	
			450	18.0	16.8	3.9	
Municipal waste	1	Loam	188	23.2 (21.7)	11.2 (10.9)	12.0 (10.8)	10
Sewage sludge	1	Silt loam	22.5	19.8 (18.0)	6.3 (5.8)	13.5 (12.2)	8
			56[a]	20.3	6.8	13.5	
			56[b]	24.4	7.5 (5.5)	16.9 (12.4)	
			89.6	20.8	6.0	14.8	

Note: Values in parentheses are for control areas.

[a] Rototill method of incorporation.

[b] Disk method of incorporation.

the sludge may have resulted in an increase in water retention at potentials less than −1500 kPa because of the coarse-textured soil's low inherent specific surface. Contrasting results were obtained for a clay loam soil that had four annual manure applications.[14] There were significant differences in water retention properties at saturation, but not at −1500 kPa.

Similar results regarding a lack of differences in available water were obtained for a field study using two silt loam soils and a sandy loam soil.[8] Although water retention increased at −33 and −1500 kPa, there were no significant differences in plant available water although 56 metric tons/ha of digested sewage sludge had been added to the sandy loam and one silt loam, and 89.6 metric tons/ha had been added to the second silt loam soil. An explanation was offered that an increase had occurred in the large pore space volume, but that the medium pore space volume had remained the same. In a study using feedlot manure applications to a clay loam soil, plant-available water was found to decrease with increasing rates of manure application.[17] It has been suggested that organic matter additions will likely increase plant-available water only for very sandy soils with low available water, although the quantity of the organic matter addition necessary to increase the available water might not be economically feasible.[24,25] This hypothesis was not supported by a recent study on a loamy sand soil in which the specific surface increased with increasing sewage sludge compost amendments and contributed to the water retention of the soil.[26] Application rates ranged from 33 Mg/ha which was too low to supply the N requirements for a crop to 268 Mg/ha which would be considered excessive. A regression equation was presented with a high coefficient of determination ($R^2 = 0.98$) in which surface area ($m^2\ kg^{-1}$) = 0.000293 × organic matter ($g\ kg^{-1}$) + 0.00387. Differences in water retention at −1500 kPa would be directly associated with increases in the surface area.

Using regression equations, Khaleel et al. found that approximately 80% of the observed variations in water-holding capacity at field capacity and wilting point for nine different soils of varying texture could be explained using the percent sand content and the relative increase in soil organic C.[1] They concluded that organic C amendments increase water-holding capacity for fine-textured soils more at field capacity than at the wilting point while the reverse relationship seems to hold for coarse-textured soils. If the organic carbon amendment increases the water-holding capacity at both field capacity and wilting point, there are likely to be minimal changes in the amount of plant-available water.

D. HYDRAULIC CONDUCTIVITY

The volume of fluid (Q) that moves through a soil per unit cross-sectional area (A) and time (t) is directly proportional to the total potential gradient ($\Delta\Psi$) which drives the fluid flow and indirectly proportional to the length (L) of the soil column through which the fluid moves. The volume of flow per unit area and time is referred to as the flux or flux density (J) (i.e., $J = Q/(A*t)$. Hydraulic conductivity (K) is the proportionality factor relating the flux density to the total potential gradient and the soil column length. Darcy's equation relates these factors for steady-state water flow within soils (i.e., $J = -K*(\Delta\Psi/L)$.

Hydraulic conductivity is normally taken to be a constant for a given water content within a given soil. When the soil is saturated, K is at a maximum value and is referred to as the saturated hydraulic conductivity (K_{sat}). When the water content within the soil is less than saturation (i.e., not saturated), K is referred to as the unsaturated hydraulic conductivity (K_{unsat}). Hydraulic conductivity usually decreases exponentially as the soil water content decreases from a saturated state to a near-dry state. When water content changes with soil depth or with time, transient conditions exist within the soil and Darcy's equation must be combined with an equation of continuity to describe accurately water movement within soils. However, for a conceptual discussion regarding the impacts of organic amendments on water flow, the basic concepts presented should be sufficient for an understanding of the interactions caused by waste additions to soils.

Fluid flow proceeds through the pore space or void areas within a soil and within the water films which coat soil particles. Coarse-textured soils with a large proportion of pore space in macropores tend to have higher rates of K_{sat} and lower rates of K_{unsat} than do fine-textured soils. Fine-textured soils with a large proportion of pore space in mesopores and micropores tend to have higher rates of K_{unsat} and lower rates of K_{sat} than coarse-textured soils.

The data concerning the effects of organic amendments on soil hydraulic conductivity are not nearly as well documented as the previously discussed soil properties. Hydraulic conductivity by its very nature is a highly variable soil property.[27] When organic amendments were added to soils, "extreme variations" have been noted in the resultant hydraulic conductivity measurements.[1] It has been noted that organic amendments may exhibit water repellency during wetting which could also influence water transmission into and within the soil.[11,12]

In a study on a coarse-textured sandy soil, K_{sat} increased with increased sludge application rates and was attributed to an increase in the total pore space.[12] Unsaturated hydraulic conductivity decreased with increased sludge application rates. It was suggested that higher K_{sat} would increase the intake of water during heavy rainstorms and that the lower K_{unsat} would decrease water losses due to evaporation. This observation was partially substantiated by a study that indicated evaporation was reduced by increased rates of manure application.[14]

In a laboratory study comparing raw and digested sludge amendments on a silt loam soil, K_{sat} values after 27 days of incubation for digested sludge-soil mixtures > raw sludge-soil mixtures > control.[18] The K_{sat} had rapidly decreased for all mixtures after 54 days, but the digested sludge-soil mixtures continued to have higher values of K_{sat}. Lower K_{sat} values for the raw sludge-soil mixtures were attributed to large amounts of mycelial growth found within the soil crust and throughout the soil mass. The K_{sat} values returned to levels near the control after 79 days.

In an 11-year study on a clay loam soil, annual manure applications resulted in increased values of K_{sat}, although the values exhibited considerable variability.[13] It was suggested that a one-time application of manure was not as effective as small annual or biennial applications in maintaining increased K_{sat} levels.

Although the limited data suggest that organic amendments increase K_{sat}, some studies have reported no change or decreases in K_{sat} as a result of organic amendments.[1] The K_{sat} of a silty clay loam soil was reported to have decreased by 25% the year after addition of 580 metric tons/ha of cattle manure.[15] After 2 years of manure applications, the K_{sat} values were reported to be extremely variable and ranged from 0.2 to 52 cm/h per unit hydraulic gradient. No significant changes in K_{sat} were determined for a clay loam soil that had been subjected to 12 annual applications of feedlot manure ranging up to 180 Mg/ha.[17] Weil and Kroontje found that a perched water table formed at a 20-cm depth within a clay loam soil during spring and early summer and was attributed to a water-repellent plowsole layer of partially decomposed poultry manure.[11]

The literature concerning the effects of organic amendments on hydraulic conductivity is limited and indicates high variability in the measurements that is partially attributed to the inherent variability associated with hydraulic conductivity quantification in field soils.[1,2,28] Because of these limitations, it has been difficult to conduct a quantitative analysis using regression equations to predict the effects of organic amendments on hydraulic conductivity.[1]

E. INFILTRATION

Infiltration is the movement of water through the soil surface and into the soil profile. The rate of infiltration into most soils is usually initially high and then exponentially decreases until some constant rate is obtained. Factors that may influence the infiltration rate include initial water content of the soil, the soil surface conditions including the macroporosity associated with the soil structure, the time from onset (potential gradient effects, surface sealing, and swelling effects), impeding subsurface layers, and the K_{sat} as related to steady-state infiltration values. Infiltration is also a highly variable soil property and may be influenced by organic amendments since the amended soil may have effects on soil surface seal formation. If organic amendments improve soil aggregation and porosity, then an indirect, favorable effect should be observed on infiltration. It was reported as early as 1937 that manure applications would increase infiltration more than a corn-corn-oats-clover rotation.[29]

Field infiltration was initially greater for the control than for areas amended with 900 metric/ha of sludge.[12] After the first hour of water application the infiltration rates were the same and the differences were attributed to the initial water repellency of the sludge. The water repellency of a partially decomposed plowsole layer of poultry manure was cited as the cause of very low spring infiltration rates (0.02 cm/h vs. 12.1 cm/h for the controls).[11] The water repellency of the manure was thought to have diminished by the following fall and much higher infiltration rates (57 cm/h) were attributed to increased earthworm activity in the manured soils.

In a field study in which digested sewage sludge was applied to two silt loam soils and a sandy loam soil, there were no statistically significant differences in infiltration from the sludge applications although infiltration was generally 1.6 to 3.7 times greater than the control areas which received no sludge.[8] The lack of statistical differences was attributed to the heterogeneity in the infiltration measurements. The results were in general agreement with a study that reported slight increases in infiltration rate from digested sewage sludge applications on a silt loam soil and a sandy loam soil.[23] A more recent study indicated that there were no significant changes in the infiltration rate for a clay loam soil after 12 annual applications of manure, although the variability associated with the infiltration measurements was large.[16]

Conditions favorable for infiltration (i.e., poor surface crust formation and a large amount of water-stable aggregates) were observed for heavily manured areas in a field study on a silty clay loam soil that had three annual applications of cattle manure.[30] The rates of water entry after 3 h of irrigation during the third year of the experiment were 0.9, 2.8, 2.6, and 3.5 cm/h for the 0, 90, 180, and 360 metric/ha application rates, respectively.

F. SOIL THERMAL PROPERTIES

There is limited information concerning the use of organic amendments and resultant changes in soil thermal properties. Average thermal conductivity values are 0.6, 7, and 21 mcal cm^{-1} sec^{-1} $°C^{-1}$ for organic matter, clay minerals, and quartz, respectively.[31] Average heat capacity values are 0.6 cal cm^{-3} $°C^{-1}$ for organic matter and 0.48 cal cm^{-3} $°C^{-1}$ for clay minerals and quartz. The addition of organic amendments would be expected to result in lower values of thermal conductivity and slightly higher values of specific heat. An important consideration would be the higher water-holding capacity associated with the incorporation of organic amendments which would be associated with higher values of specific heat.

In a 2-year field study in which sewage sludge was annually applied to a coarse-textured sandy soil, thermal conductivity decreased at any given water content as the rate of sludge application increased.[12] Although the specific heat of the sludge-soil and soil were roughly equivalent when dry, there was less temperature fluctuation in the sludge soil. This difference was attributed to the lower thermal conductivity and the higher specific heat of the sludge soil since on any given day the sludge soil, which had been amended with 900 metric tons/ha of sludge, had a higher water content than the control.

G. SOIL STRENGTH

Although a few studies have examined the effects of organic amendments on soil strength using a variety of methods, the results are not conclusive. As previously discussed, the incorporation of organic amendments promotes improved aggregation and aggregate stability and would, therefore, be expected to result in a more friable soil with decreased values of soil strength.

Unconfined compression strength of a silt loam soil decreased as application rates of shredded municipal refuse and sewage sludge increased.[9] In a different field study, the modulus of rupture decreased as manure application rate increased for a silty clay loam soil that had three annual manure applications.[15] The reduced modulus of rupture was attributed to less cohesion among the soil particles because of the increase in organic matter that resulted from the manure amendments. Addition of sewage sludge compost significantly reduced penetrometer resistance in a loamy sand soil with significantly lower resistances obtained with increasing rates of amendment.[26]

Unfortunately, the effects of adding organic amendments to soils have also produced some contrary results, indicating further study may be warranted. Soil strength determinations made on a clay loam soil equilibrated to –150 kPa water potential after four annual applications of feedlot manure indicated no significant differences due to the amendments, although up to 268 metric tons/ha of manure had been used.[14] No significant differences in soil penetration were determined for a silt loam soil that had been subjected to annual applications of 40 tons/acre of manure for more than 20 years.[7]

IV. EFFECTS OF ORGANIC AMENDMENTS ON SOIL CHEMICAL PROPERTIES

It was indicated in a review article in 1987 that "only scant" information was available in the literature on chemical and biological processes occurring in the sludge-soil system and that further research was necessary to clearly define the processes.[2] Unfortunately, the information that has been published in the interim does not add greatly to our knowledge base concerning the effects of organic amendments on soil chemical properties. Although it is still difficult to draw firm conclusions, it may be possible to make some generalizations concerning the soil cation exchange capacity and oxidation-reduction potential following the additions of organic amendments.

A. CATION EXCHANGE CAPACITY (CEC)

The addition of an organic amendment to soil is likely to increase the soil CEC simply from the additive effect of the organic matter and the high CEC associated with organic matter (1000 to 2000 mmol(–)/kg organic matter depending on soil pH). It was reported for a field study in which digested sewage sludge was applied to two silt loams and a sandy loam soil that CEC increases were attributed to the sludge organic matter additions.[8] The increases in organic C content and CEC of the sludge soil were highly correlated. The increase in CEC was attributed to the addition of the sludge which had a CEC of approximately 1250 mmol/kg of sludge and was based on the organic matter having a CEC of 2500 mmol/kg. No appreciable change in pH was attributed to sludge treatments and, therefore, pH changes were not considered to be partially responsible for the increased CEC following sludge addition.

It is also thought that the increases in CEC observed following organic amendment additions may be temporary and will decrease as the organic materials decompose. In a field study on a silt loam soil that had been amended with sewage sludge, the CEC increased nearly threefold for the highest sludge addition (240 metric tons/ha).[32] It was observed that the CEC did not change much during the first season after the initial increase. However, after 2 years, the CEC associated with the higher sludge application rates had decreased considerably while the CEC associated with the lower rates had decreased less. Approximately 1.5 years after application, the sludge-soil mixtures still had CEC values significantly greater than the control. Longer-term monitoring was not conducted to determine if the CEC would ultimately return to levels observed prior to the additions of the sludge amendment.

B. OXIDATION-REDUCTION POTENTIAL (Eh OR pe)

The addition of a large organic amendment would be expected to change the oxidation-reduction potential equilibrium even in a well-oxidized soil. In a field study using both sewage sludge and sludge compost amendments on a silt loam soil, the Eh decreased rapidly to 100 mV after addition of 240 metric tons/ha of sludge and reached –200 mV at 40 days. There was an unexplainable increase in Eh to a peak of 500 mV after 5 days following addition of 240 metric tons/ha of the sludge compost. Following this rapid increase, the Eh of the compost-soil mixture decreased more slowly than the sludge-soil mixture and reached a minimum value of 0 mV at 40 days. Organic materials in wastes act as electron donors in both chemical reactions and microbial respiration, so C-rich waste would be expected to make soils more electron rich and reducing, as illustrated by the above research.

There are also indications that, when organic amendments are added to a soil with high levels of Fe(III) and Mn(III,IV) oxides, the temporary reductions in Eh may lead to the solubility of the cations following reductive dissolution of the oxides, such that phytotoxic conditions for plant growth may result. Miller et al. reported on an incubation study in which swine manure amendments of 100 mt ha^{-1} were applied to three soils with varying levels of total Fe and Mn.[33] The manure additions caused immediate reductions in Eh ranging from 175 to 325 mV by day 4. The minimum Eh levels were considered to be related to the initial Fe and Mn contents of the three soils. The soil pH increased immediately upon the manure additions (original manure slurry pH = 7.8) from the mid 5s to the 7 to 8 range and continued to increase for the 29 days of the study. The increased pH was attributed to the initial slurry pH and the consumption of protons during the reduction processes within the soils. It was felt the heavy applications of the sludge may have resulted in phytotoxic levels of iron and especially manganese. It was recommended that it is important to aerate soils adequately that have been amended with high levels of organic wastes (possibly using frequent tillage) so as to encourage oxidizing conditions and help avoid potential phytotoxicity conditions.

V. CLOSING COMMENTS

The addition of organic C amendments to soils is likely to increase soil aggregation and stability, decrease bulk density, and increase water-holding capacity.[12,14] While it is generally thought that the addition of waste amendments to soil may result in improved soil hydraulic properties such as infiltration and hydraulic conductivity, the situation should be evaluated depending upon the nature of the soil-waste interactions.

The organic components applied in sewage sludge, manure, or other waste amendments are not considered to directly affect water transmission properties within soils with only a few exceptions such as the plugging of the soil surface layer following liquid sewage sludge applications.[2] The organic components are considered to indirectly affect water-transmission properties through effects on aggregation and porosity. It should be noted that some researchers have reported that the water-repellency nature of the sludge may have direct effects on water-transmission properties, although these effects have generally been reported to last from an hour to a few months.[11,12]

The effects of organic amendments on soil chemical properties are very complex and difficult to summarize for a general set of conditions. Increases in the soil's CEC should result simply from the addition of the organic matter within the amendment which has a characteristically high cation exchange capacity. The additions of large amounts of organic amendments are likely to create reducing conditions within soils that may persist for time intervals of several days to several weeks. The reducing conditions may also create potential phytotoxic conditions for plant growth because of increased solubilities of soil mineral oxides.

ACKNOWLEDGMENT

Contribution No. 9034 and Scientific Article No. A-7713 of the Maryland Agriculture Experiment Station and the Department of Agronomy, University of Maryland, College Park.

REFERENCES

1. Khaleel, R., Reddy, K.R., and Overcash, M.R., Changes in soil physical properties due to organic waste applications: a review, *J. Environ. Qual.*, 10(2), 133, 1981.
2. Metzger, L. and Yaron, B., Influence of sludge organic matter on soil physical properties, in *Advances in Soil Science*, Stewart, B.A., Ed., Springer-Verlag, New York, 1987, 141.
3. Catroux, G., L'Hermite, P., and Suess, E., Eds., *The Influence of Sewage Sludge Application on Physical and Biological Properties of Soils*, D. Reidel, Dordrecht, The Netherlands, 1983, 253.
4. MacRae, R.J. and Mehuys, G.R., The effect of green manuring on the physical properties of temperate-area soils, *Adv. Soil Sci.*, 3, 71, 1985.
5. Powers, W.L., Wallingford, G.W., and Murphy, L.S., Research Status on Effect of Land Application of Animal Wastes, EPA-660/2-75-010, USEPA, Washington, D.C., 1975.
6. Hall, J.E. and Coker, E.G., Some effects of sewage sludge on soil physical conditions and plant growth, in *The Influence of Sewage Sludge Application on Physical and Biological Properties of Soil*, Catroux, G., L'Hermite, P., and Suess, E., Eds., D. Reidel, Dordrecht, The Netherlands 1983, 43, 253.
7. Klute, A. and Jacob, W.C., Physical properties of sassafras silt loam as affected by long-time organic matter additions, *Soil Sci. Soc. Am. Proc.*, 14, 24, 1949.
8. Kladivko, E.J. and Nelson, D.W., Changes in soil properties from application of anaerobic sludge, *J. Water Pollut. Control Fed.*, 51(2), 325, 1979.
9. Mays, D.A., Terman, G.L., and Duggan, J.C., Municipal compost: effects on crop yields and soil properties, *J. Environ. Qual.*, 2(1), 89, 1973.
10. Webber, L.R., Incorporation of nonsegregated, noncomposted solid waste and soil physical properties, *J. Environ. Qual.*, 7(3), 397, 1978.
11. Weil, R.R. and Kroontje, W., Physical condition of a Davidson clay loam after five years of heavy poultry manure applications, *J. Environ. Qual.*, 8(3), 387, 1979.
12. Gupta, S.C., Dowdy, R.H., and Larson, W.E., Hydraulic and thermal properties of a sandy soil as influenced by incorporation of sewage sludge, *Soil Sci. Am. J.*, 41, 601, 1977.
13. Mathers, A.C. and Stewart, B.A., The effect of feedlot manure on soil physical and chemical properties, *Livestock Waste: A Renewable Resource*, 159.
14. Unger, P.W. and Stewart, B.A., Feedlot waste effects on soil conditions and water evaporation, *Soil Sci. Soc. Am. Proc.*, 38, 954, 1974.
15. Tiarks, A.E., Mazurak, A.P., and Chesnin, L., Physical and chemical properties of soil association with heavy applications of manure from cattle feedlots, *Soil Sci. Soc. Am. Proc.*, 38, 826, 1974.
16. Sommerfeldt, T.G. and Chang, C., Changes in soil properties under annual applications of feedlot manure and different tillage practices, *Soil Sci. Soc. Am. J.*, 49, 983, 1985.
17. Sommerfeldt, T.G. and Chang, C., Soil-water properties as affected by twelve annual applications of cattle feedlot manure, *Soil Sci. Soc. Am. J.*, 51, 7, 1987.
18. Epstein, E., Effect of sewage sludge on some soil physical properties, *J. Environ. Qual.*, 4(1), 139, 1975.
19. Kinsbursky, R.S., Levanon, D., and Yaron, B., Role of fungi in stabilizing aggregates of sewage sludge amended soils, *Soil Sci. Soc. Am. J.*, 53, 1086, 1989.

20. Kemper, W.D. and Koch, E.J., Aggregate stability of soils from western U.S. and Canada, *USDA Tech. Bull.*, 1355, 1966, 52.
21. Strickling, E., The effect of soybeans on volume, weight and water stability of soil aggregates, soil organic matter and crop yield, *Soil Sci. Soc. Am. Proc.*, 15, 30, 1950.
22. Bouyoucos, G.J., Effects of organic matter on the water-holding capacity and wilting point of mineral soils, *Soil Sci.*, 47, 377, 1939.
23. Kelling, K.A., Peterson, A.E., and Walsh, L.M., Effect of wastewater sludge on soil moisture relationships and surface runoff, *J. Water Pollut. Control Fed.*, 49, 1698, 1977.
24. Jamison, V.C., Changes in air-water relationships due to structural improvement of soils, *Soil Sci.*, 76, 143, 1953.
25. Jamison, V.C., Pertinent factors governing the availability of soil moisture to plants, *Soil Sci.*, 81, 459, 1956.
26. Tester, C.F., Organic amendment effects on physical and chemical properties of a sandy soil, *Soil Sci. Soc. Am. J.*, 54, 827, 1990.
27. Nielson, D.R., Biggar, J.W., and Erh, K.T., Spatial variability of field-measured soil-water properties, *Hilgardia*, 42, 215, 1973.
28. Chang, A.C., Page, A.L., and Varneke, J.E., Soil conditioning effects of municipal sludge compost, *J. Environ. Eng.*, 109, 574, 1983.
29. Smith, F.B., Brown, P.E., and Russell, J.A., The effect of organic matter on the infiltration capacity of clarion loam, *J. Am. Soc. Agron.*, 29(7), 521, 1937.
30. Mazurak, A.P., Chesnin, L., and Tiarks, A.E., Detachment of soil aggregates by simulated rainfall from heavily manured soils in eastern Nebraska, *Soil Sci. Soc. Am. Proc.*, 39, 732, 1975.
31. Devries, D.A., Thermal properties of soils, in *Physics of Plant Environment*, Van Wijk, R., Ed., Elsevier/North-Holland, Amsterdam, 1963.
32. Epstein, E., Taylor, J.M., and Chaney, R.L., Effects of sewage sludge and sludge compost applied to soil on some physical and chemical properties, *J. Environ. Qual.*, 5(4), 422, 1976.
33. Miller, W.P., Martens, D.C., and Zelazny, L.W., Effects of manure amendment on soil chemical properties and hydrous oxides, *Soil Sci. Soc. Am. J.*, 49, 856, 1985.

CHAPTER 9

Fly Ash

J. J. Bilski, A. K. Alva, and K. S. Sajwan

I. INTRODUCTION

The intensive use of coal in the U.S. for electrical power generation will increase the amount of coal residue that needs to be disposed of without causing any environmental problems. Because coal residue contains a variety of potentially hazardous substances, improper disposal and management could cause considerable environmental impacts. Ash residue is produced during the combustion of solid fuel. Depending on the design of the furnace, ashes can be carried with the flue gas, which is called fly ash (FA), or deposited underneath the boiler (bottom ash).[1] In coal-fired power plants, the characteristics of FA and bottom ash are similar. They are mixtures of inorganic constituents of the burned fuel and material that is not completely combusted. Physically, FA occurs as very fine spherical particles, having an average diameter of <10 μm. During an average coal combustion process, about 12% of the combusted fuel could become ashes.[2] The nature of coal ashes has made them excellent material for making Portland cement, mix-

0-87371-859-3/95/$0.00+$.50

ing concrete, and stabilizing soils. The annual production of coal combustion waste, as of 1983, was 75 million tons.[3] This production is expected to double by the year 2000.[4] Emission of sulfur dioxide to the atmosphere from coal-fired electric power plants is a major concern with regard to acid precipitation. This can be minimized by installing a desulfurization system which is commonly referred to as sulfur dioxide scrubbing unit. In this process, lime slurry is used to capture sulfur dioxide into a solid waste as gypsum, which is commonly referred to as flue-gas desulfurization (FGD) gypsum. The annual production of FGD gypsum, as of 1983, was 5 million tons,[3] which is expected to quadruple by the year 2000.[4]

According to the data released by the National Ash Association, on the national level, only 20% of ash produced is being utilized.[5] Therefore, a substantial amount is being accumulated in landfills and surface impoundments. The accumulation of coal combustion by-products (CCBP) in landfills raises environmental concerns because of potential leaching and contamination of groundwater aquifers by several constituents from these waste materials.

The concentrations of various mineral elements present in CCBP and the leachability of these various constituents have been investigated.[4] Based on an exhaustive literature search, the ranges in concentration of various elements as reported by Rai[4] are shown in Table 1. The data indicate wide variation in concentrations of most elements. This is in part due to variations in the coal used by the utility plants depending on the origin of the coal. In general, these CCBPs are high in Ca, Mg, Fe, K, Si, and S concentrations.

The objective of this review was to discuss effects of FA applications to soils. We have discussed the effects of FA applications on chemical and physical properties of the soil, and on microbial activity and resulting effects on plants grown on soils amended with FA. The emphasis of this chapter is to review the beneficial effects on crop growth and production, as well as potential adverse environmental impact of FA when applied to agricultural soils.

II. PRODUCTION CHEMISTRY OF FLY ASH

A characterization of flue gas emissions is important to the development of technology for reducing flue gas emissions and for minimizing health and environmental problems associated with flue gas release.

It is widely known that FA particles emitted from coal-fired plants contain several toxic trace metals.[6–9] The submicron particles, which show the greatest enrichment for many toxic trace elements, are of particular concern since they are not efficiently removed by modern particle collection devices. Furthermore, a major fraction of toxic trace elements is concentrated in the particle size, range 0.5 to 10 μm, which is inhaled and deposited in the human body. According to the data released by the U.S. Department of Health,[10] smaller particles of size fraction 0.5 to 1.0 μm are deposited primarily in the lungs, while larger particles of >1.0 μm are deposited primarily in the nasal and bronchial regions of the human respiratory system.[11]

Klein et al.[11] analyzed the concentrations of a number of elements in coal, FA, slag, and combustion gases from a large, cyclone-fed power plant. Mass balance calculations showed that the sampling and analyses done by the investigators were generally adequate to describe the flow of 37 trace elements through the power plant. Most Hg, Cl, Br, and some Se were discharged to the atmosphere as gases. As, Cd, Cu, Ga, Pb, Sb,

Table 1 Concentrations of Elements in Utility Wastes, Coal, and Soils[4]

	Element Concentrations in Different Materials					
Element (units)	Fly Ash	Bottom Ash	FGD Sludge	Oil Ash	Coal	Soils
Aluminum (wt%)	0.1–20.85	3.05–18.5	0.64–9.69	0.001–8.73	0.43–3.04	1.0–30
Antimony (µg/g)	0.8–131	<10	15.5	3–1,072	0.2–14	0.6–10
Arsenic (µg/g)	2.3–6,300	0.02–168	0.8–53.1	2.5–10,000	0.5–106	1–93
Barium (µg/g)	1–13,800	109–9,360	<25–2,280	148–1,000	150–250	70–3,000
Boron (µg/g)	10–5,000	1.5–513	42–530	0.5–600	1.2–356	2–150
Cadmium (µg/g)	0.1–130	<10	0.06–25	<11	0–6.5	0.01–0.7
Calcium (wt%)	0.11–22.30	0.22–24.10	0–34.50	0.01–33	0.5–2.67	0.7–50
Chlorine (µg/g)	13–1,720	<100–2,630	<150–8,970	200–10,000	0–5,600	20–900
Chromium (µg/g)	3.6–900	<0.2–5,820	1.6–180	10–4,390	0–610	1–1,000
Copper (µg/g)	14–2,200	3.7–932	6–340	10–130,000	1.8–185	2–300
Fluorine (µg/g)	0.4–610	2.5–104	266–1,017	1–5	10–295	10–4,000
Iron (wt%)	1.0–27.56	0.4–20.10	0.13–13.80	0.01–52.10	0.32–4.32	0.7–55.0
Lead (µg/g)	3–2,120	0.4–1,082	0.25–290	10–100,000	4–218	2–200
Magnesium (wt%)	0.04–7.72	0.2–4.8	1.99	0.06–45	0.1–0.25	0.06–0.6
Manganese (µg/g)	25–3,000	56–1,940	37.3–312	113–1,170	6–181	20–3,000
Mercury (µg/g)	0.005–12	0.005–4.2	0.005–6	<1	0.01–1.6	0.01–0.3
Molybdenum (µg/g)	1.2–236	0.84–443	<4.0–52.6	<20–779	0–73	0.2–5
Nickel (µg/g)	1.8–4,300	<10–2,939	<5–145	100–180,000	0.4–104	5–500
Nitrogen (µg/g)	250–3,300	No data	No data	No data	12,000	200–4,000
Potassium (wt%)	0.17–6.72	0.26–3.3	0.001–1.70	0.04–3.65	0.02–0.43	0.04–3.5
Selenium (µg/g)	0.2–134	0.08–14	<2–162	7.6–500	0.4–8	0.1–4
Silicon (wt%)	1.02–31.78	5.10–31.20	0.27–17.70	0.05–17.00	0.58–6.09	23–35
Silver (µg/g)	0.04–36	<9.9	<8.3	<9.9	0.04–0.08	0.01–5
Sodium (wt%)	0.01–7.10	0.08–4.13	0.02–5.53	0.1–46.5	0–0.2	0.075–1.0
Strontium (µg/g)	30–7,600	170–6,440	70.8–2,990	51.7–920	100–150	7–1,000
Sulfur (wt%)	0.04–6.44	<0.04–7.40	0.08–22.80	1.27–17.3	0.38–5.32	0.003–1.0
Vanadium (µg/g)	12–1,180	12–537	<50–261	10–460,000	0–1,281	0.7–500
Zinc (µg/g)	14–3,500	3.8–1,796	7.7–612	40–100,000	0–5,600	10–300

Se, and Zn were quite concentrated in FA compared to the slag, and were more concentrated in the ash discharged through the stack than in that collected by the precipitator. Al, Ba, Ca, Ce, Co, Eu, Fe, Hf, K, La, Mg, Mn, Rb, Sc, Si, Sr, Ta, Th, and Ti showed little preferential partitioning between the slag and the FA, with a slight tendency for a higher accumulation in slag than in FA. The partitioning of such elements as Cr, Na, Ni, and V was similar between the slag and FA.

Campbell et al.[12] conducted a detailed investigation to determine the amount of various elements in FA. They separated FA particles into various size fractions and determined the concentrations of 43 elements by X-ray fluorescence (XRF), atomic absorption (AA), and instrumental neutron activation (INAA). There was generally a good agreement between the three analytical methods with respect to concentrations of elements. Based on the concentration profiles, the elements were divided into distinct groups. The elements examined appeared to show three distinct classes of partitioning behavior. Class I: twenty elements — Al, Ba, Ca, Ce, Co, Eu, Fe, Hf, K, La, Mg, Mn, P, Rb, Sc, Si, Sm, Sr, Th, and Ti — are readily incorporated into the slag. These elements are partitioned about equally between the inlet FA and slag, with no apparent tendency to concentrate in the outlet FA. The Class I elements remain in the condensed state and are not volatilized in the combustion zone, but instead form a melt of uniform composition that becomes both FA and slag.

Class II: elements such as As, Cd, Cu, Ga, Pb, Sb, Se, and Zn are poorly incorporated into slag and are concentrated in FA as these elements are volatilized during combustion.

Class III: Hg, Cl, and Br remain completely in the gas phase. The other elements, examined by Campbell et al.,[12] failed to exhibit definite trend of partitioning. Their characteristics were intermediate between those of the elements listed under classes I and II.

The concentration of volatile elements is usually expected to decrease with increasing particle size because of the larger surface area-to-volume ratio of the smaller particles.

The area of greatest environmental concern is reduction of FA amounts during the combustion process. One of the suggestions[13] was to use more widely fluidized-bed combustion systems instead of conventional systems. Fluidized-bed combustion systems operate at 1650°F compared to 2700°F required for the conventional systems. In comparison with conventional coal-fired systems, fluidized-bed combustion seems to offer significant potential for reducing trace element emissions. However, data are lacking in several areas, particularly with regard to the emission by fluidized-bed combustion system of elements such as As, Be, Cl, Cr, Pb, Hg, Se, U, and V. Nevertheless, experimental data indicate that particulate loading of FA might be overwhelmingly decreased due to the usage of conventional control devices such as cyclones, electrostatic precipitators, or fabric filters.

Recent studies done by Menon et al.[14] showed that the concentrations of water-extractable transition metals such as Zn, Mn, Cd, Cu, and Ni as well as the B and S content of FA collected from a coal-fired power plant equipped with an electrostatic precipitator were lower than those of respective elements in water extracts of other ashes from power plants not equipped with electrostatic precipitators.

Some research has been done to determine neutralization capacity of FA materials.[6,15–17] The neutralization capacity refers to "titratable alkalinity" and is expressed in equivalents of hydronium ion (H_3O^+) consumed per unit weight of FA. The most common technique to determine neutralization capacity is by repeated acid titration, either for long periods of time or at elevated temperatures. The alkalinity of FA has been attributed to soluble compounds of Ca, Na, and possibly Mg.[9] Theis and Wirth[17] examined several factors associated with FA that were thought to be responsible for the degree of acidity or alkalinity of a FA suspension. They stated that the properties that appeared to be most responsible were oxalate-extractable (amorphous) Fe and water-soluble Ca. The Fe was the acid-forming component, while Ca was the base-forming component. They found that 3:1 ratio of Fe:Ca was a rough delineation of the ultimate acidic or basic nature of the ash.

The neutralization of acid by FA is a relatively slow process that mainly involves the particle surfaces.[6] Electron micrographs have shown that FA is comprised of spherical glass-like particles with Ca inclusions concentrated within the surface of the spheres. Elseewi et al.[18] showed that solution chemistry of FA is largely dependent on the dissolution pattern of CaO from an aluminosilicate matrix.

According to Hodgson et al.,[6] neutralization capacity was negatively correlated with concentrations of Si and Fe and positively correlated with concentrations of Ca and Mg. The ratio of Ca:Fe was more highly correlated to neutralization capacity than the concentration of either Ca or Fe. For the three sources of FA used in the study, neutralization capacity was predicted accurately by using concentrations of Ca and Mg within the pH range of 5 to 8. The ratios mentioned above were applicable to the normal soil pH range. However, the high correlation between Ca:Fe ratio and neutralization capacity of FA has been obtained using high Ca containing FA. The applicability of the above correlation to a wide range of FA sources could be questionable.

Due to the availability of large quantities of FA and the presence of high concentrations of Ca and Mg in most FA sources, FA appears to be a suitable soil amendment for liming purposes and to enhance Ca and Mg contents in the soil.[4,6,19] However, it is important to note that, above pH 9 or below 4, these materials do release high levels of soluble Al. Considerable caution should be exercised in determining the rate of application of FA as a liming material.

The differences in FA chemical composition have been investigated by Menon et al.[20] The pH, electrical conductivity, and concentration of selected elements were measured in the water extracts of five coal FA samples collected from Savannah River Site (SRS) and one from South Carolina Electric and Gas (SCE&G) plant. This work was intended to study the differences in the physicochemical properties of SRS-FA samples relative to those of a reference sample, i.e., SCE&G-FA, and to make FA-amended composts for agricultural use. Similar analyses were also performed in water extracts of a commercial organic manure "Gotta Grow", that was composted with one of the FA samples (SRS) in different proportions. All FA samples were equilibrated with deionized distilled water for 5 days (10 g of FA with 100 mL of water). The concentrations of ten metals (K, Na, Ca, Mg, Zn, Mn, Cu, Cd, Mo, and Ni) were determined in the filtrate by flame-AA spectroscopy. The results showed that FA samples used in this study differed considerably in pH, electrical conductivity, and elemental composition and that transition metals appear to bind more tightly on smaller particles than on larger ones. In general, the fraction with particle size greater than 500 μm in diameter yielded greatest concentrations of these elements in water extract. Results of the studies on the equilibration of FA samples with water show that the pH, conductivity, and the concentrations of Ca and K increased with extraction time and reached a plateau after 4 to 5 days. Accordingly, for the analysis of water-soluble cations and anions in FA, the samples should be equilibrated with water for at least 5 days.

The concentrations of water-extractable transition metals such as Zn, Mn, Cd, Cu, and Ni, as well as the B and S content of FA collected from SRS power plant, equipped with an electrostatic precipitator, were smaller than the concentrations of respective elements in water extracts of the FA from the SCE&G plant (without electrostatic precipitator).

After the studies of physicochemical properties of water extracts of FA-amended composts, authors concluded that commercial organic manures such as Gotta Grow are too rich in several mineral elements to be mixed with FA. Authors suggested that low-grade or home-made organic composts should be investigated as a possible choice for making FA-amended composts.

In summary, the composition of FA varies considerably depending on the origin of coal used for power generation. In general, FA is rich in Ca and Mg, therefore, can be used as a soil amendment for liming purpose and to enhance the availability of Ca and Mg. The distribution of trace elements is dependent on the particle size. The concentrations of trace elements are greater in smaller particle-size materials.

III. EFFECTS OF FLY ASH (FA) ON SOIL CHEMICAL PROPERTIES AND PLANT GROWTH

Adriano et al.[21] studied the effects of coal ash on soil chemical properties, and growth and mineral nutrition of corn (*Zea mays* L.) and bush bean (*Phaseolus vulgaris* L.) seedlings in a greenhouse trial. The ash used was a mixture of FA and bottom ash. Troup sandy loam (Grossarenic Paleudults) soil was used in the study. The ash was mixed with soil to give 5, 10, and 20% ash by weight. Soil analysis was performed after the termination of the experiment (i.e., following 5 weeks of growth). The soil was extracted by 1 M NH_4OAc or 0.1 M HCl and concentrations of cations and micronutrients were determined. Application of FA increased the concentrations of extractable K, Ca, Mg, Cu, Fe, Mn, and Zn in the soil. An increase in rate of application of FA increased the concentrations of these elements in the soil.

The study showed no beneficial effects of application of FA on the growth of corn or bush bean seedlings regardless of rates of application. The application of FA increased the concentrations of K and Fe in corn shoot and those of K, Ca, Mg, S, and Cu in bush bean shoot. The growth of seedlings of both of these species improved markedly in soil fertilized with N, P, and K without the addition of FA. The bean seedlings exhibited characteristic B toxicity symptoms, while the corn seedlings showed P-deficiency symptoms in FA-amended soils. Concentrations of B in the plant tissue or in the soil were not measured in the above experiment. However, B uptake by the bean seedlings must have increased in FA-amended soil as evident from the toxicity symptoms. Boron in FA is sufficiently soluble. Indeed, at low rates of application, FA can be an effective source of B for crop plants grown in soils deficient in B.[22] Mulford and Martens[23] showed that the availability of B from FA is very similar to that from $Na_2B_4O_7 \cdot 10\ H_2O$ when these amendments were added at rates to supply a given quantity of B. The availability of B is much greater from unweathered acidic FA as compared to that from weathered and leached FA.

Mulford and Martens[23] conducted a greenhouse experiment to investigate the response of alfalfa (*Medicago sativa* L.) to application of FA. Total B content measured by carmine procedure[24] in 17 FA samples used in the study varied from 236 to 618 ppm.

The uptake of B by alfalfa seedlings increased proportionately with an increase in rate of FA application. In most cases, an increase in B content was associated with a decrease in yield. Such adverse effects of B on crop yields were also reported in several studies.[25–27] Although high rates of application of FA are not desirable, it appears that FA could be an effective source of plant-available B if applied at moderate rates since the B in FA is easily available to plants.

Aitken and Bell[28] tested the uptake of B by French bean (*P. vulgaris* L.) and rhodes grass (*Chloris gayana* L.) and its phytotoxicity in an Australian FA. The ashes were untreated, leached, or adjusted to pH 6.5 and subsequently leached. The ash was mixed (0, 15, 30, 70, or 100% by weight) with a sandy loam soil. Although the available water capacity of the soil was substantially increased by FA addition, incorporation of large proportions of untreated FA resulted in poor plant growth primarily due to B toxicity. When pH of the FA was adjusted to 6.5 and subsequently leached prior to addition to the soil, plants did not show B toxicity. There were great differences in B toxicity symptoms between the two plant species tested. Rhodes grass appeared to tolerate higher B content in the growing medium than the French bean. The uptake of B was lower by the former as compared to the latter crop species.

Menon et al.[14] showed that the maximum dry matter yields of corn (*Z. mays* L.) and sorghum (*Sorghum vulgare* L.) corresponded to the increased levels of K, Ca, and N and decreased levels of B when plants were grown in 20 to 40% FA-amended compost. More recently, Menon et al.[29] investigated the effects of amending the soil with composted grass clippings or a mixture of composted grass clippings and 20% FA vs. unamended soil on the growth of mustard (*Brassica juncea* L.), collards (*B. oleracea* L.), string beans (*P. vulgaris* L.), bell pepper (*Capsicum frutescents* L.), and eggplant (*Solanum melongena* L.). The FA-amended compost was found to be effective in enhancing the dry matter yield of collard and mustard by 378 and 348%, respectively, but string beans, bell pepper, and eggplant did not show any significant increase in dry matter yield. The above-ground biomass of the latter three plant species showed toxic levels of B, which was attributed for the poor plant growth.

Scanlon and Duggan[30] studied the growth and uptake of various elements of eight woody plant species which included European black alder (*Alnus glutinosa* L.), sweet birch (*Betula lenta* L.), sycamore (*Platanus occidentalis* L.), sawtooth oak (*Quercus acutissima* L.), cherry olive (*Elaeagnus multiflora ovata* L.), autumn olive (*E. umbellata* L.), silky dogwood (*Cornus amomum* L.), and gray dogwood (*C. racemosa* L.) grown on slopes of FA mounds in eastern Tennessee. The plantings were done either without soil cover or with 10-cm cover of gravelly clay subsoil.

At the end of three growing seasons, the percentage of survival varied from 9 for sweet birch to 91 for cherry olive. The soil cover had very little affect on the survival of various tree species. Both the olive species had survival percentage ≥84%. These species showed no toxicity symptoms or imbalance in elemental concentrations. Compared to the subsoil used as soil cover, FA contained much greater levels of As, Cr, Cu, Ni, and Pb. However, only B content of the foliage of tree species was greater in the case of trees grown on FA without the soil cover as compared to that with soil cover. Boron concentrations in plants are shown in Figure 1. This clearly illustrates the availability of B from FA; therefore, if applied at moderate rates, FA could be a source of plant available B.

Plank and Martens[31] investigated the effects of FA (Fort Martin and Big Sandy) application to Tatum silt loam (clayey, mixed, thermic Typic Hapludult; pH = 5.5) on the

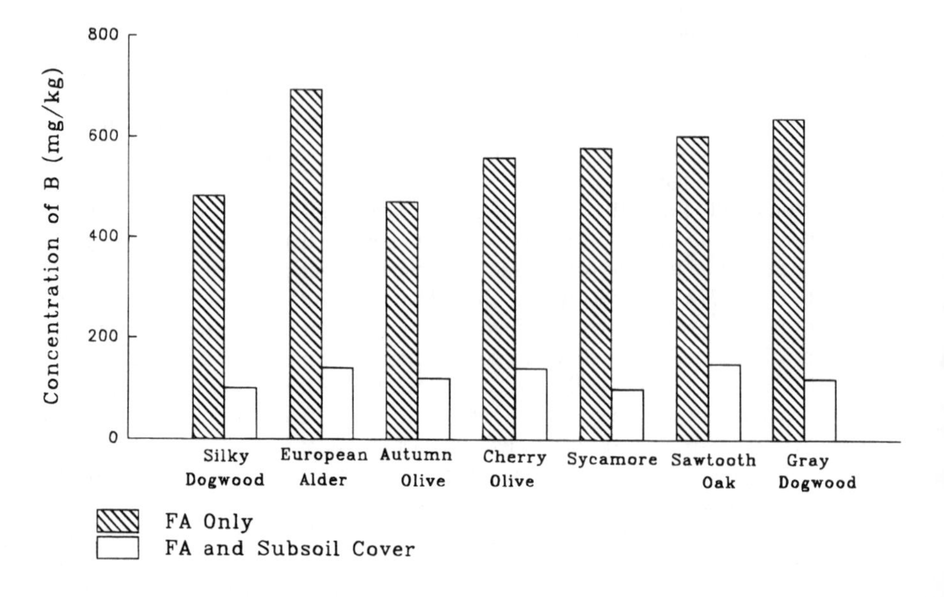

Figure 1 Concentration of B (mg/kg on dry-weight basis) in the foliage of woody plants grown on FA without and with soil cover (second season crop).[30]

availability of B to alfalfa (*Medicago sativa* L. cv. Williamsburg). Total B content of the FA sources (as measured by the procedure described by Hatcher and Wilcox[24]) varied from 232 to 370 ppm. FA was applied at rates equivalent to 1.7 and 3.4 kg B/ha based on total B content. The yield of alfalfa over 3 years increased significantly in FA-amended soil as compared to that in unamended soil due to increased availability of B. The magnitude of yield increase in FA-amended soil was similar to that in $Na_2B_4O_7 \cdot 10\ H_2O$ amended soil at equivalent rate of B. Therefore, the study demonstrated that the availability of B from FA was similar to that from $Na_2B_4O_7 \cdot 10\ H_2O$.

Studies have also been conducted to examine the Mo availability from a number of FA sources. Doran and Martens[32] determined the concentration of Mo (by the method as described by Reisenauer[33]) in a number of FA sources from 15 power plants from Virginia, Kentucky, Ohio, Tennessee, Alabama, Illinois, Montana, Minnesota, and West Virginia which varied from 5.6 to 39.3 ppm. The neutralizing capacity of these FA samples varied from 0.01 to 3.74 meq H_3O^+/g. Doran and Martens[32] further investigated the availability of Mo from two FA sources containing 5.6 and 39.3 ppm total Mo to alfalfa (*M. sativa* L. cv. Williamsburg) using an acidic Groseclose silt loam (clayey, mixed, mesic, Typic Hapludult; pH 5.2). The two sources of FA and reagent grade $Na_2MoO_4 \cdot 2\ H_2O$ were applied at rates equivalent to 0.12, 0.25, and 0.50 mg Mo/kg soil. Initial top growth of 88 days was clipped and discarded as this period was considered as equilibration period. Top growth was further clipped on 46 and 94 days after the equilibration period. The top yield as well as Mo uptake increased with an increase in rate of FA application. The beneficial effects of FA application on alfalfa growth were partly attributed to an increase in soil pH. At the highest rate of FA application, soil pH increased from 5.2 to 5.6.

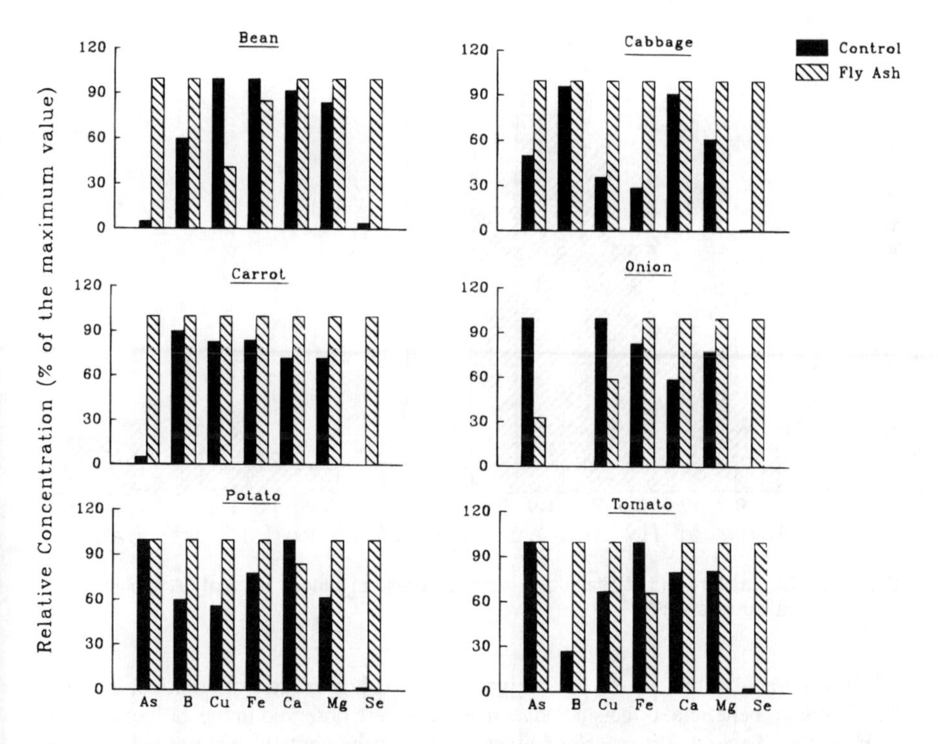

Figure 2 The concentration of some elements in edible portions of crops grown in pots on soil without or with 10% (weight basis) FA amendment (in percent of the highest value).[34]

Cary et al.[34] examined the uptake of several elements by vegetables and millet grown on FA-amended soil. FA used was from the Milliken station located in Lansing, NY. Soil used was an Arkport fine sandy loam (pH = 5.5; cation exchange capacity = 13.5 $cmol_c$ kg^{-1}). The test crops included bush bean (*Phaseolus vulgaris* L.), cabbage (*Brassica oleracea* L.), carrot (*Daucus carota* L.), Japanese millet (*Echinochloa crusgalli* L.), sweet Spanish onion (*Allium cepa* L.), potato (*Solanum tuberosum* L.), and tomato (*Lycopersicon esculentum* L.). FA was applied at a 10% rate on weight basis. As compared to the concentration of elements in the soil, the FA had greater concentrations of As, Ca, Cr, Cu, Fe, and Mg. Application of FA increased the concentration of B, Fe, Mo, and Ca in most of the crop species evaluated in this study.

The increase of elemental concentration as a result of FA addition is not consistent among all crop species as indicated in Figure 2. However, the concentrations of Se and B are consistently greater in almost all crop species grown in FA-amended soil as compared to the unamended soil.

Elseewi et al.[35] studied the availability of S from FA (Mojave Power Station, Bullhead City, Nevada) to alfalfa (*M. sativa* L.) and bermuda grass (*Cynodon dactylon* L.) in a pot experiment. FA was mixed with soil at 0.25, 0.5, 1.0, 2.0, 4.0, and 8.0% on a dry-weight basis. Dry matter yields of both crop species increased with increasing addition of FA up to 4% of the soil as shown in Figure 3. Most yield increase was obtained in

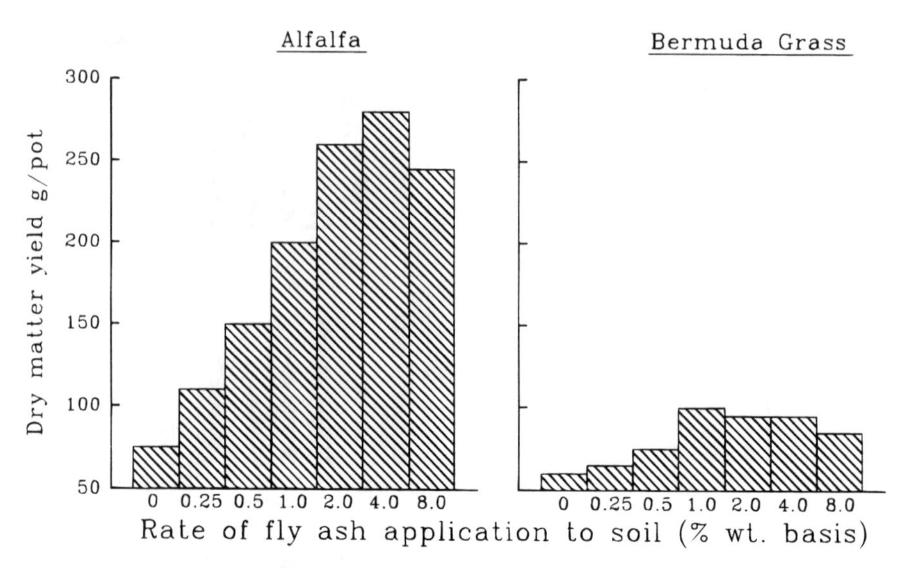

Figure 3 Dry matter yield of alfalfa and bermuda grass at various rates of FA additions to a calcareous soil.[35]

soil incorporated with 1 to 2% FA (weight basis). At higher rates of addition, although there were no beneficial effects no adverse effects were observed in the case of both the crop species. Favorable response of plants to FA amendments was attributed to an increase in S availability. The concentrations of S in plants grown in soil without FA addition were 0.057 and 0.036% for alfalfa and bermuda grass, respectively. In soils amended with FA at 8% on a dry-weight basis, S concentration increased to 0.23% in alfalfa and to 0.52% in bermuda grass.

In a parallel experiment, Elseewi et al.[35] compared the response of turnip (*B. rapa* L.) and white clover (*Trifolium repens* L.) to varying rates of S addition at 25, 50, and 100 mg/kg either as gypsum or as FA. Dry matter yield of turnip and white clover increased significantly when 25 mg/kg S was applied either as gypsum or as FA. Further increase in FA addition had marginal effects on the dry matter yield of both the crop species. On the basis of S uptake, it appeared that S availability from FA was very similar to that from gypsum.

Elseewi et al.[18] investigated the long-term effects of FA amendment on soil and plant composition as well as plant growth. FA from a western U.S. coal source was added to a calcareous soil and an acid soil at rates ranging up to and including 8% by weight. The treated soils were cropped using a native desert plant species followed by barley (*Hordeum vulgare* L.). The experiment was conducted in a greenhouse. FA application increased pH and EC of soils and also the contents of Ca, Mg, Na, B, and SO_4 in soil. Boron and SO_4 showed relatively low initial release which increased substantially during cropping. The addition of FA to soil improved dry weight of shoots of the desert species and the yield of barley. The results of plant species showed positive response with regard to concentrations of S and Mn as a result of FA amendment to soil. The response with regard to concentration of Ca, B, and Na varied among soils and plants. The availability of P, Zn, Fe, and Mn to plants appeared to be decreased in FA amended soils.

Wong and Wong[36] studied the effects of FA (China Light and Power Company, Castle Peak Station, Hong Kong) incorporation to soil on germination and seedling growth of *B. parachinensis* L. or *B. chinensis* L. FA was applied at 3, 6, 12, and 30% on air-dry weight basis of the soil. FA used in this study was high in most trace metal contents (including Cd, Co, Cu, Fe, Mn, Mo, Ni, and Pb) as compared to those in the soil sample. The germination percentage increased in soil amended with 3% FA as compared to that in unamended soil. However, at 12 and 30% FA amendment, there was a decrease in percent germination of both the test crop species. The dry weight of seedlings (12 days old) did not show any positive response to FA application. On the contrary, at 12 and 30% FA amendment, there was a significant reduction in dry matter production. The toxic effects of high rates of FA application (>12% on weight basis) to plant growth was associated with a decrease in microbial activity as indicated by inhibition of microbial respiration.[37] In the studies conducted by the same authors,[38] the application of FA at rates up to 12% to the soil resulted in greater uptake of Mn and Mo by two vegetable crops (*B. parachinensis* L. and *B. chinensis* L.) as compared to that in unamended soil. Application of FA had no consistent effect on the uptake of Cu and Ni. An increase in rate of FA application decreased the concentrations of Fe and Zn. This was attributed to an increase in soil pH as a result of FA amendment.

Fail and Wochok[39] investigated the effects of FA (Wilsonville Power Plant, Alabama) on growth of soybean (*Glycine max* L. cv. Bragg). FA was applied at 70 tons/ha on a sandy clay loam soil which had a pH of 4.4 to 5.0. Plants in FA-treated plots showed superior growth rate as compared to those in the control plots. Percentage of ground area covered by the plants accounted for 35 and 10% in FA-treated and control plots, respectively. Total dry weight per plant grown on FA-treated soil was nine-fold greater than that on untreated soil. Number of nodules per plant and the number of seed pods per plant increased by 12- and 6-fold, respectively, in FA-treated compared to untreated soil. The superior growth of soybean plants in FA-treated soil was thought to be due to an increase in soil moisture-holding capacity and an improvement in soil texture.

Pawar and Dubey[40] studied the effects of FA (Birla Industrial Complex, Nagda, India) amended to soil at 5, 10, 15, 20, 30, 40, and 100% by weight, on germination of sorghum (*Sorghum vulgare* L. cv. Vidisha), maize (*Z. mays* L. cv. Sathi and Ganga-5), wheat (*Triticum aestivum* L. sp. *vulgare* cv. N-4), and gram (*Cicer ariectinum* L. cv. H-355). Incorporation of FA up to 10% of soil weight had a stimulating effect on germination of all the crops tested as compared to that in soil without FA amendment. In the case of maize (cv. Sathi), maximum germination was observed at 15 to 20% incorporation of FA.

Martens et al.[41] conducted a greenhouse trial to test the availability of K to corn (*Z. mays* L.) plants from nine FA sources. The concentration of total K (measured by the procedure as described by Kanchiro and Sherman)[42] in the FA samples varied from 0.65 to 3.19%. The FA sources were applied to provide 75 and 113 mg K/kg soil. Reagent-grade KCl was applied at 38, 75, and 113 mg K/kg soil. The dry matter yield following 21 days of growth was significantly greater in soil treated with KCl at any K rate. Application of FA at either rate significantly increased the dry matter yield compared to that in unamended soil. Similar trend of response was also found with respect to K uptake among various treatments.

Adriano et al.[43] investigated the effects of FA (Mojave Power Generation Plant, southern Nevada) amendment on Redding (Abruptic; Durixeralf; pH = 4.8) and Hanford (Typic Xerorthent; pH = 7.1) soils at 2.5 and 5% by weight basis on growth and uptake of elements by sudan grass (*S. vulgaris* L. cv. Trudan). In the acid soil (Redding), sudan grass biomass was greater at 2.5% FA amendment than the control. Biomass production decreased when FA rate was increased to 5%. In the neutral soil (Hanford), however, biomass production decreased at all rates of FA application compared to that in unamended soil. The seedlings were clipped twice at 6-week intervals. Then the pots were leached with distilled water until the electrical conductivity (EC) of the leachate was reduced substantially. The leaching was carried out to remove excessive salinity as a result of FA amendment. The leached soil was replanted and two additional clippings were taken at 6-week intervals. The biomass production of sudan grass was much greater in the second crop following leaching. This was attributed to substantial reduction in salinity by leaching the FA-amended soil. Application of FA at 5% of soil weight increased the pH from 4.8 to 8.0 and from 7.1 to 9.1 in the acidic and neutral soils, respectively. The increase in pH due to FA application resulted in a reduction in trace element concentrations which was also reported by Phung et al.[44]

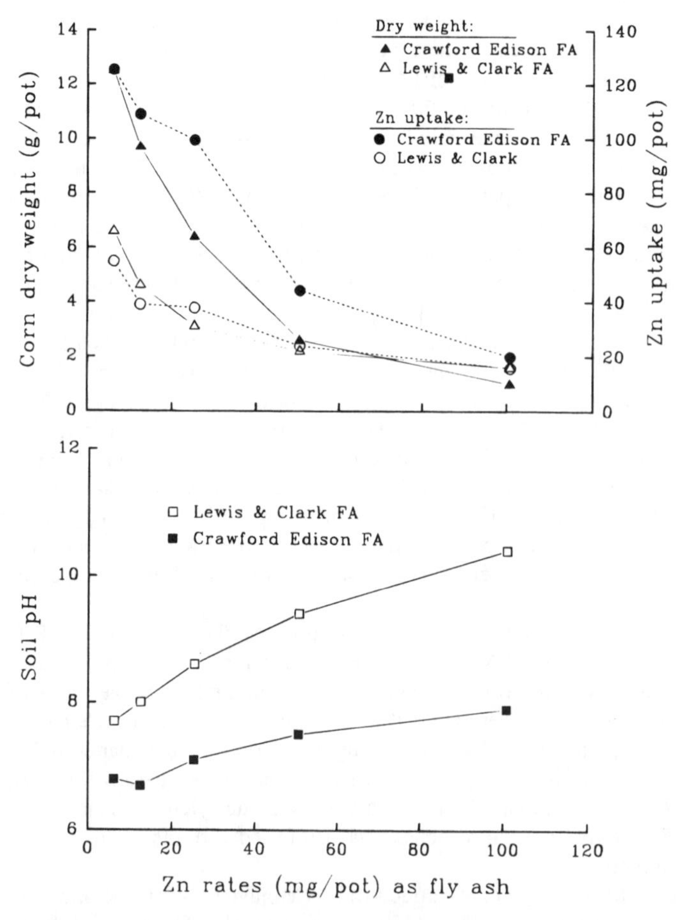

Figure 4 Effects of the application of Zn in the form of FA on the growth and Zn uptake by corn seedlings and the pH of growth medium.[45]

Schnappinger et al.[45] investigated the availability of Zn from two FA sources (Crawford Edison, Chicago, and Lewis and Clark Plant, Sidney, Montana) to corn (Z. *mays* L.) using Fredrick silt loam soil (clayey, kaolinitic, mesic, Typic Paleudult). Total Zn concentrations (following the procedure as described by Kanchiro and Sherman[42]) in the above two sources of FA were 363 and 132 ppm, respectively. FA was applied at a rate equivalent to 6.3, 12.6, 25.2, 50.4, and 100.8 mg Zn/kg soil. To compare the availability of Zn from FA with that from reagent-grade Zn source, $ZnSO_4{\cdot}7\ H_2O$ was applied at 3 and 6 mg Zn/kg soil. Application of Zn as $ZnSO_4{\cdot}7\ H_2O$ increased the dry matter production of corn seedlings as well as Zn uptake as compared to that in unamended soil. However, both dry weight and Zn uptake decreased with increasing rate of Zn application using either source of FA (Figure 4). At high rates of FA additions, soil pH increased significantly which may have been partly responsible for reduction in Zn uptake. At the lower rates of FA application, changes in soil pH were not substantial. However, Zn uptake as well as dry matter production decreased at the lower rates when compared to those in unamended soil. This suggests that certain other factors associated with FA application were responsible for the adverse effects. Although there were no data on B in plant tissue, considering readily available nature of B from FA sources,[23,31] B toxicity was suspected as the cause of decrease in dry matter production as well as Zn uptake associated with FA amendment.

Plank et al.[46] studied the effects of application of FA from Glenn Lynn Power Plant, VA, at 18, 36, 72, 108, and 144 tons/ha to Groseclose silt loam (clayey, mixed, mesic, Typic Hapludult) and Woodstown loamy fine sand (fine-loamy, siliceous, mesic, Aquic Hapludult) on yield of corn (Z. *mays* L. cv. Pioneer 3369-A). FA used in this study had a pH of 7.8 and its application at 144 tons/ha increased the soil pH from 5.8 to 6.5 and from 5.5 to 5.9, respectively, in Groseclose and Woodstown soils. Application of 18 to 144 tons/ha FA to Groseclose soil did not have a significant effect on corn yield. These applications were repeated during the following year which resulted in a cumulative rate as high as 288 tons/ha. However, neither beneficial nor adverse effects were evident at any of the FA application rates on Groseclose soil, whereas, on Woodstown soil, yields increased significantly during the second year in treatments that received a cumulative application rate of 216 and 244 tons/ha than those at the lower rates of FA application. This increase in yield was attributed to an increase in water-holding capacity of the soils at high rates of FA application.

A study by Wright[47] showed that application of FA at 144 tons/ha increased the plant available moisture in the 0- to 8-cm soil. In general, the application of FA had very little affect on the concentration of Mg, P, K, Cu, and Zn in plant tissue. The concentration of Ca increased significantly at the high rate of FA only in one of the tested soils. Increasing application of FA increased the concentration of B in plant tissue, while concentration of Mn decreased with an increase in rate of FA application. Although B concentrations were increased with increasing rate of application of FA, these concentrations never approached the toxic levels. The source of FA used in this study was weathered for 3 years; hence, a considerable portion of soluble B was removed in the leachate. Soil solution composition was measured at 15-cm depth intervals down to 60-cm depth. Annual application of 144 tons/ha FA increased the concentration of B in soil solution at all sampling depths.

Although Se in large quantities is considered as a toxic element, it is also a very important micronutrient. Many crops are deficient in Se. To study the possibility to increase Se uptake by crops, Shane et al.[48] investigated the uptake of Se by vegetables (broccoli,

B. oleracea L.; endive, *Cichorium endivia* L.; lettuce, *Lactuca sativa* L.; onion, *Allium cepa* L.; spinach, *Spinacia oleracea* L.; tomatoes, *Lycopersicum esculentum* L.; and perennial ryegrass, *Lolium multiflorum* L.) grown in FA-containing greenhouse media. The crops absorbed Se in proportion to the percentage of FA in the growing media. Broccoli, onion, and lettuce absorbed the highest concentrations. The levels of Se in ryegrass were maintained through five cuttings of the plants. The results indicate that FA may be added to growing media at a rate up to 33.6% as a source of Se but the uptake of compounds by some plant species may occur at excessive concentrations.

Wadge and Hutton[49] investigated the uptake of Cd, Pb, and Se by barley (*Hordeum vulgare* L.) and cabbage (*B. oleracea* L.) grown on soils amended with refuse incinerator FA in greenhouse experiments. Fly ash was amended at 10, 20, 30, and 40% and 5, 10, 15, and 20% on soil weight basis in barley and cabbage experiments, respectively. After 6 weeks growth, the plants were cut 2 cm above the soil surface. A second barley crop was grown on the same soil/ash amendments after all the pots were allowed to stand outside over winter for 7 months. Plant samples were digested with a mixture of concentrated nitric and perchloric acid. Concentrations of Cd and Pb in the digested samples were measured by AA spectroscopy, and Se concentration was measured by a fluorometric method. The ash amendments increased the concentrations of all three elements in the first barley crop. The concentrations approximately doubled when FA amendment was increased from 10 and 20% (on soil weight basis). The highest concentrations of Cd, Pb, and Se in barley were obtained on the soil amended with 30 to 40% FA on weight basis.

When an unweathered FA was mixed with soil at 8% by weight, there was a five- to sixfold increase in soil salinity over short period of time.[50] Similar observations were also reported by other investigators.[21,23,44,51] Soluble salt concentrations in unweathered ash deposits are very high, which may result in electrical conductivity (EC) values ≥13 $dS{\cdot}m^{-1}$. These values far exceed the levels considered to cause adverse effects for most plant species including agronomic crops (EC ≥ 4 $dS{\cdot}m^{-1}$). However, FA may be stabilized before application to reduce the impact on soil salinity. Lagooning, stockpiling, and leaching considerably reduced the detrimental effects of soluble salts and B contents associated with FA.[52]

Wadge and Hutton[49] tested the plant growth and uptake of toxic elements by barley (*Hordeum vulgare* L.) and cabbage (*B. oleraeca* L.) from the growth media containing 0, 10, 20, 30, and 40% (on weight basis) of FA. At the highest rate of FA, plant growth was impaired by FA additions and the yield was suppressed up to 30% of the control value. Weathering of the ash-amended soils reduced the detrimental effects on plant growth. For example, the yield of barley as the second crop grown in soil amended with FA (40%) increased by 70% of the control. At the same time, Cd and Se concentrations were consistently lower in the second crop. However, no such response was evident with regard to Pb concentration in barley tissue. The highest concentrations of Cd, Pb, and Se in cabbage were found in the soil amended with 20% FA and those concentrations were 147-, 18-, and 51-fold greater than the respective concentrations in cabbage grown in unamended soil. Cabbage grown in soil amended with 20% FA, following the first crop of barley, contained 72 $\mu g{\cdot}g^{-1}$ Cd as compared to 33 $\mu g{\cdot}g^{-1}$ in the first barley crop. At 20% FA amendment, Pb and Se concentrations were 43 and 1.08 $\mu g{\cdot}g^{-1}$, respectively, in barley and 16 and 0.51 $\mu g{\cdot}g^{-1}$, respectively, in cabbage.

The "availability indices" have been calculated for cabbage data,[49] expressed as the ratio of plant:soil concentration for a specific element. These indices suggest that the availability pattern for the three elements in this particular study increased in the order Pb < Se < Cd.

Lead and Cd were also the area of concern in the field experiments conducted by Schwab et al.[53] They examined the effects of manure and soil additions to the FA on the extractability and plant availability of Pb, Cd, and B. The establishment and growth of sudan grass (*Sorghum vulgare sudanese* L.), soybean (*Glycine max* L.), and a mixture of soybean plus sudan grass were studied on either unamended FA, FA-amended with composted manure (5% by volume of the top 15 cm), or composted manure (5%) plus 1% soil. Amending with manure and manure plus small amounts of soil overcame some of the negative effects of the FA and allowed plant establishment. At the time of planting, each plot was fertilized with 80 kg P ha^{-1} and 20 kg N ha^{-1}. The soil samples were extracted by ammonium bicarbonate DTPA and analyzed by inductively coupled plasma emission spectroscopy (ICPES). Plant samples were digested in nitric-perchloric acid and analyzed by ICPES. The composition of ash was characterized by measuring the concentration of various elements in a suspension of unamended ash in 0.01 M $CaCl_2$ after equilibration for 28 days. The flasks were continually aerated to ensure oxidizing conditions and to accelerate equilibration with atmospheric CO_2. The suspension was filtered prior to analysis. Activities of various constituents were calculated using the geochemical model, MINTEQA1.[54] Concentrations of Pb, Cd, and B, extracted from the FA by ammonium bicarbonate DTPA, were strongly affected by the type of FA and amendment, but were unaffected by crops. The addition of manure and manure plus soil reduced Pb, Cd, and B concentrations in growth medium. In general, the same trend was observed in Pb, Cd, and B concentrations in plant tissues. The addition of manure and manure plus soil as the amendments to FA reduced the concentrations of Pb, Cd, and B in the plant tissue.

Menon et al.[55] reported that application of FA-amended compost (20% FA) improved yields of collard greens (*B. oleracea* L.) and mustard greens (*B. juncea* L.) from 400 to 500% compared to that on bare soil control. The FA-amended composts in these experiments (20% FA) were mixed with sandy loam soil in the ratio 1:3. The soil treated with FA-amended compost showed greater concentrations of K, Ca, Mg, S, Zn, and B compared to those in the control soil.

Schwab[56] investigated the growth and elemental concentration of sudan grass (*S. vulgare* L. *sudanese*) and soybean (*G. max* L.) which were seeded directly into the ash. Treatments consisted of western and eastern FA, mixtures of it, and of FA amended with 5% composted manure or with 5% composted manure plus 1% soil. Soybean plants in all treatments showed symptoms of severe B toxicity, but sudan grass grown in FA

Table 2 Concentrations of Selected Trace Elements in Leaf Tissue in Two Tree Species Growing on Coal Fly Ash Basins

		Element Concentration (mg/kg)					
Species	**Site**	**Mn**	**As**	**Cr**	**Ni**	**Pb**	**Se**
Sweetgum	Control	1027	<0.08	1.10	3.38	<5.00	0.08
	Wet basin	109	0.20	<1.00	9.44	5.09	1.88
	Dry basin	7	0.18	1.30	1.48	<5.00	1.09
Sycamore	Control	213	<0.08	<1.00	1.95	<5.00	0.15
	Wet basin	50	0.29	<1.00	8.28	6.47	3.75
	Dry basin	12	0.11	<1.00	2.37	<5.00	0.90

From Reference 57.

amended with compost exhibited no symptoms of B toxicity. Concentrations of P, K, Mg, Zn, Cu, Mo, Cd, B, and Pb extracted by the NH_4HCO_3-DTPA soil test were correlated with the respective concentrations in the plants.

Carlson and Adriano[57] tested the growth and elemental content of trees (sweetgum, *Liquidambar styraciflua* and sycamore, *Plantanus occidentalis*) grown on abandoned coal FA basins. Both basins, "wet" and "dry", had been abandoned for approximately 20 years. The wet basin (pH = 5.58) originally received precipitator ash in ash-water slurry, while the dry basin (pH = 8.26) received both precipitator and bottom ash in dry form. Trees were established voluntarily on the wet basin following abandonment but were planted on the dry basin and a control upland site (Fuquay sand; loamy, siliceous; pH = 4.99). Tree height and diameter were measured for the above-listed trees prior to sampling for mineral analysis. The dried samples of stem and leaf tissue were wet digested in a nitric-perchloric acid mixture and concentrations of As, Cr, Ni, Pb, Se, V, and Mn were determined by ICPES. Trees from the wet basin exhibited elevated concentrations of trace elements in comparison to the trees grown on control site while the dry basin trees exhibited reduced concentrations of these elements. The concentration of Mn in trees grown on the control site was considerably greater than that in trees grown on the ash basin (Table 2). Differences in concentrations of trace elements in the leaves among the sites can generally be explained by differences in substrate trace concentrations of the respective elements and/or substrate pH. Trees from the wet ash basin generally had the highest concentrations of trace elements and these trees attained the greatest height and trunk diameter, suggesting that the elevated concentrations of trace elements in the wet basin substrate did not adversely affect the growth of these two tree species. The greater height and trunk diameter of the trees on wet basin was partly attributed to greater water-holding capacity of the substrate on this site.

Tree response to FA was also investigated in the studies conducted on poorly drained sandy soil of a small watershed in north-central Florida.[58] The results of these experiments showed beneficial effects of FA amendment on the growth of Australian pine seedlings (*Casuarina cunninghamiana* L.). Growth of the seedlings on soil amended with 56 tons unweathered FA/acre was about twofold that on the soil without FA amendment. Pine growth on the FA-amended soil after 18 months was also superior, averaging 1.9 m height vs. 1.1 m on the control. The main side effect of the FA treatment was an increase in soil pH by one unit. A similar liming effect was evident for runoff waters. The use of FA as a soil amendment for fuel wood production with nitrogen-fixing Australian pine was promising.

Amendment of soil with FA may also have an impact on pesticide persistence in soil. Albanis and Pomonis[59] tested the influence of FA (raw material lignite from thermal station; pH = 11.2) on the fate of atrazine and alachlor in soil under corn cultivation soil. The experiments were conducted under environmental conditions of loannina region in northwestern part of Greece. Fly ash was applied at 15% by weight of soil to a depth of 0 to 10 cm at the time of sowing. Analyses of soil samples 240 days following application of FA showed that the addition of FA decreased the half life of atrazine from 54 to 47 days and of alachlor from 25 to 22 days.

Research on the influence of FA on plants is often limited to plant growth and mineral composition. Very little is known on the effects of FA on plant organic components. Anderson et al.[60] examined nine glucosinolates in rutabaga (*Brassica napis* L.) grown on soil over a FA landfill (0.5- to 1.0-m thick soil layer) and on normal clay and silt loam soil. Progoitrin (2-hydroxybutyl-3-enyl glucosinolate) and neoglucobrassicin (1-

methoxy-3-undolylmethyl glucosinolate) were the most abundant glucosinolates found. Progoitrin and three other minor glucosinolates were present in greater amounts in rutabaga grown in natural soil than those in rutabaga grown in FA-amended soil. However, the latter contained greater amounts of glucobrassicin (3-indolyl-methyl glucosinolate) and neoglucobrassicin than the former.

Borkowski and Beresniewicz[61] studied the possibility of using FA as a multicomponent fertilizer for tomato cultivation. FA from brown coal produced near the town of Konin in Poland was rich in Ca (18 to 30%) and Mg (2 to 4%), along with concentrations of other elements as follows: 1 to 2% Fe, 0.2% Mn, 0.03% B, and approximately 20 mg/kg of each Zn, Mo, and Cu. Growth of tomato plants increased by 17% when 10 g FA was amended with 220 g growth medium compared to that on unamended growth medium. However, application of 40 g of FA retarded growth and yield of tomato plants. Such result might be explained by the high pH, over 7.0, and very high concentration of Mg, over 700 mg/kg.

Beresniewicz and Nowosielski[62] compared the fertilizing effects of various rates of FA having acid-neutralizing ability comparable to 15% CaO with the effect of the usage of the limestone (acid-neutralizing ability comparable to 45% CaO). The treatments included (1) FA at 5000 kg/ha, (2) limestone at 5000 kg/ha + FA at 10,000 kg/ha, (3) limestone at 10,000 kg/ha, (4) FA at 15,000 kg/ha, (5) limestone at 15,000 kg/ha, (6) FA at 20,000 kg/ha, and (7) limestone at 20,000 kg/ha. Application of FA increased the yield of vegetables (onion, celery, and cabbage) despite its low acid-neutralizing ability. The effect of lime on the soil pH lasted 4 years. At equivalent rates of addition, limestone increased soil pH more than FA. The available Mg content was greater in soils amended with FA than lime and a reverse trend was true with regard to Ca content. The increase in the sum of bases and sorption capacity of soils was greater in soils amended with limestone than FA.

In summary, most plant response studies are directed toward evaluation of plant growth and elemental composition when grown in soils amended with FA or FA plus an organic compost at various rates. There were some attempts to establish vegetation directly on FA landfill with or without soil cover. Plant species differed considerably with respect to survivability. Although FA at a moderate rate can be an effective amendment for supplying B nutrition, at excess rates B toxicity may occur on certain crop species. The availability of B is dependent on the stage of weathering of the FA. In addition, research has also shown that FA is a good source of Mo. Studies on alfalfa and bermuda grass have shown a distinct advantage of FA amendment due to an increased S availability.

Application of FA at high rates (>12% of soil on weight basis) demonstrated some adverse affects on microbial activity and in turn on plant growth.

Application of FA often results in considerable salinity problems. This can be overcome by leaching the FA soil prior to planting.

Most favorable responses to FA amendment were reported in acidic soils partly as a result of the liming property of FA. In neutral to alkaline soils, the response is often poor and may result in micronutrient problem due to a pH increase well over 7.

Some studies have also indicated an increase in soil moisture-holding capacity of the soil as a result of FA amendment. Plant response to FA amendments in terms of micronutrient and toxic trace elements is largely dependent on the composition of these elements in the FA which is a function of origin of the coal.

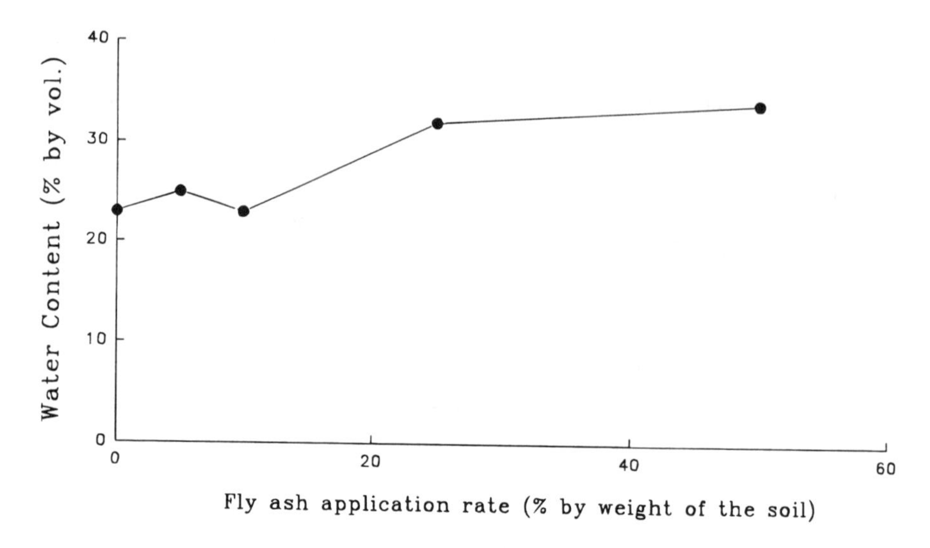

Figure 5 Water content (percent by volume at 20 cbar) of FA-amended soils (means across soil types).[65]

IV. INFLUENCE OF FLY ASH ON SOIL PHYSICAL PROPERTIES

While FA-soil chemical interactions have been thoroughly studied, the influence of FA on the soil physical properties is not fully understood. The amendment of soil with FA may potentially affect soil texture,[63,64] bulk density,[50,65] moisture relationships,[66–68] pH,[50,69-72] and soluble salts content.[66,73,74] The available-water-holding capacity (AWC) of soils of different texture has been shown to be positively correlated with the percentage of fine sand (0.2 to 0.02 mm) and organic carbon and negatively correlated with the percentage of coarse sand (0.2 to 2.0 mm).[75] This indicates that the addition of FA to soils containing only a small proportion of fine sand may increase the AWC of such soils. This hypothesis was confirmed in laboratory experiments, in which rates of FA equivalent of 0 to 753 Mg/ha were mixed with various soils.[68] There was a linear relationship between AWC and the rate of FA applied in a number of soils tested. Similar relationships were also obtained in experiments carried out under field conditions[76] using sandy loam and coarse sand amended with FA at 0 to 502 Mg/ha. The effects of FA on both soils resulted mainly from the progressive increase in moisture retained at the upper limit of available water. The results of moisture-release characteristics showed that the additions of FA increased the amount of readily available water in the soil.

Chang et al.[65] stated that the texture of FA resembles that of a silt and loam soil. High water-holding capacity of FA is probably due to the porous nature of some particles. Chang et al.[65] examined the effects of FA amendment, at 0 to 50% by volume, on water-holding capacity of five California soils of various texture. Water-holding capacity of all soils increased significantly in soils amended with FA at rates >10% compared to the unamended soils (Figure 5). A moisture release study showed that FA application of <10% did not change the characteristics of the moisture release curve. Unfortunately, from the agronomical point of view, although the addition of FA changed the water-holding capacity and the characteristics of water release curves, it did not change appreciably the amount of water available to the plant root system.

Small quantities of FA amendment did not appear to affect water-release characteristics of studied soils. When FA is added to correct trace element deficiency in soils, the amount added seldom exceeds 2.5% FA by volume. With this amount, the impact on water-holding capacity and/or release would certainly be minimal.

Application rates of FA >25% to the soil increased the passage of water through the soil profile as measured by hydraulic conductivity of soil cores.[65] The reduction in bulk density and modules of rupture as a result of application of FA in large quantities should also be considered as favorable to most agricultural soils.

In summary, application of FA generally increased the soil moisture-holding capacity. This is partly due to the particle size as well as porosity of FA particles. However, addition of FA at small quantities aimed to correct micronutrient deficiency problems did not affect the moisture-holding capacity of the soils.

V. FLY ASH AND SOIL MICROBIAL ACTIVITY

The addition of FA to soils could interfere with microbially mediated processes of organic matter decomposition and the cycling of nutrients such as C, N, S, and P in the biosphere.[77,78] Soil microbial activity is difficult to measure because of the diversity of the microbial populations and the difficulty to relate microbial numbers to microbial activity such as separation of microorganisms, microbial counts, and enzyme activities. However, several parameters can adequately represent the extent of microbial growth.

Arthur et al.[79] reported the effects of amending a soil with an acidic FA (pH 4.9) and alfalfa plants on microbial CO_2 evolution. FA amendments at 400 to 700 tons/ha rates significantly reduced CO_2 evolution as compared to that at the lower rates (0 and 100 tons/ha). These results suggested that low rates of FA amendment had minimal affects on the microbial activity.

The effects of various amounts of FA amendments on soil microbial activity were studied by Wong and Wong[37] and Pitchel and Hayes.[78] Wong and Wong[37] demonstrated that the effects of unweathered, alkaline FA on soil microbial respiration was greater in a sandy soil than in a sandy loam. It could be attributed to the higher cation exchange capacity and clay and organic matter contents in the sandy loam compared to the sandy soil.[80] Inhibition of microbial activity by FA was attributed to the effects of FA on pH and electrical conductivity.[81] This effect was apparent due to the fact that microorganism colonization increased as pulverized FA was aged.

Application of FA to soil at rates of more than 100 tons/ha usually decreased the nitrification process.[1,82] Numerous investigations have indicated that a decrease in soil microbial activity was due to the toxicity of trace elements (Cd, Cr, and Zn) present in FA on soil microorganisms.[77,78,83,84]

The effects of acidic FA on the microbial activity was attributed to toxic concentrations of trace elements.[79] However, El-Mogazi et al.[85] have expressed doubts that, even in the case of a large FA addition to the soil, a closed FA-soil disposal system would become sufficiently mixed to inhibit microbial activity or cause major changes in soil microflora.

An "ecological" dose 50% (EcD_{50}) concept was adapted to quantify the effects of FA on microbial respiration in the soil ecosystem, defined as the concentration of a toxicant that inhibits a microorganism-mediated ecological process by 50%.[77] Application of FA at rates greater than 10 to 12% of soil (on weight basis) seriously inhibited soil microbial activity in most soils.[37] Thus, to avoid any serious impact on soil microbial activities, FA application rates greater than 10 to 12% should be avoided.

Modification of poor soil conditions by crop residues, municipal wastes, and compost may even alleviate the toxic effects of FA.[43,81] Organic amendments would probably increase the organic matter and cation exchange capacity, especially of the sandy soil, and result in greater microbial activity compared to that in unamended soil. However, according to Townsend and Gillham's[86] investigations, weathering, which reduces the content of all solutes, would be very important for enhancement of microbial activity either before or after application.

Pitchel and Hayes[78] have demonstrated a marked reduction in microbe numbers following the application of FA and sludge. Application of FA at rates >5% of soil weight decreased microbe numbers. However, when FA was applied in combination with sludge, the inhibitory effects of FA on microbial activity were alleviated. This is in part due to substantial numbers of microorganisms contained in the sludge.[87] Enzyme activity in the soil usually decreased with increased rate of FA application.[78,88]

The usual method of FA disposal is transport of FA as a slurry to a settling basin with removal of water by surface drainage.[89] Klubek et al.[90] showed that the abundance of microorganisms and microbial activity increase with age of the ash basin and with establishment of a plant community. The microbial diversity also increases as the plant community on the ash basin matures. The reclamation and utilization of coal ash basins require the input of organic amendments, with low C:N ratio, to assure the availability of C and N, and the enhancement of an active microflora. The establishment of a mature plant community can be achieved more readily.

Little information is available on the effects of coal ash on the abundance of soil microorganisms and microbial activity. Although some research has been done on the influence of FA on soil microbial activity, further research is necessary to investigate the affects of FA on nitrogen transformation and other microbial processes in the soil.

In summary, FA affects on microbial activity was negligible at low rates of additions. The adverse effects of FA on the microbial activity appeared to be due to FA affects on soil pH, electrical conductivity, and the loading of toxic trace metals such as Cd, Cr, and Zn. Application of FA should not exceed 10 to 12% (on weight basis) in order to avoid any adverse effects on soil microbial activities.

Microbial populations and their activity in FA landfills depend on age of the ash basin and establishment of plant species. Addition of organic amendments with low C:N ratio enhances microbial activity.

VI. FLY ASH FOR TREATMENT OF POLLUTED WATERS

The utilization of FA for water and wastewater treatment has received little attention. Most FA has certain inherent properties that make it uniquely suited for applications in this important area of public concern.[91] FA might be utilized in the treatment of specific polluted waters and sludges. There are considerable purgative effects of FA addition on polluted lake waters, acid mine wastes, and biological sludge. The most significant FA properties are adsorptive capacity and settleability. Because of its relatively large surface area per unit volume and carbon content, FA is an effective adsorbent ma-

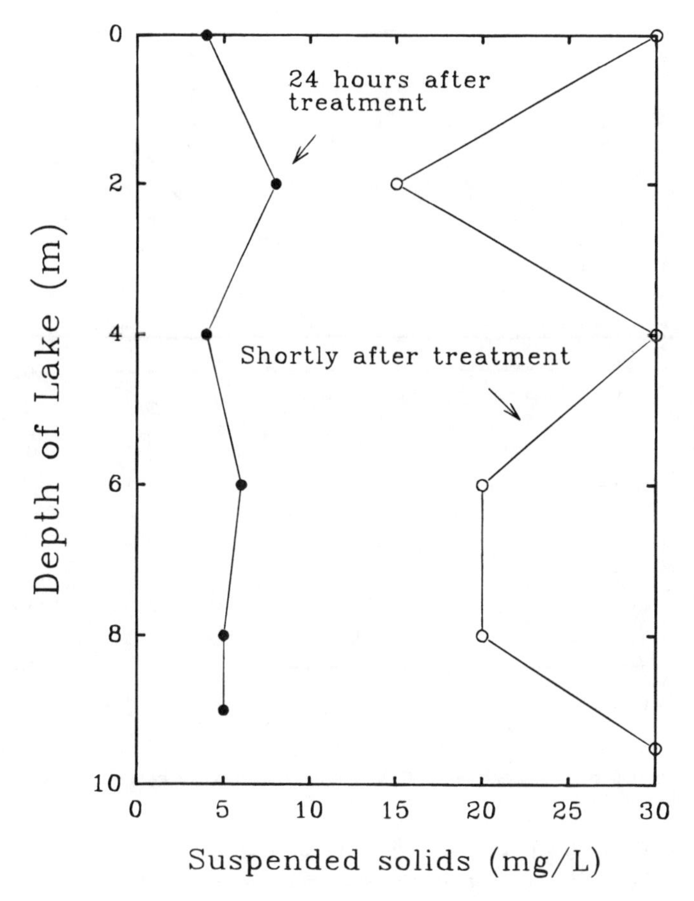

Figure 6 Changes in concentration of suspended solids in lake water following the addition of FA.[91]

terial capable of removing many organic contaminants (both dissolved and suspended) from polluted waters.[92,93] The effectiveness of FA as an adsorbent material depends on the amount of carbon it contains. FA is also capable of removing undesirable organic contaminants which are responsible for the odor and bad taste of water and the contaminations caused by algae and bacteria.

Most FA contains certain materials that are water soluble, i.e., $CaSO_4$, CaO, sodium, and potassium hydroxides.[9] The release of these chemicals from FA added to water can be very beneficial in terms of the neutralization of acid wastes.[92] Hydroxide ions as hydroxides of Ca, Na, and K released to water serve to increase water pH and also increase the concentration of Ca. Increased concentration of Ca may contribute to removal of P, which is one of the most common water pollutants, by forming almost insoluble hydroxyapatite $Ca_{10}(PO_4)_6(OH)_2$. Addition of FA to isolated lake water columns[91] resulted in significant reduction of inorganic P, suspended matter, color, and organic contaminants in the lake water. Figures 6 and 7 show FA additions to the lake water columns resulted in reduction in concentrations of suspended material and phosphate in the water.

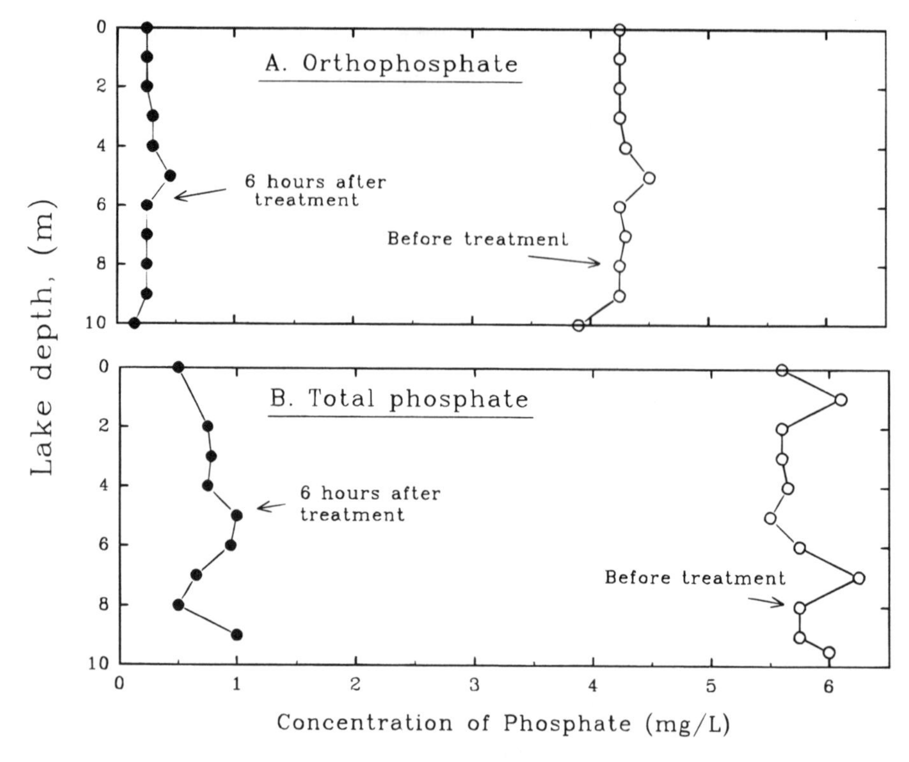

Figure 7 Reduction in concentration of phosphate in lake water column following the addition of FA.[91]

In this study, FA at 10 g/L was slowly added (in dry state) at the surface of an isolated column of 9.75 m. The concentration of suspended material in the column, which averaged 20 mg/L prior to the addition of FA, decreased to less than 5 mg/L within 24 h following the addition of FA. At the above rate of addition of FA, the removal of phosphate from the water ranged from 88 to 95% (Figure 7). The FA settled at the bottom of isolated water columns developed a physical barrier at the mud-water interface, which impaired the release of bottom pollutants into the overlaying water during the periods of intense oxygen and thermal stratification of the lake.

FA is a particulate material which has specific gravity similar to that of sand and it is capable of settling under quiescent conditions. The rate of settling velocity of a discrete particle having a size in the range of FA can be approximated by Stokes' law.[92,93] FA, however, is known to settle appreciably faster than the rate calculated by Stokes' law. It indicates that FA in certain proportions might also be useful as a factor enhancing chemical coagulation and settling of turbid waters.

Mott and Weber[94] reported that FA has significant capacity for sorption of low molecular weight organic contaminants from aqueous solution, the capacity being related to the carbon content and to the other properties specific to the FA tested. Transient diffusion experiments verified that incorporation of FA into soil-bentonite cutoff barriers can impart significant capacity for contaminant retardation of these barriers. The performance of a soil-bentonite cutoff barrier amended with high-C FA suggests that such amendments can provide significant improvement in performance with respect to retardation of contaminant breakthrough. FA appears to be useful in the reduction of dangerous water pollutants such as Pb.

Yadava et al.[95] studied the removal of Pb from contaminated water by application of FA as a function of the contact time, concentration, temperature, and pH of the adsorbate-adsorbent system. An empirical model has been tested to understand the kinetics of Pb removal at different concentrations. The amount of Pb removed reached 88.1% in water samples containing Pb in concentrations ranging from 2 to 6 mg/L, at 30°C and pH 6.4. The complete removal of Pb was achieved only in water containing 2 mg/L. The experiments were carried out by agitating 1.0 g of FA with 50 mL of Pb-contaminated water. Removal of Pb from water by adsorption on FA increased with time and adsorption maxima was attained at 110 min. The equilibrium time was independent of Pb concentration and of solution temperature. The adsorption capacity of FA for Pb decreased with the increase in the solution temperature from 20 to 40°C. In water containing 2 mg/L Pb at 30°C, an increase in solution pH from 3.0 to 6.4 increased the removal of Pb from 13.85 to 88.10%. The extent of Pb removal by FA decreased when solution pH was increased to 8.0. Accordingly, the authors concluded that FA could be successfully used to reduce Pb concentration in waters and wastewaters below the permissible limit. However, the extent of Pb decontamination by FA additions to contaminated water is dependent on several factors, such as temperature, pH of the solution, and rate of FA additions, etc.

In summary, relatively large surface area per unit volume and carbon content of FA make it an effective adsorbent material capable of removing many organic contaminants from polluted waters and sludges. High Ca content in FA helps removal of P from polluted waters by forming insoluble hydroxyapatite. FA has a tendency to settle at a fast rate. Therefore, it can be used to enhance chemical coagulation and settling turbid waters. Addition of FA has shown to effectively decrease Pb content in polluted waters.

VII. FLY ASH AND ANIMAL NUTRITION

The utilization of FA as a soil amendment may be either beneficial[96] or toxic[97] to animals fed by plants grown on such soils. It depends mainly on the amount of various elements absorbed by the plants and transferred to the food chain and into the body of animals. Although there is a great need to evaluate the response of animals to changes in various micronutrient contents as a result of growing plants in soils amended with FA, much of the research was done only on Se response as a result of FA amendment of soils used for growing animal feed crop species. Since 1959, Se has been recognized as a protective factor against white muscle disease (WMD), a debilitating myopathy in lambs and calves. In Se-deficient areas, Se supplements to animals decreased the incidence of WMD and also minimized reproductive problems such as retained placentas and low fertility.[98]

In the experiments conducted by Finkelstein and Oldfield,[96] coal FA was applied to the soil deficient in Se in amounts of 224 Mg/ha, equivalent to 4% FA (weight basis) to a depth of 19 cm soil. This level was considered effective to maintain a constant soil pH and supplying presumably adequate level of Se for optimum uptake by young orchard grass (*Dactylis glomerata*). The concentration of Se in plants appeared to be high enough to provide livestock with desirable amounts of Se in the diet without causing any Se toxicity problems. There is a great concern that crops planted or voluntarily growing on a media with excessive amounts of FA may absorb heavy metals in toxic amounts, which may result in heavy-metal toxicity to animals or humans if such plants are ingested.[99,100] Deep-rooted crops growing over soil-capped FA landfills might absorb heavy metals directly from the FA below the soil cover. Shallow-rooted crops can also absorb heavy metals by contacting FA moved upward into the soil cap by earthworm activity or by absorption of such elements as dissolved ions moved upward by capillary movement.

It is more probable that grasses and legumes rather than vegetables would be planted or voluntarily grown over soil-capped FA landfills. It has been shown[101] that sweet clover voluntarily grown on a FA landfill without soil cover absorbed an extremely high concentration of Se. When the crop was fed to goats and sheep,[102] high concentrations of Se were deposited in animal tissues and excreted in milk. Legumes have also been shown to absorb elevated concentrations of Se when grown on soil-covered FA landfills.

Stoewsand et al.[103] studied the movement of Se in soil-plant-animal food chain using rutabaga (*Brassica napus*), grown on FA landfill with soil cover, fed to rats. Rutabagas were peeled, freeze-dried, ground, and incorporated at 25% into the rats' diet. The rats were fed for 13 weeks, then Se concentration was determined in kidney, liver, and muscle by wet ashing the samples with nitric and perchloric acid. Concentration of Se was analyzed by the fluorescence procedure of Olson.[104]

The concentration of Se in rutabaga grown on amended soil control and FA was 0.17 and 4.4 ppm, respectively. The latter Se concentration is considered near toxic dietary level for monogastric animals, but not for ruminants.[101] The concentrations of Se in various tissues of rats fed with rutabaga grown on FA were significantly greater than those of the rats fed with rutabaga grown on soil without FA amendment. However, the animals did not exhibit any symptoms of Se toxicity during the feeding study. Relative liver weights of the rats fed by rutabaga grown on FA significantly increased in size. This may indicate an adaptive response related to microsomal enzyme induction to the ingestion of natural compounds present in FA-grown rutabaga.[105]

Stoewsand et al.[97] grew winter wheat on a FA landfill. It was then harvested and incorporated as 60% of a complete diet to Japanese quail (*Coturnix coturnix japonica*), followed by analysis of plant and avian tissue for residual total Se. Measurements on eggshell thickness and hepatic microsomal enzyme activity were also conducted. The eggshell thickness increased in the quail fed with high-Se diet, but hepatic microsomal enzyme activity remained unchanged. The concentrations of Se were markedly greater in all tissues of quail fed the wheat grown on FA compared to those of the quail fed on wheat grown without FA. The highest concentrations of Se were found in livers and kidneys of the quail fed with high Se wheat. These concentrations were often greater than 5 ppm, which is considered as potentially toxic for humans.[106] Although elevated by the presence of Se in seleniferous food, the presence of Se in quails' heart and muscle did not exceed 5 ppm Se. It indicates that muscle of farm animals fed with high seleniferous plants, after Se content determination, might be considered useful for the purpose of human nutrition.

The amount of FA added to a soil is an important factor for determining possible adverse effects of FA with regard to heavy metals in the plant-animal-human food chain. Excessive uptake of toxic elements by plants and resulting heavy metal problems to animals fed by these plants usually occurred when the plants were grown totally on FA or soils amended with excessive quantity of FA. The uptake and translocation into the food chain of the toxic elements from FA also depends on the species and cultivars of plants, animal species, and part of the animal body used in human nutrition.

In summary, much of the research on animal nutrition as related to FA amendment is on Se nutrition. Application of 224 Mg/ha FA to a Se-deficient soil increased the Se content of orchard grass to levels adequate for supplying Se to animals fed on that grass. The amount of FA added to a soil is an important factor for determining possible adverse effects of FA with regard to heavy metals in the plant-animal-human food chain.

VIII. FLY ASH EFFECTS ON GROUNDWATER QUALITY

Leachate from solid waste disposal sites may contaminate water supply. Possible pollution from coal FA comes from toxic metals and is of increasing concern to the agencies responsible for regulating water quality. These problems were also the area of some research activity, although coal leachate migration through the soil rooting zone and subsequent impact on plants have received little attention.

According to Carlson,[107] the chemical and physical factors that affect the quality and concentration of heavy metals in leachates from FA landfills include the frequency, intensity, duration, pH, rainfall, S content, temperature, and microorganism activity. Important hydrologic parameters that must also be considered include the hydraulic conductivity, hydraulic gradients, leachability, and attenuation capabilities of the underlying and down-gradient soils.

Although Carlson[107] investigated the migration of leachates from a reject coal pile (in South Carolina), these results are highly applicable to the leachate migration from FA landfill. The water table in the site studied was near the surface (<60 cm) and water movement was primarily to the west, toward the Savannah River. Acid leachate migration from the reject coal pile has contaminated the water table aquifer with sulfate (up to 22,200 mg/L) and metals (Fe up to 9560 mg/L, Al up to 1110 mg/L). The shallow plume of contaminated groundwater was migrating westward at a calculated velocity of 6.3 to 10.4 m/year. Leachate migration has also produced a highly acidic (pH down to 2.2) and highly saline (EC to 11.8 dS/m) rooting zone and was most likely responsible for the stressed and dead vegetation in the study area.

Cherry and Guthrie[108] analyzed the concentrations of ten toxic elements in the effluent and drainage system of a coal ash basin and determined the major modes of removal of these elements from drainage water. The experiment was conducted near a coal-fired power plant of the Savannah River Project, Aiken, South Carolina, over a period of 2 years. Samples of water, benthic sediment, macro-invertebrates, aquatic plants, and vertebrates were collected for analysis of metals by neutron activation analysis (NAA) in accordance with the procedure described by Guthrie and Cherry.[109] In these samples, the concentrations of Ti, Mn, Cu, Cr, Zn, As, Se, Co, Cd, and Hg were measured. During the first 12 months of this period, the basin was essentially filled and little settling of ash occurred. In the remaining 12 months, dredging was completed, adequate settling occurred, and most of the effluent turbidity was removed. The partitioning of most of the toxic elements was greater in sediment and biota than in water. Five of the ten metals examined (Mn, Cu, As, Zn, and Se) were biomagnified by at least one biotic component. Plants had high accumulations of Ti, Mn, As, and Hg; invertebrates had high accumulations of Co, Hg, Cu, Cr, Cd, and As; and vertebrates greatly biomagnified Se and Zn. The streamlined biotic community of the system accomplished major removal from the effluent. The magnitude of bioaccumulation of Ti, Mn, Zn, As, Se, Cd, and Hg was increased during the period of adequate settling in the basin.

Sandhu et al.[110] compared the leachability of Cd, Ni, Cr, and As in ash disposal basins of different ages located at the U.S. Department of Energy (DOE) Savannah River site and showed that, among the heavy metals examined, Ni was the most mobile and Cd was the least mobile.

The evidence of toxic metal concentrations in coal FA effluent and the movement of these metals through components of the environment suggest that further investigations are necessary to assure adequate precautions against surface and groundwater contamination.

The necessity to undertake precautions against water contamination as a result of FA usage has also been suggested by other investigators.[9,58,111,112] Dreesen et al.[111] compared the levels of trace elements extracted from FA and levels found in effluent waters from a FA landfill of a coal-fired power plant located in Fruitland, New Mexico. Trace elements in the precipitator ash were extracted with nitric acid, hydrochloric acid, citric acid, distilled water, and ammonium hydroxide. Effluent water at this plant was sampled to assess the elevation of trace element concentration compared with intake water. Fluoride was measured in the samples with an Orion 94-09A specific ion electrode and a Corning Model 112 digital volt meter. The B content was determined by thermal neutron capture prompt X-ray analysis. The trace element (As, Be, Cd, Cr, Mo, Se, V, and Zn) concentrations were determined by flameless-AA spectroscopy (Perkin-Elmer Model 306), and the concentration of Cu was determined by neutron activation. The results showed a positive correlation between those elements most extractable by water (B, F, Mo, and Se) or acid (As, B, Cd, F, Mo, and Se), and those elements that are most elevated in effluent waters (As, B, F, Mo, and Se).

There are two important factors that influence the distribution of trace elements in the extracts and effluents from FA landfills.[113] One is the surface predominance of trace elements on ash particles making them more available for mobilization. These elements include Mo, F, Se, B, As, and Cd, and all are extractable in acids. The second factor is that anionic species would remain soluble in alkaline environments while the metallic cations would be precipitated. Therefore, molybdate, borate, fluoride, selenate, and possibly arsenate, chromate, and vandanate would occur in soluble forms, whereas, Be, Cd, Cu, and Zn would be either precipitated after dissolution as a result of an increase in pH or may remain not extractable. Thus, an important consideration is whether the trace elements in the ash are soluble in the acidic environment of the venturi scrubbers or further downstream in the ash disposal system where more alkaline conditions are found. It appears that Mo, F, Se, B, and As are the elements most significantly elevated in effluent waters, and are of prime interest for future studies of soluble contaminants from coal ash in alkaline environments.

Shannon and Fine[9] investigated cation solubility of FA. They determined the concentrations of Ca, Na, Mg, and Fe in water extracts of four lignite FA from three Northern Great Plains mines. The concentrations of Mg, Na, and Fe were analyzed by AA spectroscopy and that of Ca was determined by EDTA titration. The major fraction of soluble cationic material released was represented by Ca and Na, although considerable variability existed among the four FA studied with respect to absolute concentration of Ca or Na and relative distribution of these two basic cations. Mg and Fe generally were solubilized at rates roughly one half of that for Ca and Na. As the ratio of water:FA was increased, the solubility of Ca, Mg, and Fe increased, while that of Na decreased.

The level of SO_4 was greater in dust collector ash than in electrostatic precipitation ash. Approximately 50% of the cations solubilized by water from dust collector ashes appeared to exist as SO_4. It is also necessary to note that water-soluble constituents of a typical lignite FA from the Northern Great Plains is 8.55%, while that of a typical bituminous FA is only 2.51%. This indicates that the use of FA as a soil amendment should follow its detailed chemical analysis.

Rohrman[112] noted that, from a pollution control standpoint, P, N, B, and the radioactive elements found in ash are of particular importance. The possible presence of Hg is also of concern. Though there are reports of 0.008 to 0.021% Hg in ash from West Virginia coal, no Hg was detected in samples taken from the Ohio Basin stations. Factors that determine the extent of water pollution from FA include (1) effectiveness of sluicing operations, (2) ratio of FA to other ash, (3) composition of the coal being burned, and (4) temperature of combustion.[112]

Solubility of FA in water is one of the most vital factors that determine the extent of water pollution from FA. Analysis of samples from 12 power plants showed 4 to 7% of the ash was soluble in distilled water; pH of these solutions varied from 6.2 to 11.5.[112]

At present, if FA is applied at moderate amounts as a soil amendment (2 to 5% of soil), the potential contamination of groundwater by toxic elements is not a concern.[112]

In summary, groundwater contamination by heavy metals in the FA depends on the leachate migration, soil characteristics, and depth of groundwater in landfill sites. An evaluation of leachability of Cd, Ni, Cr, and As in an ash disposal basin revealed that Ni was the most mobile while Cd was the least mobile. In general, the water-extractable elements in FA were most commonly found in elevated levels in effluent waters. In agricultural use, at present Fa is applied at moderate rates; therefore, contamination of groundwater by the constituents in FA is not a major concern at this time.

IX. FLY ASH AS SOIL-MAKING MATERIAL

In most soil science handbooks, there are similar definitions of soil. According to Stefferud,[114] soil is defined as (1) the natural medium for the growth of land plants, (2) a dynamic natural body of surface of the earth in which plants grow, composed of mineral and organic materials and living forms, and (3) the collection of natural bodies occupying parts of the earth's surface that supports plants and that have properties due to the integrated effect of climate and living organisms acting upon parent material.

Although spoil banks and coal refuse piles may vary widely in composition, both represent potential soil-making material.[115] Referring to the definitions of soil, it is noted that the growth of plants and living organisms is an essential part of the soil. Since many spoil banks and coal refuse piles lie barren for many years, it may raise the question about the ability of the addition of FA to catalyze the soil-building process and make it possible to establish and support life on these barren wastes.

Soil formation is the result of many factors and FA aids this process by eliminating or modifying those spoil and refuse properties that restrict plant growth. The physical and chemical effects that can be derived from using FA for mine spoil reclamation or in agriculture can be attributed to its source and how it is produced. FA is produced as a powdery residue when the coal is pulverized and burned in the boilers of large electric power-generating stations. The ash then contains elements that were taken up by plants millions of years ago. Although many of these elements have been identified and reported in the literature, primary interest is centered on Ca, Mg, K, P, and S and trace elements such as B, Zn, Cu, Mn, Mo, and their reactivity with soil and availability to plants.

According to the studies done by Martens,[22] most samples of FA were alkaline in reaction with neutralization from 0.04 to 3.37 meq H_3O^+/g. As a comparison, the neutralization value of $CaCO_3$ is 20 meq H_3O^+/g. Because the availability of all nutrients required by plants is affected either directly or indirectly by soil pH, it is reasonable to assume that FA application to soils influence the nutrient availability. These studies also showed that all of the samples of FA contained more B than usually occurs in soils and some of the FA samples were more rich in P and Zn than soil. High concentrations of B in FA could be beneficial, provided that B from FA particles is released slowly over a period of time.

Mixing large quantities of FA with spoil produces physical changes that may enhance plant survival and growth. Bulk density of the mixtures decreased by the additions of the lightweight amendment. Decreased bulk density results in greater pore volume, greater moisture availability, and higher air capacity, and hence better conditions for root penetration and growth.

Adding FA also modifies the particle-size distribution of the spoil because nearly all of the ash consists of fine sand and silt.[22] In the cases of muddy and clay soils, the modification results in the formation of a medium-textured soil. Soils of this texture usually have increased the pore space volume and moisture-holding capacity. Equally important is the establishment of plants on the treated spoils, which in turn provide organic matter to help form soil-enriching humus. The most successful plants used on the FA landfill sites were grasses such as festuce (*Festuca arundinacea* L.), red top grass (*Agrostis alba* L.), orchard grass (*Dactylis glomerata* L.), rye grass (*Lolium perenne* L.), and also plant species such as birdsfoot (*Phalaris canariensis*), yellow sweet clover (*Trifolium agrarium* L.), and several species of trees and shrubs.

Before a site is treated with FA, the spoil or refuse should be surveyed and sampled. Soil pH analysis of these materials mixed with varying amounts of FA determines the application rate. The analysis of micronutrients is also necessary to avoid the extensive input of potentially hazardous and toxic elements into the plant growth medium.

The results of laboratory, greenhouse, and field investigations conducted by Capp and Gillmore[115] showed that FA, under most conditions, can be treated as a safe and valuable amendment to plant growth media. Greenhouse studies proved that application of FA to soil increased the availability of B, Mo, P, K, and Zn to plants. Field trials with alfalfa showed an increase in yield with the application of FA. The yield increases were attributed to correction of B deficiency. Field experiments with two applications of 160 Mg/ha of weathered FA on clay loam and silt loam soils did not adversely affect the growth of corn plants. Results indicated that relatively high amounts of FA may be disposed on agricultural soils.

X. FLY ASH AND SEWAGE SLUDGE MIXTURE

Due to the large quantities of sewage sludge produced, there has been much interest in their possible use as a soil amendment to agricultural soils producing edible crops. However, the presence of toxic elements in certain FA and sewage sludge is of major concern. It has been reported that Se in FA[116] and Cd in sludge[117–119] are two elements of most concern regarding toxicity and absorption into edible crops. Selenium was taken up in large amounts by plants grown in soils amended with FA, while Cd, Cu, Ni, and Zn accumulated in plants grown in soils amended with sludge and B accumulated in plants grown in soils amended with both.[15,99,120,121]

Whereas a number of papers have been published on the growth of crops on soil amended with FA or sludge, few papers have been published on the use of FA and sludge mixed together and applied to soil simultaneously for plant growth. The information in this regard is lacking.

A pioneering study on FA-sewage sludge mixture as a soil amendment was reported by Adriano et al.[43] Thereafter, Elfving et al.[122] also demonstrated the potential benefit of FA-sewage sludge mixtures to various edible vegetable crops, grain millets, and apple seedlings. Adriano et al.[43] further reported that FA showed promise as a sludge companion amendment, possibly as an alternative for limestone. It was also observed that, by applying FA-sewage sludge mixtures, the Cd accumulation in plants could be substantially reduced and the weathering of FA prior to cropland disposal will overcome salinity problems associated with the use of unweathered FA while maintaining its liming potential. The application of FA along with sewage sludge to mine spoils for the establishment of deep-rooted legumes has also been reported by Singh et al.[123]

A recent study by Oyler[124] on the applications of FA-sewage sludge mixtures indicated that a 1:1 ratio of sludge to FA was found to enhance the performance of certain important herbaceous species when reclaiming a metal-contaminated site near a zinc smelter. The beneficial effects of the application of sewage sludge in combination with lime has also been observed to be very effective to coal refuse material in maintaining pH and in reducing water-soluble Al and Fe, KCl extractable Al, and total acidity,[125] its ability to supply N, P, and K to growing plants,[126] and in reducing SO_4^{2-} percolating.[127]

With our efforts to survey some key research work done during the past on utilization of FA or sewage sludge, it can be concluded that the utilization of FA or sewage sludge as a source for plant nutrients as well as an alkaline medium to reclaim acidic soils and acid mine lands has been well-documented. However, the information on the use of organic wastes and FA together as a compost mixture is clearly lacking.

In summary, combining FA with sewage sludge as a mixed amendment is in the interest of enrichment of FA with organic amendment. However, in the above mixture, the primary concern from the toxicity point of view is Se in FA and Cd in sludge. Use of an amendment containing 1:1 ratio of FA:sludge was very effective for reclaiming a contaminated soil near a zinc smelter.

XI. CONCLUSIONS

The beneficial aspects of application of FA at low to moderate rates on plant growth and yield are an encouragement to further explore the disposal of FA on agricultural soils. FA disposal on agricultural soils would be unsuccessful if adverse effects on crop production and/or environment were as striking compared to the beneficial effects. The utilization of FA in agriculture, despite many promising results, also yields environmental problems that need to be solved. Developing appropriate solutions to these problems will increase the acceptance of FA disposal on agricultural soils. For the safe disposal of FA on agricultural soils, we need to develop appropriate rates of FA application, minimize adverse environmental problems, and determine suitable plant species or cultivars that can thrive on FA-amended soils. It will be necessary to consider differences in chemical properties of various sources of FA with regard to their chemical reactions in soils. Differences in tolerances of plant species, and eventually cultivars among selected species, to soluble salt damage and to specific nutrient deficiencies and toxicities will also have to be taken into account. Further research on FA-soil-plant relationships should be conducted to develop the models of mobility and uptake of elements from soil amended with FA.

ACKNOWLEDGMENT

Florida Agricultural Experiment Station Journal Series No. R-03361.

REFERENCES

1. Carlson, C.L. and Adriano, D.C., Environmental impact of coal combustion residues, *J. Environ. Qual.*, 22, 227,, 1993.
2. Hecht, N.L. and Duvall, D.S., Characterization and Utilization of Municipal and Utility Sludges and Ashes, Environ. Prot. Technol. Series, EPA-670/2-75. U.S. Environmental Protection Agency, Washington, D.C., 1975.
3. Murarka, J.P., Inorganic and Organic Constituents in Fossil Fuel Combustion Residues, Vol. 1, Res. Project 2485-8, Electric Power Research Institute, Columbus, OH, 1987.
4. Rai, D., Inorganic and organic constituents in fossil fuel combustion residues, Vol. 2, Res. Project 2485-8, Electric Power Research Institute, Columbus, OH, 1987.
5. Adriano, D.C., Page, A.L., Elseewi, A.A., Chang, A.C., and Stranghan, J., Utilization and disposal of fly ash and other coal residues in terrestrial ecosystems: a review, *J. Environ. Qual.*, 9(3), 333, 1980.
6. Hodgson, L., Dyer, D., and Brown, D.A., Neutralization and dissolution of high-calcium fly ash, *J. Environ. Qual.*, 11(1), 93, 1982.
7. Davison, R.L., Natusch, D.F.S., and Wallace, J.K., Trace elements in fly ash, *Environ. Sci. Technol.*, 8(13), 1107, 1974.
8. Bern, J., Residues from power generation: processing, recycling and disposal: land application of waste materials, *Soil Conserv. Soc. Am.*, 226, 1976.
9. Shannon, D.G. and Fine, L.O., Cation solubilities of lignite fly ashes, *Environ. Sci. Technol.*, 8(12), 1026, 1974.
10. United States Department of Health, Education and Welfare, National Air Pollution Boards, Pub. No. AP-49, Air Quality Criteria for Particulate Matter, Washington, D.C., January 1969.
11. Klein, D.H., Andren, A., Carter, J.A., Emery, J.F., Feldman, C., Fulkerson, W., Lyon, W.S., Ogle, J.C., Taimi, Y., Van Hook, R., and Bolton, N., Pathways of thirty-seven trace elements through coal-fired power plant, *Environ. Sci. Technol.*, 9(10), 973, 1975(b).
12. Campbell, J.A., Laul, J.C., Nielson, K.K., and Smith, R.D., Separation and chemical characteristics of finely-sized fly-ash particles, *Anal. Chem.*, 50(8), 1032, 1978.
13. Fennely, P.F., Klemm, H., Hall, R., and Durocher, D., Coal burns cleaner in a fluid bed, *Environ. Sci. Technol.*, 11(3), 244, 1977.

14. Menon, M.P., Ghuman, G.S., James, J., and Chandra, K., Effect of coal fly ash-amended composts on the yield and elemental uptake by plants, *J. Environ. Sci. Health*, A27(4), 1127, 1992.
15. Furr, A.K., Kelly, W.C., Bache, C.A., Gutenmann, W.H., and Lisk, D.J., Multielement uptake by vegetables and millet grown in pots on fly ash amended soil, *J. Agric. Food Chem.*, 24(4), 885, 1976.
16. Plank, C.O. and Martens, D.C., Amelioration of soils with fly ash, *J. Soil Water Conserv.*, 28, 177, 1973.
17. Theis, T.L. and Wirth, J.L., Sorptive behavior of trace metals on fly ash in aqueous systems, *Environ. Sci. Technol.*, 11, 1096, 1977.
18. Elseewi, A.A., Straughan, B.R., and Page, A.L., Sequential cropping of fly ash-amended soils: effects on soil chemical properties and yield and elemental composition of plants, *Sci. Total Environ.*, 15, 247, 1980.
19. Van Volins, P., Arthur, M.F., and Tolle, D.A., Field and Microcosm Investigations of the Effects of Atmospheric Deposition from Fossil Fuel Combustion, 2nd Annual Progress Report, Electric Power Research Institute, Columbus, OH, 1981.
20. Menon, M.P., Ghuman, G.S., and Chandra, J.J.K., Physico-chemical characterization of water extracts of different fly ashes and fly-ash amended composts, *Water, Air and Soil Pollut.*, 50, 343, 1990.
21. Adriano, D.C., Woodford, T.A., and Ciravolo, T.G., Growth and elemental composition of corn and bean seedlings as influenced by soil application of coal ash, *J. Environ. Qual.*, 7(3), 416, 1978.
22. Martens, D.C., Availability of plant nutrients in fly ash, *Compost Sci. J.*, 15(6), 15, 1971.
23. Mulford, F.R. and Martens, D.C., Response of alfalfa to boron in flyash, *Soil Sci. Soc. Am. Proc.*, 35, 296, 1971.
24. Hatcher, J.T. and Wilcox, L.W., Colorimetric determination of boron using carmine, *Anal. Chem.*, 22, 567, 1950.
25. Cope, F., The development of a soil from an industrial waste ash, *Int. Soc. Soil Sci. Trans. Comm.*, 4, 859, 1962.
26. Hodgson, D.R. and Holliday, R., The agronomical properties of pulverized fuel ash, *Chem. Ind. Pollut.*, 66, 785, 1966.
27. Holliday, R., Hodgson, D.R., and Townsend, W.N., Plant growth on fly ash, *Nature*, 181, 1079, 1968.
28. Aitken, R.L. and Bell, L.C., Plant uptake and phytotoxicity of boron in Australian fly ashes, *Plant Soil*, 84(2), 245, 1985.
29. Menon, M.P., Sajwan, K.S., Ghuman, G.S., James, J., and Chandra, K., Fly ash-amended compost as a manure for agricultural crops, *J. Environ. Sci. Health*, 28, 2167, 1993.
30. Scanlon, D.H. and Duggan, J.C., Growth and element uptake of woody plants on fly-ash, *Environ. Sci. Technol.*, 3(13), 311, 1979.
31. Plank, C.O. and Martens, D.C., Boron availability as influenced by application of fly ash to soil, *Soil Sci. Soc. Am. Proc.*, 38, 974, 1974.
32. Doran, J.W. and Martens, D.C., Molybdenum availability as influenced by application of fly ash to soil, *J. Environ. Qual.*, 1(2), 186, 1972.
33. Reisenauer, H.M., Molybdenum, in *Methods of Soil Analysis, Agronomy, Part 2*, Vol. 9, Black, C.A., Ed., American Society of Agronomy, Madison, WI, 1965, 1050.

34. Cary, E.E., Gilbert, M., Bache, C.A., Gutenmann, W.H., and Lisk, D.J., Elemental composition of potted vegetables and millet grown on hard coal bottom ash-amended soil, *Bull. Environ. Contam. Toxicol.*, 31, 418, 1983.
35. Elseewi, A.A., Bingham, F.T., and Page, A.L., Availability of sulfur in fly ash to plants, *J. Environ. Qual.*, 7(1), 69, 1978.
36. Wong, M.H. and Wong, J.W.C., Germination and seedling growth of vegetable crops in fly ash-amended soils, *Agric. Ecosys. Environ.*, 26, 23, 1989.
37. Wong, M.H. and Wong, J.W.C., Effects of fly ash on soil microbial activity, *Environ. Pollut.*, 40, 127, 1986.
38. Wong, J.W.C. and Wong, M.H., Effects of fly ash on yield and elemental composition of two vegetables, *Brassica parachinensis* and *Brassica chinensis*, *Agric. Ecosys. Environ.*, 30, 251, 1990.
39. Fail, J.L. and Wochok, Z.S., Soybean growth on fly ash amended strip mine spoils, *Plant Soil*, 48, 473, 1977.
40. Pawar, K. and Dubey, P.S., Germination behaviours of some important crop species in flyash incorporated soils, *Prog. Ecol.*, 10, 295, 1988.
41. Martens, D.C., Schnappinger, M.G., and Zelanzy, L.W., The plant availability of potassium in fly ash, *Soil Sci. Soc. Am. Proc.*, 34, 453, 1970.
42. Kanchiro, Y. and Sherman, G.D., Fusion with sodium carbonate for total elemental analysis, *Agronomy*, 9(2), 952, 1965.
43. Adriano, D.C., Page, A.L., Elseewi, A.A., and Chang, A.C., Cadmium availability to sundangrass grown on soils amended with sewage sludge and fly ash, *J. Environ. Qual.*, 11, 197, 1982.
44. Phung, H.T., Lund, L.J., and Page, A.L., Potential use of fly ash as a liming material, in *Environmental Chemistry and Cycling Processes*, Adriano, D.C. and Brisbin, I.L. Eds., CONF-760429, U.S. Department of Commerce, Springfield, VA, 504, 1978.
45. Schnappinger, M.G., Martens, D.C., and Plant, C.O., Zinc availability as influenced by application of fly ash to soil, *Environ. Sci. Technol.*, 9(3), 258, 1975.
46. Plank, C.O., Martens, D.C., and Hallock, D.L., Effect on soil application of fly ash chemical composition and yield of corn (*Zea mays* L.) and on chemical composition of displaced soil solutions, *Plant Soil*, 42, 467, 1975.
47. Wright, J.D., Jr., The Effect of Fly Ash on Selected Physical Characteristics of the Ap Horizons of Three Virginia Soil Series, M.S. thesis, Virginia Polytech Institute and State University, Blacksburg, 1972.
48. Shane, B.S., Littman, C.B., Essick, L.A., Gutenmann, W.H., Doss, G.J., and Lisk, D.J., Uptake of selenium and mutagens by vegetables grown in fly ash containing greenhouse media, *J. Agric. Food Chem.*, 36(2), 328, 1988.
49. Wadge, A. and Hutton, M., The uptake of cadmium, lead and selenium by barley and cabbage grown on soils amended with refuse incinerator fly ash, *Plant Soil*, 96, 407, 1986.
50. Page, A.L., Elseewi, A.A., and Stranghaun, I.R., Physical and chemical properties of fly ash from coal-fired plants with reference to environmental impacts, *Residue Rev.*, 71, 83, 1979.
51. Elseewi, A.A., Bingham, F.T., and Page, A.L., Growth and mineral composition of lettuce and Swiss chard grown on fly ash amended soils, in *Environmental Chemistry and Processes*, Adriano, D.C. and Brisbin, I.L., Eds., CONF-6760429, U.S. Department Commerce, Springfield, VA, 1978a, 568.

52. Townsend, W.N. and Hodgson, D.R., Edaphological problems associated with deposits of pulverized fuel ash, in *Ecology and Reclamation of Devastated Land*, Vol. 1, Hutuik, R.J. and Davis, G., Eds., Gordon and Breach, London, 1973, 45.
53. Schwab, A.P., Tomecek, M.B., and Ohlenbush, P.D., Plant availability of lead, cadmium and boron in amended coal ash, *Water, Air, Soil Pollut.*, 297, 57, 1991.
54. Brown, D.D. and Allison, Y.D., MINTEQA1, an Equilibrium Metal Speciation Model, Report No. EPA/600/3-87/012, Environmental Research Laboratory, NTIS PB 88-144167, U.S. Environmental Protection Agency, Athens, GA, 1987.
55. Menon, M.P., Sajwan, K.S., Ghuman, G.S., James, J., and Chandra, K., Elements in coal and coal combustion residues and their potential for agricultural crops, in *Trace Elements in Coal and Coal Combustion Residues*, Keefer, R.F. and Sajwan, K.S., Eds., Lewis Publishers, Chelsea, MI, 1993, 259.
56. Schwab, A.P., Extractable and plant concentrations of metals in amended coal ash, in *Trace Elements in Coal and Coal Combustion Residues*, Keefer, R.F. and Sajwan, K.S., Eds., Lewis Publishers, Chelsea, MI, 1993, 185.
57. Carlson, C.L. and Adriano, D.C., Growth and elemental content of two tree species growing on abandoned coal fly ash basins, *J. Environ. Qual.*, 20, 581, 1991.
58. Riekerk, M., Coal-ash effects on fuel wood production and runoff water quality, *Southern J. Appl. Forest.*, 2, 99, 1983.
59. Albanis, T.A. and Pomonis, P.J., The influence of fly ash on atrazine and alachlor persistence in corn cultivation soil, *Toxicol. Environ. Chem.*, 33(3/4), 181, 1991.
60. Anderson, Y.L., Lisk, D.J., and Stoewsand, G.S., Glucosinolates in rutabaga grown in soil capped over coal fly ash, *J. Food Sci.*, 55(2), 556, 1990.
61. Borkowski, J. and Beresniewicz, A., Brown coal ash as a multicomponent fertilizer for protected tomato cultivation, *Bull. Veg. Crops Res. Work*, 88, 107, 1989 (in Polish).
62. Beresniewicz, A. and Nowosielski, O., Comparison of the fertilizing effect of brown coal ash with that of limestone on the yields of vegetables and the soil properties, *Soil Sci. Ann.*, 37(4), 141, 1986 (in Polish).
63. Plass, W.T. and Capp, J.P., Physical and chemical characteristics of surface mine spoil treated with fly ash, *J. Soil Water Conserv.*, 29(3), 119, 1974.
64. Jones, C.C. and Amos, D.F., Physical changes in Virginia soils resulting from additions of high rates of fly ash, Proc. Fourth Int. Ash Utilization Symposium, St. Louis, 1976.
65. Chang, A.C., Lund, L.J., Page, A.L., and Warunke, J.E., Physical properties of fly ash-amended soils, *J. Environ. Qual.*, 6(3), 267, 1977.
66. Capp, J.P., Power plant fly ash utilization for land reclamation in the eastern United States, in *Reclamation of Drastically Disturbed Lands*, Schaller, F.W. and Sutten, P., Eds., American Society of Agronomists, Madison, WI, 1976, 339.
67. Barber, G., Land reclamation and environmental benefits of ash utilization, in *Ash Utilization*, Escand, W.R. and Spencer, J.D., Eds., U.S. Department of Interior, Bureau of Mines, Washington, D.C., 1973, 246.
68. Salter, P.J. and Williams, J.B., Effects of pulverized fly ash on the moisture characteristics of soil, *Nature*, 215, 1157, 1967.
69. Bell, T.J. and Unger, I.A., Factors affecting the establishment of national vegetation on a coal strip mine spoil bank in Southeastern Ohio, *Am. Midl. Nat.*, 4(10), 19, 1981.
70. Spooner, A.E., Brown, D.A., and Clark, M.D., Fly Ash in an Agricultural Environment, Semi-annual report of Arkansas Power and Light, Little Rock, 1981, 36.

71. Jastraw, J.D., Zimmerman, C.A., Dovork, A.J., and Hinchman, D.R., Comparison of Lime and Fly Ash as Amendment to Acidic Coal Mine Refuse: Growth Responses and Trace Elements of Two Grasses, Annual Report, Argonne National Laboratory, Argonne, IL, 1979.
72. Adams, L.M., Capp, J.P., and Gilmore, D.W., Coal mine spoil and refuse bank reclamation with power plant fly ash compost, *Science*, 13(6), 20, 1972.
73. Scholt, L.F., Vocke, R.W., Beskid, N.J., Ness, D.D., Derickson, W.K., Siskind, B., Knight, M.J., and White, W.S., Handling of Combustion and Emission Abatement Wastes from Coal-Fired Power Plant: Implications of Fish and Wildlife Resources, U.S. Department of Interior, Fish and Wildlife Services, Washington, D.C., 1980, 196.
74. Ciravolo, T.G. and Adriano, D.C., Utilization of coal ash by crops under greenhouse conditions, in *Ecology and Resources Development*, Wali, M., Ed., Pergamon Press, Elmsford, NY, 1979, 958.
75. Salter, P.J., Berry, G., and Williams, J.B., The influence of texture on the moisture characteristics of soils. III. Quantitative relationships between particle size composition and available-water capacity, *J. Soil Sci.*, 17, 93, 1966.
76. Salter, P.J., Webb, D.S., and Williams, J.B., Effects of pulverized fuel ash on the moisture characteristics of coarse-textured soils and on crop yields, *J. Agric. Sci. Cambridge*, 77, 53, 1971.
77. Babich, H.and Stotzky, G., Environmental factors that influence the toxicity of heavy metal and gaseous pollutants to microorganisms, *Crit. Rev. Microbiol.*, 8, 99, 1983.
78. Pitchel, J.R. and Hayes, J.M., Influence of fly ash on soil microbial activity and populations, *J. Environ. Qual.*, 19, 593, 1990.
79. Arthur, M.F., Zwick, T.C., Tolle, D.A., and Van Voris, P., Effects of fly ash on microbial CO_2 evolution from an agricultural soil, *Water Air Soil Pollut.*, 22, 209, 1984.
80. Doelman, P. and Haanstra, L., Effects of lead on the decomposition of organic matter, *Soil Biol. Biochem.*, 11, 481, 1979.
81. Elliot, L.F., Tittemore, D., Papendick, R.T., Cochran, V.L., and Bezidicek, D.F., The effect of Mount St. Helen's ash on soil microbial respiration and numbers, *J. Environ. Qual.*, 1, 164, 1982.
82. Cervelli, S.G., Petruzelli, A., Perna, A., and Menicagli, R., Soil nitrogen and ash utilization: a laboratory investigation, *Agrochimica*, 30, 27, 1986.
83. Ruhling, A. and Tyler, G., Heavy metal pollution and decomposition, *Oilcos*, 24, 402, 1973.
84. Spaulding, B.P., Effects of divalent metal chlorides on respiration and extractable enzymatic activities of Douglas-fir needle litter, *J. Environ. Qual.*, 8, 105, 1979.
85. El-Mogazi, D., Lisk, D.J., and Weinstein, L.H., A review of physical, chemical and biological properties of fly ash and effects on agricultural ecosystems, *Sci. Total Environ.*, 74, 1, 1988.
86. Townsend, W.N. and Gillham, W.F., Pulverized fuel ash as a medium for plant growth, in *Ecology and Resources Degradation and Renewal*, Chadwick, M.J. and Goodman, G.T., Eds., Blackwell Scientific, Oxford, 1975, 287.
87. Alexander, M.A., *Introduction to Soil Microbiology*, John Wiley & Sons, New York, 1977.
88. Speir, T.W. and Ross, D.J., Soil phosphate and sulphatase, in *Soil Enzymes*, Burns, R.G., Ed., Academic Press, New York, 1978, 198.
89. Alberts, J.J., Newman, M.C., and Evans, D.W., Seasonal variations of trace elements in dissolved and suspended loads for coal ash ponds and pond effluents, *Water Air Soil Pollut.*, 26, 111, 1985.

90. Klubek, B., Carlson, C.L., Oliver, J., and Adriano, D.C., Characterization of microbial abundance and activity from three coal ash basins, *Soil Biol. Biochem.*, 24(11), 1119, 1992.
91. Tenney, M.W. and Echelberger, W.F., Jr., Fly Ash Utilization in the Treatment of Polluted Waters, U.S. Department of the Interior, Washington, D.C., 1970.
92. Fair, G.M., Geyer, J.C., and Okun, D.A., Water and wastewater engineering, in *Water and Wastewater Treatment and Disposal*, Vol. 2, John Wiley & Sons, New York, 1968.
93. Strates, B.S., Newman, W.F., and Lerinskas, G.L., The solubility of bone mineral. II. Precipitation of near neutral solutions of calcium and phosphate, *J. Phys. Chem.*, 61, 279, 1957.
94. Mott, H.V. and Weber, W.J., Sorption of low molecular weight organic contaminants by fly ash: considerations for the enhancement of cutoff barrier performance, *Environ. Sci. Technol.*, 26(6), 1234, 1992.
95. Yadava, K.P., Tyagi, B.S., and Singh, V.N., Fly-ash for the treatment of water enriched in lead, *J. Environ. Sci. Health*, 24(7), 783, 1989.
96. Finkelstein, E. and Oldfield, Y.E., Industrial wastes as soil amendments for orchard grass as a selenium source for grazing livestock, *Am. Soc. Animal Sci., Western Section*, 34, 176, 1983.
97. Stoewsand, G.S., Gutenmann, W.H., and Lisk, D.J., Wheat grown on fly ash: high selenium uptake and response when fed to Japanese quail, *J. Agric. Food Chem.*, 26(3), 757, 1978.
98. Harrison, J.H., Hancock, D.D., and Conrad, H.R., Selenium deficiency and avarian function in dairy cattle, *Fed. Proc.*, 41(3), 786, 1982.
99. Furr, A.K., Parkinson, T.F., Elfving, D.C., Gutenmann, W.H., Pakkala, J.S., and Lisk, D.J., Elemental content of apple millet and vegetables grown in pots of neutral soil amended with fly ash, *J. Agric. Food Chem.*, 27, 135, 1979.
100. Gutenmann, W.H., Pakkala, J.S., Churey, D.J., Kelly, W.C., and Lisk, D.J., Arsenic, boron, molybdenum, and selenium in successive cuttings of forage crops field grown on fly ash amended soil, *J. Agric. Food Chem.*, 27, 1393, 1979.
101. Furr, A.K., Parkinson, T.F., Heffron, C.L., Reid, J.T., Haschek, W.M., Gutenmann, W.H., Bache, C.A., and Lisk, D.J., Elemental content of tissues and excreta of lambs, goats, and kids fed white sweet clover growing on fly ash, *J. Agric. Food Chem.*, 26, 847, 1978.
102. Weinstein, L.H., Osmeloski, J.F., Rutzke, M., Bers, A.O., McCahan, J.B., Bache, C.A., and Lisk, D.J., Elemental analysis of grasses and legumes growing on soil covering coal fly ash landfill sites, *J. Food Safety*, 9, 291, 1989.
103. Stoewsand, G.S., Anderson, Y.L., Weinstein, L.H., Osmeloski, J.F., Gutenmann, W.H., and Lisk, D.J., Selenium in tissues of rats fed rutabagas grown on soil covering a coal fly ash landfill, *Bull. Environ. Contam. Toxicol.*, 44, 681, 1990.
104. Olson, O.E., Fluorometric analysis of selenium in plants, *J. Agric. Food Chem.*, 27, 1393, 1969.
105. Crampton, R.F., Gray, T.Y.B., Grasso, P., and Parke, D.V., Long-term studies of chemically induced liver enlargement in the rat. I. Sustained induction of microsomal enzymes with absence of liver damage on feeding phenobarbitone or butylated hydroxytoluene, *Toxicology*, 7, 289, 1977.
106. Rosenfeld, J. and Beath, O.A., *Selenium — Geobotany, Biochemistry, Toxicity and Nutrition*, Academic Press, New York, 1969.
107. Carlson, C.A., Subsurface leachate migration from a reject coal pile in South Carolina, *Water, Air Soil Pollut.*, 53, 345, 1990.

108. Cherry, D.S. and Guthrie, R.K., Toxic metals in surface waters from coal ash, *Water Res. Bull.*, 13(6), 1228, 1977.
109. Guthrie, R.K. and Cherry, D.S., Pollutant removal from coal ash basin effluent, *Water Resour. Bull.*, 2, 20, 1976.
110. Sandhu, S.S., Mills, G.L., and Sajwan, K.S., Leachability of Ni, Cd, Cr, and As from coal ash impoundments of different ages on the Savannah River Site, in *Elements in cool combustion residues*, Keefer, R.F. and Sajwan, K.S., Eds., Lewis Publishers, Chelsea, MI, 1993, 165.
111. Dreesen, D.R., Gladney, E.S., Owens, J.W., Perkins, B.L., Wienke, C.L., and Wangen, L.E., Comparison of levels of trace elements extracted from fly ash and levels found in effluent waters from a coal-fired power plant, *Environ. Sci. Technol.*, 11(10), 1017, 1977.
112. Rohrman, F.A., Analyzing the effect of fly ash on water pollution, *Environ. Manage.*, August, 76, 1971.
113. Theis, T.L., The Potential Trace Metal Contamination of Water Resources through the Disposal of Fly Ash, 2nd Natl. Conf. on Complete Water Reuse, Chicago, May 4–8, 1975.
114. Stefferud, A., Soil, in *The Yearbook of Agriculture*, U.S. Department of Agriculture, Washington, D.C., 1957, 767.
115. Capp, J.P. and Gillmore, D.W., Soil making potential of power plant fly ash in mined-land reclamation, Proc. 3rd Int. Ash Utilization Symposium, March 13–14, 1973, Association and Bureau of Mines, Pittsburgh, PA, 1973.
116. Gutemmann, W.H., Elfving, D.C., Valentino, D.I., and Lisk, D.J., Trace element absorption on soil amended with soft-coal fly ash, *BioCycle*, 22, 42, 1981.
117. Page, A.L. and Bingham, F.T., Cadmium residues in the environment, *Residue Rev.*, 48, 1, 1973.
118. Hinseley, T.D., Jones, R.L., and Ziegler, E.L. Effects on corn by application of heated anaerobically digested sludge, *Compost Sci.*, 13, 26, 1972.
119. Jones, R.L., Hinseley, T.D., Ziegler, E.L., and Tyler, J.L., Cadmium and zinc contents of corn leaf and grain produced by sludge-amended soil, *J. Environ. Qual.*, 4, 509, 1975.
120. Furr, A.K., Parkinson, T.F., Hinrichs, R.A., van Campen, D.R., Bache, C.A., Gutenmann, W.H., St. John, L.E., Jr., Pakkala, I.S., and Lisk, D.J., National survey of elements and radiochemistry in fly ashes — absorption of elements by cabbage grown in fly ash-soil mixtures, *Environ. Sci. Technol.*, 11, 1194, 1977.
121. Furr, A.K., Parkinson, T.F., Gutenmann, W.H., Pakkala, I.S., and Lisk, D.J., Elemental content of vegetables, grains and forages field-grown on fly ash amended soil, *J. Agric. Food Chem.*, 26, 357, 1978.
122. Elfving, D.C., Bache, C.A., Gutenmann, W.H., and Lisk, D.J., Analyzing crops grown on waste-amended soils, *BioCycle*, 22, 44, 1981.
123. Singh, R.N., Keefer, R.F., Ghazi, H.E., and Horvath, D.J., Evaluation of plant growth and chemical properties of mine soils following application of fly ash and organic waste materials, in *The Challenges of Change*, Halow, J.S. and Corey, J.N., Eds., 6th Int. Ash Utilization Symposium Proceedings, Morgantown, WV, 1982.
124. Oyler, T.A., Evaluation of sludge/fly ash mixtures and herbaceous species performance to reclaim metals-contaminated site near a zinc smelter, *Agron. Abstr.*, 32, 1987.

125. Pietz, R.I., Carlson, C.R., Jr., Peterson, J.R., Zenz, D.R., and Lue-Hing, C., Application of sewage sludge and other amendments to coal refuse material. I. Effects on chemical composition, *J. Environ. Qual.*, 18, 164, 1989.
126. Pietz, R.I., Carlson, C.R., Jr., Peterson, J.R., Zenz, D.R., and Lue-Hing, C., Application of sewage sludge and other amendments to coal refuse material. II. Effects on revegetation, *J. Environ. Qual.*, 18, 169, 1989.
127. Pietz, R.I., Carlson, C.R., Jr., Peterson, J.R., Zenz, D.R., and Lue-Hing, C., Application of sewage sludge and other amendments to coal refuse materials. III. Effects on percolate water composition, *J. Environ. Qual.*, 18, 174, 1989.

CHAPTER 10

Phosphogypsum and Other By-Product Gypsums

Isabelo S. Alcordo and Jack E. Rechcigl

0-87371-859-3/95/$0.00+$.50

I. INTRODUCTION

A. GENERAL USES OF BY-PRODUCT GYPSUMS AS SOIL AMENDMENT

Gypsums ($CaSO_4.xH_2O$) are used in agriculture either as sources of Ca and S for crops or as soil conditioners to improve certain physicochemical properties of problem soils.[1,2] They are available either as mined gypsum or as industrial by-products. Two relatively pure by-product gypsums are phosphogypsum[3] and the commercial-grade flue gas desulfurization gypsum called FGD gypsum.[4,5] Like mined gypsum, these by-products can be used as (1) sources of S and Ca for crops, (2) soil ameliorants for Al toxicity and subsoil acidity and infertility, (3) soil ameliorants for sodic and nonsodic dispersive soils, (4) soil conditioners for hard-setting clay soils and hardpans, (5) bulk carriers for micronutrients and low-analysis fertilizers,[1] and for phosphogypsum in particular, (6) in modifying certain cation:Ca ratios in soils[6] and in reducing NH_3-N losses from urea fertilizers[7] and farm manures.[8] Some FGD wastes such as the fluidized bed combustion (FBC) residues, while containing substantial amounts of gypsum, are used primarily to correct soil acidity in place of limestone and dolomite.[9,10]

B. WORLD PRODUCTION OF SOME BY-PRODUCT GYPSUMS AND THEIR PHYSICAL AND CHEMICAL PROPERTIES

By-product gypsums of all grades are produced by three distinct processes: (1) reaction between sulfuric acid and a Ca-containing raw material, (2) scrubbing of SO_2 from a flue gas stream using limestone or lime, and (3) neutralization of waste sulfuric acid, again, using limestone or lime. Examples of by-products produced by the first process are phosphogypsum[3] and the anhydrite fluorogypsum.[11,12] The by-products in the desulfurization of a flue gas stream vary greatly in quality. The wet flue gas desulfurization process with forced oxidation (FGD-FO) to oxidize completely SO_3-S to SO_4-S produces a relatively pure FGD gypsum.[4,5] The dry FBC process produces a FBC residue consisting of spent sorbent materials, mostly Ca and/or Mg oxides, with substantial amounts of $CaSO_4$.[13] The wet/dry spray-dry absorption (SDA) process[14–16] produces SDA residues that are mostly calcium sulfite ($CaSO_3 . {}^1\!/\!_2H_2O$) and spent Ca sorbent materials with minor amounts of $CaSO_4$.[17] Titanogypsum, kevlar gypsum, citrogypsum, and gypsums from metal pickle liquors are examples of gypsums produced by direct neutralization of waste sulfuric acid using limestone or lime. Except for some impurities in trace amounts, these by-products are relatively pure gypsum.[11,18]

1. Phosphogypsum

The co-product in wet-acid production of phosphoric acid from rock phosphate is produced according to the reaction:[2,19]

$$Ca_{10}(PO_4)_6F_2 + 10\ H_2SO_4 + 20\ H_2O \rightarrow 6\ H_3PO_4 + 10\ CaSO_4 \ .\ 2\ H_2O + 2\ HF$$

The three basic conventional processes used in wet-acid manufacture of phosphoric acid are the dihydrate, the hemihydrate, and the hemidihydrate processes. For each megagram (Mg) of P_2O_5 produced, about 4.9, 4.3, 4.9 Mg of phosphogypsum is coproduced by the dihydrate, hemihydrate, and hemidihydrate processes, respectively.[3] Based on recent data on world production of phosphoric acid, annual world production of phosphogypsum is estimated at about 125 million Mg. With only about 4% of the world's phosphogypsum production being used in agriculture and in gypsum board and cement industries,[20] some 120 million Mg of phosphogypsum are accumulating annually, most of which are piled in stacks.[21] Some are stored in abandoned quarries[22] or, in certain countries, dumped into waterways.[23]

Some of the world's major producers of phosphogypsum are the U.S.,[21] Russia,[24] Canada,[25] Japan,[26,27] India,[28] the Netherlands,[23] and Australia[22] with annual productions estimated at 40.0, 23.5, 4.0, 3.0, 2.8, 2.0, and 0.9 million Mg, respectively. Production in 1988 in West Germany was estimated at more than 0.8 million Mg. During the same year, nine European countries were reported to have used 0.9 million Mg in their cement industry.[29] Based on annual P_2O_5 production data (in million Mg), China (2.36), Brazil (1.48), Turkey (0.58), South Africa (0.50), Tunisia (0.49), South Korea (0.47), New Zealand (0.35), and Morocco (0.33) should be on the list of major phosphogypsum producers.[30]

In the U.S. 63 stacks in 12 states have been identified[21] with an estimated inventory of about 7.7 billion Mg as of 1981.[31] In another 20 years, another 800 million Mg are expected to be added to the inventory.[21] Florida, as the major phosphoric acid producer in the U.S., produces 32 million Mg of phosphogypsum annually which are stockpiled in 20 stacks.[32] Phosphogypsum inventory in Florida as of 1992 has been reported at more than 600 million Mg.[33]

In 1988 the U.S. used about 220,000 Mg of phosphogypsum as soil conditioner or fertilizer, a mere 0.5% of total annual production estimated at 40 million Mg. Main user states are California, Georgia, North Carolina, Arizona, Florida, and Virginia.[34,35] Because of environmental concerns, to be discussed later, phosphogypsum has not been used as raw material for the various gypsum-based industries.

Phosphogypsum is primarily $CaSO_4 . 2H_2O$ with small amounts of rock phosphate, sand, and clay. Table 1 shows the major chemical components of phosphogypsum from several countries as reported by various authors.[22,26,29,36–39] Particle-size distribution clusters around 0.05 mm in diameter. When moist, it is very friable with a silt-like feel. When fully air-dry, unpelleted phosphogypsum is blown off easily like dust during handling as, for example, when applying it to agricultural land. Although gypsum itself is a neutral salt, phosphogypsum is highly acidic with pH in water ranging from >2 to <5, mainly due to acid impurities, such as sulfuric, phosphoric, hydrofluoric, and fluosilicic acids.[40]

2. Flue Gas Desulfurization (FGD) Gypsum

Flue gas produced from the burning of fossil fuels contains both carrier gases and pollutants. Typical compositions of flue gases from different types of fossil fuels are given in Table 2.[41] Most of the industrialized countries have passed legislations to control emissions of oxygenated sulfur and nitrogen gases (SO_x, NO_x), and other pollutants into the atmosphere.

The state of the art of FGD processes in the U.S. as of 1975 was reviewed by Rosenberg.[42] Since then, the FGD-FO (*in situ*) process, first demonstrated by GE Environmental Systems in 1980, has become the industry standard and is used worldwide to produce mostly commercial grade FGD gypsum. The chemical reactions for the process are given as follows:

$$SO_2 + H_2O \rightarrow H_2SO_3 \qquad \text{(absorption)}$$

$$H_2SO_3 + CaCO_3 \rightarrow CaSO_3 + CO_2 + H_2O \qquad \text{(neutralization)}$$

$$Ca^{++}(aq) + SO_3^{=}(aq) + 1/2\ O_2(aq) \rightarrow CaSO_4 \qquad \text{(forced oxidation)}$$

$$CaSO_4 + 2H_2O \rightarrow CaSO_4 . 2H_2O \qquad \text{(crystallization)}$$

In round numbers, the removal of 1 kg of SO_2 requires 2 kg of $CaCO_3$ and produces 3 kg of gypsum. It is estimated that a typical 500-MW power plant burning coal containing 3% S will generate 15 Mg SO_4 h^{-1}.[5]

The production and consumption of FGD gypsum by some industrialized countries as of 1988 were reviewed by Makansi and Ellison.[11] Japan, the first to legislate emission control of dust, SO_2, NO_x, and CO_2 into the atmosphere, first produced FGD gypsum in 1970. Since then, stricter emission control regulations resulted in large increases in production, from 8000 Mg in 1970 to 1.02 million Mg in 1975. Production doubled from 1975 to 1980 to 2.09 million Mg and remained stable at around 2 million Mg from 1980 to 1990. The various by-product gypsums produced in Japan are utilized mostly in the construction industries. Japan continues to import not only mined but also by-product gypsums from several countries.[27,43]

Table 1 Typical Major Chemical Constituents of Phosphogypsum from Several Countries and Those of Mined Gypsum (% dry weight basis)

	Country					
Analysis	**Australia[22] [a]**	**Iraq[37]**	**Japan[26]**	**Sweden[29]**	**U.S. (Florida)[b]**	**Mined Gypsum[36]**
CaO	30.30–32.90	32.90	30.40	32.10	25.10–31.10	31.20–34.70
SO_3	43.00–45.20	44.90	43.50	45.20	31.90–42.00	44.20–46.70
SiO_2	0.01–5.00	0.45	4.05	0.40	3.20–17.70	0.10–2.50
Al_2O_3	0.01–0.03	1.05	0.11	—	0.19–0.57	<0.01–0.13
Fe_2O_3	0.01–0.06	0.40	0.04	0.02	0.00–0.14	0.04–0.31
MgO	0.01–0.04	0.46	0.01	0.01	0.00–0.20	0.03–0.70
Na_2O	0.03–0.35	—	0.08	0.10	0.02–0.61	—
K_2O	0.01–0.05	—	0.16	0.03	0.00–0.01	—
F	0.82–1.30	0.60	0.24	0.20	0.20–0.80	—
P_2O_5	0.28–0.80	0.18	0.29	0.30	0.50–3.70	—
Crystalline H_2O	19.20–20.60	19.20	19.00	19.30	14.40–18.80	15.90–20.20

[a] Ranges were from five samples from five producers.
[b] Ranges were from five samples from five phosphogypsum stockpiles.

In the Netherlands, all coal-fired power stations had been reported to have been fully provided with FGD units producing about 200,000 Mg of FGD gypsum in 1990. Annual production is expected to increase to 400,000 Mg by 1995. Target users are the construction industries.[17]

In Germany, federal regulations to control emissions of SO_2 enacted in 1983 led to the retrofitting of power plants with FGD units. By 1989 SO_2 emission by power plants was reduced to less than 15% of that emitted in 1982. In 1990 Germany produced 2.5 and 1.4 million Mg FGD gypsum from its hard coal-fired and lignite-fired power stations, respectively.[41] In the same year, the FGD units supplied the German gypsum-based industries with by-product gypsum. A large portion of FGD gypsum and gypsum products were exported to Belgium, Denmark, Luxembourg, the Netherlands, and Switzerland. With the reunification of East and West Germany, annual production of FGD gypsum in the country is expected to increase by another 1 to 2 million Mg.[44]

The production of FGD gypsum in the U.S. was the off-shoot of the Clean Air Act of 1970. The Act had for its purpose the setting of National Ambient Air Quality Standards (NAAQS). The current set of standards for ambient air was promulgated in April 1971. The primary standards placed the annual arithmetic mean limits on SO_2 at 0.080 mg m^{-3} or 0.03 mg kg^{-1} and NO_x at 0.100 mg m^{-3} or 0.05 mg kg^{-1}.[45] These gases are responsible for the so-called acid rain. The effect of acid rain on the soil environment had been reviewed by Rechcigl and Sparks.[46] One of the goals of the Clean Air Act Amendments of 1990 is to reduce SO_2 and NO_x by 10 and 2 million Mg by the year 2000, respectively, relative to 1980 levels. Another goal, at full implementation, is to remove some 25 million Mg of pollutants from the ambient air.[47] With increased production of solid wastes from various sources, the U.S. enacted into law the Resource Conservation and Recovery Act (RCRA) in 1976, also known as P.L. 94–580, amending the Solid Waste Disposal Act of 1965 and the Resource Recovery Act of 1970. The law deals with the management of solid and hazardous wastes and encourages energy and resource recovery.[48] Under this law, wastes from coal combustion such as bottom ash, fly ash, slag, and FGD gypsum were exempted from federal classification as "hazardous wastes", although subject to regulation as solid wastes. A new ruling from the U.S. Environmental Protection

Table 2 Typical Composition of Flue Gases from Different Types of Fossil Fuels

	Fossil Fuel			
Composition	**Hard Coal**	**Lignite**	**Oil**	**Natural Gas**
		Carrier Gases (% by vol)		
N_2	70–80	60–78	75–77	70–72
CO_2	11–15	11–15	11–14	7–10
O_2	4–7	4–7	1–5	1– 5
H_2O	3–8	8–24	6–11	1–22
		Pollutants (mg kg^{-1})		
SO_2	510–1700	100–2700	340–1700	<30
NO	200–1600	100–500	340–1000	50–1000
HCl	30–125	12–60	—	—
HF	5–60	1–2	—	—
CO	100–580	100–680	4–16	—

From Reference 41.

Agency (USEPA) to govern the disposal of FGD and other coal combustion wastes may be due sometime in the near future.

Presently, the predominant wet SO_2-scrubbing system in the U.S. is the so-called "conventional wet scrubbing" using magnesium-enhanced lime or limestone without forced oxidation. The process produces a FGD waste that is primarily calcium sulfite and gypsum. In some instances, SO_3-oxidation inhibitor is added to reduce gypsum scaling. The waste is usually mixed with fly ash (1:1) and 5% lime for stabilization before disposing in landfills.[4] The USEPA estimated that in 1984 some 14.5 million Mg of FGD wastes was produced in the U.S. By the year 2000 FGD waste production may reach 45.5 million Mg annually.[49] Total coal combustion wastes production such as bottom, fly, and coal gasification ashes, waste FGD gypsum, and FBC residue is projected to reach more than 150 million Mg annually by the year 2000.[50]

Commercial-grade FGD gypsum is produced by only a few electric utilities in the U.S. The Electric Power Research Institute (EPRI) reported in 1989 that, of the 68 utilities operating 129 electric-generating units equipped with wet scrubbers, only 10 utilities representing 16 units practiced forced oxidation to produce commercial-grade FGD gypsum. However, only three of the utilities reported commercial utilization of FGD gypsum produced.[51] Presently, the U.S. produces between 600,000 and 700,000 Mg of commercial-grade FGD gypsum, less than 50% of which are used in board manufacture, cement production, and in agriculture. However, in view of the new acid rain legislation that requires reduction of SO_2 by 10 million Mg annually, as much as 28 million Mg of FGD gypsum may be produced. Conservative estimates place this future commercial-grade FGD gypsum production from 5 to 10 million Mg annually.[52]

The existing wet scrubbers in the U.S. are located primarily in the region of the Ohio valley, the midwest, and Texas. Other scrubbers are also found in Florida, Arizona, and other western states. The utility that is believed to be the largest source of FGD gypsum in the U.S. is the Tennessee Valley Authority's (TVA's) Paradise Steam Plant via its Chemico FGD-equipped Units 1 and 2 in Drakesboro, Kentucky. Other sources are the Big Bend Plant of Tampa Electric and St. John's River Plant of Jacksonville Electric Utilities, Florida; Martin Lake Plant of Texas Utilities and Deepwater Cogen Plant of AES Utility, Texas; Sims Plant of the Grand Haven Utility, Michigan; and Bailey Plant of Northern Indiana Public Service Company, Indiana.[11]

Canada, which introduced legislation in early 1980 to reduce SO_2 emissions from coal-fired plants by 50% by 1994, has no operational wet FGD systems at present. However, two plants under construction by the New Brunswick Power and the Ontario Hydro utilities would have a total production capacity of 450,000 Mg FGD gypsum annually.[52]

The $CaSO_4.xH_2O$ contents in commercial-grade FGD gypsum could range from 91 to 99%,[11,53,54] with typical particle-size distribution clustering around 0.04 to 0.05 mm.[55,56] Depending on the production process, pH could range from 3 to 9, but most FGD gypsum's pHs are in the range of 6.0 to 8.5 determined by using 10 g gypsum in 100 mL water. Total water-soluble salts are normally no more than 1000 mg kg^{-1}.[11] Typical major chemical components of forced oxidation FGD gypsum are given in Table 3.[27,57,58]

3. *Fluidized Bed Combustion (FBC) and Spray-dry Absorption (SDA) Materials*

The FBC materials are produced by high-temperature S-removal atmospheric pressure (Ap) or pressurized (P) fluidized bed combustion (Ap-FBC or P-FBC) process. The process may be "once-through," that is nonregenerative or regenerative.[59] In the FBC system, coal is burned in a fluidized bed of granular limestone or dolomite which acts as sorbent for SO_2 released when S-containing coal is burned. To maintain the reactivity of the sorbent, fresh material is continuously fed to the bed. At the same time the spent material is continuously removed to keep the depth of the bed constant.[13] Such a system produces large quantities of FBC waste. It is estimated that a 1000-MW power plant using a FBC system without regeneration of the residue would produce about 1800 Mg of FBC material each day.[59,60] Another process by which FBC residue is produced is furnace sorption injection (FSI)[61–63] whereby an alkaline material such as limestone or a mixture of alkaline material and coal is fluidized in a jet of air and injected into or around the flame in the furnace. The main constituents of FBC materials are CaO, $CaSO_4$, and $CaSO_3$ with typical concentrations given in Table 4. Particle size distribution of the materials clusters around 1 to 2 mm (60%) with 30% in the <1-mm range.[64]

The SDA desulfurization process uses CaO which when mixed with water produces a $Ca(OH)_2$ slurry. This slurry is injected into the stream of hot flue gas, instead of into the flame, causing the $Ca(OH)_2$ to react with SO_2 to form a dry SDA residue that is mostly a mixture of $CaSO_3.{}^1\!/_2H_2O$ and CaO. It is, therefore, not a true gypsum until fur-

Table 3 Typical Chemical Analyses of FGD Gypsum

	FGD Samples (% by weight)			
Composition	U.S[57]	U.S.[57]	Germany[57]	Japan[27a]
CaO	32.20	34.00	32.70	32.28–32.76
SO_3	46.60	41.40	45.90	45.73–47.00
SiO_2	0.01	0.68	0.09	0.26–0.36
Al_2O_3	0.05	0.11	0.02	0.03–0.15
Fe_2O_3	0.02	0.05	0.04	0.03–0.08
MgO	0.22	0.29	0.11	<0.01–0.20
Na_2O	—	—	—	0.04–0.05
K_2O	0.01	0.01	0.01	0.01–0.02
F	<0.01	<0.06	<0.04	—
P_2O_5	0.01	0.04	0.01	—
Crystalline H_2O	20.80	19.20	20.70	19.76–19.94

[a] Ranges from four samples from different producers using limestone as absorbent.

Table 4 Mineralogical Analysis of FBC and SDA Materials

Constitutent	Samples (% by weight)	
	FBC[64]	SDA[17a]
$CaSO_4 \cdot 2H_2O$	—	4.5–10.0
$CaSO_4$	52.00	—
$CaSO_3 \cdot 1/2H_2O$	0.60	50.0–62.0
$CaCO_3$[b]	—	0.6–12.5
$MgCO_3$[b]	—	2.4–18.9
CaO[c]	33.00	32.9–46.9
MgO[c]	0.80	0.4–1.7
$CaCl_2$	—	0.6–7.2
NaCl	0.30	—
P_2O_5	0.02	—
Fe_2O_3	4.50[d]	0.4–1.8
Al_2O_3	—	0.8–5.4
SiO_2 + insolubles	7.00	1.4–11.1

[a] Ranges from four different sources.
[b] Values may range from trace to major constituent in FBC materials.
[c] Values depend on whether limestone or dolomite was used as sorbent in the FBC plants.

ther oxidized. The typical concentrations of the major constituents of SDA materials are given in Table 4. In West Germany, the combined production of FSI and SDA materials was reported to be 0.60 million Mg in 1989, and it is expected to increase to 1.8 and 4.3 million Mg by 1995 and 2000, respectively.[65]

II. METHODS OF APPLICATION, MOVEMENT, AND REACTIONS IN SOILS

A. METHODS OF APPLICATION

By-product gypsums are surface applied. For tilled crops, the by-product is mixed with the soil to a depth of 15 cm. In established pasture, by-product gypsum is applied to the pasture before the start of the rainy or growing season. In orchards, surface application or "within row cap" with shallow mixing to a depth of 2 to 4 cm is normally used.[64] For sodic or saline soil amelioration, the gypsum material may be dissolved in irrigation water.[66–68] Application rates depend on the purpose for which the soil amendment is applied to the soil as well as on the application frequency.

B. SOLUBILITY, SOLUTION PROPERTIES, AND SOLUBILITY RATES OF BY-PRODUCT GYPSUMS

The movement of soil amendments in the soil profile is determined by several factors, the most important of which are the solubility and solubility rate of the amendment and the reaction of the various dissolution products of interest with the soil constituents.

Samples of a Florida phosphogypsum dissolved at a consistent solubility of 2.6 and 4.3 g L^{-1} in water and Mehlich I solution (0.025 M HCl + 0.0125 M H_2SO_4), respectively, with the aqueous solution having an electrical conductivity (E_c) of 0.21 S m^{-1} and a pH of 5.2.[69] Natural or mined gypsum has a solubility of 2.41 g L^{-1} water at 25°C.[70] Bolan et al.[71] found the solubilities of FGD gypsums from the Netherlands, Illinois, and Florida to range from 2.62 to 2.67 g L^{-1}. The solubilities of phosphogypsum from Florida and mined gypsum from Britain were 2.75 and 2.54 g L^{-1}, respectively. The E_c of the saturated solutions of the various gypsum sources ranged from 0.19 for mined gypsum to 0.20 S m^{-1} for phosphogypsum. Saturated solution pH's ranged from a high of 6.9 to 7.4 for mined and FGD gypsums to a low 4.8 for phosphogypsum.

The main S constituents of FBC and SDA residues are anhydrite ($CaSO_4$) and Ca sulfite ($CaSO_3.{}^1/_2H_2O$), respectively. Natural anhydrite has a solubility of 2.1 g L^{-1} in water at 25°C, while that of Ca sulfite is 0.043 g L^{-1} water at 18°C, or about 2% that of $CaSO_4$ materials.[70] At a ratio of 1:1 (solid:water) by weight, the pH of FBC materials could go as high as 12.5 with an E_c of 1.46 S m^{-1}.[72]

Since solubility determines the levels of saturation of the soil moisture or soil water with the dissolution products Ca and SO_4 ions, the low solubility of $CaSO_3.{}^1/_2H_2O$ in SDA materials would make them poor sources of Ca and S for agriculture use. The sulfites may also be harmful to plants, by producing H_2S gas under anaerobic conditions[73] and increasing oxygen demand under aerobic conditions,[74] unless applied far in advance of planting. Clark et al.,[75] in a greenhouse study, has shown that mixing an FGD by-product containing 169 g SO_3-S kg^{-1} with soil at rates greater than 1% by weight reduced the root and shoot dry matter of maize.

The presence of other ions in the solvent may affect the solubility of the amendment. Marshall and Slusher[76] showed that the solubility of dihydrate gypsum increased in the presence of NaCl in the range of up to 3 molal. Keren and Shainberg,[77] using three sources of gypsum, analytical and mined gypsum and phosphogypsum, showed that total Ca concentrations in solutions of 0.05 and 0.10 *N* NaCl were 1.3 and 1.5 times over the Ca concentration in distilled water, respectively. Bolan et al.[71] showed that the solubilities of various sources of gypsum were much lower in 0.01 *M* $CaCl_2$ than in deionized water which they attributed to common ion effect.

The dissolution of gypsum has been described as a first-order chemical reaction.[77,78] The first-order dissolution coefficient (K_1) is given by the equation $K_1 = -\{\ln[1 - (C_t/C_s)]\}/t$, where C_t and C_s are the gypsum concentrations in the solution at time t and at saturation, respectively. Bolan et al.[71] reported that the dissolution data of various sources of gypsum in water were best fitted to a second-order kinetic equation $C_t/C_s)/(C_s - C_t) = k_2t$. In the presence of soil (5 g soil, 0.8 g gypsum in 50 mL water or 0.01 M $CaCl_2$ solution), the dissolution data fitted best the first-order kinetic equation.

The greater the value of K, the greater is the solubility rate of the material. Keren and Shainberg[77] determined $K_{1phosphogypsum}$ to be 198 and 58 and $K_{1mined\ gypsum}$ 19 and 6 ($\times 10^4$ s^{-1}), for fragment size ranges of 1.0- to 2.0- and 4.0- to 5.7-mm diameter, respectively. The corresponding $K_{1phosphogypsum}/K_{1mined\ gypsum}$ ratios were 10.4 and 9.7 for the two fragment size ranges. The actual dissolution rates of phosphogypsum and mined gypsum for the two fragment size ranges are given in Figure 1. The dissolution ratios (meq $Ca_{phosphogypsum}L^{-1}$/meq $Ca_{mined\ gypsum}L^{-1}$) were in the order of 3.0 for the smaller and 5.0 for the larger fragments at t = 2.5 min. These ratios appear to increase still at smaller values of t for both ranges of fragment sizes. Bolan et al.[71] reported the K_1 in water for powdered phosphogypsum, FGD, and mined gypsum to be 13, 18, and 9 ($\times 10^2$ min^{-1}), respectively, giving a much smaller $K_{1phosphogypsum}/K_{1mined\ gypsum}$ ratio than that reported by Keren and Shainberg.[77] The actual dissolution rates of the various sources of gypsum in water and in $CaCl_2$ solution with and without the presence of soil materials are given in Figure 2.

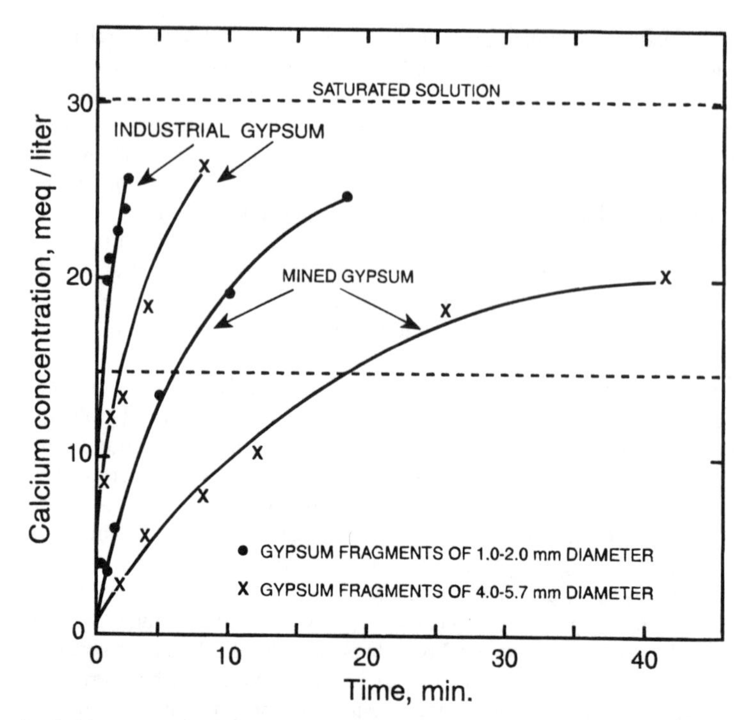

Figure 1 Calcium concentration in water vs. time plots for industrial and mined gypsum samples of two fragments sizes. (From Keren, R. and Shainberg, I., *Soil Sci. Soc. Am. J.*, 45, 103, 1981. With permission.)

Based on the solubility rate it is expected that, for a given soil type and amount of rainfall, more Ca and SO_4 from by-product gypsums will move deeper into the soil profile and at a much faster rate than from mined gypsum. Where gypsum is to be applied with the irrigation water, the contact time between the flowing water and the bed of gypsum to achieve a desired Ca and SO_4 ions concentration may be reduced by as much as ten times if phosphogypsum were used in place of mined gypsum, using Keren and Shainberg's K values. Also, the high solubility rates of by-product gypsums make them more effective ameliorants for Al toxicity, subsoil acidity, and infertility than lime materials, and ideal for sodic and saline soils reclamation.

When FBC materials are used as sources of Ca for plants, the Ca from the gypsum constituent would be more readily available than the Ca from the lime constituents in these residues. The solubility of natural calcite and aragonite ($CaCO_3$) is in the order of 0.015 g L^{-1} water which is less than 1% of the solubility of the $CaSO_4$ materials. The solubility of CaO in water is 1.31 g L^{-1} or about 50% that of $CaSO_4$.[70]

C. REACTIONS AND MOVEMENT OF $CaSO_4$ AND LIME CONSTITUENTS IN SOILS

Soils with 1:1 clay minerals have been shown to adsorb more SO_4 than those with 2:1 clay. Kamprath et al.[79] found that SO_4 adsorption in soils was directly related to SO_4 concentration in the solution and sorption decreased as pH increased from 4 to 6.

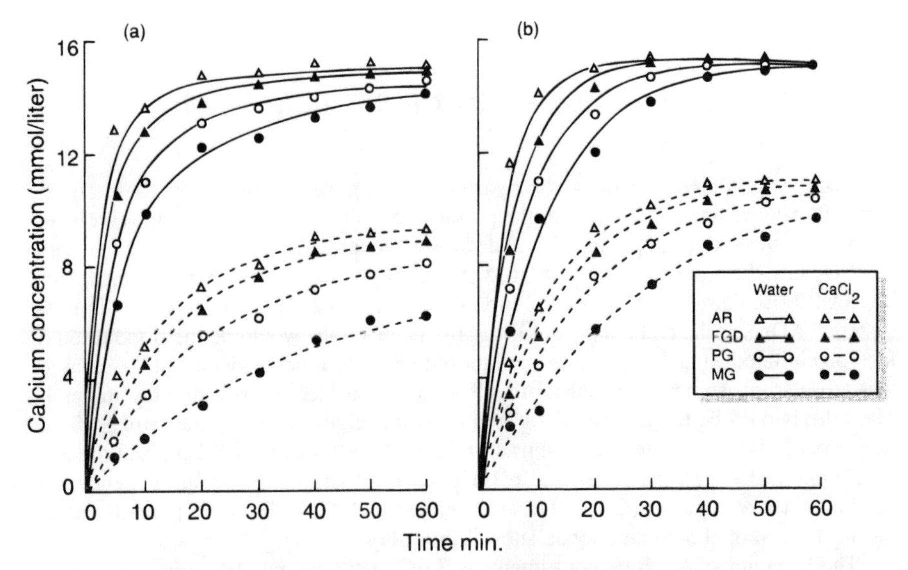

Figure 2 Dissolution (Ca dissolved) rate of phosphogypsum (PG), analytical (AR), flue gas (FGD), and mined (MG) gypsum in water and in 0.01 *M* $CaCl_2$ in the (a) absence and (b) presence of soil. (From Bolan, N.S., Syers, J.K., and Sumner, M.E., *J. Sci. Food Agric.*, 57, 527, 1991. With permission.)

The kinds of anions present in the soil may also affect SO_4-sorption. For all soils studied, increasing PO_4 concentration in the solution reduced the amount of SO_4 adsorbed by the soils.[79] Chao,[80] studying 26 inorganic and organic anions, found that PO_4 and F reduced SO_4 sorption in soils tested by 44 and 30%, respectively. Hydroxyl and HCO_3 anions increased the solution pH and consequently decreased SO_4 adsorption. Chao also found that the effect of NO_3 and SiO_4 on SO_4 sorption appeared to be pH-independent throughout the pH range from 4 to 6. The affinity of SO_4 in a ligand exchange in soils was determined to be at least ten times that of non-specifically adsorbed NO_3 and Cl.[81]

The kind of cation associated with SO_4 also affected SO_4 sorption. Chao et al.,[82] in an equilibration study, found that SO_4 sorption capacity of soils increased in the order of $CaSO_4 > K_2SO_4 > (NH_4)_2SO_4 > Na_2SO_4$. With respect to the cations saturating the exchange complex, the order was Al > Ca > K. As the pH of the solution approached neutrality, SO_4 sorption decreased regardless of the saturating cation. The greater the exchangeable Al, sesquioxide, and amorphous materials in the soil the greater was the observed pH effect.

The adsorption of SO_4 in acid soils plays the key role in amelioration of subsoil acidity. Chang and Thomas[83] proposed a comprehensive mechanism to account for SO_4 sorption in soils. Assuming a homoionic Al-saturated clay coated with hydrated oxides of Fe and Al (R), two opposing reactions were proposed:

1. Al and/or Fe hydrolysis:

$$\text{Clay-Al}_x + yH_2O + yK^+ \rightarrow \text{Clay-Al}_x(OH)_y^{yK} + yH^+$$

2. Ligand exchange:

$$\text{Clay-R}_x(\text{OH})_y + \text{SO}_4{}^{2=} \rightarrow \text{Clay-R}_x\left[(\text{OH})_{y-z}(\text{SO}_4)_z\right] + z\text{OH}^-$$

According to these reactions, the hydrolysis of exposed and/or adsorbed Al and Fe at the clay edges and/or surfaces, in the presence of salt, a K-salt, for example, releases H ions into the solution causing Al-induced (or Fe-induced) acid conditions. In the case of a SO_4-salt, the more negatively charged SO_4 readily replaces OH in a ligand exchange. The OH then neutralizes H in the solution causing a so-called "self-liming" within the system.[84,85] One of the end results predicted by the reactions would be the dissolution of face-precipitated Al and Fe, freeing the blocked permanent negative charges — charges that arose from isomorphous substitution[86] — at the surfaces of silicate clay minerals. The other would be the release of Al ions from the edges of the crystal lattice of silicate clays and exposing new ones, thus sustaining the adsorption of SO_4 ions at the positively charged edges, increasing the pH-dependent[87,88] and ionic-strength-dependent[89,90] cation exchange capacity (CEC). These increases in CEC, with proper pH control and fertilization, should help ameliorate subsoil infertility.

The behavior of the lime constituents in FBC materials may be anticipated by the reactions of pure limestone or dolomite in the soil. Metzger[91] and Brown and Munsell[92] reported that 10 to 14 years was required for surface-applied lime materials to increase soil pH to a depth of 15 cm, primarily due to the low solubility and mobility of limestone or dolomite in soils. Limestone, reacting with the soil moisture, supplies the soil with both OH and HCO_3. These, however, may be neutralized immediately by H at the soil surface. Any increase in pH in the surface soil may also increase Ca adsorption due to the extra negative charges generated in the amphoteric soil constituents.[84] Neutralization and the adsorption of Ca at the surface keep the ameliorative effect of pure lime limited to the surface.[84,93] Not only is subsoil acidity not checked by surface application of pure lime, but the practice also fails to supply Ca to the deeper horizons to alleviate subsoil infertility. However, in conjunction with $CaSO_4$, also present in FBC materials, the lime constituents in the impure by-product gypsum should correct not only surface soil acidity but help ameliorate subsoil acidity and infertility as well. Korentajer et al.[94] found that the combination of lime and gypsum applied to acid soils resulted in a greater leachability of the SO_4 ions. Morelli et al.[95] reported that a mixture of lime and gypsum applied to a Latosol resulted in a better distribution of Ca and Mg in the soil horizons, increased base saturation, and reduced exchangeable Al. A combination of 4 Mg lime with 2 Mg gypsum ha^{-1} resulted in the highest sugarcane (*Saccharum officinarum* L.) stalk production.

D. CONCENTRATIONS AND SOLUBILITIES OF SOME TRACE ELEMENT CONSTITUENTS AND THEIR REACTIONS IN SOILS

Nine trace elements, namely, B, Cl, Cu, Fe, Mn, Mo, Ni, Na, and Zn, are commonly accepted as essential micronutrients for higher plants. For animal life, Davies[96] listed 14 essential micro-elements, eight of which are specific to animals, namely, As, Co, Cr, F, I, Se, Si, and V, while six are essential to both animal and plant, namely, Cu, Fe, Mo, Mn, Ni, and Zn. In moderate amounts, some of these elements may become phytotoxic

Table 5 Total Macro- and Microelements in Phosphogypsum (PG), FGD gypsums, FBC residues, and Their Ranges in Soils

Element	PG[a]	FGD[b]	FBC[c]	Soil[d]
		Major Plant Nutrients (g kg^{-1})		
Ca	179.0–235.0[e]	227.9–242.9	240.0–460.0	<7.0–>40.0[102]
Mg	0.0–2.8	0.2–2.7	5.0–12.0	0.5–5.0[103]
K	0.0–1.7	0.1	0.5–8.0	—
P	7.9–16.1	<0.1–0.2	0.4–0.5	0.4–3.0
S	127.6–180.8	165.6–186.4	72.0–140.0	0.1–1.5
		Other Major Constituent		
Al	2.0±0.5	0.1–1.64	4.0–20.0	14.0–40.0
		Minor Plant Nutrients (mg kg^{-1})		
Fe	860–1,000	140–1,260	800–16,000	14,000–40,000
Mn	25±14	3.4–15.8	210–685	200–3,000
Mo	2.2–11.0	0.35–1.7	0.12–0.28	0.2–5
B	0.0–3.0	—	95–170	2.0–100
Cu	<82±9.6	0.6–3.2	12.0–19	2.0–100
Zn	<340±21	2.9–<7	29.0–105	10.0–300
Ni	0.0–2.0	—	13.0–29	5.0–500
Na	520±79	27–729	679–7,050[g]	—
Cl	150±4.7	<20–2,347	—	
		Trace Animal Nutrients (mg kg^{-1})		
As[f]	0.76–0.94	0.14–1.74	<0.1–0.3[100]	8.4–17[g]
Co	0.58±0.15	0.05–0.46	35.0[g]	0.4–400
Cr[f]	6±1.40	2.20–11	9.0–23	5.0–1,000
F	2000–8,000	95–729	—	—
I	0.90–3.80	<0.30–1.7	—	—
Se[f]	0.72–2.10	<1.00–5.2	0.16–0.58	0.1–2
V	1.8–4.0	1.57–9.06	—	11.0–63[h]
		EPA "Toxicity Index" metals (mg kg^{-1})		
Ag	<1.3±0.64	—	<10.0[99]	0.18[i]
Ba	<210±24	2.5–7.0	40.0[100]	—
Cd	3–4	—	0.50	0.01–0.70
Hg	0.28–0.40	0.04–0.36	—	0.1–1.71[j]
Pb	2–13	—	1.5–7.5	2.0–200

[a] From References 38 and 39 unless otherwise indicated.
[b] From Reference 57; ranges in five FGD gypsum samples.
[c] From Reference 98 unless otherwise indicated.
[d] From Reference 101 unless otherwise indicated.
[e] Range from Table 1.
[f] Among the USEPA "toxicity index" metals.
[g] From Reference 104; range in Scottish, English, and Welsh soils.
[h] From Reference 104; range in German soils.
[i] From Reference 104; mean in English and Welsh soils.
[j] From Reference 104; range in Scottish soils.

to plants and animals. Even in trace amounts, As, Cr, as well as Ag, Ba, Cd, Hg, Pb, and Se are of environmental concern.[97] Table 5 shows the concentrations of the major and trace elements in phosphogypsum,[38,39] FGD gypsum,[57] and FBC residues.[98–100] Their concentrations in certain soils are also presented as a reference.[101–104]

None of the by-products contain sufficient quantities of minor plant nutrients to supply the needs of agricultural crops, if applied alone in moderate amounts. Among the trace elements essential to animals yet harmful in excess amounts, only F in phosphogypsum appears to present some concern. Fluoride, present originally in rock phosphate, may be found in relatively high concentrations in phosphogypsum. Phosphogypsums from Australia,[22] India,[28] and Florida[38] were reported to contain from 11 to 13, 5 to 40, and 2 to 8 g F kg^{-1} of phosphogypsum, respectively. Another concern is the presence of heavy metals, particularly the so-called EPA "toxicity index" metals.[97] Except for Ag, Ba, and Cd in phosphogypsum, their concentrations were no larger than the range reported in soils. It has been suggested that the heavy-metal contaminants detected in phosphogypsum are likely to be predominantly compounds of Ca or SO_4.[39] The major and minor plant nutrients and F in phosphogypsum have been found to be soluble in acid solution (0.025 *M* HCl + 0.0125 *M* H_2SO_4). Their solubilities were correlated to the solubility of phosphogypsum itself. In water, Ca, P, K, and F were moderately soluble and Cu and Mn were highly soluble. Iron and Al were relatively insoluble in water.[69]

III. SPECIFIC USES AND RATES OF APPLICATION

A. SOURCE OF SULFUR AND CALCIUM FOR CROPS

Sulfur is essential to plant growth. In general, plants contain as much S as P, the usual range being from 0.2 to 0.5% on a dry-weight basis. Sulfur is as important as N as a constituent of the amino acids cysteine, cystine, and methionine in proteins that account for 90% of S in plants. It is also needed in the formation of oil in crops such as peanut (*Arachis hypogaea* L.), soybean (*Glycine max* [L.] Merr.), flax (*Linum usitissimum*), and rapeseed (*Brassica campestris*).[105]

Calcium is also important to plant life, with concentration ranging from 0.2 to 1.0% in plant tissues. Calcium deficiency shows itself in the failure of terminal buds and apical tips of roots to develop. Lack of Ca also results in general breakdown of membrane structures, with resultant loss in retention of cellular diffusible compounds. Disorders in the storage tissues of fruits and vegetables frequently indicate Ca deficiency.[105]

Sulfur deficiencies in soils are being reported with greater frequency throughout the world. The major reason is the increasing use of S-free high-analysis fertilizers.[106,107] There is no doubt that these deficiencies could become more severe as more countries adopt FGD of their coal-fired power stations.

Comprehensive reviews of the S status of soils in the U.S. by region as well as in other parts of the world have been presented by Tabatabai.[108] Blevins,[109] based on other studies and an ongoing program called "Test for S" conducted since 1982, indicated that the number of states reporting S deficiency in the U.S. has progressively increased from 13 in 1962, to 36 in 1986, and 48 in 1991. Only Nevada and Utah had no report of S deficiency. The recent studies by Raun and Barreto[110] to determine the response of corn (*Z. mays*, L.) to gypsum in six countries in Central America showed localized response to S in Guatemala, Panama, and El Salvador. A study by Tandon[111] on crop response to S in India showed that nearly half of over 40 crops fertilized with S under field conditions responded to S fertilization. Average increases in yields ranged from 16 to 41%.

The need for Ca by plants may be readily satisfied by liming materials such as calcitic and dolomitic limestone. However, lime application in large amounts on certain soils can be detrimental to plant growth. Kamprath,[112] in a review of the effect of lime on Oxisols and Ultisols, reported that lime application that raised the soil pH to 7 resulted in reduced rates of water infiltration, reduced availability of P, B, Mn, and Zn, and reduced growth of sudangrass (*Sorghum vulgare* var. sudanese L.), corn, and soybean. Therefore, for certain soils that require large amounts of Ca to support commercially viable crop yields, or for crops that need large amounts of readily soluble Ca such as peanut, a Ca source other than lime, such as the gypsum materials, is necessary. The response of cereal crops, grain legumes, sugarcane, fruits and vegetables, and forage crops to phosphogypsum and mined gypsum applied as source of S and Ca has been reviewed by Alcordo and Rechcigl[1] and Shainberg et al.[2]

The higher the application rates of any soil amendment containing hazardous impurities, the greater is the concern for environmental pollution. Tandon[111] reported that S uptake by crops ranges from 5 to 46 kg S ha^{-1}. Hence, the rate of S application as plant nutrient is normally very moderate. Experimental annual rates to determine the response of most crops to S seldom reach 90 kg S ha^{-1}, with increments between levels ranging from 5 to 20 kg ha^{-1}. When a gypsum material is used as a source of S, the maximum annual rate seldom exceeds 0.5 Mg ha^{-1}, whether mined or FGD gypsum or phosphogypsum. Peanuts possess a unique nutritional requirement in that supplemental Ca must be supplied to the "peg", a modified stem that penetrates the soil surface to form the pod or nut.[113] Gascho and Alva[114] reported that, as a source of Ca for peanut, phosphogypsum is normally applied at the annual rates of 0.5 to 1.0 Mg ha^{-1}. This range of application is also applicable to sugarcane.[115] Golden[116] used 2.5 Mg phosphogypsum per hectare in Louisiana. At low annual rates of application of by-product gypsums, environmental concerns may not be justified. However, rates as high as 11.2 Mg ha^{-1} phosphogypsum and 22.4 Mg ha^{-1} fluorogypsum have been applied to alligator clay soils in Louisiana with significant increases in sugarcane stalk and sugar yields.[117,118] When annual rates of application are as high as 5 or more Mg ha^{-1} environmental concerns as to the effect of major and trace constituents of by-product gypsums on plant tissue, groundwater, soil, and the ambient atmosphere in the case of ^{226}Ra-bearing by-products become valid issues.

B. AMELIORANT FOR AL TOXICITY AND SUBSOIL ACIDITY AND INFERTILITY

Jackson[119] grouped the acid or acid-forming constituents in soils as (1) free mineral acids such as H_2SO_4, (2) organic acids, and (3) active Al and Fe. Soils dominated by each of these soil acids are (1) the sulfate soils of Thailand,[120] (2) the virgin Spodosols of Florida,[121,122] and (3) the Oxisols and Ultisols of the tropics and subtropics.[123] It is with the last group of acid soils that gypsum materials have found application as a subsoil acidity ameliorant.

Acid soils typically have a pH of <5 in water and about 4 in salt at the surface to a depth of 1 m.[2] They constitute from 40 to 50% of potentially arable highly weathered soils,[124] estimated at more than 800 million ha worldwide.[125] They are located primarily in the tropics and subtropics where intense chemical weathering occurs due to high temperature and rainfall. Under such conditions the bases that are highly soluble are leached to greater depths than are Al and Fe, leaving behind soil horizons enriched in active Al and Fe which, in the late stage, may be compacted by their oxides and hydrous oxides. As the deeper horizon becomes more acidic, more Al and Fe are solubilized from primary minerals and from their secondary oxides and hydrous oxides.[126] This, in turn, increases the Al and Fe saturation of the exchange complex of the colloidal soil constituents, making the subsoil infertile. When Al saturation of the exchange capacity exceeds 60%, appreciable amounts of Al^{3+} get into the soil solution[127] which could build up to levels toxic to plant roots. Toxicity due to excess Al may inhibit root penetration and proliferation as is frequently observed in the highly weathered acidic soils of the southeastern U.S.[128–132] Adams and Moore,[133] investigating root growth in subsoil horizons of Coastal Plain soils in the U.S., found not only Al toxicity in the argillic horizons (Bt) but even a more prevalent Ca deficiency in both eluvial and illuvial horizons wherein Ca saturations were $\leq 17\%$.

For soils with serious subsoil acidity and infertility, surface application of lime may not be the practical answer to the problem.[91,92] Heavy application of lime to overcome low solubility and downward mobility may prove deleterious to the physicochemical properties of these soils.[112] Deep placement of lime which requires the use of heavy machinery has not found widespread application even in advanced countries because of the cost involved. In developing countries, where most of the problem exists, the cost of the heavy equipment needed is simply prohibitive, making deep lime placement unrealistic. The use of gypsum and gypsum by-products, alone or in combination with other chemical or mechanical treatments to enhance their ameliorative efficiency, appears to be the most practical approach to the worldwide problem of subsoil acidity and infertility. A comprehensive review of the theoretical basis for the use of gypsum materials to alleviate subsoil acidity, subsoil infertility, and Al toxicity has been done by Alcordo and Rechcigl.[1]

Application rates of gypsum materials for subsoil acidity amelioration are relatively high. In Georgia, Hammel et al.[134] found that 35 Mg mined gypsum per hectare incorporated to a depth of 15 cm increased soybean grain yield by 26% in the second year and corn silage yield by 35% in the third year. The 2-year average grain soybean and the third year corn silage yields in gypsum-treated plots were 38 and 36% higher, respectively, than in one where the subsoil was loosened and mixed down to 1 m. The study showed that chemical modification of the acid and infertile soil profile is more effective than physical modification in improving soybean and corn yields in these soils. Rechcigl et al.[135] in Virginia and Odom[136] in Alabama reported that application of gypsum at 13 and 10 Mg ha^{-1}, respectively, increased alfalfa (*Medicago sativa*, L.) forage yields. Farina and Channon[137] reported that gypsum at 10 Mg ha^{-1} applied to corn on a limed, strongly acidic Plinthic Paleudult soil increased grain yield by an average of 19% over that of the limed control over a 3-year period. Progressive reduction in the level of exchangeable Al was accompanied by increased subsoil Ca, Mg, and SO_4-S. Water pH increased markedly in the zone of maximum SO_4-sorption/precipitation. Effects of gypsum on subsoil root development of corn were striking by the fourth cropping season. In Louisiana, Caldwell et al.[138] reported that by-product gypsum applied on acid Gigger silt loam in 1986 not only increased cotton yields in 1987, 1988, and 1989 but also increased exchangeable Ca at the depths of 30, 45, and 60 cm during the first and third year, respectively. In lime-treated plots, increases in exchangeable Ca were mainly at the top 15 cm and none at 30 cm. Sumner,[85] using mined gypsum and phosphogypsum applied at 5 to 10 Mg ha^{-1} incorporated into the soil in several field experiments on a range of soils in southeastern U.S., concluded that there were no differences between the two gypsum sources based on crop responses and soil reactions. Highly significant and economically profitable yield responses were obtained for alfalfa, corn, soybean, cotton (*Gossypium hirsutum* L.), and peaches (*Prunas persica* L.). Mined gypsum and phosphogypsum application enhanced root penetration and proliferation in the subsoil, where previous conditions often prevented root growth.

Considering the very high rates of applications required in subsoil acidity amelioration, the environmental aspects associated with the use of by-product gypsums cannot be ignored.

C. AMELIORANT FOR SODIC AND NONSODIC DISPERSIVE SOILS

In the arid regions of the world evapotranspiration exceeds rainfall. Soluble salts in the water table move upward in the soil profile instead of downward as occurs in regions of acid soils. In such regions the dominant soil clay mineral is montmorillonite, but illite and vermicullite are also common.[2] The soils in these regions may contain various soluble salts and exchangeable Na at levels that interfere with the growth of most crops. When the E_c of the saturation extracts of these soils exceeds 0.2 S m^{-1} at 25°C with a Na adsorption ratio (SAR = $Na/[Mg + Ca]^{1/2}$) >15, the soil is classified as saline sodic or simply sodic. The saturation extracts of ordinary saline soils have the same E_c value as a sodic but with a SAR of <15.[139] Because of the Na ions on the exchange complex of the colloidal fraction, these soils are highly dispersive in water of low salt concentration. The dispersed clay particles normally clog the soil pores during rain or irrigation, reducing or completely stopping water infiltration, and form hard crusts at the surface upon drying, making seedling emergence difficult. Poor water infiltration rate (IR) during rain or irrigation and hard surface crusting upon drying are the two major problems that need to be ameliorated if sodic soils are to be used for crop production.

The studies of Hilgard[140] on the toxicities of Na_2CO_3, NaCl, and Na_2SO_4 to plants and those of Kelley and Brown[141] on the reclamation of so-called black alkali (sodic) and white alkali (saline) soils in California established gypsum as the material of choice for sodic and saline soil reclamation. Later work by the staff of the Salinity Laboratory of the U.S. Department of Agriculture (USDA)[142] led to the recommendation that for each $mmol_c$ of Na that needs to be replaced and leached from a 15-cm depth, 2.016 Mg of $CaSO_4$ or 0.358 Mg S are required per hectare.[139] Theoretical models also have been formulated to predict gypsum and leaching requirements of Na-affected soils on the basis of kinetic[143,144] or equilibrium chemistry.[145,146] Mined gypsum has been traditionally used for sodic soil reclamation, but the use of FGD gypsum or phosphogypsum may prove to be more efficient than using mined gypsum because of their high solubility rates.[71,77]

The rates of mined gypsum or by-product gypsum application for sodic soil reclamation are rather high. In California, phosphogypsum is applied at an initial rate of 22.4 Mg ha^{-1} which is mixed with the soil to a depth of 15 cm, followed by 11.2 Mg ha^{-1} every second year thereafter.[147] In India, initial rates as high as 32 Mg ha^{-1} of phosphogypsum had been used for sodic soil reclamation, despite high F content.[28]

Nonsodic dispersive soils have also been shown to benefit from the application of gypsum materials at rates between 3 and 10 Mg ha^{-1}. The surface soils of the red brown earths of Australia whose SAR values are less than the U.S. standard for sodic soils were observed to disperse spontaneously under field conditions.[148] Sims and Rooney[149] reported substantial improvement in the physical condition of these soils when gypsum was applied at 2.24 to 4.5 Mg ha^{-1} on the surface of fallowed land in the spring of the year before cropping. Loveday's[150] work, which established a general relationship between crust strength, hydraulic conductivity (HC), and seedling emergence for a range of dispersive soils, led to the implementation of the recommendation on the use of gypsum on hard-setting wheat soils in Australia.[151]

Some soil series in the southeastern U.S. also have highly dispersive surface soils,[152–154] despite their high kaolinite and sesquioxides contents.[2,155] In Georgia, Miller[156] reported that surface-applied 5 Mg ha^{-1} phosphogypsum doubled the final IR of a Typic Hapladults and reduced soil loss by 30 to 50%. Miller[157] concluded that increases in water intake and reductions in runoff and soil loss in runoff were primarily the effects of ionic strength. The high E_c (0.05 to 0.13 S m^{-1}) in the runoff maintained by phosphogypsum kept soil clays flocculated thereby reducing crusting.[156]

Agassi[158] using five soils from Israel, showed that plots where phosphogypsum was spread over the surface maintained the highest IR compared to those where phosphogypsum was mixed with the surface soil or the control. The IR of phosphogypsum-treated plots remained high while that of the control approached zero when the cumulative "rainfall" exceeded 60 mm. Agassi[159] also demonstrated under field conditions that phosphogypsum at 5 to 10 Mg ha^{-1} spread over the surface effectively reduced soil erosion and runoff in wheat fields at various ranges of rainfall. Kazman et al.[160] showed that phosphogypsum at 5 Mg ha^{-1} kept a high IR on a Natanya sandy loam soil with exchangeable sodium percentage (ESP) of 11.6 under a simulated rainfall of 26 mm h^{-1} and even beyond a cumulative rainfall of 70 mm. At this ESP, the IR of the control approached zero at 30 mm of cumulative rainfall.

D. AMELIORANT FOR SUBSOIL HARDPANS AND HARD-SETTING CLAY SOILS

Physical barriers in the soil profile such as natural hardpans, dense textural B horizons, and tillage pans formed by heavy farm machinery may limit root penetration and proliferation.[161] Radcliffe et al.[162] showed that cone index values, a measure of mechanical impedance, were lower on gypsum-treated plots that had been cropped for several years than on the fallowed plots. They concluded that gypsum increased subsoil root activity which, in turn, reduced subsoil mechanical impedance. A review by Sumner et al.[163] presented mechanical impedance and aggregate stability data that clearly demonstrated that both mined gypsum and phosphogypsum contributed to conditions that helped improve penetration of subsoil hardpans by roots in highly weathered soils. They suggested that gypsum materials affected subsoil hardpans by influencing flocculation and aggregation, thus leading to better root penetration and proliferation. These, in turn, further improved subsoil aggregation.

In Asia, most, if not all, of the lowland rice soils are heavy clay soils that are puddled during land preparation to permit hand transplanting of seedlings. Puddling physically destroys soil aggregates. Reduction during continuous flooding solubilizes the cementing agents of Fe oxides and hydrous oxides further breaking down the finer aggregates.[164–167] In areas where rainfall or irrigation water is not sufficient to support a summer lowland rice crop, rice fields are left idle because of extreme difficulty in preparing them for a dryland crop. Fields cake and deep cracks open on the soil surface forming large and hard lumps. When mixed with the puddled topsoil at the last puddling, phosphogypsum or mined or FGD gypsum should induce fine cracking at the surface with friable soil lumps due to ionic strength effect of the dissolved gypsum and the nonplastic property of the undissolved gypsum particles. In Australia, gypsum is dissolved in the first irrigation water to obtain flocculation rather than Ca saturation.[66] Gypsum applied in such a manner was reported to result in a more friable surface soil, increased rate of water entry, and increased yields due to better crop establishment. Rates of application ranged from 0.56 to 4.48 Mg ha^{-1}. Collings[25] reported that phosphogypsum is also being used to ameliorate heavy clay soils in Canada.

E. BULK CARRIER FOR MICRONUTRIENTS AND LOW-ANALYSIS FERTILIZERS AND OTHER POTENTIAL USES

Phosphogypsum, where readily available, is a potential bulk carrier for micronutrients and low analysis fertilizer formulations. Pelletized phosphogypsum, enriched with micro- and macronutrients, has shown promise for use with urea and sulfate of potash magnesia as pelletizing agents.[168]

Phosphogypsum mixed with urea at 2.3 times the weight of the latter reduced NH_3-N loss by 85%.[7] Da Gloria et al.[8] reported that ordinary superphosphate was effective in reducing NH_3-N losses from chicken manure, but not agricultural gypsum. An amount of ordinary superphosphate three times greater than that indicated by the stoichiometric calculation was needed to reduce NH_3-N losses by 90%. Acidity in phosphogypsum and in ordinary superphosphate may have played a definite role in reducing NH_3 volatilization. It may also be noted that ordinary superphosphate fertilizers contain considerable quantities of phosphogypsum.

Another use of phosphogypsum is indicated by Alva,[6] who used phosphogypsum to effect a differential leaching of cations from a soil, thus changing the cation:Ca ratios. The use of phosphogypsum to control cation:Ca ratios such as Mg:Ca ratio to meet specific requirements of crops without affecting soil pH has not been fully explored.

F. FBC MATERIALS AS SUBSTITUTES FOR AGRICULTURAL LIME

Soil acidity could be a major growth-limiting factor in crop production. Soil acidity or H ion activity per se is apparently not injurious to plants at the pH values usually observed in soils. Plants showing adverse effects when growing in acid soils grow normally in nutrient solutions at the same pH values.[169] One of the adverse effects of soil acidity appears to be in reducing metal ion uptake by plants due to competition for carrier sites, so that a higher-solution concentration of a given ion is necessary to support a given rate of absorption at lower pH values.[170,171] Another is direct toxicity of some elements such as Mn[172] and Al.[173-179] A recent comprehensive review of soil acidity and plant growth has been given by Robson.[180]

The lime components in FBC residues such as CaO and MgO make these materials ideal substitutes for agricultural limestone or dolomite. The total lime content, expressed as the neutralizing potential of the material compared to an equal amount of ground $CaCO_3$ or agricultural limestone, ranges from 31 to 100%, averaging 60%.[7] Several studies indicated that the best agricultural use of FBC materials is as a lime source for croplands,[9,181–184] orchards,[72,185–190] and pastures.[191] The application rates are largely determined by the lime requirement of the soil and the lime content of the FBC materials, as well as their heavy metal contents. A manual on the use of FBC materials has been prepared by the USDA, Agricultural Research Service (ARS). The manual took into consideration soil loading of heavy metals contained in FBC materials.[10]

Rates of 3.8 to 19.9 Mg FBC residues per hectare applied to red clover (*Trifolium pratense* L.), tall fescue (*Festuca arundinacea* Schreb.), buckwheat (*Fagopyrum sagittatum* Gilib.), and oat (*Avena sativa* L.) resulted in increased tissue Mg and S in all species and corrected soil pH as effectively as $Ca(OH)_2$ without any adverse effect on plant growth.[182] Rates of 12.5 and 25.0 Mg ha^{-1} increased plant growth of apple (*Malus domistica*) seedlings, increased tissue Ca, and reduced potentially high soil Mn levels.[188] Because of the high rates of application of FBC materials as lime substitutes, it is necessary to consider the levels of the heavy metals added to the soil and their potential levels in the edible crop tissues.

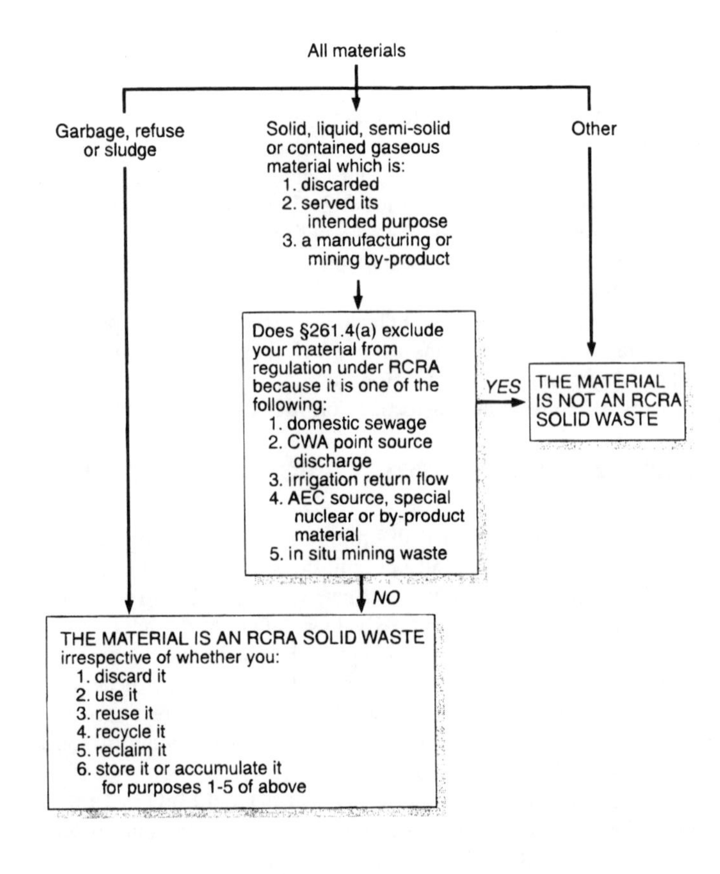

Figure 3 Definition of a solid waste.[192]

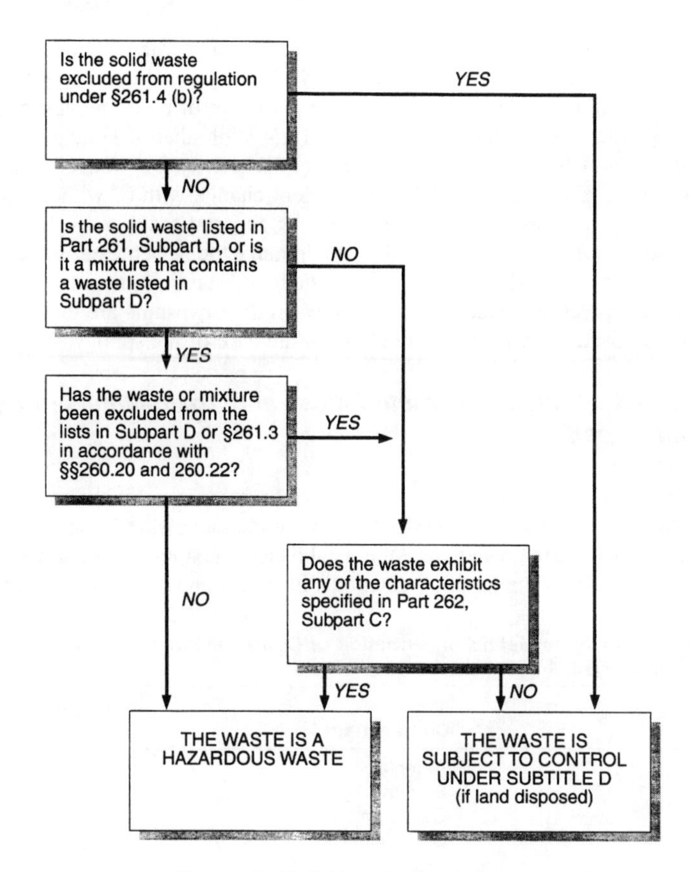

Figure 4 Definition of a hazardous waste.[192]

IV. ENVIRONMENTAL ASPECTS

A. BY-PRODUCT GYPSUMS AND THE USEPA HAZARDOUS WASTE CRITERIA

The USEPA has summarized the determination of a solid waste and a hazardous waste (Figures 3 and 4).[192] A "solid waste" is defined as any discarded material that is not excluded by section 261.4(a) of 40 CFR, Part 261, or that is not excluded by variance granted under sections 260.30 and 260.31 of 40 CFR, Part 260. A solid waste that is also a "hazardous waste" is one that is not excluded from regulation as a hazardous waste under Section 261.4(b) of 40 CFR, Part 261.[97] Under 40 CFR, Part 261, Subparts C and D, a hazardous waste may fall under one or more categories (hazard codes in parenthesis), namely, ignitable (I), corrosive (C), reactive (R), toxicity characteristic (E), acute hazardous (H), or toxic (T). The defining criteria are determined according to USEPA-approved analytical procedures.

On the basis of the physical and chemical properties of by-product gypsums the materials are not "ignitable". They do not ignite through friction, absorption of moisture, or spontaneous chemical changes. They are not "corrosive" wastes, defined as aqueous or dry solid wastes that have pHs of ≤2.0 or ≥12.5 determined according to USEPA Method 5.2 of Section 260.11 of 40 CFR.[192] The pH range of phosphogypsum is >2 to <5; that of FBC materials is generally less than 12.5, while that of FGD gypsums falls in the middle of the USEPA values. On the basis of reactivity, by-product gypsums are also not "reactive". They do not exhibit any violent change with or without explosion or form potentially explosive mixtures with water or generate toxic gases, fumes, or vapors. They also do not contain materials that would make them "acute hazardous" waste. The FBC and SDA materials, however, that contain substantial amounts of CaO may generate heat upon wetting. More pertinent to by-product gypsums are the toxicity criteria and the radioactivity content, particularly in the case of phosphogypsum.

1. The USEPA "Toxicity Characteristics" and Toxic Metals in By-Product Gypsums

A solid waste exhibits "toxicity" or "toxicity characteristics" if the extract from a waste sample contains any of the USEPA "toxicity characteristics" contaminants at a concentration equal to or greater than the established maximum contamination level (MCL) listed in Table 6. The leaching potentials of mined and by-product gypsums and

Table 6 Maximum Toxic Metal Concentration of Contaminants for Toxicty Characteristics Determination

USEPA HW No.[a]	Contaminant	Regulatory Level (mg L^{-1})
D004	Arsenic	5.0
D005	Barium	100.0
D006	Cadmium	1.0
D007	Chromium	5.0
D008	Lead	5.0
DOO9	Mercury	0.2
D010	Selenium	1.0
D011	Silver	5.0

[a] Hazardous waste number.

From Reference 97.

Table 7 Leaching Potentials (mg L^{-1}) of Phosphogypsum (PG), FGD Gypsum, FBC Residues, and Mined Gypsum for Toxic Metals in Relation to the National Primary Drinking Water Standards (NPDWS)

Toxic Metal	Gypsum Source				
	PG[39]	FGD[59]	FBC[59]	Mined[59]	NPDWS[193]
As	0.013	0.058–0.069	<0.04	0.001	0.05
Ba	0.200	<1.00	<1.00	<1.00	1.00
Cd	0.010	<0.01	<0.01	<0.01	0.01
Cr	0.040	<0.05	0.09	<0.05	0.05
Pb	0.010	<0.04	<0.04	<0.04	0.05
Hg	0.001	0.001–0.0013	0.0005	0.0011	0.002
Se	0.003	0.017–0.080	0.001	<0.001	0.01
Ag	0.060	<0.04	<0.04	<0.04	0.05

Table 8 Radionuclide Contents (pCi g^{-1}) of Phosphogypsum (PG), FGD Gypsum, and FBC Residues

Nuclide	PG[38]	FGD[194]	FBC[a]
^{226}Ra	9.60–42.00[b]	0.03–0.13	0.59–0.96
^{210}Po	24.30[c]	—	0.04–0.11
^{210}Pb	30.78[c]	—	8.00
^{234}U	1.47–2.44	—	0.20–0.66
^{235}U	0.13–0.17	—	0.20–0.40
^{238}U	1.61–2.41	—	0.38–0.68
^{227}Th	0.47–0.61	—	—
^{228}Th	0.03–0.06	—	0.32–0.47
^{228}Ra	—	—	—
^{230}Th	1.95–3.91	—	0.47–0.54
^{232}Th	0.07–0.13	0.02–0.09	0.19–0.34

[a] Range in combustion bed materials and spent bed solids in EXXON miniplant FBC samples.[99]
[b] Range in nine phosphogypsum stacks sampled to a maximum depth of 30 m at 3-m interval.

the National Primary Drinking Water Standards (NPDWS)[193] for each toxic metal are given in Table 7. Based on the leaching potentials for these toxic metals, the various by-product gypsums covered in this review do not appear to fall under the USEPA "toxicity characteristic" or the "toxicity" classification.

2. *Radioactivity as a USEPA Hazardous Waste Criterion*

Another criterion that could render solid wastes hazardous is radioactivity. The various radionuclide contaminants found in by-product gypsums are given in Table 8.[38,99,194] To be excluded from the list of hazardous wastes, the average radionuclide concentration must be less than 5 pCi ^{226}Ra g^{-1} of solid waste, or less than 10 μCi for any single discrete source.[195] Table 8 shows that FGD gypsum and FBC materials have relatively low radioactivity compared to the hazardous waste standards. For these reasons, flue gas control wastes generated from the combustion of coal or other fossil fuels in the ordinary process of generating electricity are classified as "not hazardous" under Section 261.4(b).91[97] Their disposal, however, is subject to regulation as RCRA solid wastes.

B. RADIONUCLIDES AND RADIOACTIVITY IN PHOSPHOGYPSUM

The major environmental concern in the use of phosphogypsum even as a soil amendment is radioactivity, particularly one produced from marine-deposited rock phosphate or phosphorites. In the manufacture of phosphoric acid, 86% of the ^{238}U goes with the phosphoric acid while 80% of the ^{226}Ra remains with the phosphogypsum.[35,196] The higher the radioactivity content in the raw material, the higher is the level in the phosphogypsum.

1. *A Brief Review of Radioactivity and Radionuclides*

Three long-lived radionuclides are naturally present in rocks in both primary and secondary geologic formations. These are ^{232}Th, ^{235}U, and ^{238}U. In the process of their ra-

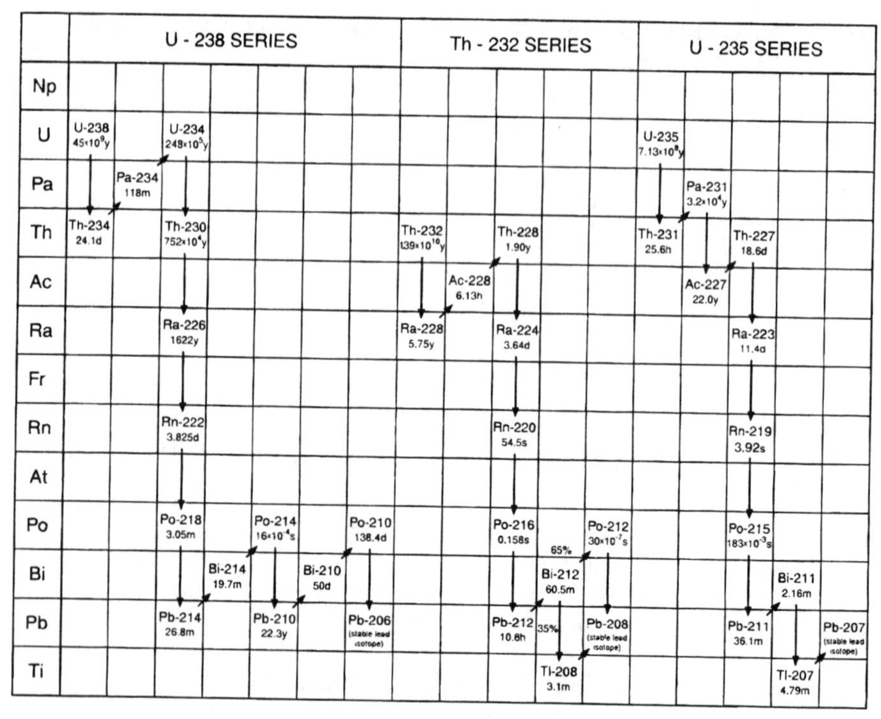

Figure 5 The three naturally occurring radioactive decay series.[197]

dioactive decay, these radionuclides generate their respective series of cascading shorter-lived radionuclides, each ending with a stable isotope of ^{208}Pb, ^{207}Pb, and ^{206}Pb for ^{232}Th, ^{235}U, and ^{238}U series, respectively (Figure 5). Under a "closed system" and over time, a state is reached when so-called parents and daughters in the whole decay series or in an isolated fraction thereof will have equal radioactivity. At such a state, the decay series is said to have reached a secular or radioactive equilibrium. A system in equilibrium may be disequilibrated should a daughter or parent be moved into or out of the system at a rate that is significant relative to the lifetime of the daughter and over a distance that is significant relative to the size of the system.[197]

In the radioactive decay of parents and daughters, one of two kinds of particles is expelled from the nucleus. An alpha decay expels a particle labeled alpha, consisting of two protons and two neutrons and carrying a 2^+ charge. Its expulsion from the nucleus results in a reduction in atomic weight by 4 units and atomic number by 2. Alpha particles have kinetic energies ranging between 3 and 9 million electron volts (MeV), 931 MeV being equal to 1 amu (atomic mass unit). A beta decay expels a beta particle, weighing 1/7500th of the alpha particle and carrying a single charge, which may be negative (electron from a neutron) or positive (positron from a proton). In a negative or positive beta decay, there is no change in the mass number of the atom, but a change in mass does occur with the release of decay energy carried off by the particles.[198]

The expulsion of these particles from the nucleus may or may not be accompanied by an emission of high-energy electromagnetic radiations called gamma rays. However, very frequently an alpha or beta decay leaves the nucleus in a state of excitation, and this excitation energy is released by gamma-ray emission. Like the alpha particles, gamma rays are emitted with discrete energies. Having no mass or charge, gamma rays possess great penetrating power.[198]

The stability of these decaying radionuclides is expressed in half-life. A half-life is defined as the time it takes for exactly one-half of the initial number of atoms to decay to the daughter product. The likelihood of decay for any one single atom has been found to be statistically random and can be related to its probability of decay or decay constant in a unit of time. This makes it logical to express the concentration of radionuclides in radioactivity or activity units — disintegration per second (dps)[199] — rather than in mass units. Consistent with the above concept, the concentration of a given radionuclide is normally expressed in Becquerel (1 Bq = 1 dps) or Curie (1 Ci = 3.7×10^{10} dps) or pCi (1 pCi = 1 Ci $\times 10^{-12} = 3.7 \times 10^{-2}$ dps; hence, 1 pCi g^{-1} = 37 Bq Kg^{-1}).

2. Geochemistry of Parent Radionuclides and Nuclide Daughters

Uranium Series — The three naturally occurring isotopes of U are ^{238}U, ^{235}U, and ^{234}U (Figure 5), the most abundant being ^{238}U, the main source of contaminant in phosphoric acid manufacture. Uranium ions exist in several valence states. In the most reduced valence state (U^{4+}), U ion is quite immobile and behaves as a large lithophile cation. In this form, U may substitute for Ca. In the hexavalent state (U^{6+}), U as the uranyl $(UO_2)^{2+}$ ion is relatively mobile and readily forms stable complexes with carbonate in the pH range 4 to 10.[200] Complexes are also readily formed with phosphate, sulfate, fluoride, and silicate ions.[197]

In solid phase, U may be found as a necessary constituent of some minerals such as uraninite (UO_2) or coffinite ($USiO_4$). As an impurity in some minerals, it may occupy a limited number of lattice or interstitial sites such as in calcite or phosphorite. It is also found as an adsorbed or coprecipitated constituent in Fe oxyhydroxides and in organic materials such as in organic-rich shale.[199] Rogers and Adams[201] reported that the Chattanooga shale of the eastern U.S. contained an average U concentration of 79 mg kg^{-1} compared with non-uraniferous shale which averaged about 2 mg kg^{-1}. Many coals and lignites may also be enriched with U. Szalay[202] reported that U concentrations in uraniferous bioliths ranges from 50 to 200 mg kg^{-1}. If coal or lignite mined from such uraniferous deposits is used as fuel in power plants, the scrubbing of SO_2 from the flue gas using limestone may result in the production of FGD gypsum with significant amounts of radionuclides as contaminants.

Uranium as U^{4+}, being similar in ionic size as Ca^{2+}, is believed to substitute for Ca^{2+} in the apatite lattice in phosphorites.[203] In marine phosphorites,. the U concentration is appreciable, ranging up to about 400 mg kg^{-1} for some samples from South Carolina which are stratigraphically equivalent to the Florida deposits. The median U concentration in Florida land pebble has been reported to range from 100 to 200 mg kg^{-1}.[204] Natural contamination of groundwater and ambient atmosphere may result when marine phosphorite deposits are exposed to subaerial conditions. Weathering may disrupt the radioactive equilibrium within these deposits and may cause the release of radiogenic species into the groundwater or into the ambient atmosphere.[199] Mining of these deposits causes similar disruption in the radioactive equilibrium.

Thorium Series — There are three Th isotopes with long half-life but only ^{228}Th comes from the ^{232}Th decay series, whereas the longer-lived ^{230}Th daughter emanates from ^{238}U.

The other three Th isotopes, ^{234}Th, ^{231}Th, and ^{227}Th, are all short lived with half-life in days (Figure 5). All Th ions are tetravalent (Th^{4+}). They form sparingly soluble organic and inorganic complexes. In view of its low solubility, Th activities in groundwater are reported to be very low.[197] In common igneous rocks, the ratio between Th and U (Th to U) has been reported to be between 3 and 4. In ocean-floor phosphorites, the concentration of Th tends to be less than 5 mg kg^{-1} compared to that of U which is generally about 100 mg kg^{-1}.[199]

Radium Daughter Decay — Radium is found as a daughter in all three decay series, namely, U-238, Th-232, and U-235, all in the oxidation state of 2 (Ra^{2+}). The element is chemically similar to Ba and, to a lesser degree, to Ca. It is quite mobile in Cl compounds or complexes, while moderate amounts of sulfate ion inhibit mobility due to its tendency to coprecipitate as $RaSO_4$ along with $BaSO_4$. Also, whereas $RaCl_2$ is soluble in water, the solubility of $RaSO_4$ is only 2.0×10^{-5} g L^{-1} of water at 25° C.[70] As a daughter of both U and Th, Ra is often found in interstitial sites in crystals replacing Ca as in carbonates.[199]

The 1976 USEPA National Interim Primary Drinking Water (NIPDW) regulations placed a MCL for ^{226}Ra at 3 pCi L^{-1} and 5 pCi L^{-1} for both ^{226}Ra and ^{228}Ra.[205] The MCL proposed in 1991 are 20 pCi L^{-1} for ^{226}Ra as well as for ^{228}Ra. The proposed MCL for U is 30 pCi L^{-1} or 0.020 mg L^{-1}.[206] A reconnaissance water sampling in central and northern Florida found that 19% of the water samples had ^{226}Ra activities equal to or in excess of 3 pCi L^{-1}. The maximum recorded was 90 pCi L^{-1}.[207]

Radon Daughter Decay — ^{226}Ra decays to give off ^{222}Rn gas, which is of great environmental concern due to its association with various kinds of cancer not only from inhalation but also through ingestion of Rn-enriched drinking water.[208–210] Three Rn isotopes are generated, one in each decay series, all with very short half-life, namely, 3.9 s for ^{219}Rn, 54.5 s for ^{220}Rn, and 3.8 days for ^{222}Rn. Being a noble gas, Rn is relatively unreactive with other atoms but is highly soluble in water. The proposed drinking water MCL for ^{222}Rn is 300 pCi L^{-1}.[206] Radon in municipal drinking water may have less than 100 pCi L^{-1}. Well waters may have as much as 25,000 pCi L^{-1}.[211] In some cases, ^{222}Rn in natural water may exceed 4.5×10^5 pCi L^{-1}.[199] When ingested ^{222}Rn has been shown to have an unexpectedly long biologic residence time.[200] Also, dissolved ^{222}Rn tends to be degassed from the water and contributes significant amounts to indoor radon.[212–213] ^{222}Rn decays to produce the longer-lived ^{210}Pb.

Lead Daughter Decay — The radioactive isotopes of Pb are ^{212}Pb from ^{232}Th, ^{214}Pb and ^{210}Pb from ^{238}U, and ^{211}Pb from ^{235}U. Lead salts are relatively insoluble in water. Except for ^{210}Pb which has a half-life of 22.1 years all other Pb isotopes have half-life in matter of hours.

Polonium Daughter Decay — Polonium isotopes are generated by the three decay series with ^{216}Po and ^{212}Po coming from ^{232}Th; ^{218}Po, ^{214}Po, and ^{210}Po from ^{238}U; and ^{215}Po and ^{211}Po from ^{235}U. Polonium is highly reactive and is rapidly adsorbed. Although it is a very rare element, its concentration in some groundwater could be high enough to be of concern. Except for ^{210}Po, which has a half-life of 138 days, all other Po isotopes have a half-life in the order of seconds and minutes. ^{210}Po decays to produce another important radionuclide ^{210}Pb. Each of the primary parent decay series ends with its respective stable Pb isotope (Figure 5).

Table 9 Radionuclide Concentration Ratios Relative to ^{226}Ra Concentration of a Reference Phosphogypsum

Radionuclide	Concentration (pCi g^{-1})
^{226}Ra	1.000
^{210}Po	1.040
^{210}Pb	1.400
^{228}Th	0.133
^{228}Ra	0.133
^{230}Th	0.187
^{232}Th	0.123
^{234}U	0.120
^{235}U	0.005
^{238}U	0.110

From Reference 214.

3. Radiation Exposure and Health Effects (Risk) Assessment

The three ways by which one is exposed to ionizing radiation are (1) direct exposure as in the case of gamma rays, (2) inhalation in the case of ^{222}Rn gas and radionuclide-contaminated particulates, and (3) ingestion in the case of drinking water or eating food contaminated with various radionuclides. Risk assessments for such exposures are evaluated for three population types: (1) the individual, (2) the critical population group (CPG), and (3) the general population. The radionuclides concentrations in a reference phosphogypsum relative to ^{226}Ra (Table 9) provide the basis for the risk estimates which are given in terms of ^{226}Ra concentration.[214]

Radiation dose is directly related to radioactivity exposure or intake. The quantity "dose equivalent" is used to express the time-integrated energy deposited per unit mass of a tissue volume weighted by factors that account for the biological effectiveness of the different kinds of radiation. The expected cumulative effect of a given exposure or intake on an organ or tissue projected to the next 50 years is referred to as the "committed dose equivalent" for that organ or tissue. Because the various radionuclides are not distributed uniformly throughout the body, there is a nonuniform irradiation of the various organs and tissues in the body. The quantity "committed effective dose" has been introduced to compare cases of different dose distribution. It is the uniform whole-body dose expected to present the same risk as the actual nonuniform irradiation.[215]

Table 10 Dose Conversion Factors (DCF)

Nuclide	Inhalation DCF (mrem pCi^{-1})^a	Ingestion DCF (mrem pCi^{-1})^a	Direct Gamma DCF (mrem year^{-1}) (pCi m^{-2})$^{-1}$
^{226}Ra	8.6×10^{-3}	1.3×10^{-3}	1.67×10^{-4}
^{210}Po	9.4×10^{-3}	1.9×10^{-3}	8.55×10^{-10}
^{210}Pb	1.4×10^{-2}	5.4×10^{-3}	0
^{228}Th	3.4×10^{-1}	4.0×10^{-4}	3.37×10^{-4}
^{228}Ra	4.8×10^{-3}	1.4×10^{-3}	9.04×10^{-5}
^{230}Th	3.3×10^{-1}	5.5×10^{-4}	8.88×10^{-8}
^{232}Th	1.6×10^{0}	2.7×10^{-3}	6.56×10^{-8}
^{234}U	1.3×10^{-1}	2.8×10^{-4}	8.00×10^{-8}
^{235}U	1.2×10^{-1}	2.5×10^{-4}	6.41×10^{-8}
^{238}U	1.2×10^{-1}	2.7×10^{-4}	1.67×10^{-5}

^{a} 50-year committed dose equivalent from 1 year of intake (uptake).

From Reference 214.

Table 11 Risk Conversion Factors

Nuclide	Inhalation Risk (pCi^{-1})	Ingestion Risk (pCi^{-1})	Direct Gamma Risk ($(pCi\ m^{-2})^{-1}$)
^{226}Ra	2.8×10^{-9}	9.4×10^{-11}	5.7×10^{-11}
^{210}Po	2.4×10^{-9}	1.4×10^{-10}	2.9×10^{-16}
^{210}Pb	1.4×10^{-9}	5.5×10^{-10}	0
^{228}Th	7.2×10^{-8}	1.3×10^{-11}	4.8×10^{-11}
^{228}Ra	5.8×10^{-10}	7.0×10^{-11}	3.1×10^{-11}
^{230}Th	2.9×10^{-8}	2.3×10^{-11}	2.7×10^{-14}
^{232}Th	2.9×10^{-8}	2.1×10^{-11}	2.0×10^{-14}
^{234}U	2.5×10^{-8}	7.5×10^{-11}	2.4×10^{-14}
^{235}U	2.3×10^{-8}	7.3×10^{-11}	5.5×10^{-12}
^{238}U	2.2×10^{-8}	7.4×10^{-11}	7.23×10^{-13}

Note: 70-year lifetime risk of fatal cancer from 1 year of exposure.

From Reference 214.

Table 12 Radon Risk Conversion Factors

Exposure Scenario	Inhalation Risk[a] ($(pCi\ m^{-3})^{-1}$)
Indoor	4.4×10^{-8}
Outdoor	4.4×10^{-9}

[a] 70-year lifetime risk of fatal cancer from 1 year of exposure to ^{220}Rn and ^{222}Rn daughters.

From Reference 214.

Radiation exposure, in units of Bq or pCi, is converted to dose equivalent — in units of Sievert (1 Sv = 1 equivalent Joule/kg) or rem (1 rem = 0.01 Sv; mrem = 1 rem × 10^{-3})[215] — using the appropriate dose conversion factors (DCFs) for ingested or inhaled particles and for gamma rays (Table 10). Radiation exposures are converted directly into health risk estimates, expressed as 70-year lifetime risk of fatal cancer from 1 year of exposure, using risk conversion factors (RCFs) (Tables 11 and 12).[214]

4. Health Risks Associated with Use of Phosphogypsum in Agriculture

The USEPA, in its effort to determine the applicability of phosphogypsum for agriculture, has begun to evaluate the health risks involved in using phosphogypsum for six scenarios or field uses. The various parameters assumed for the scenarios are summarized in Table 13. For all scenarios, health risks were evaluated for

1. Agricultural worker spending 2000 h/year at the site and exposed to direct gamma radiation, without allowance for shielding, and inhaling air at the rate of 8000 $m^3/year^{-1}$ with the outside air containing an average dust load of 500 mg m^{-3}.
2. On-site individual living in a house constructed on a site and working at the same site previously used for agriculture, being exposed to direct gamma radiation, indoor Rn inhalation, and using water from a contaminated well.
3. The CPG which includes individuals who might be exposed to the highest doses, as a result of normal daily activities, through inhalation of contaminated dust and

Table 13 Parameters Assumed by USEPA to Assess Health Risks of Phosphogypsum (PG) for Seven Agricultural Uses or Scenarios

Parameter	Scenario						
	1[a]	2[a]	3[a]	4[a]	5[b]	6[b]	7[c]
Type of soil	Clay	Sand	Clay	Sand	Clay	Sand	
^{226}Ra in PG (pCi g^{-1})	3–26	3–26	3–26	3–16	3–26	3–26	3–60
Receptor[d]	1,2,3,4[d]	1,2,3,4	1,2,3,4	1,2,3,4	1,2,3,4	1,2,3,4	1
Distance (m)[e]	100 and 890	100 and 890	100 and 6,440	100 and 6,440	100 and 1,000	100 and 1,000	100
Initial PG applied	—	—	—	—	17,600	17,600	
PG rate (lb $acre^{-1}$)	1,460[f]	1,460[f]	4,470[f]	4,470[f]	8,800[g]	8,800[g]	4,995–10,000[f]
Period applied (years)[h]	100	100	100	100	100	100	100
Tillage depth (cm)	22	22	46	46	30	30	23
Acres per farm	138	138	1,000	1,000	556	556	

[a] Phosphogypsum as source of S and Ca for crops, 1 and 2 at low rate, 3 and 4 at high rate.
[b] Phosphogypsum for sediment control for eroded and leached soils.
[c] Evaluates effect of phosphogypsum containing a range of ^{226}Ra concentrations at various rates of application to the on-site person exposed to direct gamma radiation and radon inhalation.
[d] Populations exposed: 1 — agricultural worker, 2 — on-site individual, 3 — member of a critical population group (CPG), 4 — off-site individual.
[e] Distance from the edge or boundary of the field treated with phosphogypsum.
[f] Applied biennially at rates indicated.
[g] Applied biennially after the initial rate.
[h] Risks evaluated for various pathways of contamination after 100 years of sustained biennial application of phosphogypsum.

From Reference 214.

Table 14 Risk Assessment Results for Scenario 7 — Radon Exposure Risks[a] to the On-Site Individuals as a Function of Phosphogypsum Application Rate and ^{226}Ra Concentration

Application Rate (lb acre^{-1})[b]	^{226}Ra Concentration in Phosphogypsum (pCi g^{-1})						
	3	7	15	20	30	45	60
500	1.0×10^{-7}	2.4×10^{-7}	5.1×10^{-7}	6.8×10^{-7}	1.0×10^{-6}	1.6×10^{-6}	2.1×10^{-6}
1,000	2.1×10^{-7}	4.8×10^{-7}	1.0×10^{-6}	1.4×10^{-6}	2.1×10^{-6}	3.1×10^{-6}	—
1,500	3.1×10^{-7}	7.5×10^{-7}	1.6×10^{-6}	2.1×10^{-6}	3.1×10^{-6}	—	—
2,500	5.1×10^{-7}	1.2×10^{-6}	2.6×10^{-6}	3.4×10^{-6}	—	—	—
5,000	1.0×10^{-6}	2.4×10^{-6}	5.1×10^{-6}	—	—	—	—
10,000	2.1×10^{-6}	4.8×10^{-6}	1.0×10^{-5}	—	—	—	—

[a] Lifetime risk from 1 year of exposure.
[b] 1.12 x lb acre^{-1} = kg ha^{-1}.

From Reference 214.

Table 15 Risk Assessment Results for Scenario 7 — External Gamma Risks[a] to the On-Site Individuals as a Function of Phosphogypsum Application Rate and ^{226}Ra Concentration

Application Rate (lb acre^{-1})[b]	^{226}Ra Concentration in Phosphogypsum (pCi g^{-1})						
	3	7	15	20	30	45	60
500	1.1×10^{-7}	2.6×10^{-7}	5.7×10^{-7}	7.5×10^{-7}	1.1×10^{-6}	1.7×10^{-6}	2.3×10^{-6}
1,000	2.3×10^{-7}	5.3×10^{-7}	1.1×10^{-6}	1.5×10^{-6}	2.3×10^{-6}	3.4×10^{-6}	—
1,500	3.4×10^{-7}	7.9×10^{-7}	1.7×10^{-6}	2.3×10^{-6}	3.4×10^{-6}	—	—
2,500	5.7×10^{-7}	1.3×10^{-6}	2.8×10^{-6}	3.8×10^{-6}	—	—	—
5,000	1.1×10^{-6}	2.6×10^{-6}	5.7×10^{-6}	—	—	—	—
10,000	2.3×10^{-6}	5.3×10^{-6}	1.1×10^{-5}	—	—	—	—

[a] Lifetime risk from 1 year of exposure.
[b] $1.12 \times$ lb acre^{-1} = kg ha^{-1}.

From Reference 214.

Table 16 Risk Assessment Results for Scenario 7 — Total Risks[a] to the On-Site Individuals as a Function of Phosphogypsum Application Rate and ^{226}Ra Concentration

Application Rate	^{226}Ra Concentration in Phosphogypsum (pCi g^{-1})						
(lb $acre^{-1}$)[b]	3	7	15	20	30	45	60
500	2.2×10^{-7}	5.1×10^{-7}	1.1×10^{-6}	1.4×10^{-6}	2.2×10^{-6}	3.3×10^{-6}	4.4×10^{-6}
1,000	4.4×10^{-7}	1.0×10^{-6}	2.2×10^{-6}	2.9×10^{-6}	4.4×10^{-6}	6.5×10^{-6}	—
1,500	6.5×10^{-7}	1.4×10^{-6}	3.3×10^{-6}	4.4×10^{-6}	6.5×10^{-6}	—	—
2,500	1.1×10^{-6}	2.5×10^{-6}	5.4×10^{-6}	7.2×10^{-6}	—	—	—
5,000	2.2×10^{-6}	5.1×10^{-6}	1.1×10^{-5}	—	—	—	—
10,000	4.4×10^{-6}	1.0×10^{-5}	2.2×10^{-5}	—	—	—	—

[a] Lifetime risk from 1 year of exposure.
[b] $1.12 \times$ lb $acre^{-2}$ = kg ha^{-1}.

From Reference 214.

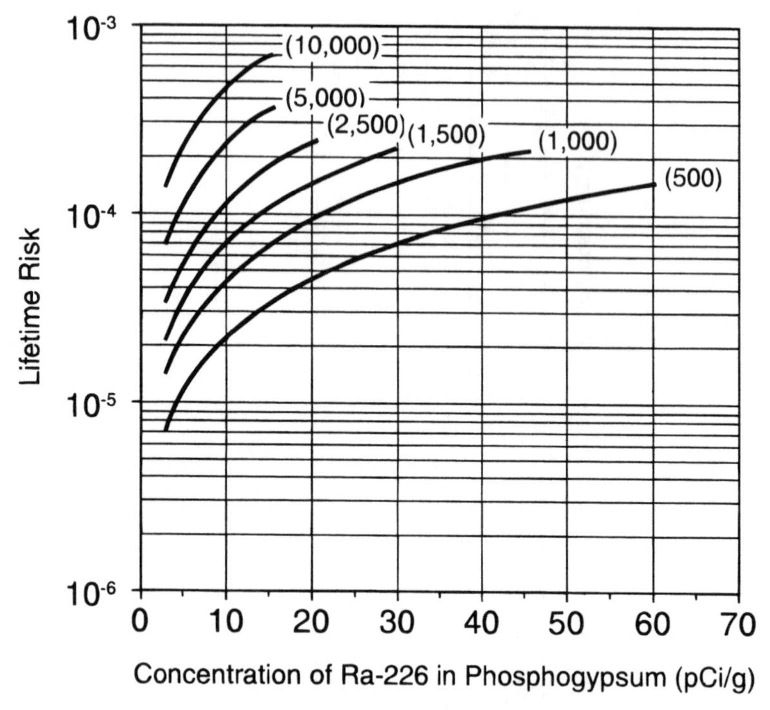

Figure 6 Risk assessment results for scenario 7 — ^{222}Rn exposure risks to the on-site individual as a function of ^{226}Ra content of phosphogypsum for six application rates (lb $acre^{-1}$ in parenthesis; $1.12 \times$ lb $acre^{-1}$ = kg ha^{-1}).[214]

ingestion of drinking water from a contaminated well, from foodstuffs contaminated by well water, and foodstuffs grown on fertilized soil.

4. Off-site individual who can be exposed by ingestion of water contaminated via the groundwater pathway and ingestion of river water contaminated by surface runoff.[214]

Lifetime maximum individual risks (MIRs) from 1 year of exposure were determined using the PATHRAE[216] dose assessment model. The results from such a model showed that, for each of the agricultural scenarios, the highest doses and risks resulted from external gamma exposure and from indoor Rn inhalation to the on-site individual. The results also indicated that, the larger the rate of application and the higher the radionuclide concentration, the greater the risks. A seventh scenario to determine the risks to the on-site individual from various combinations of application rates and ^{226}Ra concentrations in phosphogypsum due to Rn inhalation, gamma radiation, and to both was then evaluated. The lifetime risks for scenario 7 from 1 year of exposure are given in Tables 14 to 16. The risks for a 70-year continuous exposure are presented in Figures 6 to 8. Assuming 100 years of biennial phosphogypsum application at 3 Mg ha^{-1} (2700 lb $acre^{-1}$), conversion from agricultural to residential use after 100 years, and a lifetime risk of 1×10^{-4} as a presumptively safe limit, USEPA came up with the final rule on the use of phosphogypsum for agriculture such that the MIR for a 70-year continuous exposure does not exceed 3×10^{-4}. Thus, the USEPA determined that phosphogypsum

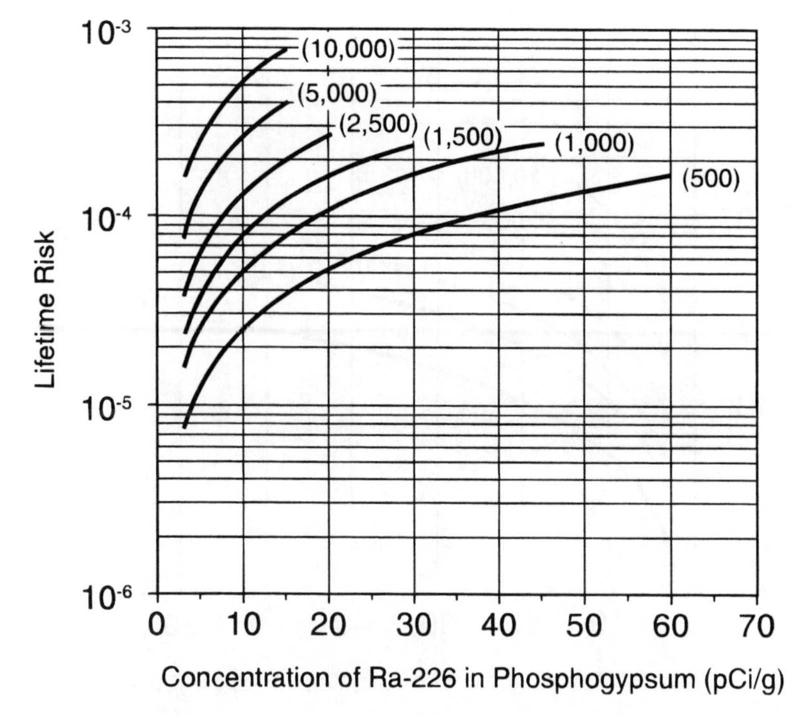

Figure 7 Risk assessment results for scenario 7 — external gamma exposure risks to the on-site individual as a function of ^{226}Ra content of phosphogypsum for six application rates (lb $acre^{-1}$ in parenthesis; 1.12 × lb $acre^{-1}$ = kg ha^{-1}).[214]

containing 10 pCi ^{226}Ra g^{-1} or less may be used for agriculture in the U.S. (Figure 9).[217] The USEPA's final rule, however, is being contested by the Fertilizer Institute which argued that the USEPA dose-risk model and parameter selections overestimated dose and risk by a factor of 2.5. The institute suggested that the USEPA "consider allowing the application of phosphogypsum containing 26 pCi g^{-1} ^{226}Ra to continue for an extended period of time (e.g., 5 years) so that further studies can determine the fate of ^{226}Ra in agricultural soils and verify or disprove the USEPA model of ^{226}Ra fate in soils."[218]

For applications in other countries where MIRs other than the USEPA's are acceptable and using phosphogypsum with ^{226}Ra content higher than 30 pCi g^{-1}, the use of Figure 8 should prove invaluable.[214]

V. IMPACT OF FGD GYPSUM AND FBC MATERIALS ON PLANT TISSUE, SOIL, AND GROUNDWATER

A. IMPACT OF FGD GYPSUM AS SOIL AMENDMENT

The quality of FGD gypsums generated by wet process with forced oxidation normally approaches, if not exceeds, that of mined gypsum.[59] These by-products of the electric power industry are produced to meet commercial-grade requirements to substitute

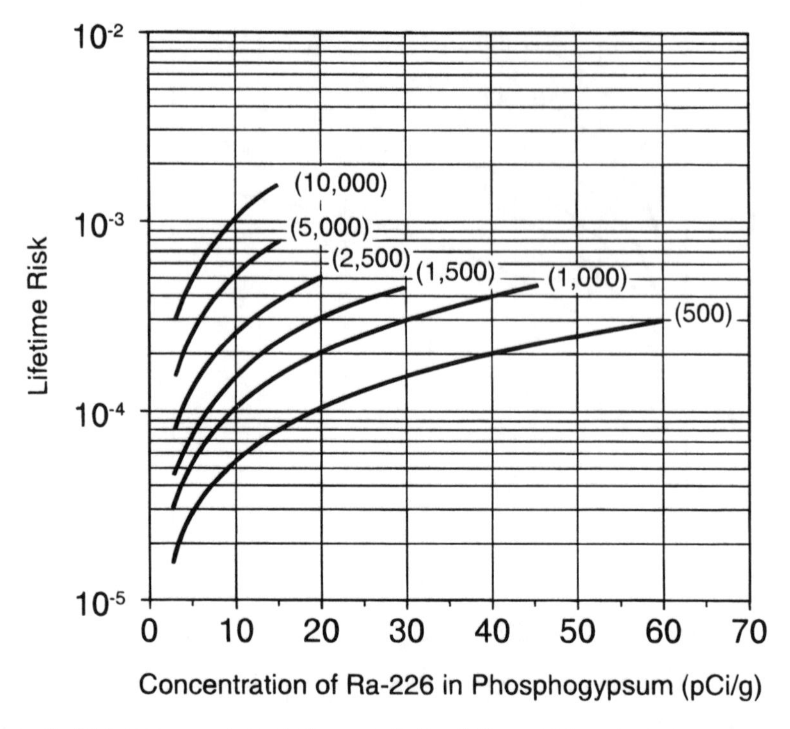

Figure 8 Risk assessment results for scenario 7 — ^{222}Rn and external gamma exposure risks to the on-site individual as a function of ^{226}Ra content of phosphogypsum for six application rates (lb $acre^{-1}$ in parenthesis; $1.12 \times$ lb $acre^{-1}$ = kg ha^{-1}).[214]

for mined gypsum. No work has been done on their impact on the environment when used as soil amendments as their effects on soil, water, and plant tissue are not expected to be different from those of mined gypsum.

B. IMPACT OF FBC MATERIALS AS SOIL AMENDMENT

1. Effect on Plant Tissue Quality and Soil

The main use of FBC materials as a soil amendment is as a substitute for agricultural lime. For FBC materials with low lime content,[100] they may have to be applied at three or more times the rate of pure $CaCO_3$ to achieve the desired soil pH. This would make the application rates relatively high.

The main concern in the use of FBC materials as soil amendments is not radioactivity but the quantities of trace elements and heavy metals. Based on soil column studies in the greenhouse, Sidle et al.[219] reported that of the heavy metals analyzed (Cd, Cr, Cu, Ni, Pb, Sr, and Zn) in 14 leachings from acid soil columns treated with FBC residue, none appeared to move through the column. Barnhisel and Thom[184] compared the effect of FBC material and $CaCO_2$ on the germination and growth of alfalfa and soybean under greenhouse conditions and on the chemical composition of both soils and plants. Rates used, in megagrams per hectare, were 4 and 8 for $CaCO_3$ and 8, 16, and 32 for FBC

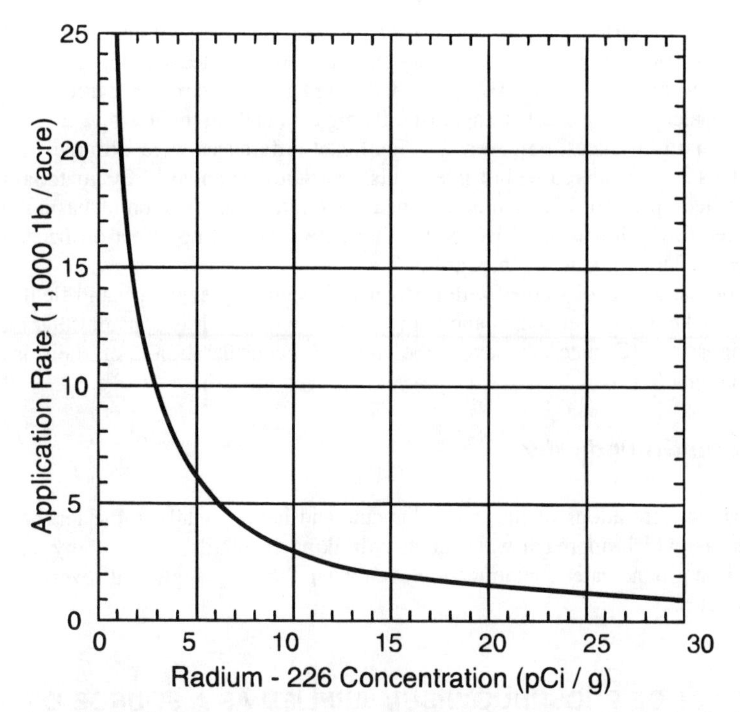

Figure 9 Application rates (lb $acre^{-1}$ in parenthesis; 1.12 × lb $acre^{-1}$ = kg ha^{-1}) of phosphogypsum as a function of ^{226}Ra concentration for lifetime risk of 3×10^{-4}.[214]

materials. The results showed that excessive rates of FBC material adversely affected the germination of alfalfa seeds but not that of soybean. The high rates also decreased the levels of Mg in the plant tissue. Results for heavy metals indicated that Pb, Cd, Cu, Ni, and Cr concentrations in alfalfa herbage and soybean grain were not related to treatments. Also, the concentrations of most of the metals were below the detection limits using inductively coupled plasma (ICP) analysis. The levels of Al, Mn, Zn, and Mo in plant tissues were at or near levels considered excessive when FBC material was applied as a top dressing, but not when it was uniformly mixed with the soil. Boron concentrations in soybean grain at the third harvest were higher with FBC material (25.0 to 26.1 mg kg^{-1}) than with $CaCO_3$ (13.3 to 14.8 mg kg^{-1}). In alfalfa, B concentrations were consistently higher in FBC-treated (42 to 54 mg kg^{-1}) than in $CaCO_3$-treated plants (18 to 19 mg kg^{-1}).

In field experiments, Stout et al.[9] used FBC materials and $Ca(OH)_2$ as lime sources to compare their effect on yield and on the elemental composition of snap beans (*Phaseolus vulgaris* L.), broccoli (*Brassica oleracea*, Botrytis Group), lettuce (*Lactuca sativa* L.), and beets (*Beta vulgaris* L.). The results showed no significant difference between FBC and $Ca(OH)_2$ treatments on the levels of the major nutrients (Ca, Mg, K, P, S, and N) nor in the trace elements and heavy metals (Na, Fe, Zn, Mn, Cu, Al, Ni, Sr, Cr, Cd, Pb, and Co) in all crops, with the exception of Zn in lettuce. Zinc concentration in lettuce was higher with FBC material than with $Ca(OH)_2$, with 40 and 36 mg kg^{-1}, respectively. Trace elements and heavy-metal concentrations in corn leaves were no different in the two treatments, but Na and Cu in corn grains were higher with FBC material than with $Ca(OH)_2$, with 64 and 50 mg kg^{-1} for Na and 3.19 and 2.68 for Cu. Application of FBC material resulted in higher Na in soybean leaves, 358 against 339 mg kg^{-1}, but lower Cd in grains, 0.09 against 0.15 mg kg^{-1}. Oats fertilized with FBC material had elevated Cd in grain, 0.10 against 0.04 mg kg^{-1}, and Zn in the straw, 25 against 21 mg kg^{-1}. In all instances, however, the significant differences were extremely small and the values are considered within safe levels. In a study on apples, FBC material and gypsum applied up to 16 times the determined basic rate of application (1 basic rate = 115 g Ca/tree for both sources; FBC containing 28% Ca), had no effect on fruit flesh levels of N, K, Mn, Fe, Cu, Al Zn, and Pb.[187]

Based on the results of studies with FBC materials under greenhouse and field conditions on uptake of trace elements and heavy metals by crops, it is apparent that metal concentrations in FBC materials have to be taken into consideration upon their use as soil amendments.[10]

2. *Effect on Groundwater*

The low concentrations of the trace elements and heavy metals in the leachate of FBC materials, which approach water quality drinking standards, rule out any significant impact on groundwater contamination even when these are applied at several times the $CaCO_3$ rates.

VI. IMPACT OF PHOSPHOGYPSUM APPLIED AS A SOURCE OF S AND CA TO FORAGE GRASSES GROWING ON A FLORIDA SPODOSOL SOIL

The main environmental concerns in the use of phosphogypsum as soil amendments are the toxic metals, F, and radionuclides impurities. In an environmental study, Rechcigl et al.[220] used a Florida phosphogypsum as source of S and Ca on established bahiagrass (*Paspalum notatum* Flugge) and annual ryegrass (*Lolium multiflorum* Lam.) on a tilled land (*sandy, siliceous, hyperthermic Aeric Haplaquods*) applied at the rates of 0, 0.4, 2, and 4 Mg ha^{-1}. The 0.4 Mg rate was applied annually for 3 years and the higher rates applied only initially to duplicate 0.1 ha plots. Runoff and groundwater at depths of 60 and 120 cm were sampled for pH, E_c, F, radionuclides, and toxic metals determinations. Soil and plant tissue samples were also collected for radionuclides and heavy metal analyses. Radon gas was determined both as soil surface Rn flux and as ambient atmospheric Rn. The following discussions are limited to the bahiagrass results, unless indicated otherwise.

A. IMPACT OF TOXIC METALS AND F ON THE SOIL, SURFICIAL GROUNDWATER, AND PLANT TISSUE

Baker and Chesnin[101] have discussed the main concerns with the toxic and potentially toxic elements in soils. Recent reviews on the geobiochemistry of these elements and other heavy metals have been published by Adriano.[221] A more general review by several authors has been published by Alloway.[222]

Table 17 Chemical Analyses of Phosphogypsum (PG) Used in the Study and of Other PG Samples from Florida PG Stacks[38,39]

	Concentration (unit)		
Elements	**U.S. Agric. Chem. Corp.[a]**	**Pembroke Laboratory[a]**	**Average in Florida PG[a]**
	Plant Nutrient (%)		
Major			
Calcium (Ca)	24.5	25.6	26.2
Sulfur (S)	19.8	—[b]	19.5
Phosphorus (P)	—[b]	0.3	0.7
Minor (mg kg^{-1})			
Iron (Fe)	690.0	460.0	860–1000 ± 300–600
Sodium (Na)	260.0	260.0	520 ± 79
Potassium (K)	38.0	110.0	200–230 ± 83–94
Molybdenum (Mo)	—[b]	17.0	2.2–11 ± 1.4–2.2
Magnesium (Mg)	16.0	13.0	<940 ± <27
Boron (B)	—[b]	<10.0	<3.0
Zinc (Zn)	6.0	6.2	<340 ± 21
Copper (Cu)	0.6	2.1	<82 ± <9.6
Manganese (Mn)	—[b]	1.9	25 ± 14
Chloride (Cl)	—[b]	—[b]	<150 ± <4.7
Nickel (Ni)	—[b]	—[b]	<2.0
Phytotoxic			
Fluorides (F)	4300	—[b]	5000
Aluminum (Al)	—[b]	1100	2000 ± 540
USEPA toxic metals (TM) for "toxicity characteristic" (mg kg^{-1})(TC) determination			
Barium (Ba)	46.0	45.0	<210 ± <24
Arsenic (As)	5.0	5.0	0.76–0.9 ± 0.26–0.3
Silver (Ag)	2.0	<0.2	<1.3 ± <0.64
Cadmium (Cd)	0.7	1.1	3.4–4.0
Selenium (Se)	<0.05	1.6	0.7–2.1 ± 0.44–0.72
Mercury (Hg)	—[b]	<0.01	0.3–0.4 ± 0.25–0.28
Lead (Pb)	—[b]	4.0	2.0–13.0
Chromium (Cr)	—[b]	2.9	6.0 ± 1.4

[a] Moisture-free basis.
[b] Not determined in the analysis.

The chemical constituents of the phosphogypsum used by Rechcigl et al.[220] in the study and their averages for Florida phosphogypsums are given in Table 17. The measured USEPA leaching potentials for toxic metals in several Florida phosphogypsums and their estimates in the phosphogypsum used in the study are given in Table 18. The projected increases in concentrations of the toxic metals due to application of 4 Mg phosphogypsum ha^{-1} at the top 0- to 15-cm soil depth and actual determined values are given in Table 7. From the data, the toxic metals in phosphogypsum do not appear to present any environmental concerns to the soil. In water, assuming that the total leachable amounts are evenly mixed all at once with 0.30 m × 10,000 m^2 of groundwater, the projected elevation of toxic metal concentrations are again negligible relative to the national primary drinking water standards (Table 17). In another ongoing study by the authors,[223] toxic metals in bahiagrass regrowth and hay forages from pastures amended with 0, 10, and 20 Mg phosphogypsum per hectare were shown to be not different for Cd, Cr, Hg, Ni, Pb, and Se. Mays and Mortvedt[224] showed that Cd contents of corn,

Table 18 USEPA Toxic Metals (TM) Concentrations in Leachate for "Toxicity Characteristics" (TC) for Solid Wastes,[97] Their Leachability, Concentrations, and Leaching Potentials (LP) in PG Samples from Florida PG Stacks,[39] and the Estimated LP in PG Used in the Study

	EPA TC metals							
Sample	**As**	**Ba**	**Cd**	**Cr**	**Pb**	**Hg**	**Se**	**Ag**
	USEPA TC for solid waste (mg L^{-1})							
	5.0	100	1.0	5.0	5.0	0.2	1.0	5.0
	LP of Florida PG Stack Samples							
Stack A 1	0.02	0.2	0.01	0.04	0.01	0.000	0.002	0.08
2	0.02	0.2	0.01	0.05	0.03	0.000	0.003	0.07
3	0.03	0.0	0.01	0.07	0.01	0.000	0.004	0.01
Stack B 1	0.01	0.0	0.01	0.01	0.00	0.001	0.003	0.04
2	0.01	0.2	0.01	0.01	0.00	0.001	0.005	0.04
Stack C 1	0.02	0.2	0.02	0.02	0.00	0.001	0.005	0.04
2	0.01	0.4	0.03	0.05	0.00	0.001	0.003	0.10
Stack D	0.01	0.0	0.01	0.11	0.01	0.001	0.003	0.09
Stack E	0.01	0.2	0.03	0.03	0.03	0.001	0.004	0.06
Stack F	0.00	0.1	0.01	0.01	0.04	0.001	0.002	0.05
Stack G	0.01	0.3	0.01	0.01	0.01	0.001	0.002	0.04
Stack H	0.02	0.3	0.01	0.05	0.00	0.001	0.003	0.04
Stack I	0.02	0.2	0.01	0.02	0.01	0.004	0.002	0.07
Average stack LP	0.02	0.2	0.01	0.04	0.01	0.001	0.003	0.06
	Leachability of TM (mg kg^{-1}) in PG Samples from Florida PG Stacks							
Total TM in PG[a]	0.85	105	0.59	6.00	1.30	0.34	1.40	0.67
Amount Leached	0.24	3.2	0.28	0.80	0.36	0.01	0.06	0.96
% leachable	28	3	47	13	28	4	4	—[b]
	Estimated LP of PG Used in the Study (mg L^{-1})							
Estimates[c]	0.12	0.09	0.02	0.02	0.03	<0.001	0.003	0.30

[a] Averages for samples from PG stacks A1, B1, E, F, and H.
[b] Indeterminate (total in PG and amount leached were 0.67 and 0.96 mg kg^{-1}, respectively, giving a leachability >100 %).
[c] [(TM in PG used)/(TM in PG stacks)] × average stack LP.

wheat, and soybean grains fertilized with 22 and 112 Mg phosphogypsum ha^{-1} containing 0.23 mg Cd kg^{-1} were not different from those of the control. Thus, it is safe to conclude that application of phosphogypsum at agricultural rates does not present any short- or long-term hazards to the soil, water, or crop due to toxic metal impurities.

Fluoride is a cumulative poison. Intake of small quantities over an extended period may lead to the accumulation of F in animals to toxic levels.[225] Several investigators have found that adverse effects of chronic F toxicosis on bones and teeth can result in leg stiffness, lameness, reduced feed intake, a decline in general health, and greatly reduced animal productivity.[226–228] Doses of F that cause acute toxicity have been found to cause gastroenteritis, muscular weakness, chronic convulsion, pulmonary congestion, and respiratory and cardiac failure.[228] Based on dietary F tolerances developed by the National Research Council, dietary F should not exceed 40 mg kg^{-1} for beef and dairy

Table 19 Projected Increases In and Measured Concentrations of Toxic Metals in a Florida Spodosol Soil Cropped to Forages Grasses and Amended with PG Applied as a Source of S and Ca 1 Year after Application

Parameter	EPA "Toxicity Characteristics" Metals							
	As	Ba	Cd	Cr	Pb	Hg	Se	Ag
Total added[a] to the soil from 4.0 Mg PG ha^{-1}: (g ha^{-1})	20.0	184.0	4.4	11.6	16.0	<0.04	6.4	8.0
Projected[b] increase in the top 0–15 cm soil from 4.0 Mg PG ha^{-1} (mg kg^{-1} soil)	0.010	0.092	0.002	0.006	0.008	<0.0001	0.003	0.004
Measured concentrations in the top 0–15 cm soil: (mg kg^{-1} soil) Mg PG ha^{-1}								
0.0	—	—	0.005a[c]	0.005a	0.130a	0.010a	0.000	—
0.4	—	—	0.005a	0.010a	0.105a	0.010a	0.000	—
2.0	—	—	0.000a	0.010a	0.140a	0.005a	0.000	—
4.0	—	—	0.020a	0.010a	0.145a	0.000a	0.000	—

[a] See Table 17.
[b] Based on a 2×10^6 kg soil ha^{-1}.
[c] Means with same letter are not different at $p \leq 0.05$.

cattle, 60 mg kg^{-1} for horses, and 60 and 150 mg kg^{-1} for breeding ewes and feeder lambs, respectively. Some data on bovines indicate that long-term feeding of diets containing more than 30 mg F kg^{-1} may lead to fluorosis within a period of 2 to 3 years.[229] Suttie[228] reported that forages in areas free of F contamination will generally contain 5 to 10 mg F kg^{-1} on dry matter basis. Forages subjected to atmospheric F pollution may show a tenfold variation in F content from season to season, tending to be lower during the early part of summer and higher in the fall.

Keerthisinghe et al.[230] showed that subterranean clover (*Trifolium subterraneum*) receiving NaF at 25, 50, and 100 mg F kg^{-1} soil had F contents in the plant tissue of 33,

Table 20 Fluoride (F) Content in Regrowth and Hay Bahiagrass Forage from a Pasture Growing on a Florida Spodosol Soil Amended with Phosphogypsum (PG) as Source of S and Ca, Averaged by Year and Over a 3-Year Period, 1990–1992 (mg F kg^{-1})

Forage	Treatment Mg PG ha^{-1}	1990	1991	1992	Means
Regrowth	0.0	8.8a	6.3b	7.6a	7.2a
	0.4	8.0a	6.8ab	8.3a	7.7a
	2.0	10.2a	7.3ab	7.7a	7.9a
	4.0	9.7a	9.8a	7.9a	8.6a
Hay	0.0	9.5a	7.4a	2.4a	5.8a
	0.4	7.5a	6.9a	2.6a	5.3a
	2.0	5.5a	7.6a	2.5a	5.2a
	4.0	7.5a	7.0a	2.6a	5.4a

Note: Means with same letter are not different at $p \leq 0.05$.

Table 21 Electrical Conductivity (E_c), Fluoride (F) Content, and pH of Runoff and Groundwater Sampled at 60- and 120-cm Depths from a Florida Spodosol Soil Cropped to Bahiagrass and Amended with Phosphogypsum (PG) as Source of S and Ca, Averaged by Year and Over a 3-Year Period, 1990–1992

Depth	Treatment (Mg PG ha^{-1})	1990	1991	1992	Means
			$s\ m^{-1}$		
Runoff	0.0	0.03a	0.03a	0.02ab	0.03a
	0.4	0.03a	0.02a	0.03a	0.03a
	2.0	0.05a	0.03a	0.01b	0.03a
	4.0	0.06a	0.05a	0.03a	0.04a
60 cm	0.0	0.02a	0.03b	0.02a	0.02b
	0.4	0.03a	0.04b	0.03a	0.03b
	2.0	0.04a	0.04b	0.02a	0.03b
	4.0	0.07a	0.12a	0.04a	0.08a
120 cm	0.0	0.02a	0.03b	0.02b	0.02a
	0.4	0.03a	0.04ab	0.04b	0.04a
	2.0	0.03a	0.11a	0.06ab	0.06a
	4.0	0.03a	0.07ab	0.09a	0.05a
			mg F L^{-1}		
Runoff	0.0	0.22a	0.09a	0.17c	0.15c
	0.4	0.27a	0.13a	0.56a	0.35a
	2.0	0.42a	0.13a	0.29bc	0.25b
	4.0	0.69a	0.12a	0.41ab	0.35a
60 cm	0.0	0.11c	0.14a	0.23a	0.15b
	0.4	0.14bc	0.15a	0.43a	
0.20ab					
	2.0	0.17ab	0.33a	0.83a	0.36a
	4.0	0.22a	0.30a	0.71a	0.36a
120 cm	0.0	0.10a	0.08bc	0.16a	0.11a
	0.4	0.21a	0.14ab	0.28a	0.21a
	2.0	0.19a	0.27a	0.35a	0.25a
	4.0	0.20a	0.22ab	0.32a	0.23a
			pH		
Runoff	0.0	5.4a	5.3ab	4.9a	5.2a
	0.4	6.3a	5.9a	5.9a	6.0a
	2.0	5.5a	5.3ab	5.4a	5.4a
	4.0	5.9a	4.3b	5.0a	4.9a
60 cm	0.0	4.58a	4.83a	4.45a	4.65a
	0.4	4.61a	4.57a	4.63a	4.60a
	2.0	5.08a	5.56a	5.30a	5.33a
	4.0	4.68a	4.90a	5.17a	4.88a
120 cm	0.0	5.20a	4.85a	5.10a	5.08a
	0.4	5.80a	5.25a	5.48a	5.58a
	2.0	5.22a	4.48a	4.85a	4.93a
	4.0	4.91a	5.28a	3.97a	4.81a

Note: Means with same letter are not different at $p \leq 0.05$.

52, and 73 mg F kg^{-1} dry matter, respectively, across three rates of P. The F contents in subterranean clover receiving phosphogypsum and gypsum at rates of 2.5, 5, and 10 Mg ha^{-1} ranged from 19 to 22 mg F kg^{-1}. In a pot experiment conducted at the Ona Research Center,[231] F content of bahiagrass increased dramatically with F application in excess of 125 mg F kg^{-1} soil, reaching a maximum of 250 for NaF and 50 mg kg^{-1} for F in phosphate plant waste water.

Rechcigl et al.[220] used phosphogypsum containing 0.43% F. At 4 Mg ha^{-1} rate, this would raise the F content of the soil by 8.6 mg F kg^{-1} to the depth of 15 cm. The mean F concentrations in bahiagrass regrowth and hay tissues for the individual years and for the 3-year period are given in Table 20. The 4 Mg ha^{-1} rate increased F content of regrowth plant tissue over the control only in the second year (1991). Treatment effects were not significantly different when averaged over the 3-year period. The treatments had no effect on tissue F in the mature (hay) forage, in individual years, and when averaged over the 3-year period.[220]

The effects of phosphogypsum on surficial groundwater F, E_c, and pH are given in Table 21. The table shows that 0.4 Mg ha^{-1} phosphogypsum applied annually and 4 Mg applied initially increased mean F concentrations in runoff in 1992 and when averaged over the 3-year period. Treatments of 2 and 4 Mg ha^{-1} applied initially increased F concentrations over the control during the first year (1990) in water samples collected at 60-cm depth and when averaged over the 3-year period. Fluoride concentrations in water samples collected at 120-cm depth were not affected by the treatments when averaged over the 3-year period. In instances when phosphogypsum application increased F concentration in water, the magnitude of the F values remained substantially below the NPDW MCL of 4 mg F L^{-1}.[195]

The E_c of runoff was not affected by treatments when effects were averaged over the 3-year period. Application of 4 Mg ha^{-1} increased E_c of groundwater sampled at 60 cm depth in 1991 and when averaged over the 3-year period. The effect of 4 Mg ha^{-1} phosphogypsum on water sampled at 120-cm depth was significant over the control only in the third year but not when averaged over the 3-year period. All E_c values were substantially below the upper range of 0.15 S m^{-1} for potable water in the U.S.[205]

Phosphogypsum up to 4 Mg ha^{-1} had no statistically measurable effect on the pH of runoff and subsurface water.

B. IMPACT OF RADIONUCLIDES IN PHOSPHOGYPSUM

1. Effect on Soil Radionuclide Concentrations

The ^{226}Ra concentrations in U.S. soils by individual measurements ranged from 0.23 to 4.2 pCi g^{-1}. The state average, as determined by gamma spectroscopy, ranged from 0.65 in Alaska to 1.5 pCi g^{-1} in Kentucky, Nevada, New Mexico, and Ohio. The average concentrations for the whole U.S. were 1.1, 0.98, 1.0 pCi g^{-1} for ^{226}Ra, ^{232}Th, and ^{238}U, with a 0.77 correlation coefficient between ^{226}Ra and ^{238}U, indicating roughly a radioactive equilibrium in most soil samples.[232]

The USEPA standards for contaminated soils in any processing area is 5 pCi g^{-1} in excess of background level from the surface to 15 cm below and 15 pCi g^{-1} averaged over 15-cm thick layers of soil more than the 15 cm below the surface.[233] For agricultural land use, the National Council on Radiation Protection and Measurement (NCRPM) suggested concentration limits for ^{226}Ra and ^{210}Pb at 40 and 20 pCi g^{-1}, respectively.[234]

Table 22 ^{226}Ra, ^{210}Pb, and ^{210}Po Concentrations in a Florida Spodosol Soil Cropped to Bahiagrass and Amended with Phosphogypsum (PG) as Source of S and Ca, Averaged over a 3-Year Period, 1990–1992, By Depth

Treatment (Mg PG ha^{-1})	Depth (cm)					
	0–15	15–30	30–45	45–60	60–75	75–90
			pCi ^{226}Ra g^{-1}			
0.0	0.55a	0.67a	0.56a	0.52a	0.55a	0.75a
0.4	0.46a	0.43a	0.58a	0.58a	0.75a	0.70a
2.0	0.65a	0.63a	0.43a	0.58a	0.53a	0.35a
4.0	0.58a	0.80a	0.40a	0.47a	0.48a	0.48a
			pCi ^{210}Pb g^{-1}			
0.0	0.61a	0.55a	0.42a	0.32a	0.45a	0.42a
0.4	0.59a	0.28a	0.22b	0.42a	0.48a	0.33a
2.0	0.63a	0.43a	0.38b	0.37a	0.60a	0.34a
4.0	0.59a	0.30a	0.22b	0.08a	0.28a	0.32a
			pCi ^{210}Po g^{-1}			
0.0	0.53a	0.28a	0.23a	0.24a	0.21a	0.23b
0.4	0.58a	0.68a	0.75a	0.64a	0.40a	0.32b
2.0	0.90a	0.32a	0.44a	0.49a	0.31a	0.44b
4.0	0.79a	0.45a	0.23a	0.64a	0.79a	1.07a

Note: Means with same letter are not different at $p \leq 0.05$.

^{210}Po and ^{210}Pb, the longer-lived decay products of ^{222}Rn, are largely formed in the atmosphere, but their individual half-life is sufficiently long that they have separate existence from their Rn parent.[235]

^{226}Ra, ^{210}Po, and ^{210}Pb are the most significant radionuclides in phosphogypsum (Table 11). Hence, the study and review were limited to these radionuclides as they impact on bahiagrass and the immediate environment. Mortvedt[236,237] has reviewed several other radionuclides as they impact on agriculture. Miller and Sumner[238] also considered other radionuclides in their study to determine the impacts of phosphogypsum on soils. Mullins and Mitchell[239] reported that application of phosphogypsum, with 21 and 0.008 pCi g^{-1} ^{226}Ra and ^{210}Po, respectively, at 0, 0.145, and 0.295 Mg ha^{-1} to winter wheat had little or no effect on ^{226}Ra and ^{210}Po in soil. Values for ^{226}Ra in a Benndale soil series ranged from 0.10 to 0.23 for the control and 0.08 to 0.18 pCi g^{-1} for the phosphogypsum-treated plots. In a Dothan soil series, ^{226}Ra in control plots ranged from 0.23 to 0.35 and 0.18 to 0.33 in treated plots. Values for ^{210}Po in both soil series were below the analytical detection limit. Miller and Sumner,[238] using phosphogypsum containing 17.7, 2.4, and 17.7 pCi g^{-1} ^{266}Ra, ^{210}Po, and ^{210}Pb, respectively, applied at 10 Mg ha^{-1}, found that there was no detectable elevated levels of any radionuclide in the topsoil or subsoil, 5 years after application. Mays and Mortvedt[224] incorporated phosphogypsum containing 25 pCi g^{-1} ^{226}Ra into a soil (coarse-silty, siliceous, thermic Glossic Fragiudults) at 0, 22, and 112 Mg ha^{-1} to determine the effects of disposing phosphogypsum on agricultural land. The soil was sampled after a corn-wheat-soybean crops sequence. Their results showed that phosphogypsum at 112 Mg ha^{-1} elevated ^{226}Ra in the soil from 0.94 to 1.98 pCi g^{-1} at the 0- to 15-cm depth but had no effect at the lower depths.

The ^{226}Ra, ^{210}Po, and ^{210}Pb concentrations in phosphogypsum used in the bahiagrass-phosphogypsum environmental study were 18.1, 24.3, and 30.8 pCi g^{-1}, respectively. The effects of phosphogypsum application to a bahiagrass pasture on soil radionuclides are given in Table 22. There was no measurable effect of phosphogypsum application on the levels of ^{226}Ra, ^{210}Pb, and ^{210}Po analyzed by depth at 15-cm intervals over the whole 90-cm soil profile except at the depth of 75 to 90 cm.

2. *Effect on Plant Tissue Radionuclide Concentrations*

Simon and Ibrahim[240] reviewed ^{226}Ra uptake by plants. Radionuclide uptake is influenced by the plant species, soil properties, and by the chemical form of radionuclide. This uptake may be described in terms of the "concentration ratio" or CR (CR = concentration in dry plant tissue/concentration in oven-dry soil). When plant tissue is expressed in fresh-weight basis, the ratio is termed "soil-to-plant-transfer factor".[241] It is normally assumed that a linear relationship exists between the radionuclides concentration in the soil and in the tissue of the plant grown on it. The National Council on Radiation Protection and Measurements (NCRPM)[234] cited a study of 11 types of root and leafy vegetables grown on soil contaminated with U tailings in which a linear relationship was observed between ^{226}Ra concentration in the plant tissue and in the soil. Lindekin and Coles[242] reported that uptake of ^{226}Ra from soil with 0.5 pCi g^{-1} by broccoli and turnips (*Brassica rapa*) were lower by a factor of 2 in soil with 5200 mg Ca kg^{-1} compared to another soil with 3100 mg Ca kg^{-1}. Mislevy et al.[243] showed that biomass plants growing on two phosphatic clay soils had different levels of ^{226}Ra after 1 year. Elephant grass (*Pennisetum purpureum* L. "PI 300086") and leucaena (*Leucaena leucocephala* [Lam.] De Wet) contained the lowest concentrations averaging 0.09 and 0.13 pCi g^{-1}. Desmodium (*Desmodium cinerascens* A. Gray) with 0.37 pCi g^{-1} had the highest ^{226}Ra content. On the average, plants grown on mined phosphatic clay containing 24.3 pCi g^{-1} had 0.23 pCi g^{-1} in their tissue while those grown on unmined soil containing 0.24 pCi g^{-1} had 0.04 pCi g^{-1}. Thus, mined phosphatic clay that had 101 times more radioactivity than the unmined soil increased radioactivity in the plant tissue only by six times. The mean CR for mined and unmined soils for these biomass crops were 0.009 and 0.170, respectively. These data indicate not only that uptake may not be linear[244] but also that plants may take up ^{226}Ra only to a certain limit. Guidry et al.,[245] in an exhaustive study of the relationships among food plant species, soil properties, and levels of radionuclides, reported that for ^{226}Ra strong positive relationships exist between ^{226}Ra in crop tissue and ^{226}Ra in soil, soil pH, and soil organic matter content. A weak negative relationship was found for CEC, Mg, and Ca. Food ^{210}Pb levels were also positively correlated with soil ^{210}Pb, but there were no clear-cut relationships between the soil properties previously mentioned and food ^{210}Pb as well as between food ^{210}Po and soil ^{210}Po. Mays and Mortvedt[224] showed that phosphogypsum from 0, 22, and 112 Mg ha^{-1} had no effect on the concentrations of ^{226}Ra in corn, wheat, and soybean grains. Corn, wheat, and soybean fertilized with phosphogypsum at 112 Mg ha^{-1} were determined to contain 0.001, 0.022, and 0.0.01 pCi g^{-1} compared with the control with 0.001, 0.012, and 0.027 pCi g^{-1}, respectively. Mullins and Mitchell[239] reported that ^{226}Ra and ^{210}Po concentrations in wheat forage fertilized with phosphogypsum were no different from those of control samples. In Florida, Myhre et al.[246] reported that phosphogypsum applied at rates up to 2.24 Mg ha^{-1} had no effect on ^{226}Ra concentration in citrus fruit juice. The highest concentration observed in a sample was 3.0 pCi L^{-1} while the highest mean was 1.72 pCi L^{-1}. Fruit juice from trees grown on Myakka soil had the highest overall mean concentration of 1.22 compared to 0.52 and 0.69 pCi L^{-1} ^{226}Ra on Oldsmar and Smyrna soils, respectively. There was no relationship between ^{226}Ra content in fruit juice and phosphogypsum application rates.

Table 23 **^{226}Ra, ^{210}Pb, and ^{210}Po Concentrations in Regrowth and Hay Bahiagrass Forage from a Pasture Growing on a Florida Spodosol Soil Amended with Phosphogypsum (PG) as Source of S and Ca, Averaged by Year and Over a 3-Year Period, 1990–1992**

Treatment (Mg PG ha^{-1})	1990	1991	1992	Means[a]
	Regrowth Forage			
	pCi ^{226}Ra g^{-1}			
0.0	0.01a[b]	0.04a	0.13a	0.05b
0.4	0.02a	0.04a	0.17a	0.06b
2.0	0.01a	0.06a	0.13a	0.05b
4.0	0.03a	0.09a	0.23a	0.10a
	pCi ^{210}Pb g^{-1}			
0.0	0.20a	0.50ab	2.53a	0.93a
0.4	0.13b	0.73a	2.13a	0.93a
2.0	0.14b	0.62ab	2.10a	0.87a
4.0	0.13b	0.46c	2.22a	0.81a
	pCi ^{210}Po g^{-1}			
0.0	0.04a	0.52a	0.10b[c]	0.33a
0.4	0.10a	0.73a	0.10b	0.48a
2.0	0.19a	0.51a	0.10b	0.38a
4.0	0.22a	0.37a	0.20a	0.30a
	Hay Forage			
	pCi ^{226}Ra g^{-1}			
0.0	0.03b	0.03a	—[d]	0.03c
0.4	0.10a	0.13a	—	0.11a
2.0	0.03b	0.10a	—	0.06bc
4.0	0.09a	0.09a	0.20	0.09ab
	PCi ^{210}Pb g^{-1}			
0.0	0.80a	0.75a	2.29a	1.33a
0.4	0.95a	0.85a	3.05a	1.45a
2.0	0.75a	0.90a	2.20a	1.28a
4.0	0.87a	0.40a	2.75a	1.23a
	PCi ^{210}Po g^{-1}			
0.0	0.73a	0.27a	—[d]	0.57a
0.4	0.62a	0.59a	0.20a	0.61a
2.0	0.50a	0.78a	0.10b	0.61a
4.0	0.87a	0.85a	0.10b	0.86a

[a] Statistics computed based solely on harvests where each treatment had at least one detected value associated with it.
[b] Means with same letter(s) are not different at $p \leq 0.05$.
[c] Significance due to MSE being very small.
[d] Not detected in samples.

Table 23 shows the radionuclide contents of bahiagrass in regrowth and mature forages which were determined for 3 years. The 3-year means showed that application of 4 Mg phosphogypsum appeared to have increased ^{226}Ra in bahiagrass regrowth and hay forages but not ^{210}Pb and ^{210}Po. Miller and Sumner,[238] using 10 Mg ha^{-1} phosphogypsum, also found no difference in ^{226}Ra uptake by alfalfa grown in phosphogypsum-treated plots and in the control.

Table 24 **^{226}Ra, ^{210}Pb, and ^{210}Po Concentrations in Runoff and Groundwater Sampled at 60- and 120-cm Depths from a Florida Spodosol Soil Cropped to Bahiagrass and Amended with Phosphogypsum (PG) as Source of S and Ca, Averaged by Year and over a 3-Year Period, 1990–1992**

Depth	Treatment (Mg PG ha^{-1})	1990	1991	1992	Means
		pCi ^{266}Ra L^{-1}			
Runoff	0.0	0.11b[a]	0.20c	0.27a	0.19a
	0.4	0.13b	0.35b	0.50a	0.37a
	2.0	0.07b	0.40a	0.20a	0.19a
	4.0	0.42a	0.35b	0.47a	0.42a
60 cm	0.0	—	0.40a	—	0.40a
	0.4	—	0.60a	—	0.60a
	2.0	—	0.55a	—	0.55a
	4.0	—	0.70a	—	0.70a
120 cm	0.0	0.69a	0.86a	0.73a	0.80a
	0.4	1.35a	1.36a	0.97a	1.23a
	2.0	0.93a	1.59a	0.87a	1.13a
	4.0	1.12a	1.11a	1.75a	1.33a
		pCi ^{210}Pb L^{-1}			
Runoff	0.0	0.05a	1.00a	0.50ab	0.45a
	0.4	0.17a	1.10a	1.10a	0.87a
	2.0	0.00a	0.60a	0.60ab	0.36a
	4.0	0.41a	0.60a	0.20b	0.40a
60 cm	0.0	—	0.45b	—	0.45b
	0.4	—	1.05b	—	1.05b
	2.0	—	0.80b	—	0.80b
	4.0	—	2.50a	—	2.50a
120 cm	0.0	0.31a	0.57a	0.45a	0.42b
	0.4	0.36a	0.86a	0.55a	0.60ab
	2.0	0.34a	1.63a	1.13a	1.03a
	4.0	0.25a	1.02a	1.03a	0.77ab
		pCi ^{210}Po L^{-1}			
Runoff	0.0	0.37a	0.40a	0.50a	0.44a
	0.4	0.25a	1.30a	0.23a	0.54a
	2.0	0.29a	0.60a	0.23a	0.31a
	4.0	0.41a	0.40a	0.27a	0.35a
60 cm	0.0	—	0.35a	—	0.35a
	0.4	—	0.25a	—	0.25a
	2.0	—	0.35a	—	0.35a
	4.0	—	0.40a	—	0.40a
120 cm	0.0	1.13ab	1.51a	1.93a	1.30a
	0.4	0.50b	0.79a	0.15a	0.55a
	2.0	0.57b	0.94a	0.55a	0.72a
	4.0	1.28a	1.01a	0.53a	0.98a

[a] Means with same letter are not different at $p \leq 0.05$.

3. Effect on Surficial Groundwater Radionuclide Concentrations

Radionuclide concentrations in groundwater collected from a bahiagrass pasture fertilized with phosphogypsum are given in Table 24. Runoff from the 4 Mg-treated plots showed higher ^{226}Ra than from all other plots in the first year. In the second year, runoff from phosphogypsum-treated plots showed higher ^{226}Ra concentrations than from the control. The 3-year average, however, indicated that phosphogypsum treatments had no effect on ^{226}Ra concentrations in runoff. There was also no difference in the concentrations of ^{210}Po and ^{210}Pb in runoff in individual years and when averaged over 3 years.

Water samples at 60-cm depth were collected only during the second year (1991). There was no effect of treatments on ^{226}Ra and ^{210}Po, but 4 Mg ha^{-1} phosphogypsum appeared to have elevated ^{210}Pb concentration (Table 24). The data, however, are very limited and not conclusive. Miller and Sumner,[238] in a leaching study using 30-cm long undisturbed soil columns and 10 Mg ha^{-1} phosphogypsum rate, found that ^{226}Ra was significantly leached from a sandy but not from a clayey soil. Total amount leached was 5 pCi or 3.5% of the 142 pCi initially added in phosphogypsum. No measurable amounts of ^{210}Po and ^{210}Pb were detected in the leachates.

Water samples collected at 120-cm depth showed no significant difference related to phosphogypsum rates in ^{226}Ra, ^{210}Po, and ^{210}Pb concentrations (Table 24). The ^{226}Ra concentrations in all water samples collected over the years at all depths were extremely low compared to the proposed MCL of 20 or the present standard of 3 pCi L^{-1}. The implicit MCLs for ^{210}Pb and ^{210}Po are 1.0 and 15.0 pCi L^{-1}, respectively (Roessler, C.E., personal communication).

4. Effect on Soil Rn Flux, Gamma Radiations, and Ambient Atmospheric Rn

When ^{222}Rn and its progenies are inhaled, they may deposit in the respiratory system where they can irradiate the bronchial and lung tissues. Exposure to ^{222}Rn and its progenies at high concentrations can cause lung cancer as it does among uranium miners. Cumulative exposure to low concentrations for long periods may present a risk of lung cancer.[247]

Table 25 ^{222}Rn Surface Flux (pCi m^{-2} s^{-1}) Over a Florida Spodosol Soil Cropped to Bahiagrass Pasture and Amended with Phosphogypsum (PG) as Source of S and Ca, Averaged by Year and Over a 3-Year Period, 1990–1992

Treatment (Mg PG ha^{-1})	1990	1991	1992	Means
0.0	0.04a	0.03a	0.02a	0.03a
0.4	0.03a	0.03a	0.03a	0.03a
2.0	0.04a	0.05a	0.03a	0.04a
4.0	0.03a	0.04a	0.04a	0.04a

Note: Means with same letter are not different at $p \leq 0.05$.

Table 26 Gamma Radiation (uR h^{-1}) Over a Florida Spodosol Soil Cropped to Bahiagrass and Amended with Phosphogypsum (PG) as Source of S and Ca, Averaged by Year and Over a 3-Year Period, 1990–1992

Treatment (Mg PG ha^{-1})	1990–1991	1991–1992	1992	Means
0.0	4.29a	5.28a	5.48a	4.81a
0.4	4.54a	5.06a	5.45a	5.09a
2.0	4.55a	5.37a	5.35a	5.04a
4.0	4.71a	5.08a	5.23a	4.99a

Note: Means with same letter are not different at $p \leq 0.05$.

Table 27 Ambient Atmospheric ^{222}Rn (pCi L^{-1}) Over a Florida Spodosol Soil Cropped to Bahiagrass and Amended with Phosphogypsum (PG) as Source of S and Ca, Averaged by Year and Over a 3-Year Period, 1990–1992

Treatment (Mg PG ha^{-1})	1990–1991	1991–1992	1992	Means
0.0	0.22b	0.22a	0.20a	0.20a
0.4	0.27a	0.26a	0.19a	0.24a
2.0	0.19c	0.24a	0.20a	0.21a
4.0	0.20c	0.21a	0.21a	0.20a

Note: Means with same letter are not different at $p \leq 0.05$.

The concentration of Rn progeny is measured by a unit called working level (WL; WL $\times$ 10^{-3} = mWL) using air grab samples according to the procedure of Rolle.[248] One WL is the concentration of Rn progeny in equilibrium with 200 pCi L^{-1} of ^{222}Rn in air[244] or any combination of short-lived Rn decay products in 1 L of air that will result in the emission of alpha particles with a potential total energy of 13 $\times$ 10^4 MeV.[233]

Radon progeny concentrations inside a weighing room, 3.6 $\times$ 7.2 m, where about 0.5 Mg phosphogypsum was stored, were measured in the morning after Rn gas was allowed to accumulate overnight. A mean Rn progeny concentration of 12 mWL, six times higher than that of the air outside, was recorded. Field measurements of the WL at the workers' breathing zone during a 3-h application of phosphogypsum by hand broadcast showed a mean concentration of 0.10 mWL.[220] These measurements show that agricultural workers handling phosphogypsum in the field are exposed to no more radioactivity than they are in their homes which in Florida is about 4 mWL indoors.[247]

Three radiation parameters were measured over the bahiagrass plots treated with phosphogypsum. ^{222}Rn flux (pCi m^{-2} s^{-1}) is the amount of ^{222}Rn gas emanating from the ground surface. It is measured using a large-area activated charcoal canister according to USEPA guidelines.[249,250] The ambient atmospheric ^{222}Rn (pCi L^{-1} of air) and the gamma radiation (uR h^{-1}) were measured using electret ion chambers (EICs).[251–253] For reference, individual measurements of gamma radiation in the U.S. ranged from <1 to 34 uR h^{-1}. State averages vary from 3.3 in Texas to 14 uR h^{-1} in Colorado, Nevada, and Wyoming.[232] The value in Florida ranged from 5 to 10 uR h^{-1}.[254] Radon soil surface flux measurements have been reported to vary by a factor of 250, ranging from 0.005 to 1.41 with an average of 0.43 pCi m^{-2} s^{-1}.[255,256] The individual values of ambient atmospheric ^{222}Rn in the U.S. measured using EICs ranged from 0.0 to 1.11 pCi L^{-1} with a median concentration of 0.39 pCi L^{-1}. Mean annual outdoor values ranged from 0.16 to 0.57 pCi L^{-1}.[257]

The results of soil Rn flux, gamma radiation, and ambient atmospheric Rn measurements over the bahiagrass pasture fertilized with phosphogypsum are given in Tables 25 through 27, respectively. The tables show that phosphogypsum up to 4 Mg ha^{-1} had no statistically measurable effect on Rn flux, gamma radiation, and ambient atmospheric ^{222}Rn, measured several times during a year period, on the annual means and when averaged over 3 years. Soil ^{222}Rn flux values were about one tenth that of the reported national average of 0.43 pCi m^{-2} s^{-1}, but were within the range of individual values cited earlier for reference. A flow-through method using electret ion chambers to measure soil Rn flux confirmed the relatively low values found in the study.[223] The gamma radiation and the ambient atmospheric Rn values were also all within the values reported.[255–257]

VII. CONCLUSIONS

The toxic metals and fluorides in industrial by-product gypsums do not appear to constitute environmental hazards to surficial groundwater, soil, and crop tissue at rates used in agriculture. The USEPA maximum limit of 10 pCi ^{226}Ra g^{-1} phosphogypsum for agricultural use appears to be too low. Almost all studies show little or no statistically measurable radiological effect in crops, soil, surficial groundwater, and the ambient atmosphere from phosphogypsum containing twice as much ^{226}Ra as the USEPA limit. There is the need to work with phosphogypsum with higher radioactivity and at high application rates until statistically measurable effects are achieved in order to base risk assessments on measured or empirical data.

ACKNOWLEDGMENTS

We thank the Florida Institute of Phosphate Research (FIPR), Bartow, Florida, for the research grant that made the writing of this review possible. We also express our appreciation to Drs. G. D. Nifong of FIPR, Rosa M. C. Muchovej of University of Florida, R. F. Korcak, and K. D. Ritchey of USDA-ARS in Beltsville, Maryland, and Beckley, West Virginia, respectively, for their critical review of the paper.

REFERENCES

1. Alcordo, I.S. and Rechcigl, J.E., Phosphogypsum in agriculture: a review, in *Advances in Agronomy*, Vol. 49, Sparks, D.L., Ed., Academic Press, San Diego, 1993, 55.
2. Shainberg, I., Sumner, M.E., Miller, W.P., Farina, M.P.W., Pavan, M.A., and Fey, M.V., Use of gypsum on soils: a review, in *Advances in Soil Science*, Vol. 9, Stewart, B.A., Ed., Springer-Verlag, New York, 1989, 1.
3. Kouloheris, A.P., Chemical nature of phosphogypsum as produced by various wet process phosphoric acid processes, in *Phosphogypsum*, Borris, D.P. and Boody, P.W., Eds., Florida Institute of Phosphate Research, Bartow, FL, 1980, 7.
4. Mosher, R.E., Glover, R.L., and Colley, J.D., Current and future trends in gypsum-producing FGD systems by U.S. power utilities, in *New Frontiers for Byproduct Gypsum*, McAdie, H.G., Compiler, ORTECH International, Ontario, 1988, 83.
5. Saleem, A., GE's worldwide experience with IFO based gypsum producing FGD systems, Proc. 2nd Int. Conf. FGD and Chemical Gypsum, ORTECH International, Ontario, 1991.
6. Alva, A.K. and Gascho, G.J., Differential leaching of cations and sulfate in gypsum amended soils, *Commun. Soil Sci. Plant Anal.*, 22, 1195, 1991.
7. Bayrakli, F., Ammonia volatilization losses from different fertilizers and effect of several urease inhibitors, CaCl2 and phosphogypsum on losses from urea, *Fert. Res.*, 23, 147, 1990.
8. da Gloria, N.A., Barretto, M.C.V., Moraes, C.J., and Mattiazzo-Prezotto, M.E., Avaliação do gesso e de alguns fosfatos como inibidores da volatilização de amonia de estercos, *R. Bras. Ci. Solo, Campinas*, 15, 297, 1991.

9. Stout, W.L., Menser, H.A., Hern, J.L., and Bennett, O.L., Fluidized bed combustion waste in food production, in *Solid Waste Research and Development Needs for Emerging Coal Technologies*, American Society of Civil Engineers, New York, 1979, 170.
10. Stout, W.L., Hern, J.L., Korcak, R.F., and Carlson, C.W., *Manual for Applying Fluidized Bed Combustion Residue to Agricultural Lands*, USDA, Agricultural Research Service, 1988.
11. Makansi, J. and Ellison, W., Worldwide progress in the utilization of by-product gypsum, in *New Frontiers for Byproduct Gypsum*, McAdie, H.G., Compiler, ORTECH International, Ontario, 1988, 1.
12. Gaynor, J.C., Differences between chemical and natural gypsum, in *New Frontiers for Byproduct Gypsum*, McAdie, H.G., Compiler, ORTECH International, Ontario, 1988, 143.
13. Henschel, D.B., The EPA R&D program to assess the solid residue from the fluidized-bed combustion process, in *Solid Waste Research and Development for Emerging Coal Technologies*, American Society of Civil Engineers, New York, 1979, 37.
14. Faber, P.S. and Livengood, C.D., Characterization of an industrial spray dryer at Argonne National Laboratory, Project Summary, 8th Symp. on Flue Gas Desulfurization, Industrial Environmental Research Laboratory, Research Triangle Park, NC, 1984, 12.
15. Fortune, O., Bechtel, T.F., Puska, E., and Arello, J., Design and initial operation of the spray dry FGD system at the Marquette Michigan Board of Light and Power Shiras #3 plant, Project Summary, 8th Symp. on Flue Gas Desulfurization, Industrial Environmental Research Laboratory, Research Triangle Park, NC, 1984, 12.
16. Gustke, J.M., Wayne, E., and Morgan, M.D., Overview and evaluation of two years of operation of the Riverside spray dryer system, Project Summary, 8th Symp. on Flue Gas Desulfurization, Industrial Environmental Research Laboratory, Research Triangle Park, NC, 1984, 12.
17. Bloem, P.J.C. and Sciarione, B.J.G., Application of spray-dry products in building materials, Proc. 2nd Int. Conf. FGD and Chemical Gypsum, ORTECH International, Ontario, 1991, 29.1.
18. Hammer, E.J. and Bennett, R., Industrial end-uses for by-product gypsum, in *New Frontiers for Byproduct Gypsum*, ORTECH International, Ontario, 1988, 189.
19. Becker, P., *Phosphates and Phosphoric Acid*, Marcel Dekker, New York, 1983, 6.
20. Lin, K.T., Lai, C.I., Ghafoori, N., and Chang, W.F., High-strength concrete utilizing industrial by-product, in *Proc. 3rd Int. Symp. in Phosphogypsum*, Vol. 2, Chang, W.F., Ed., Florida Institute of Phosphate Research, Bartow, FL, 1990, 456.
21. USEPA, Environmental Impact Statement for Proposed NESHAPS for Radionuclides — Background Information Document, Vol. 2 (Draft), USEPA, Washington, D.C., 1989.
22. Beretka, J., The current state of utilization of phosphogypsum in Australia, in *Proc. 3rd Int. Symp. Phosphogypsum*, Vol. 2, Florida Institute of Phosphate Research, Bartow, FL, 1990, 394.
23. van der Sloot, H.A., and de Groot, A.J., Environmental aspects of the utilization of pellets prepared from fosfogypsum, coal fly ash and cement, in *Contaminated Soil*, Assink, J.W. and van den Brink, W.J., Eds., Martinus Nijhoff, Dordrecht, The Netherlands, 1985, 919.
24. Novikov, A.A., Klassen, P.V., and Evenchik, S.D., The status and trends of phosphogypsum utilization in the USSR, in *Proc. 3rd Int. Symp. Phosphogypsum*, Vol. 2, Florida Institute Phosphate Research, Bartow, FL, 1990, 594.

25. Collings, R.K., Phosphogypsum in Canada, in *Phosphogypsum*, Borris, D.P. and Boody, P.W., Eds., Florida Institute of Phosphate Research, Bartow, FL, 1980, 583.
26. Miyamoto, M., Phosphogypsum utilization in Japan, in *Phosphogypsum*, Borris, D.P. and Boody, P.W., Eds., Florida Institute of Phosphate Research, Bartow, FL, 1980, 535.
27. Hara, N., Utilization of FGD gypsum for building materials in Japan, in *New Frontiers for Byproduct Gypsum*, McAdie, H.G., Compiler, ORTECH International, Ontario, 1988, 171.
28. Mishra, U.N., Use of phosphogypsum in reclamation of sodic soils in India, in *Phosphogypsum*, Borris, D.P. and Boody, P.W., Eds., Florida Institute of Phosphate Research, Bartow, FL, 1980, 223.
29. Erlendstadt, G., The Salzgitter technology for the refining of by-product gypsum: experience in Norway, in *New Frontiers for Byproduct Gypsum*, McAdie, H.G., Compiler, ORTECH International, Ontario, 1988, 123.
30. BSC, Inc., *World Fertilizer Plant List and Atlas*, 8th ed., British Sulphur Corporation, London, 1984.
31. USEPA, Report to Congress: Wastes from the Extraction and Beneficiation of Metallic Ores, Phosphate Rock, Asbestos, Overburden from Uranium Mining, and Oil Shale, USEPA, Washington, D.C., 1985.
32. Chang, W.F., *Reclamation, Reconstruction and Reuse of Phosphogypsum for Building Materials*, Florida Institute of Phosphate Research, Bartow, FL, 1987.
33. McFarlin, R.F., Current and anticipated research priorities for the utilization and management of phosphate process by-products and wastes, in *Phosphate Fertilizers and the Environment*, Schultz, J.J., Ed., International Fertilizer Development Center, Muscle Shoals, AL, 1992, 179.
34. TVA, Commercial Fertilizers, TVA/NFCD-85/5, Bulletin Y-207, Tennessee Valley Authority, Muscle Shoals, AL, 1988.
35. USEPA, Diffuse NORM Waste, Waste Characterization and Risk Assessment (Draft), Office of Radiation Programs, Washington, D.C., 1991, B-2-15.
36. Cole, H.H., *The Gypsum Industry in Canada*, Canada Department of Mines Mine Branch Report No. 714, F.A. Acland, 2 Ottawa, 1930.
37. Khalil, N.F, Alnuaimi, N.M., and Mustafa, M.H., Utilization of Phosphogypsum in Iraq, in *Proc. 3rd Int. Symp. Phosphogypsum*, Vol. 2, Chang, W.F., Ed., Florida Institute of Phosphate Research, Bartow, FL, 1990, 402.
38. May, A. and Sweeney, J., Assessment of environmental impacts associated with phosphogypsum in Florida, in *Phosphogypsum*, Borris, D. P. and Boody, P.W., Eds., Florida Institute Phosphate Research, Bartow, FL, 1980, 415.
39. May, A. and Sweeney, J., *Evaluation of Radium and Toxic Element Leaching Characteristics of Florida Phosphogypsum Stockpiles, Report of Investigation 8776*, Bureau of Mines, Tuscaloosa Research Center, AL, 1983.
40. Nifong, G.D., Environmental Aspects of Phosphogypsum (mimeo notes), Florida Institute of Phosphate Research, Bartow, FL, September 1988.
41. Umlauf, J., FGD in West Germany with emphasis on the SHU process, in *Proc. 2nd Int. Conf. FGD and Chemical Gypsum*, ORTECH International, Ontario, 1991, 15.1.
42. Rosenberg, H.S., Engdahl, R.B., Genco, G.M., Bloom, S.G., Ball, D.A., and Oxley, J.H., Status of Stack Gas Technology for SO2 Control (Final report, Part II), Electric Power Research Institute, Palo Alto, CA, 1975.
43. Venta, G.J., Utilization of chemical gypsum in Japan, in *Proc. 2nd Int. Conf. FGD and Chemical Gypsum*, ORTECH International, Ontario, 1991, 3.1.

44. Stein, V., FGD-gypsum in the United Germany: trends of demand and supply, in *Proc. 2nd Int. Conf. FGD and Chemical Gypsum*, ORTECH International, Ontario, 1991, 7.1.
45. Louis, T. and Buonicore, A.J., *Air Pollution Control Equipment*, Vol. 1, CRC Press, Boca Raton, FL, 1988, 5.
46. Rechcigl, J.E. and Sparks, D.L., Effect of acid rain on the soil environment: a review, *Commun. Soil Sci. Plant Anal.*, 16, 653, 1985.
47. Stebbins, L.H., *Summary and Outline: Clean Air Act Amendments of 1990*, Missimer & Associates, Jacksonville, FL, 1992, 328.
48. Carnes, S., The siting and institutional impacts of RCRA, in *Solid Waste Research and Development for Emerging Coal Technologies*, American Society of Civil Engineers, New York, 1979, 31.
49. USEPA, *Wastes from the Combustion of Coal by Electric Utility Power Plants*, USEPA, Washington, DC, 1988.
50. Carlson, C.L. and Adriano, D.C., Environmental impacts of coal combustion residues, *J. Environ. Qual.*, 22, 227, 1993.
51. Steffan, P. and Golden, D., FGD gypsum utilization: survey of current practices and assessment of market potential, in *Proc. 2nd Int. Conf. FGD and Chemical Gypsum*, ORTECH International, Ontario, 1991, 4.1.
52. Thayer, A.G., The potential impact of desulphogypsum on the North American gypsum board industry, in *Proc. 2nd Int. Conf. FGD and Chemical Gypsum*, ORTECH International, Ontario, 1991, 2.1.
53. Makkinejad, N., Bohm, H., and Niess, T., The variables affecting the production of commercial grade gypsum from the FGD process, in *Proc. 2nd Int. Conf. FGD and Chemical Gypsum*, ORTECH International, Ontario, 1991, 11.1.
54. Sekhar, N., Evolution and engineering aspects of gypsum production, dewatering, handling and storage in wet flue gas desulphurization process, in *Proc. 2nd Int. Conf. FGD and Chemical Gypsum*, ORTECH International, Ontario, 1991, 16.1.
55. Feeney, S., Downs, B., and Novak, J., In-situ forced oxidation retrofit at Michigan South Central Power Agency's Endicott Station, in *Proc. 2nd Int. Conf. FGD and Chemical Gypsum*, ORTECH International, Ontario, 1991, 10.1.
56. Wilhelm, J.H. and Stone, W.W., Techniques of gypsum dewatering, in *Proc. 2nd Int. Conf. FGD and Chemical Gypsum*, ORTECH International, Ontario, 1991, 13.1.
57. Kocman, V., Modern methods for the analysis of flue gas and by-product gypsums, in *New Frontiers for Byproduct Gypsum*, McAdie, H.G., Compiler, ORTECH International, Ontario, 1988, 155.
58. Barber, J.W., Hudgens, B.A., Byers, C.D., and Nathan, V.R., Characterization of synthetic gypsum, in *Proc. 2nd Int. Conf. FGD and Chemical Gypsum*, ORTECH International, Ontario, 1991, 26.1.
59. Keairns, D.L., Sun, C.C., Peterson, C.H., and Newby, R.A., Fluid-bed combustion and gasification solids disposal, in *Solid Waste Research and Development for Emerging Coal Technology*, American Society of Civil Engineers, New York, 1979, 92.
60. Ruth, L.A., Regeneration of CaSO4 in FBC, in *Proc. 4th Int. Conf. Fluidized-Bed Combustion*, The MITRE Corp., McLean, VA, 1975, 425.
61. Chughtai, M.Y. and Michelfelder, S., Direct desulfurization through additive injection in the vicinity of the flame, in *Project Summary: 8th Symp. Flue Gas Desulfurization*, Industrial Environmental Research Laboratory, Research Triangle Park, NC, 1984, 5.

62. McElroy, M.W., Review of EPRI research on furnace sorbent injection SO2 control, in *Project Summary: 8th Symposium on Flue Gas Desulfurization*, Industrial Environmental Research Laboratory, Research Triangle Park, NC, 1984, 5.
63. Ross, G. and Finlay, P.G., Environment Canada's program for the management of byproducts from thermal power emission control systems, in *New Frontiers for Byproduct Gypsum*, ORTECH International, Ontario, 1988, 55.
64. Korcak, R.F., Fluidized bed material applied at disposal levels: effects on an apple orchard, *J. Environ. Qual.*, 17, 469, 1988.
65. Taubert, U., Occurrence and recycling of FGD gypsum in Europe especially in the Federal Republic of Germany, in *Proc. 2nd Int. Conf. FGD and Chemical Gypsum*, ORTECH International, Ontario, 1991, 1.1.
66. Davidson, J.L. and Quirk, J.P., The influence of dissolved gypsum on pasture establishment on irrigated sodic clays, *Aust. J. Agric. Res.*, 12, 100, 1961.
67. Keisling, T.C., Rao, P.S.C., and Jessup, R.E., Pertinent criteria for describing the dissolution of gypsum beds in flowing water, *Soil Sci. Soc. Am. J.*, 42, 234, 1978.
68. Kemper, W.D., Olsen, J., and deMooy, C.J., Dissolution rates of gypsum in flowing water, *Soil Sci. Soc. Am. Proc.*, 39, 458, 1975.
69. Alcordo, I.S. and Rechcigl, J.E., The use of solubility curves as a method to characterize phosphogypsum and its constituents, *Commun. Soil Sci. Plant Anal.*, 23, 2595, 1992.
70. Weast, R.C., Ed., *Handbook of Chemistry and Physics*, 61st ed., CRC Press, Boca Raton, FL, 1981, B-72.
71. Bolan, N.S., Syers, J.K., and Sumner, M.E., Dissolution of various sources of gyspum in aqueous solutions and in soil, *J. Sci. Food Agric.*, 57, 527, 1991.
72. Korcak, R.F., Effects of applied sewage sludge compost and fluidized bed material on apple seedling growth, *Commun. Soil Sci. Plant Anal.*, 11, 571, 1980.
73. Adams, D.F. and Farwell, S.O., Sulfur gas emissions from stored flue gas desulfurization sludges, *J. Air Pollut. Control Assoc.*, 31, 557, 1981.
74. Santhanam, C.J., Lunt, R.R., Johnson, S.L., Cooper, C.B., Thayer, P.S., and Jones, J.W., Health and environmental impacts of increased generation of coal ash and FGD sludges, *Environ. Health Perspect.*, 33, 131, 1979.
75. Clark, R.B., Zeto, S.K, Ritchey, K.D., Wendell, R.R., and Baligar, V.C., Effects of coal flue gas desulfurization by-products and calcium-sulfite, -sulfate and -carbonate on maize grown in acid soil, *Plant Soil*, in press.
76. Marshall, W.L. and Slusher, R., Thermodynamics of calcium sulfate dihydrate in aqueous sodium chloride solutions, 0–110°, *J. Phys. Chem.*, 70, 4015, 1966.
77. Keren, R. and Shainberg, I., Effect of dissolution rate on the efficiency of industrial and mined gypsum in improving infiltration of a sodic soil, *Soil Sci. Soc. Am. J.*, 45, 103, 1981.
78. Barton, F.M. and Wilde, N.M., Dissolution rates of polycrystalline samples of gypsum and orthorhombic corms of calcium sulphate by rotating disc method, *Trans. Faraday Soc.*, 67, 3590, 1971.
79. Kamprath, E.J., Nelson, W.L., and Fitts, J.W., The effect of pH, sulfate and phosphate concentrations on the adsorption of sulfate by soils, *Soil Sci. Soc. Am. Proc.*, 20, 463, 1956.
80. Chao, T.T., Anion effects on sulfate adsorption by soils, *Soil Sci. Soc. Am. Proc.*, 28, 581, 1964.
81. Gebhardt, H. and Coleman, N.T., Anion adsorption by allophanic tropical soils. II. Sulfate adsorption, *Soil Sci. Soc. Am. Proc.*, 38, 259, 1974.

82. Chao, T.T., Harward, M.E., and Fang, S.C., Cationic effects on sulfate adsorption by soils, *Soil Sci. Soc. Am. Proc.*, 27, 35, 1963.
83. Chang, M.L. and Thomas, G.W., A suggested mechanism for sulfate adsorption by soils, *Soil Sci. Soc. Am. Proc.*, 27, 281, 1963.
84. Reeve, N.G. and Sumner, M.E., Amelioration of subsoil acidity in Natal Oxisols by leaching of surface-applied amendments, *Agrochemophysica*, 4, 1, 1972.
85. Sumner, M.E., *Gypsum as an Ameliorant for the Subsoil Acidity Syndrome*, Florida Institute of Phosphate Research, Bartow, FL, 1990.
86. de Villiers, J.M. and Jackson, M.L., Cation exchange capacity variations with pH in soil clays, *Soil Sci. Soc. Am. Proc.*, 31, 473, 1967.
87. Hue, N.V., Adams, F., and Evans, C.E., Sulfate retention by an acid BE horizon of an Ultisol, *Soil Sci. Soc. J.*, 49, 1196, 1985.
88. Alva, A.K., Sumner, M.E., and Miller, W.P., Reactions of gypsum or phosphogypsum in highly weathered acid subsoils, *Soil Sci. Soc. Am. J.*, 54, 993, 1990.
89. van Raij, B. and Peech, M., Electrochemical properties of some Oxisols and Alfisols of the tropics, *Soil Sci. Soc. Am. Proc.*, 36, 587, 1972.
90. Keng, J.K.C. and Uehara, G., Chemistry, mineralogy and taxonomy of Oxisols and Ultisols, *Soil Crop Sci. Soc. Fl. Proc.*, 33, 119, 1974.
91. Metzger, W.H., Distribution of fertilizer residues in the soil after fourteen years of a fertilizer experiment with alfalfa, *J. Am. Soc. Agron.*, 26, 620, 1934.
92. Brown, B.A. and Munsell, R.I., Soil acidity at various depths as influenced by time since application, placement, and amount of limestone, *Soil Sci. Soc. Am. Proc.*, 3, 217, 1938.
93. Pearson, R.W., Childs, J., and Lund, Z.F., Uniformity of limestone mixing in acid subsoils as a factor in cotton root penetration, *Soil Sci. Soc. Am. Proc.*, 37, 727, 1973.
94. Korentajer, L., Byrnes, B.H., and Hellums, D.T., Effect of liming and leaching on the sulfur-supplying capacity of soils, *Soil Sci. Soc. Am. J.*, 47, 525, 1983.
95. Morelli, J.L., Dalben, A.E., Almeida, J.O.C., and Dematte, J.L.I., Calcario e gesso na produtividade da cana-de-asucar e nas caracteristicas quimicas de um Latossolo de textura media alico, *R. Bras. Ci. Solo, Campinas*, 16, 187, 1992.
96. Davies, B.E., Trace metals in the environment: retrospect and prospect, in *Biogeochemistry of Trace Metals*, Adriano, D.C., Ed., Lewis Publishers, Chelsea, MI, 1992, 1.
97. USEPA, *Identification and Listing of Hazardous Waste*, 40 CFR, Part 261, 7-1-92 Edition, USEPA, Washington, D.C., 1992, 27.
98. Hern, J.L., Stout, W.L., Sidle, R.C., and Bennett, O.L., Characterization of fluidized bed combustion waste composition and variability as they relate to disposal on agricultural lands, in *Proc. 5th Int. Conf. Fluidized-Bed Combustion*, Washington, D.C., 1977, 833.
99. Robinson, J.M., Kindja, R.J., Young, C.W., Hall, R.R., and Fennelly, P., *Environmental Aspects of Fluidized-Bed Combustion*, USEPA, Washington, D.C., 1981.
100. Stone, R. and Kahle, R.L., *Environmental Assessment of Solid Residues from Fluidized-Bed Fuel Processing* (final report), USEPA, Washington, D.C., 1987.
101. Baker, D.E. and Chesnin, L., Chemical monitoring of soils for environmental quality and animal and human health, *Adv. Agron.*, 27, 1975, 305.
102. Haby, V.A., Russelle, M.P., and Skogley, E.O., Testing soils for potassium, calcium, and magnesium, in *Soil Testing and Plant Analysis*, Westerman, R.L., Ed., Soil Science Society of America, Madison, WI, 1990, 191.
103. Mengel, K. and Kirkby, E.A., Magnesium, in *Principles of Plant Nutrition*, International Potash Institute, Worblaufen-Bern, Switzerland, 1978, 411.

104. Angelone, M. and Bini, C., Trace elements concentrations in soils and plants of Western Europe, in *Biogeochemistry of Trace Metals*, Adriano, D. C., Ed., Lewis Publishers, Chelsea, MI, 1992, 19.
105. Tisdale, S.L., Nelson, W.L., and Beaton, J.D., *Soil Fertility and Fertilizers*, 4th ed., Macmillan, New York, 1985.
106. Jordan, H.V., *Sulfur as a Plant Nutrient in the Southern United States*, USDA Tech. Bull. No. 1297, U.S. Government Printing Office, Washington, D.C., 1964.
107. Morris, R.J., The importance of sulphur in agriculture — an overview, in *Proc. Int. Symp. on Sulphur In Agricultural Soils*, Portch, S. and Hussain, Sk. G., Eds., The Bangladesh Agricultural Research Council and The Sulphur Institute, Dhaka, Bangladesh, 1986, 1.
108. Tabatabai, M.A., Ed., *Sulfur in Agriculture*, American Society of Agronomy/Crop Science Society of America/Soil Science Society of America, Madison, WI, 1986.
109. Blevins, Z., How important is sulphur in increasing farm production?, in *Proc. 41st Annu. Meet. Fertilizer Industry Round Table*, The Fertilizer Industry Round Table, Glen Arm, MD, 1991, 74.
110. Raun, W.R. and Barreto, H.J., Maize grain yield response to sulphur fertilization in *Central America, Sulphur Agric.*, 16, 26, 1992.
111. Tandon, H.L.S., Sulphur in Indian agriculture, update 1992, *Sulphur Agric.*, 16, 20, 1992.
112. Kamprath, E.J., Potential detrimental effects from liming highly weathered soils to neutrality, *Soil Crop Sci. Soc. Fl. Proc.*, 31, 200, 1971.
113. Cox, F.R., Adams, F., and Tucker, B.B., Liming, fertilization, and mineral nutrition, in *Peanut Science and Technology*, Pattee, H.E. and Young, C.T., Eds., American Peanut Research and Education Society, Yoakum, TX, 1982, chap. 6.
114. Gascho, G.J. and Alva, A.K., Beneficial effects of gypsum for peanut, in *Proc. 3rd Int. Symp. Phosphogypsum*, Chang, W.F., Ed., Florida Institute of Phosphate Research, Bartow, FL, 1990, 376.
115. Gosnell, J.M. and Long, A.C., A sulphur deficiency in sugarcane, *South Afr. Sugar Tech. Assoc.*, 43, 26, 1969.
116. Golden, L.E., *Results from By-product Gypsum Outfield Tests with Sugarcane, 1975–1982*, Report of Projects, Dept. of Agron., La. Agr. Exp. Stn., Baton Rouge, LA, 1982.
117. Buselli, E.M., Gypsum Effects on an Alligator Clay Soil and Sugarcane in Louisiana, M.S. thesis, Louisiana State University, Baton Rouge, 1988.
118. Breithaupt, J.A., Effect of By-Product Gypsum on Yield and Nutrient Content of Sugarcane and Soil Properties, M.S. thesis, Louisiana State University, Baton Rouge, 1989.
119. Jackson, M.L., Aluminum bonding in soils: a unifying principle in soil science, *Soil Sci. Soc. Am. Proc.*, 27, 1, 1963.
120. Parkpian, P., Pongsakul, P., and Sangtong, P., Characteristics of acid soils in Thailand, a review, in *Plant-Soil Interactions at Low pH*, Wright, R.J., Baligar, V.C., and Murrmann, R.P., Eds., Kluwer Academic, Dordrecht, The Netherlands, 1991, 397.
121. Carlisle, V.W. and Fiskell, J.G.A., Relationship of acidity to other soil properties in certain Weston, Leon and Blanton profiles on three marine terraces, *Soil Crop Sci. Soc. Fl. Proc.*, 22, 92, 1962.
122. Pettry, D.E., Carlisle, V.W., and Caldwell, R.E., Spodic horizons in selected Leon and Immokalee soils, *Soil Crop Sci. Soc. Fl. Proc.*, 25, 160, 1965.

123. Kamprath, E.J., Exchangeable aluminum as a criterion for liming leached mineral soils, *Soil Sci. Soc. Am. Proc.*, 34, 252, 1970.
124. Sanchez, P.A., Advances in the management of Oxisols and Ultisols in tropical South America, in *Proc. Int. Seminar on Soil Environment and Fertility Management of Intensive Agriculture*, Society of the Science of Soil and Manure, Tokyo, 1977, 535.
125. Orvedal, A.C. and Ackerson, K.T., *Agricultural Soil Resources of the World* (mimeo), USDA, Soil Conservation Service, Washington, D.C., 1972.
126. Magistad, O.C., The aluminum content of the soil solution and its relation to soil reaction and plant growth, *Soil Sci.*, 20, 181, 1925.
127. Nye, P., Craig, D., Coleman, N.T., and Ragland, J.L., Ion exchange equilibrium involving aluminum, *Soil Sci. Soc. Am. Proc.*, 25, 14, 1961.
128. Rios, M.A. and Pearson, R.W., R. W., The effect of some chemical environmental factors on cotton root behavior, *Soil Sci. Soc. Am. Proc.*, 28, 232, 1964.
129. Adams, F. and Lund, Z.F., Effect of chemical activity of soil solution aluminum on cotton root penetration of acid subsoils, *Soil Sci.*, 101, 193, 1966.
130. Pearson, R.W., Soil environment and root development, in *Plant Environment and Efficient Water Use*, Pierre, W. H., Kirkham, D., Pesek, J., and Shaw, R., Eds., American Society of Agronomy, Madison, WI, 1966, 95.
131. Adams, F., Pearson, R.W., and Doss, B.D., Relative effects of acid subsoils on cotton yields in field experiments and on cotton roots in growth-chamber experiments, *Agron. J.*, 59, 453, 1967.
132. Soileau, J.M. and Engelstad, O.P., Cotton growth in an acid fragipan subsoil. I. Effects of physical soil properties, liming, and fertilization on root penetration, *Soil Sci. Soc. Am. Proc.*, 33, 915, 1969.
133. Adams, F. and Moore, B.L., Chemical factors affecting root growth in subsoil horizons of Coastal Plain soils, *Soil Sci. Soc. Am. Proc.*, 47, 99, 1983.
134. Hammel, J.E., Sumner, M.E., and Shahandeh, H., Effect of physical and chemical profile modification on soybean and corn production, *Soil Sci. Soc. Am. J.*, 49, 1508, 1985.
135. Rechcigl, J.E., Edmisten, K.L., Wolf, D.D., and Reneau, R.B., Jr., Response of alfalfa grown on acid soil to different chemical amendments, *Agron. J.*, 80, 515, 1988.
136. Odom, J.W., Alfalfa response to gypsum, boron and subsoiling on an acid Ultisol, in *Agronomy Abstracts*, American Society of Agronomy, Madison, WI, 1991, 296.
137. Farina, M.P.W. and Channon, P., Acid subsoil amelioration. II. Gypsum effects on growth and subsoil chemical properties, *Soil Sci. Soc. Am. J.*, 52, 175, 1988.
138. Caldwell, A.G., Hutchinson, R.L., Kennedy, C.W., and Jones, J.E., Effect of rates of lime and by-product gypsum on movement of calcium and sulfur into an acid soil, in *Agronomy Abstracts*, American Society of Agronomy, Madison, WI, 1990, 264.
139. Kilmer, V.J., Ed., *Handbook of Soil and Climate in Agriculture*, CRC Press, Boca Raton, FL, 1982.
140. Hilgard, E.W., *Soils — Their Formation, Properties, Composition, and Relation to Climate and Plant Growth* in the Humid and Arid Regions, Macmillan, New York, 1907.
141. Kelly, W.P. and Brown, S.M., Principles governing the reclamation of alkali soils, *Hilgardia*, 8, 149, 1934.
142. Richards, L.A., *Diagnosis and Improvement of Saline and Alkali Soils*, USDA Agricultural Handbook No. 60, USDA, Washington, D.C., 1954.
143. Melamed, D., Hanks, R.J., and Willardson, L.S., Model of salt flow in soil with a source-sink term, *Soil Sci. Soc. Am. Proc.*, 41, 29, 1977.

144. Glas, T.K., Klute, A., and McWhorter, D.B., Dissolution and transport of gypsum in soils. I. Theory, *Soil Sci. Soc. Am. Proc.*, 43, 265, 1979.
145. Dutt, G.R., Terkeltoub, R.W., and Rauschkolb, R.S., Prediction of gypsum and leaching requirements for sodium-affected soils, *Soil Sci.*, 114, 93, 1972.
146. Tanji, K.K., Donnen, L.D., Ferry, G.V., and Ayers, R.S., Computer simulation analysis on reclamation of salt-affected soil in San Joaquin Valley, California, *Soil Sci. Soc. Am. Proc.*, 36, 127, 1972.
147. Lindekin, C.L., Radiological considerations of phosphogypsum utilization in agriculture, in *Phosphogypsum*, Borris, D.P., and Boody, P.W., Eds., Florida Institute Phosphate Research, Bartow, FL, 1980, 401.
148. Rengasamy, P., Greene, R.S.B., Ford, G.W., and Mehanni, A.H., Identification of dispersive behavior and the management of red brown earths, *Aust. J. Soil Sci.*, 22, 413, 1984.
149. Sims, H.J. and Rooney, D.R., Gypsum for difficult clay wheat growing soils, *J. Dept. Agric., Victoria*, 63, 401, 1965.
150. Loveday, J., Recognition of gypsum-responsive soils, *Aust. J. Soil Res.*, 12, 87, 1974.
151. Howell, M., Gypsum use in the wheat belt, *J. Agric. West Aust.*, 28, 40, 1987.
152. Hendrickson, B.H., Barnett, A.P., Carreker, J.R., and Adams, W.E., *Runoff and Erosion Control Studies on Cecil Soil in the Southern Piedmont*, USDA-Agricultural Research Service Tech. Bull. No. 1281, U.S. Government Printing Office, Washington, D.C., 1963.
153. Reicosky, D.C., Cassel, D.K., Blevins, R.L., Gill, W.R., and Naderman, G.C., Conservation tillage in the Southeast, *J. Soil Water Conserv.*, 32, 13, 1977.
154. Chiang, S.C., Radcliffe, D.E., Miller, W.P., and Newman, K.D., Hydraulic conductivity of three Southeastern soils as affected by sodium, electrolyte concentration, and pH, *Soil Sci. Soc. Am. J.*, 51, 1293, 1987.
155. Perkins, H.F., Owen, V., Hammel, J.E., and Price, E.A., Soil Characteristics of the Plant Science Farm of the University of Georgia College Experiment Station, University of Georgia Exp. Stn. Bull. No. 287, 1982.
156. Miller, W.P., Infiltration and soil loss of three gypsum-amended Ultisols under simulated rainfall, *Soil Sci. Soc. Am. J.*, 51, 1314, 1987.
157. Miller, W.P., *Use of Gypsum to Improve Physical Properties and Water Relations in Southeastern Soils*, Florida Institute of Phosphate Research, Bartow, FL, 1989.
158. Agassi, M., Morin, J., and Shainberg, I., Infiltration and runoff control in the semi-arid region of Israel, *Geoderma*, 28, 345, 1982.
159. Agassi, M., Shainberg, I., and Morin, J., Infiltration and runoff in wheat fields in the semi-arid region of Israel, *Geoderma*, 36, 263, 1985.
160. Kazman, Z., Shainberg, I., and Gal, M., Effect of low levels of exchangeable sodium and applied phosphogypsum on the infiltration rate of various soils, *Soil Sci.*, 135, 184, 1983.
161. Bowen, H.D., Alleviating mechanical impedance, in *Modifying the Root Environment to Reduce Crop Stress*, Arkin, G.F., and Taylor, H.M., Eds., ASAE Monograph No. 4, American Society of Agricultural Engineers, St. Joseph, MI, 1981, 21.
162. Radcliffe, D.E., Clark, J.L., and Sumner, M.E., Effect of gypsum and deep-rooting perennials on subsoil mechanical impedance, *Soil Sci. Soc. Am. J.*, 50, 1566, 1986.
163. Sumner, M.E., Radcliffe, D.E., McCray, M., and Clark, R.L., Gypsum as an ameliorant for subsoil hardpans, *Soil Technol.*, 3, 253, 1990.
164. Kawaguchi, K. and Matsuo, Y., Movements of active oxide in dry paddy soil profiles, 6th Int. Cong. Soil Science Transactions, Paris, 1956, R-9, 533.

165. Takai, Y. and Kamura, T., The mechanism of reduction in waterlogged paddy soil, *Folia Microbiol.*, 11, 304, 1966.
166. Alcordo, I.S., The Flow Properties of Al-montmorillonite, Fe-montmorillonite, and Soil Clay Suspension and Their Changes with Anaerobic Reduction, Doctoral thesis, Department of Agronomy, University of Illinois, Champaign-Urbana, IL, 1968.
167. Harmsen, K. and van Breemen, N., A model for the simultaneous production and diffusion of ferrous iron in submerged soils, *Soil Sci. Soc. Am. Proc.*, 39, 1063, 1975.
168. Hunter, A.H., *Use of Phosphogypsum Fortified with Other Selected Elements as a Soil Amendment on Low Cation Exchange Soils*, Florida Institute of Phosphate Research, Bartow, FL, 1989.
169. Couto, W., Soil pH and Plant Productivity, in *CRC Handbook of Agricultural Productivity*, Vol. 1, Rechcigl, M., Ed., CRC Press, Boca Raton, FL, 1982, 71.
170. Jacobson, L., Overstreet, R., King, H.M., and Handley, R., A study of potassium absorption by barley roots, *Plant Physiol.*, 25, 639, 1950.
171. Nielson, T.R. and Overstreet, R., A study of the role of the hydrogen ion in the mechanism of potassium absorption by excised barley roots, *Plant Physiol.*, 30, 303, 1955.
172. Fried, M. and Peach, M., The comparative effects of lime and gypsum upon plants grown on acid soils, *J. Am. Soc. Agron.*, 38, 614, 1946.
173. Ligon, W.S. and Pierre, W.H., Soluble aluminum studies. II. Minimum concentrations of aluminum found to be toxic to corn, sorghum, and barley in culture solutions, *Soil Sci.*, 34, 307, 1932.
174. Vlamis, J., Acid soil infertility as related to soil-solution and solid-phase effects, *Soil Sci.*, 75, 383, 1953.
175. Matsumoto, H., Hirasawa, E., Torikai, H., and Takahashi, E., Localization of absorbed aluminum in pea root and its binding to nucleic acids, *Plant Cell Physiol.*, 17, 127, 1976.
176. Matsumoto, H. and Morimura, S., Repressed template activity of chromatin of pea roots treated with aluminum, *Plant Cell Physiol.*, 21, 951, 1980.
177. Munns, D.N. and Franco, A.A., Soil constraints to legume production, in *Biological Nitrogen Fixation for Tropical Agriculture*, Graham, P.H. and Harris, S.C., Eds., Centro Internacional de Agricultura Tropical, Cali, Columbia, 1982, 135.
178. de Carvalho, M.M., Edwards, D.G., Asher, C.J., and Andrew, C.S., Effects of aluminum on nodulation of two Stylosanthes species grown in nutrient solution, *Plant Soil*, 64, 141, 1982.
179. Horst, W.J., Wagner, A., and Marschner, H., Effect of aluminum on root growth, cell-division rate and mineral element contents in roots of Vigna unguiculata genotypes, *Z. Pflanzenphysiol.*, 109, 95, 1983.
180. Robson, A.D., Ed., *Soil Acidity and Plant Growth*, Academic Press, Sydney, 1989.
181. Stout, W.L., Sidle, R.C., Hern, J.L., and Bennett, O.L., Effects of fluidized bed combustion waste on the Ca, Mg, S, and Zn levels in red clover, tall fescue, oat, and buckwheat, *Agron. J.*, 71, 662, 1979.
182. Bennett, O.L., Agricultural applications: environmental aspects of AFBC, in *Source Book on Fluidized Bed Combustion*, Massachusetts Institute of Technology, Technology Division, Bedford, MA, 1983.
183. Bennett, O.L., Hern, J.L., and Perry, H.D., Agricultural uses of atmospheric fluidized bed combustion residue (AFBCR) — a seven year study, Proc. 2nd Annu. Pittsburgh Coal Conference, 1985, 558.
184. Barnishel, R.I. and Thom, W.O., Utilization of Fluidized Bed Combustion Waste, Final report, Tennessee Valley Authority, Chattanooga, TN, 1987.

185. Korcak, R.F., Fluidized bed material as a calcium source for apples, *HortScience*, 14, 163, 1979.
186. Korcak, R.F., Fluidized bed material as a lime substitute and calcium source for apple seedlings, *J. Environ. Qual.*, 9, 147, 1980.
187. Korcak, R.F., Effectiveness of fluidized bed material as a calcium source for apples, *J. Am. Soc. Hort. Sci.*, 107, 1138, 1982.
188. Wrubel, J.J., Jr., Korcak, R.F., and Childers, N., Orchard studies utilizing fluidized bed material, *Commun. Soil Sci. Plant Anal.*, 13, 1071, 1982.
189. Korcak, R.F., Wrubel, J.J., and Childers, N., Peach orchard studies utilizing fluidized bed material, *J. Plant Nutr.*, 7, 1597, 1984.
190. Edwards, J.H., White, A.W., and Bennett, O.L., Fluidized bed combustion residue as an alternative liming material and Ca source, *Commun. Soil Sci. Plant Anal.*, 16, 621, 1985.
191. Smedley, K.O., Fontenot, J.P, and Allen, V.G., Effects of fluidized bed combustion residue application to reclaimed land on yield and composition of forage and performance of grazing steers, Proc. Int. Grassland Symposium, Japan, 1985, 1055.
192. USEPA, *An Overview of Subtitle C Regulations*, 40 CFR, Part 260, 7-1-92 ed., USEPA, Washington, D.C., 1992, chap. 1.
193. USEPA, *National Primary Drinking Water Regulations*, 40 CFR, Part 141, 7-1-92 ed., USEPA, Washington, D.C., 1992, 589.
194. Hoeksema, H.W. and van der Brugghen, F.W., Minimizing trace element concentrations in FGD gypsum, in *Proc. 2nd Int. Conf. FGD and Chemical Gypsum*, ORTECH International, Ontario, 1991, 8.1.
195. USEPA, Proposed rules, part 250 — hazardous waste guidelines and regulations, *Fed. Reg.*, 43 (243), 58954, 1978.
196. Roessler, C.E., Control of radium in phosphate mining, beneficiation and chemical processing, in *The Environmental Behaviour of Radium*, Vol. 2, International Atomic Energy Agency, Vienna, 1990, 269.
197. Osmond, J.K., Cowart, J.B., Humphreys, C.L., and Wagner, B.E., Radioelement Migration in Natural and Mined Phosphate Terrains (final report), Florida Institute of Phosphate Research, Bartow, FL, 1985.
198. Choppin, G.R., *Nuclei and Radioactivity*, W.A. Benjamin, New York, 1964.
199. Cowart, J.B. and Burnett, W.C., The distribution of uranium and thorium decay-series radionuclides in the environment, in *Symp. Proc. Naturally Occurring Radionuclides in Agricultural Products*, Hanlon, E.A., Ed., University of Florida, Gainesville, 1991, 3.
200. Langmuir, D., Uranium solution-mineral equilibrium at low temperatures with applications to sedimentary ore deposits, *Geochim. Cosmochim. Acta*, 42, 547, 1978.
201. Rogers, J.J.W. and Adams, J.A.S., Uranium, in *Handbook of Geochemistry*, Wedepohl, K.H., Ed., Springer-Verlag, New York, 1969, 92-A-1.
202. Szalay, A., Accumulation of uranium and other micrometals in coal and organic shales and the role of humic acids in these geochemical enrichments, *Kungl. Svenska Vetenskapakademien*, 5, 23, 1969.
203. Altschuler, Z.S., *Geochemistry of Uranium in Apatite and Phosphorite*, U.S. Geological Survey, Washington, D.C., 1958.
204. Menzel, R.G., Uranium, radium, and thorium content in phosphate rocks and their possible radiation hazard, *J. Agric. Food Chem.*, 16, 231, 1968.
205. APHA/AWWA/WPCP, *Standard Methods for Examination of Water and Wastewater*, 16th ed., American Public Health Association, American Waterworks Association, Water Pollution Control Federation, Washington, D.C., 1985, 652.

206. USEPA, National primary drinking water regulations; radionuclides; proposed rule, *Fed. Reg.*, Part II, 33050, 1991.
207. Irwin, C.A. and Hutchinson, C.B., *Reconnaissance Water Sampling for Ra-226 in Central and Northern Florida*, U.S. Geological Survey, Washington, D.C., 1976.
208. Lyman, G.H., Lyman, C.G., and Johnson, W., Association of leukemia with radium groundwater contamination, *JAMA*, 454, 621, 1985.
209. Gosink, R.A., Baskaran, M., and Holleman, D.F., Radon in the human body from drinking water, *Health Physics*, 59, 919, 1990.
210. Mose, D.G., Mushrush, G.W., and Chrosniak, C.E., A comparison between radon in water and the incidence of cancer: a case study from northern Virginia and southern Maryland, USA, in *Symp. Proc. Naturally Occurring Radionuclides in Agricultural Products*, Hanlon, E.A., Ed., University of Florida, Gainesville, 1991, 213.
211. USEPA, *Radon Reduction Methods, a Homeowner's Guide*, 2nd ed., USEPA, Washington, D.C., 1987.
212. USEPA, *Removal of Radon from Household Water*, USEPA, Washington, D.C., 1987.
213. USEPA, *Radon Reference Manual*, USEPA, Washington, D.C., 1987.
214. USEPA, Potential Uses of Phosphogypsum and Associated Risks, Background Information Document, USEPA, Washington, D.C., 1992.
215. Roessler, C.E., Bolch, W.E., Birky, B., and Roessler, G.S., Dose estimation and risk assessment for naturally-occurring radionuclides in agricultural products, in *Symp. Proc. on Naturally Occurring Radionuclides in Agricultural Products*, Hanlon, E.A., Ed., University of Florida, Gainesville, 1991, 194.
216. USEPA, *PATHRAE-EPA: A Performance Assessment Code for the Land Disposal of Radioactive Wastes, Documentation and Users Manual*, USEPA, Washington, D.C., 1987.
217. USEPA, National emission standards for hazardous air pollutants; national emission standards for radon emissions from phosphogypsum stacks, *Fed. Reg.*, 57(107), 23305, 1992.
218. The Fertilizer Institute, *Review of the Phosphogypsum Final Rule and Background Information Document*, The Fertilizer Institute, Washington, D.C., 1992.
219. Sidle, R.C., Stout, W.L., Hern, J.L., and Bennett, O.L., Solute movement from fluidized bed combustion waste in acid soils and mine spoil columns, *J. Environ. Qual.*, 8, 236, 1979.
220. Rechcigl, J.E., Alcordo, I.S., Roessler, C.E., and Littell, R.C., Influence of Phosphogypsum on Forage Yield and Quality, and on the Environment in a Typical Florida Spodosol Soil (final report), Florida Institute of Phosphate Research, Bartow, FL, 1993.
221. Adriano, D.C., Ed., *Biogeochemistry of Trace Metals*, Lewis Publishers, Chelsea, MI, 1992.
222. Alloway, B.J., Ed., *Heavy Metals in Soils*, John Wiley & Sons, New York, 1990.
223. Rechcigl, J.E., Alcordo, I.S., Roessler, C.E., Littell, R.C., and Alva, A.K., Impact of Phosphogypsum on Radon Emissions and on Radioactivity and Heavy Metals in Soil, Groundwater, and Bahiagrass Forage, Progress Report, December 1994, Florida Institute of Phosphate Research, Bartow, FL, 1993.
224. Mays, D.A. and Mortvedt, J.J., Crop response to soil applications of phosphogypsum, *J. Environ. Qual.*, 15, 78, 1986.
225. Campbell, J.R. and Lasley, J.F., *The Science of Animals that Serve Mankind*, McGraw-Hill, New York, 1969, 439.
226. Hobbs, C.S., Moorman, R.P., Jr., Griffith, J.M., West, J.L., Merriman, G.M., Hansard, L.S., and Chamberlain, C.C., Fluorosis in Cattle and Sheep, Tennessee University Agric. Exp. Stn. Bull. No. 235, 1954.

227. Allcroft, R., Burns, K.N., and Hebert, C.N., Fluorosis in cattle. II. Development and alleviation: experimental studies, in *Animal Diseases Surveys Report No. 2*, Ministry of Agriculture, Fisheries and Food, Her Majesty's Stationery Office, London, 1965.
228. Suttie, J.W., Effects of Fluoride on livestock, *J. Occup. Med.*, 19, 40, 1977.
229. Church, D.C., *Digestive Physiology and Nutrition of Ruminants*, Vol. 2, O & B Books, Corvallis, OR, 1979, 356.
230. Keerthisinghe, G., McLaughlin, M.J., and Freney, J.R., Use of gypsum, phosphogypsum and fluoride to ameliorate subsurface acidity in a pasture soil, in *Plant-Soil Interactions in Low pH*, Wright, R.J., Baligar, V.C., and Murrmann, R.P., Eds., Kluwer Academic, Dordrecht, The Netherlands, 1991, 509.
231. IFAS, Position Statement and Supporting Reports Regarding the Proposed Farmland Industries Phosphate Fertilizer Complex Near the Ona Agricultural Research Center, Institute of Food and Agricultural Sciences, University of Florida, Ona, 1980.
232. Myrick, T.E., Berven, B.A., and Haywood, F.F., *State Background Radiation Levels: Results of Measurements Taken During 1975–1979*, Oak Ridge National Laboratory, Oak Ridge, TN, 1981.
233. USEPA, *Health and Environmental Protection Standards for Uranium and Thorium Mill Tailings*, 40 CFR, 7-1-92 ed., USEPA, Washington, D.C., 1992, chap. 1.
234. NCRPM, Exposures from the Uranium Series with Emphasis on Radon and its Daughters, NCRP Report No. 77, National Council on Radiation Protection and Measurements, Bethesda, MD, 1984.
235. NCRPM, Exposure of the Population in the United States and Canada from Natural Background Radiation, NCRP Report No. 94, National Council on Radiation Protection and Measurements, Bethesda, MD, December 1987.
236. Mortvedt, J.J., Plant and soil relationships of uranium and thorium decay series radionuclides, in *Symp. Proc. Naturally Occurring Radionuclides in Agricultural Products*, Hanlon, E.A., Ed., University of Florida, Gainesville, 1991, 56.
237. Mortvedt, J.J., The radioactivity issue — effects on crops grown on mined phosphate lands, P-fertilized soils, and phosphogypsum-treated soils, in *Phosphate Fertilizers and the Environment*, Schultz, J.J., Ed., International Fertilizer Development Center, Muscle Shoals, AL, 1992, 271.
238. Miller, W.P. and Sumner, M.E., Impacts from Radionuclides on Soil Treated with Phosphogypsum (final report), Agronomy Department, University of Georgia, Athens, 1992.
239. Mullins, G.L. and Mitchell, C.C., Jr., *Use of Phosphogypsum to Increase Yield and Quality of Annual Forages*, Florida Institute of Phosphate Research, Bartow, FL, 1990.
240. Simon, S.L. and Ibrahim, S.A., Biological uptake of radium by terrestrial plants, in *The Environmental Behaviour of Radium*, Vol. 1, Tech. Rep. Series No. 310, International Atomic Energy Agency, Vienna, 1990, 545.
241. Till, J.E. and Meyer, H.R., *Radiological Assessment*, U.S. Nuclear Regulatory Commission, Washington, D.C., 1983.
242. Lindekin, C.L and Coles, D.G., The radium-226 content of agricultural gypsum, Proc. Symp. Public Health Aspects of Radioactivity in Consumer Products, U.S. Regulatory Commission Report NUREG/CP-0001, Washington, D.C., 1977, 369.
243. Mislevy, P., Blue, W.G., and Roessler, C.E., Productivity of clay tailings from phosphate mining. I. Biomass crops, *J. Environ. Qual.*, 18, 95, 1989.

244. Simon, S.L. and Ibrahim, S.A., The plant/soil concentration ratio for calcium, radium, lead, polonium: evidence from non-linearity with reference to substrate concentration, *J. Environ. Radioactivity*, 5, 123, 1987.
245. Guidry, J.J., Roessler, C.E., and McClave, J.T., Radioactivity in Foods Grown on Mined Phosphate Lands (final report), Florida Institute Phosphate Research, Bartow, FL, 1990.
246. Myhre, D.L., Martin, H.W., and Nemec, S., Yield, 226Ra concentration, and juice quality of oranges in groves treated with phosphogypsum and mined gypsum, in *Proc. 3rd Int. Symp. Phosphogypsum* (*suppl.*), Florida Institute Phosphate Research, Bartow, FL, 1990, 11.
247. FIPR, *Radiation and Your Environment: A Guide to Low-level Radiations for Citizens of Florida*, Florida Institute of Phosphate Research, Bartow, FL.
248. Rolle, R., Rapid working level monitoring, *Health Physics*, 22, 233, 1972.
249. USEPA, National emission standards for hazardous air pollutants; radionuclides; final rule and notice of reconsideration, *Fed. Reg.*, Part II, EPA 40 CFR Part 61, 54(240), 51674, 1989.
250. Hartley, J.N. and Freeman, H.D., Radon Flux Measurements on Gardinier and Royster Phosphogypsum Piles Near Tampa and Mulberry, Florida, Eastern Environmental Radiation Facility, Montgomery, AL, 1985.
251. Rechcigl, J.E., Alcordo, I.S., Roessler, C.E,. and Stieff, L., Methods of measuring ambient atmospheric radon and radon surface flux associated with phosphogypsum treatment of a Florida Spodosol soil, *Commun. Soil Sci. Plant Anal.*, 23, 2581, 1992.
252. Fjeld, R.A., Montague, K.J., Haapala, M.H., and Kotrappa, P., Field test of electret ion chambers for environmental monitoring, *Health Physics*, 66(2), 147, 1994.
253. Price, J.G., Rigby, J.G., Christensen, L., Hess, R., LaPointe, D.D., Ramelli, A.R., Desilets, M., Hopper, R.D., Kluesner, T., and Marshall, S., Radon in outdoor air in Nevada, *Health Physics*, 66(4), 433, 1994.
254. Roessler, C.E., Radiological aspects of phosphogypsum, in Proc. Symp. Natural Radiations and Technologically Enhanced Natural Radiation in Florida, Health Physics Society (Florida chapter), Winter Haven, FL, 1987, 320.
255. Wilkening, M.H., Clements, W.E., and Stanley, D., Radon-222 flux measurements in widely separated regions, in *Natural Radiation in the Environment II*, Adams, J.A.S., Lowder, W.M., and Gesell, T.F., Eds., U.S. Energy Research and Development Administration Report, 1972, National Technical Information Service, Springfield, VA, 1972.
256. NCRPM, Control of Radon in Houses, NCRP Report No. 103, National Council on Radiation Protection and Measurements, Bethesda, MD, 1989.
257. Hopper, R.D., Levy, R.A., Rankin, R.C., and Boyd, M.A., National ambient radon study, in Proc. Int. Symp. Radon and Radon Protection, Philadelphia, April 1991.

CHAPTER 11

Agronomic Practices in Relation to Soil Amendments and Pesticides

Thanh H. Dao and Paul W. Unger

0-87371-859-3/95/$0.00+$.50

I. INTRODUCTION

A. ENHANCED PUBLIC AWARENESS OF QUALITY OF THE ENVIRONMENT

Technological advances in agriculture have contributed extensively to annual bountiful harvests and a great diversity of agricultural products of high quality in the U.S. Since the Great Depression of the 1930s, rapid progress has occurred in crop and animal germplasm improvement, farm machinery engineering, soil conservation, fertility amendments and enhancement techniques, and development of effective pest control chemicals.

Soil amendments have been used to increase land productivity and efficiency of providing food, fiber, and shelter for an increasing population working outside of agriculture since antiquity. In modern agricultural production systems, attainment of the most economic yields and a reasonable return for land and labor investments are achieved through the addition of many physical off-farm inputs such as improved seeds, commercial fertilizers, pesticides, and irrigation. It is not surprising to find an ever-increasing specialization of U.S. farming systems, a high production efficiency, and a steadfast abundance of products that are achieved by a dwindling farm population.

In the process, U.S. agriculture has drastically changed the structure and functions of the natural landscape. The ecological balance has been modified on about a quarter of a billion of cultivated hectares, not considering the environment that surrounds them. Public concern about the environment has become widespread and its roots may be traced back to the days of *Silent Spring*.[1] Today, both agriculturists and the nonagricultural public share concern for the environment. The challenge is to seek management alternatives that mitigate environmental degradation while maintaining agricultural productivity and profitability. Another challenge to both groups is to arrive at a mutual understanding of objectives and suitable solutions that preclude overreaction and restrictive regulation that may put agricultural industries in noncompetitive positions in an ever-increasing competitive global economy.

B. CONTAMINATION OF SURFACE AND GROUNDWATER RESOURCES

1. Surface Water Contamination

In the 1960s and 1970s, numerous instances of silting of lakes and reservoirs, dams, navigable waters, and estuaries made news headlines and drew public attention to the ever-increasing problem of soil erosion. Erosion is the source of 99% of the total suspended solids in U.S. waterways. An estimated 5 billion tons of soil reaches the nation's waterways and surface water impoundments each year.[2] Sediments reaching waterways clog streams and drainage channels, reduce the storage capacity of reservoirs and lakes, and cause deterioration of aquatic and terrestrial wildlife habitats. Water transportation systems are affected as navigable channels and harbors have to be dredged at annual costs of over $300 million.

Detection of agricultural and industrial contaminants in streams, lakes, and many other surface bodies of water also provide cause for public concern and outcry. In addition to turbidity caused by sediment, eutrophication of lakes and many public recreational surface water bodies have been traced to agricultural practices as well as to urban use of household chemicals. For example, the Great Lakes were heavily polluted from municipal and industrial discharges during the 1960s and 1970s until a massive cleanup effort was mounted by both the U.S. and Canada to restore some of their past natural splendor. The marine life and shellfish industries of the Chesapeake Bay have been threatened during the past decades by discharges from states that surround the bay[3] and from thermal pollution of nuclear power plants dotting the coastal zone.

Numerous management practices have been suggested to cope with nonpoint source pollution of surface waters from agricultural activities.[4,5] Sediment discharges have been successfully brought under control with widespread adoption of landscape-slope modification techniques, and the use of water-diversion structures such as terraces, grass waterways, and contour farming practices. Nutrient management is greatly improved by extensive reliance on soil testing, and a broader understanding of crop nutrient requirements. Integrated pest management is widely practiced with improved understanding of the biology of pests, their life cycle and that of vectors, and the availability of effective pesticides. However, recent national monitoring efforts reveal a number of hot spots around the nation. A recent finding of elevated concentrations of atrazine [6-chloro-*N*-ethyl-*N*′-(1-methylethyl)-1,3,5-triazine-2,4-diamine], alachlor [2-chloro-*N*-(2,6-diethylphenyl)-*N*-(methoxymethyl)acetamide], and a number of other pesticides in the Mississippi River basin serves as a reminder that the problem is still prevalent for that river and its tributaries.[6,7] The situation has serious implications for agriculture. The incident also points to our inability to cope with seasonal variations of contaminant levels, and to understand the fate of transient spikes during contaminant transport in rivers and streams, their impact on public water suppliers, and the ability of the suppliers to comply with the Safe Drinking Water Act regulations.

2. Groundwater Contamination

The general distress over groundwater contamination is growing because of increasing evidence of chemical intrusion in water bodies underlying areas of agricultural activities. It has been thought that the soil has an immense capacity to filter out waste by-products to clarify contaminated water. Only in the early 1980s did we begin to realize the scope of the groundwater contamination problem.[8–10]

Groundwater contamination is a more complex problem than pollution of surface water because it is more difficult to detect and to rectify. Agriculture accounts for about 70% of groundwater usage. Although most of the volume is for crop irrigation, groundwater is the principal source of drinking water for our rural population. The presence of synthetic chemicals in drinking water is being detected in national surveys of public and private wells and reservoirs. Prominent sources of the chemicals include hazardous waste landfills, surface impoundments, lagoons, underground storage tanks, septic tanks, well injection sites, chemical spills, concentrated livestock operations, and agricultural fields due to extensive use of fertilizers and pesticides.

The U.S. Environmental Protection Agency is completing a survey of 101 pesticides, 25 pesticide metabolites, and nitrate in 1350 statistically selected drinking water wells and community water systems across the U.S. to assess the extent of the problem as mandated by Congress.[10] Samples from the majority of wells contain concentrations lower than the health advisory levels. Approximately 10.4% of community water system wells and 4.2% of rural domestic water wells in the U.S. contain at least one of the surveyed chemicals. Over half of those public and private drinking water wells contain nitrate above the reporting limit of 0.15 mg L^{-1}, and 1.2% of community water system wells and 2.4% domestic water wells have nitrate above the maximum contaminant level of 10 mg N L^{-1}. Although rigorous data analysis is still underway, the goals of the survey are to develop knowledge of the chemical transport pathways in relation to soil and landscape factors and to obtain an understanding of how the magnitude of the contamination is associated with patterns of use and management practices in place at the soil surface. Data from groundwater monitoring studies may become a pesticide registration requirement. As such, they may provide the basis for a new regulatory approach to control environmental contamination by chemicals that have properties indicative of a propensity to leach. This would be in addition to current assessments of human risk, ecological risk, environmental exposure, and environmental fate.

C. WIND EROSION: AIR QUALITY IN SEMI-ARID AND ARID ENVIRONMENTS

Wind erosion in the U.S. causes movement of an estimated 2 billion tons of soil each year.[2] Agricultural crops are most vulnerable to damage by blowing soil particles after emergence and early seedling growth.[11] Upon settling on tree or crop plant leaves, dust impairs gaseous exchange and photosynthetic efficiency of the plant. Windblown soil depletes the productivity of arable lands. Dust impairs air quality and worsens respiratory ailments and allergies of man and animals. It also permeates homes, offices, businesses, and factories causing significant human suffering, time loss, and equipment abrasion and premature failures. Blowing dust reduces visibility, causing travel difficulty and traffic accidents.

Wind erosion is a serious problem in arid and semi-arid regions and in sandy soil areas of the U.S. Cultivated coarse-textured soils of the Atlantic Coast Flatwoods, the southern Coastal Plains, the Great Lakes regions, and the Great Plains are prone to wind erosion.[12] Droughts and subsequent wind erosion damage crops and the land. For example, severe droughts were common for the Great Plains for centuries and the region was labeled as "the Great American Desert" on most maps made before 1860. The Southern Plains have suffered from prolonged droughts about once every 20 years since their settlement in the 1880s. However, intermittent periods of wet weather sustained settlement of the region. In the summer of 1931, another long drought began and persisted for 7 years. Notorious dust storms became widespread 2 years later and were frequent until the fall of 1938. Millions of acres were damaged by wind erosion, with many soils of the region losing between 5 to 30 cm of topsoil during these Dust Bowl years. Since the 1930s, much of the damaged lands have been converted to grassland and cropland with good soil and water conservation practices to minimize wind erosion.

D. CHANGING REGULATORY PHILOSOPHY AND STRATEGIES

There are several federal and numerous state statutes that provide oversight authority to federal, state, and local governments to regulate soil amendment uses, mitigate their off-site contamination, and prevent human exposure. They include the following federal laws that serve as foundations of all local and state regulations affecting agriculture:

The Federal Insecticide, Fungicide, and Rodenticide Act (FIFRA), as amended, provides the authority to require registration of new pesticides, review and reregister existing pesticides, and regulate the marketing and use of pesticides. The goal is to ensure that, when used according to label directions, pesticides do not pose unreasonable risks to human health (the Delaney's clause of the Federal Food, Drug, and Cosmetic Act) or the environment.

The 1987 Water Pollution Control Act (Clean Water Act), in particular Section 314 (clean lakes), Section 319 (nonpoint sources), and Section 320 (national estuaries), provides regulatory authority for controlling point sources of pollution through a permit program based on treatment and controlled effluent release and a nonpoint source program based on land management plans.

The Safe Drinking Water Act of 1987 requires the establishment of maximum contaminant levels (MCL) for environmental contaminants in public drinking water systems that may have adverse health effects. Guidelines are developed for meeting drinking water standards for all public water suppliers. The act also mandates the development of nonregulatory health advisory levels on contaminants for which MCLs have not been established.

The Coastal Zone Management Act of 1972, as amended, establishes coastal zones as areas of special economic and ecological significance. The Act provides for comprehensive consideration of land and water uses of these areas to mitigate physical and biological degradation of the coastal environment.

There is increasing recognition that pollution control programs are failing at the local levels in spite of the aforementioned legislation. The presence of synthetic chemicals associated with agricultural use in the environment is confirmed by the results of numerous air quality monitoring studies,[13] lakes, streams, rivers,[6] and drinking water well surveys.[12]

Pinpointing the sources of contamination is difficult because contaminants can reach the atmosphere and groundwater from diffuse sources other than normal field application practices. At least in quasi-point source cases, controllable mechanisms can be implemented. They include regulatory requirements such as curtailing accidental spills at manufacturing sites by using double containment structures, preventive measures to minimize back-siphoning into wells, direct channeling into unprotected or damaged well heads at mixing/loading sites, improper disposal practices, and overflow from unprotected stockyards of agricultural chemical distribution centers and dealerships.

There are increasing efforts to target "leachers" as priority candidates for regulation and develop an understanding of how the contamination is associated with patterns of use and management practices in place at the soil surface. The evidence would become the basis for developing contamination mitigation strategies. Knowledge of vulnerable regions would be used to target monitoring efforts and restrict use and conditions of use. The USEPA Pesticide and Nitrogen Management strategies have been drafted to chart a changing regulatory direction for the future. Owing to constraints imposed by political and social factors, which are further affected by fiscal pressures and limits of growth of federal spending, two ideological frameworks have emerged as guiding principles of a national strategy of protection of the environment.

1. The concept of differential protection, as the public has come to understand that not all resources have equal societal values. There is a recognition of the need for a flexible approach to the protection of the environment. Priorities must be established for environmental protection based on the greatest opportunities to reduce risks to society, human health, the natural ecosystem, and the economy of the locality, the region, and/or the nation.
2. The pollution prevention concept. This approach to environmental protection is an alternative to the approach of containment of contaminants or waste by-products once they have been released into the environment and of remediation once contamination has occurred. The emphasis is on source reduction where use of toxic materials would be reduced. Recycling is widely encouraged to minimize wastes and promote wise use of natural resources.

Many economic incentives have been suggested for putting the burden of cost of pollution remediation directly on the responsible party. There are deepening concerns that the cross-compliance approach has had marginal access at the local level.[14] That approach primarily depended on the deterrence of direct civil and criminal litigation, voluntary participation, and economic benefits gained from commodity support programs. It is conceivable that the evolving coordination of environmental policies among federal agencies will foster the development of coordinated and successful state water resource management programs. Most states have a legal framework that includes some avenues to address agricultural contamination of surface and groundwater resources. However, few states have developed preventive programs as they have to bear fiscal responsibility and huge costs for administering such programs.[15]

Considerable uncertainty exists about the human health effects of the levels of toxic contaminants detected in surface and groundwater. The uncertainty is a driving factor in increasing public pressure for stiffer regulation of agricultural amendments, regardless of societal benefits that may be foregone if the public perceives that the quality of their supply of drinking water and water for recreational use is being compromised. The results may have a dramatic effect on the availability of types of agricultural practices, crops, and livestock that may be used or produced in particular regions of the country, irrespective of specific climate, soil, and landscape factors.

II. EFFECTS OF CROPPING PRACTICES

A. PRODUCTION STRATEGIES AND SOIL AMENDMENTS

Much research has been done on the fate of fertilizers, micronutrients, pesticides, municipal sludge, agricultural and industrial by-products, and other amendments in major agrosystems during the past 4 decades. Soil amendments are released into a very complex natural environment in which a multitude of physical, chemical, and biological processes interact at all conceivable scales of measurement (Figure 1). A summary of specific physical and chemical properties of soil amendments that control their behavior and fate in the environment are represented in Figure 2, while those of agricultural practices are outlined in Figure 3. This research has shown that we can responsibly intervene with reason and wisdom gained from experience and scientific knowledge to reduce the undesirable effects of soil amendments and pesticides on nontargeted environments. Best management practices (BMPs) have been formulated on a continuing basis to cope with agricultural non-point source pollution of surface water resources to meet the challenges of natural resource conserva-

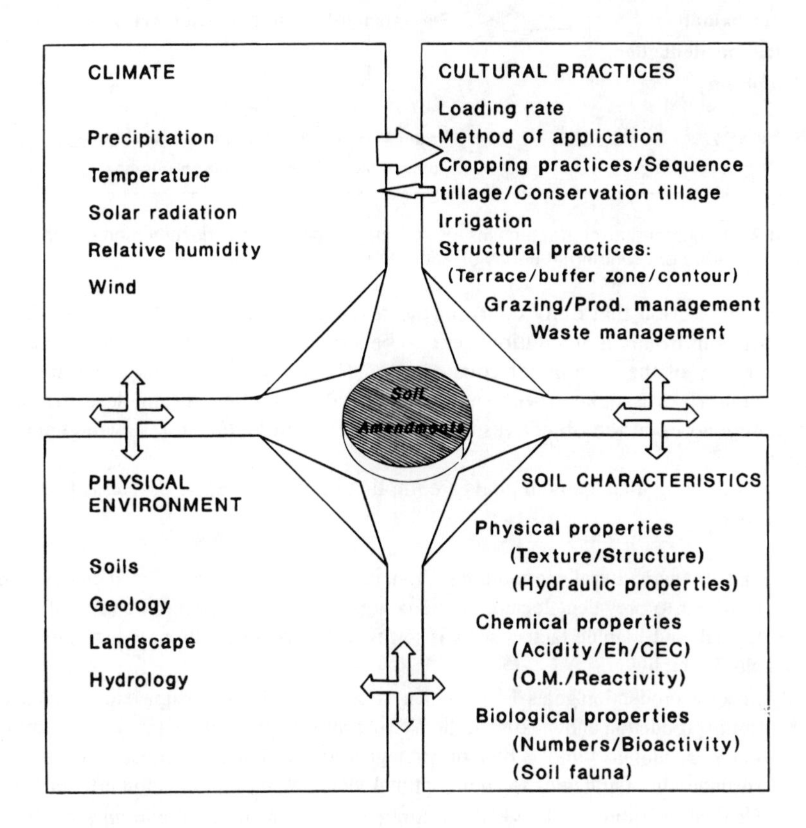

Figure 1 Interrelationships among major factors affecting the behavior and fate of soil amendments in the natural environment.

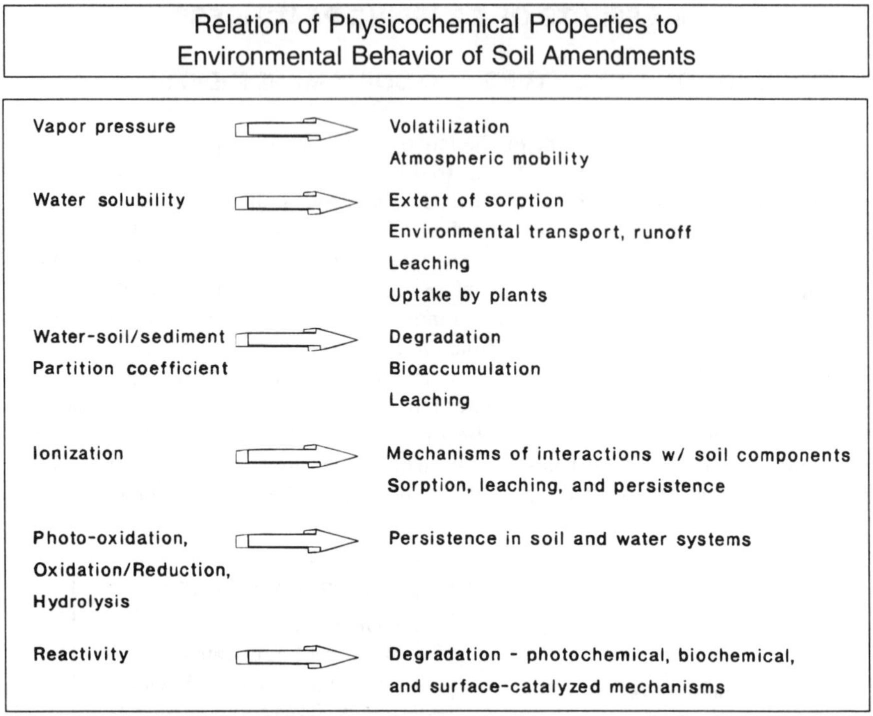

Relation of Physicochemical Properties to
Environmental Behavior of Soil Amendments

Property		Behavior
Vapor pressure	→	Volatilization Atmospheric mobility
Water solubility	→	Extent of sorption Environmental transport, runoff Leaching Uptake by plants
Water-soil/sediment Partition coefficient	→	Degradation Bioaccumulation Leaching
Ionization	→	Mechanisms of interactions w/ soil components Sorption, leaching, and persistence
Photo-oxidation, Oxidation/Reduction, Hydrolysis	→	Persistence in soil and water systems
Reactivity	→	Degradation - photochemical, biochemical, and surface-catalyzed mechanisms

Figure 2 Fundamental characteristics of soil amendments and the environment that influence environmental behavior and fate processes.

tion and protection that daily confront the resource managers and producers around the world. Environmental solutions tend to be agrosystem specific for the climate and physiography of the area under consideration. Therefore, an accurate picture of environmental risks associated with each agricultural production system described above would depend upon the objectives of the producer and his/her total management approach.

The following management goals are implicit in major U.S. agrosystems that are designed for

1. Specific agricultural commodities. Production practices are developed and used in response to prevalent local climatic conditions and specific environmental, biological, and edaphic factors such as soil type, landscape position, topography, and physiography.
2. Specific production goals. Management strategies are either designed to attain maximum production and maximize short-term profitability (maximal yield goals, maximum economic yield goals) or strategies are designed to enhance long-term sustainability (strategies to attain optimal yield, increase production efficiency).
3. General environmental objectives. Strategies seek to insure environmental compatibility, to control soil erosion, the eutrophication of lakes and reservoirs, and to improve environmental quality.

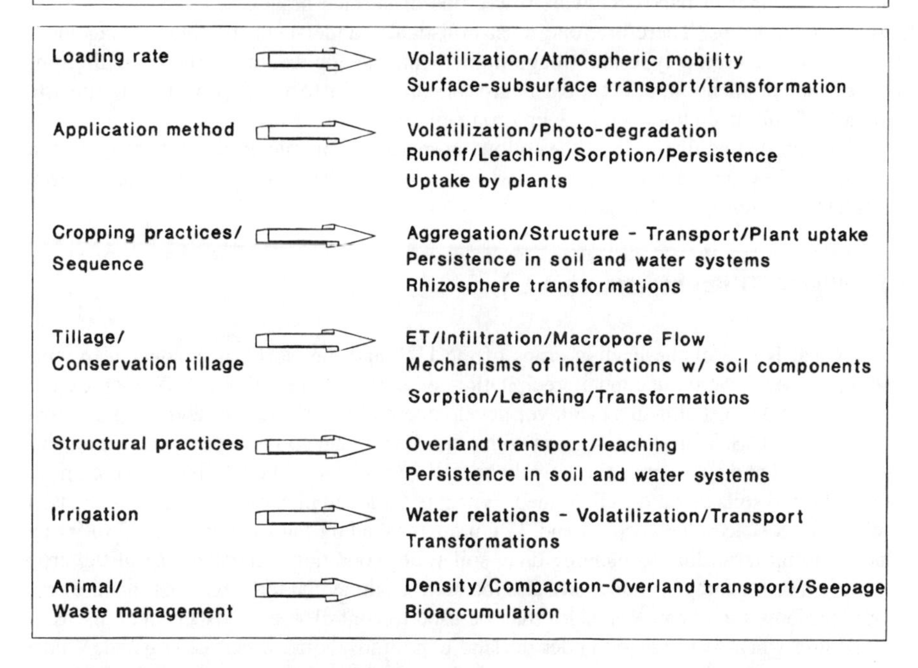

Figure 3 Agronomic practices and their effects on basic soil processes and the environmental fate of soil amendments.

In addition, we must consider whether the production system is purely animal agriculture, purely plant agriculture, or a combination of the two types in order to categorize the nature of agricultural management systems. In the latter case, there is a distinction between practices that are employed when cropping and livestock production are in separate operations or in a mixed agricultural production enterprise that utilizes the same land for the dual purpose of cropping and livestock production to maximize return from that land. Lastly, an important intrinsic character of a management system is how water for production is managed since water management practices vary with crops, soils, location, climate, and production objective.

B. CHARACTERISTICS OF MAJOR AGRONOMIC PRODUCTION SYSTEMS

Existing cropping systems represent the integration of a basic understanding of plant growth and development, water and nutrient requirements, and hydrologic and climatological processes to optimize plant population density, spacing, fertilization, pest management, and other cultural practices into functional production systems. Each commodity implies specific sets of management practices. Cultural practices such as planting geometry, crop rotation, and land management (seedbed preparation, cultivation, nutrient management, weed control measures) are tailored to the kind of crop or crop sequences.

Each variation in cultural practices has a modifying effect on the environmental behavior of soil amendments. For example, the off-site effects of fertilizers would differ whether they are applied in bands for wide-row crops or surface broadcast for narrow-row or solid-seeded crops. Therefore, one must holistically understand the total management context, i.e., the crop type, the tillage implement, the application methods, or the timing of such application, etc., for each agrosystem in order to assess properly the specific impact of soil amendments on the environment.

We can distinguish at least the following major agronomic crops that are grown in the U.S. They include small grains (wheat, oats, rye, barley), corn, soybean, cotton, sorghum, potatoes, and forages.[16]

1. *Wheat* (Triticum aestivum L.)

Wheat is one of the leading crops of the U.S. and the world. It is adapted to temperate regions where the annual precipitation averages between 250 to 1700 mm, except for warm and moist climates that favor development of wheat diseases and lodging. Most wheat in the U.S. is grown in regions with rainfall less than 750 mm/year and where winters are cold and summers are relatively hot. Wheat is best adapted to fertile medium- to fine-textured soils that are well drained. Wheat is drilled in plowed, disked, and harrowed soil in rows spaced between 15 and 35 cm apart. Seeding rates range from 33 to 90 kg ha^{-1}, varying according to planting date, soil water conditions, intended use of the crop (for grain or for forage production), and location. Commercial fertilizers and manure provide nutrients for wheat. Pesticides may be used to control weeds, insects, and diseases.

While wheat is produced under dryland conditions, some wheat is irrigated. Wheat is usually grown as a monoculture, although it is often seen doublecropped with summer crops in the more humid eastern and southeastern states. Conservation tillage is also used by many producers to maintain surface crop residues for soil protection against erosion and for conserving soil water. For example, the Palouse region is a very fertile winter wheat-producing area in the Pacific Northwest. However, it also has the most rapidly eroding landscape, with annual erosion rates ranging from 200 to 450 Mg/ha during the winter months due to partially frozen loess soils and high amounts of winter precipitation. Soil erosion reduces crop yields and impairs soil productivity and environmental quality. The collaborative research and education efforts of the STEEP (solutions to environmental and economic problems) program have made extensive advances in conservation practices to curb the erosion problem over the past 15 years.

In the southern Great Plains, the conventional intensive culture of winter wheat caused annual sediment discharges of 3.2 to 15.9 Mg/ha, exceeding the tolerance values "T" of several benchmark soils of the region.[17] Mean annual concentrations of dissolved N and P ranged from 1.1 to 1.9 mg N/L and 0.1 to 0.4 mg P/L, respectively. While these annual discharge levels are relatively low, in comparison to other natural or geologic sources of nitrate, for example, they would be further reduced by the adoption of conservation tillage methods. Crop residues in conservation tillage systems have been suggested to be potential sources of discharged nutrients. Surface residue application may enhance the transport of phosphorus (0.6 to 12.2 kg/ha) in surface runoff, compared to soil incorporation. However, crop residues were found to also act as a temporary storage medium for nitrogen and herbicides. Weathered wheat straw and corn stover residues have been observed to be a significant mechanism of immobilization and retardation of chemical transport to the subsurface of lands under conservation tillage. The extent of carbon sequestration in conservation tillage will be further discussed under the topic of soil management practices.

2. *Corn* (Zea mays L.)

Corn is another major economic crop in many areas of the U.S. and the world. Corn has a wide range of vegetative types and is adapted to a wide range of environmental conditions. The region of largest production in the U.S. (the Corn Belt) has a mean summer temperature of 25 to 27°C and a frost-free season of 140 days. The annual precipitation ranges between 600 and 1000 mm. Corn is usually planted about mid-May in a well-prepared seedbed. Planting is 5 to 8 cm deep in 0.75- to 1-m spaced rows. The average number of plants ranges between 30 and 40 thousand per hectare, depending upon local soil water and precipitation amount and distribution. Much of the corn is cultivated continuously with intensive fertilization and chemical insect and disease control methods.

Conservation tillage methods such as ridge-tillage and no-tillage have been increasingly used in the Corn Belt states to maintain surface crop residues and control soil erosion. Water quality problems are extensive in the Midwest, arising primarily from the monocultures of corn and soybean. The widespread detection of nitrate-N and selected pesticides in underlying aquifers and in drinking water wells of the region[10] has prompted ongoing changes in cultural methods and in nutrient management practices that include a strong consideration for plant nutrient sufficiency concepts, cover crop N credits, and manure nutrient credits. Optimizing fertilizer placement near the seed, for conservation tillage systems in particular, has been necessary to ensure the availability and use-efficiency of nutrients to growing seedlings while minimizing the hazards of osmotic and/or toxicity stress caused by soil amendments.

Rotations used in the Corn Belt include a 3-year rotation of corn, small grains, and a legume cover crop such as clover (*Trifolium* spp.), vetch (*Vicia* spp.), alfalfa (*Medicago sativa* L.), or a legume-grass mixture. Winter annual cover crops, particularly forage legume cover crops, are often used to provide not only soil cover but also fixed N to the corn crop. Cover crop management is critical as their growth affects stored water and its availability to the following corn crop. However, storage and plant use of precipitation occurring during the corn-growing season would be enhanced. The desiccated cover crop residues can increase infiltration rates and decrease evaporation losses, conserving water for optimal corn growth and development.

3. *Soybean* (Glycine max L.)

The climatic requirements for soybean are about the same as those for corn. However, soybean is a short-day plant and is sensitive to photoperiod. Soybean is grown primarily in the midwest and south regions of the U.S. in short rotation with corn, cotton, and small grains or in place of corn as a full season crop. The general cultural practices for soybeans are similar to those of corn. Soybeans are inoculated with *Rhizobium* spp. and are best planted in May or early June in a well-prepared seedbed in 0.75- to 1-m spaced rows. Early control of weeds is critical due to somewhat slow early growth of soybean seedlings. Soybeans are harvested for the beans, as hay in the immature stage or as silage when grown in combination with corn or other crops such as cowpeas (*Vigna unguiculata* L.), sorghum sudan [*Sorghum bicolor sudanense* (L.) Moench], or small grains.

Soybean is produced as a monoculture in many areas of the eastern half of the U.S. However, soils are susceptible to erosion following soybean production due to low amounts of surface residues that remain after crop harvest. Sediment and associated particulate nutrients pose water quality risks to the region's streams and rivers[6] and underground aquifers. In recent years, doublecropping soybean and small grains (i.e., wheat, rye) has become widespread in the southeastern U.S., in place of the monocropped soybean system. The practice maximizes land use and crop production. Planting soybean into standing crop residues produces yields equal or better than those under conventional monocroppped soybean practices. The year-round cropping system maintains soil cover and has controlled soil erosion. Other winter cover crops such as clovers (*Trifolium* spp.), common chickweed (*Stellaria media* L.), Canada bluegrass (*Poa compressa* L.), and downy brome (*Bromus tectorium* L.) can significantly increase soil cover by 30 to 50% during the wet period of late spring and early summer. Mean annual dissolved ammonium- and nitrate-N and P were reduced by an average 37, 75, and 37% when compared to a bare soil control.[18]

4. *Cotton* (Gossypium hirsutum L.)

Cotton is primarily found in the humid South Atlantic states, the South, the Southwest, and in the irrigated regions of southern California. Climatic conditions favorable for cotton are mean summer temperature of 25°C, a frost-free season of 180 to 200 days, and a minimum of 500 mm of precipitation. As cotton stalk destruction is important for the control of the *Heliothis* complex, the land usually is clean-tilled and formed into raised beds, ridges, and furrows for planting and irrigation. Cotton produces satisfactory yields over a wide range of plant populations. Populations of 50 to 150 thousand plants per hectare are used on 0.75-m-spaced rows in South Texas and on 1-m-spaced rows in the humid mid-South. Cotton is harvested after a chemical treatment to defoliate the plants, which reduces boll rotting on lower branches, eliminates dampness, and reduces fiber staining. On the Texas High Plains, some cotton is harvested after freezing weather causes leaf drop and plant drying.

In recent years, some cotton has been planted into residues of previous grain crops or cover crops, with the residues providing protection against wind erosion. Winter cereals have been used as cover crops that are killed with herbicides about 3 weeks prior to cotton planting. Strip tillage or in-row tillage and direct planting of cotton in the killed cover provided protection of the soil surface and seedlings against erosion. Small grain cover crops have been the preferred choice over leguminous species such as clovers, hairy vetch (*Vicia villosa* L.), cicer milkvetch (*Astralagus cicer* L.), or lespedeza (*Lespedeza stipula* L.) because of production economics and plant residue persistence. Decomposing small grain cover crop residues have been observed to reduce leaching of residual nitrate in soil and, in fact, may well increase N fertilizer requirement of cotton by about 30 kg/ha compared to winter fallow without cover crops. Nutrient placement with respect to the seeds thus becomes critical in such a system and in reduced tillage methods as weather conditions also contributes to the relative response of cotton to applied inputs and other soil amendments.

5. ***Grain sorghum*** **(Sorghum bicolor [L.] Moench)**

Grain sorghum is adapted to semiarid regions of the U.S. and the world. It is more drought tolerant than other major grain crops and is well suited to regions characterized by inadequate water supplies to meet the evapotranspiration demand, erratic distribution of seasonal precipitation, and highly variable precipitation and surface water supplies from year to year. Sorghum in the U.S. is grown primarily in the southern states with the major production area being the southern and central Great Plains. Only about 17% of sorghum in the U.S. is irrigated.[19]

In the central and southern Great Plains, sorghum usually is planted in May or June because soil temperatures above about 16°C are needed for rapid seed germination and seedling establishment. Typical plant populations are about 100 thousand plants per hectare on dryland and two to three times that amount with irrigation. Adequate plant nutrients are required for achieving good yields with commercial fertilizer and manure applications. One or more applications of pesticides may be required to control insects on sorghum, depending on location and severity of the outbreak.

Sorghum may be planted on clean-tilled soil, but is often planted under reduced-tillage conditions where some or all residues of the previous crop remain on the surface. Sorghum is usually grown in sequence with wheat in a 3-year rotation, i.e., wheat-sorghum-fallow (WSF) that has expediently met the need to control wind erosion. The conservation of stored soil water has greatly stabilized dryland production of sorghum in semiarid climates. Conservation tillage also reduces N and P transport in runoff compared to conventional tillage practices. Annual sediment and total N and P losses average 8877, 7.28, and 2.5 kg/ha/year with conventional sorghum cropping practices.[20] However, these soluble P concentrations generally exceed limits associated with accelerated eutrophication of surface waters (0.01 mg P/L). Sediment and nutrient discharges can be potentially reduced by a factor of 10 or greater with conservation tillage or no-tillage practices (281, 0.76, and 0.28 kg/ha/year).

A soybean-sorghum rotation also has been suggested as a cropping sequence to profit from fixed N by the grain legume crop because grain sorghum yields are often higher when rotated to other crops. Rotational effects include a disrupted pest/vector cycle as well as the granulating effect of different root systems and the production of a wide array of organic substrates that promote biological activity and diversity of the rhizosphere. In the humid southeast, crimson clover (*T. incarnatum* L.) and hairy vetch cover crops can potentially reduce N fertilizer needs by 99 to 123 kg/ha, minimizing reliance upon agricultural chemical inputs and environmental risks.

6. ***Potatoes*** **(Solanum tuberosum L.)**

Potatoes are produced primarily in cool climates where mean summer temperatures are about 18°C or lower. Day length, cool temperatures, and optimal fertility favor tuber formation. Potatoes are grown on soil types that range from sandy loams to silt loams with a pH of 5.0 to 5.5. Some potatoes are also grown on peat. Commercial potato production relies heavily on use of commercial fertilizers. Application rates range from 120 kg N, 70 to 90 kg P, 100 kg K ha^{-1} in Maine to 300 kg N, 35 kg P, and 100 to 120 kg K ha^{-1} in the Columbia River basin of Washington.[21]

Potato quality and starch content are highly dependent on nutrient management. Potatoes also respond well to green-manure crops and animal manure applications. Plowing is often performed to incorporate the legume residues or manure. Several diseases affect the foliage and tubers of potatoes. Fumigants are used to control soil nematodes prior to planting seed potatoes. Oftentimes resistant varieties are used. Insecticides are used to control vector insects, aphids, and the Colorado potato beetle that is one of the most destructive pests of potato. Planting rates average 1800 to 2500 kg ha^{-1} in rows spaced 0.9 m apart. A uniform water supply is critical to the production of tubers that are well formed and uniform in size. Timely irrigation is used to supply water to blooming potato plants, while tubers are developing. Potatoes are dug mechanically after the vines are killed to stop growth and hasten maturity.

Clean seedbed preparation methods leaves the soil surface nearly bare and prone to wind and water erosion. Winter wheat and Austrian winter peas (*Pisum sativum* subsp. *arvense* L.) have been used as cover crops that are killed prior to potato planting. The method of combining subsoiling, fertilizer banding, cover crops, and planting in a narrow-tilled strip provides an acceptable erosion abatement technique as well as a precision placement of nutrients to meet plant needs for optimal potato growth and development.

7. Forages

Annual and perennial grasses, legumes, and grass-legume mixtures are widely grown throughout the U.S. (an estimated 28 million hectares in 1990), providing forage for grazing animals or hay for later consumption by animals.[16] Forages are often intercropped or double-cropped with major grain crops such as corn, soybeans, and small grains. When hay is baled and removed from the land, nutrients are removed also. Hence, good forage production relies heavily on good plant nutrient management. Commercial fertilizers, animal manures, municipal wastewater, and sewage sludge are common nutrient sources for forage crops. When forages are grazed, nutrients removed by animals are recycled to the land through the manure, although the distribution is highly spatially variable.

Land application of agricultural by-products, food processing wastewater, and municipal wastewater and sludge allows for recycling of plant nutrients, although their contents are highly variable depending upon factors such as the type of raw materials and treatment process. However, concerns over land operations such as odor, metal contents, pathogens, and land suitability and availability have curbed the interest for land spreading of these materials on agricultural fields. Irrigation with wastewater remains a popular nutrient recycling and water renovation practice in forage production and forestry ecosystems. Over 57 to 71% of applied N can be assimilated by reed canarygrass (*Phalaris arundinacea* L.), field corn, and a large portion of the potassium and phosphorus requirements of forage grasses can be met by the practice.[22] Careful attention to water and nutrient loading rates must be exercised to avoid runoff and nutrient discharge from the application sites as well as deep percolation losses. The impact of grazing livestock on forage production and the environment will be considered later under the topic of animal management practices.

C. DRYLAND PRODUCTION PRACTICES

Water is vital for agricultural production, and agronomic yields are reduced more frequently by water shortage than by any other input and environmental factor. Although water table management and drainage are required for humid areas of the mideast and southeast U.S., rainfed agricultural production systems require improved technologies for increasing precipitation-use efficiency of crops, pasture, and range plants. There are many attempts to modify plant water requirements via germplasm enhancement and selection of crops and cultivars with high water-stress tolerance. Crop enhancement is also focused on increasing rooting depth and depth of soil water extraction.

Water management practices are linked to soil management practices that are designed to maintain soil water at levels sufficient to allow crop growth and production. Management practices are designed to increase water infiltration and storage of precipitation and snowmelt, and to reduce runoff and soil water evaporation in dryland farming systems in the Midwest and the Great Plains states (snow catchment, surface residues management, standing stubble, furrow-diking, and fallow).

Under dryland and rainfed conditions, weather prediction may play a significant role in overall water management for crop production. Timely and accurate predictions may allow the producers to plan tillage, seeding, and harvesting operations, and to adjust the timing of applying soil amendments and agricultural chemicals for their most effective utilization. For example, with accurate predictions, producers can avoid applying soil amendments prior to a major precipitation event, thus reducing the potential for wash-off from plant and soil surfaces, discharge from the field, or deep leaching into the soil. On the other hand, fertilizers applied with incoming drought conditions would remain unused by the crop, and excess nitrogen, phosphorus, or potassium remains available for later transport from the field or through the soil.

D. IRRIGATED PRODUCTION PRACTICES

Irrigation of cropland and pastures is practiced to reduce the risks of production caused by low or poorly distributed precipitation or to remove salts from the root zone in arid regions of the western states. Nationwide, 15% of cropland is irrigated, but that land area produced $30 billion or about 51% of total crop sales in 1987.[16] Timely irrigation is needed to supply optimum amounts of water at the most critical growth stages of crops such as vegetables, fruits and nuts, rice, and specialty crops. For example, water is provided in large quantities to cotton in California, Arizona, New Mexico, and parts of Texas to meet daily evapotranspiration that averages 5.1 mm day^{-1}, and to maintain adequate soil water content in the root zone.

Overall, irrigated agriculture uses about 40% of all freshwater withdrawals in the U.S. A wide range of water delivery systems is used for irrigation. They include surface/gravity, subirrigation, sprinkler, and drip systems. Irrigation management practices are designed to efficiently use available water, improve scheduling to meet the plant water demands, and, in all cases, improve application uniformity. Considerably more water is withdrawn from streams, reservoirs, and aquifers than is utilized by plants because typical irrigation water use efficiency is estimated at about 60 to 70%. Practices to reduce evaporation and seepage losses from reservoirs and from conveyance structures would conserve large quantities of water. Localized or wide-scale water shortages of varying intensity and duration recently experienced in areas of west Texas and southern California reemphasize the need for conservation and new technologies to achieve effective and efficient use of this resource.

Deficit irrigation (limited irrigation, irrigation of alternate rows) is practiced to conserve scarce resources in some semiarid farming regions. Sprinkler irrigation systems offer the greatest flexibility in terms of the rate and time of water application. However, sprinkler irrigation on sloping land may result in major runoff losses when the systems are improperly operated or designed. Many high-pressure systems are costly to operate and maintain and are being replaced by medium- and low-pressure systems to reduce energy costs. The low-energy precision application system (LEPA) has been developed to enhance water-distribution uniformity and water-use efficiency of crops. It has the capability to apply irrigation water and agricultural chemicals in precise amounts and locations on the field. Hence, fertigation and chemigation techniques are further refined to allow precision management of inputs. Timely applications at low rates can reduce leaching losses. Small quantities and repeated applications of soluble nutrients such as N and pest control chemicals may be metered to meet crop needs or pest intensity, thus avoiding overloading the system and reducing the potential for off-site dispersion and transport.

Best management practices in such crop and water management systems for soil amendment usage involve careful timing of water, nutrient, and pesticide applications in order to protect surface and groundwater resources. Water and chemical management practices must be optimized to cope with combinations of amendment properties and field characteristics that are conducive to potential leaching such as high water solubility, low soil-water partition coefficient, relatively long half-life, high water table, soil permeability, and groundwater recharge rates.

The potential effects of irrigation management practices on environmental quality and water contamination vary with the type of water delivery system being used. Deep percolation and transport of water and dissolved solutes are more likely with surface/gravity systems than with drip or sprinkler systems. Furthermore, significant quantities of irrigation water may escape as runoff from the field when containment structures and systems for recycling tailwater into the irrigation system are not in use. Use of such systems can prevent contamination of surface water bodies. Surge-flow and cablegation systems which allow application of high volumes of water in a relatively short time achieve soil sealing and uniform water application on the field[23] and lower the potential for deep percolation and the high volume of tailwater runoff that is common when surface irrigation systems are used.

Quantity and timing of irrigation have a direct effect on transport of soil amendments. Decisions about when to irrigate and how much water to apply are routinely made throughout the growing season. Improper scheduling can result in applying too much or too little water. Overapplication may lead to deep percolation or runoff losses of water and soil-applied amendments. Existing approaches and emerging technologies to mitigate off-site impacts include use of (1) furrow diking in conjunction with the LEPA system to reduce runoff; (2) tillage systems that retain surface residues and thus enhance infiltration under some conditions; and (3) other runoff-retarding practices such as contouring, land leveling, and terracing.

III. EFFECTS OF SOIL MANAGEMENT PRACTICES

A. LANDFORMING PRACTICES

Conventional management approaches to control nonpoint source contaminants rely on structural practices. These include terraces, grass waterways, vegetative buffer or filter strips, tile drains, sediment detention basins, and other structures to intercept, reroute, or retain water and thus modify the transport of contaminants by runoff water.

Most structural controls can be easily installed, but they are costly and usually installed on a voluntary cost-sharing basis. They also require frequent maintenance. However, the management practices in place prior to the installation of structures are still used. Therefore, for the most part, these structural practices are only a substitute for implementing real management changes from established practices that are at the root of the environmental pollution problems.

B. TILLAGE SYSTEMS

1. Types

Inversion or Clean Tillage — Clean tillage is the farming practice that often involves one primary tillage operation to bury surface residues, control weeds, and loosen the soil, and one or more secondary operations for additional weed control and seedbed preparation. The primary operation often is with an implement (e.g., a moldboard plow) that inverts the tillage layer and leaves the surface virtually devoid of crop residues. While such tillage buries residues, controls weeds, and loosens the soil, it leaves the surface bare and subject to the abrasive action of falling raindrops, which may destroy surface soil aggregates, cause surface sealing, and thus increase runoff and erosion.

Runoff may transport sediments and agricultural chemicals from the land, thus potentially polluting surface water. Runoff of nutrients, industrial and agricultural by-products, municipal wastewater and sludge, and pesticides from application sites has been studied intensively for the past two decades in a coordinated effort to control nonpoint source pollution. The processes by which chemicals enter runoff and streamflow are complex. As the intensity and duration of precipitation increase, water flow within a field becomes concentrated because of tillage patterns, planting geometry, and field slope. Plant nutrients, metals, and pesticides are entrained in the runoff stream by diffusion and turbulent convection as dissolved solutes in the aqueous phase. They are also transported while sorbed on suspended soil particles or sediments that have been detached by the impact energy of raindrops or by abrasion and turbulent flow. In general, the potential for runoff is greater with clean tillage than with conservation tillage (see next section), which involves management of crop residues on the soil surface.

Conservation Tillage — In developing practices to control soil erosion, we observe that these same practices have the potential for reducing nonpoint source pollution. Conservation tillage practices are attractive because they allow us to achieve many environmental objectives while improving the economics of farming. Conservation tillage systems, by definition, encompass practices that retain a minimum of 30% surface cover by residues after crop planting to control water erosion. To control wind erosion, surface residues equivalent to 1100 kg ha^{-1} of small grain residues are required.[24] The use of conservation tillage systems for soil erosion control has increased in the U.S. in recent years for most agrosystems. Based on the planted area devoted to conservation tillage for the 1968 to 1986 period, projections of adoption rates for the year 2000 and 2010 lie in the 52 to 80% and 63 to 83% ranges, respectively.[25] The 1988 national survey of tillage practices has revealed a U.S. total of about 36 million ha managed under conservation tillage.

In some conservation tillage systems, avoidance of mechanical soil disturbance other than that needed for opening a narrow slit in the soil for seed placement retains most plant residues on the surface after harvest of plant parts of economic value. This practice (no-tillage) leaves cultivated land under permanent cover, an ideal condition similar to that found on native rangeland. The plant residues can effectively decrease the velocity and volume of runoff water. Major pollutants of lakes and streams are sediments and the synthetic chemicals they carry in overland water flow or by wind erosion. The reduction of sediment losses from cultivated fields results in improved air and surface water quality. However, because runoff is reduced, more water infiltrates the soil. If this water is not utilized by the above-ground vegetation or lost by evaporation, the potential for leaching of chemicals through the soil and, hence, groundwater contamination is increased.

Use of selective herbicides in lieu of tillage to control weeds makes many mechanical machine-intensive operations unnecessary. In addition to the reduction in topsoil loss that is not always visible, the net results of adopting conservation tillage practices are savings in fuel, equipment needs, wear, and tear, and labor. Conservation tillage practices can prevent or greatly reduce pollution from nonpoint sources, thus helping to meet the Conservation Provisions of the 1985 Food Security Act and the conservation compliance and environmental protection requirements of the 1990 Farm Bill.

2. Equipment

A wide variety of tillage equipment is available for achieving almost any desirable soil condition. Included are moldboard, disk, and lister plows that invert the plow layer; rotary tillers and disks (one-way, tandem, offset) that thoroughly mix the plow layer; and sweep, blade, chisel, and field cultivator implements that loosen the plow layer without inversion or thorough mixing.

Inversion implements effectively loosen the plow layer, control existing weeds, and bury crop residues, but secondary tillage is usually required for subsequent weed control and final seedbed preparation. Implements for secondary tillage include disks, harrows, sweep plows, and field cultivators. Inversion tillage results in a bare soil surface, thus increasing the potential for greater runoff and transport of amendments from the field than where some residues remain on the surface. Mixing implements effectively loosen soil and control existing weeds, but leave some crop residues on the surface. Some additional tillage is usually needed for additional weed control and seedbed preparation. Retention of some residues on the surface helps to reduce runoff and amendment transport.

Tillage that only loosens the plow layer without inverting or thoroughly mixing it retains most crop residues on the soil surface. Even repeated operations with loosening implements often retain sufficient residues on the soil surface to reduce runoff and amendment transport. However, tillage that only loosens the soil sometimes does not result in effective weed control. In general, sweeps, blades, and field cultivators, which undercut the surface to sever weed roots, provide good weed control when the soil is relatively dry and when sufficient time elapses for the plants to die before precipitation occurs. When the soil is wet, weed control with these implements is usually poor. Chisel implements usually also result in poor weed control.

Various types of equipment are available for amendment (fertilizers, pesticides, lime, sewage sludge, fly ash, etc.) application to soils, with the appropriate type being dependent on the amendment to be applied. Dry materials can be broadcast on the surface or placed in soil behind chisel, disk, hoe, or slide openers. Manure is broadcast on the surface, and liquid or slurried materials may be applied to the surface or injected into the soil behind suitable openers. Surface-applied materials often are incorporated with soil by subsequent tillage, but may remain on the surface until moved into the soil by precipitation when a no-tillage system is used. Anhydrous ammonia is injected into soil through chisel-type openers. Most herbicides and insecticides are applied by ground sprayers or by airplane.

3. Effects on Soil Properties

Many factors influence the effect of tillage methods on soil properties. Included are implement used, tillage depth, and speed of operation. Soil texture, organic matter content, and water content also play an important role in determining rate and extent of the alterations. Soil properties having the most effect on the fate of amendments are soil bulk density, aggregation, hydraulic properties, pH, cation/anion exchange capacity, and organic matter content (Figure 1).

As the adoption of alternative production methods increases for most crops, many of the known transformation processes affecting nutrient, metal, and pesticide fate in the environment are influenced by the physical, chemical, and biological alterations in soil properties induced by the reduction or absence of mechanical tillage. Although considerable knowledge has been generated in the past decade on conservation tillage practices, it has become apparent that many soil processes require many years to establish a new equilibrium upon reduction of tillage intensity or elimination of tillage. The effects of management practices on the fate of soil amendments in the new microenvironment of conservation tillage systems must be viewed in light of the length of time these practices have been adopted since the last soil disturbance and the extent of tillage.

The major changes in surface ecology included (1) physical changes such as macroporosity, bulk density, water content, water infiltration and storage, and attenuated soil temperature fluctuations; (2) chemical changes in soil properties such as soil pH, and organic matter characteristics and content; and (3) changes in numbers and kinds for soil organisms and indigenous crop residues' organisms. These are fundamental changes; thus, one must understand the context of the physical, chemical, and biological changes to properly assess their specific impact on soil amendments.

Primary differences in the surface microenvironment between conventional tillage and alternative practices are summarized below.

Physical Changes — Tillage operations significantly alter the ecological balance both above and below ground. Tillage operations mix soil and incorporate organic debris into the soil, depending upon the degree and type of tillage implement used. In contrast, incorporation of surface residues into no-tillage soil occurs more slowly through wetting and drying, freezing and thawing, and the action of soil-inhabiting macroorganisms and soil-borne microorganisms.

A comparison of soil bulk density of Pachic and Udic Paleustolls under three tillage management systems showed that, less than 2 months after planting winter wheat, the plowed soil reconsolidated and had highest soil densities in the surface 0- to 0.3-m depth. Meanwhile, small differences disappeared with increasing depth between subtilled and no-tillage soils.[26] A temporal variation of soil density is expected with the plowed treat-

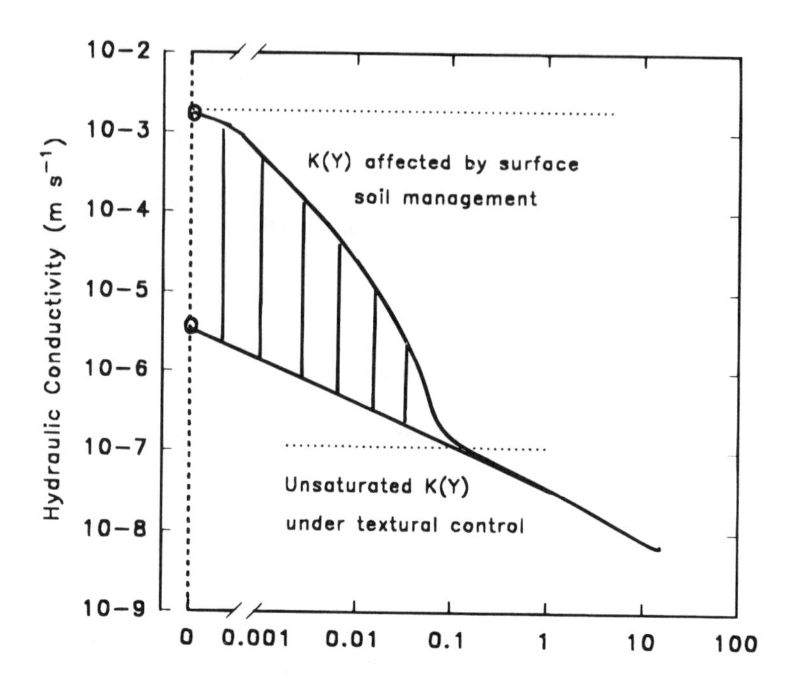

Figure 4 Effects of surface soil management practices on the relationship between hydraulic conductivity and water pressure relationship in a soil medium.

ment, with low soil density values observed soon after mechanical tillage is performed. However, the soil settles under the weight of its mass. It deforms from wetting and drying,[27] the impact of raindrops of subsequent precipitation, and the extent of soil aggregate stability. In general, runoff increases as soil density increases. Consequently, maintaining a soil in a loosened, noncompacted condition is essential for reducing runoff and, hence, transport of potential contaminants to surface waters. On the other hand, no-tillage soil bulk density shows less temporal variability and, for example, is lower during the majority (70 to 80%) of the growing season of winter wheat, compared to that of a plowed soil.[26] The lower density is the result of partial incorporation of crop residues into the surface soil and organic debris of the previous crop's root system remaining undisturbed while building up soil organic matter content and an elaborate channel structure. There are fine cracks, convoluted fine root channels, and small-diameter wormholes present in the soil. Although these biopores represent less than 1% of the bulk soil volume, roots of subsequent crops utilize and colonize these pathways of low resistance.[28] The well-developed structure and a greater macroporosity also are caused by increased activity of arthropods and earthworms, as compared to conventionally tilled soil. All these biological factors may contribute to the increased porosity and consequently improve hydraulic properties of the continuously no-tillage soils (Figure 4).[26,29,30]

Conservation tillage is conducive to maintaining soil organic matter contents at levels higher than those of clean tillage. With no-tillage, organic matter is highest at the surface, which helps stabilize surface aggregates, minimizes surface sealing, and maintains favorable water infiltration rates. In structured soils, water flow takes place in macropores along the ped faces or biopores such as wormholes or old root channels, thereby bypassing the bulk of the soil. Macropore flow manifests as a pattern of spikes, or fingers, moving water deeper than predicted by Darcian flow. With reduced runoff, however, there is a greater potential for leaching of amendments to groundwater. The leaching of nonreactive chemicals deviates from piston flow behavior or from chromatographic models for reactive solutes through a heterogenous soil medium. Preferential or facilitated transport of nitrate and soil-applied herbicides[31] has been suggested to explain the large differences in the expected and actual transport of bromacil [5-bromo-6-methyl-3-(1-methylpropyl)-2,4-^{1}H,^{3}H, pyrimidinedione] and napropamide [*N,N*-diethyl-2-(1-naphthalenyloxy) propanamide] through a structured clay soil.

The residue layer in conservation tillage systems also provides a mulching effect and decreases soil water evaporation from reduced- and no-tillage soils.[32] Evaporation from a bare soil surface is a three-stage process. Stage 1 is an energy-limiting stage and occurs at the potential evaporation rate (pan evaporation). The other two stages are limited by water flow to the soil surface.[33] Stage 1 evaporation rates are significantly reduced with increasing surface residues (0 to 3500 g m^{-2}).[34] Thus, surface soil layers of tilled soils fluctuate widely and are drier than those of no-tillage soil in periods of high potential evaporation throughout the year.[26] Moist soil conditions (at about 60% water-holding capacity) are favorable for enhanced microbial activity and biochemical reactions in soil. Coupled with the altered temperature regime of no-tillage soils, these conditions are conducive to enhanced rate of biological degradation of soil-borne pesticides, microbial mineralization-immobilization of nutrients, or the mineralization of sludge or animal manure.[31]

The surface residue layer also has a mulching effect and attenuates temperature fluctuations.[36,37] The primary mechanism of the attenuation is the change in radiant energy

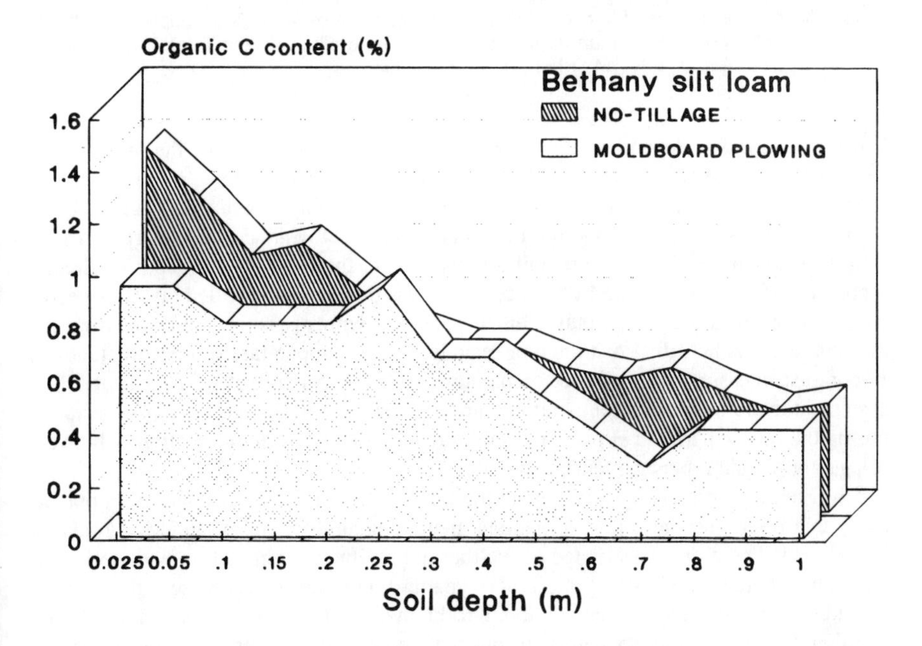

Figure 5 Effect of tillage methods on soil organic carbon deposition as a function of soil depth after 8 years of continuous surface residue management. (From Dao, T.H., *Research Trends in Agricultural Science, Soil Science, Council Sci. Res. Integration*, Trivandrum, India, 1, 9, 1993. With permission.)

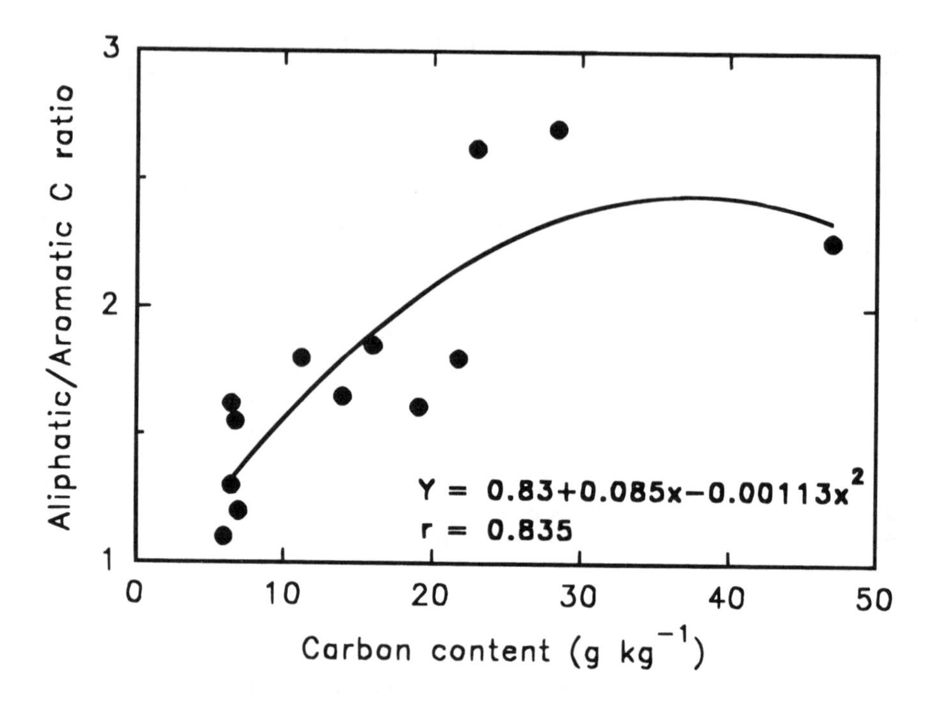

Figure 6 Characteristics of humic substances extracted from 0- to 15-cm depth of no-tillage soils. (From Stearman, G.K., Lewis, R.J., Tortorelli, L.J., and Tyler, D.D., *Soil Sci. Soc. Am. J.*, 53, 744, 1989. With permission.)

balance. Therefore, depending upon the cropping system and time of the year, the effects of tillage on soil temperature will vary in magnitude and in trend, to be either beneficial or limiting to plant processes. The effects of cropping systems and indirectly the amount of soil surface coverage by the unharvested plant residues are illustrated by the following examples. Under corn, soil temperature in the 0- to 30-cm depth is affected more by surface residues and by surface roughness than any other tilled layer property.[36] However, when these temperatures are normalized with respective daily maxima and minima, there is no effect of tillage or residue on soil temperature diurnal fluctuations. In contrast, measurements of soil temperature at 0- and 5-cm depths in a wheat-fallow system indicate maximum temperatures are highest with no-tillage (standing residues) during the fallow period.[37] However, minimum temperatures differ only slightly between tillage and residue management practices.

Soil temperature changes influence the kinetics of soil processes and microbial activity that affect nitrogen transformations, metal availability, and pesticide degradation in the field. Therefore, knowledge of the thermal regime in conjunction with soil-water fluxes will help us understand microbial dynamics and rates of processes affecting soil amendment availability and degradation. This knowledge also enhances our understanding of other factors such as root growth, nutrient uptake, and rhizosphere transformation processes that are strongly influenced by daily soil temperature and water fluctuations.

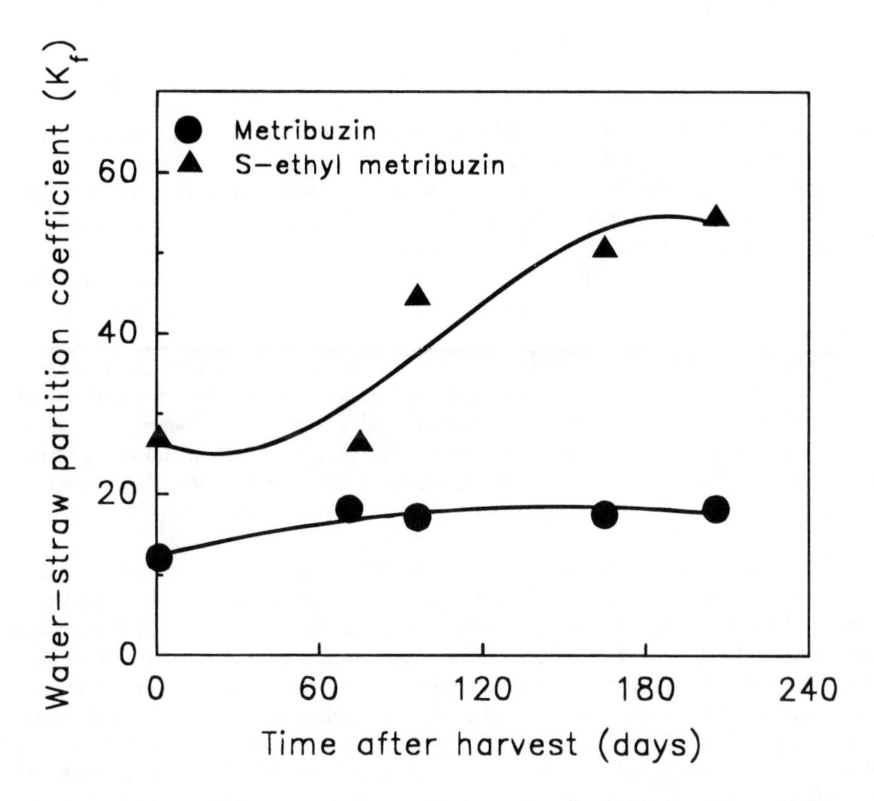

Figure 7 Retention characteristics of fresh and field-weathered winter wheat (*Triticum aestivum* L.) straw for metribuzin and S-ethyl metribuzin herbicides. (After Reference 44.)

Chemical Changes — A shift from conventional tillage to no-tillage methods results in gradual stratification of soil properties, resembling the ecosystem of a native pasture. Selected nitrogen fertilization practices and the lack of soil mixing without tillage may lead to increases in surface soil acidity.[38] Such changes affect the retention and persistence of polarizable and basic herbicides,[39] the soil- and solution-phase equilibria of metals, and, hence, their availability in soil and sludge-amended soil.[40]

The greatest impact of the residue mulch on the surface of reduced/no-tillage soil is the addition of soluble carbon to the surface soil layers.[41,42] Direct seeding and leaving wheat stubble on the soil surface increase soil organic carbon content (Figure 5) as well as its composition (Figure 6) about 4 to 5 years after conversion from intensive tillage to no-tillage management. Soil organic matter influences numerous soil conditions, including density, aggregation, and soil-amendment interactions. In general, soil density decreases and aggregate stability increases as organic matter content increases. Soil aggregation involves the bonding together of soil particles. Soils with stable aggregates at the surface resist aggregate breakdown and surface sealing during rainstorms, thus tending to reduce runoff. Soils with sufficiently large dry stable aggregates (>0.84 mm) also resist wind erosion, thus minimizing contamination of the atmosphere. Some amendments adhere to organic matter. Thus, if sufficient organic matter is present and prevented from transport in runoff, leaching of the amendments should be reduced. Elevated levels of soil organic carbon tend to increase sorption of organic pesticides in upper depths of no-tillage soils. For example, soil-water partition coefficients of metribuzin [4-amino-6-(1,1-dimethylethyl)-3-(methylthio)-1,2,4-triazin-5(4*H*)-one] in no-tillage soils are approximately 1.5 to 5 times greater than those of plowed soils. The increased deposition of organic carbon from the crop residues, as well as the mulch itself (Figure 7),[44] jointly decrease the availability of metribuzin in soil water and its transport to the subsurface of fields under conservation tillage. As the carbon-rich soil-residue interface is also the most microbially active zone, the herbicide is subjected to enhanced degradation.[42,45]

There is an increase in pH-dependent exchange capacity with the increased carbon deposition. Metal availability in sludge-amended soils is reduced in such a carbon-rich medium.[40,46] Organic constituents in municipal sludge and animal manure form complexes with metals such as cadmium, copper, zinc, etc., restricting metal bioavailability.[47] However, manganese complexation by organic ligands has resulted in phytotoxicity as the process may account for greater than 75% of the soil solution metal concentrations.[48,49]

Conservation tillage practices are effective alternative practices to reduce particulate losses of plant nutrients, in particular nitrogen and phosphorus that are associated with runoff.[50] Phosphorus fertilization practices tend to induce accumulation in surface soil layers in relatively insoluble inorganic forms. If erosion is controlled, very little is lost as particulate phosphorus, which is a very potent agent of eutrophication of surface waters. Generally, phosphorus is not a significant groundwater contaminant, unless the water table is high or it enters aquifers via direct conduits, as in karst landscapes. Surface crop residues, upon weathering, may be a source of phosphorus.[51,52] Therefore, as with any nutrient input, care is needed to ensure minimal runoff entrainment with adequate soil erosion control.

Biological Changes — Tillage affects soil water, air, temperature, and the cycling of carbon and nitrogen in soil, thereby regulating microbial activity and numbers. Incorporated or decomposing surface residues of plants provide a growth substrate that is converted into microbial biomass and soil organic matter. Differences in rates of conversion occur between soil management practices. In a Colorado site, total N in the 0- to 10-cm depth has declined to 50, 68, and 73% of an adjacent native sod after 16 years of cultivation with plowing, stubble-mulch, and no-tillage management.[53] Concurrently, soil microbial biomass and CO_2 respiration were highest in the no-tillage treatment compared to the stubble-mulch or plowed treatments. No-tillage soil biomass carbon has increased from 27 to 83% greater than that under conventional tillage.[54] A stratification with soil depth has also been observed in C, N, P, and S contents, paralleled by the stratification in microbial numbers as well as classes of microorganisms.[55] Organic amendments such as manure and sludge also increase microbial numbers and activity.[35]

With respect to metabolism of pesticide or organic amendment, evidence of biodegradation kinetics following a second-order rate law with respect to microbial cell density has been reported for simple substrates such as methanol, phenol, and t-butyl alcohol.[56] More accurate descriptions of pesticide fluxes are necessary to develop management practices that will give us the control needed to tailor efficient pest control programs while preventing undesirable off-site impacts.

C. SOIL-SPECIFIC CROP MANAGEMENT

Unique crop management practices are better adapted or even essential for specific soil conditions. Soil conditions that strongly affect management options include texture and surface slope. Options affected by texture include crops to be grown, irrigation amount and frequency, and amendment applications.

Most grain, oilseed, and fiber crops are adapted to soils of all textures, but tuber and groundnut crops (e.g., potato and peanut) are better adapted to coarser-textured soils (loams to sands). Coarse-textured soils, however, have low water- and nutrient-holding capacities. Therefore, they must be irrigated or fertilized more frequently with smaller amounts to avoid excessive percolation of water or leaching of nutrients.

Topography has a major effect on runoff and, hence, soil erodibility and amendment transport. On slightly sloping soils (up to 3%), most management practices are adapted, except for conditions noted above. However, on steeper slopes, some tillage and related practices are better adapted than others. Conservation tillage, especially no-tillage, is an adaptable management option that permits cropping of steeply sloping soils without subjecting them to excessive runoff and erosion, and, hence, transport of dissolved and adsorbed amendments.

Recent developments of real-time sensing technology for nutrient (i.e., nitrate), soil properties (i.e., organic matter content, pH), standing vegetation, in particular weeds, and the quantitative description of field spatial variability mark the beginning of an era of prescription farming.[57] The possibilities are abundant; the emerging technologies such as computer-controlled application technology and precise geo-referenced positioning of farming equipment with global-positioning systems[58] allow specific customizing of rates of inputs such as fertilizers, pesticides, and irrigation water at every location on a field, matched to actual plant needs, soil-holding capacity, or pest density. Spatially correlated data of soil characteristics can be encoded on microprocessors as digital maps, or remotely sensed pest spatial density to vary input dosage and blends as the application/delivery equipment moves across the field. The net benefits to the environment of such an approach are (1) the reduction of inputs, hence, minimizing off-site impact, and (2) minimizing yield loss caused by underapplication, hence, maximizing profitability.

IV. EFFECTS OF ANIMAL MANAGEMENT PRACTICES

Livestock are effective converters of forage into high-quality proteins that are important to human nutrition and health. As such, livestock also have a significant role in the diversification of farming enterprises. They add stability to farm income by making farm receipts less vulnerable to price volatility of commodity markets. Animal agriculture accounts for over 52% of the gross agricultural receipts in the U.S. ($89.6 billion) in 1990.[16]

Livestock production is practiced on large ranches that are common in the Great Plains states and western states or on small farms in the Southeast. These enterprises involve either diffuse operations on pastures and rangelands with herds of various sizes or concentrated feedlots where annual animal concentrations of 30 to 50 thousand head of livestock are common. Dairy and swine operations each involve various numbers of animals, with the animals generally confined to a relatively small area or even totally confined as in some swine operations. Most poultry operations involve many thousands of birds in confined quarters.

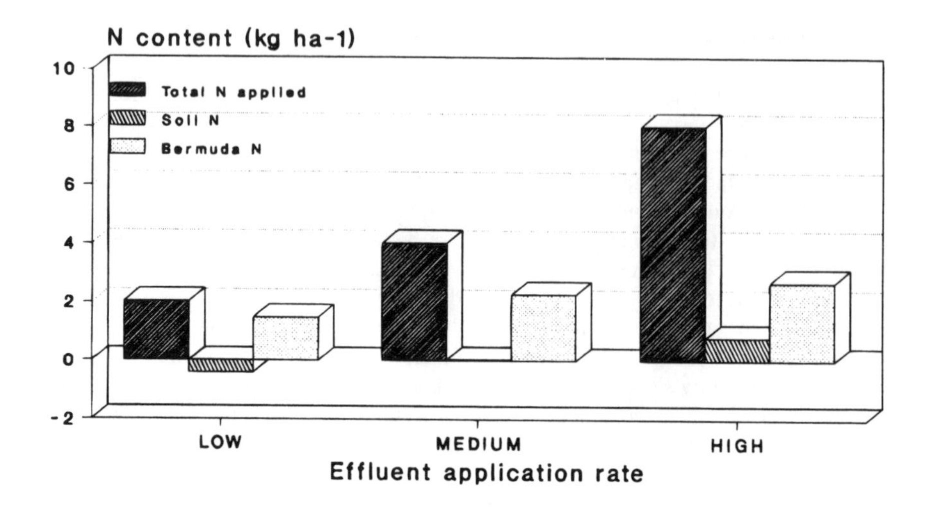

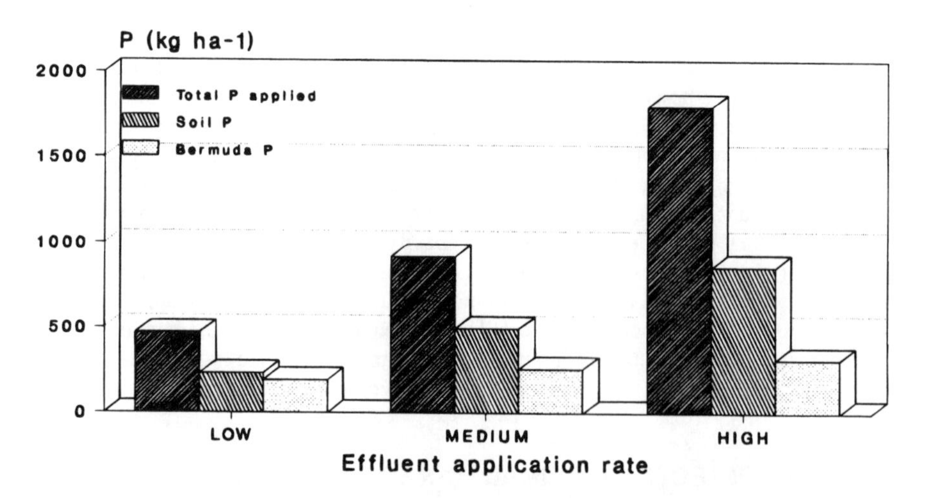

Figure 8 Nutrient recovery in soil and by bermudagrass (*Cynodon dactylon* L.) from three levels of swine lagoon wastewater applications to bermudagrass. (From King et al., *J. Environ. Qual.*, 14, 14, 1985. With permission.)

Livestock and poultry manure may be a major source of excess nutrients, in particular nitrate, salts, organic matter, antibiotics, pathogenic organisms, and other contaminants of the atmosphere and surface and ground waters. Subsurface hydrogeology, landscape, containment management, and weather all contribute to the likelihood of discharge to surface waters and of groundwater contamination. Nitrate leaching from animal manure storage such as open unpaved cattle feedlots,[59,60] dairy operations,[61] poultry litter,[62] runoff detention ponds, liquid manure lagoons and stockpiles, and from land applications of manures poses a real threat to water resources.[63,64] Seepage of effluent from livestock feed and silage storage areas may pose similar risks to the surrounding environment. In many cases, the relationship between animal production practices and those for handling animal wastes and environmental contamination potential have been recognized. A number of technologies are available to mitigate adverse impacts of the various livestock and poultry production systems to soil and water resources. We will distinguish the following three major categories of animal production practices to explore the opportunities to reduce environmental impacts of each of the three groups.

A. MIXED CROPPING-LIVESTOCK MANAGEMENT SYSTEMS

Mixed management systems are categorized to imply varying levels of interactions that exist between crop and animal production. These sectors mutually influence each other, although they may be physically separate. In such a case, the link is economic as one sector may finance or facilitate the flow of production inputs such as capital, labor, and time to the other sector.

1. Separate Feed and Livestock Production Practices

Producing animal feed includes activities ranging from growing forages and grains for on-farm use to support the production of a handful of animals to raising them as cash crops to supply feedlots, and other highly concentrated livestock production enterprises such as dairy operations.

Feed grain crops such as corn and sorghum and other food grains and fiber crops (i.e., soybean, cotton) from which processed agricultural by-products are produced and used as animal feed are managed as described previously. With respect to forages, the annual production of all hay is about 135 million metric tons, which has been valued at $11.1 billion in 1990. That amount supplied roughage to about 76 million animal units during the same year.[16] Although normally stored in sheds, temporary stacks of harvested forage and uncovered hay bales that dot the rural landscape lose nutrients upon precipitation and weathering. The effluent generates concentrated plumes that leach and disperse in the subsurface environment. Similarly, silage sorghum, corn, legumes, and grass hay are often stored partially fermented in silos or other sheltered and closed facilities. During storage, these wet feeds lose water that drains to the surrounding soil. Effluent composition varies with the feed, and may be acidic and high in soluble nutrients. Collection structures for effluent would minimize pollution risks to surface water impoundments, streams, and to groundwater.

Forage crops are frequently irrigated with municipal wastewater to enhance water disposal and dry matter accumulation. These production practices allow the scavenging of nutrients contained in wastewaters as well as clarifying the waterstream prior to reentry into natural streams and rivers (Figure 8). Oftentimes, the needs for water disposal strain the limits of hydraulic load of the land application systems. System failure results in runoff, particularly during the cold winter months when evapotranspiration is low and the soil may freeze. There is the need to cope with the changing soil hydraulic characteristics during the lifetime of the disposal systems in addition to management of water-loading rates.[66] Reductions in infiltration and the ensuing decline in system capacity are correlated with particulate and anoxic conditions that build up in the soil. These changes modify the transformation processes affecting nutrients, and other soil amendments. For example, nitrogen loss pathways are shifted to atmospheric emissions of denitrification products. Therefore, volatilization may become a proportionately larger pathway for nutrient removal than absorption by vegetative surfaces until the advent of soil drainage remediation such as the addition of gypsum.

2. Integrated Cropping-Livestock Production Systems

Effective integration of cropping and livestock systems exists in ancient agriculture as evidenced from archeological finds. Today, we can observe very successful mixed production systems on Amish or Mennonite farms in parts of the U.S. and in many developing countries.

The basic system relies on the cycling of resources, where the outputs of one sector provide the production inputs of the other sector. For example, large livestock production enterprises are generally associated with crop production land areas of the Corn Belt. Animal manure is cycled into crop production for soil improvement and to reduce reliance on commercial fertilizers. The recycling practice also lowers the potential for pollution from livestock wastes.

Grazing of winter cereals, in particular winter wheat during its vegetative growth stage, is a common practice in a six-state region in the southern Great Plains. The area includes parts of Colorado, New Mexico, Nebraska, Kansas, Oklahoma, and Texas, and is known as the winter wheat pasture area. Wheat is produced for the dual purpose of providing winter forage for grazing ruminant livestock and for grain. The vegetative growth phase of the crop provides a production input for the animal production sector. The combined economic return from wheat includes both that of the grain crop and those of animal products. Such a production system requires different management practices than a crop grown strictly for grain.

Although mutual influence occurs in such a mixed system, we know more about how to manage the cropping sector to sustain productivity as it directly influences the feed supplies to the animal sector than we seem to know about the effects of livestock management on the crop production sector. Animal impacts include the effects of devegetation, erratic deposition of manure, animal traction, and soil trampling which in turn influence the physical, chemical, and biological characters of the agroecosystem. For example, the nutrient flow between plant-soil-animal compartments, plant species composition and population dynamics in response to grazing, plant and soil water relations, and soil hydraulic and landscape hydrological properties are all modified by livestock grazing and traction.

The impact of animal traction on the stability of the cropping sector is unclear. In these systems, livestock are released in wheat pastures as soon as feasible. It is not uncommon to find animals in the field when wheat is in the two- to three-leaf stage, particularly in years with dry autumns. When livestock trample the soil where they browse or get their water, they cause substantial increase in soil bulk density when compared to that of nongrazed lands.[67,68] Trampling degrades soil structure by the churning action of hooves under a load that also depends upon animal species. Animal traffic causes shallower compaction of soils than that induced by farming equipment. However, the area trampled by livestock may be more extensive.

Trampling occurs for extended periods of time under varied climatic and soil conditions. Initial stocking rates range from 350 to 900 kg initial live-weight per hectare in the fall. In the spring, animal density may reach 2500 to 3000 kg ha^{-1}. Bulk densities of 1.4 to 1.6 Mg m^{-3} are commonly measured in hoof prints left by animals weighing between 140 to 230 kg. By the end of the grazing period, the surface soil is compacted and massive in the top 0.10 to 0.15 m. Compacted soil restricts water, air, and plant root penetration, and thereby restricts vegetation growth. Reduced water infiltration into compacted soil often resulted in overland flow and runoff from grazing lands. As a result, amendments may be transported by runoff to surface water supplies. Producers who use intensive tillage continue to use the plow to disrupt the compacted soil after livestock grazing. Chiseling is often recommended to producers who practice conservation tillage methods.[68] Nonetheless, grazing management must be optimized to keep defoliation rates appropriate to the landscape, vegetation, and soil conditions of the site. Maintenance of appropriate vegetative biomass and cover insure that plant vigor, regrowth capacity, and nutrient and water use efficiency are maintained high and thus prevent soil degradation and erosion.

Land treatment with animal manures can restore good soil physical conditions on grazing lands as well as croplands. However, in either case, proper credits must be made for residual N. Precise knowledge of crop requirements and yield goals, nutrient characteristics of the amendment source, and timing of field application in relation to weather conditions may minimize contamination risks to surface and underground water supplies. Soil incorporation or subsurface injection of liquid sources as opposed to surface application minimizes environmental risks from the use of livestock wastes. There is an acute need for developing optimized land application guidelines. New less labor-intensive technologies are also needed to quantify the magnitude of volatilization and seepage losses from storage areas, holding ponds, and land applications as a function of

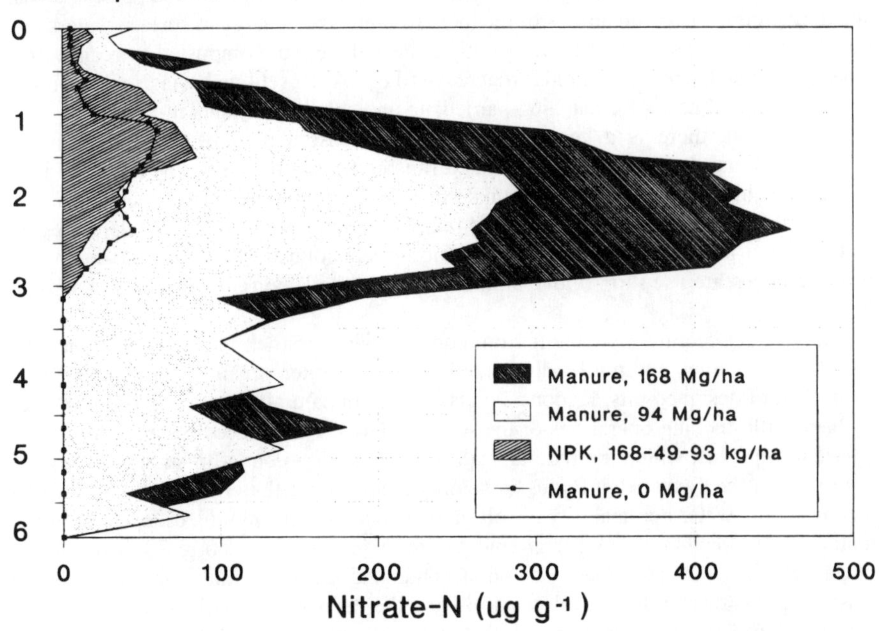

Figure 9 Nitrate-N distribution in soil following 5 years of cattle and poultry manure applications to a Rhodic Paleudult. (From Cooper, J.R., Reneau, R.B., Jr., Kroontje, W., and Jones, G.D., *J. Environ. Qual.*, 13, 189, 1984. With permission.)

climate variables, structure design, and hydrogeology. The eventual knowledge gained may lead to its enhanced use as a nutrient resource in agronomic production as opposed to a waste material to be disposed of.

Likewise, the use of municipal wastewater, sludge, and solid wastes on croplands may be handled in a similar manner. Proper credits of soil N and nutrient and toxic metal contents for sludge should receive particular attention in establishing application rates.[69] This is particularly important for range plants that do not have a high input requirement to preserve the ecological balance of the rangelands. Soil incorporation, when feasible, should be practiced to minimize nutrient overload and volatile losses, as well as minimizing direct ingestion by grazing livestock and wildlife.

B. CONCENTRATED LIVESTOCK AND POULTRY PRODUCTION SYSTEMS

A large population of livestock and poultry exists on numerous small farms and ranches in the Southeast. Predominant species include dairy cattle, beef cattle, swine, broilers, turkeys, and laying hens. Large quantities of animal manures are produced for a region where cropland and grassland are in short supply for land application. Nutrients, feed and antibiotic-derived metals,[70] salts, and enteric pathogens of livestock[71] are potential threats to surface and groundwater resources of this area of high rainfall.[62]

In the northeast, much of the farmland is sloping, often bordered by small streams. The growing season is relatively short and the ground is snow covered and frozen for long periods of the year. During the winter, livestock are kept in confined holding areas, and they produce large quantities of manure. To prevent discharge to surface waters, effluent and solid wastes must be frequently collected, stored, composted to reduce volume, and applied to prepared fields. Vegetative filter strips,[72] settling basins, and detention ponds provide a means for catching particulate and soluble nutrients in runoff.[73,74]

In contrast, there is a general trend toward consolidation of ownership in the Southwest. Large holding and concentrated finishing cattle feedlots exist owing to land availability. In an active feedlot, the soil surface develops a compacted manure-soil layer that restricts the infiltration of water and retards the leaching of salts, metals, nitrogen, and other nutrients of livestock manure.[75] Appropriate collection structures and containment practices are used to remove animal wastes while leaving intact this surface-seal layer.

Most states require that runoff from concentrated livestock feeding operations and dairies be contained and not be discharged directly to enter streams. Containment may be in custom-designed pits, lagoons, ponds, etc., or in naturally occurring lakes as used by beef cattle-feeding operations of the southern Great Plains. Feedlots are constructed adjacent to playas, which are natural shallow lakes with no outlet to streams. They are put to use in the Texas High Plains as holding areas for feedlot effluent and runoff collection.[76] Low water permeability of soil in the playa bottoms minimizes leaching of potential contaminants to the underground aquifer, which is about 60 to 70 m below the playa bottom. Other man-made detention ponds and lagoons containing animal liquid waste and wastewater also develop a seal from the deposition of colloidal organic matter and microbial activity and growth that clog soil pores along the sides and bottom of the structures.[74,77] Runoff from stored manure must also be contained, usually by the same structures as above. Runoff collected in playas generally evaporates because potential evaporation greatly exceeds precipitation in the region where playas exist. Runoff collected in pits and lagoons often is spread on cropland or pastures where plants utilize the nutrients.

Rates of seepage and leaching from cattle feedlot surfaces and natural playas are usually low (<10^{-8} m s^{-1}). However, Miller[76] has observed seepage rates ranging between 10^{-8} and 2.1×10^{-7} m s^{-1} for selected feedlots in the High Plains of Texas. Nitrate leaching from animal manure has been found under feedlots and manure stockpiles and adjacent to the runoff-containment structures (Figure 9). In addition to overall management, the rate of leaching is greatly affected by soil conditions and precipitation. Leaching is most likely on permeable soils when precipitation is relatively high. Leaching is slight on the slowly permeable soils of West Texas where precipitation is limited.

In addition to leaching losses, ammonia volatilization from surface-applied manure slurry accounts for 19 to 80% of total ammoniacal N in manure.[78] Fluxes of 12 kg N ha^{-1} h^{-1} are observed within the first hours after application,[79] indicating that more than half of the total ammonia losses from cattle slurry may occur during the first day after application. Volatile loss may be drastically reduced if slurries are injected into the soil or acidified prior to land application.[79,80] Where cultivation is consistent with soil management practices, incorporation immediately after application, even to a shallow depth, would drastically reduce the loss of the more volatile components. Malodorous components of livestock manure can be controlled by similar subsurface injection or by drying either by weather conditions conducive to drying or the use of drying agents. Sulfur-containing compounds are formed under the anoxic or low oxidation-reduction (0- to 200-mV) conditions that occur in soils treated with high rates of beef cattle and swine manure slurries.

C. PERMANENT PASTURE AND RANGE MANAGEMENT PRACTICES

Extensive livestock grazing operations in the U.S. are distributed across sparsely vegetated rangelands and forested lands of the western states in comparison to the more intensively managed pastures of the eastern and southeastern regions of the U.S. For the more fragile environments, livestock production may be the best agricultural enterprise that fits the landscape, particularly for erosion-prone lands. This requires proper matching of farming and ranching systems to the physical characteristics and limitations of the lands, and dictates careful management of the grazing animals. Range vegetation is a diverse plant community that is in equilibrium with the soil, water, temperature, and other climatic factors of the site. Severe ecological damage has been done to rangeland in the past century by misuse and ignorance of their fragility. Although there has been a general improvement in range conditions, about 60% of the nation's rangelands are still in fair to poor condition.[2] Thus, the most appropriate management approach to these natural ecosystems would be to judiciously control the frequency and intensity of grazing and the time of year they could be put to use. Good management not only would improve forage availability for livestock and wildlife grazing, and preserve and improve wildlife habitat, but may also enhance the quantity and quality of water that originates from rangeland watersheds.

Manure production and distribution in these extensive animal production enterprises may vary by two to three orders of magnitude as animal density varies from 10 to 5600 kg live-weight per hectare, depending on the climate, soils, topography, and grazing management across the nation. Cycling of nutrients and organic matter from deposited animal wastes promotes biological activity and slowly builds structure and productivity of the fragile soil ecosystem. Although animal excreta is deposited in a highly variable manner on grazed lands,[81,82] ruminant livestock return about 75 to 90% of the N ingested in consumed forages to the soil.[83,84] The spatially variable deposition can be concentrated to certain parts of the field such as shade areas, traffic paths, and drinking areas. Elevated concentrations of extractable phosphorus and potassium are observed near water sources and extend from 10 to 20 m into intensively managed pastures.[85] Net transfer of nutrients by grazing animals from the total area of a pasture or range to a small proportion of the total area (about 3 to 6%) can pose potential off-site risks to bordering or underlying water resources.

Deleterious effects of grazing animals to streams and water impoundments can be mitigated by restricting their access and maintaining adequate vegetative buffer or riparian zones. Vegetative filter strips are used as a living barrier to filter sediment, particulate matter, nutrients, and animal excreta and associated microorganisms from runoff and wastewaters.[72,86–89] Upon entering a vegetative filter strip, suspended materials in runoff are effectively removed by deposition, filtration, absorption, adsorption, leaching, and volatilization.

Whether filter strips are effective tools to remove soluble contaminants is uncertain. On one hand, nitrate-N is filtered and immobilized by riparian zones of an agricultural watershed.[90] On the other hand, filter strips retain only 15% of ammonia-N from dairy barnyard runoff and N retention is lowest during snowmelt periods.[91] A computer simulation of the performance of vegetative filter strips for 230 actual field conditions, using the Chemicals, Runoff, and Erosion from Agricultural Management Systems (CREAMS) model, indicates that 10 to 80% of particulate matter and sediment-associated chemicals are removed by filter strips.[92] The effectiveness of a filter strip is dependent upon soil slope, slope profile shape, and vegetation density. However, the simulation showed that entrainment of dissolved nutrients and pesticides is not affected by a filter strip.[92] It seems that the residence time of the contaminant in a vegetative filter zone plays an important role in the zone's efficiency for removing soluble nutrients. How much is removed will depend on how fast contaminants pass through the vegetative barrier. For example, mass retention may be high for small and intermittent water applications, whereas the percentage of retention of soluble contaminant declines as hydraulic loading rate increases.

V. MODELING THE OFF-SITE EFFECTS OF AGRONOMIC PRACTICES

A. SYSTEM INTEGRATION TO EVALUATE FARMING SYSTEMS AND DEVELOP BEST MANAGEMENT PRACTICES

Mathematical representations of processes acting on farm inputs serve a fundamental need to integrate multidisciplinary aspects that characterize a particular agrosystem. The number of variables and relevant factors are substantial and the interactions between the various aspects are complex. This leads to an increasing reliance on computers and computer models to integrate the individual pieces of information into a comprehensive understanding of the fate of soil amendments in agriculture. During the last decade, substantial advances have been made in computer models to predict behavior of particulate, chemical, and biological contaminants in the environment. By the same token, models are tools for simulating the fate of a new amendment or new uses of an existing one to project the likely ecological impact of such a use.

Computer models are also used to make large-scale assessment of agricultural management systems and to develop best management practices. They can pinpoint specific knowledge gaps, given the vast number of combinations of soil amendment types and forms, soil, crop, and climate to assist producers, researchers, and resource managers in designing effective management strategies. Alternative decisions can be made based on the relative merits of various management options. The short-term benefit can be maximized when an optimal economic yield is achieved with a minimum input of amendments for the highest levels of crop growth and pest control. The long-term benefit to the environment and society would be the sum total of optimized farming systems that are based on informed analyses rather than primarily based on experience and economics.

B. SELECTED MANAGEMENT MODELS

A detailed consideration of existing models is precluded because they are numerous and each is intended for a specific use. They represent a wide range of temporal and spatial scales. It is left to the user to find and apply the appropriate model to one's application. In this discussion, reference is limited to models that are sensitive to changes in management practices. Further reference to model assumptions, simplifications, model purpose, structure, and data requirements can be found in excellent past reviews.[93–96]

Oftentimes, it is neither feasible nor practical to employ a screening model that is broad or a comprehensive research model that is data intensive to guide management of soil amendments at a particular site. Yet, there are attempts at streamlining research models in such a manner that inputs are easily understood and more readily available, and that model output is organized as to be readily interpreted to guide water management and amendment applications under field conditions.

The CREAMS model may be used to make relative comparisons of soil and chemical contaminant loads from alternative management practices on field-scale areas.[97] The model has been successfully applied to a number of production systems to compare the effects of management practices for corn[98] or watershed land treatments and vegetative filter strip performance.[92] CREAMS has been modified to address one-dimensional distribution of agricultural chemicals and is called GLEAMS (Groundwater Loading Effects of Agricultural Management Systems).[99]

The Erosion Productivity Impact Calculator (EPIC) has been developed to correlate productivity loss to soil erosion over long periods of time. The model incorporates hydrology, weather, erosion, plant growth, nutrients, temperature, and tillage, and makes an economic analysis of erosion loss and the sustainability of production systems.[100]

The Simulation of Production and Utilization of Rangelands (SPUR) model addresses both the field- and basin-scale responses to environmental factors and plant and animal management practices, and calculates the economic impacts of the management decisions of these resources.[101]

Nitrogen, Tillage, and Residue Management (NTRM) is a comprehensive model that was developed to evaluate practices of management of nitrogen sources in conventional and conservation tillage systems.[102]

The Nitrogen Leaching and Economic Analysis Package (NLEAP) has been developed to estimate N loss as mitigated by climate, soils, and agronomic management practices.[103] The model allows projection of economic impact of N management practices and environmental contamination of aquifers. Field assessments of NLEAP provided reasonable estimates of post-harvest soil nitrate levels for irrigated crops along the South Platte River.[104]

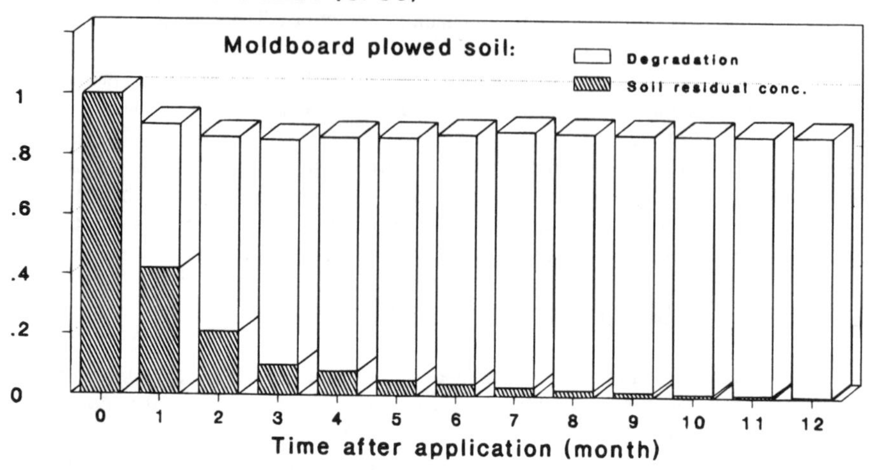

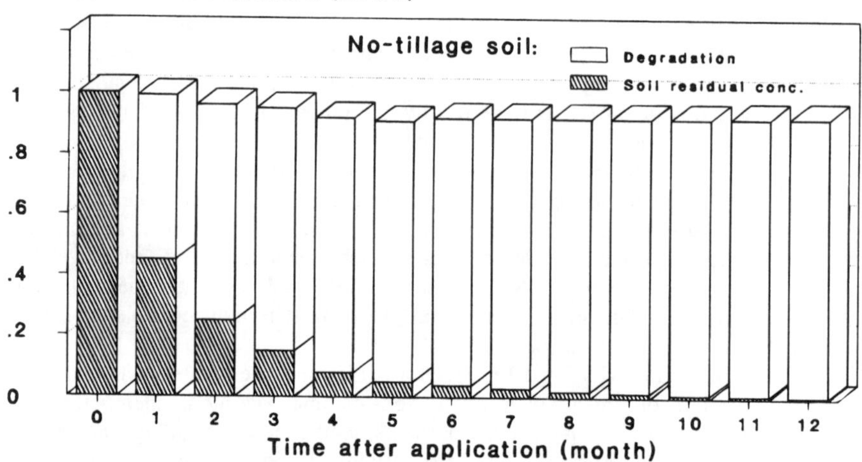

Figure 10 Pesticide Root Zone Model (PRZM) simulation of the fate of annual applications of metribuzin to plowed and no-tilled Bethany silt loam under winter wheat: effect of tillage method on herbicide degradation in the 0- to 1.5-m depth. (From Dao, T.H., *Research Trends in Agricultural Science, Soil Science*, Council Sci. Res. Integration, Trivandrum, India, 1, 9, 1993. With permission.)

DRASTIC is a qualitative tool to evaluate the potential vulnerability of any hydrogeological setting.[105] It is an acronym that stands for seven parameters thought to be the most important factors controlling groundwater contamination potential: depth to water (D), recharge (R), aquifer material (A), soil (S), topography (T), impact of vadose zone (I), and the aquifer hydraulic conductivity (C). Each factor is assigned a weight that describes its importance in the contamination process to arrive at an overall numerical

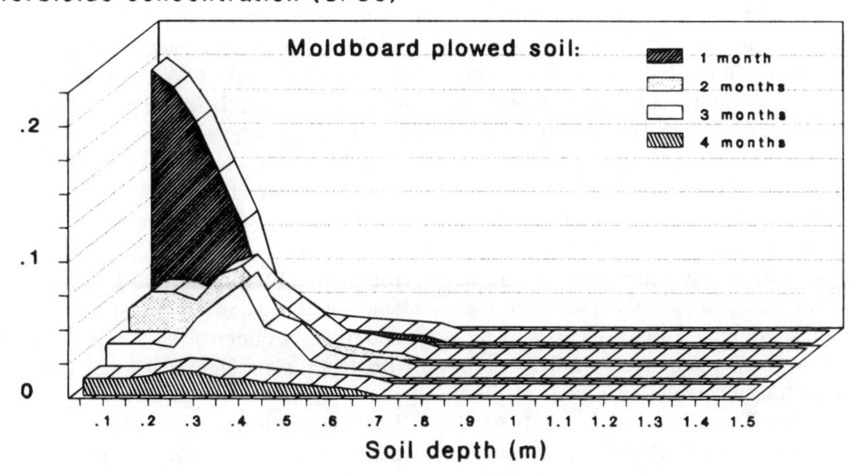

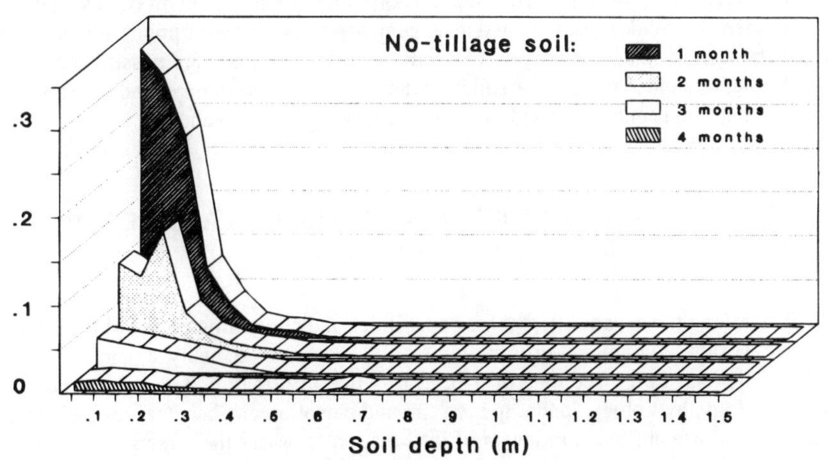

Figure 11 Pesticide Root Zone Model (PRZM) simulation of the fate of annual applications of metribuzin to plowed and no-tilled Bethany silt loam under winter wheat: temporal and soil distribution of residual herbicide in the 0- to 1.5-m depth. (From Dao, T.H., *Research Trends in Agricultural Science, Soil Science*, Council Sci. Res. Integration, Trivandrum, India, 1, 9, 1993. With permission.)

index. A ranking method and nonparametric statistics are used to compare relative groundwater contamination potentials.

The Pesticide Root Zone Model (PRZM) has been developed by the USEPA[106,107] to assess the transient transport of pesticides in the root zone as its name implies. The model uses a simplified water balance based on the Soil Conservation Service curve number system to describe surface management practices. PRZM allows optional estimation routines for some model parameters such as soil property changes with time and kinetic rates of pesticide degradation other than first-order reaction rates. The model has been used with limited success in an assessment of the ability of residue management practices in attenuating the impact of climate variability and management-induced soil alterations on the field behavior of metribuzin (Figures 10 and 11).[42,108] A new version is under development and will incorporate simultaneous accumulation of parent chemicals and metabolites and calculation of volatilization loss (Carsel, personal communication).

The Leaching Estimates and Chemistry Model [LEACHM-(N, P)] is a research process-based model of water and distinct modules for nitrogen and pesticide fates, plant uptake, and transformations in the unsaturated zone.[109] The water balance is based on a complex iteration of the Richards equation. A field assessment of transformation and transport of the nematicide aldicarb has been performed with LEACHM.[110] The results shows general agreement between measured and simulated soil water and total residues concentrations. Discrepancies may be attributed to descriptions of hydraulic conductivity, water content, and suction relationship with depth, plant uptake of water, and pesticide transformation kinetics.

The Root Zone Water Quality Model (RZWQM) has been developed by the USDA-ARS to compare agricultural chemical management practices. The model utilizes an expert system approach to describe the movement of water, including macropore flow, the environmental fate, and transport of nutrients and pesticides at the field-scale, primarily in the vertical dimension.[111] The model incorporates plant growth processes, soil chemical, nutrient, pesticide processes, and management processes. Descriptions of management activities such as tillage, irrigation, cultivation, and their timing are possible because algorithms are included to describe soil bulk density changes with time, and rainfall events. Field evaluation is underway and provisions are being made to address two-dimensional transport processes.

C. GEOGRAPHIC INFORMATION SYSTEMS (GIS) AND RESOURCE AND ENVIRONMENTAL MANAGEMENT

Employing the expanding graphic capabilities and computing power of the computer science world, geographical information systems can encode, analyze, and display geographically based digitized data of natural resource information. Georeferenced data of soils, geology, hydrology, landscape terrain, and elevation can be superimposed on agricultural land use attributes in layers of information to assess the effects of management options and the impact of alternative decisions. For example, GIS can provide a rapid means to assess potential environmental problem areas, to formulate compliance planning, and to monitor changes by performing regional and global modeling. The analytical results may be in the form of descriptive reports, tables, or colored and textured maps at any desired scale. For example, the vulnerability of groundwater to chemical contamination of a unique region of the nation, i.e., the Pearl Harbor watershed, Hawaii[112] can be constructed from layers of information about soils, topography, land cover, roads, streams, and field boundaries.

There are several commercial GIS software packages such as ARC/INFO, MAP-INFO, and INTERGRAPH. In the public domain, an interactive Unix-based GIS software, called Geographic Resources Analysis Support System (GRASS), has been developed by the U.S. Army Corps of Engineers, Construction Engineering Research Laboratory (USA-CERL), and has for some time seen wide applications in resource man-

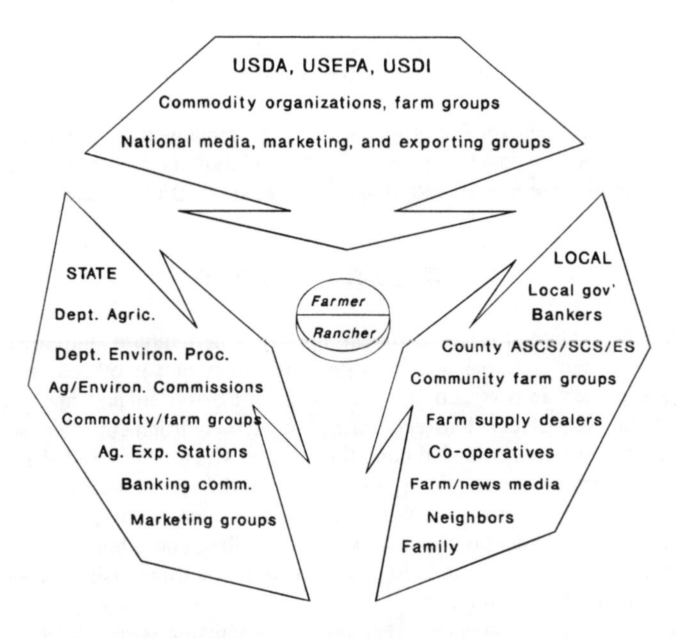

Figure 12 Institutional and peer groups that may influence the production and environmental decision-making process on the farm/ranch.

agement by the Corps of Engineers (i.e., hydrologic investigations of the Pinon Canyon, Las Animas County, Colorado).[113] GRASS is used by university departments of geology, geography, anthropology, archeology, and landscape architecture in site mapping as well as by a large number of federal agencies and private sector groups. In agriculture and natural resource management, GRASS has been applied to digitize soil surveys, and to perform natural resource inventory and analyses (i.e., NRI SOILS-5), farm or ranch conservation planning, watershed planning and monitoring, and water quality assessment at county, regional, and national levels.[114]

The most fundamental data requirements of GIS are the spatially coordinated soil, geology, and hydrography databases at all levels of detail. Their unavailability is the major impediment to a wide use in natural resource conservation and environmental planning. The Soil Conservation Service is in the process of updating and digitizing county soil surveys for the nation. The USEPA has been maintaining a digital file of surface water networks of the U.S. (REACH) based on USGS hydrography data. Digitized hydrologic features and land surface elevation data exist at several scales and resolutions.[115] Aquifer boundaries and hydraulic properties are found in the USGS WATSTORE database.[116] Due to the high level of activities, there are reported differences in encoding standards in addition to the completeness of the process across the country that would impede the task of assembling the information into a national system. There is also the matter of public accessibility to the various databases. Nonetheless, the "open architecture" approach to GIS should be promoted as it would allow the combination or addition of new databases and new modular computer models, expert systems, and decision-aid tools to the core code, thus forming a powerful information management system. The practice would allow inclusion of local details such as weather, soils, landscape, geology, cropping practices, and animal production practices as well as integrating these details into more regional or national databases for resource management and environmental protection strategies.

The validation process for GIS outputs, however, remains the central and continuing issue of debate as it is in any other means of simulation modeling of agroecosystem performance, the management of farm inputs, and environmental impacts.

VI. A CLOSING THOUGHT

The behavior and fate of soil amendments applied to agricultural lands are controlled by climate, soil, tillage, livestock and/or cropping practices, etc., in other words, the total management approach. In perspective, each strategy is agrosystem and site specific. The agronomic practices in effect extensively define the microenvironment where soil amendments will reside, and the ecological conditions under which they may behave and disperse as time progresses. Oftentimes, bioactivity, pathways of transformations, and cross-media transport are modified as we change from one practice to an alternative practice. A predominant fate process may yield to another, and a holistic evaluation of the total farming system alone can help us gain a comprehensive insight of amendment behavior and fate in the environment.

Nonetheless, a set of strategies to agricultural production is to a large extent producer specific. This human factor is the true unknown variable in the environmental impact equation. We may be able to model the physical, chemical, and biological processes of crop or animal growth, climate changes, and the physical environment, but we cannot yet predict the subjectivity of the human operator. How he/she chooses to farm and makes each individual operational decision will dictate how much, if any, soil amendments will be dispersed in the environment.

The challenge is to understand the thought process and the decision-making process, as well as the motives of the producer. Germane institutional and peer spheres of influence are complex (Figure 12). Fortunately, as we have gleaned from suggested solutions and insights presented throughout the previous sections, there are a great number of technology tools and system-oriented innovations to aid the producer in choosing options that help him/her stay in business and in compliance with current environmental laws and regulations. The long-term benefits to the environment and society will be derived from a willingness to change, and will be equal to the sum total of optimized farming systems.

REFERENCES

1. Carson, R.L., *Silent Spring*, Fawcett, Greenwich, CT, 1962.
2. U.S. Department of Agriculture, *The Second RCA Appraisal. Soil, Water, and Related Resources on Non-Federal Land in the U.S. Analysis of Condition and Trends*, U.S. Government Printing Office, Washington, D.C., 1989.
3. U.S. Environmental Protection Agency, *Chesapeake Bay Program: Findings and Recommendations*, U.S. EPA, Philadelphia, 1983.
4. Novotny, V. and Chesters, G., *Handbook of Non-Point Pollution: Sources and Management*, van Nostrand Reinhold, New York, 1981.

5. Wauchope, R.D., The pesticide content of surface water drainage from agricultural fields: a review, *J. Environ. Qual.*, 7, 459, 1978.
6. Goolsby, D.A., Coupe, R.C., and Markovchick, D.J., Distribution of Selected Herbicides and Nitrate in the Mississippi River and its Major Tributaries, April through June 1991, U.S. Geological Survey, Water Resour. Invest. Rep. 91-4163. U.S. Government Printing Office, Washington, D.C., 1991.
7. U.S. Department of Agriculture, Atrazine Task Group, USDA Working Group on Water Quality, Atrazine in Surface Waters. A Report to the USDA Working Group on Water Quality, U.S. Government Printing Office, Washington, D.C., 1992.
8. Cohen, C., Eiden, C., and Lorber, M.N., Monitoring groundwater for pesticides, in *Evaluation of Pesticides in Groundwater*, Garner, W.Y. et al., Eds., ACS Symposium Series 315, American Chemical Society, Washington, D.C., 1986, 170.
9. U.S. Environmental Protection Agency, Pesticides in Groundwater Database. 1988 Interim Report, U.S. EPA, Office Pesticide Programs, Washington, D.C., 1988.
10. U.S. Environmental Protection Agency, National Survey of Pesticides in Drinking Water Wells. Phase I Report, U.S. EPA 570/9-90-015, Washington, D.C., 1990.
11. Lyles, L. and Woodruff, N.P., Abrasive action of windblown soil on plant seedlings, *Agron. J.*, 52, 533, 1960.
12. Soil Conservation Service, USDA, Facts about Wind Erosion and Dust Storms on the Great Plains, Publication leaflet No. 394, U.S. Government Printing Office, Washington, D.C., 1961.
13. U.S. Environmental Protection Agency, National Air Quality and Emission Trend Reports, U.S. EPA, Research Triangle Park, NC, 1990.
14. Alber, D.G. and Shortle, J.S., Cross compliance and water quality protection, *J. Soil Water Conserv.*, 44, 453, 1989.
15. Office of Technology Assessment, U.S. Congress, Beneath the Bottom Line: Agricultural Approaches to Reduce Agrichemical Contamination of Groundwater, OTA-F-418, U.S. Government Printing Office, Washington, D.C., 1990.
16. U.S. Department of Agriculture, *Agricultural Statistics*, 1991. U.S. Government Printing Office, Washington, D.C., 1991.
17. Smith, S.J., Sharpley, A.N., Naney, J.W., Berg, W.A., and Jones, O.R., Water quality impacts associated with wheat culture in the Southern Plains, *J. Environ. Qual.*, 20, 244, 1991.
18. Zhu, J.C., Gantzer, C.J., Anderson, S.H., Albert, E.E., and Beuselinck, P.R., Runoff, soil, and dissolved nutrient losses from no-till soybean with winter cover crops, *Soil Sci. Soc. Am. J.*, 53, 1210, 1989.
19. Krieg, D. R. and Lascano, R. J., Sorghum, in *Irrigation of Agricultural Crops*, Stewart, B.A. and Nielsen, D. R., Eds., Agronomy 30, American Society of Agronomy, Madison, WI, 1990, 719.
20. Sharpley, A.N., Smith, S.J., Williams, J.R., Jones, O.R., and Coleman, G.A., Water quality impacts associated with sorghum culture in the Southern Plains, *J. Environ. Qual.*, 20, 239, 1991.
21. Hammond, M.W., Cost analysis of variable fertility management of P and K for potato production in central Washington, presented at Workshop on Research and Development Issues in Soil-Specific Crop Management, Minneapolis, MN, April 14–16, 1992.
22. Smith, J.H. and Peterson, J.R., Recycling of nitrogen through land application of agricultural, food processing, and municipal wastes, in *Nitrogen in Agricultural Soils*, Stevenson, F.J., Ed., (Agronomy 22), American Society of Agronomy, Madison, WI, 1982, chap. 21.

23. Miller, D.E., Aarstad, J.S., and Evans, R.G., Control of furrow erosion with crop residues and surge-flow irrigation, *Soil Sci. Soc. Am. J.*, 51, 421, 1987.
24. Conservation Technology Information Center (CTIC), Tillage definitions, *Conserv. Impact*, 8, 7, 1990.
25. Schertz, D.L., Conservation tillage: an analysis of acreage projections in the U.S., *J. Soil Water Conserv.*, 43, 256, 1988.
26. Dao, T.H., Tillage and winter wheat residue management effects on soil water infiltration and storage, *Soil Sci. Soc. Am. J.*, 57, 1586, 1994.
27. Mapa, R.B., Green, R.E., and Santo, L., Temporal variability of soil hydraulic properties with wetting and drying subsequent to tillage, *Soil Sci. Soc. Am. J.*, 50, 1133, 1986.
28. Ehlers, W., Observations on earthworm channels and infiltration on tilled and untilled loess soil, *Soil Sci.*, 119, 242, 1975.
29. Edwards, W.M., Norton, L.D., and Redmond, C.E., Characterizing macropores that affect infiltration into non-tilled soil, *Soil Sci. Soc. Am. J.*, 52, 483, 1988.
30. Clothier, B.E., Root zone processes and water quality: the impact of management, in *Proc. Int. Symp. Water Quality Modeling of Agricultural Non-Point Sources*, DeCoursey, D.G., Ed., USDA, ARS-81, Logan, UT, 1990, 659.
31. White, R.E., Dyson, J.S., Gerstl, A., and Yaron, B., Leaching of herbicides through undisturbed cores of a structured clay soil, *Soil Sci. Soc. Am. J.*, 50, 277, 1986.
32. Unger, P.W. and Parker, J.J., Evaporation reduction from soil with wheat, sorghum, and cotton residues, *Soil Sci. Soc. Am. J.*, 40, 938, 1976.
33. Idso, S.B., Reginato, R.J., Jackson, R.D., Kimball, B.A., and Nakayama, F.S., The three stages of drying of a field soil, *Soil Sci. Soc. Am. J.*, 38, 831, 1974.
34. Steiner, J.L., Tillage and surface residue effects on evaporation from soils, *Soil Sci. Soc. Am. J.*, 53, 911, 1989.
35. Fraser, D.G., Doran, J.W., Sahs, W.W., and Lesoing, G.W., Soil microbial populations and activities under conventional and organic management, *J. Environ. Qual.*, 17, 585, 1988.
36. Gupta, S.C., Larson, W.E., and Allmaras, R.R., Predicting soil temperatures and soil heat flux under different tillage-surface residue conditions, *Soil Sci. Soc. Am. J.*, 47, 1212, 1984.
37. Unger, P.W., Residue management effects on soil temperature, *Soil Sci. Soc. Am. J.*, 52, 1777, 1988.
38. Blevins, R.L., Thomas, G.W., and Cornelius, P.L., Influence of no-tillage and nitrogen fertilization on certain soil properties after 5 years of continuous corn, *Agron. J.*, 69, 383, 1977.
39. Cheng, H.H., Picloram in soil: extraction and mechanism of adsorption, *Bull. Environ. Contam. Toxicol.*, 6, 28, 1971.
40. Bell, P.F., James, B.R., and Chaney, R.L., Heavy metal extractability in long term sewage sludge and metal salt-amended soils, *J. Environ. Qual.*, 20, 481, 1991.
41. Unger, P. W., Organic matter, nutrient, and pH distribution in no- and conventional-tillage semiarid soils, *Agron. J.*, 83, 186, 1991.
42. Dao, T.H., Role of management systems in groundwater protection strategies: effects of soil and water conservation practices on field behavior of herbicides, *Research Trends in Agricultural Science, Soil Science*, Vol. 1, Council Sci. Res. Integration, Trivandrum, India, 9, 1993.

43. Stearman, G.K., Lewis, R.J., Tortorelli, L.J., and Tyler, D.D., Characterization of humic acid from no-tilled and tilled soils using carbon-13 nuclear magnetic resonance, *Soil Sci. Soc. Am. J.*, 53, 744, 1989.
44. Dao, T.H., Field decay of wheat straw and its effects on metribuzin and S-ethyl metribuzin sorption and elution from crop residues, *J. Environ. Qual.*, 20, 203, 1991.
45. Dao, T.H., Behavior and subsurface transport of agrochemicals in conservation systems, in *Groundwater Water Quality and Agricultural Practices*, Fairchild, D.M., Ed., Lewis Publishers, Chelsea, MI, 1987, chap. 13.
46. McBride, M.B., Transition metal bonding in humic acid: an ESR study, *Soil Sci.*, 126, 200, 1978.
47. Chaney, R.L., Crop and food chain effects of toxic elements in sludges and effluents, in *Recycling Municipal Sludges and Effluents on Land*, National Association of State Universities and Land-Grant Colleges, Washington, D.C., 1973, 129.
48. Hue, N.V., A possible mechanism for manganese phytotoxicity in Hawaiian soils amended with a low manganese sewage sludge, *J. Environ. Qual.*, 17, 473, 1988.
49. Hue, N.V., Silva, J.A., and Arifin, R., Sewage sludge-soil interactions as measured by plant and soil chemical composition, *J. Environ. Qual.*, 17, 384, 1988.
50. Gilliam, J.W. and Hoyt, G.D., Effect of conservation tillage on fate and transport of nitrogen, in *Effects of Conservation Tillage on Groundwater Quality*, Logan, T.J. et al., Eds., Lewis Publishers, Chelsea, MI, 1987, 217.
51. Schreiber, J.D. and McDowell, L.L., Leaching of nitrogen, phosphorus, and organic carbon from wheat straw residues. I. Rainfall intensity, *J. Environ. Qual.*, 14, 251, 1985.
52. Sharpley, A.N. and Smith, S.J., Mineralization and leaching of P from soil incubated with surface-applied and incorporated crop residue, *J. Environ. Qual.*, 18, 101, 1989.
53. Follett, R.F. and Schimel, D.S., Effect of tillage practices on microbial biomass dynamics, *Soil Sci. Soc. Am. J.*, 53, 1091, 1989.
54. Staley, T.E., Edwards, W.M., Scott, C.L., and Owens, L.B., Soil microbial biomass and organic component alterations in a no-tillage chronosequence, *Soil Sci. Soc. Am. J.*, 52, 998, 1988.
55. Doran, J.W., Soil microbial and biochemical changes associated with reduced tillage, *Soil Sci. Soc. Am. J.*, 44, 765, 1980.
56. Hickman, T.G. and Novak, J.T., Relationship between subsurface biodegradation rates and microbial density, *Environ. Sci. Technol.*, 23, 525, 1989.
57. Carr, P.M., Carlson, G.R., Jacobsen, J.S., Nielsen, G.A., and Skogley, E.O., Farming by soil, not fields: a strategy for increasing fertilizer profitability, *J. Prod. Agric.*, 4, 57, 1991.
58. Tyler, D.A., Positioning technology (GPS), presented at Workshop on Research and Development Issues in Soil-Specific Crop Management, Minneapolis, MN, April 14–16, 1992.
59. Smith, S.J., Mathers, A.C., and Stewart, B.A., Distribution of N forms in soil receiving cattle feedlot wastes, *J. Environ. Qual.*, 9, 215, 1980.
60. Cooper, J.R., Reneau, R.B., Jr., Kroontje, W., and Jones, G.D., Distribution of nitrogenous compounds in a Rhodic Paleudult following heavy manure application, *J. Environ. Qual.*, 13, 189, 1984.
61. Adriano, D.C., Pratt, P.F., and Bishop, S.E., Nitrate and salt in soils and groundwaters from land disposal of dairy manure, *Soil Sci. Soc. Am. Proc.*, 35, 759, 1971.

62. Liebhardt, W.C., Golt, C., and Tupin, J., Nitrate and ammonium concentrations of groundwater resulting from poultry manure applications, *J. Environ. Qual.*, 8, 211, 1979.
63. McCalla, T.M., Use of animal manure wastes as a soil amendment, *J. Soil Water Conserv.*, 29, 213, 1974.
64. Chang, C., Sommerfeldt, T.G., and Entz, T., Soil chemistry after eleven annual applications of cattle feedlot manure, *J. Environ. Qual.*, 20, 475, 1991.
65. King, L.D., Westerman, P.W., Cummings, G.A., Overcash, M.R., and Burns, J.C., Swine lagoon effluent applied to "coastal" bermudagrass. II. Effects on soil, *J. Environ. Qual.*, 14, 14, 1985.
66. Sigrist, R.L., Soil clogging during subsurface wastewater infiltration as affected by effluent composition and loading rate, *J. Environ. Qual.*, 16, 181, 1987.
67. Warren, S.D., Nevill, M.B., Blackburn, W.H., and Garza, N.E., Soil responses to trampling under intensive rotation grazing, *Soil Sci. Soc. Am. J.*, 50, 1336, 1986.
68. Dao, T.H., Morrison, J.E., Jr., and Unger, P.W., Soil compaction and bearing strength in residue management systems, in *Crop Residue Management To Reduce Erosion and Improve Soil Quality in the Southern Plains*, Stewart, B.A. and W.C. Moldenhauer, Eds., USDA, Conserv. Res. Rept. No. 37, U.S. Government Printing Office, Washington, D.C., 1994, 40.
69. U.S. Environmental Protection Agency, *Process Design Manual. Land Application of Municipal Sludge*, U.S. EPA-625/1-83-016, Cincinnati, OH, 1983.
70. Payne, G.G., Martens, D.C., Kornegay, E.T., and Linderman, M.D., Availability and form of copper in three soils following eight annual applications of copper-enriched swine manure, *J. Environ. Qual.*, 17, 740, 1988.
71. Shere, B.M., Miner, J.R., Moore, J.A., and Buckhouse, J.C., Indicator bacterial survival in stream sediments, *J. Environ. Qual.*, 21, 591, 1992.
72. Schwer, C.B. and Clausen, J.C., Vegetative filter treatment of dairy milkhouse wastewater, *J. Environ. Qual.*, 18, 446, 1989.
73. Edwards, W.M., Owens, L.B., and White, R.K., Managing runoff from a small paved beef feedlot, *J. Environ. Qual.*, 12, 281, 1983.
74. Miller, M.H., Robinson, J.B., and Gillham, R.W., Self-sealing of earthen liquid manure storage ponds, I. A case study, *J. Environ. Qual.*, 14, 533, 1985.
75. Mielke, L.N., Swanson, N.P., and McCalla, T.M., Soil profile conditions of cattle feedlots, *J. Environ. Qual.*, 3, 14, 1974.
76. Miller, W.D., *Infiltration Rates and Groundwater Quality beneath Cattle Feedlots, Texas High Plains*, Water Quality Office, U.S. EPA Water Pollution Control Res. Series No. 16060 EGS 01/71, Washington, D.C., 1971.
77. Rowsell, J.G., Miller, M.H., and Groenevelt, P.H., Self-sealing of earthen liquid manure storage ponds. II. Rate and mechanism of sealing, *J. Environ. Qual.*, 14, 539, 1985.
78. Sommer, S.G. and Olesen, J.E., Effects of dry matter content and temperature on ammonia loss from surface-applied cattle slurry, *J. Environ. Qual.*, 20, 679, 1991.
79. Pain, B.F. and Thompson, R.B., Ammonia volatilization from livestock slurries applied to land, in *Nitrogen in Organic Wastes Applied to Soils*, Hansen, J.A. and Henriksen, K., Eds., Academic Press, San Diego, 1989, 202.
80. Hoff, J.D., Nelson, D.W., and Sutton, A.L., Ammonia volatilization from liquid swine manure applied to cropland, *J. Environ. Qual.*, 10, 90, 1981.

81. Peterson, R.G., Lucas, H.L., and Woodhouse, W.W., Jr., The distribution of excreta by freely grazing cattle and its effect on pasture fertility. I. Excretal distribution, *Agron. J.*, 48, 440, 1956.
82. Afzal, M. and Adams, W.A., Heterogeneity of soil N in pasture grazed by cattle, *Soil Sci. Soc. Am. J.*, 56, 1160, 1992.
83. Wilkinson, S.R. and Lowry, R.W., Cycling of mineral nutrients in pasture ecosystems, in *Chemistry and Biochemistry of Herbage*, Vol. 2, Butler, G.W. and Bailey, R.W. Eds., Academic Press, New York, 1973, 247.
84. Ball, R., Keeney, D.R., Theobald, P.W., and Nes P., N balance in urine-affected areas of a New Zealand pasture, *Agron. J.*, 71, 309, 1979.
85. West, C.P., Mallarino, A.P., Wedin, W.F., and Marx, D.B., Spatial variability of soil chemical properties in grazed pastures, *Soil Sci. Soc. Am. J.*, 53, 784, 1989.
86. Young, R.A., Huntrods, T., and Anderson, W., Effectiveness of vegetated buffer strips in controlling pollution from feedlot runoff, *J. Environ. Qual.*, 9, 483, 1980.
87. Dickey, E.C. and Vanderholm, D.H., Vegetative filter treatment of livestock feedlot runoff, *J. Environ. Qual.*, 10, 279, 1981.
88. Soil Conservation Service, USDA, *National Handbook of Conservation Practices*, U.S. Government Printing Office, Washington, D.C., 1984.
89. Jacobs, T.C. and Gilliam, J.W., Riparian losses of nitrate from agricultural drainage waters, *J. Environ. Qual.*, 14, 472, 1985.
90. Lowrance, R.R., Todd, R.L., and Asmussen, L.E., Nutrient cycling in an agricultural watershed. I. Phreatic movement, *J. Environ. Qual.*, 13, 22, 1984.
91. Schellinger, G.R. and Clausen, J.C., Vegetative filter treatment of dairy barnyard runoff in cold regions, *J. Environ. Qual.*, 21, 40, 1992.
92. Nicks, A.D., Williams, R.D., Krider, J.N., and Lewis, J.A., Simulation of filter strip effectiveness for non-point pollution control, presented at the Environmentally-Sound Agriculture Conference, Orlando, FL, April 16–18, 1991.
93. DeCoursey, D.G., *Proc. Int. Symp. Water Quality Modeling of Agricultural Non-Point Sources*, Parts I and II, USDA, ARS-81, Logan, UT, 1990.
94. Jones, R.L. and Hanks, R.J., Review of unsaturated zone leaching models from a user's perspective, in *Proc. Int. Symp. Water Quality Modeling of Agricultural Non-Point Sources*, DeCoursey, D.G., Ed., USDA, ARS-81, Logan, UT, 1990, 129.
95. Wagenet, R.J. and Rao, P.S.C., Modeling pesticide fate in soils, in *Pesticides in the Soil Environment: Processes, Impacts, and Modeling*, Cheng, H.H., Ed., Soil Sci. Soc. Am. Book Series 2, Soil Science Society of America, Madison, WI, 1990, chap. 10.
96. Leonard, R.A., Movement of pesticides into surface waters, in *Pesticides in the Soil Environment: Processes, Impacts, and Modeling*, Cheng, H.H., Ed., Soil Sci. Soc. Am. Book Series 2, Soil Science Society of America, Madison, WI, 1990, chap. 9.
97. Knisel, W.G., CREAMS: A Field-Scale for Chemicals, Runoff, and Erosion from Agricultural Management Systems, USDA Conservation Res. Rep. 26, U.S. Government Printing Office, Washington, D.C., 1980.
98. Crowder, B.M., Pionke, H.B., Epp, D.J., and Young, C.E., Using CREAMS and economic modeling to evaluate conservation practices: an application. *J. Environ. Qual.*, 14, 428, 1985.
99. Leonard, R.A., Knisel, W.G., and Still, D.A., GLEAMS: groundwater loading effects of agricultural management systems, *Trans. Am. Soc. Agric. Eng.*, 30, 1403, 1987.

100. Jones, C.A., Dyke, P.T., Williams, J.R., Kiniry, J.R., Benson, V.W., and Griggs, R.H., EPIC: an operational model for evaluation of agricultural sustainability, *Agric. Sys.*, 37, 341, 1991.
101. Wight, J.R. and Skiles, J.W., SPUR: Simulation of Production and Utilization of Rangelands. Documentation and User Guide, ARS-63, USDA-ARS, Boise, ID, 1987.
102. Shaffer, M.J. and Larson, W.E., NTRM, a Soil-Crop Simulation Model for Nitrogen, Tillage, and Crop Residue Management, USDA Conservation Res. Rep. 34-1, U.S. Government Printing Office, Washington, D.C., 1987.
103. Shaffer, M.J., Wylie, B.K., Follett, R.F., and Bartling, P.N.S., Using climate/weather data with NLEAP model to manage soil fertility and nitrate leaching, *Agron. Abstr.*, 23, 1992.
104. Wylie, B.K., Shaffer, M.J., and Brodahl, M.K., Validation of nitrate leaching and economic analysis package (NLEAP) using residual soil nitrate from irrigated croplands in Colorado, *Agron. Abstr.*, 24, 1992.
105. Dean, J.D., Jowist, P.P., and Donigan, A.S., Jr., *Leaching Evaluation of Agricultural Chemicals (LEACH) Handbook*, EPA-600/3-84-068. U.S. Government Printing Office, Washington, D.C., 1984.
106. Carsel, R.F., Smith, C.N., Mulkey, L.A., Dean, J.D., and Jowise, P.P., *User's Manual for the Pesticide Root Zone Model: Release 1*, U.S. EPA-600/3-84-109, U.S. Government Printing Office, Washington, D.C., 1984.
107. Carsel, R.F., Jones, R.L., Hansen, J.H., Lamb, R.L., and Anderson, M.P., A simulation procedure for groundwater quality assessment of pesticides, *J. Contamin. Hydrol.*, 2, 125, 1988.
108. Dao, T.H., Managing variability of climate and soil characteristics in conservation tillage systems: effects on field behavior of herbicides, presented at the Workshop on Research and Development Issues in Soil-Specific Crop Management, Minneapolis, MN, April 14–16, 1992.
109. Wagenet, R.J. and Hutson, J.L., *LEACHM: Leaching Estimates and Chemistry Model*, Continuum Vol. 2, Water Research Institute, Cornell University, Ithaca, NY, 1987.
110. Wagenet, R.J. and Hutson, J.L., Predicting the fate of non-volatile pesticides in the unsaturated zone, *J. Environ. Qual.*, 15, 315, 1986.
111. U.S. Department of Agriculture, Root Zone Water Quality Model. Version 1.0, Technical documentation, USDA-ARS Great Plains Systems Res. Tech. Rep. 2, Ft. Collins, CO, 1992.
112. Oki, D.S. Miyahira, R.N., Green, R.E., Giambelluca, T.W., Lau, L.S., Mink, J.F., Schneider, R.C., and Little, D.N., Assessment of the Potential for Groundwater Contamination Due to Proposed Urban Development in the Vicinity of the U.S. Navy Waiawa Shaft, Pearl Harbor, HI, Water Resources Res. Center, Special Report 03.02.90, University of Hawaii at Manoa, Honolulu, 1990.
113. Jahn, P.C., Investigating Pinon Canyon groundwater, GRASSCLIPPINGS Newsletter, *J. Geogr. Info. Sys.*, 6, 1, 1994.
114. Mitchell, K.M. and Pike, D.R., The use of a geographic information system to develop a comprehensive pesticide management plan, *Weed Sci. Soc. Am. Abstr.*, 78 (No. 234), 33, 1993.
115. National Cartographic Information Center (NCIC), Digital Cartographic and Geographic Data, U.S. Geological Survey, Reston, VA, 1985.
116. Baker, C.H., Jr. and Foulk, D.G., WATSTORE User's Guide, Groundwater File. Open-File Report 75-589, U.S. Geological Survey, Reston, VA, April 1984.

CHAPTER 12

Genetically Engineered Microbial Amendments in Soils

J. L. Wiebers, R. L. Hill, and J. S. Angle

I. INTRODUCTION

The rapidly evolving field of molecular biology has resulted in the production of genetically engineered microbes (GEMs) containing desirable agronomic and environmental attributes. Most GEMs that will be released into soil are bacteria that have been modified to express the production of foreign and unique proteins. For example, bacteria have been engineered to degrade a variety of recalcitrant organic materials.[1] Organisms containing the metabolic capability to degrade organic contaminants can then be used as a simple and inexpensive method to bioremediate contaminated soils. Engineered bacteria are also now available that can deliver toxic pesticides to specific populations of pests.[2] The use of GEMs in this manner may reduce the need for application of synthetic pesticides to entire ecosystems and the undesirable environmental consequences associated with the use of pesticides. Other traits such as enhanced ability to fix atmospheric nitrogen and the prevention of frost damage to crops have also been incorporated into the genome of bacteria that may find use in environmental and agricultural systems.[3,4]

0-87371-859-3/95/$0.00+$.50

We are on the verge of widespread practical use of recombinant organisms. There are currently a large number of recombinant organisms awaiting regulatory approval to be used in the field. Experience over the past 15 years has shown that these organisms generally only affect the specific target population and have little effect on nontarget species or the whole of the ecosystem. Most recombinant organisms have been weakened or disarmed so as to limit their exposure and activity in the ecosystem. Despite the consensus that most GEMs will have limited impact upon the environment, each potential field release must be considered on a case-by-case basis. Unforeseen environmental consequences and effects on nontarget organisms require extensive evaluation prior to granting permission for a specific release.

Several other concerns must be addressed prior to the release of a GEM into the soil. Transfer of recombinant genes into competent indigenous organisms is a concern since this would result in the incorporation of recombinant genes into bacteria adapted for survival in soil.[5,6] This transfer could potentially give the recipient organism a competitive advantage and thus affect the balance of the ecosystem. A second concern is whether the GEM will survive outside the laboratory for an extended period of time.[7] Although most GEMs are altered so that survival is restricted, unknown environmental influences could affect the potential for survival. A third concern, and one that has yet to receive extensive consideration, is whether the GEM will be mobile within the soil ecosystem and therefore lost to other environments, such as ground or surface water.[8] Runoff of a GEM from soil into surface water could pose a threat to a variety of nontarget organisms. For example, many bacteria have been engineered that produce *Bacillus thuringiensis* (Bt) toxins and these organisms will soon be applied to soil to control specific insect pests. If this organism were to be lost from the soil via runoff and into surface bodies of water, potential effects on nontarget aquatic invertebrates could result. A reduction in aquatic invertebrates would ultimately have an effect upon the entire aquatic ecosystem. Leaching of GEMs through the soil profile and into groundwater is another undesirable consequence that could be associated with a field release. Contamination of drinking water supplies with recombinant organisms is a serious concern and one that would restrict the use of these organisms. On the other hand, if the purpose of a field release was to remediate a contaminated aquifer, then it would be desirable for organisms to leach through the soil and into groundwater. Unless contact is made between a GEM and the contaminant, remediation is not possible. It is therefore important to understand the potential for leaching of a GEM through the soil and the processes that affect leaching.

Leaching of recombinant bacteria through the soil profile is obviously a serious concern and one that must be addressed when considering the safety of a potential field release. Organisms may have the potential to rapidly move through the soil profile if they are small in size or are surrounded by a non-reactive cell wall. Alternatively, large cells surrounded by thick polysaccharide capsules may move only a few centimeters into soil. Since the regulatory process for an impending release requires that all potential safety concerns be addressed prior to formal approval, each organism must be studied on a case-by-case basis. While an examination of physiological traits and simple survival and competition data may be readily obtained, an assessment of the potential for leaching is complicated and expensive. This assessment is often beyond the capability of the institution seeking approval for a release.

As an alternative to conducting studies for each organism in each soil, a modeling approach can be used to predict potential problems. To date, however, most models used to predict movement of microbes through soil have failed to be validated due to the extreme differences between soluble contaminants for which the models were designed and the particulate nature of microorganisms. In addition, microbes in soil may either grow or die off during the leaching process. Adsorption to soil colloids and organic matter is also significantly different for bacteria compared to chemical contaminants. The focus of the current chapter is therefore to examine some of the more common models that may be used to assess the movement of microbes through soil.

II. THE BACTERIA TRANSPORT PROCESS

Bacteria are transported in the soil both vertically and horizontally as a result of water movement.[9] The bacteria transport process is best described as a function of the combined forces of convection, dispersion, and retention. The effects of each of these forces on bacteria transport are a function of the soil environment. Properties that influence bacterial transport within soil include the volumetric water content of the soil, soil porosity, structure, vegetative growth on and within the soil, and pH. These properties exhibit increasing variation with increased field size. In general, larger fields are less homogeneous and more difficult to characterize than smaller fields. There is also much variation of physical properties between different fields. This variance makes it difficult to accurately predict the fate of microorganisms in the natural environment where local variation may significantly influence bacterial transport. Models designed to predict bacterial transport generally consider the effects of soil characteristics on the convection, dispersion, and retardation of bacteria within soil systems.

A. CONVECTION

Convection is the process by which nonsorbing substances move through the soil at an average rate equal to the average velocity of the flowing solution. Convection is sometimes used interchangeably with advection to describe transport in soils. Advection is defined as the process by which contaminants are transported by the soil solution, while convection is more accurately used to describe the movement of heat through a system.[9,10] Although advection is probably a more appropriate term to describe solute and particle movement, convection will be used in this discussion to conform with its use in the soil science literature.

Convection in soils is a function of the input flux and the hydraulic conductivity of the soil. Nonreactive, soluble tracers such as bromide or chloride are often used to determine convective velocity in soils.[10,11] Tracers are added either as a pulse or a step input to the soil and measurements of the tracer concentrations in the effluent are used to calculate the average velocity of the soil solution. Tracers are commonly used in laboratory soil columns maintained under steady-state flow since such conditions considerably simplify the subsequent transport analysis. Research has also shown that predictions of solute transport based on steady-state flow models are comparable to those based on transient flow models.[12,13]

The pore size distribution and the range of pore shapes within a soil affects the fluid velocity through the soil profile by creating a distribution of velocities related to the distribution of pore sizes and shapes.[14] In a soil with a homogeneous random pore distribution, the range of velocities is represented by a mean velocity. Preferential flow occurs in soils with a bimodal pore size distribution where a tracer or a particle may flow preferentially through large pores and bypass the bulk of the soil matrix. This phenomena is observed in soils where the breakthrough curves exhibit considerable tailing.[15]

Convection is the primary bacterial transport mechanism.[16] Studies have shown that bacteria travel longer distances in soils with large pore sizes, maintained under heavy flux conditions where the convective processes predominate. Smith et al. reported "the degree of macropore flow influences the rate of movement of water or other non-interacting solutes, but it determines the extent of *Escherichia coli* transport."[17] In general, coarse-textured soils have a greater percentage of macropores than do fine-textured soils; therefore, greater bacterial movement is achieved in coarse-textured soils.[8,18,19] The interrelationship between the quantity of water within a soil and the soil's pore size distribution is reflected by the energy status or matric potential of the soil water. The effect of matric potential on the active movement of *Pseudomonas aeruginosa* was studied by Griffin and Quail, who concluded that a critical volume of water is required for movement of bacteria.[20] Bitton et al. found that bacterial movement through soil columns did not occur when water content was at or below field capacity.[21] They reported that bacteria did not move with the water in sandy soils when the water content was less than 15%. Wong and Griffin surveyed available data regarding the effect of matric potential on bacteria movement and found that it is improbable that bacteria will move at matric potentials approaching –500 cm of water.[22] Germann et al. used the kinematic wave approximation to evaluate the transport of *E. coli* in the vadose zone.[23] This study concluded that the macropore moisture content must be greater than 0.015 (m^3/m^3) and macropore conductance must be greater than 0.1 (m/s) for significant transport of microbes.

Bacteria have exhibited greater movement when macropores are saturated. A study evaluating three different genera of bacteria and their movement in soil at different matric potentials found all genera of bacteria moved significantly longer distances at a matric potential of –50 cm than at –150 cm.[22] McCoy and Hagedorn noted that, under saturated conditions, preferential flow affected the transport of microorganisms by allowing them to bypass interaction with the bulk of the soil matrix.[24] This observation may explain why some studies have reported that bacteria exhibited movement through soils at rates faster than rates observed for conservative tracers.[25] Bacteria seemingly move through soils in continuous liquid pathways that become more tortuous as the air phase enters the flow channel during conditions extenuating toward unsaturated flow. Berry and Hagedorn suggested that unsaturated flow conditions are not conducive for bacterial transport due to the increased tortuosity of the flow path.[8]

Other parameters that influence macropore flow and therefore may have a pronounced effect on convection include bulk density, soil structure, and vegetative growth. In a study using packed soil columns filled with a sieved Ede loamy sand, van Elsas et al. found significantly greater movement of bacterial cells in soils with lower bulk densities.[26] The effect of bulk density on the convective movement of bacteria in soils is likely related to the increase in macroporosity and the improvement in soil structure associated with decreased bulk density. The importance of soil structure on solute transport processes within soils must be considered when planning experiments designed to study microbial transport. It has been noted that cell transport through mixed or sieved soils is insignificant relative to transport through most structured soils.[17,27] Undisturbed soil columns are more likely to be representative of natural field soils than packed columns because the structure and the macroporosity of the field is better maintained. At the same time, undisturbed soil microcosms may be particularly useful for evaluating the persistence and gene transfer of GEMs due to their easy replication and ability to control the effects of single soil factors on bacterial movement.[28] It is difficult or impossible to obtain accurate estimations of the natural heterogeneity of transport processes found within field soils when using laboratory columns; therefore, results obtained in laboratory experiments may not be representative of field transport variability.[26] Studies using larger columns may give more realistic results than those studies using smaller columns. Predictive models based on column experiments should be validated by field experiments when possible. The presence of plant roots or other biological channels may provide additional flow paths for bacteria movement. The presence of roots in the soil profile has been observed to increase the downward movement of *P. fluorescens*, *Bradyrhizobium japonicum*, and *P. putida*.[26,29-31] The effect of plant roots may be more important for transport in heavier textured soils with more developed plants. A recent study reported that plant roots had no effect on transport in a loamy sand that had been planted with young plants.[18]

B. DISPERSION

Dispersion is the process by which a contaminant is distributed as it flows through the soil. Hydrodynamic dispersion (also referred to as mechanical mixing) results from the mixing that occurs within a fluid during tortuous convective movement of the fluid as it travels through different soil pores. Diffusion results from the mixing that occurs due to differences in potential gradients. When attempting to mathematically describe transport processes in soils, a dispersion coefficient is commonly used that jointly considers both diffusion and hydrodynamic dispersion.[10,11,32] Dispersion is primarily a function of the motion of the soil solution, except at very low velocities where diffusion may also have a substantial effect.[11,33] The form of the dispersion coefficient is highly dependent on the nature of the porous medium.[34] Dispersion is affected by the physical characteristics of the soil including grain size, nonuniformity, pore size distribution, and structural features.[9] In general, the amount of dispersion observed in a soil is directly proportional to the degree of soil structure within that soil. Cassel et al. reported that greater dispersion was observed in undisturbed soil cores than in disturbed soil cores.[35] Preferential flow channels can be created in structured soils where there is a great diversity of pore sizes in the soil.[11] The dispersion coefficient is representative of the heterogeneity of the soil.[36] For large-scale field experiments, heterogeneity usually increases with field size.

The dispersivity is a proportionality coefficient that describes the relationship between the dispersion coefficient and the mean pore water velocity as follows:

$$D = \lambda v \tag{1}$$

where λ (cm) is the dispersivity, D (cm^2 day^{-1}) is the dispersion coefficient, and v (cm day^{-1}) is the mean pore water velocity.[11] Wierenga and van Genuchten have shown that, although the dispersion coefficient is an increasing function of the mean pore water velocity, dispersivity is largely unaffected by changes in velocity.[37]

In column experiments, the dispersion coefficient has been reported to be linearly related to the length of the column, provided that the column length exceeds 10 cm.[38] Therefore, it has been suggested that it is important to report the depth of evaluation with any reported dispersion coefficient. Contrary results were reported by Wierenga and van Genuchten, who analyzed the movement of a tritium tracer in small (30-cm length) and large (6-m length) soil columns.[37] They did not find any significant increase in the dispersion coefficient with depth; however, they did report that the dispersivity values were five times greater in the longer columns.

The relationships between the dispersion coefficient (D), velocity (V), and column length (L) are often described by the use of Peclet numbers (P), where P = (VL)/D. Peclet numbers were used by Pfannkuch to explain five regimes of dispersion observed for solid particles as outlined by Rose:[34]

1. If P is less than 0.3, the system operates under pure molecular diffusion.
2. If P is between 0.3 and 5, magnitudes of convection and diffusion are equal.
3. If P is greater than 5 and less than 1000, convection dominates diffusion; diffusion is minimal.
4. If P is between 1000 and 150,000, there is pure convection, provided Darcy's law is valid. There is no diffusion.
5. If P is greater than 150,000, there is pure convection, but Darcy's law is not valid; inertia and turbulence are present. This situation is unlikely to occur in soils except under conditions where macropore flow dominates.

Under conditions where diffusion is negligible compared to hydrodynamic dispersion, all displacements through the same column should be adequately described by the same Peclet number.[39] Discrepancies between Peclet numbers of tracers used in identical columns may arise if one tracer is subject to kinetic, nonequilibrium adsorption while the other tracer is not subject to the same adsorptive process.[39]

The effect of dispersion on bacterial movement may be influenced by bacterial size. Fontes et al.[40] observed that dispersion had a greater influence on the movement of larger cells than on the movement of smaller cells. This observation was attributed to the fact that the movement of the larger cells is restricted to macropores that are known to have longer distances between pore intersections.

C. RETENTION

Corapcioglu and Haridas identified three retention mechanisms for bacteria in soils: mechanical filtering, sedimentation, and adsorption. Mechanical filtering and adsorption are most effective for the retention of bacteria in soils.[24,41,42] Viruses are primarily retained in soils by adsorption, since their small sizes are difficult to filter.[9]

Mechanical filtering accounts for bacteria removed from the liquid phase by exclusion from flow paths in the soil matrix due to their relatively large sizes. Germann and Douglas reported that pores that transport particles have diameters at least an order of magnitude greater than the particles that are transported.[43] This restrictive process called "straining" is a major inhibitor of bacterial transport. One method used to eliminate the effects of straining is to study bacterial movement in sandy soils with pores and pore necks large enough to allow bacteria to move easily.[44] This method may restrict the process of filtering in the short term, but over time the growth of microorganisms in the soil may clog pores and cause filtering to occur. Bacterial accumulations on grain surfaces form clusters called dendrides. The effects of straining increase with dendridic growth.[45] Bacterial clogging of pores has been observed under both aerobic and anaerobic conditions. A recent study conducted by Vandevivere and Baveye reported that strictly aerobic bacteria are able to reduce the saturated hydraulic conductivity of a sand by up to four orders of magnitude.[46] The presence of organic material in the soil may increase the filtration ability of the soil by creating a "biological mat".[47]

A second mechanism of bacterial retention in soils is sedimentation. Sedimentation is gravitational deposition on soil particles and occurs when bacteria have a different density than the soil solution. There are differing opinions about the importance of sedimentation in bacteria transport. Some studies report that sedimentation is not a major factor for bacteria retention.[36,41] These studies view microbial transport as a process involving the movement of individual microorganisms. Corapcioglu and Haridas suggested that some small-sized bacteria are neutrally buoyant and, therefore, do not settle.[41] The average density for microorganisms is approximately 1 g/cm^3, which is also the average density of water; therefore, sedimentation should not occur with average bacteria. Matthess and Pekdeger suggested that sedimentation is not an important force influencing bacteria transport since the repulsing surface forces between microbial cells are greater than the kinetic energy of a small particle transported by the groundwater flow.[36] Sedimentation becomes increasingly important in studies considering bacteria as aggregate particles rather than as distinct individuals. Theoretically, this may be a more appropriate perspective for studying bacteria transport because microorganisms typically adsorb to small particles, aggregate with themselves, or bioflocculate in solution.[48] Hagedorn et al.[49] reviewed the results of several studies on the transport of bacteria through the solum in relation to land application of domestic waste waters and concluded that sedimentation of bacterial clusters occurs throughout the zone of saturated flow. Tan et al.[19] observed that bacterial sedimentation and aggregation did not occur in either distilled water or in 25 m*M* $CaCl_2$ (calcium chloride).

A third mechanism of bacterial retention in soils is adsorption. Adsorption is the adhesion of microorganisms to soil particles and may affect bacterial transport in three ways. First, adsorption may affect the movement of bacteria within the soil by retention of the bacteria that are adsorbed by large soil particles. Second, adsorption may affect the decay rate of the bacteria by protecting adsorbed bacteria from inactivation. Third, adsorption may enhance bacteria transport by enabling bacteria adsorbed to very small particles to travel with the particles with less danger of immobilization from adsorption to larger particles. This third concept was proposed by Gannon et al.,[27] who found that bacteria motility and the lower-bound estimated value of the adsorption coefficient were inversely correlated. The effect of this third type of adsorption on bacteria transport has not been fully explained.

The adsorption potential of soils is affected by various soil physical and chemical properties such as texture, pH, bulk density, and soil moisture. Tan et al.[19] reported that bacteria in the fine soil fraction were retained near the soil surface partially because of the greater adsorptive capacity of the fine fraction. Many studies have attributed greater bacteria movement in coarse-textured soils to the low adsorptive capacity of soils containing less clay.[21,24] Peterson and Ward[50] reported that the adsorption of bacteria is probably not a significant retention factor for sand, loamy sand, or sandy loam soils. The increased coarseness of the soil material may decrease the available surface area of the soil and, therefore, provide fewer sites for bacteria adsorption.[44]

The adsorption capacity of a soil increases with decreasing pH below 8.[41] Therefore, decreasing the pH may increase bacteria retention. Increasing the salt content of the soil solution may also increase the adsorption capacity of a soil due to double-layer compression.[19] In general, both microbial and mineral particles are negatively charged. These charges create a long-range repulsive force determined by the surface charge density of the microorganisms and the mineral particles. The repulsing electrostatic forces are stronger than the van der Waals forces that function between interactively attracted particles; therefore, negatively charged particles stay in suspension and are not easily adsorbed.[36] Additionally, these repulsing forces may help microbes achieve greater motility through anion exclusion.[9] Anion exclusion is the process by which negatively charged microbial particles are pushed to the center of the pore where the average velocity is greater than the velocity of the solution moving closer to the solid surfaces.[11] The range of the repulsive force depends on the electrolyte concentration of the soil solution and the type of counter-ion species in solution.[8] Divalent ions are more effective in reducing repulsive forces than monovalent ions. Particles suspended in weaker electrolyte concentrations experience less adsorption and, therefore, are able to travel longer distances. This phenomena was observed by Tan et al. in an experiment where bacteria moved significantly further in the same sand when suspended in distilled water than when suspended in either 3- or 25-m*M* $CaCl_2$ or 75-m*M* KCl solutions.[44] Bacteria also moved significantly further in the 3-m*M* $CaCl_2$ solution than in the 25-m*M* $CaCl_2$ solution.

Efforts to predict adsorption based on soil properties or microbial characteristics have not been successful.[10] Adsorption isotherms describe the relationship between the amount of substance adsorbed and the concentration of substance in solution at a given temperature.[11] Two general types of adsorption isotherms are the Langmuir and the Freundlich isotherms. The Freundlich isotherm assumes the soil surface has different types of adsorption sites; therefore, adsorption increases with increasing solution concentration. The Langmuir isotherm assumes the adsorbed concentration increases linearly at low concentrations and approaches a constant at high concentrations. Virus adsorption has been successfully described by the linear Freundlich isotherm.[51] Tan et al.[44] used isotherms in a model designed to predict bacteria movement in short coarse sand and fine sand-packed columns under unsteady, unsaturated flow conditions over very short time periods (i.e., 2 to 30 min). Following infiltration of the initial input concentration, columns were sectioned and sampled for microbial concentrations. Retardation for bacteria suspended in distilled water was described using a linear adsorption isotherm, while retardation for bacteria suspended in 25-m*M* $CaCl_2$ was described using a nonlinear Freundlich adsorption isotherm. The authors indicated two problems were encountered when using this approach. First, the recovery of bacteria decreased with increasing adsorption, perhaps due to inadequate time provided for bacteria desorption. Second, it was difficult to determine the adsorption isotherm for bacteria leached with $CaCl_2$ through coarse sand because the adsorbed bacteria concentration increased at an unusually rapid rate with increasing solution concentration. Concerns have been raised that batch adsorption isotherms may not adequately explain the adsorption of bacteria that occurs in soil columns because the isotherms fail to account for the kinetic processes that may influence bacterial sorption in soil systems.[52] Nonequilibrium adsorption models may offer a more accurate method for characterizing the adsorption-desorption process in bacterial transport.[47,52] This hypothesis is in agreement with Corapcioglu and Haridas, who reported that bacterial retention caused by straining, sedimentation, and adsorption is a rate-controlled reaction for bacteria, and an equilibrium-controlled reaction for viruses.[45] Nonequilibrium models may include parameters that account for the kinetic adsorption processes involved in transport.

III. MATHEMATICAL TRANSPORT MODELS

Existing theories describing the movement of bacteria in soils are incomplete and tests of existing models against laboratory data are few.[40,48] Additional research on the kinetics of bacterial transport is required before a completely successful process-based predictive model of bacterial movement can be formulated.[9,26,43,53] Several researchers have presented theoretical bacteria transport models involving the parameters previously described, but commonly conclude that insufficient data exist for conclusive analysis.[41,50,53]

A. STOCHASTIC MODELS

Many bacterial transport studies attempt to predict the rate of bacterial movement through soils using transport models. Transport models are limited in their abilities to predict bacterial transport in soils due to the variability of the flow field resulting from the nonuniform distribution of particles in suspension.[48] The convection-dispersion equation that is commonly used to describe solute transport in porous media is also limited in its ability to accurately describe the effects of local variation of soil physical properties on transport.[11] Some researchers have suggested that bacterial transport may be best described using stochastic models that account for the variable velocities observed in soils using a statistical treatment of transport parameters.[10,53] This approach was used by Peterson and Ward to determine the impact of various parameters in a bacterial transport model.[50] The stochastic approach was particularly useful in their study as a tool to determine the probability of bacteria transport beyond the 120-cm soil depth under various input conditions. One criticism of stochastic models is that they are only useful for predicting movement at specific sites under specific conditions.

B. THE CONVECTION-DISPERSION EQUATION (CDE)

The CDE describes the overall transport process by quantifying the relationship between various transport parameters including mean pore water velocity, dispersion, and retardation. The general equation commonly used to describe the one-dimensional transport of a solute in a uniform soil under steady-state flux conditions was first presented by Lapidus and Amundson as follows:

$$\frac{\delta C}{\delta t} = D\frac{\delta^2 C}{\delta X^2} - V\frac{\delta C}{\delta X} \tag{2}$$

where C is the concentration (meq/cm^3), D is the dispersion coefficient (cm^2/day), V is the mean pore water velocity (flux divided by the volumetric water content; cm/day), X is the distance (cm), and t is the time (days).[54] This equation has been the foundation for analysis for much of the transport research in soil physics.[12,37,55–57,59] It has been shown that, except for small values of V/D, all solutions of this equation predict nearly symmetrical concentration distributions.[15] The term tailing is used to describe nonsymmetrical distributions. For conditions when solutes interact with the soil, this equation may be written as follows:

$$R\frac{\delta C}{\delta t} = D\frac{\delta^2 C}{\delta X^2} - V\frac{\delta C}{\delta X} \tag{3}$$

Where R represents the retardation factor. The R term reduces to one for non-reactive solutes and is defined as follows:

$$R = 1 + \frac{\rho k}{\theta} \tag{4}$$

where ρ is the soil bulk density (g cm^{-3}), θ is the volumetric water content (cm^3 cm^{-3}), and, if the adsorption process is linear, k (cm^3 g^{-1}) is the empirical distribution coefficient.[32,60,61]

Parameter estimation techniques typically use a nonlinear least-squares minimization procedure to fit data obtained from column tracer experiments to estimate the values of coefficients.[32,39,62] Least-squares computer techniques are among the most accurate and most convenient of methods available for determining D and R.[63] It is important to hold a calculated known value for either R or V constant while estimating the other parameters because these two coefficients are highly interdependent.[32] Best fitted values of D and R are obtained when Equation 3 is augmented with supplemental conditions characterizing the initial concentration of the system and the boundary conditions. This is especially important for short column experiments where the Peclet number is small or large-scale field experiments with large dispersivities.[63] If the Peclet number is greater than about 20, the fitted value of D becomes roughly independent of the analytical solution used for determination.[63] Van Genuchten and Alves[64] have published a compilation of different solutions for the CDE equation under a variety of boundary conditions.

Breakthrough curves plotting the pore volumes of effluent against the relative concentration graphically illustrate the transport processes occurring. The height of the breakthrough curve increases with increasing dispersivity.[60] Increasing the value of R shifts the curve to the right.[60] Decreasing the value of R (perhaps due to anion exclusion) causes the curve to shift left. Increasing the dispersivity increases the width of the breakthrough curve.[40] Normally breakthrough curves peak over one pore volume. Tailing can be an indication of preferential flow.

Three general models based on Equation 3 are used to describe the transport process in a porous medium: the linear equilibrium model, the two-site/two-region kinetic nonequilibrium model, and the one-site kinetic nonequilibrium model. One of the primary differences between these models is the method of evaluating R or, more specifically, k (see Equation 4).

The linear equilibrium model assumes that adsorption is a linear, instantaneous process described as a linearized isotherm by van Genuchten:

$$s = kc \tag{5}$$

where s is the adsorbed concentration, c is the solution concentration, and k is an empirical distribution constant that may be predicted using batch isotherms.[56,57] The assumption that adsorption is instantaneous may not be valid for flow experiments; therefore, applications of linear models to field experiments may lead to serious underpredictions of field leaching potential.[65]

Linear equilibrium models have not adequately described the solute transport process in situations where bimodal porosity leads to two-region flow or in situations where the adsorption process is controlled by either one-site or two-site kinetic nonequilibrium adsorption processes. For example, a herbicide study by Davidson and McDougal[66] observed considerable tailing in the effluent curve that was not predicted by the use of a time-dependent adsorption coefficient in a convective-dispersive equation. Research indicates that the adsorption processes of bacteria and many chemicals are not adequately described by a single linear equilibrium approach.[8,21,52,56–58,61,67–71] Under nonequilibrium conditions, either a two-site or a one-site kinetic adsorption mechanism may be used to better describe the adsorption process.[56,57] The two-site sorption theory assumes a combined effect of two types of adsorption sites: (type-1) instantaneous linear adsorption and (type-2) time-dependent, nonlinear adsorption sites.[56,57,61] Models designed to describe transport using this theory are referred to in the literature as two-site kinetic adsorption models. Adsorption on both type-1 and type-2 sites at equilibrium is described by Parker and van Genuchten in the following equations:[32]

$$s_1 = k_1 c = Fkc \tag{6}$$

$$s_2 = k_2 c = (1 - F)kc \tag{7}$$

where the subscripts 1 and 2 refer to type-1 and type-2 sites, and F is the fraction of all sites occupied by type-1 sorption sites. Total adsorption at equilibrium is defined as

$$s = s_1 + s_2 = kc \tag{8}$$

Recall that type-1 sites are always at equilibrium; therefore, the adsorption rate for type-1 sites may be defined as follows:

$$\frac{\delta s_1}{\delta t} = Fk\frac{\delta c}{\delta t} \tag{9}$$

The adsorption rate for the type-2 kinetic nonequilibrium sites may be described by the following linear and reversible rate equation:

$$\frac{\delta s_2}{\delta t} = \alpha(k_2 c - s_2) \tag{10}$$

where α is a first-order rate coefficient (day^{-1}). These equations can be organized into the following transport model described by van Genuchten and Wagenet with an additional coefficient representing microbial decay (μ) as follows:

$$\frac{\delta s_2}{\delta t} = \alpha\left[(1 - F)\,kc - s_2\right] - \mu_{s2} s_2 \tag{11}$$

$$\frac{\delta\,(\theta + F\rho k)\,c}{\delta t} = \frac{\delta}{\delta}x\left(\theta D\frac{\delta c}{\delta x} - qc\right) - \alpha\rho\left[(1 - F)\,kc - s_2\right] - \theta\mu_{liq}c - F\rho k\mu_{s1}c \tag{12}$$

For a detailed description of the analytical solution for this model see Reference 61. Two-site models may give an improved description of bacteria retention in soils because they consider both the instantaneous sorption of bacteria due to electrochemical surface interactions between the bacteria and mineral surfaces and the kinetic, time-dependent retention observed for bacteria in some studies. Berry and Hagedorn[8] reported that the process of bacterial cell adhesion itself can be described using a two-site model. Two-site/two-region models have successfully described solute transport in a variety of experiments.[68–71]

Alternatively, one could use a one-site kinetic nonequilibrium model to describe bacteria transport. This approach assumes that the adsorption of bacteria is kinetically controlled and not instantaneous. Therefore, all adsorption sites in these models are considered type-2 sites. The following one-site kinetic nonequilibrium model has been described by van Genuchten and Wagenet:

$$\frac{\delta s}{\delta t} = \alpha \left(kc - s\right) - \mu_s s \tag{13}$$

$$\frac{\delta c}{\delta t} + \frac{\rho}{\theta}\frac{\delta s}{\delta t} = D\frac{\delta^2}{\delta x^2} - v\frac{\delta c}{\delta x} - \mu_{liq}c - \frac{\rho\mu_s}{\theta}s \tag{14}$$

where μ describes first-order decay.[61] Because the one-site model is a special case of the two-site model, analytical solutions for the two-site model are also valid for the one-site model.[61] A one-site kinetic nonequilibrium model was used by Hornberger et al.[52] to describe bacterial transport. They concluded that adsorption of bacteria in columns is kinetically controlled rather than instantaneous because bacteria were strongly retained in the soil column, but no retardation in breakthrough time was observed.

Nonequilibrium models may give improved descriptions of the transport processes occurring for contaminant concentrations exhibiting nonsymmetrical breakthrough curves. This seems to often occur in microbial transport studies. Wollum and Cassel[67] observed considerable tailing in a column displacement experiment studying the movement of streptomycetes. Bitton et al.[21] also observed considerable tailing while studying the movement and retention of an encapsulated form of *Klebsiella aerogenes* in saturated soil columns. In a study examining the movement of bromide and labeled bacteria through a contaminated sandy aquifer, Harvey and Garabedian[72] found that little difference in accuracy was obtained when the kinetic, nonequilibrium approach was used instead of the linear, instantaneous approach. This conclusion was attributed to the fact that there was little retardation observed in the experiment.

IV. MODELS CONSIDERING POPULATION DYNAMICS

Validated transport models considering the effect of population dynamics on transport are difficult to find. Models may be simplified by assuming that bacterial growth is limited or insignificant. Experiments are typically conducted using microorganisms that are in a resting state, or at low temperatures, or over short time periods to allow models to operate without die-off or growth rate coefficients.[27,44,52] These models are limited in their use to predict microbial concentrations under normal field conditions over time.

Some models use growth and death-rate coefficients from the literature and not from experiments conducted under the site conditions used in the model.[10] This type of approach underestimates the importance of accurate death and growth rate coefficients. One-dimensional transport models Sumatra-1 and WORM were used to evaluate potential transport results of GEMs released to the environment in several hypothetical scenarios.[39,63,73] The studies compared the results of both models run with and without the first-order production and zero-order decay coefficients. Results indicated that the presence of decay constants reduced the relative predicted concentration by up to five orders of magnitude at the 40-cm depth after 60 days.[53] Peterson and Ward[50] performed a variety of Monte Carlo simulations to determine the probability of bacterial movement under four different input parameters. Their results showed that bacteria are at least twice as likely to reach or surpass the 120-cm soil depth when the die-off and distribution coefficients are not included in the transport equation.

Models of microbial survival and plasmid transfer have used simple exponential growth and decay mass action approaches.[74,75] The inactivation of microorganisms has been accurately described as a first-order reaction; however, the die-off coefficient has been observed to be a highly variable parameter, ranging several orders of magnitude for any given type of bacteria due to the effects of environmental factors on bacterial die-off. Crane and Moore[47,76] reviewed data examining die-off rates. They found that effects of physical and climatic factors on bacteria die-off rates have not been quantitatively defined for two reasons. First, the effects of many factors, such as temperature and pH, are nonlinear. Second, many experiments do not have a sufficiently complete database to evaluate the effects of important, interrelated factors.

Several models may be used to describe microbial transport under steady-state flow in a one-dimensional homogeneous system. Where adsorption can be described by a linear isotherm (Equation 5), microbial transport in soils can be described by the following linear equilibrium adsorption model:

$$R\frac{\delta c}{\delta t} = D\frac{\delta^2 c}{\delta x_2} - v\frac{\delta c}{\delta x} - \mu c + \gamma \tag{15}$$

where

$$\mu = \mu_w + \frac{\mu_s \rho k}{\theta} \tag{16}$$

and

$$\gamma = \gamma_w + \frac{\gamma_s \rho k}{\theta} \tag{17}$$

define μ (day^{-1}) and γ (cfu/day) as the first-order decay and zero-order production terms, respectively, each represented as a component contribution of both the liquid and the solid phases.[32] An analytical solution for this equation was presented by van Genuchten.[57] Coefficients for Equation 15 can be estimated using a least-squares minimization parameter estimation program, such as CXTFIT.[32] Best values are obtained for coefficients by limiting the number of unknown coefficients fitted to the data. This is particularly important for coefficients that are mutually dependent, such as V and R, as well as for coefficients that have been observed to have significant interactions, such as μ and γ.[32]

Several analytical solutions for Equation 15 have been presented by van Genuchten, who reported that Equation 15 becomes easier to solve mathematically when the zero-order production term (γ) becomes zero, and all solutions presented may be used without further modification.[57] Modifications are necessary for the limiting case when only the first-order decay term (μ) becomes zero.

Recall that Equations 11 and 12 can be used to model microbial transport using the two-site kinetic nonequilibrium model. An analytical solution of the two-site kinetic nonequilibrium model has been presented by van Genuchten and Wagenet.[61] Six independent dimensionless parameters are introduced as follows: P (Peclet number), R (retardation factor), a site-partitioning coefficient β, a rate coefficient ω, and two dimensionless degradation coefficients, μ_{liquid} and μ_{sorbed}. When $\mu_{liquid} = \mu_{sorbed}$ these coefficients may be represented by a single degradation rate term, μ. R is the retardation factor previously defined in Equation 4. Other dimensionless parameters are defined as follows:

$$P = \frac{vL}{D} \tag{18}$$

where L represents the column length, v is the mean pore water velocity, and D is the dispersion coefficient.

$$\beta = \frac{\theta + F\rho k}{\theta + \rho k} = \frac{R_m}{R} \tag{19}$$

$$\omega = \frac{\alpha\,(1 - \beta)\,RL}{v} \tag{20}$$

where α is a first-order rate coefficient. A program modification of the CXTFIT code named CXT4 was designed for parameter estimation of the two-site non-equilibrium model with degradation.[61] This program utilizes a least-squares minimization procedure to optimize degradation parameters in the general two-site/two-region model.[61,77] The same parameters listed above are used in the CXT4 program.

The one-site nonequilibrium model is a special case of the two-site nonequilibrium model. A one-site model designed to consider first-order decay is defined by Equations 13 and 14. The one-site model assumes that all sorption sites are type-2 sites; therefore, F (the fraction of type-1 sites) is reduced to zero.[56,57] The dimensionless variables used for the one-site model are the same as those used for the two-site model with the following exceptions as explained by van Genuchten:

$$\beta = \frac{1}{R} \tag{21}$$

$$\omega = \frac{\alpha\,(R - 1)\,L}{v} \tag{22}$$

Recall that the one-site model is a special case of the two-site model and, therefore, analytical solutions for the two-site model may be directly applied to the one-site model.[56,57]

V. MODEL COMPARISONS TO PREDICT GEM TRANSPORT

Wiebers compared the use of the one-site kinetic nonequilibrium, two-site kinetic nonequilibrium, and linear equilibrium transport models to predict the movement of a GEM (*Pseudomonas aeruginosa*) through undisturbed, loamy sand soil columns.[78] Transport parameters were optimized for each model using least-squares minimization techniques. Two separate column studies were performed. An independent model validation was performed where transport parameters derived from the first study were used to predict GEM movement in the second study. A dependent validation was also performed where the mean pore water velocity and dispersion coefficient from the second study were used to predict GEM movement in the second study. All models estimated GEM transport utilizing a population degradation coefficient.

Microbial data is commonly log-transformed prior to analysis to facilitate comparisons between population numbers. Transport models typically evaluate contaminant movement over time based on an initial input and use transport parameters based on least-squares minimization procedures. Log-transformed data do not allow equal distribution of estimation errors in the least-squares procedure. Furthermore, log-transformed values cannot be used as a basis for data reduction based on input values since the scale of reduction is not linear. Data were, therefore, reduced by dividing observations of bacteria in a particular depth interval by the input population.

Wiebers determined that the one-site kinetic nonequilibrium model was the most accurate of the three transport models for predicting GEM transport using the independent validation.[78] The linear equilibrium model consistently overpredicted GEM populations both in the depth intervals and over time. The two-site kinetic nonequilibrium model was not able to accurately predict GEM movement using the V and D transport parameters determined from the initial model calibration in the first study.

The predictions of the two-site nonequilibrium model were greatly improved when the transport parameters from the second study were used in a dependent validation. Two-site model predictions of relative population over time were more accurate than the other two models and gave more accurate predictions of the relative GEM population over time than it did over depth. The linear model overpredicted GEM populations consistently for both the dependent and independent model validations. The predictions of the one-site kinetic nonequilibrium model were similar for both the independent and dependent model calibrations. The one-site nonequilibrium model was considered to be the best model to use when describing GEM transport because consistently good results were obtained for both the independent and dependent validations.

VI. CLOSING COMMENTS

The selection and use of a predictive model to estimate the leaching potential of a GEM will likely be specific for the soil and microorganism under consideration. Preliminary research using simple column systems of coarse-textured soils has indicated that one-site kinetic nonequilibrium models may offer advantages in predicting GEM transport for a limited number of specific organisms under the specified conditions. Whether these models will offer the best predictions when applied to field soils of varying textures and for a broad range of microorganisms remains to be seen.

Population degradation coefficients and transport parameters determined using least-squares minimization procedures and used to model GEM die-off and GEM transport may give results that are biased. Least-squares minimization procedures attempt to determine parameters and coefficients in a manner to minimize the values of residual estimation errors. Since a GEM is likely to be introduced into the soil at a high initial input value and will likely decay at an exponential rate, use of the least-squares minimization procedure to estimate degradation coefficients describing GEM populations within the soil will likely overpredict GEM populations following the rapid decay in the first few days after introduction. Similar overprediction is also likely to occur when estimating transport parameters since larger data values following GEM introduction will have a greater effect on the residual estimation errors.

Steps should be taken to reduce overprediction when estimating population degradation coefficients and transport parameters by placing more weight on the microbial populations observed in the later stages of monitoring. It should be emphasized that within a few days after introduction, the GEM population observed in soils is likely to be quite small relative to the GEM input population, however, the total GEM population numbers within the soil may still be quite large. When evaluating the potential hazards related to the leaching of microorganisms in soils, the situation where only one GEM leaches into the ground water may present a potential hazard.

ACKNOWLEDGMENT

Contribution No. 9033 and Scientific Article No. A-7712 of the Maryland Agricultural Experiment Station and the Department of Agronomy, University of Maryland, College Park.

REFERENCES

1. Doyle, J.D., Short, K.A., Stotzky, G., King, R.J., and Seidler, R.M., Ecologically significant effects of *Pseudomonas putida* genetically engineered to degrade 2,4-dichlorophenoxyacetate on microbial populations and processes in soil, *Can. J. Microbiol.*, 37, 682, 1991.
2. McGaughey, W.H., Insect resistance to the biological insecticide *Bacillus thuringiensis, Science*, 229, 193, 1985.
3. Sharples, F.E., Ecological aspects of hazard identification for environmental uses of genetically engineered organisms, in *Risk Assessment in Genetic Engineering*, Levin, M.A. and Strauss, H.S., Eds., McGraw-Hill, New York, 1990.
4. Naimon, J.S., Using expert panels to assess risks of environmental applications: a case study of the 1988 Frostban risk assessment, in *Risk Assessment in Genetic Engineering*, Levin, M.A. and Strauss, H.S., Eds., McGraw-Hill, New York, 1990.
5. Glew, J.G., Angle, J.S., and Sadowsky, M.J., *In vivo* transfer of plasmid R68.45 from *Pseudomonas aeruginosa* to the indigenous population in non-sterile soil, *Microbial Releases*, 1, 237, 1993.
6. Smit, E., van Elsas, J.D., van Veen, J.A., and deVos, W.M., Detection of plasmid transfer from *Pseudomonas fluorescens* to indigenous bacteria in soil by using bacteriophage R2f for donor counterselection, *Appl. Environ. Microbiol.*, 57, 3482, 1991.
7. Tiedje, J.M., Colwell, R.R., Grossman, Y.L., Hodson, R.E., Lenski, R.E., Mack, R.N., and Regal, P.J., The planned introduction of genetically engineered organisms: ecological considerations and recommendations, *Ecology*, 70, 298, 1989.
8. Berry, D.F. and Hagedorn, C., Soil and groundwater transport of microorganisms, in *Assessing Ecological Risk of Biotechnology*, Ginzburg, L.B., Ed., Butterworth-Heinemann, Boston, 1991.

9. Gerba, C.P., Yates, M.V., and Yates, S.R., Modeling microbial transport in the subsurface: a mathematical discussion, in *Modeling the Environmental Fate of Microorganisms*, Hurst, C.J., Ed., American Society for Microbiology, Washington, D.C., 1991, 77.
10. Yates, M.V. and Yates, S.R., Modeling microbial transport in the subsurface, in *Modeling the Environmental Fate of Microorganisms*, Hurst, C.J., Ed., American Society for Microbiology, Washington, D.C., 1991, 48.
11. Jury, W.A., Gardner, W.R., and Gardner, W.H., *Soil Physics*, 5th ed., John Wiley & Sons, New York, 1991, 218.
12. Wierenga, P.J., Solute distribution profiles compared with steady-state and transient water movement models, *Soil Sci. Soc. Am. J.*, 41, 1050, 1977.
13. Destouni, G., Applicability of the steady state flow assumption for solute advection in field soils, *Water Resour. Res.*, 27, 2129, 1991.
14. Biggar, J.W. and Nielsen, D.R., Miscible displacement and leaching phenomenon, in *Irrigation of Agricultural Lands*, Hagan, R.M., Haise, R.R., and Edminster, T.W., Eds., (Agronomy 11), American Society of Agronomy, Madison, WI, 1967, 254.
15. van Genuchten, M.Th. and Wierenga, P.J., Mass transfer studies in sorbing porous media. I. Analytical solutions, *Soil Sci. Soc. Am. J.*, 40, 473, 1976.
16. Peterson, T.C. and Ward, R.D., Bacterial transport in coarse soils beneath on-site wastewater treatment systems, Colorado State University Exp. Stat. Tech. Bull. No. TB87-4, Colorado State University, Fort Collins, 1987.
17. Smith, M.S., Thomas, G.W., White, R.E., and Ritonga, D., Transport of *Escherichia coli* through intact and disturbed soil columns, *J. Environ. Qual.*, 14, 87, 1985.
18. Trevors, J.T., van Elsas, J.D., van Overbeek, L.S., and Starodub, M., Transport of genetically engineered *Pseudomonas flourescens* strain through a soil microcosm, *Appl. Environ. Microbiol.*, 56, 401, 1990.
19. Tan, Y., Bond, W.J., Rovira, A.D., Brisbane, P.G., and Griffin, D.M., Movement through soil of a biological control agent *Pseudomonas fluorescens, Soil Biol. Biochem.*, 23, 821, 1991.
20. Griffin, D.M. and Quail, G., The movement of bacteria in moist particulate systems, *Aust. J. Biol. Sci.*, 21, 579, 1968.
21. Bitton, G., Lhahav, N., and Henis, Y., Movement and retention of *Klebsiella aerogenes* in soil columns, *Plant Soil*, 40, 373, 1974.
22. Wong, P.T.W. and Griffin, D.M., Bacterial movement at high matric potentials — in artificial and natural soils, *Soil Biol. Biochem.*, 8, 215, 1976.
23. Germann, P.F., Smith, M.S., and Thomas, G.W., Kinematic wave approximation to the transport of *Escherichia coli* in the vadose zone, *Water Resour. Res.*, 23, 1281, 1987.
24. McCoy, E.L. and Hagedorn, C., Transport of resistance-labeled *Escherichia coli* strains through a transition between two soils in a topographic sequence, *J. Environ. Qual.*, 9, 686, 1980.
25. Keswick, B.H., Wang, D., and Gerba, C.P., The use of microorganisms as ground water tracers: a review, *Ground Water*, 20, 142, 1982.
26. van Elsas, J.D., Trevors, J.T., and Overbeek, L.S., Influence of soil properties on the vertical movement of genetically-marked *Pseudomonas fluorescens* through large soil microcosms, *Biol. Fertil. Soils*, 10, 249, 1990.
27. Gannon, J.T., Mingelgrin, U., Alexander, M., and Wagenet, R.J., Bacterial transport through homogeneous soil, *Soil Biol. Biochem.*, 23. 1155, 1991.

28. Bentjen, S.A., Fredrickson, J.K., van Voris, P., and Li, S.W., Intact soil-core microcosms for evaluating the fate and ecological impact of the release of genetically engineered microorganisms, *Appl. Environ. Microbiol.*, 55, 198, 1989.
29. Parke, J.L., Moen, R., Roviraa, A.D., and Bowen, G.D., Soil water flow effects the rhizosphere distribution of a seed-borne biological control agent, *Pseudomonas fluorescens, Soil Biol. Biochem.*, 18, 583, 1986.
30. Trevors, J.T. and Berg, G., Conjugal RP4 transfer between pseudomonads in soil and recovery of RP4 plasmid dna from soil, *Syst. Appl. Microbiol.*, 11, 223, 1989.
31. Madsen, E.L. and Alexander, M., Transport of *Rhizobium* and *Pseudomonas* through soil, *Soil Sci. Soc. Am. J.*, 46, 557, 1982.
32. van Genuchten, M.Th. and Parker, J.C., Boundary conditions for displacement experiments through short laboratory soil columns, *Soil Sci. Soc. Am. J.*, 48, 703, 1984.
33. Skopp, J. and Gardner, W.R., Miscible displacement: an interacting flow region model, *Soil Sci. Soc. Am. J.*, 56, 1680, 1992.
34. Rose, D.A., Hydrodynamic dispersion in porous materials, *Soil Sci.*, 123, 277, 1977.
35. Cassel, D.K., Krueger, T.H., Schroer, F.W., and Norum, E.B., Solute movement through disturbed and undisturbed soil cores. *Soil Sci. Soc. Am. Proc.*, 38, 36, 1974.
36. Matthess, G. and Pekdeger, A., Survival and transport of pathogenic bacteria and viruses in ground water, in *Ground Water Quality*, Ward, C.H., Giger, W., and McCarty, P.L., Eds., John Wiley & Sons, New York, 1985, 472.
37. Wierenga, P.J. and van Genuchten, M.Th., Solute transport through small and large unsaturated soil columns, *Ground Water*, 27, 35, 1989.
38. Gupta, S.K. and Pandey, R.N., Effect of column length on hydrodynamic dispersion coefficient during solute movement, *J. Indian Soc. Soil Sci.*, 29, 388, 1981.
39. van Genuchten, M.Th., Wierenga, P.J., and O'Connor, G.A., Mass transfer studies in sorbing porous media. III. Experimental evaluation with 2,4,5-T, *Soil Sci. Soc. Am. J.*, 41, 278, 1977.
40. Fontes, D.E., Mills, A.L., Hornberger, G.M., and Herman, J.S., Physical and chemical factors influencing transport of microorganisms through porous media, *Appl. Environ. Microbiol.*, 57, 2473, 1991.
41. Corapcioglu, D.K. and Haridas, A., Transport and fate of microorganisms in porous media: a theoretical investigation, *J. Hydrol.*, 72, 149, 1984.
42. Harvey, R.W., Parameters involved in modeling movement of bacteria in groundwater, in *Modeling the Environmental Fate of Microorganisms*, Hurst, C.J., Ed., American Society for Microbiology, Washington, D.C., 1991, 25, 178.
43. Germann, P.F. and Douglass, L.A., Comments on 'Particle Transport Through Porous Media' by Laura M. McDowell-Boyer, James R. Hunt, and Nicholas Sitar, *Water Resour. Res.*, 23, 1697, 1987.
44. Tan, Y., Bond, W.J., and Griffin, D.M., Transport of bacteria during unsteady, unsaturated soil water flow, *Soil Sci. Soc. Am. J.*, 56, 1331, 1992.
45. Corapcioglu, M.Y. and Haridas, A., Microbial transport in soils and groundwater: a numeric model, *Adv. Water Res.*, 8, 188, 1985.
46. Vandevivere, P. and Baveye, P., Saturated hydraulic conductivity reduction caused by aerobic bacteria in sand columns, *Soil Sci. Soc. Am. J.*, 56, 1, 1992.
47. Yates, M.V. and Yates, S.R., Modeling microbial fate in the subsurface environment, *Crit. Rev. Environ. Control*, 17, 307, 1988.
48. McDowell-Boyer, L., Hunt, J.R., and Sitar, N. Particle transport through porous media, *Water Resour. Res.*, 22, 1901, 1986.

49. Hagedorn, C., McCoy, E.L., and Rahe, T.M., The potential for groundwater contamination from septic effluents, *J. Environ. Qual.*, 10, 1, 1981.
50. Peterson, T.C. and Ward, R.D., Development of a bacterial transport model for coarse soils, *Water Res. Bull.*, 25, 349, 1989.
51. Lance, J.C., Gerba, C.P., and Wang, D.S., Comparative movement of different enteroviruses in soil columns, *J. Environ. Qual.*, 11, 347, 1982.
52. Hornberger, G.M., Mills, A.L., and Hermann, J.S., Bacterial transport in porous media: evaluation of a model using laboratory observations, *Water Resour. Res.*, 28, 915, 1992.
53. Dickinson, R.A., Problems with using existing transport models to describe microbial transport in porous media, in *Modeling the Environmental Fate of Microorganisms*, Hurst, C.J., Ed., American Society for Microbiology, Washington, D.C., 1991, 21.
54. Lapidus, L. and Amundson, N.R., Mathematics of adsorption in beds. VI. The effects of longitudinal diffusion in ion exchange and chromatographic columns, *J. Phys. Chem.*, 56, 984, 1952.
55. Skopp, J. and Gardner, W.R., Miscible displacement: an interacting flow region model, *Soil Sci. Soc. Am. J.*, 56, 1680, 1992.
56. van Genuchten, M.Th., Non-Equilibrium Transport Parameters from Miscible Displacement Experiments, Res. Rep. 119, U.S. Salinity Laboratory Riverside, CA, 1981a.
57. van Genuchten, M.Th., Analytical solutions for chemical transport with simultaneous adsorption, zero-order production and first-order decay, *J. Hydrol.*, 49, 213, 1981b.
58. Rao, P.S.C., Davidson, J.M., Jessup, R.E., and Selim, H.M., Evaluation of conceptual models for describing non-equilibrium adsorption-desorption of pesticides during steady-flow in soils, *Soil Sci. Soc. Am. J.*, 43, 22, 1979.
59. Biggar, J.W. and Nielsen, D.R., Miscible displacement. II. Behavior of tracers, *Soil Sci. Soc. Am. Proc.*, 26, 126, 1962.
60. Wierenga, P.J., Solute transport, in *Future Developments in Soil Science Research: A Collection of SSSA Golden Anniversary Contributions Presented at Annual Meeting in New Orleans, LA 30 November–5 December, 1986*, Boersma, L.L., Ed., Soil Science Society of America, Madison WI, 1986, 23.
61. van Genuchten, M.Th. and Wagenet, R.J., Two-site/two-region models for pesticide transport and degradation: theoretical development and analytical solutions, *Soil Sci. Soc. Am.*, 53, 1303, 1989.
62. Elprince, A.M. and Day, P.R., Fitting solute breakthrough equations to data using two adjustable parameters, *Soil Sci. Soc. Am. J.*, 41, 39, 1977.
63. van Genuchten, M.Th. and Wierenga, P.J. Solute dispersion coefficients and retardation factors, in *Methods of Soil Analysis, Part I. Physical and Mineralogical Methods*, 2nd ed., American Society of Agronomy, Soil Science Society of America, Madison, WI, 1986, 1025.
64. van Genuchten, M.Th. and Alves, W.J., Analytical Solutions of the One-Dimensional Convective-Dispersive Solute Transport Equation, U.S. Department of Agriculture, Tech. Bull. No. 1661, 1982, 151.
65. Rao, P.S.C. and Jessup, R.E., Sorption and movement of pesticides and other toxic organic substances in soils, in *Chemical Mobility in Soils*, Kral, D.M., Ed., Soil Sci. Soc. Am. Spec. Pub. No. 11, Soil Science Society of America and American Society of Agronomy, Madison, WI, 1983, 183.

66. Davidson, J.M. and McDougal, J.R., Experimental and predicted movement of three herbicides in a water-saturated soil, *J. Environ. Qual.*, 2, 428, 1973.
67. Wollum, A.G. II and Cassel, D.K., Transport of microorganisms in sand columns, *Soil Sci. Soc. Am. J.*, 42, 72, 1978.
68. Gamerdinger, A.P., Lemley, A.T., and Wagenet, R.J., Non-equilibrium sorption and degradation of three 2-chloro-s-triazine herbicides in soil-water systems, *J. Environ. Qual.*, 2, 815, 1991.
69. Selim, H.M. and Amacher, M.C., A second-order approach for modeling solute retention and transport in soils, *Water Resour. Res.*, 24, 2061, 1988.
70. De Camargo, O.A., Biggar, J.W., and Nielsen, D.R., Transport of inorganic phosphorus in an alfisol, *Soil Sci. Soc. Am. J.*, 43, 884, 1979.
71. Cameron, D.R. and Klute, A., Convective-dispersive solute transport with a combined equilibrium and kinetic adsorption model, *Water Resour. Res.*, 13, 183, 1977.
72. Harvey, R.W. and Garabedian, S.P., Use of colloid filtration theory in modeling movement of bacteria through a contaminated sandy aquifer, *Environ. Sci. Technol.*, 25, 178, 1991.
73. Versar, Inc., Ambient Exposures to Recombinant Microorganisms Intentionally Released to Municipal and Pulp and Paper Industry Wastewaters, Draft final report prepared for Exposure Assessment Branch, Exposure Evaluation Division, Office of Toxic Substances, U.S. Environmental Protection Agency, under EPA Contract no. 68-02-4254, Task No. 45, Washington, D.C., September 30, 1987.
74. Knudsen, G.R., Walter, M.V., Porteous, L.A., Prince, V.J., Armstrong, J.L., and Seidler, R.J., Predictive model of conjugative plasmid transfer in the rhizoshpere and phyllosphere, *Appl. Environ. Microbiol.*, 54, 343, 1988.
75. Levin, B.R., Stewart, F.M., and Rice, V.A., The kinetics of conjugative plasmid transmission: fit of a simple mass action model, *Plasmid*, 2, 247, 1979.
76. Crane, S.R. and Moore, J.A., Modeling enteric bacterial die-off: a review, *Water, Air, Soil Pollut.*, 27, 411, 1986.
77. Gamerdinger, A.P., Wagenet, R.J., and van Genuchten, M. Th., Application of two-site/two-region models for studying simultaneous non-equilibrium transport and degradation of pesticides, *Soil Sci. Soc. Am. J.*, 54, 957, 1990.
78. Wiebers, J.L., Comparison of Prediction Models to Estimate Genetically Engineered Microbe Transport, M.Sc. thesis, University of Maryland, College Park, 1993, 110.
79. USEPA, Report of the Biotechnology Science Advisory Committee, USEPA, Washington, D.C., 1987.

Index